T0182978

Lecture Notes in Computer Science 9042

More information about this series at http://www.springer.com/series/7407

Alexander Gelbukh (Ed.)

Computational Linguistics and Intelligent Text Processing

16th International Conference, CICLing 2015 Cairo, Egypt, April 14–20, 2015 Proceedings, Part II

Editor
Alexander Gelbukh
Centro de Investigación en Computación
Instituto Politécnico Nacional
Mexico DF
Mexico

ISSN 0302-9743 ISSN 1611-3349 (electronic)
Lecture Notes in Computer Science
ISBN 978-3-319-18116-5 ISBN 978-3-319-18117-2 (eBook)
DOI 10.1007/978-3-319-18117-2

Library of Congress Control Number: 2014934305

LNCS Sublibrary: SL1 – Theoretical Computer Science and General Issues

Springer Cham Heidelberg New York Dordrecht London

Printed on acid-free paper

Springer International Publishing AG Switzerland is part of Springer Science+Business Media (www.springer.com)

Preface

CICLing 2015 was the 16th Annual Conference on Intelligent Text Processing and Computational Linguistics. The CICLing conferences provide a wide-scope forum for discussion of the art and craft of natural language processing research, as well as the best practices in its applications.

This set of two books contains four invited papers and a selection of regular papers accepted for presentation at the conference. Since 2001, the proceedings of the CICLing conferences have been published in Springer's *Lecture Notes in Computer Science* series as volumes 2004, 2276, 2588, 2945, 3406, 3878, 4394, 4919, 5449, 6008, 6608, 6609, 7181, 7182, 7816, 7817, 8403, and 8404.

The set has been structured into 13 sections representative of the current trends in research and applications of Natural Language Processing:

- Lexical Resources
- Morphology and Chunking
- Syntax and Parsing
- Anaphora Resolution and Word Sense Disambiguation
- Semantics and Dialogue
- Machine Translation and Multilingualism
- Sentiment Analysis and Emotions
- Opinion Mining and Social Network Analysis
- Natural Language Generation and Summarization
- Information Retrieval, Question Answering, and Information Extraction
- Text Classification
- Speech Processing
- Applications

The 2015 event received submissions from 62 countries, a record high number in the 16-year history of the CICLing series. A total of 329 papers (second highest number in the history of CICLing) by 705 authors were submitted for evaluation by the International Program Committee; see Figure 1 and Tables 1 and 2. This two-volume set contains revised versions of 95 regular papers selected for presentation; thus, the acceptance rate for this set was 28.9%.

In addition to regular papers, the books feature invited papers by:

- Erik Cambria, Nanyang Technical University, Singapore
- Mona Diab, George Washington University, USA
- Lauri Karttunen, Stanford University, USA
- Joakim Nivre, Uppsala University, Sweden

who presented excellent keynote lectures at the conference. Publication of full-text invited papers in the proceedings is a distinctive feature of the CICLing conferences.

Table 1. Number of submissions and accepted papers by topic[1]

Accepted	Submitted	% accepted	Topic
19	51	37	Emotions, sentiment analysis, opinion mining
19	56	34	Text mining
17	65	26	Arabic
17	58	29	Information extraction
17	49	35	Lexical resources
15	53	28	Information retrieval
14	35	40	Under-resourced languages
12	45	27	Semantics, pragmatics, discourse
11	40	28	Clustering and categorization
11	33	33	Machine translation and multilingualism
10	29	34	Practical applications
8	37	22	Social networks and microblogging
8	21	38	Syntax and chunking
7	17	41	Formalisms and knowledge representation
7	23	30	Noisy text processing and cleaning
5	21	24	Morphology
4	12	33	Question answering
4	10	40	Textual entailment
3	9	33	Natural language generation
3	8	38	Plagiarism detection and authorship attribution
3	13	23	Speech processing
3	21	14	Summarization
3	12	25	Word sense disambiguation
2	10	20	Computational terminology
2	8	25	Co-reference resolution
2	16	12	Named entity recognition
2	9	22	Natural language interfaces
1	1	100	Computational humor
1	15	7	Other
1	11	9	POS tagging
0	7	0	Spelling and grammar checking

[1] As indicated by the authors. A paper may belong to more than one topic.

Furthermore, in addition to presentation of their invited papers, the keynote speakers organized separate vivid informal events; this is also a distinctive feature of this conference series.

With this event, we continued with our policy of giving preference to papers with verifiable and reproducible results: in addition to the verbal description of their findings given in the paper, we encouraged the authors to provide a proof of their claims in electronic form. If the paper claimed experimental results, we asked the authors to make available to the community all the input data necessary to verify and reproduce these results; if it claimed to introduce an algorithm, we encourage the authors to make the algorithm itself, in a programming language, available to the public. This additional

Table 2. Number of submitted and accepted papers by country or region

Country or region	Authors	Papers[2]		Country or region	Authors	Papers[2]	
	Subm.	Subm.	Accp.		Subm.	Subm.	Accp.
Algeria	6	2.5	0.5	Lebanon	2	1	–
Argentina	2	0.5	0.5	Malaysia	3	3	
Australia	4	1.00	1.00	Mexico	13	5.95	1.08
Belgium	7	2	1	Morocco	16	6	–
Brazil	19	8.5	2.5	Myanmar	3	1.5	–
Canada	20	9.5	6	The Netherlands	2	1	–
Chile	5	1	–	New Zealand	4	2	1
China	37	15.05	4.3	Nigeria	5	2	1
Colombia	6	2	1	Oman	1	1	–
Czech Rep.	11	3.33	2	Peru	6	3.5	0.5
Egypt	89	42.8	13.33	Philippines	2	1	–
Estonia	1	1	–	Poland	3	2	–
Ethiopia	2	1	–	Portugal	8	3	2
Finland	6	1.75	1.75	Qatar	1	1	–
France	30	13.05	2.58	Romania	7	5	1
Georgia	1	1	–	Russia	11	7.33	1.67
Germany	15	6.83	1.33	Saudi Arabia	10	4.42	0.5
Greece	3	1	–	Serbia	3	1	–
Hong Kong	10	3.7	2.7	Singapore	1	1	1
Hungary	2	2	2	South Africa	1	0.83	0.5
India	98	57	15	Spain	16	6.92	3.92
Iran	1	1	–	Sri Lanka	7	3	0.67
Iraq	2	1.33	0.5	Sweden	2	0.67	–
Ireland	4	1.33	1.33	Switzerland	4	2.25	1.25
Israel	6	2	–	Tunisia	57	23.58	4
Italy	9	3.33	0.33	Turkey	29	14.42	2.83
Japan	17	7.25	1.25	Ukraine	2	0.33	–
Jordan	1	0.5	–	UAE	4	2.83	0.5
Kazakhstan	4	1	1	UK	23	10.75	3.58
Korea, South	1	0.5	–	USA	36	13.78	6.58
Kuwait	1	0.17	0.17	Vietnam	3	1.67	–
				Total:	705	329	95

[2] By the number of authors: e.g., a paper by two authors from the USA and one from UK is counted as 0.67 for the USA and 0.33 for UK.

electronic material will be permanently stored on the CICLing's server, www.CICLing.org, and will be available to the readers of the corresponding paper for download under a license that permits its free use for research purposes.

In the long run, we expect that computational linguistics will have verifiability and clarity standards similar to those of mathematics: in mathematics, each claim is accompanied by a complete and verifiable proof, usually much longer than the claim itself; each theorem's complete and precise proof—and not just a description of its general idea—is made available to the reader. Electronic media allow computational linguists to provide material analogous to the proofs and formulas in mathematic in full length—

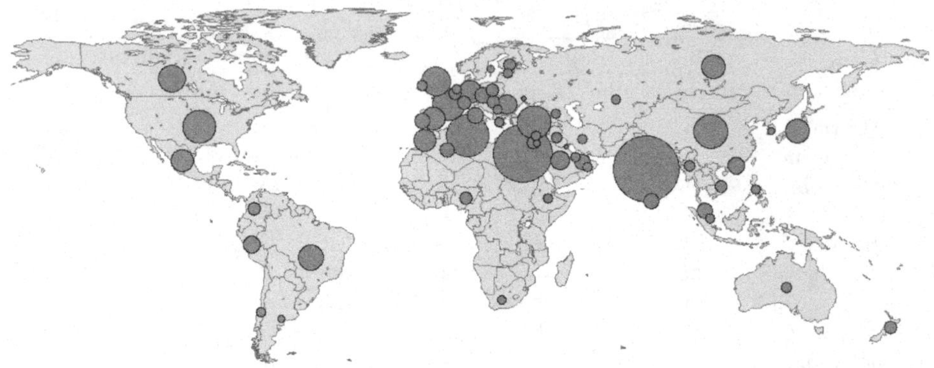

Fig. 1. Submissions by country or region. The area of a circle represents the number of submitted papers.

which can amount to megabytes or gigabytes of data—separately from a 12-page description published in the book. More information can be found on http://www.CICLing.org/why_verify.htm.

To encourage providing algorithms and data along with the published papers, we selected three winners of our Verifiability, Reproducibility, and Working Description Award. The main factors in choosing the awarded submission were technical correctness and completeness, readability of the code and documentation, simplicity of installation and use, and exact correspondence to the claims of the paper. Unnecessary sophistication of the user interface was discouraged; novelty and usefulness of the results were not evaluated—instead, they were evaluated for the paper itself and not for the data.

In this year, we introduced a policy of allowing—and encouraging—papers longer than the 12-page limit included in the fee. The reason was an observation that longer papers tend to be more complete and useful for the reader. In contrast, when a restrictive page limit is enforced, very often the authors have to omit important details; to the great frustration of the readers, this usually renders the whole paper largely useless because the presented results cannot be reproduced. This our observation was strongly confirmed by the fact that all four papers selected for the Best Paper Awards, especially the winners of the first two places, were much over the usual 12-page limit imposed by other conferences.

The following papers received the Best Paper Awards, the Best Student Paper Award, as well as the Verifiability, Reproducibility, and Working Description Awards, correspondingly:

Best Paper 1st Place: *Automated Linguistic Personalization of Targeted Marketing Messages Mining User-generated Text on Social Media*, by Rishiraj Saha Roy, Aishwarya Padmakumar, Guna Prasaad Jeganathan, and Ponnurangam Kumaraguru, India;

Best Paper 2nd Place: *Term Network Approach for Transductive Classification*, by Rafael Geraldeli Rossi, Solange Oliveira Rezende, and Alneu de Andrade Lopes, Brazil;

Best Paper 3rd Place: *Building Large Arabic Multi-domain Resources for Sentiment Analysis*, by Hady ElSahar and Samhaa R. El-Beltagy, Egypt;

Best Student Paper:[1] *Translation Induction on Indian Language Corpora using Translingual Themes from Other Languages*, by Goutham Tholpadi, Chiranjib Bhattacharyya, and Shirish Shevade, India;

Verifiability 1st Place: *Domain-specific Semantic Relatedness from Wikipedia Structure: A Case Study in Biomedical Text*, by Armin Sajadi, Evangelos E. Milios, and Vlado Keselj, Canada;

Verifiability 2nd Place: *Translation Induction on Indian Language Corpora using Translingual Themes from Other Languages*, by Goutham Tholpadi, Chiranjib Bhattacharyya, and Shirish Shevade, India;

Verifiability 3rd Place: *Opinion Summarization using Submodular Functions: Subjectivity vs Relevance Trade-off*, by Jayanth Jayanth, Jayaprakash S., and Pushpak Bhattacharyya, India.[2]

The authors of the awarded papers (except for the Verifiability award) were given extended time for their presentations. In addition, the Best Presentation Award and the Best Poster Award winners were selected by a ballot among the attendees of the conference.

Besides its high scientific level, one of the success factors of CICLing conferences is their excellent cultural program in which all attendees participate. The cultural program is a very important part of the conference, serving its main purpose: personal interaction and making friends and contacts. The attendees of the conference had a chance to visit the Giza Plateau with the Great Pyramid of Cheops and the Sphinx—probably the most important touristic place on Earth; The Egyptian Museum, the home to the largest collection of Pharaonic or ancient Egyptian relics and pieces; and the Old Cairo, to mention only a few most important attractions.

In this year we founded, and held in conjunction with CICLing, the First Arabic Computational Linguistics conference, which we expect to become the primary yearly event for dissemination of research results on Arabic language processing. This is in accordance with CICLing's mission to promote consolidation of emerging NLP communities in countries and regions underrepresented in the mainstream of NLP research and, in particular, in the mainstream publication venues. With founding this new conference, and with the very fact of holding CICLing in Egypt in a difficult moment of its history, we expect to contribute to mutual understanding, tolerance, and confidence between the Arabic world and the Western world: the better we know each other the more lasting peace between peoples.

[1] The best student paper was selected from among papers of which the first author was a full-time student, excluding the papers that received Best Paper Awards.

[2] This paper is published in a special issue of a journal and not in this book set.

I would like to thank all those involved in the organization of this conference. In the first place, these are the authors of the papers that constitute this book: it is the excellence of their research work that gives value to the book and sense to the work of all other people. I thank all those who served on the Program Committee of CICLing 2015 and of the First Arabic Computational Linguistics conference, the Software Reviewing Committee, Award Selection Committee, as well as additional reviewers, for their hard and very professional work. Special thanks go to Ted Pedersen, Savas Yıldırım, Roberto Navigli, Manuel Vilares Ferro, Kenneth Church, Dafydd Gibbon, Kais Haddar, and Adam Kilgarriff for their invaluable support in the reviewing process.

I would like to thank Prof. Tarek Khalil, president of Nile University (NU), for welcoming CICLing at NU. I also want to cordially thank the conference staff, volunteers, and members of the Local Organization Committee headed by Prof. Samhaa R. El-Beltagy and advised by Prof. Hussein Anis. In particular, I am very grateful to Ms. Aleya Serag El-Din for her great effort in coordinating all the aspects of the conference. I wish to thank the Center for Informatics Science for all the support they have provided. I am deeply grateful to the administration of the Nile University for their helpful support, warm hospitality, and in general for providing this wonderful opportunity of holding CICLing in Egypt. I am also grateful to members of the Human Foundation for their great effort in planning the cultural program. I acknowledge support from Microsoft Research, the project CONACYT Mexico–DST India 122030 "Answer Validation through Textual Entailment," and SIP-IPN grant 20150028.

The entire submission and reviewing process was supported for free by the EasyChair system (www.EasyChair.org). Last but not least, I deeply appreciate the Springer staff's patience and help in editing these volumes and getting them printed in very short time—it is always a great pleasure to work with Springer.

March 2015 Alexander Gelbukh

Organization

CICLing 2015 was hosted by the Nile University, Egypt and was organized by the CICLing 2015 Organizing Committee in conjunction with the Nile University, the Natural Language and Text Processing Laboratory of the Centro de Investigación en Computación (CIC) of the Instituto Politécnico Nacional (IPN), Mexico, and the Mexican Society of Artificial Intelligence (SMIA).

Organizing Chair

Samhaa R. El-Beltagy	Nile University, Egypt

Organizing Committee

Samhaa R. El-Beltagy	Chair	Nile University, Egypt
Hussein Anis	Senior Adviser	Nile University, Egypt
Aleya Serag El-Din	Principal co-coordinator	Nile University, Egypt
Yasser Nasr	Finance	Nile University, Egypt
Sameh Habib	Facilities and Logistics	Nile University, Egypt
Hoda El-Beleidy	Accommodation	Nile University, Egypt
Mariam Yasin Abdel Ghafar	Cultural activities	The Human Foundation, Egypt
Yasser Bayoumi	Engineering	Nile University, Egypt
Mahmoud Gabr	IT	Nile University, Egypt
Layla Al Roz		Nile University, Egypt
Hady Alsahar		Nile University, Egypt
Mohamed Fawzy		Nile University, Egypt
Amal Halby		Nile University, Egypt
Muhammad Hammad		Nile University, Egypt
Talaat Khalil		Nile University, Egypt
Yomna El-Nahas		Nile University, Egypt

Program Chair

Alexander Gelbukh	Instituto Politécnico Nacional, Mexico

Program Committee

Ajith Abraham	Machine Intelligence Research Labs (MIR Labs), USA
Rania Al-Sabbagh	University of Illinois at Urbana-Champaign, USA
Marianna Apidianaki	LIMSI-CNRS, France
Alexandra Balahur	European Commission Joint Research Centre, Italy
Sivaji Bandyopadhyay	Jadavpur University, India

Srinivas Bangalore	AT&T Labs-Research, USA
Leslie Barrett	Bloomberg, LP, USA
Roberto Basili	University of Rome Tor Vergata, Italy
Núria Bel	Universitat Pompeu Fabra, Spain
Anja Belz	University of Brighton, UK
Pushpak Bhattacharyya	IIT Bombay, India
António Branco	University of Lisbon, Portugal
Nicoletta Calzolari	Istituto di Linguistica Computazionale – CNR, Italy
Nick Campbell	TCD, Ireland
Michael Carl	Copenhagen Business School, Denmark
Violetta Cavalli-Sforza	Al Akhawayn University, Morocco
Niladri Chatterjee	IIT Delhi, India
Kenneth Church	IBM, USA
Dan Cristea	A.I.Cuza University of Iași, Romania
Walter Daelemans	University of Antwerp, Belgium
Samhaa R. El-Beltagy	Cairo University, Egypt
Michael Elhadad	Ben-Gurion University of the Negev, Israel
Anna Feldman	Montclair State University, USA
Alexander Gelbukh (Chair)	Instituto Politécnico Nacional, Mexico
Dafydd Gibbon	Universität Bielefeld, Germany
Gregory Grefenstette	Inria, France
Eva Hajičová	Charles University, Czech Republic
Sanda Harabagiu	University of Texas at Dallas, USA
Yasunari Harada	Waseda University, Japan
Karin Harbusch	University of Koblenz and Landau, Germany
Ales Horak	Masaryk University, Czech Republic
Veronique Hoste	Ghent University, Belgium
Nancy Ide	Vassar College, USA
Diana Inkpen	University of Ottawa, Canada
Aminul Islam	Dalhousie University, Canada
Guillaume Jacquet	JRC, Italy
Doug Jones	MIT, USA
Sylvain Kahane	Université Paris Ouest Nanterre La Défense and CNRS/Alpage, Inria, France
Alma Kharrat	Microsoft Research, USA
Adam Kilgarriff	Lexical Computing Ltd, UK
Philipp Koehn	University of Edinburgh, UK
Valia Kordoni	Humboldt University Berlin, Germany
Leila Kosseim	Concordia University, Canada
Mathieu Lafourcade	LIRMM, France
Krister Lindén	University of Helsinki, Finland
Bing Liu	University of Illinois at Chicago, USA
Elena Lloret	University of Alicante, Spain
Bente Maegaard	University of Copenhagen, Denmark
Cerstin Mahlow	University of Stuttgart, IMS, Germany

Suresh Manandhar	University of York, UK
Sun Maosong	Tsinghua University, China
Diana McCarthy	University of Cambridge, UK
Alexander Mehler	Goethe-Universität Frankfurt am Main, Germany
Rada Mihalcea	University of Michigan, USA
Evangelos Milios	Dalhousie University, Canada
Jean-Luc Minel	Université Paris Ouest Nanterre La Défense, France
Dunja Mladenic	Jožef Stefan Institute, Slovenia
Marie-Francine Moens	Katholieke Universiteit Leuven, Belgium
Masaki Murata	Tottori University, Japan
Preslav Nakov	Qatar Computing Research Institute, Qatar Foundation, Qatar
Roberto Navigli	Sapienza Università di Roma, Italy
Joakim Nivre	Uppsala University, Sweden
Kjetil Nørvåg	Norwegian University of Science and Technology, Norway
Attila Novák	Pázmány Péter Catholic University, Hungary
Kemal Oflazer	Carnegie Mellon University in Qatar, Qatar
Constantin Orasan	University of Wolverhampton, UK
Ekaterina Ovchinnikova	KIT, Karlsruhe and ICT, University of Heidelberg, Germany
Ivandre Paraboni	University of São Paulo – USP/EACH, Brazil
Maria Teresa Pazienza	University of Rome, Tor Vergata, Italy
Ted Pedersen	University of Minnesota Duluth, USA
Viktor Pekar	University of Birmingham, UK
Anselmo Peñas	Universidad Nacional de Educación a Distancia, Spain
Stelios Piperidis	Institute for Language and Speech Processing, Greece
Octavian Popescu	FBK-IRST, Italy
Marta R. Costa-Jussà	Institute For Infocomm Research, Singapore
German Rigau	IXA Group, Universidad del País Vasco / Euskal Herriko Unibertsitatea, Spain
Fabio Rinaldi	IFI, University of Zürich, Switzerland
Horacio Rodriguez	Universitat Politècnica de Catalunya, Spain
Paolo Rosso	Technical University of Valencia, Spain
Vasile Rus	The University of Memphis, USA
Horacio Saggion	Universitat Pompeu Fabra, Spain
Patrick Saint-Dizier	IRIT-CNRS, France
Franco Salvetti	University of Colorado at Boulder and Microsoft Inc., USA
Rajeev Sangal	Language Technologies Research Centre, India
Kepa Sarasola	Euskal Herriko Unibertsitatea, Spain
Roser Sauri	Pompeu Fabra University, Spain

Hassan Sawaf	eBay Inc., USA
Satoshi Sekine	New York University, USA
Bernadette Sharp	Staffordshire University, UK
Grigori Sidorov	Instituto Politécnico Nacional, Mexico
Vivek Kumar Singh	Banaras Hindu University, India
Vaclav Snasel	VSB-Technical University of Ostrava, Czech Republic
Efstathios Stamatatos	University of the Aegean, Greece
Josef Steinberger	University of West Bohemia, Czech Republic
Jun Suzuki	NTT, Japan
Stan Szpakowicz	SITE, University of Ottawa, Canada
Juan-Manuel Torres-Moreno	Laboratoire Informatique d'Avignon / Université d'Avignon et des Pays de Vaucluse, France
George Tsatsaronis	Technical University of Dresden, Germany
Dan Tufiș	Institutul de Cercetări pentru Inteligență Artificială, Academia Română, Romania
Olga Uryupina	University of Trento, Italy
Renata Vieira	Pontifícia Universidade Católica do Rio Grande do Sul, Brazil
Manuel Vilares Ferro	University of Vigo, Spain
Aline Villavicencio	Universidade Federal do Rio Grande do Sul, Brazil
Piotr W. Fuglewicz	TiP Sp. z o. o., Poland
Bonnie Webber	University of Edinburgh, UK
Savaş Yıldırım	Istanbul Bilgi University, Turkey

Software Reviewing Committee

Ted Pedersen	University of Minnesota Duluth, USA
Florian Holz	Universität Leipzig, Germany
Miloš Jakubíček	Lexical Computing Ltd, UK, and Masaryk University, Czech Republic
Sergio Jiménez Vargas	Universidad Nacional, Colombia
Miikka Silfverberg	University of Helsinki, Finland
Ronald Winnemöller	Universität Hamburg, Germany

Award Committee

Alexander Gelbukh	Instituto Politécnico Nacional, Mexico
Eduard Hovy	Carnegie Mellon University, USA
Rada Mihalcea	University of Michigan, USA
Ted Pedersen	University of Minnesota Duluth, USA
Yorick Wilks	University of Sheffield, UK

Additional Referees

Naveed Afzal
Rodrigo Agerri
Saad Alanazi
Itziar Aldabe
Iñaki Alegria
Haifa Alharthi
Henry Anaya-Sánchez
Vít Baisa
Francesco Barbieri
Janez Brank
Jorge Carrillo de Albornoz
Dave Carter
Jean-Valère Cossu
Victor Manuel Darriba Bilbao
Elnaz Davoodi
Claudio Delli Bovi
Bart Desmet
Steffen Eger
Luis Espinosa Anke
Tiziano Flati
Ivo Furman
Dimitrios Galanis
Dario Garigliotti
Mehmet Gençer
Rohit Gupta
Afli Haithem
Ignacio J. Iacobacci
Milos Jakubicek
Magdalena Jankowska
Assad Jarrahian
Kirill Kireyev
Vojtech Kovar
Majid Laali
John Lowe
Andy Lücking
Raheleh Makki Niri
Shachar Mirkin
Alexandra Moraru
Andrea Moro
Mohamed Mouine
Mohamed Outahajala
Michael Piotrowski
Alessandro Raganato
Arya Rahgozar
Christoph Reichenbach
Francisco José Ribadas-Pena
Alvaro Rodrigo
Francesco Ronzano
Pavel Rychly
Armin Sajadi
Ehsan Sherkat
Janez Starc
Yasushi Tsubota
Tim vor der Brück
Rodrigo Wilkens
Tuğba Yıldız

First Arabic Computational Linguistics Conference

Program Chairs

Alexander Gelbukh	Instituto Politécnico Nacional, Mexico
Khaled Shaalan	The British University in Dubai, UAE

Program Committee

Bayan Abu-shawar	Arab Open University, Jordan
Hanady Ahmed	Qatar University, Qatar
Hend Al-Khalifa	King Saud University, Saudi Arabia
Mohammed Attia	Al-Azhar University, Egypt

Aladdin Ayesh	De Montfort University, UK
Khalid Choukri	ELDA, France
Samhaa R. El-Beltagy	Cairo University, Egypt
Ossama Emam	ALTEC, Egypt
Aly Fahmy	Cairo University, Egypt
Alexander Gelbukh (Chair)	Instituto Politécnico Nacional, Mexico
Ahmed Guessoum	University of Science and Technology Houari Boumediene, Algeria
Nizar Habash	New York University Abu Dhabi, UAE
Kais Haddar	MIRACL Laboratory, Faculté des sciences de Sfax, Tunisia
Lamia Hadrich Belguith	MIRACL Laboratory, Tunisia
Sattar Izwaini	American University of Sharjah, UAE
Mark Lee	University of Birmingham, UK
Sherif Mahdy Abdou	RDI, Egypt
Farid Meziane	University of Salford, UK
Herman Moisl	Newcastle University, UK
Ahmed Rafea	American University in Cairo, Egypt
Allan Ramsay	University of Manchester, UK
Mohsen Rashwan	Cairo University, Egypt
Paolo Rosso	Technical University of Valencia, Spain
Nasredine Semmar	CEA, France
Khaled Shaalan (Chair)	The British University in Dubai, UAE
William Teahan	Bangor University, UK
Imed Zitouni	IBM Research, USA

Additional Referees

Sherif Abdou
Nora Al-Twairesh
Feryal Haj Hassan
Harish Tayyar Madabushi

Website and Contact

The webpage of the CICLing conference series is http://www.CICLing.org. It contains information about past CICLing conferences and their satellite events, including links to published papers (many of them in open access) or their abstracts, photos, and video recordings of keynote talks. In addition, it contains data, algorithms, and open-source software accompanying accepted papers, in accordance with the CICLing verifiability, reproducibility, and working description policy. It also contains information about the forthcoming CICLing events, as well as contact options.

Contents – Part II

Sentiment Analysis and Emotion Detection

Opinion Mining and Social Network Analysis

Best Paper Award, First Place:

Natural Language Generation and Text Summarization

Information Retrieval, Question Answering, and Information Extraction

Text Classification

Speech Processing

Applications

Erratum

Contents – Part I

Grammar Formalisms and Lexical Resources

Invited Paper:

Invited Paper:

Morphology and Chunking

Syntax and Parsing

Anaphora Resolution and Word Sense Disambiguation

Semantics and Dialogue

Invited Paper:

Verifiability Award, First Place:

Machine Translation and Multilingualism

Best Student Paper Award;
Verifiability Award, Second Place:

Sentiment Analysis and Emotion Detection

The CLSA Model: A Novel Framework for Concept-Level Sentiment Analysis

Erik Cambria[1], Soujanya Poria[1], Federica Bisio[2], Rajiv Bajpai[1], and Iti Chaturvedi[1]

[1] School of Computer Engineering, Nanyang Technological University, Singapore
[2] DITEN, University of Genoa, Italy
{cambria,sporia,rbajpai,iti}@ntu.edu.sg, federica.bisio@edu.unige.it
http://sentic.net

Abstract. Hitherto, sentiment analysis has been mainly based on algorithms relying on the textual representation of online reviews and microblogging posts. Such algorithms are very good at retrieving texts, splitting them into parts, checking the spelling, and counting their words. But when it comes to interpreting sentences and extracting opinionated information, their capabilities are known to be very limited. Current approaches to sentiment analysis are mainly based on supervised techniques relying on manually labeled samples, such as movie or product reviews, where the overall positive or negative attitude was explicitly indicated. However, opinions do not occur only at document-level, nor they are limited to a single valence or target. Contrary or complementary attitudes toward the same topic or multiple topics can be present across the span of a review. In order to overcome this and many other issues related to sentiment analysis, we propose a novel framework, termed concept-level sentiment analysis (CLSA) model, which takes into account all the natural-language-processing tasks necessary for extracting opinionated information from text, namely: microtext analysis, semantic parsing, subjectivity detection, anaphora resolution, sarcasm detection, topic spotting, aspect extraction, and polarity detection.

1 Introduction

Concept-level sentiment analysis is a natural-language-processing (NLP) task that has recently raised growing interest both within the scientific community, leading to many exciting open challenges, as well as in the business world, due to the remarkable benefits to be had from financial market prediction. The potential applications of concept-level sentiment analysis, in fact, are countless and span interdisciplinary areas such as political forecasting, brand positioning, and human-robot interaction.

For example, Li et al. [54] implemented a generic stock price prediction framework and plugged in six different models with different analyzing approaches. They used Harvard psychological dictionary and Loughran-McDonald financial sentiment dictionary to construct a sentiment space. Textual news articles were then quantitatively measured and projected onto such a sentiment space. The models' prediction accuracy was evaluated on five years historical Hong Kong Stock Exchange prices and news articles and their performance was compared empirically at different market classification levels.

A. Gelbukh (Ed.): CICLing 2015, Part II, LNCS 9042, pp. 3–22, 2015.
DOI: 10.1007/978-3-319-18117-2_1

Rill et al. [90] proposed a system designed to detect emerging political topics in Twitter sooner than other standard information channels. For the analysis, authors collected about 4 million tweets before and during the parliamentary election 2013 in Germany, from April until September 2013. It was found that new topics appearing in Twitter can be detected right after their occurrence. Moreover, authors compared their results to Google Trends, observing that the topics emerged earlier in Twitter than in Google Trends.

Jung and Segev [49] analyzed how communities change over time in the citation network graph without additional external information and based on node and link prediction and community detection. The identified communities were classified using key term labeling. Experiments showed that the proposed methods can identify the changes in citation communities multiple years in the future with performance differing according to the analyzed time span.

Montejo-Raez et al. [66] introduced an approach to sentiment analysis in social media environments. Similar to explicit semantic analysis, microblog posts were indexed by a predefined collection of documents. In the proposed approach, performed by means of latent semantic analysis, these documents were built up from common emotional expressions in social streams.

Bell et al. [5] proposed a novel approach to social data analysis, exploring the use of microblogging to manage interaction between humans and robots, and evaluating an architecture that extends the use of social networks to connect humans and devices. The approach used NLP techniques to extract features of interest from textual data retrieved from a microblogging platform in real-time and, hence, to generate appropriate executable code for the robot. The simple rule-based solution exploited some of the 'natural' constraints imposed by microblogging platforms to manage the potential complexity of the interactions and to create bi-directional communication.

All current approaches to sentiment analysis focus on just a few issues related to processing opinionated text, the most common being polarity detection. However, there are many NLP problems that need to be solved –at the same time– to properly deconstruct opinionated text into polarity values and opinion targets. Detecting a polarity from a document without deconstructing this into specific aspects, for example, is pointless as we may end up averaging positive and negative polarity values associated to different product features. Moreover, we may have the best polarity detection tool on the market but, if this is unable to recognize sarcasm, it could infer a completely wrong polarity.

To this end, we propose the CLSA model (Fig. 1) as reference framework for researchers willing to take a more holistic and semantic-aware approach to sentiment analysis, which also applies to the multimodal realm [82]. The main contributions of the proposed model are that (a) it promotes the analysis of opinionated text at concept-, rather than word-, level and (b) it takes into account all the NLP tasks necessary for extracting opinionated information from text. The rest of the paper consists of an overview of the CLSA model (Section 2), followed by a description of each constituting module of the model, namely: microtext analysis (Section 3), semantic parsing (Section 4), subjectivity detection (Section 5), anaphora resolution (Section 6), sarcasm detection (Section 7), topic spotting (Section 8), aspect extraction (Section 9), and polarity detection (Section 10). Section 11, finally, offers concluding remarks.

2 Model Overview

Sentiment analysis is a 'suitcase' research field that contains many different areas, not only related to computer science but also to social sciences, e.g., sociology, psychology, and ethics. The CLSA model focuses on the computational foundations of sentiment analysis research to determine eight key NLP tasks or modules, that are necessary for the correct interpretation of opinionated text, namely:

1. Microtext analysis, for normalizing informal and irregular text (Section 3)
2. Semantic parsing, for deconstructing natural language text into concepts (Section 4)
3. Subjectivity detection, for filtering non-opinionated or neutral text (Section 5)
4. Anaphora resolution, for resolving references in the discourse (Section 6)
5. Sarcasm detection, for detecting sarcastic opinions and flip their polarity (Section 7)
6. Topic spotting, for contextualizing opinions to a specific topic (Section 8)
7. Aspect extraction, for deconstructing text into different opinion targets (Section 9)
8. Polarity detection, for detecting a polarity value for each opinion target (Section 10)

3 Microtext Analysis

Due to the exponential growth of social media, an increasing number of applications, e.g., Web mining, Internet security, cyber-issue detection, and social media marketing, need microtext feature selection and classification.

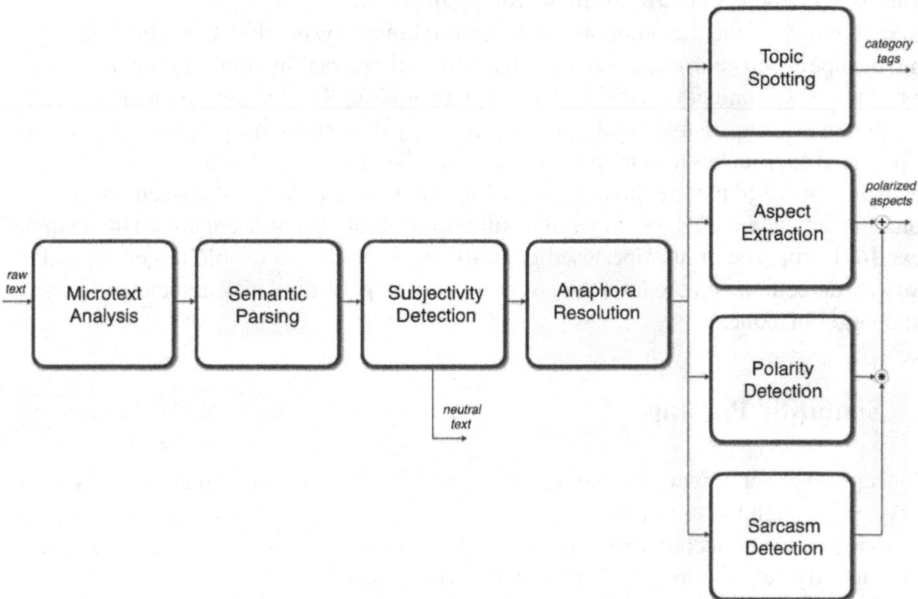

Fig. 1. The CLSA model

Existing sentiment resources developed on non-microblogging data, in fact, turn out to be very inaccurate on informal text [52]. Some of the fundamental characteristics of microtext are a highly relaxed spelling and the reliance on abbreviations, acronyms, and emoticons [89]. This causes problems when trying to apply traditional NLP tools and techniques, e.g., information extraction, automated summarization, and text-to-speech, which have been developed for conventional English text. It could be thought that a simple find-and-replace pre-processing on the microtext would solve that problem. However, the sheer diversity of spelling variations makes this solution impractical; for example, a sampling of Twitter studied in [57] found over 4 million out-of-vocabulary words. Moreover, new spelling variations are created constantly, both voluntarily and accidentally.

The challenge of developing algorithms to correct the non-standard vocabulary found in microtexts is known as text message normalization. The first step in tackling this challenge is to realize that the number of different spelling variations may be massive but they follow a small number of simple basic strategies [57], such as `abbreviation` and `phonetic substitution`. In [111], authors used web blogs to create a corpus for sentimental analysis and exploited emoticons as mood indicators. They used support vector machine (SVM) and conditional random field (CRF) learners to classify sentiments at sentence-level and investigated various strategies to infer the overall sentiment of documents. Later, many other works proposed to do the same using Twitter sentiment [46,72] and buzz [46,74,72]. More recent studies [52,28,65,24] exploit Twitter-specific features such as emoticons, hashtags, URLs, @symbols, capitalizations, and elongations to better detect polarity from microblogging text.

While most of the literature on Twitter sentiment analysis refers to supervised learning, unsupervised approaches [101,75] have recently gained increasing popularity. This is because less number of training data are available for Twitter sentiment analysis and it is practically impossible to train the system every time new data come in. Usually, unsupervised approaches to sentiment analysis involve the creation of a sentiment lexicon in an unsupervised manner first, and then the detection of polarity of unseen text using a function dependent on the number of positive and negative words contained in the input text. [47] proposed an unsupervised graph-based approach to enable target-dependent polarity detection, i.e., the inference of positive or negative polarity associated to a specific target in a query.

4 Semantic Parsing

Concept-level sentiment analysis [12,13,14,85] focuses on the semantic analysis of text [39] through the use of web ontologies or semantic networks, which allow the aggregation of the conceptual and affective information associated with natural language opinions. By relying on large semantic knowledge bases, such approaches step away from the blind use of keywords and word co-occurrence count, but rather rely on the implicit features associated with natural language concepts. Unlike purely syntactical techniques, concept-based approaches are able to detect also sentiments that are expressed in a subtle manner, e.g., through the analysis of concepts that do not explicitly convey any emotion, but which are implicitly linked to other concepts that do so.

The bag-of-concepts model can represent semantics associated with natural language much better than bags-of-words [17]. In the bag-of-words model, in fact, a concept such as `cloud computing` would be split into two separate words, disrupting the semantics of the input sentence (in which, for example, the word `cloud` could wrongly activate concepts related to `weather`). Concept extraction is one of the key steps of automatic concept-level text analysis. [19] used domain specific ontologies to acquire knowledge from text. Using such ontologies, the authors extracted 1.1 million common sense knowledge assertions.

Concept mining is useful for tasks such as information retrieval [87], opinion mining [16], text classification [112]. State-of-the-art approaches mainly exploit term extraction methods to obtain concepts from text. These approaches can be classified into two main categories: linguistic rules [22] and statistical approaches [114,1]. [114] used term frequency and word location and, hence, employed a non-linear function to calculate term weighting. [1] mined concepts from the Web by using webpages to construct topic signatures of concepts and, hence, built hierarchical clusters of such concepts (word senses) that lexicalize a given word. [32] and [103] combined linguistic rules and statistical approaches to enhance the concept extraction process.

Other relevant works in concept mining focus on concept extraction from documents. Gelfand et al. [35] have developed a method based on the Semantic Relation Graph to extract concepts from a whole document. They used the relationship between words, extracted from a lexical database, to form concepts. Nakata [70] has described a collaborative method to index important concepts described in a collection of documents. [81] proposed an approach that uses dependency-based semantic relationships between words. [86] used a knowledge-based concept extraction method relying on a parse graph and a common-sense knowledge base, coupled with a semantic similarity detection technique allowing additional matches to be found for specific concepts not present in the knowledge base.

Lexico-syntactic pattern matching is also a popular technique for concept extraction. [40] extracted hyponomy relations from text from Grolier's Encyclopedia by matching four given lexico-syntactic patterns. Her theory explored a new direction in the field of concept mining. She claimed that existing hyponomy relations can be used to extract new lexical syntactic patterns. [58] and [59] used the "is-a" pattern to extract Chinese hyponymy relations from unstructured Web corpora and obtained promising results.

5 Subjectivity Detection

Subjectivity detection aims to automatically categorize text into subjective or opinionated (i.e., positive or negative) versus objective or neutral and is hence useful to analysts in government, commercial and political domains who need to determine the response of the people to different events [98,9]. Linguistic pre-processing can be used to identify assertive sentences that are objectively presented and remove sentences that are mere speculations and, hence, lack sentiments [69]. This is particularly useful in multi-perspective question-answering summarization systems that need to summarize different opinions and perspectives and present multiple answers to the user based on opinions derived from different sources.

Previous methods used well established general subjectivity clues to generate training data from un-annotated text [91]. In addition, features such as pronouns, modals, adjectives, cardinal numbers, and adverbs have shown to be effective in subjectivity classification. Some existing resources contain lists of subjective words, and some empirical methods in NLP have automatically identified adjectives, verbs, and N-grams that are statistically associated with subjective language. However, several subjective words such as 'unseemingly' occur infrequently, consequently a large training dataset is necessary to build a broad and comprehensive subjectivity detection system.

While there are several datasets with document and chunk labels available, there is a need to better capture sentiment from short comments, such as Twitter data, which provide less overall signal per document. Hence, in [91], authors used extraction pattern learning to automatically generate linguistic structures that represent subjective expressions. For example, the pattern 'hijacking' of $< x >$, looks for the noun 'hijacking' and the object of the preposition $< x >$. Extracted features are used to train state-of-the-art classifiers such as SVM and Naïve Bayes that assume that the class of a particular feature is independent of the class of other features given the training data [108].

Sentence-level subjectivity detection was integrated into document-level sentiment detection using minimum cuts in graphs where each node is a sentence. The graph cuts try to minimize the classification error of a baseline classifier, e.g., Naïve Bayes, over sentences. The contextual constraints between sentences in a graph could lead to significant improvement in polarity classification [76]. On the other hand, bag-of-words classifiers represent a document as a multi set of its words disregarding grammar and word order. They can work well on long documents by relying on a few words with strong sentiments like 'awesome'. However, distributional similarities of words, such as co-occurrence matrix and context information, are unable to capture differences in antonyms. This is a problem typical of sentiment analysis, as semantic similarity and affective similarity are often different from each other, e.g., `happy` and `sad` are two similar concepts in a semantic sense (as they are both emotions) but they are very different in an affective sense as they bear opposite polarities.

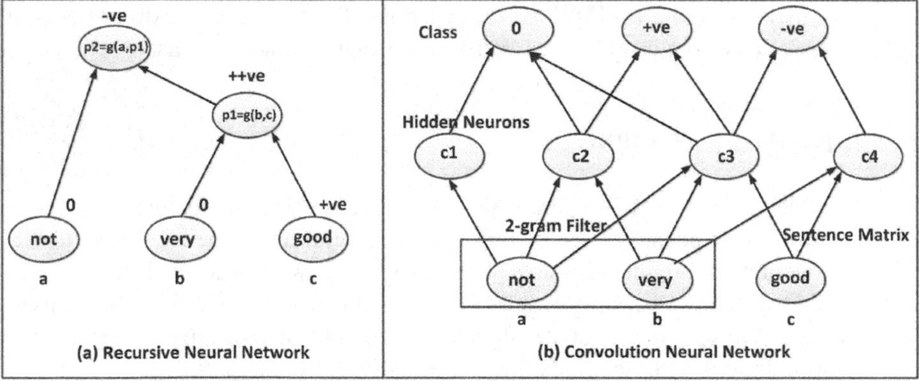

Fig. 2. Example of subjectivity detection using (a) Recursive NN (b) Convolution NN

Several works have explored sentiment compositionality through careful engineering of features or polarity shifting rules on syntactic structures. However, sentiment accuracies for binary positive/negative classification for single sentences has not exceeded 80% for several years. When including a third 'neutral' class, the accuracy falls down to only 60%. It can be seen that many short n-grams are neutral while longer phrases are well distributed among positive and negative subjective sentence classes. Therefore, matrix representations for long phrases and matrix multiplication to model composition are being used to evaluate sentiment.

In such models, sentence composition is modeled using deep neural networks such as recursive auto-associated memories [50,36]. Recursive neural networks (RNN) predict the sentiment class at each node in the parse tree and try to capture the negation and its scope in the entire sentence. In the standard recursive neural network, each word is represented as a vector and it is first determined which parent already has all its children computed. Next, the parent is computed via a composition function over child nodes. In the RNN matrix, the composition function for long phrases depends on the words being combined and hence is linguistically motivated.

However, the number of possible composition functions is exponential. Hence, [93] introduced a recursive neural tensor network that uses a single tensor composition function to define multiple bilinear dependencies between words. Fig. 2 (a) illustrates the state space of a recursive neural network. Parent feature vectors are computed in a bottom-up manner by combining child nodes using composition function g. The polarity at each node is determined using its feature vector and a baseline classifier. Fig. 2 (b) illustrates the state space of a convolution neural network, where the input are feature vectors of words and the hidden neurons use convolution filters to detect patterns. Each neuron in the output layer corresponds to a single polarity type.

6 Anaphora Resolution

Anaphora can be defined as the presupposition that points back to some previous item. The pointing back reference is called anaphor and the entity to which it refers is its antecedent. The process of determining the antecedent of an anaphor is called anaphora resolution. Anaphora resolution is an open NLP challenge that needs to be tackled in many domains, including machine translation, summarization, question-answering systems and sentiment analysis. In machine translation, for example, anaphora must be resolved to disambiguate pronouns and develop a reliable machine translation algorithm. Current machine translation systems usually do not go beyond sentence level, thus failing a complete discourse understanding. Automatic text summarization systems, instead, need anaphora resolution for the selection of meaningful sentences in text. Because such systems often select salient sentences based on the words or concepts these contain, in fact, they may miss key sentences that use anaphoric expressions to provide important information about previously mentioned topics.

There are various types of anaphora. The most widespread ones are: pronominal anaphora, which is realized by anaphoric pronouns; adjectival anaphora, realized by anaphoric possessive adjectives; and one-anaphora, the anaphoric expression is realized by a "one" noun phrase (Fig. 3).

Pronominal anaphora: Lucy went to cinema. She was happy.
Adjectival anaphora: Lucy went to cinema. Her boyfriend bought the tickets.
One-anaphora: I broke my phone. I have to buy a new one.

Fig. 3. Different types of anaphora

When resolving anaphora, some constraints must be respected:

- Number agreement: it is necessary to distinguish between singular and plural references.
- Gender agreement: it is necessary to distinguish between male, female, and neutral genders.
- Semantic consistency: it is assumed that both the antecedent clause and the one containing the anaphora are semantically consistent.

Grammatical, syntactic or pragmatic rules have been widely used in the literature to identify the antecedent of an anaphor. Hobbs' algorithm [41] searches parse trees (i.e., basic syntactic trees) for antecedents of a pronoun. After searching the trees, it checks the number and gender agreement between a specified pronoun and its antecedent candidates. The method proposed by Lappin and Leass [53], termed Resolution of Anaphora Procedure (RAP), is a discourse model in which potential referents have degrees of salience. In particular, authors try to solve pronoun references by finding highly salient referents compatible with pronoun agreement features. [51] proposed a modified version, which is based on part-of-speech tagging with a shallow syntactic parse indicating grammatical rules. The centering theory [11] is based on the presence and the consequent searching of a focus, or center, of the discourse and on the assumption that subsequent pronouns have the strong tendency to refer to it. In [20], a distance metric function is introduced to calculate the similarity between two noun phrases.

The Anaphora Matcher proposed by [30] embeds semantic knowledge in anaphora resolution, by means of the lexical database WordNet, used to acquire semantic information about words in sentences. However, the algorithm still focuses on words and not on concepts, thus losing the possibility to connect a pronoun to a general, multi-word concept (e.g., 'Lucy went to cinema. It was amazing', the concept `go to the cinema` cannot be related to the pronoun using only words). An alternative to the syntactical constraints is represented by the statistical approach introduced by [26]. In order to match the anaphor, the model uses the statistical information represented by the frequencies of patterns obtained from a selected corpus to find the antecedent candidate with the highest frequency. CogNIAC (COGnition eNIAC) [3] solves the association of pronouns with limited knowledge and linguistic resources. It achieves high precision for some pronouns.

In [63], the input is checked against agreement and for a number of so-called antecedent indicators. Candidates are assigned scores by each indicator and the candidate with the highest score is returned as the antecedent. [4] offers an evaluation environment for comparing anaphora resolution algorithms. [64] presents the system MARS, which operates in fully automatic mode and employs genetic algorithms to achieve optimal performance. In [55], the resolution of anaphora is achieved by employing both the WordNet ontology and heuristic rules. The percentage of correctly resolved anaphora reaches almost 80%.

7 Sarcasm Detection

Sarcasm is always directed at someone or something. A target of sarcasm is the person or object against whom or which the ironic utterance is directed. Targets can be the sender himself, the addressee or a third party (or a combination of the three). The presence of sarcastic sentences may completely change the meaning of the whole review, therefore misleading the interpretation of the review itself.

While the use of irony and sarcasm is well studied from its linguistic and psychologic aspects, sarcasm detection is still represented by very few works in the computational literature. [79] suggested a theoretical framework in which the context of sentiment words shifts the valence of the expressed sentiment. This is made on the assumption that, though most salient clues about attitude are provided by the lexical choice of the writer, the organization of the text also provides relevant information for assessing attitude. To this end, authors described how the base attitudinal valence of a lexical item is modified by lexical and discourse context and propose a simple implementation for some contextual shifters. In [100], a semi-supervised algorithm for sarcasm identification in product reviews is proposed. The authors proposed a set of pattern-based features to characterize sarcastic utterances, combined with some punctuation-based features. The experiments were performed on a dataset of about 66,000 Amazon reviews, and a precision of 77% and recall of 83.1% were obtained in the identification of sarcastic sentences. [29] extended this approach to a collection of 5.9 million tweets and 66,000 product reviews from Amazon, obtaining F-scores of 0.78 and 0.83, respectively.

Their algorithmic methodology is based on patterns. In order to extract such patterns automatically, they classified words into high-frequency words and content words. After filtering these patterns, other generic features are added: sentence length in words, number of "!" and "?" characters in the sentence, number of quotes in the sentence, and number of capitals words in the sentence. Then, they employed a k-nearest neighbors (kNN)-like strategy for the classification task. In [56], a classifier is trained for the detection of Dutch tweets, by exploiting the use of intensifiers such as hyperbolic words, which are able to strengthen the sarcastic utterance. [38] proposed a method for the identification of sarcasm in Twitter, where each message is codified based on lexical and pragmatic factors, the former including unigrams and dictionary-based factors, the latter combining positive and negative emoticons and tweet replies. The authors then employed and compared performances of SVM and logistic-regression machines used for the classification task.

8 Topic Spotting

Topic spotting or auto-categorization is about classifying or tagging a piece of text with one or more category labels. Unlike topic modeling, topic spotting does not focus on clustering the words of a large text corpus into set of topics but rather giving a context to the input text. In other words, topic spotting is more similar to short text conceptualization than topic modeling (which is inapplicable to short texts). Let us consider the multi-word statement `score grand slam`, taken as an unit, it is obviously related to tennis, however word-by-word passes on very surprising semantics. Correspondingly, in a bag-of-words model, the expression `get withdrawn` would not convey the meaning `withdraw from a bank`. Psychologist Gregory Murphy began his highly acclaimed book [68] with the statement "Concepts are the glue that holds our mental world together". Still, Nature magazine book review calls it an understatement because "Without concepts, there would be no mental world in the first place"[8].

Undoubtedly, the ability to conceptualize is a defining characteristic of humanity. We focus on conceptualizing from texts or words. For example, given the word "India," a person will form in his mind concept such as a country or region. Given two words, "India" and "China," the top concepts may shift to an Asian country or a developing country, etc. Given yet another word, "Brazil," the top concept may change to BRIC or an emerging market and so forth (Fig. 4). Besides generalizing from instances to concepts, human beings also form concepts from descriptions. For example, given the words 'body', 'smell' and 'color', the concept 'wine' comes to our mind. Certainly, instances and descriptions may get mixed up, for example, we conceptualize 'apple' and 'headquarters' to a company but 'apple', 'smell' and 'color' to a fruit.

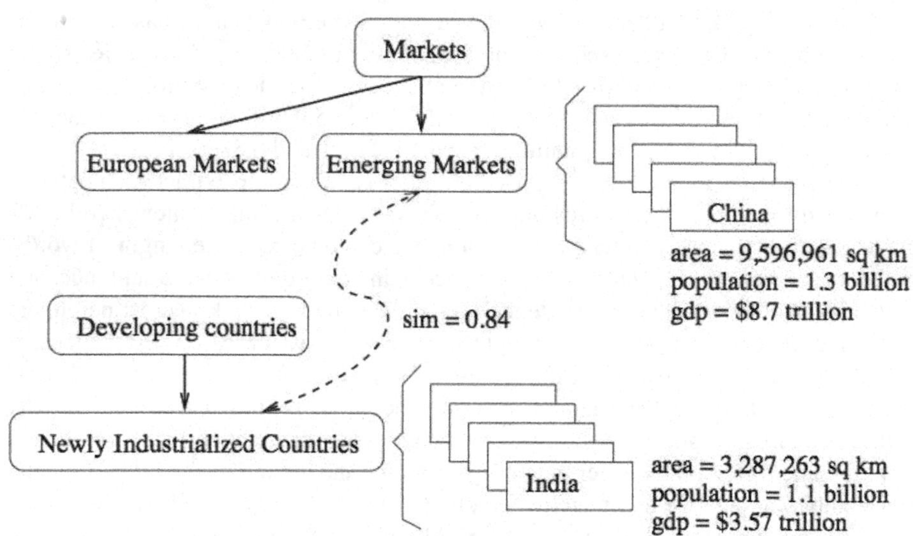

Fig. 4. Context-based short text conceptualization

The question is whether machines can do it. Much work has been devoted to the topic discovery from a text. The task of classifying the textual data that has been culled from sites on the World Wide Web is both difficult and intensively studied [25,48,71]. [96] proposed a bag-of-words model to classify tweets into set of generic classes. As classification classes they considered "News", "Events", "Opinions", "Deals", and "Private Messages". Though their method does not deal with topic spotting, the classification method is surely helpful to spot topics in tweets. An unsupervised method was proposed by [21] to leverage topics at multiple granularity. In [27], broad and generic Twitter categories based on the topics are described. All of these approaches, however, are only good at classifying tweets within some limited, generic and pre-defined topics.

In order to accurately detect topics from tweets, a concept-based approach is necessary. Recently, Wang et al. [105] presented a novel framework to classify any given short text to general categories. The method relies on a bag-of-concept model and a large taxonomy, it learns a concept model for each category, and conceptualizes short text to a set of relevant concepts. Similar approaches were proposed by [95], who used a large probabilistic knowledge base and Bayesian inference mechanism to conceptualize words in short text, and [18], who merged common and common-sense knowledge for topic spotting in the context of open-domain sentiment analysis.

9 Aspect Extraction

In opinion mining, different levels of analysis granularity have been proposed, each one having its advantages and drawbacks. Aspect-based opinion mining [43,31] focuses on the relations between aspects and document polarity. Aspects are opinion targets, i.e., the specific features of a product or service that users like or dislike. For example, the sentence "The screen of my phone is really nice and its resolution is superb" expresses a positive polarity about the phone under review. More specifically, the opinion holder is expressing a positive polarity about its *screen* and *resolution*; these concepts are thus called opinion targets, or aspects. It is important to identify aspects because reviewers may express opposite polarities about different aspects in the same sentence. Without such an identification, sentences like "I love the touchscreen of iPhone6 but the battery lasts too little" may be categorized as neutral because the average polarity is null when, in fact, the reviewer is very positive about one aspect but very negative about another.

The task of identifying aspects in a given opinion is called aspect extraction. Aspect extraction from opinions was first studied by Hu and Liu [43]. They introduced the distinction between explicit and implicit aspects, but only dealt with the former. They used a set of rules based on statistical observations. Hu and Liu's method was later improved by Popescu and Etzioni [80] and by Blair-Goldensonh [6]. [80] assumes the product class is known in advance. Their algorithm detects whether a noun or noun phrase is a product feature by computing PMI between the noun phrase and the product class. Poria et al. [83] proposed a set of linguistic rules to extract aspect terms, e.g., `speaker` and employed a knowledge base technique to extract the aspect category, e.g., `sound`. Scaffidi et al. [92] presented a method that uses language model to identify product features, under the assumption that product features are more frequent in product reviews than in general natural language text.

Topic modeling has been widely used as a basis to perform extraction and grouping of aspects [44,23]. In the literature, two models have been considered: pLSA [42] and LDA [7]. Both models introduce a latent variable 'topic' between the observed variables 'document' and 'word' to analyze the semantic topic distribution of documents. In topic models, each document is represented as a random mixture over latent topics, where each topic is characterized by a distribution over words. The LDA model defines a Dirichlet probabilistic generative process for document-topic distribution; in each document, a latent aspect is chosen according to a multinomial distribution, controlled by a Dirichlet prior α. Then, given an aspect, a word is extracted according to another multinomial distribution, controlled by another Dirichlet prior β.

Some existing works employing these models include the extraction of global aspects (such as the brand of a product) and local aspects (such as the property of a product) [99], the extraction of key phrases [10], the rating of multi-aspects [106] and the summarization of aspects and sentiments [61]. [113] employed Maximum-Entropy to train a switch variable based on POS tags of words and use it to separate aspect and sentiment words. [62] added user feedback into LDA as a response variable connected to each document. In [60], a semi-supervised model was proposed. DF-LDA [2] also represents a semi-supervised model, which allows the user to set must-link and cannot-link constraints. A must-link means that two terms must be in the same topic, while a cannot-link means that two terms cannot be in the same topic.

[107] proposed two semi-supervised models for product aspect extraction, based on the use of seeding aspects. Within the category of supervised methods, [45] employed seed words to guide topic models to learn topics of specific interest to a user, while [106] and [67] employed seeding words to extract related product aspects from product reviews.

10 Polarity Detection

Polarity detection is the most popular sentiment analysis task. In fact, many research works even use the terms 'polarity detection' and 'sentiment analysis' interchangeably. This is due to the definition of sentiment analysis as the NLP task that aims to classify a piece of text as positive or negative. As discussed before, however, there are several other tasks that need to be taken into account in order to correctly infer the polarity associated with one or more opinion targets in informal short text. Existing approaches to polarity detection can be grouped into four main categories: keyword spotting, lexical affinity, statistical methods, and concept-level approaches.

Keyword spotting is the most naïve approach and probably also the most popular because of its accessibility and economy. Polarity is inferred after classifying text into affect categories based on the presence of fairly unambiguous affect words like 'happy', 'sad', 'afraid', and 'bored'. Elliott's Affective Reasoner [33], for example, watches for 198 affect keywords, e.g., 'distressed' and 'enraged', plus affect intensity modifiers, e.g., 'extremely', 'somewhat', and 'mildly'. Other popular sources of affect words are Ortony's Affective Lexicon [73], which groups terms into affective categories, and Wiebe's linguistic annotation scheme [109].

Lexical affinity is slightly more sophisticated than keyword spotting as, rather than simply detecting obvious affect words, it assigns arbitrary words a probabilistic 'affinity' for a particular emotion. For example, 'accident' might be assigned a 75% probability of being indicating a negative affect, as in 'car accident' or 'hurt by accident'. These probabilities are usually trained from linguistic corpora [110,97,94,88].

Statistical methods, such as latent semantic analysis (LSA) and SVM, have been popular for polarity detection from text and have been used by researchers on projects such as Goertzel's Webmind [37], Pang's movie review classifier [78], and many others [77,102,104]. By feeding a machine learning algorithm a large training corpus of affectively annotated texts, it is possible for the systems to not only learn the affective valence of affect keywords, but such a system can also take into account the valence of other arbitrary keywords (like lexical affinity), punctuation, and word co-occurrence frequencies. However, statistical methods are generally semantically weak, meaning that, with the exception of obvious affect keywords, other lexical or co-occurrence elements in a statistical model have little predictive value individually. As a result, statistical classifiers only work with acceptable accuracy when given a sufficiently large text input. So, while these methods may be able to detect polarity on the page or paragraph level, they do not work well on smaller text units such as sentences.

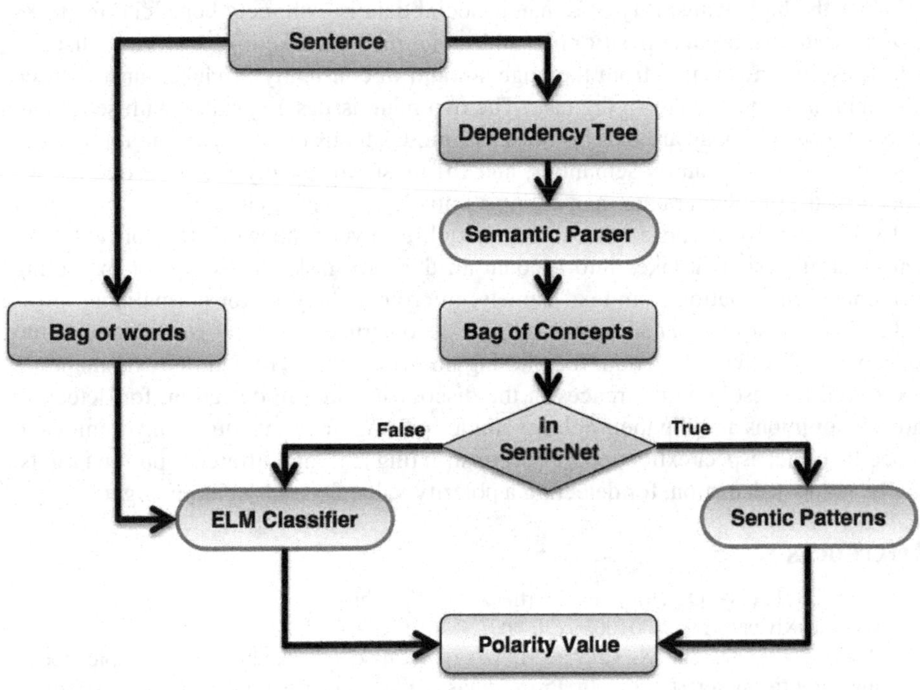

Fig. 5. Sentic patterns

Concept-based approaches focus on a semantic analysis of text through the use of web ontologies [34] or semantic networks [15], which allow grasping the conceptual and affective information associated with natural language opinions. By relying on large semantic knowledge bases, such approaches step away from the blind use of keywords and word co-occurrence count, but rather rely on the implicit meaning/features associated with natural language concepts. Unlike purely syntactical techniques, concept-based approaches are able to detect also sentiments that are expressed in a subtle manner, e.g., through the analysis of concepts that do not explicitly convey any emotion, but which are implicitly linked to other concepts that do so.

Besides these four categories, there are hybrid frameworks for polarity detection that propose an ensemble of two (or more) of the above-mentioned approaches, e.g., sentic patterns [84], which employ both statistical methods and concept-based techniques to infer the polarity of short texts (Fig. 5).

11 Conclusion

Sentiment analysis is a research field germane to NLP that has recently raised growing interest both within the scientific community, leading to many exciting open challenges, as well as in the business world, due to the remarkable benefits to be had from financial market prediction.

While the high availability of dynamic social data is extremely beneficial for tasks such as branding, product positioning, and social media marketing, the timely distillation of useful information from the huge amount of constantly produced unstructured information is a very challenging task. The two main issues associated with sentiment analysis research today are that (a) most techniques focus on the syntactic representation of text, rather than on semantics, and (b) most works only focus on one or two aspects of the problem, rather than taking a holistic approach to it.

To this end, we proposed the CLSA model, a novel framework for concept-level sentiment analysis that takes into account all the NLP tasks necessary for extracting opinionated information from text, namely: microtext analysis, for normalizing informal and irregular text; semantic parsing, for deconstructing natural language text into concepts; subjectivity detection, for filtering non-opinionated or neutral text; anaphora resolution, for resolving references in the discourse; sarcasm detection, for detecting sarcastic opinions and flip their polarity; topic spotting, for contextualizing opinions to a specific topic; aspect extraction, for deconstructing text into different opinion targets; finally, polarity detection, for detecting a polarity value for each opinion target.

References

1. Agirre, E., Ansa, O., Hovy, E., Martínez, D.: Enriching very large ontologies using the www. arXiv preprint cs/0010026 (2000)
2. Andrzejewski, D., Zhu, X., Craven, M.: Incorporating domain knowledge into topic modeling via dirichlet forest priors. In: Proceedings of the 26th Annual International Conference on Machine Learning, pp. 25–32. ACM (2009)
3. Baldwin, B.: Cogniac: high precision coreference with limited knowledge and linguistic resources. In: Proceedings of a Workshop on Operational Factors in Practical, Robust Anaphora Resolution for Unrestricted Texts, pp. 38–45. Association for Computational Linguistics (1997)

4. Barbu, C., Mitkov, R.: Evaluation tool for rule-based anaphora resolution methods. In: Proceedings of the 39th Annual Meeting on Association for Computational Linguistics, pp. 34–41. Association for Computational Linguistics (2001)
5. Bell, D., Koulouri, T., Lauria, S., Macredie, R., Sutton, J.: Microblogging as a mechanism for human-robot interaction. Knowledge-Based Systems 69, 64–77 (2014)
6. Blair-Goldensohn, S., Hannan, K., McDonald, R., Neylon, T., Reis, G.A., Reynar, J.: Building a sentiment summarizer for local service reviews. In: Proceedings of WWW 2008 Workshop on NLP in the Information Explosion Era (2008)
7. Blei, D.M., Ng, A.Y., Jordan, M.I.: Latent dirichlet allocation. The Journal of Machine Learning Research 3, 993–1022 (2003)
8. Bloom, P.: Glue for the mental world. Nature 421, 212–213 (2003)
9. Bonzanini, M., Martinez-Alvarez, M., Roelleke, T.: Opinion summarisation through sentence extraction: An investigation with movie reviews. In: Proceedings of the 35th International ACM SIGIR Conference on Research and Development in Information Retrieval SIGIR 2012, pp. 1121–1122 (2012)
10. Branavan, S.R.K., Chen, H., Eisenstein, J., Barzilay, R.: Learning document-level semantic properties from free-text annotations. Journal of Artificial Intelligence Research 34(2), 569 (2009)
11. Brennan, S.E., Friedman, M.W., Pollard, C.J.: A centering approach to pronouns. In: Proceedings of the 25th Annual Meeting on Association for Computational Linguistics, pp. 155–162. Association for Computational Linguistics (1987)
12. Cambria, E.: An introduction to concept-level sentiment analysis. In: Castro, F., Gelbukh, A., González, M. (eds.) MICAI 2013, Part II. LNCS, vol. 8266, pp. 478–483. Springer, Heidelberg (2013)
13. Cambria, E., Fu, J., Bisio, F., Poria, S.: AffectiveSpace 2: Enabling affective intuition for concept-level sentiment analysis. In: AAAI, Austin, pp. 508–514 (2015)
14. Cambria, E., Gastaldo, P., Bisio, F., Zunino, R.: An ELM-based model for affective analogical reasoning. Neurocomputing 149, 443–455 (2015)
15. Cambria, E., Hussain, A.: Sentic Computing: A Common-Sense-Based Framework for Concept-Level Sentiment Analysis. Springer, Cham (2015)
16. Cambria, E., Hussain, A., Havasi, C., Eckl, C., Munro, J.: Towards crowd validation of theUK national health service. In: WebSci, Raleigh (2010)
17. Cambria, E., Schuller, B., Liu, B., Wang, H., Havasi, C.: Knowledge-based approaches to concept-level sentiment analysis. IEEE Intelligent Systems 28(2), 12–14 (2013)
18. Cambria, E., Song, Y., Wang, H., Howard, N.: Semantic multi-dimensional scaling for open-domain sentiment analysis. IEEE Intelligent Systems 29(2), 44–51 (2014)
19. Cao, C., Feng, Q., Gao, Y., Gu, F., Si, J., Sui, Y., Tian, W., Wang, H., Wang, L., Zeng, Q., et al.: Progress in the development of national knowledge infrastructure. Journal of Computer Science and Technology 17(5), 523–534 (2002)
20. Cardie, C., Wagstaff, K., et al.: Noun phrase coreference as clustering. In: Proceedings of the 1999 Joint SIGDAT Conference on Empirical Methods in Natural Language Processing and Very Large Corpora, pp. 82–89 (1999)
21. Chen, M., Jin, X., Shen, D.: Short text classification improved by learning multi-granularity topics. In: IJCAI, pp. 1776–1781. Citeseer (2011)
22. Chen, W.L., Zhu, J.B., Yao, T.S., Zhang, Y.X.: Automatic learning field words by bootstrapping. In: Proc. of the JSCL, vol. 72, Tsinghua University Press, Beijing (2003)
23. Chen, Z., Liu, B.: Mining topics in documents: standing on the shoulders of big data. In: Proceedings of the 20th ACM SIGKDD International Conference on Knowledge Discovery and Data Mining, pp. 1116–1125. ACM (2014)

24. Chikersal, P., Poria, S., Cambria, E.: SeNTU: Sentiment analysis of tweets by combining a rule-based classifier with supervised learning. In: Proceedings of the International Workshop on Semantic Evaluation (SemEval 2015) (2015)
25. Cohen, W.W., Hirsh, H.: Joins that generalize: Text classification using whirl. In: KDD, pp. 169–173 (1998)
26. Dagan, I., Itai, A.: Automatic processing of large corpora for the resolution of anaphora references. In: Proceedings of the 13th Conference on Computational Linguistics, vol. 3, pp. 330–332. Association for Computational Linguistics (1990)
27. Dann, S.: Twitter content classification. First Monday 15(12) (2010)
28. Davidov, D., Tsur, O., Rappoport, A.: Enhanced sentiment learning using twitter hashtags and smileys. In: Proceedings of the 23rd International Conference on Computational Linguistics: Posters, pp. 241–249. Association for Computational Linguistics (2010)
29. Davidov, D., Tsur, O., Rappoport, A.: Semi-supervised recognition of sarcastic sentences in twitter and amazon. In: Proceedings of the Fourteenth Conference on Computational Natural Language Learning, pp. 107–116. Association for Computational Linguistics (2010)
30. Denber, M.: Automatic resolution of anaphora in english. Eastman Kodak Co. (1998)
31. Ding, X., Liu, B., Yu, P.S.: A holistic lexicon-based approach to opinion mining. In: Proceedings of First ACM International Conference on Web Search and Data Mining (WSDM 2008), pp. 231–240. Stanford University, Stanford(2008)
32. Du, B., Tian, H., Wang, L., Lu, R.: Design of domain-specific term extractor based on multi-strategy. Computer Engineering 31(14), 159–160 (2005)
33. Elliott, C.D.: The Affective Reasoner: A Process Model of Emotions in a Multi-Agent System. PhD thesis, Northwestern University, Evanston (1992)
34. Gangemi, A., Presutti, V., Reforgiato, D.: Frame-based detection of opinion holders and topics: a model and a tool. IEEE Computational Intelligence Magazine 9(1), 20–30 (2014)
35. Gelfand, B., Wulfekuler, M., Punch, W.F.: Automated concept extraction from plain text. In: AAAI 1998 Workshop on Text Categorization, pp. 13–17 (1998)
36. Glorot, X., Bordes, A., Bengio, Y.: Domain adaptation for large-scale sentiment classification: A deep learning approach. In: Proceedings of the Twenty-eight International Conference on Machine Learning, ICML (2011)
37. Goertzel, B., Silverman, K., Hartley, C., Bugaj, S., Ross, M.: The Baby Webmind project. In: AISB, Birmingham (2000)
38. González-Ibáñez, R., Muresan, S., Wacholder, N.: Identifying sarcasm in twitter: a closer look. In: Proceedings of the 49th Annual Meeting of the Association for Computational Linguistics: Human Language Technologies: Short Papers, vol. 2, pp. 581–586. Association for Computational Linguistics (2011)
39. Harlambous, Y., Klyuev, V.: Thematically reinforced explicit semantic analysis. International Journal of Computational Linguistics and Applications 4(1), 79–94 (2013)
40. Hearst, M.A.: Automatic acquisition of hyponyms from large text corpora. In: Proceedings of the 14th Conference on Computational Linguistics, vol. 2, pp. 539–545. Association for Computational Linguistics (1992)
41. Hobbs, J.R.: Resolving pronoun references. Lingua 44(4), 311–338 (1978)
42. Hofmann, T.: Probabilistic latent semantic indexing. In: Proceedings of the 22nd Annual International ACM SIGIR Conference on Research and Development in Information Retrieval, pp. 50–57. ACM (1999)
43. Hu, M., Liu, B.: Mining and summarizing customer reviews. In: Proceedings of the ACM SIGKDD International Conference on Knowledge Discovery & Data Mining, pp. 168–177 (2004)
44. Hu, Y., Boyd-Graber, J., Satinoff, B., Smith, A.: Interactive topic modeling. Machine Learning 95(3), 423–469 (2014)

45. Jagarlamudi, J., Daumé III, H., Udupa, R.: Incorporating lexical priors into topic models. In: Proceedings of the 13th Conference of the European Chapter of the Association for Computational Linguistics, pp. 204–213. Association for Computational Linguistics (2012)
46. Jansen, B.J., Zhang, M., Sobel, K., Chowdury, A.: Twitter power: Tweets as electronic word of mouth. Journal of the American Society for Information Science and Technology 60(11), 2169–2188 (2009)
47. Jiang, L., Yu, M., Zhou, M., Liu, X., Zhao, T.: Target-dependent twitter sentiment classification. In: Proceedings of the 49th Annual Meeting of the Association for Computational Linguistics: Human Language Technologies, vol. 1, pp. 151–160. Association for Computational Linguistics (2011)
48. Joachims, T.: Text categorization with support vector machines: Learning with many relevant features. In: Nédellec, C., Rouveirol, C. (eds.) ECML 1998. LNCS, vol. 1398, Springer, Heidelberg (1998)
49. Jung, S., Segev, A.: Analyzing future communities in growing citation networks. Knowledge-Based Systems 69, 34–44 (2014)
50. Kalchbrenner, N., Grefenstette, E., Blunsom, P.: A convolutional neural network for modelling sentences. CoRR, abs/1404.2188 (2014)
51. Kennedy, C., Boguraev, B.: Anaphora for everyone: pronominal anaphora resoluation without a parser. In: Proceedings of the 16th Conference on Computational Linguistics, vol. 1, pp. 113–118. Association for Computational Linguistics (1996)
52. Kouloumpis, E., Wilson, T., Moore, J.: Twitter sentiment analysis: The good the bad and the omg! In: ICWSM, vol. 11, pp. 538–541 (2011)
53. Lappin, S., Leass, H.J.: An algorithm for pronominal anaphora resolution. Computational linguistics 20(4), 535–561 (1994)
54. Li, X., Xie, H., Chen, L., Wang, J., Deng, X.: News impact on stock price return via sentiment analysis. Knowledge-Based Systems 69, 14–23 (2014)
55. Liang, T., Wu, D.-S.: Automatic pronominal anaphora resolution in english texts. In: ROCLING (2003)
56. Liebrecht, C.C., Kunneman, F.A., van den Bosch, A.P.J.: The perfect solution for detecting sarcasm in tweets# not. In: ACL (2013)
57. Liu, F., Weng, F., Wang, B., Liu, Y.: Insertion, deletion, or substitution?: normalizing text messages without pre-categorization nor supervision. In: Proceedings of the 49th Annual Meeting of the Association for Computational Linguistics: Human Language Technologies: Short Papers, vol. 2, pp. 71–76. Association for Computational Linguistics (2011)
58. Liu, L., Cao, C., Wang, H.: Acquiring hyponymy relations from large chinese corpus. WSEAS Transactions on Business and Economics 2(4), 211 (2005)
59. Liu, L., Cao, C.-G., Wang, H.-T., Chen, W.: A method of hyponym acquisition based on "isa" pattern. Journal of Computer Science, 146–151 (2006)
60. Lu, Y., Zhai, C.: Opinion integration through semi-supervised topic modeling. In: Proceedings of the 17th International Conference on World Wide Web, pp. 121–130. ACM (2008)
61. Lu, Y., Zhai, C.X., Sundaresan, N.: Rated aspect summarization of short comments. In: Proceedings of the 18th International Conference on World Wide Web, pp. 131–140. ACM (2009)
62. Mcauliffe, J.D., Blei, D.M.: Supervised topic models. In: Advances in Neural Information Processing Systems, pp. 121–128 (2008)
63. Mitkov, R.: Robust pronoun resolution with limited knowledge. In: Proceedings of the 36th Annual Meeting of the Association for Computational Linguistics and 17th International Conference on Computational Linguistics, vol. 2, pp. 869–875. Association for Computational Linguistics (1998)

64. Mitkov, R., Evans, R., Orăsan, C.: A new, fully automatic version of mitkov's knowledge-poor pronoun resolution method. In: Gelbukh, A. (ed.) CICLing 2002. LNCS, vol. 2276, pp. 168–186. Springer, Heidelberg (2002)
65. Mohammad, S.M., Kiritchenko, S., Zhu, X.: NRC-Canada: Building the state-of-the-art in sentiment analysis of tweets. In: Proceedings of the Second Joint Conference on Lexical and Computational Semantics (SEMSTAR 2013) (2013)
66. Montejo-Raez, A., Diaz-Galiano, M., Martinez-Santiago, F., Urena-Lopez, A.: Crowd explicit sentiment analysis. Knowledge-Based Systems 69, 134–139 (2014)
67. Mukherjee, A., Liu, B.: Aspect extraction through semi-supervised modeling. In: Proceedings of the 50th Annual Meeting of the Association for Computational Linguistics: Long Papers, vol. 1, pp. 339–348. Association for Computational Linguistics (2012)
68. Murphyp, G.L.: The big book of concepts. MIT Press (2002)
69. Murray, G., Carenini, G.: Subjectivity detection in spoken and written conversations. Natural Language Engineering 17, 397–418 (2011)
70. Nakata, K., Voss, A., Juhnke, M., Kreifelts, T.: Collaborative concept extraction from documents. In: Proceedings of the 2nd Int. Conf. on Practical Aspects of Knowledge management (PAKM 1998). Citeseer (1998)
71. Nigam, K., McCallum, A.K., Thrun, S., Mitchell, T.: Text classification from labeled and unlabeled documents using em. Machine Learning 39(2-3), 103–134 (2000)
72. O'Connor, B., Balasubramanyan, R., Routledge, B.R., Smith, N.A.: From tweets to polls: Linking text sentiment to public opinion time series. In: ICWSM, vol. 11, pp. 122–129 (2010)
73. Ortony, A., Clore, G., Collins, A.: The Cognitive Structure of Emotions. Cambridge University Press, Cambridge (1988)
74. Pak, A., Paroubek, P.: Twitter as a corpus for sentiment analysis and opinion mining. In: LREC (2010)
75. Paltoglou, G., Thelwall, M.: Twitter, myspace, digg: Unsupervised sentiment analysis in social media. ACM Transactions on Intelligent Systems and Technology (TIST) 3(4), 66 (2012)
76. Pang, B., Lee, L.: A sentimental education: Sentiment analysis using subjectivity summarization based on minimum cuts. In: Proceedings of the 42nd Annual Meeting of the Association for Computational Linguistics (ACL 2004) (2004)
77. Pang, B., Lee, L.: Seeing stars: Exploiting class relationships for sentiment categorization with respect to rating scales. In: ACL, pp. 115–124. Ann Arbor (2005)
78. Pang, B., Lee, L., Vaithyanathan, S.: Thumbs up? Sentiment classification using machine learning techniques. In: EMNLP, Philadelphia, pp. 79–86 (2002)
79. Polanyi, L., Zaenen, A.: Contextual valence shifters. In: Computing Attitude and Affect in text: Theory and Applications, pp. 1–10. Springer (2006)
80. Popescu, A.-M., Etzioni, O.: Extracting product features and opinions from reviews. In: Proceedings of Conference on Empirical Methods in Natural Language Processing (EMNLP 2005), pp. 3–28 (2005)
81. Poria, S., Agarwal, B., Gelbukh, A., Hussain, A., Howard, N.: Dependency-based semantic parsing for concept-level text analysis. In: Gelbukh, A. (ed.) CICLing 2014, Part I. LNCS, vol. 8403, pp. 113–127. Springer, Heidelberg (2014)
82. Poria, S., Cambria, E., Hussain, A., Huang, G.-B.: Towards an intelligent framework for multimodal affective data analysis. Neural Networks 63, 104–116 (2015)
83. Poria, S., Cambria, E., Ku, L.-W., Gui, C., Gelbukh, A.: A rule-based approach to aspect extraction from product reviews. In: COLING, Dublin, pp. 28–37 (2014)
84. Poria, S., Cambria, E., Winterstein, G., Huang, G.-B.: Sentic patterns: Dependency-based rules for concept-level sentiment analysis. Knowledge-Based Systems 69, 45–63 (2014)

85. Poria, S., Gelbukh, A., Cambria, E., Hussain, A., Huang, G.-B.: EmoSenticSpace: A novel framework for affective common-sense reasoning. Knowledge-Based Systems 69, 108–123 (2014)
86. Rajagopal, D., Cambria, E., Olsher, D., Kwok, K.: A graph-based approach to commonsense concept extraction and semantic similarity detection. In: WWW, pp. 565–570. Rio De Janeiro (2013)
87. Ramirez, P.M., Mattmann, C.A.: Ace: improving search engines via automatic concept extraction. In: Proceedings of the 2004 IEEE International Conference on Information Reuse and Integration, IRI 2004, pp. 229–234. IEEE (2004)
88. Rao, D., Ravichandran, D.: Semi-supervised polarity lexicon induction. In: EACL, Athens, pp. 675–682 (2009)
89. Read, J.: Using emoticons to reduce dependency in machine learning techniques for sentiment classification. In: Proceedings of the ACL Student Research Workshop, pp. 43–48. Association for Computational Linguistics (2005)
90. Rill, S., Reinel, D., Scheidt, J., Zicari, R.: Politwi: Early detection of emerging political topics on twitter and the impact on concept-level sentiment analysis. Knowledge-Based Systems 69, 14–23 (2014)
91. Riloff, E., Wiebe, J.: Learning extraction patterns for subjective expressions. In: Proceedings of the 2003 Conference on Empirical Methods in Natural Language Processing, pp. 105–112 (2003)
92. Scaffidi, C., Bierhoff, K., Chang, E., Felker, M., Ng, H., Jin, C.: Red opal: product-feature scoring from reviews. In: Proceedings of the 8th ACM Conference on Electronic Commerce, pp. 182–191. ACM (2007)
93. Socher, R., Perelygin, A., Wu, J.Y., Chuang, J., Manning, C.D., Ng, A.Y., Potts, C.: Recursive deep models for semantic compositionality over a sentiment treebank (2013)
94. Somasundaran, S., Wiebe, J., Ruppenhofer, J.: Discourse level opinion interpretation. In: COLING, Manchester, pp. 801–808 (2008)
95. Song, Y., Wang, H., Wang, Z., Li, H., Chen, W.: Short text conceptualization using a probabilistic knowledgebase. In: IJCAI, Barcelona (2011)
96. Sriram, B., Fuhry, D., Demir, E., Ferhatosmanoglu, H., Demirbas, M.: Short text classification in twitter to improve information filtering. In: Proceedings of the 33rd International ACM SIGIR Conference on Research and Development in Information Retrieval, pp. 841–842. ACM (2010)
97. Stevenson, R., Mikels, J., James, T.: Characterization of the affective norms for english words by discrete emotional categories. Behavior Research Methods 39, 1020–1024 (2007)
98. Tang, D., Wei, F., Yang, N., Zhou, M., Liu, T., Qin, B.: Learning sentiment-specificword embedding for twitter sentiment classification (2014)
99. Titov, I., McDonald, R.: Modeling online reviews with multi-grain topic models. In: Proceedings of the 17th International Conference on World Wide Web, pp. 111–120. ACM (2008)
100. Tsur, O., Davidov, D., Rappoport, A.: Icwsm-a great catchy name: Semi-supervised recognition of sarcastic sentences in online product reviews. In: ICWSM (2010)
101. Turney, P.: Thumbs up or thumbs down? Semantic orientation applied to unsupervised classification of reviews. In: ACL, Philadelphia, pp. 417–424 (2002)
102. Turney, P., Littman, M.: Measuring praise and criticism: Inference of semantic orientation from association. ACM Transactions on Information Systems 21(4), 315–346 (2003)
103. Velardi, P., Fabriani, P., Missikoff, M.: Using text processing techniques to automatically enrich a domain ontology. In: Proceedings of the International Conference on Formal Ontology in Information Systems, vol. 2001, pp. 270–284. ACM (2001)
104. Velikovich, L., Goldensohn, S., Hannan, K., McDonald, R.: The viability of web-derived polarity lexicons. In: NAACL, Los Angeles, pp. 777–785 (2010)

105. Wang, F., Wang, Z., Li, Z., Wen, J.-R.: Concept-based short text classification and ranking. In: Proceedings of the 23rd ACM International Conference on Conference on Information and Knowledge Management, pp. 1069–1078. ACM (2014)
106. Wang, H., Lu, Y., Zhai, C.: Latent aspect rating analysis on review text data: a rating regression approach. In: Proceedings of the 16th ACM SIGKDD International Conference on Knowledge Discovery and Data Mining, pp. 783–792. ACM (2010)
107. Wang, T., Cai, Y., Leung, H.-F., Lau, R.Y.K., Li, Q.: Huaqing Min. Product aspect extraction supervised with online domain knowledge. Knowledge-Based Systems 71, 86–100 (2014)
108. Wiebe, J., Riloff, E.: Creating subjective and objective sentence classifiers from unannotated texts. In: Proceedings of the 6th International Conference on Computational Linguistics and Intelligent Text Processing, pp. 486–497 (2005)
109. Wiebe, J., Wilson, T., Cardie, C.: Annotating expressions of opinions and emotions in language. Language Resources and Evaluation 39(2), 165–210 (2005)
110. Wilson, T., Wiebe, J., Hoffmann, P.: Recognizing contextual polarity in phrase-level sentiment analysis. In: HLT/EMNLP, Vancouver (2005)
111. Yang, C., Lin, K.H., Chen, H.-H.: Emotion classification using web blog corpora. In: IEEE/WIC/ACM International Conference on Web Intelligence, pp. 275–278. IEEE (2007)
112. Yuntao, Z., Ling, G., Yongcheng, W., Zhonghang, Y.: An effective concept extraction method for improving text classification performance. Geo-Spatial Information Science 6(4), 66–72 (2003)
113. Zhao, W.X., Jiang, J., Yan, H., Li, X.: Jointly modeling aspects and opinions with a maxent-lda hybrid. In: EMNLP, pp. 56–65. Association for Computational Linguistics (2010)
114. Zheng, J.H., Lu, J.L.: Study of an improved keywords distillation method. Computer Engineering 31(18), 194–196 (2005)

Building Large Arabic Multi-domain Resources for Sentiment Analysis

Hady ElSahar and Samhaa R. El-Beltagy

Center of Informatics Sciences, Nile University, Cairo, Egypt
hadyelsahar@gmail.com, samhaa@computer.org

Abstract. While there has been a recent progress in the area of Arabic Sentiment Analysis, most of the resources in this area are either of limited size, domain specific or not publicly available. In this paper, we address this problem by generating large multi-domain datasets for Sentiment Analysis in Arabic. The datasets were scrapped from different reviewing websites and consist of a total of 33K annotated reviews for movies, hotels, restaurants and products. Moreover we build multi-domain lexicons from the generated datasets. Different experiments have been carried out to validate the usefulness of the datasets and the generated lexicons for the task of sentiment classification. From the experimental results, we highlight some useful insights addressing: the best performing classifiers and feature representation methods, the effect of introducing lexicon based features and factors affecting the accuracy of sentiment classification in general. All the datasets, experiments code and results have been made publicly available for scientific purposes.

1 Introduction

In the past few years, Sentiment Analysis has been the focus of many research studies due to the wide variety of its potential applications. Many of these studies have relied heavily on available resources mostly in the form of polarity annotated datasets [15–17, 20] or sentiment lexicons such as SentiWordNet [5].

At the same time, the Arabic language has shown rapid growth in terms of its users on the internet, moving up to the 4th place in the world ranking of languages by users according to internetworldstats[1]. This, along with the major happenings in the Middle East, shows a large potential for Sentiment Analysis and consequently an urgent need for more reliable processes and resources for addressing it.

Because of that, there has been an increasing interest and research in the area of Arabic Sentiment Analysis. However, The Arabic Language remains under resourced with respect to the amount of the available datasets. This can be attributed to the fact that most resources developed within studies addressing Arabic Sentiment Analysis, are either limited in size, not publicly available or developed for a very specific domain.

Having said that, a handful of recently published work addresses the issue of availing large Arabic resources for Sentiment Analysis [4, 6, 12]. In this work, we follow in the footsteps of these, by creating large multi-domain datasets of annotated reviews

[1] http://www.internetworldstats.com/stats7.htm

A. Gelbukh (Ed.): CICLing 2015, Part II, LNCS 9042, pp. 23–34, 2015.
DOI: 10.1007/978-3-319-18117-2_2

which we publicly avail to the scientific community. The datasets cover the following domains: movies, hotels, restaurants and products and are made up of approximately 33K reviews. Furthermore we make use of each of the generated datasets to build domain specific sentiment lexicons.

We make use of the multi-domain generated lexicons to perform extensive experiments, benchmarking a wide range of classifiers and feature building methods for the task of sentiment classification. Experimental results provide useful insights with respect to the performance of various classifiers, the effect of different content representations, and the usefulness of the generated lexicons when used solely and when combined with other features. Furthermore, we study the effect of document length and richness with subjective terms on the performance of the sentiment classification task, with the aim of finding the document criteria which affects the performance of the sentiment classification the most.

2 Related Work

Building Sentiment Analysis resources for the Arabic language has been addressed by a number of researchers. For sentiment annotated corpora, Rushdi-Saleh et al. [18] presented OCA; a dataset of 500 annotated movie reviews collected from different web pages and blogs in Arabic. Although the dataset is publicly available, it is limited in size and only covers the movie reviews domain.

Abdul-Mageed & Diab [1] presented the AWATIF multi-genre corpus of Modern Standard Arabic labeled for subjectivity and Sentiment Analysis. The corpus was built from different resources including the Penn Arabic Treebank, Wikipedia Talk Pages and Web forums. It was manually annotated by trained annotators and through crowd sourcing. The dataset targets only Modern Standard Arabic (MSA) which is not commonly used when writing reviews on most websites and social media. Moreover the dataset is not available for public use.

LABR [4, 12] is a large dataset of 63K, polarity annotated, Arabic Book reviews scrapped from www.goodreads.com. On this site (GoodReads), each review is rated on a scale of 1 to 5 stars. The creators of LABR have made use of these ratings by mapping them to sentiment polarities. The dataset was then used for the tasks of sentiment polarity classification and rating classification. The large scale dataset is publicly available for use; however it only covers the domain of book reviews.

For sentiment lexica, as a part of a case study exploring the challenges in conducting Sentiment Analysis on Arabic social media, El-Beltagy et al. [7] developed a sentiment lexicon including more than 4K terms. The lexicon was semi-automatically constructed through expanding a seed list of positive and negative terms by mining conjugated patterns and then filtering them manually. El-Sahar & El-Beltagy [8] presented a fully automated approach to extract dialect sentiment lexicons from twitter streams using lexico-syntactic patterns and point wise mutual information.

More recently, SANA, a large scale multi-genre sentiment lexicon was presented [2, 3]. SANA is made up of 224,564 entries covering Modern Standard Arabic, Egyptian Dialectal Arabic and Levantine Dialectal Arabic. SANA is built from different resources including the Penn Arabic Treebank [10], Egyptian chat logs, YouTube comments, twitter and English SentiWordNet. Some of the lexicon components were built manually, others were obtained using automatic methods such as machine translation.

Various techniques were used to evaluate the generated lexicon. The lexicon is not publicly available.

3 Building the Datasets

Finding and extracting Arabic reviewing content from the internet is considered to be a hard task relative to carrying out the same task in English [18]. This is due to the smaller number of Arabic based e-commerce and reviewing websites over the internet, as well as less activity by users of these sites. Also, many Arabic speakers use the English language or Arabic transliterated in Roman characters to write their reviews. All this has had a big impact on reducing the amount of pure Arabic reviews on the internet. Fortunately, the Arabic reviewing content over the internet has recently shown a significant growth; moreover new reviewing websites have been established. In this study we make use of the available reviewing Arabic content over the internet to create multi-domain datasets reliable for the task of Sentiment Analysis.

3.1 Dataset Generation

For the automatic generation of annotated datasets, we utilize the open-source Scrapy[2] framework, which is a framework for building custom web crawlers. The datasets cover four domains as follows:

1. **Hotel Reviews (HTL):** For the hotels domain 15K Arabic reviews were scrapped from TripAdvisor[3]. Those were written for 8100 Hotels by 13K users.
2. **Restaurant Reviews (RES):** For the restaurants domain two sources were scrapped for reviews: the first is Qaym[4] from which 8.6K Arabic reviews were obtained, and the second is TripAdvisor from which 2.6K reviews were collected. Both datasets cover 4.5K restaurants and have reviews written by over 3K users.
3. **Movie Reviews (MOV):** The movies domain dataset was built out of scrapping 1.5K reviews from elcinema.com[5] covering around 1K movies.
4. **Product Reviews (PROD):** For the Products domain, a dataset of 15K reviews was scraped from the Souq[6] website. The dataset includes reviews from Egypt, Saudi Arabia, and the United Arab Emirates and covers 6.5K products for which reviews were written by 7.5K users.

Each of websites above provides for each review, the text of the review as well as a rating entered by the reviewer. The rating reflects the overall sentiment of the reviewer towards the entity s/he reviewed. So, for each review, the rating was extracted and normalized into one of three categories: positive, negative, or mixed using the same approach adopted by Nabil et al. and Pang et al. [12, 14]. To eliminate any irrelevant and re-occurring spam reviews, we eliminate all redundant reviews.

---

[2] `www.scrapy.org`
[3] `www.tripadvisor.com`
[4] `www.Qaym.com`
[5] `www.elcinema.com`
[6] `www.souq.com`

3.2 Datasets Statistics

In order to better understand the nature of the various collected datasets, a set of statistics reflecting the number of reviews each contains, the number of users who contributed to the reviews, the number of items reviewed, and the polarity distribution of the reviews, was generated for each dataset. These are presented in Table 1 and Figure 1. The total number of collected unique reviews was approximately 33K. The dataset that had the most redundancy was the PROD dataset where out of the total 14K scrapped reviews, only 5K were unique. This can be attributed to the fact that lots of reviews in this dataset consist of only one word, which increases the probability of redundancy.

As shown in Figure 1, the total the number of positive reviews is far larger than that of negative reviews in all of the datasets. The same phenomenon was also observed by the authors of LABR [4], the dataset collected for book reviews.

Figure 2 shows a box plot representing the average number of tokens per review for each dataset. As can be seen, the movie reviews in the MOV dataset are by far the longest. By examining the dataset, it was observed that this is due to the fact that reviewers tend to write long descriptions of the movie they are reviewing, within their review. On the other hand, the PROD dataset tends to have the shortest reviews. Later in this paper we investigate the effect of the length of document on the process of sentiment classification.

Table 1. Summary of Dataset Statistics

	HTL	RES#1	RES#2	MOV	PROD	ALL
#Reviews	15579	8664	2646	1524	14279	42692
#Unique Reviews	15562	8300	2640	1522	5092	**33116**
#Users	13407	1639	1726	416	7465	24653
#Items	8100	3178	1476	933	5906	19593

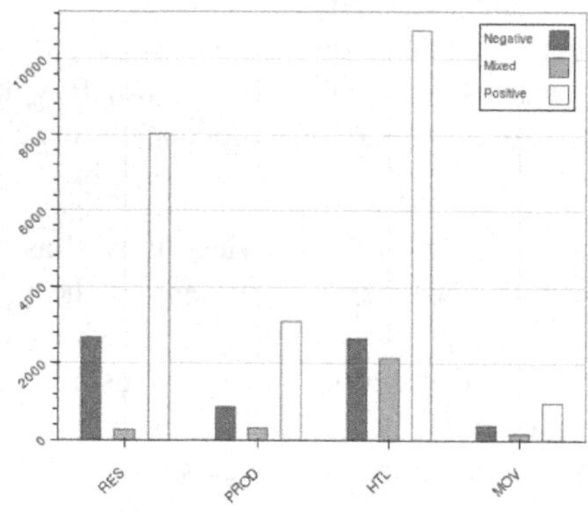

Fig. 1. Number of reviews for each class

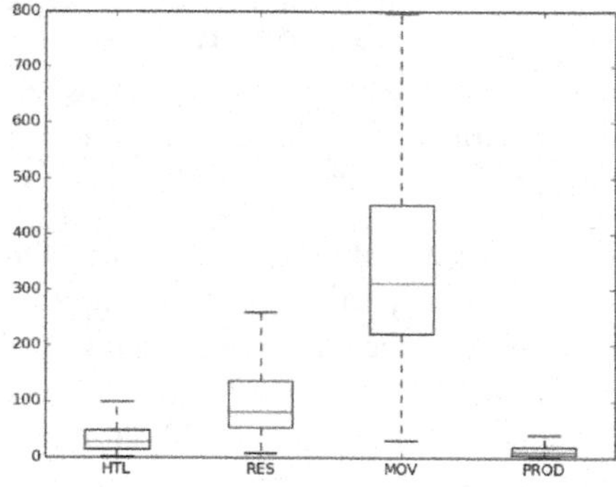

Fig. 2. Box plot showing number of tokens for each of the datasets

4 Building Lexicons

In this section we introduce a method to generate multi-domain lexicons out of the collected reviews datasets. The approach followed is a semi-supervised one, that makes use of the feature selection capabilities of Support Vector Machines [21] to select the most significant phrases contributing to accuracy of sentiment classification and is very similar to that presented by Nabil, Aly and Atiya [12].

To build a lexicon, we follow an approach that generates unigrams and bi-grams from the collected documents. For selecting the set of most significant features we utilize 1-norm Support Vector Machines [21] displayed in (1). 1-norm support vector machines use the L1 penalty $\|\beta\|_1$ calculated as shown in (2).

The L1 regularization results in sparser weight vectors than the L2 (3) regularization, in which only the top significant features will end up with weights larger than zero. Moreover, L1 regularization has proven to be superior to L2 regularization when the number of features is larger than the number of samples, or in other words when there are many irrelevant features [13], which is our case as we use all the extracted n-grams as features.

$$\arg min_{\beta,\beta_0} \sum_{i=1}^{n} [1 - y_i(x_i^T \beta + \beta_0)]_+ + \lambda \|\beta\|_1 \tag{1}$$

$$\|x\|_1 = \sum_{i=1}^{n} |x_i| \tag{2}$$

$$\|x\| = \sqrt{\sum_{i=1}^{n} |x_i|^2} \tag{3}$$

In addition to the previously generated multi-domain datasets, we make use of the LABR dataset for book reviews [4, 12] in order to generate multi-domain lexica covering the book reviews domain as well. We use each of the datasets individually and split each into two parts: 80% for training & validation (we use this as well to generate our lexicons), and 20% for testing. The aim of testing is to assess the usability of learned lexicons on classifying unseen data.

Out of the training examples, we start by building a bag of words model for each dataset where features are the set of unigrams and bigrams and values are simple word counts. Since we are interested in generating a sentiment lexicon of positive and negative terms only, we use only reviews tagged with a positive or negative class.

We use cross validation to tune the soft margin parameter C ($\lambda = 1/2C$). Higher values of C add a higher penalty for the misclassified points rather than maximizing the separation margin. So the optimization problem will lead to a larger number of selected features to reduce the misclassified errors. Lower values of C, result in smaller vectors which are more sparse, leading to a lower number of selected features which might lead to underfitting when the selected features are not enough for the classification process. The best performing classifier is the classifier with the highest accuracy with the least amount of selected features.

After this step, we rank the non-zero coefficients of the best model parameters and map them to the corresponding unigram and bigram features. Features with the highest positive value coefficients are considered to be the highest discriminative features of the documents with positive sentiment. On the other hand, n-grams which correspond to the highest negative value coefficients are considered to indicate a negative sentiment. Based on this, we automatically label the n-grams with the corresponding class label.

This process was repeated for each of the datasets. The resulting unigrams and bigrams from each, was then reviewed by two Arabic native speaker graduate students. The reviewers were asked to manually filter incorrect or irrelevant terms and to keep only those which match with their assigned label thus indicating positive or negative sentiment.

The result of this process was a set of domain specific lexicons extracted from each dataset. In addition, we combined all lexicons into one domain general lexicon; sizes of the different lexicons are shown in table 2.

Table 2. Summary of lexicon sizes

	HTL	RES	MOV	PROD	LABR	ALL
# Selected features	556	1413	526	661	3552	6708
# Manually filtered	218	734	87	369	874	1913

5 Experiments

In this section we design a set of experiments, aiming to: a) validate the usefulness of the generated datasets and lexicons, and b) to provide extensive benchmarks for different machine learning classifiers and feature building methods over the generated datasets to aid future research work. The experiments consisted of three variations which are described in the following subsections.

5.1 Dataset Setups

Experiments on each of the generated datasets were done independently. We also ran the experiments on the LABR book reviews dataset [12]. We explore the problem of sentiment classification as a 2 class classification problem (positive or negative) and a 3 class classification problem (positive, negative and mixed). We ran the experiments using 5-fold cross validation. Moreover, we re-ran our trained classifiers on the 20% unseen testing dataset to make sure that our models are not over fitted. All experiments were carried out using both balanced and unbalanced datasets, but due to paper length limitations the experiments carried out on unbalanced datasets, are documented in a separate report.

5.2 Training Features

For building feature vectors we applied several methods that have been widely utilized before in sentiment classification such as word existence, word count [17, 20] and TFIDF [18].

We also used Delta TFIDF [11]. This method is a derivative of TFIDF in which each n-gram is assigned a weight equal to the difference of that n-gram's TFIDF scores in the positive and negative training corpora as represented in (4). In this equation, $V_{t,d}$ is the Delta TFIDF value for term t in document d, $C_{t,d}$ is the number of times term t occurs in document d, P_t and N_t are the number of positive and negative labeled documents in the training set with the term t while $|P|$ and $|N|$ are the sizes of the positive and negative labeled documents in the training sets.

$$V_{t,d} = C_{t,d} * \log_2\left(\frac{|P|}{P_t}\right) - C_{t,d} * \log_2\left(\frac{|N|}{N_t}\right) \quad (4)$$

This method promises to be more efficient than traditional TFIDF, especially in the reviews domain as common subjective words like "Good", "Bad", "Excellent" are likely to appear in a large number of documents leading to small IDF values, even though these terms are highly indicative. At the same time, these terms don't reoccur frequently within the same document, as users tend to use synonyms to convey the same meaning, which overall results in smaller values of TFIDF.

Another type of feature representation was examined in which feature vectors were comprised entirely of entries from previously generated lexicons. A document is then represented by the intersection of its terms with the lexicon terms or simply by the matches in the document from the lexicon, and their count. We apply this feature representation method once by using domain specific lexicons on each of their respective datasets, and another using the combined lexicon. We refer to those feature representation methods as Lex-domain and Lex-all respectively.

The experiments examine the effect of combining feature vectors generated from Lex-domain and Lex-all, with those generated from TF-IDF, Delta-TFIDF and Count. The effect of this step is discussed in details in the next section.

5.3 Classifiers

For the training and classification tasks, experiments were done using Linear SVM, Logistic regression, Bernoulli Naive Bayes, K nearest neighbor and stochastic gradient descent. The linear SVM parameters were set using cross validation.

Combining different features, classifiers and dataset setups resulted in 615 experiments for each of the datasets. The detailed experiments results and the source code of the experiments have been made publically available for research purposes[7], but a summary of what the authors think are the most important experiments, is presented in the next sub-section.

---

[7] http://bit.ly/1wXue3C

Table 3. Ranking of clssifiers by average accuracy

Classifier	Accuracy	
	2 Classes	3 Classes
Linear SVM	**0.824**	**0.599**
Bernoulli NB	0.791	0.564
LREG	0.771	0.545
SGD	0.752	0.544
KNN	**0.668**	**0.469**

Table 4. Average accuracy associated with of each of the feature representations with and without combining lexicon based features

	Features	Lexicon	LABR	MOV	RES	PROD	HTL	Average
2 Class	Lex-domain	N/A	0.727	0.703	0.811	0.740	0.859	0.768
	Lex-all	N/A	0.746	0.739	0.826	0.732	0.868	0.782
	Count	None	0.806	0.710	0.810	0.725	0.866	0.783
		Lex-domain	0.810	0.703	0.816	0.745	0.874	0.790
		Lex-all	**0.812**	0.733	0.819	0.745	0.873	0.796
	TFIDF	None	0.739	0.552	0.761	0.723	0.730	0.701
		Lex-domain	0.786	0.723	0.819	0.751	0.876	0.791
		Lex-all	0.783	**0.743**	0.836	0.758	0.876	**0.799**
	Delta-TFIDF	None	0.739	0.535	0.745	0.694	0.746	0.692
		Lex-domain	0.771	0.704	0.831	0.752	0.884	0.789
		Lex-all	0.779	0.721	**0.846**	**0.759**	**0.887**	0.798
3 Class	Lex-domain	None	0.510	0.503	0.578	0.524	0.630	0.549
	Lex-all	None	0.529	0.491	0.607	0.494	0.649	0.554
	Count	None	0.603	0.497	0.563	0.520	0.669	0.570
		Lex-domain	0.605	0.484	0.579	0.532	0.669	0.574
		Lex-all	**0.606**	**0.526**	0.589	**0.537**	**0.671**	**0.586**
	TFIDF	None	0.546	0.348	0.513	0.473	0.575	0.491
		Lex-domain	0.578	0.520	0.581	0.536	0.653	0.574
		Lex-all	0.577	0.510	0.599	0.510	0.661	0.572
	Delta-TFIDF	None	0.527	0.340	0.471	0.442	0.549	0.466
		Lex-domain	0.555	0.503	0.588	0.531	0.656	0.566
		Lex-all	0.567	0.476	**0.606**	0.505	0.669	0.565

6 Results and Discussion

This section highlights some of the experiments performed seeking answers for the proposed research questions. We present below the results recorded by experimenting on the balanced datasets. In the detailed experiments report we also present the results for the unbalanced datasets.

6.1 Best Performing Classifiers and Features

Comparing the performance of different classifiers, we average the accuracy of each classifier over all datasets using all feature building methods; the results are shown in Table 3.

It can be observed, that both the 2 class and 3 class classification problems yielded the same ranking for best and worst classifiers. Linear SVM proved to be the best preforming classifier over all datasets scoring a significant difference than the rest of the classifiers while the worst preforming classifier was the K Nearest Neighbor. These results are very similar to those reported by many previous research works on sentiment classification and specifically the benchmarks of the LABR dataset for book reviews [4, 12].

To compare the effect of employing different feature representation methods, we calculate the accuracy of each one of them averaged over all classifiers; results are shown in Table 4. For the 2 class classification problem, the top three feature representation methods were Delta-TFIDF, TFIDF and Count, when combined with Lex-all, the feature vectors of the combined lexicon. The same three feature representation methods also ranked on top, for the 3 class classification problem.

The least performing methods were TF-IDF and Delta-TFIDF when used solely without combining with any lexicon based feature vectors, with a 10% drop in the accuracy than the top performing feature representations.

6.2 Accuracy of Lexicon Based Features Solely and Combined with Other Features

Lexicon based feature vectors of Lex-domain and Lex-all solely achieved a fair performance in comparison with the best performing features with less than a 2% drop in the average accuracy. These results were obtained using unseen test data different from the one used to build lexicons.

Given that the maximum length of the Lexicon based features Lex-Domain and Lex-all is 2K, while other feature vectors can grow up to several millions. This proves that Lexicon based features generated from a sample of the datasets can lead to much simpler classifiers.

Combining lexicon based features with other features provided large improvements on the total accuracy, with 10% in cases of TFIDF and Delta-TFIDF and 2% in case of Counts.

Using domain general features Lex-all rather than Lex-domain, doesn't show a significant difference in the overall accuracy in our case, as the length of the generated lexicons are relatively small and although they are representing multi-domains, all of them are generated from the reviews domain in which users tend to use similar language structure.

6.3 Effect of Document Length and Richness with Subjective Terms on Sentiment Classification

In order to show the effect of document length and subjectivity richness on the performance of sentiment classification, we label each of the misclassified documents out of the 2 class classification problem with its number of terms and subjectivity score. The subjectivity score is the sum of all polarities of the subjective terms in the document. To calculate this score we rely on the generated domain general lexicon to detect subjective words in a document and their polarity.

Additionally, a set of negation detection rules were crafted and used to flip the term polarity if it happened to be negated. Then, documents of similar size and subjectivity score are grouped together and the average error rate is calculated for each. The resulting error rates were used to plot the heat map shown in Fig3. From the resulting heat map, the following can be observed:

- The error rate increases as the subjectivity score tends to zero and decreases as the subjectivity grows in any of the positive or negative directions. This applies for small, mid-range and relatively long documents.
- For extremely long documents, a document's subjectivity seizes to correlate with the error rate. We find that documents longer than 1K words with very high subjectivity, achieve very high error rates. This is probably because longer documents allow more topic drifts, criticizing other subjects or having comparisons with other entities which are not handled explicitly by any of our classifiers.
- Extremely short documents by definition have a very limited number of terms and hence cannot have a high subjectivity score, which often results from matching with multiple entries in subjectivity lexicon. As a result, the majority of extremely short documents end up with high error rates.
- Finally, we find that the error rate for mid-range documents is slightly shifted to the positive side. At the same time, the maximum subjectivity scores on the positive side are higher than on the negative side which is consistent with the observation that negative terms are less frequently used than positive terms [9, 19].

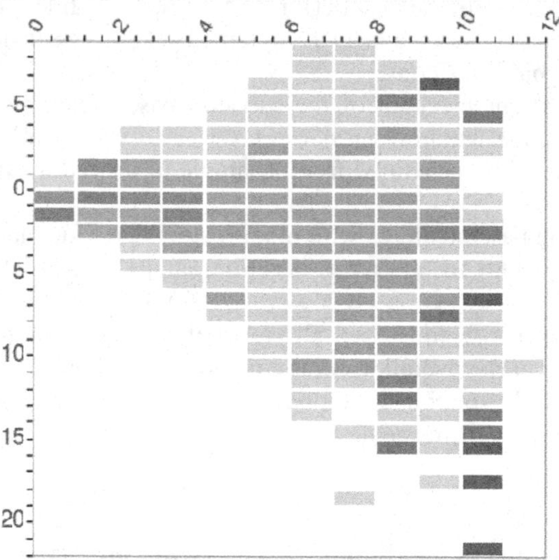

Fig. 3. Heat map showing the error rate for various document lengths and subjectivity score groups. The horizontal axis shows the log of the document lengths, while the vertical axis represents the subjectivity scores and the color gradient is the error rate (the darker the worse).

7 Conclusion and Future Work

In this study, we introduced large multi-domain datasets for Sentiment Analysis. The datasets were scrapped from multiple reviewing websites in the domains of movies, hotels, restaurants and products. Moreover we presented a multi-domain lexicon of 2K entries extracted from these datasets.

Although the generated lexicon isn't very large, the results of the experiments have shown that abstracting reviews by lexicon based features only has achieved a relatively fair performance for the task of sentiment classification.

An extensive set of experiments was performed for the sake of benchmarking the datasets and testing their viability for both two class and three class sentiment classification problems. Out of the experimental results, we highlighted that the top performing classifier was SVM and the worst was KNN, and that the best performing feature representations were the combination of the lexicon based features with the other features.

Finally according to the error analysis on the task of sentiment classification, we find that the document length and richness with subjectivity both affect the accuracy of sentiment classification, in which; sentiment classifiers tend to work better when the documents are rich with polar terms of one class, i.e., high values of subjectivity score. However, this often doesn't hold when the document length is extremely short or long.

Although the generated datasets cover multiple domains, they are all generated only from reviews. Thus, their usefulness for social media Sentiment Analysis is yet to be studied. This might include generation of additional datasets to cover cases that don't show up in the reviews domain, but common in social media like advertisements and news. This is a motivation for future research work.

References

1. Abdul-Mageed, M., Diab, M.: AWATIF: A Multi-Genre Corpus for Modern Standard Arabic Subjectivity and Sentiment Analysis. In: LREC, pp. 3907–3914 (2012)
2. Abdul-mageed, M., Diab, M.: SANA: A Large Scale Multi-Genre, Multi-Dialect Lexicon for Arabic Subjectivity and Sentiment Analysis. In: Proceedings of the Ninth International Conference on Language Resources and Evaluation (LREC 2014), pp. 1162–1169 (2014)
3. Abdul-Mageed, M., Diab, M.: Toward Building a Large-Scale Arabic Sentiment Lexicon. In: Proceedings of the 6th International Global WordNet Conference, pp. 18–22 (2012)
4. Aly, M., Atiya, A.: LABR: A Large Scale Arabic Book Reviews Dataset, pp. 494–498. Aclweb.Org. (2013)
5. Baccianella, S., et al.: SentiWordNet 3.0: An Enhanced Lexical Resource for Sentiment Analysis and Opinion Mining. In: Proceedings of the Seventh International Conference on Language Resources and Evaluation (LREC 2010), pp. 2200–2204 (2010)
6. Badaro, G., et al.: A Large Scale Arabic Sentiment Lexicon for Arabic Opinion Mining. In: ANLP 2014, pp. 176–184 (2014)
7. El-Beltagy, S., Ali, A.: Open Issues in the Sentiment Analysis of Arabic Social Media: A Case Study. In: Proceedings of 9th International Conference on Innovations in Information Technology (IIT), pp. 215–220 (2013)

8. ElSahar, H., El-Beltagy, S.R.: A Fully Automated Approach for Arabic Slang Lexicon Extraction from Microblogs. In: Gelbukh, A. (ed.) CICLing 2014, Part I. LNCS, vol. 8403, pp. 79–91. Springer, Heidelberg (2014)
9. Jerry, B., Osgood, C.: The pollyanna hypothesis. J. Verbal Learning Verbal Behav. 8(1), 1–8 (1969)
10. Maamouri, M., et al.: The penn arabic treebank: Building a large-scale annotated arabic corpus. In: NEMLAR Conference on Arabic Language Resources and Tools, pp. 102–109 (2004)
11. Martineau, J., et al.: Delta TFIDF: An Improved Feature Space for Sentiment Analysis. In: Proc. Second Int. Conf. Weblogs Soc. Media (ICWSM), vol. 29, pp. 490–497 (2008)
12. Nabil, M., et al.: LABR: A Large Scale Arabic Book Reviews Dataset. arXiv Prepr. arXiv1411.6718 (2014)
13. Ng, A.: Feature selection, L 1 vs. L 2 regularization, and rotational invariance. In: ICML (2004)
14. Pang, B., et al.: Thumbs up? Sentiment Classification using Machine Learning Techniques. In: Conf. Empir. Methods Nat. Lang. Process. (EMNLP 2002), pp. 79–86 (2002)
15. Pang, B., Lee, L.: Opinion Mining and Sentiment Analysis. In: The 42nd annual meeting on Association for Computational Linguistics, pp. 271–278 (2004)
16. Pang, B., Lee, L.: Seeing stars: Exploiting class relationships for sentiment categorization with respect to rating scales 1 (2005)
17. Pang, B., Lee, L.: Thumbs up? Sentiment classification using machine learning techniques. In: Proc. Conf. Empir. Methods Nat. Lang. Process., Philadephia, Pennsylvania, USA, July 6-7, pp. 79–86 (2002)
18. Rushdi-Saleh, M., Martin-Valdivia, T.: OCA: Opinion corpus for Arabic. J. Am. Soc. Inf. Sci. Technol. 62(10), 2045–2054 (2011)
19. Taboada, M., et al.: Lexicon-Based Methods for Sentiment Analysis (2011)
20. Turney, P.D.: Thumbs Up or Thumbs Down? Semantic Orientation Applied to Unsupervised Classification of Reviews. In: Proceedings of the 40th Annual Meeting on Association for Computational Linguistics, pp. 417–424. Association for Computational Linguistics (2002)
21. Zhu, J., et al.: 1 -norm Support Vector Machines. Advances in Neural Information Processing Systems 16(1), 49–56 (2004)

Learning Ranked Sentiment Lexicons

Filipa Peleja and João Magalhães

CITI, Departamento de Informática, Faculdade de Ciências e Tecnologia,
Universidade Nova de Lisboa, 2829-516 Caparica, Portugal
filipapeleja@gmail.com, jm.magalhaes@fct.unl.pt

Abstract. In contrast to classic retrieval, where users search factual information, opinion retrieval deals with the search of subjective information. A major challenge in opinion retrieval is the informal style of writing and the use of domain-specific jargon to describe the opinion targets. In this paper, we present an automatic method to learn a space model for opinion retrieval. Our approach is a generative model that learns sentiment word distributions by embedding multi-level relevance judgments in the estimation of the model parameters. The model is learned using online Variational Inference, a recently published method that can learn from streaming data and can scale to very large datasets. Opinion retrieval and classification experiments on two large datasets with 703,000 movie reviews and 189,000 hotel reviews showed that the proposed method outperforms the baselines while using a significantly lower dimensional lexicon than other methods.

1 Introduction

The increasing popularity of the WWW led to profound changes in people's habits. Search is now going beyond looking for factual information, and now people wish to search for the opinions of others to help them in their own decision-making [16, 17]. In this new context, sentiment expressions or opinion expressions, are important pieces of information, specially, in the context of online commerce [12]. Therefore, modeling text to find meaningful words for expressing sentiments (sentiment lexicons) emerged as an important research direction [1, 9, 13, 28].

In this work we investigate the viability of automatically generating a sentiment lexicon for opinion retrieval and sentiment classification applications. Some authors have tackled opinion retrieval by re-ranking search results with an expansion of sentiment words. For example, Zhang and Ye [28] describe how to use a generic and fixed sentiment lexicon to improve opinion retrieval through the maximization of a quadratic relation model between sentiment words and topic relevance. In contrast, Gerani et al. [9] applies a proximity-based opinion propagation method to calculate the opinion density at each point in a document. Later, Jo and Oh [13] proposed a unified aspect and sentiment model based on the assumption that each sentence concerns one aspect and all sentiment words in that sentence refer to that sentence. Finally, Aktolga and Allan [1] targeted the task of sentiment diversification in search results. The common element among these works [1, 9, 13, 28] is the use of the

A. Gelbukh (Ed.): CICLing 2015, Part II, LNCS 9042, pp. 35–48, 2015.
DOI: 10.1007/978-3-319-18117-2_3

sentiment lexicon SentiWordNet [3]. In these scenarios, the most common and quite successful approach, is the use of a pre-defined list of positive and negative words as a sentiment lexicon.

In the present work we aim at capturing sentiment words that are derived from online users' reviews. Even though we can find some examples on the literature the majority of the available work is focused on sentiment classification as positive, negative and neutral [17, 22] or joint aspect-sentiment features [13], and only a few approaches are focused on the task of automatically defining specific sentiment vocabularies [7].

A major challenge in opinion retrieval is the detection of words that express a subjective preference and, more importantly, common domain-specific sentiment words, such as jargon. However, because domain dependencies are constantly changing and opinions are not on a binary scale, capturing the appropriate sentiment words for opinion ranking can be a particularly challenging task.

In our study, we found that current state-of-the-art sentiment strategies that use static lexicons [8] were too coarse-grained. Moreover, static lexicons are too generic and are not adequate for ranking, which is central to many applications: they do not consider domain words, have fixed sentiment word weights (which are sometimes simply positive/negative or have more than one sentiment weight), and do not capture interactions between words. In this paper we aim to deliver a sentiment resource specifically designed for rank-by-sentiment tasks. Because we argue that a simple weight is not enough, we identify two main steps for building this resource: (1) the identification of the lexicon words, and (2) the inference f a word sentiment distributions.

The proposed algorithm is related to the Labeled LDA algorithm [18] and LDA for re-ranking [19]. However, a fundamental difference is that we add an extra hierarchical level to smooth sentiment word distributions across different sentiment relevance levels.

In summary, this paper makes the following contributions:

- A fully generative automatic method to learn a domain-specific lexicon from a domain-specific corpus, which is fully independent of external sources: there is no need for a seed vocabulary of positive/negative sentiment words, and unlike previous approaches, such as Hu and Liu [11], it has no explicit notion of sentiment word polarities.
- A hierarchical supervised method is used to enhance the ability of learning sentiment word distributions in specific contexts. The uncertainty that arises from the sentiment word polarities used in previous works [3, 27], are naturally mitigated in our proposal by ensembles of sentiment word distributions that co-occur in the same context.

With our method, opinion ranking is computed from sentiment word distributions. The space model for opinion ranking ensembles sentiment word distributions, thereby capturing not only generic sentiment words but also domain-specific jargon.

The formal model and its inference are detailed in section 3. In section 4 we describe the evaluation of the proposed method over two datasets on opinion retrieval and opinion classification experiments. In both cases our method outperforms or matches existing methods in the NDCG, P@30 and MAP metrics. Furthermore, we provide an in-depth analysis of the lexicon domain-specificity and diversity.

2 Learning Ranked Sentiment Lexicons

We address the problem of creating a sentiment lexicon based on user reviews without human supervision and propose to identify the sentiment words using a multi-level generative model of users' reviews. Intuitively, we use a generative probabilistic model that ties words to different sentiment relevance levels, creating a sentiment rank over the entire sentiment lexicon. The main contribution of the proposed approach is that the model infers a sentiment lexicon by analyzing users' reviews as sentiment ranked sets of documents.

Problem Formalization: consider a set of M documents $\mathcal{D} = \{d_1, \dots, d_l\}$ containing user opinions towards a given product. Each review d_i is represented by a ple (w_i, s_i), where $w_i = (w_{i,1}, \dots w_{i,N})$ a vector of N is word counts and $s_i \in \{1, \dots, R\}$ is the associated sentiment level value that quantifies the user opinion towards the product, corresponding to the user rating. Our goal is to learn a fine-grained lexicon of sentiment words that best captures the varying level of user satisfaction.

2.1 Latent Dirichlet Allocation

Latent Dirichlet Allocation (LDA) is a generative model that explores word co-occurrences at the document-level and at the level of K latent topics. For each latent topic it samples a word distribution from a prior Dirichlet distribution. LDA is a well-known method for modelling text documents with latent topics [2, 6, 15, 21]. In Figure 1 we present the smoothed LDA graphical model – refer to [4] for details. LDA model explores the probability of a sequence of words and its hidden topics and is given by

$$p(w,z) = \int p(\theta) \cdot \prod_{n=1}^{N} p(z_n \mid \theta) p(w_n \mid z_n) \, d\theta$$

where θ is the per-document topic Dirichlet$(\cdot \mid \alpha)$ distribution, z is the per-word topic assignment following a Multinomial$(\cdot \mid \theta^{(d)})$ distribution, and w corresponds to the set of words observed on each document. Our intuition is that by adding a new hierarchical level to LDA, we can embed sentiment level information in the latent topics.

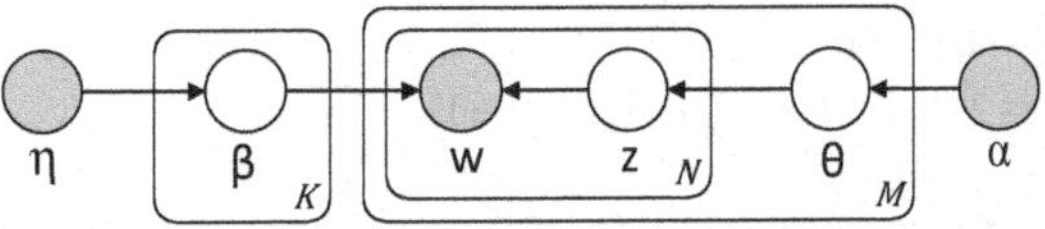

Fig. 1. LDA graphical model

2.2 Rank-LDA

The LDA method explores co-occurrences at document level, identifying the latent topics z and their respective associated words. However, our methodology is different in the sense that we extract words or pairs of words associated to a sentiment. Figure 2 presents the graphical model of the proposed Rank-LDA method. At its core, the Rank-LDA links the hidden structure (latent topics) to the sentiment level of each document. In the hidden structure a set of hidden topics are activated for each sentiment level. Hence, while LDA models a topic as a distribution over a fixed vocabulary [4], Rank-LDA computes the distribution of words over the topics that best describe an association to a sentiment. Notice that this sentiment-topic association is different from previous work [2, 16, 18, 19, 25], where LDA was used to capture the topic distributions over the words that best describe the product.

In Figure 2 the random variables α, η and π correspond to the distribution priors, β is the per-corpus topic Dirichlet($\cdot\,|\eta$) distribution, $s_i \in \{1, \ldots, R\}$ is the per-document sentiment relevance level and sw is the per-word random variable corresponding to its sentiment distributions across different sentiment levels.

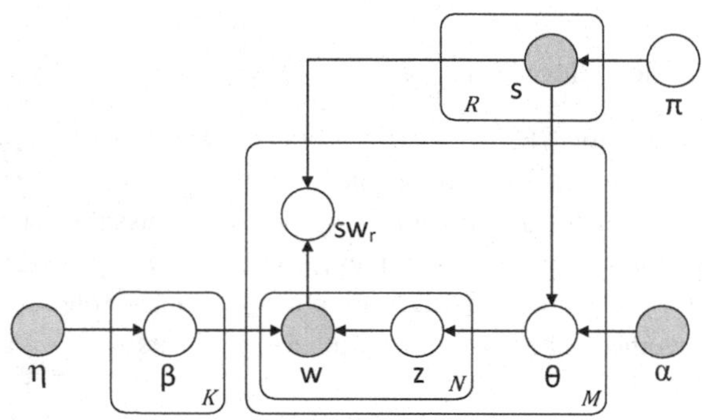

Fig. 2. Rank-LDA graphical model

ALGORITHM 1. The Rank-LDA generative process

For each topic $k \in \{1, \ldots, K\}$
  Generate $\beta_k = (\beta_{k,1}, \ldots, \beta_{k,N}) \sim \mathrm{Dir}(\cdot\,|\eta)$
For each document d:
  For each topic $k \in \{1, \ldots, K\}$
    Generate $s_k^{(d)} \in \{1, \ldots, R\} \sim \mathrm{Mult}(\cdot\,|\pi)$
  Generate $\alpha^{(d)} = L^{(d)} \cdot \alpha$
  Generate $\theta^{(d)} = (\theta_{k,1}, \ldots, \theta_{k,N}) \sim \mathrm{Dir}(\cdot\,|\alpha)$
  For each $i \in \{1, \ldots, N_d\}$:
    Generate $z_i \in \left\{\lambda_1^{(d)}, \ldots, \lambda_{M_d}^{(d)}\right\} \sim \mathrm{Mult}(\cdot\,|\theta^{(d)})$
    Generate $w_i \in \{1, \ldots, V\} \sim \mathrm{Mult}(\cdot\,|\,\beta_{z_i})$
For each sentiment word w_i:
  Compute the marginal distribution $\int p(sw_i \mid w_i, s)ds$

The proposed method is related to Labeled-LDA [18] and Supervised-LDA (sLDA) [5], but with significant differences. Firstly, we tie the latent topics to sentiment levels in our method. Therefore, the hidden topics will encode the words ranked by sentiment level and then by topic relevance. While the sLDA assumes that labels are generated by the topics and in Labeled-LDA the labels activate/de-activate the topics, which is the role of the projection matrix $L^{(d)}$. In Rank-LDA we further extend the projection matrix $L^{(d)}$ to link the topic latent variables r_k to sentiment levels $s_k^{(d)}$. Thus, the rows of the projection matrix will correspond to a set of topics associated to a given sentiment level. In particular, consider the case where we have 3 sentiment levels and 2 latent topics per sentiment level, if a given document d has a sentiment level equal to 2, then $s_k^{(d)} = (0,0,1,1,0,0)$ and the projection matrix would be:

$$\begin{pmatrix} 0\,0 & 1\,0 & 0\,0 \\ 0\,0 & 0\,1 & 0\,0 \end{pmatrix}.$$

This formalization models a document as a set of sentiment ranked words. Algorithm 1 describes the generative process of the Rank-LDA model.

2.3 Sentiment Word Distributions

A key characteristic of sentiment words is the sharing of sentiment words across different levels of sentiment, although in different proportions. The result is that word distributions over topics associate a word relevance to different sentiment levels. The distributions of Figure 3 depict the marginal distributions of each sentiment word. The distributions of interactions between sentiment words are also embedded in the hierarchical model structure (illustrated in Figure 4 – Section 3 Experiments). Figure 3 illustrates a sample of the density distribution of words per sentiment level, for some exemplificative sentiment words: *emotion*, *love*, *wonderful*, *awful*, *heart* and *terrible*. The sentiment word distributions are given by the density distribution

$$p(\mathrm{w} \mid s_i) = \int p(\theta) \cdot \prod_{n=1}^{N} p(z_n \mid \theta, s_i) p(w_n \mid z_n)\, d\theta + \tau$$

in which we compute the marginal distribution of a word given a sentiment level, over the K latent topics of the Rank-LDA model. The variable τ is a smoothing parameter set to 0.01.

The sentiment word distribution function can also be used to rank words by its positive/negative weight and to calculate a word relevance in different sentiment levels. A straightforward way of achieving this conversion is through the function

$$RLDA(w_{i,j}) = \frac{p(w|s=i) - p(w|s=j)}{min(p(w|s=i), p(w|s=j))}$$

where $p(w|s=i)$ and $p(w|s=j)$ denote the word values of sentiment relevance for word w in ratings i and j. The obtained lexicon with Rank-LDA is denoted as RLDA.

The D-TFIDF method [14] has proven to be quite successful in sentiment analysis. In this model, all term (t) frequencies are observed, for each document (d), computing

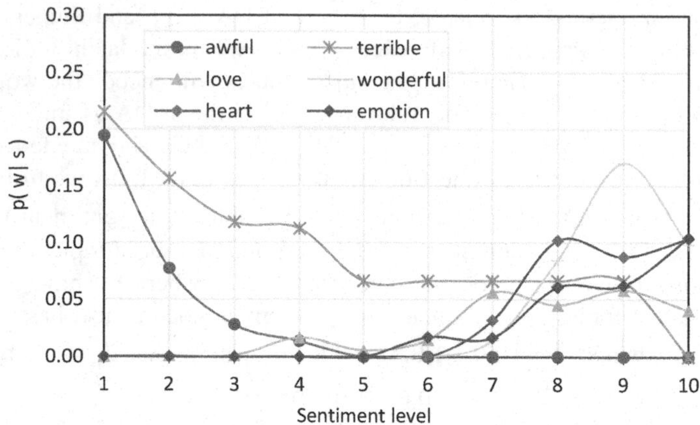

Fig. 3. The sentiment distributions of words emotion, love, wonderful, awful, heart and terrible

the respective log-odds between the number of negative (N_t) and positive documents (P_t). In addition, D-TFIDF is given by $\delta_{t,d} = tf_{t,d} \cdot log(N_t/P_t)$. Following Martineau et al [14] method we added the term sentiment log-odds in the RLDA expression. We will refer to this approach as D-RLDA.

Model Inference: Computationally, in RLDA, reviews are rated in a particular scale (usually 1 to 10 or 1 to 5). We iteratively compute a set of topic distributions per sentiment level – a review rating level – and this is repeated until all sentiment levels have been incorporated in the RLDA structure. In this hierarchical approach, rating information is imposed over the topic distributions, rendering distributions of words that enable us to identify words used to express different levels of sentiment relevance. We adapted the online variational Bayes (VB) optimization algorithm proposed by Hoffman, Bach and Blei [10] to estimate the RLDA parameters. Batch algorithms do not scale to large datasets and are not designed to handle stream data. The online VB algorithm is based on online stochastic optimization procedure.

2.4 Opinion Ranking

To rank opinions by sentiment level we implemented a ranking algorithm that minimizes the distance between the query sentiment level qs_i and the inferred sentiment level $p(s = qs_i \mid d_j)$ of each review d_j. The reviews contain multiple words and we obtained a sentiment word distribution over each document j and sentiment relevance level i – $p(\mathrm{w}|s_i)$ for each word. Hence, reviews are ranked according to

$$p(s = qs_i \mid d_j) \propto exp\left(\sum\nolimits_k \gamma_{j,k} \cdot p(w_j|s_i)\right)$$

The parameters $\gamma_{j,k}$ are optimized to minimize the expected cost between the observed rating and the inferred cost. When ranking opinions, one wishes to retrieve reviews that are close in range to the query. Thus, the squared error cost function was chosen to minimize the penalty over close sentiment levels and maximize the penalty over more distant sentiment levels.

2.5 Opinion Classifier

Starting from the word distributions we designed a straightforward approach that aims at identifying the reviews' sentiment levels. Using a binary classifier to classify reviews as positive or negative. The intuition behind this is simple: the review sentiment level is adapted to a positive versus negative viewpoint. Additionally, we have implemented two finer-grained sentiment classifiers: a multiclass classifier (one-vs-all), that aims at finding their most probable sentiment level and, a multiple Bernoulli (multilevel) classifier as it benefits from observing each sentiment level individually.

3 Experiments

In this section we describe the two experiments we carried out to assess the effectiveness of the proposed method. While the first experiment concerns opinion ranking, the other evaluates the performance of the Rank-LDA in binary, multiclass and multilevel sentiment classification task. The evaluation metrics P@5, P@30, NDCG and MAP were used for the opinion ranking experiment; and precision, recall and F1 in the classification experiment.

3.1 Datasets and Baselines

IMDb-Extracted: This dataset contains over 703,000 movie reviews. Reviews are rated in a scale of 1 to 10. We crawled this dataset because most of the existing review datasets either lacked the rating scale information, targeted multiple domains, did not capture cross-item associations or were limited to small numbers.

TripAdvisor: This dataset contains 189,921 reviews, and each review is rated in a scale of 1 to 5. This dataset was made available by Wang et al. [25]. The dataset was split into 94,444 documents for training and 95,477 documents for testing purposes.

We further analyze our method to understand its general characteristic and possible utility in sentiment applications. Thus, we compare our lexicon with three well-known sentiment lexicons, a LDA topic model and a method to build polarity lexicons based on a graph propagation algorithm:

- **SentiWordNet [8]:** this lexicon was built with a semi-automatic method where some manual effort was used to curate some of the output.
- **MPQA [27]:** this lexicon provides a list of words that have been annotated for intensity (weak or strong) in the respective polarity – positive, negative or neutral. The lexicon was obtained manually and an automatic strategy is employed afterwards.
- **Hu-Liu [11]:** this lexicon contains no numerical scores. Based on the premise that misspelled words frequently occur in users' reviews these words are deliberately included in the lexicon.

- **D-TFIDF [14]:** full vocabulary baselines. D-TFIDF combines TFIDF with a weight that measures how a word is biased to a dataset.
- **LLDA [18]:** Labeled LDA is a topic model that constrains LDA by defining one-to-one correspondence between LDA's latent topics and user tags. In the present work, tags will correspond to user ratings.
- **Web GP [24]:** A method based on graph propagation algorithms to construct polarity lexicons from lexical graphs.

Our methods are: Rank-LDA (RLDA), as described in section 2.2 and section 2.3; D-**RLDA** which uses **D-TFIDF** weighting scheme adapted to RLDA method; and, following the strategy described in section 2.3, we have also computed Rank-LLDA (**RLLDA**) and Rank-Web GP (**RWGP**).

Table 1. Number of words in lexicons built from IMDb and TripAdvisor datasets

	IMDb	**TripAdvisor**
RLDA/D-RLDA	9,510	4,936
RLLDA	55,428	15,086
RWGP	1,406	875
D-TFIDF	367,691	123,678
LLDA	97,808	44,248
Web GP	3,647	2,261

3.2 Experiments: Opinion Ranking

In this section we present the evaluation results in a task of opinion retrieval by rating level. Table 2 shows the opinion retrieval performances. A user review is represented as a query and the relevance judgment is the rating level. The table shows that the proposed methods RLDA, D-RLDA, RLLDA and RWGP, LLDA and Web GP are consistently effective across the four evaluation metrics (P@5, P@30, MAP and NDCG).

In general, these lexicons outperform the static lexicons. However, for the TripAdvisor dataset, the metric D-TFIDF presents fairly good results. We note, in comparison with the proposed metrics, for both MAP and NDCG that the metric D-TFIDF presents equally good results. However, we would like to recall that D-TFIDF presents a weight for all words (367,691 and 123,678 words in the IMDb and TripAdvisor datasets respectively). This serves as relevant difference in comparison to all the other sentiment lexicons (Table 1). Noticeably, the D-TFIDF metric would not be as useful as the proposed approach for creating sentiment lexicons. For instance, for the TripAdvisor dataset, the words that D-TFIDF lexicon identifies as having higher positive and negative relevance are {cevant, untrained, unconcerned, enemy} and {leonor, vaporetto, unpretentions, walter}, respectively; in contrast, the D-RLDA lexicon identifies {full, great, excellent, wonderful} and {tell, call, dirty, bad}as the words having higher positive and negative relevance respectively. These examples illustrate the discriminative nature of D-TFIDF and the generative nature of D-RLDA.

LLDA [18] is a model of multi-labeled corpora that addresses the problem of associating a label (a rating, in our case) with one topic. In particular LLDA is strongly competitive with discriminative classifiers in multi-label classification tasks. However, we note that despite presenting equally good results, it requires a higher number of words to correctly perform the opinion retrieval tasks. Intuitively, the proposed task could be approximated to a topic classification task for which LLDA is more appropriate. However, LLDA is not capturing sentiment words. Indeed, similarly to D-TFIDF, it is capturing words that best describe each rating level. On the other hand, the web-derived lexicon [24], Web GP, performs at a similar level although with a considerable lower number of words – approximately 50% lower than the ones captured by RLDA. Web GP constructs a polarity lexicon using graph propagation techniques whereas the graph captures semantic similarities between two nodes. In contrast our method relies on LDA generative model to capture semantic similarities of words. Nonetheless, unlike Web GP, RLDA does not require a seed of manually constructed words to produce the lexicon. In addition, asserting the ideal number of sentiment words that are required for a sentiment lexicon can be highly challenging. As a consequence, for sentence level classification tasks, sentiment words selected by Web GP may not be enough to discriminate sentiments at sentence level.

3.3 Experiments: Opinion Classification

To evaluate the gains of using the proposed method in a supervised opinion classification task we measured the performance of the lexicons in a binary (B), multilevel (MB) and one-against-all (OAA) opinion classification task (Table 3).

The MB classifier predicts the rating that presents the highest probability in a rating range of 1 to 10 (IMDb) or 1 to 5 (TripAdvisor). MB entails a greater challenge than positive vs. negative, or vs. all, as unlike the other classifiers, the MB classifier attempts to distinguish between similar ratings [20]. In Table 3, for the IMDb dataset, we can verify that the MB classifier was outperformed by the B and OAA classifiers. Mid-range ratings represent a greater challenge than high or low ratings. We found that the TripAdvisor dataset has a lower rating range – i.e., lower uncertainty between mid-range opinions. In other words, users tend to be blunter when writing a highly positive or negative review. Obviously these mid-range reviews negatively affect the performance overall. For instance, Jo and Oh [13] have chosen to remove all ratings from borderline reviews from the classification. However, in this experiment we chose to remain as close to the real data as possible. When analyzing the results for both datasets, we see that our method has an optimal performance, consistently outperforming other lexicons or being as good as the best.

Previous observations about hotel reviews vocabulary [26] were also confirmed by our study. The vocabulary used in hotel reviews is more "contained" than the one used in movie reviews. In particular, in the latter users tend to be more creative and less concise (IMDb data). Users create longer documents discussing different topics and frequently recur to the use of synonyms to avoid boring the reader with repetition [14, 23]. This domain characteristic is reflected in the classification performance, which performs better in the domains where both the vocabulary and the documents length are more concise. Results also show that generic sentiment lexicons (e.g. SWN) can perform quite well on sentiment analysis tasks. However almost always below other finer-grained lexicons.

Table 2. Opinion ranking. P@5, P@30, MAP and NDCG for two datasets. * is the best result, the statistical significance t-test showed that the differences in retrieval results between D-RLDA and SentiWordNet are statistically different.

	IMDb				**TripAdvisor**			
	P_5	**P_30**	**MAP**	**NDCG**	**P_5**	**P_30**	**MAP**	**NDCG**
RLDA	92.00	90.67	56.33	78.17	92.00	98.67	65.34	81.34
DRLDA	90.00	91.67*	56.37	78.18	96.00	98.67	65.33	81.31
RLLDA	94.00*	91.00	55.12	77.16	100.00*	98.00	65.92	81.50
RWGP	92.00	88.67	56.64	79.21*	100.00*	98.00	64.02	81.47
Hu-Liu	82.00	76.67	43.85	72.44	92.00	90.00	55.76	78.12
MPQA	82.00	81.67	46.22	73.61	100.00*	87.33	57.99	78.95
SWN	88.00	89.00	53.52	76.77	92.00	96.00	63.70	80.89
D-TFIDF	76.00	81.67	54.72	77.01	96.00	99.33*	66.51	81.81*
LLDA	92.00	90.34	55.04	77.13	100.00*	98.67	66.61*	81.87*
Web GP	88.00	89.67	57.20*	79.11	96.00	96.00	65.36	81.83*

Table 3. Opinion classification. Precision (P), recall (R) and F1-measure for binary classification, P for multiple Bernoulli (MB) and one-against-all (OAA) for two datasets. * is the best result, significance was tested using t-test and all classifiers were all different from the baseline with a value of $p < 0.01$.

	IMDb					**TripAdvisor**				
	Binary			**MB**	**OAA**	**Binary**			**MB**	**OAA**
Method	P	R	F1	P	P	P	R	F1	P	P
RLDA	89.05	88.59	88.82	73.02	70.29	94.47	93.41*	93.94*	88.40	90.89
D-RLDA	89.87*	86.85	88.33	73.09	70.98	94.24	93.12*	93.68*	94.24	90.87
RLLDA	84.21	97.04	90.17*	73.67*	80.80*	95.73*	91.56	93.60	95.58*	91.30*
RWGP	81.61	96.63	88.48	69.39	76.61	94.90	88.78	91.73	93.52	88.40
Hu-Liu	73.46	94.82	82.78	61.43	65.87	94.52	65.11	77.11	94.52	83.12
MPQA	75.59	93.52	83.60	62.04	66.05	94.27	73.56	82.64	94.27	84.42
SWN	73.90	99.50*	84.81	68.59	68.77	94.38	91.55	92.95	94.38	88.48
D-TFIDF	91.05	85.76	88.33	70.36	73.68	94.78	92.47	93.61	94.78	90.77
LLDA	83.21	97.04	89.60	73.67*	80.62	95.79*	87.69	91.56	95.58*	91.38*
Web GP	82.44	97.12	89.18	71.53	78.53	95.53	91.46	93.45	94.85	89.85

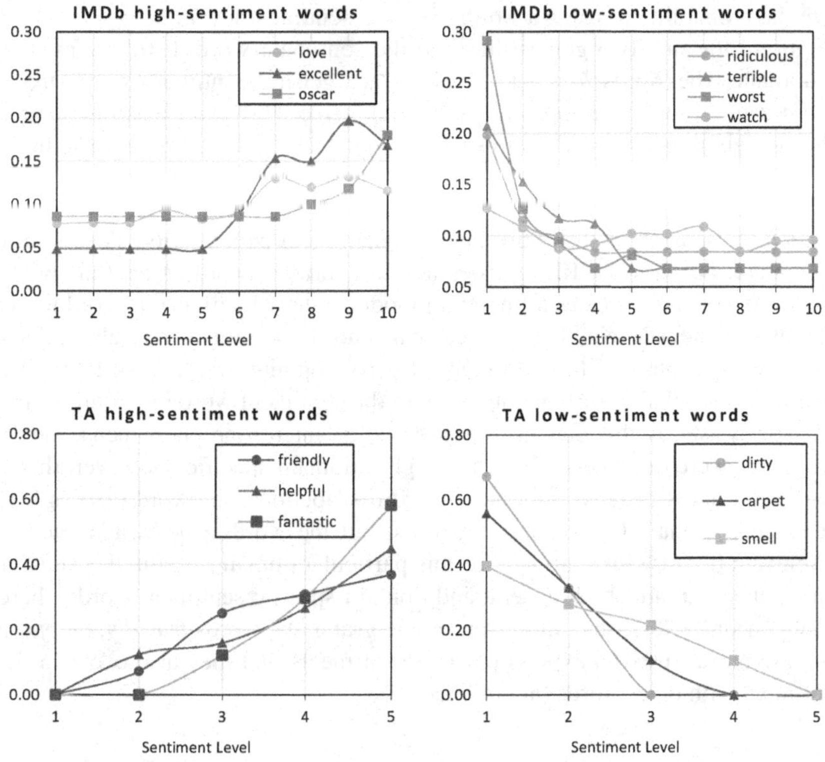

Fig. 4. Sentiment word distributions for the datasets IMDb and TripAdvisor (TA)

Martineau et al. [14] proposed the metric D-TFIDF to weight words scores. Martineau et al. found that D-TFIDF improved the classification accuracy of subjective sentences in the Pang and Lee subjectivity dataset[1]. Moreover, variants of our proposed method outperformed the static lexicons with D-TFIDF showing similar performance. We note, however, that the RLDA lexicon required 2.6% (Table 1) of the words used by D-TFIDF, this entails a very aggressive - and effective - feature selection.

3.4 Qualitative Evaluation

One of the key properties of the proposed method is the sentiment word distributions for specific domains. Rank-LDA leverages on the rating scale assigned to reviews to learn a structured and generative model that represents the entire domain. This generative quality of the model, guarantees that words are represented by probability distributions across the entire range of sentiment levels. Figure 4 depicts examples of sentiment word distributions. In these figures the conditional probability density

[1] `http://www.cs.cornell.edu/people/pabo/movie-review-data/`

functions for each word is presented. We selected a sample of sentiment words to illustrate the probability of using a word given the sentiment level.

In Figure 4 the first two graphs illustrate the sentiment word distributions for the IMDb domain. The words *love* and *excellent* are general sentiment words that are used from a mid-range to a top-level sentiment value. However, it is interesting to note that in this domain the domain-specific sentiment word *oscar* is only used to express a highly positive sentiment. On the other hand the second graph illustrates words that are mostly used to express negative sentiment. We note that the sentiment word *watch* is used across the entire range of sentiment expressivity. This is an important feature, because the RLDA does not categorize a word as neutral (or positive/negative), instead it creates a fine-grain model of how likely is this word to occur at different sentiment levels. This is a critical feature to learn more elaborate sentiment word-interactions and to build more effective opinion retrieval systems. In the third and fourth graphs we turn our attention to the sentiment word distributions in the TripAdvisor dataset. In this domain we observed an interesting phenomena: the most positive words were quite general and not highly domain-specific. However, this was not true for the most negative sentiment word distributions: the word *dirty* is highly relevant in this domain (for obvious reasons), but the words *carpet*, and *smell*, are highly relevant because they are key for this particular domain. Again, this illustrates that our method captures both general and domain-specific sentiment words, thereby generating adequate lexicons. Moreover, on the graph, the words *friendly*, *helpful* and *fantastic* are mostly used to express positive sentiments, but they also occur in negative sentiment with other words (not or but).

4 Conclusions

To tackle opinion retrieval problems we proposed the Rank-LDA method, a generative model, to learn a highly structured sentiment space model that learns a domain-specific sentiment lexicon. It characterizes words in terms of sentiment distributions and, through its latent structure, it also captures interactions between words, thereby creating joint-distributions of sentiment words. We examined the impact of the dimensionality of the hidden structure in two datasets: IMDb-Extracted dataset and TripAdvisor dataset which contained 367,691 and 123,678 different words, respectively. To assess the generalization of the proposed method, we used these datasets from two distinct domains, in opinion retrieval and classification experiments. In all experiments, the improvements of the proposed method over the baselines, were as good as, or better than, existing methods. It is important to note that the word is not assigned a fixed value but a probability distribution instead. Our approach takes advantage of a more compact representation of opinion documents to achieve those levels of performance. This balance between performance and dimensionality is due to the sentiment word distributions that model a distribution density over the entire range of sentiment relevance.

Acknowledgements. This work was partially supported by the Portuguese Agency for Innovation under project QSearch (23133) co-financed by the *Fundo Europeu de Desenvolvimento Regional* (FEDER) through the Lisbon's Operational Program.

References

[1] Aktolga, E., Allan, J.: Sentiment diversification with different biases. In: Proceedings of the 36th International ACM SIGIR, p. 593 (2013)

[2] Andrzejewski, D., Zhu, X.: Latent Dirichlet Allocation with topic-in-set knowledge. In: Proceedings of the NAACL HLT 2009 Workshop on Semi-Supervised Learning for Natural Language Processing, pp. 43–48 (2009)

[3] Baccianella, S., et al.: SentiWordNet 3.0: An enhanced lexical resource for sentiment analysis and opinion mining. In: Proc.7th International LREC (2010)

[4] Blei, D.M., et al.: Latent dirichlet allocation. J. Mach. Learn. Res. 3, 993–1022 (2003)

[5] Blei, D.M., McAuliffe, J.D.: Supervised Topic Models. In: Advances in Neural Information Processing Systems (NIPS) (2007)

[6] Chemudugunta, C., et al.: Modeling Documents by Combining Semantic Concepts with Unsupervised Statistical Learning. In: Proceedings of the 7th International Conference on The Semantic Web, pp. 229–244 (2008)

[7] Chen, L., et al.: Extracting Diverse Sentiment Expressions with Target-Dependent Polarity from Twitter. In: Proc. of the 6th AAAI/ICWSM (2012)

[8] Esuli, A., Sebastiani, F.: Sentiwordnet: A publicly available lexical resource for opinion mining. In: Proc. of the 5th International LREC, pp. 417–422 (2006)

[9] Gerani, S., et al.: Proximity-based opinion retrieval. In: Proceeding of the 33rd international ACM SIGIR, p. 403 (2010)

[10] Hoffman, M., et al.: Online learning for latent dirichlet allocation.In: Advances in Neural Information Processing Systems (NIPS) (2010)

[11] Hu, M., Liu, B.: Mining and summarizing customer reviews. In: Proceedings of the tenth ACM SIGKDD, pp. 168–177 (2004)

[12] Jakob, N., et al.: Beyond the stars: exploiting free-text user reviews to improve the accuracy of movie recommendations. In: Proceeding of the 1st CIKM Workshop on Topic-sentiment Analysis for Mass Opinion (TSA), pp. 57–64 (2009)

[13] Jo, Y., Oh, A.H.: Aspect and sentiment unification model for online review analysis. In: Proc. of the 4h ACM Web Search and Data Mining, pp. 815–824 (2011)

[14] Martineau, J., Finin, T.: Delta TFIDF: An Improved Feature Space for Sentiment Analysis. In: Proc. of the AAAI on Weblogs and Social Media (2009)

[15] Mei, Q., et al.: Topic sentiment mixture: modeling facets and opinions in weblogs. In: Proceedings of the 16th International Conference WWW, pp. 171–180 (2007)

[16] Moghaddam, S., Ester, M.: ILDA: Interdependent LDA model for learning latent aspects and their ratings from online product reviews. In: Proceedings of the 34th International ACM SIGIR, pp. 665–674 (2011)

[17] Pang, B., Lee, L.: Opinion mining and sentiment analysis. Foundations and Trends in Information Retrieval 2(1-2), 1–135 (2008)

[18] Ramage, D., et al.: Labeled LDA: A Supervised Topic Model for Credit Attribution in Multi-labeled Corpora. In: EMNLP, pp. 248–256 (2009)

[19] Song, Y., et al.: Topic and keyword re-ranking for LDA-based topic modeling. In: Proceedings of the 18th ACM CIKM, pp. 1757–1760 (2009)

[20] Sparling, E.I.: Rating: How Difficult is It? In: RecSys 2011 Proceedings of the fifth ACM Conference on Recommender Systems, pp. 149–156 (2011)
[21] Titov, I., McDonald, R.: Modeling online reviews with multi-grain topic models. In: Proc. of the 17th International Conference on WWW, pp. 111–120 (2008)
[22] Turney, P.: Thumbs up or thumbs down? Semantic orientation applied to unsupervised classification of reviews. In: Proc. of the 40th ACL, pp. 417–424 (2002)
[23] Turney, P.D.: Mining the Web for Synonyms: PMI-IR versus LSA on TOEFL. In: Proc. of the 12th European Conference Machine Learning, pp. 491–502 (2001)
[24] Velikovich, L., et al.: The Viability of Web-derived Polarity Lexicons. In: Human Language Technologies: Proceedings of the NAACL, pp. 777–785 (2010)
[25] Wang, H., et al.: Latent aspect rating analysis on review text data. In: Proc. of the 16th ACM SIGKDD, p. 783 (2010)
[26] Weichselbraun, A., et al.: Extracting and Grounding Contextualized Sentiment Lexicons. IEEE Intelligent Systems 28(2), 39–46 (2013)
[27] Wilson, T., et al.: Recognizing contextual polarity in phrase-level sentiment analysis. In: Proceedings of the Conference on HLT/EMNLP, pp. 347–354 (2005)
[28] Zhang, M., Ye, X.: A generation model to unify topic relevance and lexicon-based sentiment for opinion retrieval. In: Proceedings of the 31st Annual International ACM SIGIR, pp. 411–418 (2008)

Modelling Public Sentiment in Twitter: Using Linguistic Patterns to Enhance Supervised Learning

Prerna Chikersal[1], Soujanya Poria[1], Erik Cambria[1],
Alexander Gelbukh[2], and Chng Eng Siong[1]

[1] School of Computer Engineering, Nanyang Technological University, Singapore
[2] Centro de Investigación en Computación, Instituto Politécnico Nacional, Mexico
prerna1@e.ntu.edu.sg, {sporia,cambria,aseschng}@ntu.edu.sg,
gelbukh@cic.ipn.mx
http://sentic.net

Abstract. This paper describes a Twitter sentiment analysis system that classifies a tweet as positive or negative based on its overall tweet-level polarity. Supervised learning classifiers often misclassify tweets containing conjunctions such as "but" and conditionals such as "if", due to their special linguistic characteristics. These classifiers also assign a decision score very close to the decision boundary for a large number tweets, which suggests that they are simply unsure instead of being completely wrong about these tweets. To counter these two challenges, this paper proposes a system that enhances supervised learning for polarity classification by leveraging on linguistic rules and sentic computing resources. The proposed method is evaluated on two publicly available Twitter corpora to illustrate its effectiveness.

Keywords: Opinion Mining, Sentiment Analysis, Sentic Computing.

1 Introduction

Nowadays, an increasing number of people are using social media to express their opinions on various subjects, as a result of which a vast amount of unstructured opinionated data has become available. By analysing this data for sentiments, we can infer the public's opinion on several subjects and use the conclusions derived from this to make informed choices and predictions concerning those subjects [1]. However, due to the volume of data generated, manual sentiment analysis is not feasible. Thus, automatic sentiment analysis is becoming exceedingly popular [2].

Polarity classification is a sub-task of sentiment analysis that focusses on classifying text into positive and negative, or positive, negative and neutral. Document-level polarity classification involves determining the polarity of opinions expressed in an entire document, whereas sentence-level polarity classification involves determining the polarity of opinions expressed in a single sentence. Another level of polarity classification is aspect-based polarity classification, which involves extracting opinion targets from text and then determining the polarity of the text towards that particular target. Surveys of methods used for various levels of sentiment analysis can be found in [3,4,5].

A. Gelbukh (Ed.): CICLing 2015, Part II, LNCS 9042, pp. 49–65, 2015.
DOI: 10.1007/978-3-319-18117-2_4

Tweets are short microblogging texts containing a maximum of 140 characters. They can include multiple sentences and often contain misspelled words, slangs, URLs, elongations, repeated punctuations, emoticons, abbreviations and hashtags. These characteristics make extracting sentiment and opinions from tweets a challenge, and hence an interesting topic of research. This paper focusses on tweet-level polarity classification, which involves predicting the overall polarity of opinions expressed in a tweet. We focus on classifying the tweets into positive or negative, and ignore the neutral class.

This paper is organised as follows. Section 2 explains the motivation or need for the proposed method; Section 3 briefly discusses related research; Section 4 describes the method; Section 5 presents and analyses the experimental results obtained; finally, Section 6 concludes the paper.

2 Motivation for Our Method

Supervised learning classifiers commonly used for polarity classification rely on feature vectors extracted from the text to represent the most important characteristics of the text. Word N-grams, which are denoted by the frequencies of contiguous sequences of 1, 2, or 3 tokens in the text, are the most commonly used features for supervised sentiment analysis. While such classifiers [6,7,8] have been shown to perform reasonably well, studies such as [9], [10] and [11] show that using a "one-technique-fits-all" solution for all types of sentences is not good enough due to the diverse types of linguistic patterns found in sentences. That is, the presence of modal verbs such as "could" and "should", conjunctions such as "but" and "or" and conditionals such as "if", "until", "unless", and "in case" in a text substantially worsen the predictions of a supervised classifier.

Furthermore, supervised learning classifiers classify each tweet with a certain probability or decision (confidence) score. For a large number of tweets, the decision score predicted by a typical supervised classifier is very close to the decision boundary. This implies that the classifier is unsure about which class the tweets in question belong to and so cannot assign class labels to them with much confidence. Thus, the class labels assigned to such tweets are either completely incorrect or correct mostly by fluke.

To prove this notion, we train a Support Vector Machine (SVM) classifier using n-grams (n = 1,2,3) as features on $\approx$ 1.6 million tweets provided by [8] and test it on 1794 positive and negative tweets provided by [12] and plot the decision scores computed by the SVM in Figure 1. In the graph, we can see that frequency of misclassifications reduce as we move away from the decision boundary ($y = 0$). We find that 341 tweets out of 1794 tweets are misclassified by the SVM, however 239 out of the 341 misclassified tweets have a decision score that lies between -0.5 and $+0.5$. Thus, the SVM is simply unsure instead of being completely wrong about these 239 tweets. If we consider all the predictions of the SVM, we get a misclassification rate[1] of $\approx 19\%$. But, if we exclude all the predictions whether right (475 tweets) or wrong (239 tweets) with a decision score between -0.5 and $+0.5$, we get a misclassification rate of only $\approx 9.4\%$. This means that if we consider the classification of only those tweets that the SVM is confident about, we can say that it correctly classifies over 90% of the tweets!

[1] misclassification rate $= \frac{Number\ of\ Incorrect\ Classifications}{Total\ Number\ of\ Classifications}$

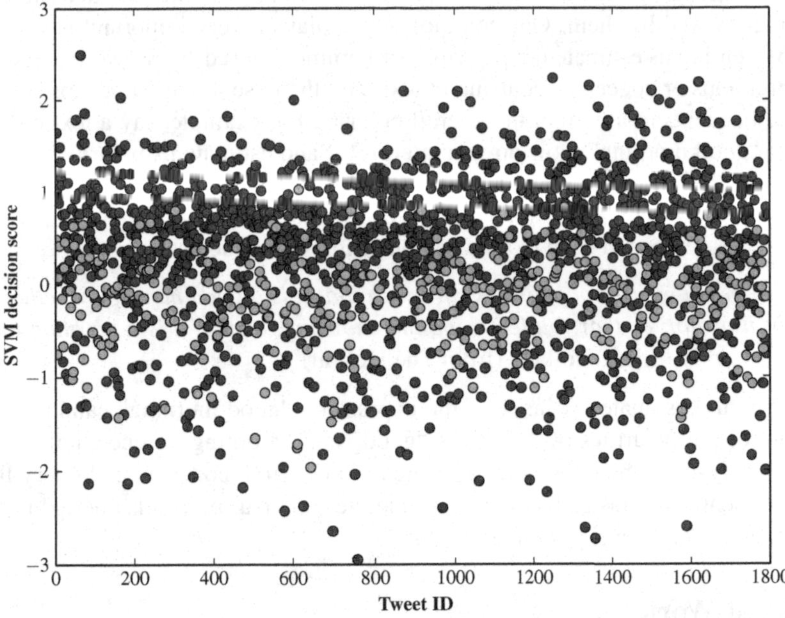

Fig. 1. SVM decision scores plotted for the classification of 1794 tweets into positive or negative using n-grams (n=1,2,3) as features. Tweets with decision score above 0 are labelled as positive, while tweets with decision score below 0 are labelled as negative. Blue = correctly classified (1453 tweets), Green = misclassified (341 tweets). 239 out of the 341 misclassified tweets have decision score between -0.5 and +0.5, implying that the SVM is simply unsure about them.

So, from the above, we can deduce that it would be beneficial to design a classifier that:

- Can handle special parts-of-speech of grammar such as conjunctions and conditionals.
- Uses a secondary (high-confidence) classifier to verify or change the classification labels of the tweets the SVM computes a very low decision or confidence score for.

To handle the special parts-of-speech of grammar, we modify the n-gram features provided as input to the classifier, based on linguistic analysis of how these parts-of-speech are used in sentences [13]. The scope of the method proposed in this paper is limited to the conjunction "but" and the conditionals "if", "unless", "until" and "in case".

Furthermore, we design an unsupervised rule-based classifier to verify or change the classification labels of the tweets the SVM computes a very low decision score for. The rules used by this classifier are based on our linguistic analysis of tweets, and leverage on sentiment analysis resources that contain polarity values of words and phrases. The primarily resource used for this purpose is SenticNet [14] – a semantic and affective resource for concept-level sentiment analysis, which basically assigns polarity values to concepts taken from a common-sense knowledge base called ConceptNet [15].

As human beings, we are able to understand the meaning of texts and determine the sentiment conveyed by them. Our common-sense plays a very important role in this process by helping us estimate the polarities of commonly used single-word and multi-word expressions or concepts occurring in text, and then use the relationships between words and concepts to ascertain the overall polarity. For example, say a text contains the phrase "good morning"; how do you interpret it and estimate its polarity? Luckily, depending on the context, our common-sense helps us deduce whether the expression "good morning" is used as a wish, as a fact, or as something else. Otherwise, without common-sense, we would need to ask each other questions such as

> *"Do you wish me a good morning, or mean that it is a good morning whether I want it or not; or that you feel good this morning; or that it is a morning to be good on?"* – J.R.R. Tolkien (from The Hobbit)

Moreover, the estimated polarity of the expression "good morning" cannot merely be the sum of the polarities of the words "good" and "morning". Hence, unlike most sentiment analysis methods, we prefer to break tweets into concepts and query those concepts in SenticNet, instead of relying completely on bag-of-words queried in lexicons containing word-level polarities.

3 Related Work

In this section, we briefly review some concepts and commonly used techniques for sentiment analysis that are relevant to the method proposed in this paper.

3.1 Supervised Learning for Sentiment Analysis

A text sample is converted to a feature vector that represents its most important characteristics. Given the feature vectors X and class labels Y for N number of training tweets, the supervised learning algorithm approximates a function F such that $F(X) = Y$. Now, in the testing phase, given feature vectors X' for T number of unlabelled tweets, the function F predicts labels Y' using $F(X') = Y'$ for each of the unlabelled tweets.

The most commonly used features for sentiment analysis are *term presence* and *term frequency* of single tokens or unigrams. The use of higher order n-grams (presence or frequency of 2,3,..,n contiguous tokens in a text) such as bigrams and trigrams is also prevalent, and allows for encoding of the tokens' positional information in the feature vector. Parts-of-speech and negation based features are also commonly used in sentiment analysis. Studies such as [16] and [17] focus on techniques used to represent negation, detect negation words, and determine the scope of negation in text.

More recent studies such as [6], [18], [19], and [20], exploit microblogging text or Twitter-specific features such as emoticons, hashtags, URLs, @symbols, capitalisations, and elongations to enhance sentiment analysis of tweets.

3.2 Unsupervised Learning and Linguistic Rules for Sentiment Analysis

Usually, unsupervised approaches for sentiment analysis such as [21] involve first creating a sentiment lexicon in an unsupervised manner, and then determining the polarity of a text using some function dependent on the number or measure of positive and negative words and/or phrases present in the text. A comparison of supervised methods and other unsupervised methods can be found in [22].

In [9], the authors define dependency-based linguistic rules for sentiment analysis, and merge those rules with common-sense knowledge, and machine learning to enhance sentiment analysis. Our proposed method is based on the idea illustrated in [9], however the linguistic rules we define are limited and not dependency based, because most dependency parsers do not perform well for microblogging texts such as tweets. Moreover, it is desirable to perform sentiment analysis of social media texts in real-time, and dependency parsers cannot be used in real-time due to the large time complexity of their algorithms. In this paper, our goal is to create a Twitter sentiment analysis classifier that classifies tweets in real-time while countering the two challenges postulated in 2.

3.3 Concept-Level Sentiment Analysis and Sentic Computing

So far sentiment analysis approaches relying on keyword spotting, word co-occurrence frequencies, and bag-of-words have worked fairly well. However, with increase in user-generated content such as microblogging text and the epidemic of deception phenomenon such as web-trolling and opinion spam, these standard approaches are becoming progressively inefficient. Thus, sentiment analysis systems will eventually stop relying solely on word-level techniques and move onto concept-level techniques. Concepts can be single-word or multi-word expressions extracted from text. Multi-word expressions are often more useful for sentiment analysis as they carry specific *semantics and sentics* [23], which include common-sense knowledge (which people acquire during their early years) and common knowledge (which people gather in their daily lives). The survey in [24] explains how Natural Language Processing research is evolving from methods based on bag-of-words to bag-of-concepts and finally on bag-of-narratives. In this paper, we define linguistic rules which rely on polarity values from a concept-level common-sense knowledge base called SenticNet [14].

4 The Proposed Method

Before analysing raw tweets for sentiments, we pre-process them. During pre-processing, all the @*<username>* references are changes to @*USER* and all the *URLs* are changed to *http://URL.com*. Then, we use the CMU Twitter Tokeniser and Parts-of-Speech Tagger [25] to tokenise the tweets and assign a parts-of-speech tag to each token. Apart from nouns, verbs, adjectives and adverbs, this tagger is also able to tag injunctions, and microblogging-specific tokens such as emoticons, hashtags, and URLs.

The proposed sentiment analysis system is illustrated in Figure 2, and is explained in detail in this section.

4.1 Emoticon Rules

Using the tokens in a tweet and the output of the tagger, we are able to find all the tokens that represent emoticons in the tweet. Since people often repeat certain punctuations to

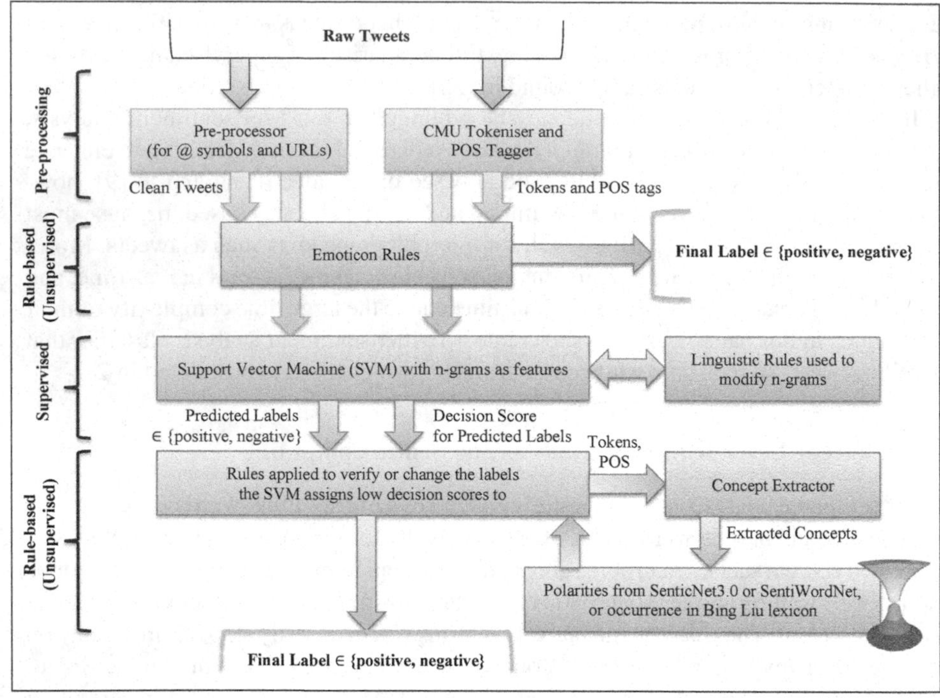

Fig. 2. Flowchart of the Proposed Twitter Sentiment Analysis System

emphasise emoticons, we remove all repeated characters from every emoticon string to obtain the bag-of-emoticons present in the tweet. Table 1 is a manually created list of usually polar emoticons along with their semantic orientation (*positive or negative*). We match emoticons in the bag-of-emoticons of the tweet to the list of positive or negative emoticons, and count the number of positive and the number of negative emoticons present in the tweet.

Table 1. Manually Created List of Positive and Negative Emoticons

Orientation	List of Emoticons
Positive	(-: , (: , =) , :) , :-) , =‘) , :‘) , :‘-) , =-d , =d , ;d , :d , :-d , ^_^ , ^_^ , :] , ^_- , ^_* , ^^
Negative	)-: ,): , =(,]: , :[, :(, :-(, >;(, >:(, :_(, d’x , :‘(, :”(, =’[, :’(, :’-(, \: , :/ , (~_~) , >__> , <(’-’)> , </3

Then, we apply the following rules to classify the tweet:

- If a tweet contains one or more positive emoticons and no negative emoticons, it is labeled as *positive*.
- If a tweet contains one or more negative emoticons and no positive emoticons, it is labeled as *negative*.
- If neither of the two rules above apply, the tweet is labeled as *unknown*.

If these emoticon-based rules label a tweet as *positive* or *negative*, we consider that label to be the final label outputted by our system. However, all tweets labelled as *unknown* by these rules are passed into the next stage in our sentiment analysis pipeline, that is the supervised learning classifier.

4.2 Support Vector Machine (SVM) with N-grams as Features

For supervised learning, we represent each tweet as a feature vector of case-sensitive n-grams (unigrams, bigrams, and trigrams). These n-grams are frequencies of sequences of 1, 2 or 3 contiguous tokens in a tweet. The TF-IDF [26] weighting scheme from information retrieval is applied to the frequency counts, and L1 regularisation is used for feature selection and dimensionality reduction. Finally, a Support Vector Machine is trained using the LIBLINEAR library [27].

To account for negation, we append the string "_NEG" to all negated tokens in a tweet. All tokens between certain negation words and the next punctuation mark are considered to be negated, as long as they are either nouns, adjectives, adverbs, or verbs. This is so because negating emoticons, hashtags or URLs would not make sense. Apart, from this, no other feature related to negation is used in the feature vector.

For this purpose, we take into account the following negation words: *never, no, nothing, nowhere, noone, none, not, havent, haven't, hasnt, hasn't, hadnt, hadn't, cant, can't, couldnt, couldn't, shouldnt, shouldn't, wont, won't, wouldnt, wouldn't, dont, don't, doesnt, doesn't, didnt, didn't, isnt, isn't, arent, aren't, aint, ain't.*

In section 4.4, the same method will be to find negated tokens in order to invert their polarity values.

4.3 Modifying N-grams According to Linguistic Rules

As mentioned in section 2, typical supervised learning methods based on n-grams perform badly on sentences containing special parts-of-speech such as conjunctions and conditionals commonly used in grammar, due to their peculiar linguistic characteristics. We theorise that one such characteristic is that a certain part of the sentence either becomes irrelevant for sentiment analysis or possesses a semantic orientation that is opposite to the sentence's overall orientation.

We analyse tweets containing the conjunction "but" and the conditionals "if", "unless", "until", and "in case", and formulate rules that should enable removal of irrelevant or oppositely oriented n-grams from the tweet's feature vector, before it is used for supervised learning.

Below are a few examples of tweets containing "but" at different syntactic positions. In each tweet, the most salient part that is the part that contributes considerably to the overall polarity of the tweet is underlined. In certain tweets however, if no salient part can be found or is ambiguous, nothing is underlined. The overall polarity of the tweet is indicated in parenthesis.

(1) @USER Tell you at our Friday lunch. Sorry for the late reply but yes we can eat somewhere on Marshall tomorrow haha *(positive)*

(2) it may have been against the minnows of FC Gomel, but a great performance from Rodger's Reds at Anfield tonight, and a great start! *(positive)*

(3) SP to support UPA, but oppose anti-people policies: Samajwadi Party on Saturday said it will continue to oppose *(negative)*

(4) Taylor Kitsch may not be a leading man, or able to open a movie, but he was quite good in The Bang Bang Club- Ryan Phillippe as well *(positive)*

(5) S/O to @USER ! I don't really know her but she seems real chill. She may not know how to spell Peyton Siva, but still follow her! *(positive)*

(6) you didnt win ABDC but you won over my heart you may not know me but imma true ICONiac by heart *(positive)*

(7) I laughed a little, but not sharing them out with anyone. Will the weather be good tomorrow for Boris Bikes? *(positive)*

(8) Gutted I'm missing the cardigan match on Saturday! But more important things to do *(negative)*

From the examples above, we observe that the part of the sentence posterior to the word "but" is usually (though not always) a better indicator of the overall polarity of the tweet, as compared to the anterior part. This premise holds true for examples (1) to (6), but does not work for a few examples such as (7) and (8).

In example (7), it is difficult to determine the most salient part of the tweet. This could be because that tweet appears to be only weakly positive, and could even be interpreted as negative if we only consider the posterior part of "but". In example (8), the most salient part of the tweet is anterior to the word "but", perhaps because the polarity of the posterior part is too subtle or even neutral. Nevertheless, in this paper, we will only focus on formulating rules that work for tweets similar to examples (1) through (6), as handling tweets similar to (7) and (8) is too difficult and requires more complex linguistic analysis which is beyond the scope of this study.

Furthermore, it is difficult to automatically determine the salient part in tweets similar to example (6), due to grammatical errors introduced by the writer of the tweet. That is, example (6) should contain 2 separate sentences, but there is no punctuation mark to separate "...my heart" and "you may not know me...", which makes it very hard for us to pick out the phrases "you won over my heart" and "imma true ICONiac by heart" as the most salient parts. Hence, to best handle such tweets, if there are more than one "but"s in the same sentence of a tweet, only the part posterior to the last occurrence of the word "but" is to be considered as the most salient.

Hence, we propose the following strategy to modify n-grams for tweets containing the conjunction "but":

1. We use the Punkt sentence tokeniser [28] to break a tweet into sentences.
2. In each sentence, we find the location of the last occurrence of the word "but"
3. We remove all tokens except the tokens posterior to (occurring after) that location. So, the modified sentence only contains tokens succeeding the last "but".
4. Once we have processed all the sentences in the tweet, we merge the modified sentences together to obtain the modified tweet.

Moving forwards, below are a few examples of tweets containing "if" at different syntactic positions. In each tweet, the most salient part that is the part that contributes considerably to the overall polarity of the tweet is underlined. In certain tweets however, if no salient part can be found or is ambiguous, nothing is underlined. The overall polarity of the tweet is indicated in parenthesis.

(1) If Gerald Green doesn't have the most hops in the league then he definitely is a strong 2nd!! *(positive)*

(2) If you're not coming to the SVSU vs. Wayne State game tomorrow, watch it on CBS College Sports or FSN Detroit. It's about to be hype! *(positive)*

(3) If the Lakers still had Jordan Farmar,Trevor Ariza,&Shannon Brown I'd be watching them ..I dont like the Lakers but they were entertaining. *(positive)*

(4) if you follow @USER ill make you my famous Oreo brownie on Sunday!!! *(positive)*

(5) Juniors playing powderpuff, if you aren't at practice tomorrow you will NOT play, it starts at 5:30pm, hope to see you there! *(negative)*

(6) @USER can you please come to SXSW in Austin in March? I've wanted to see you for years & it would be amazing if you played a show here! *(positive)*

From the above examples, we can see that as compared to "but", "if" has many more syntactic positions, such as:

(i) if <condition clause> then <consequent clause>
(ii) if <condition clause>, <consequent clause>
(iii) if <condition clause> <missing then/comma, or other> <consequent clause>
(iv) <consequent clause> if <condition clause>

According to syntax, example (1) is of type (i), example (2), (3) and (5) are of type (ii), example (4) is of type (iii), and example (6) is of type (iv). In examples (1) and (2), the most salient part of the tweet is the part that occurs after "then" or after the comma (,). Even in example (3), the part just after the first comma succeeding the "if" includes the most salient part of the tweet. However, example (3) contains both "if" and "but", which makes it harder to automatically determine the most salient part.

Moreover, in examples (4) and (5), the most salient part is not preceded by a "then" or comma, due to grammatical errors introduced by the writer or due to the informal nature of tweets. In example (6), "if" occurs in the middle of the sentence, such that even though the consequent clause usually precedes the "if" in such cases, it is hard to automatically determine the scope of the most salient part. Hence, determining the most salient part in tweets similar to (4), (5), and (6) requires more complex linguistic analysis which is beyond the scope of this study.

In this paper, we will only focus on tweets similar to (1), (2), and (3). Also, while the examples above are only limited to the conditional "if", we will also handle the conditionals "unless", "until" and "in case". For these conditionals, we consider the most salient part of the tweet to be the part that occurs after the first comma succeeding the conditional, whereas for "if" we consider the part occurring after "then" as well as the comma.

Therefore, we propose the following strategy to modify n-grams for tweets containing the conditionals "if", "unless", "until" and "in case":

1. We use the Punkt sentence tokeniser [28] to break a tweet into sentences.
2. In each sentence, we find the location of the last occurrence of the conditional ("if", "unless", "until" or "in case")
3. Then, we find the location of the first comma (and also "then" in case of "if") that occurs after the conditional.
4. We remove all tokens between the conditional and the comma/"then" including the conditional and the comma/"then". All the remaining tokens now make up the modified sentence.
5. Once we have processed all the sentences in the tweet, we merge the modified sentences together to obtain the modified tweet.

In case a tweet contains a conditional as well as the conjunction "but", only "but" rules are applied.

Finally, using the modified tweets, we create new feature vectors containing modified unigrams, bigrams, and trigrams for each tweet. These modified n-grams are then provided as input to the Support Vector Machine (SVM) specified in section 4.2, instead of the n-grams that are typically used.

4.4 Tweaking SVM Predictions Using Linguistic Rules and Sentic Computing

During training, a Support Vector Machine (SVM) approximates a hyperplane or decision boundary that best separates data points (feature vectors of samples) belonging to n different classes (feature vectors of samples = n-grams of tweets, n = 2 and class $\in$ {positive, negative} in our case). The data points that "support" this hyperplane on either sides are known as support vectors.

Each trained SVM has a scoring function that computes the decision score for each new sample, based on which the class label is assigned. The SVM decision score for classifying a sample is the signed distance from the sample's feature vector x to the decision boundary, and is given by:

$$SVM\ Decision\ Score = \sum_{m}^{i=1} \alpha_i y_i G(x_i, x) + b \quad (1)$$

where $\alpha_1, \alpha_2, \ldots \alpha_n$, and b are the parameters estimated by the SVM, $G(x_i, x)$ is the dot product in the predictor space between x and the support vectors, and m is the number of training samples.

As explained in section 2, the decision score for a large number of tweets is too low, implying that the SVM is unsure about the label it assigns to them, because their feature vector lies very close to the decision boundary. Hence, after running the supervised classifier on all the unlabelled tweets, we get the decision score computed by it for each tweet to determine the confidence of the SVM's predictions.

For tweets with an absolute decision score or confidence below 0.5, we discard the class labels assigned by the SVM and instead use an unsupervised classifier to predict their class labels. This unsupervised classification process works as follows:

1. The tweets are modified using the method describes in section 4.3, in order to take into account conjunctions and conditionals.
2. Single-word and multi-word concepts are extracted from the tweets in order to fetch their polarities from SenticNet [14]. These concepts are extracted using algorithm 1.

Algorithm 1. Given a list of *tokens* in a tweet and a list of their corresponding POS tags, this algorithm extracts a bag-of-concepts from tweets

```
token1 = []; pos1 = [];
{First, remove all stop words from the tweet tokens}
for each token, tag in tokens, pos do
   if token is NOT a stop word then
      append token to token1 and tag to pos1
   end if
end for
concepts = []
{adjacent tokens with the following POS tags² are extracted as multi-word concepts}
conceptTagPairs = [("N", "N"), ("N", "V"), ("V", "N"), ("A", "N"), ("R", "N"), ("P", "N"),
("P", "V")]
for ti in range(0, len(tokens1)) do
   token = tokens1[ti]; tag = pos1[ti];
   prevtoken = tokens1[ti-1]; prevtag = pos1[ti-1];
   token_stem = Stem(token); prevtoken_stem = Stem(prevtoken);
   {raw tokens and stemmed tokens are extracted as single-word concepts}
   append token to concepts
   append token_stem to concepts
   if (prevtag, tag) in conceptTagPairs then
      append prevtoken+" "+token to concepts
      append prevtoken_stem+" "+token_stem to concepts
   end if
end for
```

---

[2] "N" = Noun, "V" = Verb, "A" = Adjective, "R" = Adverb, "P" = Preposition.

3. Then, we query all these concepts in SenticNet in order to get their polarities. If a single-word concept is not found in SenticNet, it is queried in SentiWordNet [29], and if it is not found in SentiWordNet, it is searched for in the list of positive and negative words from the Bing Liu lexicon [30]. The number of positive and negative concepts, and the polarity of the most polar concept is noted as the tweet's most polar value. The Bing Liu lexicon only contains a list of around 2000 strongly positive and 4800 strongly negative words, and no polarity values. So, the polarity of all positive words in the Bing Liu lexicon is assumed as +1.0 while the polarity of all negative words is assumed as −1.0.
4. Based on the number of positive and negative concepts, and the most polar value occurring in the tweet, the following rules are applied to classify it:
   - If the number of positive concepts is greater than the number of negative concepts and the most polar value occurring in the tweet is greater than or equal to 0.6, the tweet is labelled as positive.
   - If the number of negative concepts is greater than the number of positive concepts and the most polar value occurring in the tweet is less than or equal to –0.6, the tweet is labelled as negative.
   - If neither of the two rules stated above apply, the tweet is labeled as unknown by the rule-based classifier, and the SVM's low confidence predictions are taken as the final output of the system.

5 Experiments and Results

We train our SVM [27] classifier on around 1.6 million positive and negative tweets provided by [8]. First, the training data is divided into 80% train and 20% validation sets, and the "c" parameter is selected as 0.4 through 10-fold cross-validation. Then, the model is trained on 100% of the training data.

Table 2. Results obtained on 1794 positive/negative tweets from the SemEval 2013 dataset

Method	Positive			Negative			Average		
	P	R	F	P	R	F	P	R	F
N-grams	90.48	82.67	86.40	61.98	76.45	68.46	76.23	79.56	77.43
N-grams and Emoticon Rules	90.62	83.36	86.84	62.99	76.65	69.15	76.80	80.00	78.00
Modified N-grams	89.95	84.05	86.90	63.33	74.59	68.50	76.64	79.32	77.70
Modified N-grams, and Emoticon Rules	90.10	84.73	87.33	64.41	74.79	69.22	77.26	79.76	78.27
Modified N-grams, Emoticon Rules, and Word-level Unsupervised Rules	91.40	86.79	89.04	68.55	77.89	72.92	79.97	82.34	80.98
Modified N-grams, Emoticon Rules, and Concept-level Unsupervised Rules	92.42	86.56	89.40	68.96	80.79	74.41	80.69	83.68	81.90

We evaluate our proposed method on two publicly available datasets – SemEval 2013 [12] test set and SemEval 2014 [12] test set. Neutral tweets are removed from each dataset, which leaves 1794 and 3584 positive/negative tweets in the SemEval 2013 and SemEval 2014 datasets respectively. Tables 2 and 3 show the results obtained on these two datasets. In these tables, each row shows the precision (P), recall (R), and F-score for the positive, and negative classes, followed by the average positive and negative precision, recall, and F-score. All values in the tables are between 0 and 100, and are rounded off to 2 decimal places. This section will focus on discussing and analysing the results shown.

In order to gauge the effectiveness of our method, we consider averaged positive and negative F-score (F_{avg}) as the primary evaluation metric, and the standard n-grams based supervised model as a benchmark. It is important to note that apart from TF-IDF weighed frequency counts of n-grams, this standard n-grams benchmark model also takes negation into account.

Table 3. Results obtained on 3584 positive/negative tweets from the SemEval 2014 dataset

Method	Positive			Negative			Average		
	P	R	F	P	R	F	P	R	F
N-grams	89.92	81.90	85.72	61.20	75.66	67.67	75.56	78.78	76.69
N-grams and Emoticon Rules	89.74	83.05	86.27	62.50	74.85	68.11	76.12	78.95	77.19
Modified N-grams	89.39	82.90	86.02	62.00	73.93	67.44	75.69	78.41	76.73
Modified N-grams, and Emoticon Rules	89.25	83.97	86.53	63.29	73.22	67.89	76.27	78.60	77.21
Modified N-grams, Emoticon Rules, and Word-level Unsupervised Rules	90.22	86.24	88.19	67.37	75.25	71.09	78.80	80.75	79.64
Modified N-grams, Emoticon Rules, and Concept-level Unsupervised Rules	90.41	86.20	88.25	67.45	75.76	71.37	78.93	80.98	79.81

On comparing the standard n-grams model with the n-grams and emoticon rules model, we can see that emoticon rules increase F_{avg} by 0.57 and 0.50 in the 2013 and 2014 datasets respectively. Comparison between the modified n-grams model, and modified n-grams and emoticon rules model also shows that emoticon rules increase F_{avg} by 0.57 and 0.48 in the two datasets respectively. Thus, this shows that the emoticon rules formulated by us significantly improve sentiment analysis.

Modifying n-grams using linguistic rules for conjunctions and conditionals increases F_{avg} by 0.27 and 0.04 in the two datasets respectively. While the increase is not very significant for the 2014 dataset, modified n-grams are still better than standard n-grams as (i) they do increase the overall F_{avg} and the increase is quite significant in the 2013 dataset, (ii) a typical Twitter corpus contains a very small percentage of tweets with such conjunctions and conditionals, and hence even a small improvement is very encouraging.

Next, we observe the results obtained by tweaking the SVM's predictions using the method specified in section 4.4. In this, we also compare the results obtained by using a bag-of-concepts model to the results obtained by using a bag-of-words (or single-word concepts only) model. We see that the F_{avg} of the bag-of-concepts model is 0.92 more than the bag-of-words model for the 2013 dataset, and 0.17 more than the bag-of-words model for the 2014 dataset. So, even though the effect of moving to concept-level sentiment analysis from word-level sentiment analysis will vary from one dataset to another, concept-level sentiment features will almost always perform better since they already include word-level sentiment features.

On comparing the results obtained by the modified n-grams and emoticon rules model with the modified n-grams, emoticon rules and concept-level unsupervised rules model, we see that tweaking the SVM's predictions using rules and sentic computing increases the F_{avg} by 3.63 and 2.6 in the two datasets respectively. Hence, this shows that the linguistic rules and sentic computing based secondary classifier proposed by us, substantially improve the result and is thus very beneficial for sentiment analysis.

Overall, our final sentiment analysis system achieves a F_{avg} score that is 4.47 units and 3.12 units higher than the standard n-grams model.

6 Conclusion and Future Work

In this paper, we describe the pipeline of a Twitter sentiment analysis system that enhances supervised learning, by using modified features for supervised learning as well as applying rules based on linguistics and sentic computing. Based on our results, we can conclude that unsupervised emoticon rules and modified n-grams for supervised learning help improve sentiment analysis. They do so by handling peculiar linguistic characteristics introduced by special parts-of-speech such as emoticons, conjunctions and conditionals. Moreover, we have shown that verifying or changing the low-confidence predictions of a supervised classifier using a secondary rule-based (high-confidence, unsupervised) classifier is also immensely beneficial.

In the future, we plan to further improve performance of our classifier [31]. We will do this by further analysing the linguistics of tweets to take into account other conjunctions such as "or", conditionals such as "assuming", or modal verbs such as "can", "could" , "should", "will" and "would". We also plan to develop more sophisticated rules to improve the classification of tweets that the supervised classifier assigns a low decision score to. Apart from deeper linguistic analysis and better rules, expanding common-sense knowledge bases such as SenticNet [14] and the use of concept based text analysis [32] can also help to boost the predictions of the unsupervised classifier, thereby improving the predictions of the whole system. The proposed approach can also be fed to a multimodal sentiment analysis framework [33][34]. Future work will also explore the use of common-sense vector space resources such as [35,36], construction of new ones [37], and extraction of aspects from the tweets [38], as well as richer n-gram [39], vector space [40], or graph-based [41] text representations.

Acknowledgements. We wish to acknowledge the funding for this project from Nanyang Technological University under the Undergraduate Research Experience on CAmpus (URECA) programme.

References

1. Li, H., Liu, B., Mukherjee, A., Shao, J.: Spotting fake reviews using positive-unlabeled learning. Computación y Sistemas 18, 467–475 (2014)
2. Alonso-Roríс, V.M., Santos Gago, J.M., Pérez Rodríguez, R., Rivas Costa, C., Gómez Carballa, M.A., Anido Rifón, L.: Information extraction in semantic, highly-structured, and semi-structured web sources. Polibits 49, 69–75 (2014)
3. Cambria, E., Schuller, B., Xia, Y., Havasi, C.: New avenues in opinion mining and sentiment analysis. IEEE Intelligent Systems 28, 15–21 (2013)
4. Pang, B., Lee, L.: Opinion mining and sentiment analysis. Foundations and Trends in Information Retrieval 2, 1–135 (2008)
5. Liu, B.: Sentiment analysis and opinion mining. Synthesis Lectures on Human Language Technologies 5, 1–167 (2012)
6. Kouloumpis, E., Wilson, T., Moore, J.: Twitter sentiment analysis: The good the bad and the omg? In: ICWSM 2011, pp. 538–541 (2011)
7. Pak, A., Paroubek, P.: Twitter as a corpus for sentiment analysis and opinion mining. In: LREC, vol. 10, pp. 1320–1326 (2010)
8. Go, A., Bhayani, R., Huang, L.: Twitter sentiment classification using distant supervision. CS224N Project Report, Stanford, pp. 1–12 (2009)
9. Poria, S., Cambria, E., Winterstein, G., Huang, G.B.: Sentic patterns: Dependency-based rules for concept-level sentiment analysis. Knowledge-Based Systems 69, 45–63 (2014)
10. Liu, Y., Yu, X., Liu, B., Chen, Z.: Sentence-level sentiment analysis in the presence of modalities. In: Gelbukh, A. (ed.) CICLing 2014, Part II. LNCS, vol. 8404, pp. 1–16. Springer, Heidelberg (2014)
11. Narayanan, R., Liu, B., Choudhary, A.: Sentiment analysis of conditional sentences. In: Proceedings of the 2009 Conference on Empirical Methods in Natural Language Processing, vol. 1, pp. 180–189. Association for Computational Linguistics (2009)
12. Nakov, P., Rosenthal, S., Kozareva, Z., Stoyanov, V., Ritter, A., Wilson, T.: SemEval-2013 task 2: Sentiment analysis in Twitter. In: Proceedings of the International Workshop on Semantic Evaluation, SemEval, vol. 13 (2013)
13. Datla, V.V., Lin, K.I., Louwerse, M.M.: Linguistic features predict the truthfulness of short political statements. International Journal of Computational Linguistics and Applications 5, 79–94 (2014)
14. Cambria, E., Olsher, D., Rajagopal, D.: SenticNet 3: a common and common-sense knowledge base for cognition-driven sentiment analysis. In: Twenty-Eighth AAAI Conference on Artificial Intelligence, pp. 1515–1521 (2014)
15. Speer, R., Havasi, C.: ConceptNet 5: A large semantic network for relational knowledge. In: The Peoples Web Meets NLP, pp. 161–176. Springer (2013)
16. Wiegand, M., Balahur, A., Roth, B., Klakow, D., Montoyo, A.: A survey on the role of negation in sentiment analysis. In: Proceedings of the Workshop on Negation and Speculation in Natural Language Processing, pp. 60–68. Association for Computational Linguistics (2010)
17. Councill, I.G., McDonald, R., Velikovich, L.: What's great and what's not: learning to classify the scope of negation for improved sentiment analysis. In: Proceedings of the Workshop on Negation and Speculation in Natural Language Processing, pp. 51–59. Association for Computational Linguistics (2010)
18. Davidov, D., Tsur, O., Rappoport, A.: Enhanced sentiment learning using twitter hashtags and smileys. In: Proceedings of the 23rd International Conference on Computational Linguistics: Posters, pp. 241–249. Association for Computational Linguistics (2010)
19. Mohammad, S.M., Kiritchenko, S., Zhu, X.: NRC-Canada: Building the state-of-the-art in sentiment analysis of tweets. In: Proceedings of the Second Joint Conference on Lexical and Computational Semantics (SEMSTAR 2013) (2013)

20. Chikersal, P., Poria, S., Cambria, E.: SeNTU: Sentiment analysis of tweets by combining a rule-based classifier with supervised learning. In: Proceedings of the International Workshop on Semantic Evaluation (SemEval 2015) (2015)
21. Turney, P.D.: Thumbs up or thumbs down?: Semantic orientation applied to unsupervised classification of reviews. In: Proceedings of the 40th Annual Meeting on Association for Computational Linguistics, pp. 417–424. Association for Computational Linguistics (2002)
22. Chaovalit, P., Zhou, L.: Movie review mining: A comparison between supervised and unsupervised classification approaches. In: Proceedings of the 38th Annual Hawaii International Conference on System Sciences, HICSS 2005, pp. 112c–112c. IEEE (2005)
23. Cambria, E., Hussain, A.: Sentic Computing: A Common-Sense-Based Framework for Concept-Level Sentiment Analysis. Springer, Cham (2015)
24. Cambria, E., White, B.: Jumping NLP curves: A review of natural language processing research. IEEE Computational Intelligence Magazine 9, 48–57 (2014)
25. Owoputi, O., O'Connor, B., Dyer, C., Gimpel, K., Schneider, N., Smith, N.A.: Improved part-of-speech tagging for online conversational text with word clusters. In: HLT-NAACL, pp. 380–390 (2013)
26. Salton, G., Buckley, C.: Term-weighting approaches in automatic text retrieval. Information Processing & Management 24, 513–523 (1988)
27. Fan, R.E., Chang, K.W., Hsieh, C.J., Wang, X.R., Lin, C.J.: LIBLINEAR: A library for large linear classification. The Journal of Machine Learning Research 9, 1871–1874 (2008)
28. Kiss, T., Strunk, J.: Unsupervised multilingual sentence boundary detection. Computational Linguistics 32, 485–525 (2006)
29. Esuli, A., Sebastiani, F.: Sentiwordnet: A publicly available lexical resource for opinion mining. In: Proceedings of LREC, vol. 6, pp. 417–422. Citeseer (2006)
30. Liu, B., Hu, M., Cheng, J.: Opinion observer: Analyzing and comparing opinions on the web. In: Proceedings of the 14th International Conference on World Wide Web, pp. 342–351. ACM (2005)
31. Schnitzer, S., Schmidt, S., Rensing, C., Harriehausen-Mühlbauer, B.: Combining active and ensemble learning for efficient classification of web documents. Polibits 49, 39–45 (2014)
32. Agarwal, B., Poria, S., Mittal, N., Gelbukh, A., Hussain, A.: Concept-level sentiment analysis with dependency-based semantic parsing: A novel approach. Cognitive Computation, 1–13 (2015)
33. Poria, S., Cambria, E., Hussain, A., Huang, G.B.: Towards an intelligent framework for multimodal affective data analysis. Neural Networks 63, 104–116 (2015)
34. Poria, S., Cambria, E., Howard, N., Huang, G.B., Hussain, A.: Fusing audio, visual and textual clues for sentiment analysis from multimodal content. Neurocomputing (2015)
35. Poria, S., Gelbukh, A., Cambria, E., Hussain, A., Huang, G.B.: EmoSenticSpace: A novel framework for affective common-sense reasoning. Knowledge-Based Systems 69, 108–123 (2014)
36. Cambria, E., Fu, J., Bisio, F., Poria, S.: Affectivespace 2: Enabling affective intuition for concept-level sentiment analysis. In: Twenty-Ninth AAAI Conference on Artificial Intelligence, pp. 508–514 (2015)
37. Vania, C., Ibrahim, M., Adriani, M.: Sentiment lexicon generation for an under-resourced language. International Journal of Computational Linguistics and Applications 5, 63–78 (2014)
38. Poria, S., Cambria, E., Ku, L.W., Gui, C., Gelbukh, A.: A Rule-Based Approach to Aspect Extraction from Product Reviews. In: Proceedings of the Second Workshop on Natural Language Processing for Social Media (SocialNLP), pp. 28–37. Association for Computational Linguistics and Dublin City University (2014)
39. Sidorov, G.: Should syntactic n-grams contain names of syntactic relations? International Journal of Computational Linguistics and Applications 5, 139–158 (2014)

40. Sidorov, G., Gelbukh, A., Gómez-Adorno, H., Pinto, D.: Soft similarity and soft cosine measure: Similarity of features in vector space model. Computación y Sistemas 18, 491–504 (2014)
41. Das, N., Ghosh, S., Gonçalves, T., Quaresma, P.: Comparison of different graph distance metrics for semantic text based classification. Polibits 49, 51–57 (2014)

Trending Sentiment-Topic Detection on Twitter

Baolin Peng[1,2], Jing Li[1,2], Junwen Chen[1,2], Xu Han[1,3], Ruifeng Xu[4], and Kam-Fai Wong[1,2,3]

[1] Dept. of Systems Engineering & Engineering Management, The Chinese University of Hong Kong, Shatin, N.T., Hong Kong
[2] MoE Key Laboratory of High Confidence Software Technologies, China
[3] Shenzhen Research Institute, The Chinese University of Hong Kong, Hong Kong
[4] Harbin Institute of Technology, Shenzhen Graduate School, China
{blpeng,lijing,jwchen,kfwong}@se.cuhk.edu.hk, xuruifeng@hitsz.edu.cn

Abstract. Twitter plays a significant role in information diffusion and has evolved to an important information resource as well as news feed. People wonder and care about what is happening on Twitter and what news it is bringing to us every moment. However, with huge amount of data, it is impossible to tell what topic is trending on time manually, which makes real-time topic detection attractive and significant. Furthermore, Twitter provides a platform of opinion sharing and sentiment expression for events, news, products etc. Users intend to tell what they are really thinking about on Twitter thus makes Twitter a valuable source of opinions. Nevertheless, most works about trending topic detection fail to take sentiment into consideration. This work is based on a non-parametric supervised real-time trending topic detection model with sentimental feature. Experiment shows our model successfully detects trending sentimental topic in the shortest time. After a combination of multiple features, e.g. tweet volume and user volume, it demonstrates impressive effectiveness with 82.3% recall and surpasses all the competitors.

Keywords: Twitter, Online social network, Trending topic detection, Sentiment analysis.

1 Introduction

Twitter, one of the most popular microblogging websites in the world, is regarded more as a new type of information source as well as news feed than an online communication platform[9]. It not only brings strong influence to our daily life, but also servers as a mirror reflecting real-life-events. On Twitter, people are now more concerned about what is happening on the world instead of what their friends or family are doing. Is Justin Bieber having a new girlfriend? How terrible the damage brought by Typhoon Haiyan to Philippines? Did Portugal or Sweden win the qualification for World Cup finals? A case study given by Sakaki[19] shows that the Twitter can act as a bursting event monitor and enable the information propagates even faster than other news media.

A. Gelbukh (Ed.): CICLing 2015, Part II, LNCS 9042, pp. 66–77, 2015.
DOI: 10.1007/978-3-319-18117-2_5

With more than 500 million registered users posting 340 million tweets per day[1], it is impossible to detect big news or important events in time manually. Additionally, Twitter provides a platform for opinion sharing and sentiment expression. On Twitter, people tends to express their true feelings and tells what they are really thinking about. One of the reasons why Twitter is regarded as valuable data resource is that it contains hundreds and thousands of opinions including discussions about social events, praises or complains about products, and so on[1]. By knowing how sentiment changes from time to time, we will have a better understanding about the evolution of topics, thus helps us determine whether a topic is trending or not. However, little real-time models for trending topic detection has taken sentiment features into consideration.

As such, we propose a real-time non-parametric trending topic detection model combining sentiment analysis methods. In experiment, the topics detected by our model are compared to the topics selected as trending topics by Twitter's own system. And it demonstrates that our system is able to detect trending topics using less time compared to current state-of-the-art real-time trending topic detection systems. Moreover, the topics detected by our model has stronger sentiment, some of which has brought strong influence to society although they maybe ignored by Twitter company at that time. Furthermore, we make a combination of multiple features and yield a compound model, which achieves highest effectiveness.

To sum up, the main contribution of this paper are mainly two folds :

- Our proposed approach utilizes time-varied sentiment to enhance the state-of-the-art real-time trending topic detection model.
- We propose a novel compound model combining multiple features, i.e. tweet volumes, sentiment and user volumes, which obtains best effectiveness and quick response.

2 Related Work

As to topic detection, topic model is one of the most well-known method. Diao et al.[5] processes a LDA model and made some adaptations for the diversity and noise in microblog. Gao et al. [6] proposed an incremental Gibbs sampling algorithm based on HDP to incrementally derive and rene the labels of clusters, thus helps to reveal the connections between topics. Though these works claims to have amazing result compared to their baseline topic model, they are not suitable to deal with streaming data thus can not be applied for real-time topic detection task. For real-time topic detection, there are three kinds of approaches[13]. The most popular one is to analyze the deviation of topics' activity relative to some baseline. Twitter monitor [12] cluster words to form topics and combine features like temporal features and social authority as an accurate description of each topic, adding to user interaction they successfully detect real-time topics over Twitter Streams. Similar methods are also given by Becker et al.[3] and Cataldi et al.[4].

[1] http://techcrunch.com/2012/07/30/analyst-twitter-passed-500m-users-in-june-2012-140m-of-them-in-us-jakarta-biggest-tweeting-city/

Nikolov[13] proposed a model based on time series classification and argue that from pattern of tweets with respect to timeline, we could determine which topic is trending. His work is attractive and claimed to be state-of-the-art but only takes tweets' amount variation into consideration and ignores the importance of sentiment. Whereas, Twitter is a platform for sentiment expression and opinion sharing, sentiment analysis and opinion mining draws particular attention in Twitter analysis. Among all the research topics in sentiment analysis, sentiment classification is perhaps most extensively studied[16] one. Its goal is to classify a subjective document as positive or negative with the help of some machine learning methods. Classical sentiment classification works mainly focus on two aspects.

One is to refine machine learning approaches and apply them to classify documents. Pang and Lee[17] are pioneers to use these classification algorithms to classify sentiment of movie reviews. In their work, Naive Bayes, Maximum Entropy model and Support Vector Machine (SVM) are applied to determine polarity of reviews as positive or negative. Back to the year of 1999, Wiebe et al.[23] came up with a method using Naive Bayes to determine whether a document is subjective or objective. Inspired by this, Pang and Lee[15] used a hierarchical classification model that treated subjectivity classification as the first step of sentiment analysis before polarity classification. Subjectivity classification with no doubt make sense based on the assumption that objective sentence imply little opinion though it maybe not correct under some special circumstances.

Another important aspect is feature selection. Six types of features are summarized as the most common features appeared in previous works [11]. Features based on terms and their frequencies are the most simple kind of feature but shown in Pang and Lee[17] to be effective using Naive Bayes and SVM as sentiment classifier. Part of speech of each word is also of importance, which is already proved effective in sentiment tagging task by Riloff et al.[18] and Wiebe et al.[22]. Sentiment words and phrases expressing positive or negative sentiment, are regarded as efficient sentiment indicators. Specifically, adjectives, adverbs, verbs and nouns are probable to be considered as sentiment words. Sentiment shifters can change sentiment orientations so that they should not be ignored.

3 Methodology

In this section, we describe how our model detect real-time sentimental topics. Firstly, sentiment scores indicating both polarity and intensity are given to tweets based on SVM sentiment classifier. Furthermore, sentimental temporal series are determined by estimating distribution of sentiment scores. Then we build a real-time trending topic detection model with sentimental features. A coarse-grained problem definition is given in the following part in Section 3.1

3.1 Problem Definition

Discovering topics from the massive social network information is a worthy task when we faces the big data scenario in social computing domain. But we still want to focus on a more specific scenario, which is sentiment sensitive and dynamic.

Taking a further considering, besides the newest events, we are arguing that the topic received heavier sentiment from users would be more tend to become a trending one. The intuition is directed, think about our everyday life, even though some news might get large volume of mentioned for it is formality (like a government document get published), but it is hardly to say it would be a hot topic if nobody have strong emotion or sentiment on it; Conversely, most of the trending topic is the exciting or controversial ones.

Taking such issues into account, and concerning the data gathering, we decide to define our problem on the Twitter platform, as Twitter provide sufficient API to manipulate the data, as well as an official streaming API, which would be stated in detail in Section 3.2

3.2 Data Collection and Preprocessing

Twitter Streaming API has accessing rate limitations for common developers, so it is impossible to reach a relative higher sampling rate. The sampling rate from the API is guaranteed to be no lower than 1% of the whole server streams according to Twitter documentation. We collected the raw tweets from Streaming API through more than 20 days, and around 10GB data was returned in every 24-hour. We totally gained more than nine hundreds of millions tweets, which are divided uniformly into three dataset for training, developing and testing respectively [2].

The trending hashtags in each hour are also provided by Twitter API, we collected them within the same period with the streaming data, and totally more than 1400 hot, or trending hashtags are retrieved. Each hashtag has a timestamp to indicate in which hour of which day it was trending.

Our goal is to decide whether a topic is trending or not. Generally speaking, the traditional way to defining a topic or event in social computing and information retrieval domain is using the bag-of-word model, which is the most simple but also most effective one. It defines the topic as a set of keywords, most of which is verb or noun, without considering the ordering of them, and practically this model could roughly depict an event already. On the other hand, Classical topic models such as LDA are not applicable here since we aim at detecting real-time topics from streaming data.

In this work, we regard hashtags consist of a word or a phrase with a hash symbol as a corse-grained topic, e.g. #londonriot, #TwitterParty,#nowplaying. Hashtags are created by Twitter users as a way to categorize tweets thus illustrates the connection among the topic. For one thing, hashtags are popular among Twitter users. Wang et al.[20] measured on a dataset with around 0.6 million randomly selected tweets and found that around 14.6% tweets have at least one hashtag. For another, real-time trending topics can be detected via trend analysis of hashtags and avoid delay brought by topic detection processing like clustering. In addition, we are not so care how the topic is described but how to find a trending one. Thus we would only choose a hashtag in Twitter to be a

[2] All the codes and data will be published and uploaded to google drive after reviewing.

topic, instead of using bag-of-word, and this setting could erase the complication on clustering keywords into a single topic.

Trending hashtags were selected from the hot topics received from Twitter as training set. We filtered out the hashtags existing for too short or too long period (less than 4 hours or longer than 24 hours), as these kinds of topics usually brought many noise and might be unstable along the time line. Also, to limit the topics lasting period within 24 hours is to avoid the periodic patterns. For the non-trending hashtags which is testing set, we would just randomly pick from the tags which were not existing in the trending list, and lasting for enough time (at least 4 hours) meanwhile. The time series would be generated with respect to each of these filtered hashtags.

To accelerate hashtag extraction and analysis, MapReduce is applied here. Each hashtag could be a key in MapReduce framework. The combination of tweet list relevant to a topic could be referred to the Reduce procedure. So we conclude that our preprocessing could be perfectly fit in the MapReduce framework. For each tweet, the mapper produces a pair of hashtag and twitter ID and the reducer combines twitter IDs relevant to each hashtag into a list.

As sentiment analysis is an important step in our work, NLP related preprocessing work such as tokenization, lemmatization, and part-of-speech annotation is required, which is implemented by Stanford Core NLP[3].

3.3 Sentimental Time Series

Before sentimental topic detection, we give sentiment scores to each tweet to indicate both the polarity and sentiment intensity. Sentiment analysis in this paper is based on SVM classifier with unigram features. According to Pang and Lee[15], unigram feature model is simple but has a good performance and when fed to SVM classifier, it achieves impressive performance and surpasses all other competitors[7]. A bit tweet set with 1600000 subjective tweets labeled by emoticons as positive or negative release by Go. et al. [7] is utilized as training data.

In many previous works, sentiment words are chosen as features in sentiment classification. However, most of these works are based on subjectivity lexicon, such as MPQA[4]. Unfortunately, subjective words in these lexicons hardly appear in tweets because Twitter users tend to use very casual language. So feature words are selected from training set, a huge corpus of tweets with emoticons, most of which can somehow regarded as subjective microblogs.

However, such a huge training set contains 794876 words and phrase but most of them are noise for sentiment analysis. According to Go et al.[7], usernames, links and repeated letters can be eliminated with the help of regular expression, thus shrinks the feature set down to 45.85%. Though they perform a good reduction to feature space, there are some other properties we can take advantage of for further feature reduction.

[3] http://nlp.stanford.edu/downloads/corenlp.shtml

[4] http://mpqa.cs.pitt.edu

Part-of-speech (POS) is considered of big significance in sentiment classification. Nouns, verbs, adjectives and adverbs are all probable sentiment indicators. Barboca and Feng[2] also mentioned top 5 features as positive polarity, negative polarity, verbs, good emoticons and upper cases based on training data. So we only keep emoticons and upper cases as well as words annotated as noun, verb, adjective or adverb.

Furthermore, word frequency in different polarity helps to evaluate its subjectivity. For example, words like happy, great, wonderful, will appear a lot more frequently in positive tweets. It also acts as a polarity indicator for Naive Bayes. In the last step of feature selection, we filter out words equally appearing in both polarity determined by the following indicator:

$$i(f) = \begin{cases} 1 & \text{if } 1\text{-}\theta < P(f|POS) < \theta \\ 0 & \text{otherwise} \end{cases}$$

and a feature f will be removed if $i(f) = 1$.With all aforementioned steps, we finally shrink features down to 8.65% of the original size.

Moreover, we'd like to define *sentiment score* in Definition 1 to illustrate both the polarity and sentiment intensity of a tweet.

Definition 1. *Sentiment score is a real value in [-1,1]: negative score indicates negative polarity while positive score denotes positive polarity. And a larger absolute value means higher degree of sentiment.*

Furthermore, SVM classifier not only determine the polarity but also helps to evaluate how subjective a tweet is. In fact, the farther the distance between tweet vector and SVM hyperplane, the more words as positive or negative sentiment indicators it contains, thus it is more subjective, or in other words, it expresses stronger sentiment. As a result, we use the score returned by *SVMlight*[8] as a sentiment score.

Specifically, we should point out that sentiment scores are relative instead of absolute. In other words, it only makes sense when sentiment of different tweet are comparing with each other. In fact, it is impossible to define exactly how strong the sentiment of a tweet is but sentiment comparison is much easier. For example, different person will have different evaluation in sentiment of the tweet 'I've done a good job'. But for tweet 'I've done a good job ' and 'I've done a great job', almost everyone will agree with the statement that the latter one has stronger positive sentiment than the previous one. As a result, we scale our sentiment scores into the interval of $[-1, 1]$ so as to fit Definition 1.

With sentiment scores given to Twitter topics, sentimental time series can be thus given. Considering that sentiment scores are relative values, a few steps mush be made in case of sentiment score's spiky or bursty distribution. Firstly, Histogram Equalization, which is a well-known method in image processing to adjust contrast using image's histogram, is applied for sentiment score smoothing due to the property of sentiment relativity. This method usually gains a higher global contrast particularly when the distribution of sentiment score is represented by close contrast values. Though it may reduce local contrast, through this adjustment, the intensities can be better distributed on the histogram and it effectively spreads out the most frequent intensity values thus increase the global contrast.

In order to calculate the temporal sentiment series, we propose a method to give a score to each time window based on the sentiment scores of tweets. First of all, we estimate the sentiment distribution in each time window by Parzen Window method with Gaussian Kernel given in the following:

$$p_n(s) = \frac{1}{nh_n} \sum_{i=1}^{n} K(\frac{s - s_i}{n}) \quad (1)$$

where $K(x) = \frac{1}{\sqrt{2\pi}} exp^{-\frac{x^2}{2}}$, n is the number of tweets, s_i indicate the sentiment score given to the i-th tweet in this time window, and h_n is the window width for sentiment score. Parzen window is non-parametric way to estimate the probability density function of a random variable inspired by histogram estimation method. And Gaussian Kernel is fundamental in data smoothing problem where inferences about the population are made, based on a finite data sample.

As a result, by calculate the expectation of sentiment in a time window, i.e. $e(t_i) = \int_{-1}^{1} sp_n(s)ds$ and we give a value $e(t_i)$ to time window t_i thus get a sequence $e(t_1), e(t_2), \cdots, e(t_n)$ as sentimental time series of a hashtag.

3.4 Real-Time Trending Topic Detection

The original time series is sharp due to rapid change on Twitter as shown in the left figure in Figure 1. In order to lower down noises and cover higher order, i.e. second order information, we applied the normalizing method mentioned in [13] to time series before Trending Topic Detection, then we can smooth the spiky time series as shown in the right figure in Figure 1. Specifically speaking, assume the $s[n]$ is the value of time series s at the index (time stamp) n and the method consists of the following steps:

- Baseline normalization: $s[n] = \frac{s[n]}{\sum_i^N s[i]}$, N is the total length of the series
- Spike normalization: $s[n] = |s[n] - s[n-1]|^\alpha$.
- Smoothing: $s[n] = \sum_(m = n - N_s mooth + 1)^n s[m]$, where the $N_s mooth$ is a parameter that controlling how smooth we want.
- Logarithmic: $s[n] = logs[n]$.

Then we use a supervised model proposed by Nikolov[13] with sentimental time series to detect trending sentimental topics.

In the model, we need to label some positive and negative instances as indicating signals. Positive instances are those hashtags known as trending ones. We use Twitter API to retrieve the trending hashtags that are suggested by Twitter system, which are partially labeled by people from Twitter, and apply them as positive sample of trending topic. In addition, some negative instances, i.e. non-trending hashtags, are also required. In addition, we randomly select some

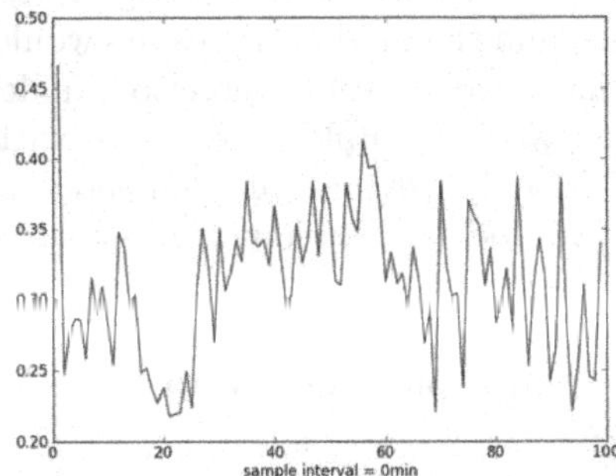

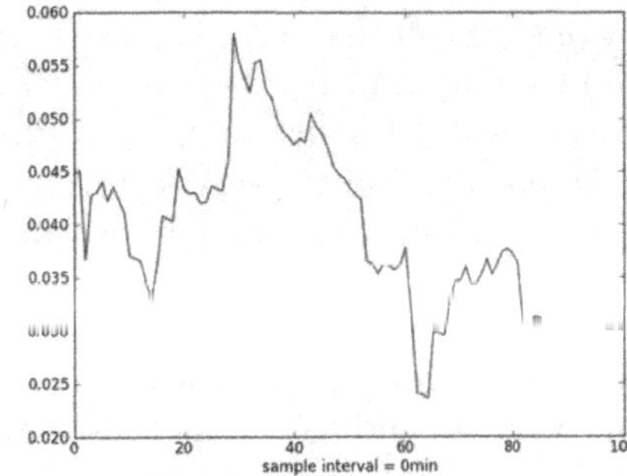

Fig. 1. Time Series Before and After Normalization

not trending hashtags to be negative samples. With positive set denoted by R_+ and negative set indicated as R_-, we are able to calculate the voting of each of these signals for an input time-series as a evaluation for trending topics:

$$R(s) = \frac{\sum_{c \in R_+} exp(-r \cdot d(s,c))}{\sum_{c \in R_-} exp(-r \cdot d(s,c))} \tag{2}$$

where r is a parameter that used to constrain system's sensitivity and $d(s_1, s_2)$ defines distance between two time series s_1 and s_2, and specifically speaking the Euclidean distance defined as following:

$$d(s_1, s_2) = \sqrt{\sum_i (s_{1i} - s_{2i})^2} \tag{3}$$

Then the value of $R(s)$ is used to determine which class the time series S belongs to, thus can tell whether a hashtag time series is trending or not.

4 Experiments

We tested our system on the test set with streaming data as previous mentioned, and would evaluate the results with two aspects: The trending prediction effectiveness and the trending detection time.

4.1 Effectiveness Evaluation

To evaluate the effectiveness of our system, we apply the (precision, recall, F-score) schema. We point out that in the context of topic detection, more emphasize should be put on recall rather than precision. As even if the system yields many actually non-trending topics, people don't want to miss the ones that are really trending. So we expect our system to guarantee high recall, which is just the case in the later experiments.

The test set contains around 400 trending topics and 300 non-trending topics. They are passed through the candidate systems, and the predicting results would be used to compare with the true values. Besides trending topic detection model with sentimental feature denoted by S-Model, we also implement model with other kinds of features like tweet volume (V-Model), which is implemented in the work of Nikolov[13] and user volume (UV-Model). Moreover, we make a linear combination of various features given in equation 4 as a compound one and integrate it into our model. Specifically in the equation below, $R_v(s)$, $R_s(s)$ and $R_u(s)$ are voting score of time series s given by our model with tweet V-Model, S-Model and UV-Model respectively.

$$R_v(s) = w_1 R_v(s) + w_2 R_s(s) + w_3 Ru(s) \tag{4}$$

We evaluate each model with precision, recall and F-score and furthermore compare it with baseline Bursting Model, i.e. Twitter Monitor[12]. The evaluating results are shown in Table 1.

Table 1. Effectiveness Evaluation of Different Models

Model	Precision	Recall	F-score
Bursting Model	60.5%	48.1%	53.6%
V-Model	65.3%	80.1%	71.9%
S-Model	57.5%	63.7%	60.4%
V+UV-Model	63.5%	75.1%	68.8%
V+S-Model	65.2%	**82.3%**	**73.3%**
V+U+S-Model	**68.3%**	78.1%	72.9%

Bursting model's performance just reaches the basic requirement line, but one thing interesting is even though the recall is low in this model, the precision is relatively good enough comparing with other models. The intuitive interpretation is that in the bursting model is stricter on judging whether a time series is trending or not, due to the bursting degree might not be spike enough to trigger the model for many actually trending time series.

Obviously, the volume feature is the most indicative features among all. Even with the classification model trained only on it, the recall is very high, just no more than 3 percentages lower than the highest recall among all. One could interpret this as the trending of time series are generally regarding the volume variation as indicator. This is also why some classic topic detection models, e.g. the bursting model, would take tweet volume series as their inputs.

The overall performance of our sentimental model is better than the baseline, with a relative higher recall, which is what we expected. We measured the hashtag set detected only by S-Model and only by V-Model with some sampled hashtags shown in table 2.

We can see that our sentimental model will pay more attention to topics with strong sentiment expression although they may not be selected as trending topics by Twitter company. For example, #TellAFeministThankYou is started by Melissa McEwan of Shakesville with the purpose to response to harassment

Table 2. Comparison of Sampled hashtags detected by V-Model and S-Model

V-Model	S-Model
#LoQueMasDeseoEs	#HappyBirthdayHarryFromLatinas
#mbv	#WaysToPissOffYourValentine
#NXZEROnoEncontro	#giornatadellamemoria
#EresLittleMonsterSi	#TuCaraMeSuena15
#PraSempreNossoEncantoPE	#WeWantMarcoLopezFor5Minutes
#RANHariBaru	#10TheBestMoviesEver
#QueremosBandaCineNoEncontro	#TellAFeministThankYou
#MJ50	#NationalSigningDay

of feminists on Twitter. It stroke American society and there were many news reports talking about this event.

When we combine multiple features together, interesting things happen. Even though user volume should have indicating function, when it was combined with the volume features the system performance even dropped; this might due to the correlation between user and tweet volume is relatively high, and when they are applied together even more noise would be created. Nevertheless, the compound model with volume and sentiment features, yields the highest F-score among all models, when the precision almost stay the same with only V-model. Yet the highest precision is achieved not by V-only or V-S model, but the compound-all model, namely applied all the V, S and U. So in general, from the overall aspect, one could expect the V-S model to be the best system, unless higher precision is required.

4.2 Detecting Time Evaluation

Since our model is a real-time model dealing with streaming data, how fast a trending topic is detected plays an important role. We also make an evaluation on the time trending topics are detected, which is illustrated by *detecting time* and defined as the following:

Since detecting time may vary from topic to topic so we calculate the average detecting time of different models and the results are shown in Table 3

What we can conclude from above table is that, the baseline, bursting model make the decision even after the Twitter company reports the trending topics. And the model our system based on can successfully detect trending topics before the report of Twitter company with whatever kind of feature. And among all kinds of features, V+S+U feature acts most slowly. And V+S-Model, which performs the best in the evaluation of effect has a not-bad performance, even faster than V-Model.

The model with the best detecting time is our S-Model. We interpret it as the sentiment information is more fierce at the beginning, and also have more stable patterns, which might give the system more confidence when it tries to give prediction when the topics start to spread. Also this property affects the performance of V+S-Mode which has better detecting time than V-Model.

Table 3. Detecting Time Evaluation of Different Models

Model	Detecting Time
Bursting Model	10.5 min
V-Model	-18 min
S-Model	**-33 min**
V+U-Model	-15 min
V+S-Model	-22 min
V+U+S-Model	-5 min

5 Conclusions

In this paper, we propose a real-time non-parametric model to detect trending topic. This work is based on a state-of-the-art supervised trending topic detection model and takes sentiment variation into consideration. In experiments, our model has the fastest response while making a descent decision. Although it does not have the best performance in effectiveness evaluation, i.e. precision, recall and F-Score, it successfully detect topics with stronger sentiment. Moreover, by a linear combination of multiple features, especially the tweet volume feature and sentimental feature, we get a model with best effectiveness and quick response.

However, there exists other kinds of features that are likely to make a contribution to the dynamic models. In addition, we could define different kinds of topics in the future, rather than the traditional concept which majorly based on the keyword bursting intuition. Moreover, the dynamic model, i.e. time series classification model might be applied to other social computing domains[14], e.g. rumour detection[10,21], along with a compound model taking several dynamic features into account at the same time.

Acknowledgement. This research is partially supported by General Research Fund of Hong Kong (417112), Shenzhen Fundamental Research Program (JCYJ20130401172046450, JCYJ20120613152557576), RGC Direct Grant (417613), National Natural Science Foundation of China No. 61370165 and Shenzhen Peacock Plan Research Grant KQCX20140521144507925.

References

1. Agarwal, A., Xie, B., Vovsha, I., Rambow, O., Passonneau, R.: Sentiment analysis of twitter data. In: LSM, pp. 30–38. Association for Computational Linguistics (2011)
2. Barbosa, L., Feng, J.: Robust sentiment detection on twitter from biased and noisy data. In: COLING(Poster), pp. 36–44. Association for Computational Linguistics (2010)
3. Becker, H., Naaman, M., Gravano, L.: Beyond trending topics: Real-world event identification on twitter. In: ICWSM (2011)
4. Cataldi, M., Di Caro, L., Schifanella, C.: Emerging topic detection on twitter based on temporal and social terms evaluation. In: MDMKDD, p. 4. ACM (2010)

5. Diao, Q., Jiang, J., Zhu, F., Lim, E.P.: Finding bursty topics from microblogs. In: ACL, pp. 536–544. Association for Computational Linguistics (2012)
6. Gao, Z.J., Song, Y., Liu, S., Wang, H., Wei, H., Chen, Y., Cui, W.: Tracking and connecting topics via incremental hierarchical dirichlet processes. In: ICDM, pp. 1056–1061. IEEE (2011)
7. Go, A., Bhayani, R., Huang, L.: Twitter sentiment classification using distant supervision. CS224N Project Report, Stanford, pp. 1–12 (2009)
8. Joachims, T.: Svmlight: Support vector machine. SVM Light Support Vector Machine http://svmlight. joachims. org/, University of Dortmund 19(4) (1999)
9. Kwak, H., Lee, C., Park, H., Moon, S.: What is twitter, a social network or a news media? In: WWW, pp. 591–600. ACM (2010)
10. Li, B., Zhou, L., Wei, Z., Wong, K., Xu, R., Xia, Y.: Web information mining and decision support platform for the modern service industry. In: ACL, pp. 97–102 (2014), http://aclweb.org/anthology/P/P14/P14-5017.pdf
11. Liu, B.: Sentiment analysis and opinion mining. Synthesis Lectures on Human Language Technologies 5(1), 1–167 (2012)
12. Mathioudakis, M., Koudas, N.: Twittermonitor: Trend detection over the twitter stream. In: SIGMOD, pp. 1155–1158. ACM (2010)
13. Nikolov, S.: Trend or No Trend: A Novel Nonparametric Method for Classifying Time Series. Ph.D. thesis, Massachusetts Institute of Technology (2012)
14. Ou, G., Chen, W., Wang, T., Wei, Z., Li, B., Yang, D., Wong, K.: Exploiting community emotion for microblog event detection. In: EMNLP, pp. 1159–1168 (2014)
15. Pang, B., Lee, L.: A sentimental education: Sentiment analysis using subjectivity summarization based on minimum cuts. In: ACL, p. 271. ACL (2004)
16. Pang, B., Lee, L.: Using very simple statistics for review search: An exploration. In: COLING (Posters), pp. 75–78 (2008)
17. Pang, B., Lee, L., Vaithyanathan, S.: Thumbs up?: Sentiment classification using machine learning techniques. In: ACL, pp. 79–86. Association for Computational Linguistics (2002)
18. Riloff, E., Wiebe, J., Wilson, T.: Learning subjective nouns using extraction pattern bootstrapping. In: HLT-NAACL, pp. 25–32. Association for Computational Linguistics (2003)
19. Sakaki, T., Okazaki, M., Matsuo, Y.: Earthquake shakes twitter users: Real-time event detection by social sensors. In: WWW, pp. 851–860. ACM (2010)
20. Wang, X., Wei, F., Liu, X., Zhou, M., Zhang, M.: Topic sentiment analysis in twitter: a graph-based hashtag sentiment classification approach. In: CIKM, pp. 1031–1040. ACM (2011)
21. Wei, Z., Chen, J., Gao, W., Li, B., Zhou, L., He, Y., Wong, K.: An empirical study on uncertainty identification in social media context. In: ACL, pp. 58–62 (2013)
22. Wiebe, J., Riloff, E.: Creating subjective and objective sentence classifiers from unannotated texts. In: Gelbukh, A. (ed.) CICLing 2005. LNCS, vol. 3406, pp. 486–497. Springer, Heidelberg (2005)
23. Wiebe, J.M., Bruce, R.F., O'Hara, T.P.: Development and use of a gold-standard data set for subjectivity classifications. In: ACL, pp. 246–253. Association for Computational Linguistics (1999)

EmoTwitter – A Fine-Grained Visualization System for Identifying Enduring Sentiments in Tweets

Myriam Munezero[1], Calkin Suero Montero[1], Maxim Mozgovoy[2], and Erkki Sutinen[1]

[1] School of Computing, University of Eastern Finland, Joensuu, Finland
{mmunez,calkins,erkki.sutinen}@uef.fi

[2] The University of Aizu, Aizu-wakamatsu, Fukushima, Japan
mozgovoy@u-aizu.ac.jp

Abstract. Traditionally, work on sentiment analysis focuses on detecting the positive and negative attributes of sentiments. To broaden the scope, we introduce the concept of *enduring sentiments* based on psychological descriptions of sentiments as enduring emotional dispositions that have formed over time. To aid us identify the enduring sentiments, we present a fine-grained functional visualization system, EmoTwitter, that takes tweets written over a period of time as input for analysis. Adopting a lexicon-based approach, the system identifies the Plutchik's eight emotion categories and shows them over the time period that the tweets were written. The enduring sentiment patterns of *like* and *dislike* are then calculated over the time period using the flow of the emotion categories. The potential impact and usefulness of our system are highlighted during a user-based evaluation. Moreover, the new concept and technique introduced in this paper for extracting enduring sentiments from text shows great potential, for instance, in business decision making.

Keywords: Sentiment Analysis, Emotions, Enduring.

1 Introduction

The ability to accurately identify and reflect the sentiments 'out there' is valuable and challenging. This ability influences real world text analysis applications such as, targeting marketing or political campaigns to users and voters with specific likes and dislikes, or identifying online antisocial behavior e.g., cyberbullying. Thus, understanding and being able to detect the development of sentiments in text is a desirable asset.

The area of natural language processing (NLP) that broadly deals with the computational treatment of opinions, feelings, emotions and subjectivity in texts is sentiment analysis (SA) [20]. Current work in SA focuses on classifying sentiments based on the polarity/valence (positive, negative, neutral) of text [20]. Some research have further explored classifying the sentiment intensity ([31]; [29]); while others have further explored extracting features such as the source and target of a

A. Gelbukh (Ed.): CICLing 2015, Part II, LNCS 9042, pp. 78–91, 2015.
DOI: 10.1007/978-3-319-18117-2_6

sentiment expressed in text [30]. Although sentiments are characterized by valence and intensity, they are in fact more complex.

Sentiments are not just momentous constructs, they are defined as "an acquired and relatively permanent major neuropsychic disposition to react emotionally, cognitively, and conatively toward a certain object (or situation) in a certain stable fashion, with awareness of the object and the manner of reacting" [2] (See section 2.6 for further discussion). Unlike brief emotional episodes, sentiments about an object are formed over time and are enduring [23]. For example, a single tweet may read "I am angry at my sister today", this statement is an emotional response to something that "the sister" has done "today", whereas the enduring sentiment towards the sister might be in fact pleasant and loving majority of the time. It is the dynamics of sentiment formation that lead to "enduring patterns of liking and disliking" of objects [26]. Research in SA has thus far concentrated on the momentous expression of feelings and emotions, and not made strides to identify the enduring sentiments (see [17] for an analysis on the differences between emotions and sentiments).

To rectify this problem, our work in this paper goes into the deeper analysis of "sentiment over time" to explore the concept of *enduring sentiment.* The proliferation of social media allows us to obtain enough user data to go deeper and analyze this concept. Hence, beyond negative and positive attributes, we can investigate changes in emotion dispositions formed over time and how these changes reflect into sentiments.

The main contribution of our work relies on the ability to extract intrinsic emotional knowledge through the social network Twitter and provide a fine-grained visualization of that knowledge. EmoTwitter presents a visual time analysis of the emotional information flow of Twitter users towards certain topics and estimates the enduring sentiment towards those topics. The current work is part of a broader project on detecting antisocial behavior from online sources, whereby identifying enduring sentiments is beneficial in the prediction of future behavior.

2 Background

2.1 Twitter Microblogging Platform

Twitter is a social microblog platform that allows people to post their views and sentiments on any subject; from new products launched, to favorite movies or music to political decisions [14]. As a microblog platform, users can only share short messages called tweets (max 140 characters) which are usually written by one person who updates it personally [14]. The content of tweets can be personal or things that a person considers of interest.

The popularity of Twitter and the vast amount of information posted through it has made it attractive for natural language analysts. The tweets are also public and hence accessible to researchers unlike most social network sites [28]. Furthermore, tweets are reliably time stamped so that they can be analyzed from a temporal perspective.

2.2 Sentiment Analysis for Twitter

SA has usually been treated as a simple classification task, classifying texts into positive and negative (and sometimes neutral) categories (see [20] for a review). Martínez-Cámara et al., [14] note that SA on tweets is no different from the analysis on long texts even though the short length of the tweets or their linguistic style that tends to be informal, with abbreviations, short hand, idioms, misspellings and incorrect use of grammar, make it difficult for NLP analysis.

Over the past two years a large number of SA programs have been developed to predict the sentiment content of texts in tweets. Normally, analyzing sentiment in tweets takes one of two approaches: lexicon-based or machine learning (ML). Lexicon-based approaches rely upon different features such as the presence of emoticons or certain words and phrases in order to determine the polarity of a sentence or document [28]. For example, Hogenboom et al., [10] made use of emoticons as features. Emoticons are a reoccurrence in tweets, and were found to be a good classification feature. Other lexicons used for sentiment classification include SentiWordNet [4] and WordNet-affect [27], both which contain words that have been labeled with their polarity orientation.

On the other hand, ML techniques involve the building of classifiers (e.g. Näive Bayes, maximum entropy and support vector machines) from instances of labeled tweets [7]. ML techniques can be employed to use several features including ones from lexicons for classification.

Our work falls within the lexicon-based approach. Advantageously, using the lexicon-based approach allows our system to handle slang words, misspellings and also keyword sets for different sets of languages like Spanish or French.

2.3 Enduring Sentiment

Murray and Morgan [18] define sentiment as "a more or less enduring disposition (predilection or readiness) in a personality to respond with a positive or negative affect to a specified entity". The word enduring is defined by Oxford as "lasting over a period of time" [19]. In Gordon [8], a similar definition of sentiment is found; sentiments are "socially constructed patterns of sensations, expressive gestures, and cultural meanings organized around a relationship to a social object, usually another person (...) or group such as a family". Broad [1] explains that a sentiment is formed when a certain object is constantly perceived or thought of by a person and, over time, the person creates a dispositional idea towards the object. This dispositional idea has corresponding emotional tendencies that are evoked whenever the person perceives, thinks about the object or any symbols related to the object. In Pang and Lee [20], a sentiment "suggests a settled opinion reflective of one's feelings", where the word 'settled' indicates something that reoccurs over time.

The definition of sentiment in SA has often been simplified to, for instance, as an explicit or implicit expression in text of the writer's positive, negative or neutral regard toward a subject [12]. However, beyond this, sentiments are enduring and that is the focus of our investigation.

Identifying the enduring sentiment could prove beneficial, for instance, in sorting reviewers by relevancy. For example, a bad review of the movie 'Sky Fall' from a person who likes action movies would have more merit for a recommendation system, than a bad review from someone who dislikes action movies. The enduring sentiment attribute can additionally be used in market analysis to identify the loyal customers; those who have liked a brand or product for a long period of time even if they have been instances of dislike.

2.4 Temporal Sentiment Visualization

Havre et al., [9] proposed an information visualization system called ThemeRiver that visualizes thematic variations over time within a large collection of documents. The "river" flows from left to right through time, changing width to depict changes in thematic strength of temporally associated documents. Colored "currents" flowing within the river narrow or widen to indicate decreases or increases in the strength of an individual topic or a group of topics in the associated documents.

Mishne and Rijke [15] also developed a system called MoodViews, which is a collection of tools for analyzing, tracking and visualizing moods and mood changes in blogs posted by LiveJournal users.

Fukuhara et al., [6] similarly to ThemeRiver, focus on the visualization of both topical and sentiment flow along within a timeline. The method accepts texts with timestamps such as Weblogs and news articles, and produces two kinds of graphs, i.e., (1) topic graph that shows temporal change of topics associated with a sentiment, and (2) sentiment graph that shows temporal change of sentiments associated with a topic.

TwitInfo, a prototype system for monitoring events on Twitter, uses a timeline graph showing the major peaks of publication of tweets about a particular topic, the most relevant tweets, and the polarity of the opinions they express [13].

Duan et al., [3] further developed three interactive widgets that are arranged together to create coordinated multiple views: the sentiment trend view showing the temporal sentiment dynamics and sentiment comparisons among different categories/topics, the chart visualization view illustrating the associated structured facets, and the snippet/document panel providing details of documents and context of sentiment. Mash-up capabilities among the three views allow the user to navigate the data set using optimal interactions.

A more recent visualization system is that from Kempter et al., [11] called EmotionWatch. It automatically recognizes emotions using the Geneva Emotion Wheel, version 2.0 [25]. They score tweets into 20 discrete emotion categories. In their research, they found out that 20 emotions are too many for users.

However in all the visualization systems above, they aggregate the sentiments and summarize the results topic-wise or polarity-wise. None of them have yet to create a visual representation illustrating the flow of emotions a single user has towards a topic over a period of time. This is necessary when analyzing the emotional dispositions that form the enduring sentiment.

Ours is the first visualization system to combine multiple views while allowing for extraction of emotions, as well as polarity and the analysis the emotions over time in order to identify the enduring sentiments.

3 EmoTwitter

For the purposes of emotional analysis of tweets we have designed and implemented an automated system that performs several actions. First, EmoTwitter accepts a Twitter user name and downloads Twitter posts written by the user (Fig. 1). The system then extracts the emotions present in the tweets, their negative and or positive attributes, and the topics in the tweets. It then produces a fine-grained visualization of the emotional information in the collection and produces a visualization of the enduring sentiment.

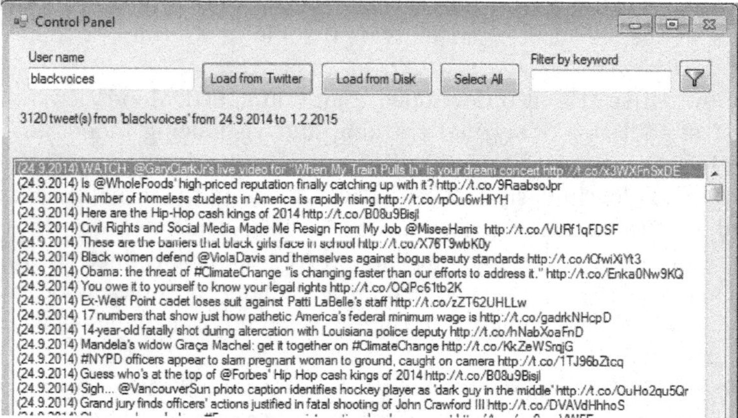

Fig. 1. View of downloaded tweets from Twitter user 'blackvoices'. The window shows the total count and time period of the downloaded tweets. It also includes a place to enter keyword and filter tweets by that particular keyword.

3.1 Downloading Tweets

It is fairly easy to obtain a collection of tweets as the posts are inherently public and a percentage of them are available for download [28]. Using the Twitter API, our software system can download a collection of past tweets of any specified user. For the subsequent analysis, we remove retweets as they might reflect emotions of Twitter users other than the user in consideration. Note, however, that the Twitter API can only return up to 3,200 of a user's most recent tweets, including retweets. A possible way to overcome this limitation would be to constantly follow certain users and store their tweets as they appear.

3.2 Extraction of Emotional Indicators

In order for a richer exploration of emotions that goes beyond the mere polarity of tweets, we extract emotions from the tweets by comparing each sentence in a tweet against the NRC word-emotion association lexicon [16]. The lexicon has been manually annotated into eight categories according to Plutchik's [21] eight basic emotions: joy, sadness, fear, anger, anticipation, surprise, disgust, and trust. The annotations also include scores for whether a word is positive or negative. Each score in the lexicon is simply a boolean marker, denoting whether the given word belongs to a given emotion category. In our calculations, when a word in a tweet matches a word in the lexicon, we mark that word with a score of 1 within the matched emotion category, and when the word does not match any word in the lexicon, we mark it with a score of 0. Currently the lexicon includes emotional annotations for 6,468 unique words. Further descriptions of the lexicon can be found in an article by Mohammad and Turney [16].

In our work, an emotion score (*eScore*) is calculated for each one of Plutchik's eight categories represented in each tweet as follows:

$$eScore_{category} = \frac{eWords_{category}}{eWords_{all}} \tag{1}$$

Where:

- $\mathrm{eWords}_{\mathrm{category}}$ is the number of words in the uploaded tweets that have nonzero emotional score for the category according to the NRC lexicon.

- $\mathrm{eWords}_{\mathrm{all}}$ is the number of words in the uploaded tweets that have nonzero emotional score for any category according to the NRC lexicon.

3.3 Data Visualization

Our system visualizes obtained scores with a variety of graphical forms. The visualization shows the following:

3.3.1 Emotion Distribution

Emotional scores, calculated over a sequence of tweets, are visualized on a radar chart that directly resembles Plutchik's wheel of emotions.

The radar chart (Fig. 2) has eight independent axes, corresponding to the individual primary emotions. For each axis, we calculate a point of average emotional score over the uploaded tweets (i.e., for each tweet, we calculated the *eScore* and then summed up all the *eScores*, and divided the resulting value by number of entries). Then these points become vertices of a filled polygon, thus providing a convenient visualization for the eight primary emotional scores.

3.3.2 Emotion Polarity

In addition, polarity attribute scores are visualized on a bar graph (see Fig. 3). Since "positive" and "negative" annotations are directly present in the NRC lexicon, in Fig. 3, we make use of this information. The visualization shows the positive and negative average values for the given range of uploaded tweets.

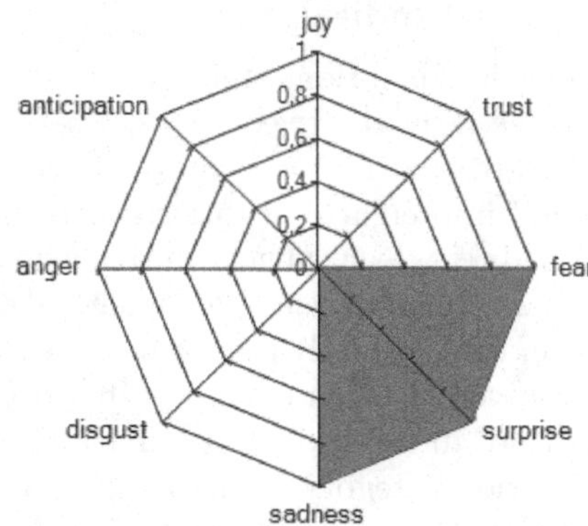

Fig. 2. Distribution of emotions within the downloaded tweets of Twitter user 'black-voices'

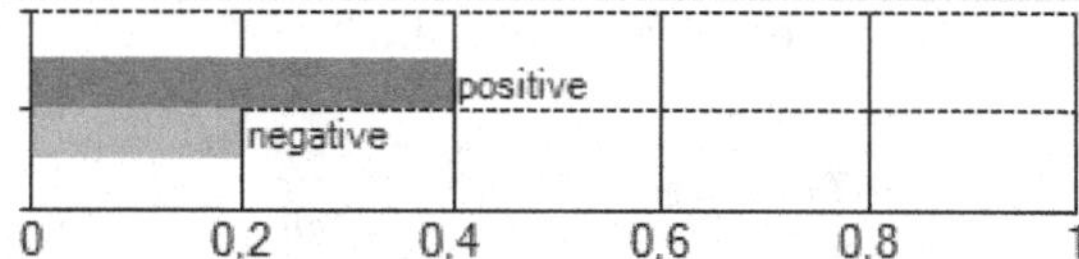

Fig. 3. Downloaded tweets' positive and negative score averages of Twitter user 'black-voices'

3.3.3 Topics

Most topics found frequently in the uploaded tweets are displayed as a word cloud, which is convenient for identifying frequent topics conveyed by one particular Twitter user (see Fig. 4). Before building a word cloud, we apply a stop-word removal procedure and Porter's stemming algorithm [22]. These steps help to focus on the linguistically significant components of text by removing common words that are not topics such as "without", "soon", "sometime", etc.

EmoTwitter further allows for the interaction with the word cloud. By clicking on any word in the word cloud, the system filters the tweets to only display those tweets talking about the clicked topic. Fig. 5 displays a screenshot of the word cloud where for instance the topic "Ferguson" has been selected.

3.3.4 Temporal Flow of Emotions

Our system also allows for the visualization of the temporal flow of emotions in the uploaded tweets. For the temporal flow-chart, a user has the choice of visualizing all the emotions or selecting a combination of emotions. Fig. 6 shows the visualization when four of the emotions are selected.

For the temporal flow-chart, we build average emotional scores for each time entry and thus obtain a calendar-like view of emotional changes in the writings. In the given samples, the entries are given for the three different dates; hence,

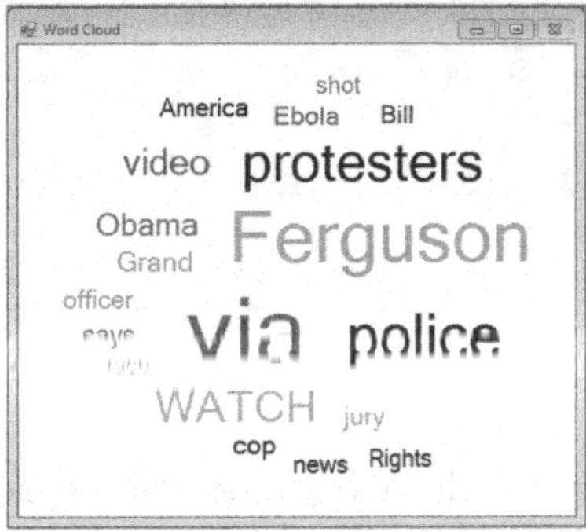

Fig. 4. Word cloud illustrating frequent topics for the Twitter user 'blackvoices'

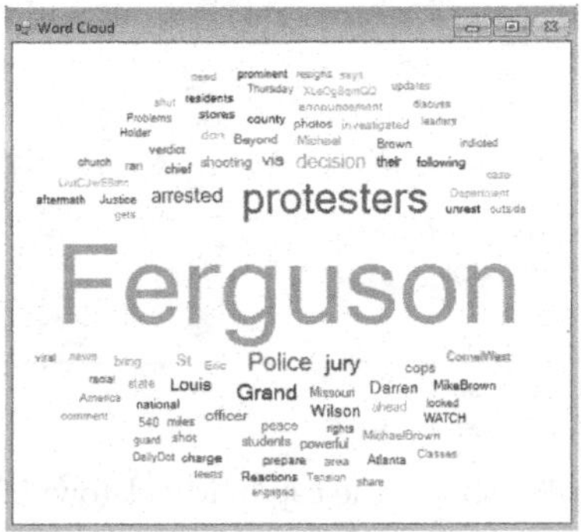

Fig. 5. Word cloud illustrating a selected topic 'Ferguson' from the topics of Twitter user 'blackvoices'

the graphs are built for all time points. The Y-axis values for the flow chart are calculated in the same manner as for the *eScore*, (i.e., number of emotional words for the given category divided by the number of emotional words).

As sentiments can be classified according to the nature of their emotional disposition or according to their objects [5], we classify the enduring sentiments according to the former as it fits our purpose of investigating the sentiments formed towards an object.

Thus, we mapped the observed emotions given in the tweets onto two broad categories of enduring sentiments, *Like* and *Dislike* with the following formula:

$$Like = Average(eScore_{joy,trust,anticipation}) \tag{2}$$

$$Dislike = Average(eScore_{anger,fear,disgust,sadness}) \tag{3}$$

Whereby, *like* includes the emotions that have a positive evaluation of the object, i.e, joy, trust and anticipation. *dislike* includes the emotions that have a negative evaluation of the object, i.e, anger, fear, disgust, and sadness.

That is to say that like and dislike are enduring tendencies to experience certain emotions whenever an object comes to mind and or in contact.

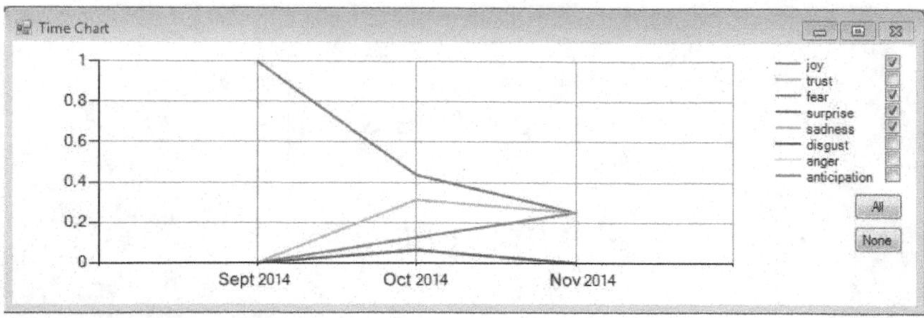

Fig. 6. Emotion distribution for the Twitter user 'blackvoices' with four emotions selected: joy, fear, surprise, and sadness

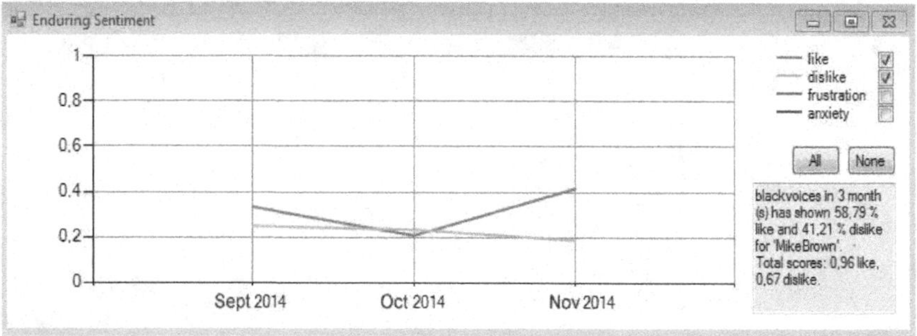

Fig. 7. Like and dislike distribution for the topic 'MikeBrown' from Twitter user 'blackvoices'

A similar graph as Fig. 6 is used to visualize the *like* and *dislike* patterns, with the option of viewing both or one at a time.

Fig. 7, illustrates that a person may experience various emotions towards a particular topic over a period of time. But the enduring pattern of like or dislike towards a topic will become visible over time. In our system, as we are at the moment only able to retrieve and analyze tweets over a couple of months, we made a summary in the enduring sentiment window to indicate percentage and total scores of like and dislike that could be indicative of the enduring sentiment targeted at a particular object. The scores are defined as follows:

$$Like(\%) = \frac{eScore(Like)}{eScore(Like) + eScore(Dislike)} \tag{4}$$

$$Dislike(\%) = \frac{eScore(Dislike)}{eScore(Like) + eScore(Dislike)} \tag{5}$$

$$TotalLike = \sum eScore(Like) \tag{6}$$

$$TotalDislike = \sum eScore(Dislike) \tag{7}$$

Hence, in Fig. 7 we see that for instance, Twitter user 'blackvoices' has shown 58.79% like, 41.21% dislike, 0.96 total like, and 0.67 total dislike towards the topic 'MikeBrown' in the period of three months.

4 Preliminary Evaluation

EmoTwitter is a visualization tool that is designed to explore the emotional and enduring sentiment content in tweets. As an exploratory visualization system, it is difficult to define appropriate evaluation metrics. Since the goal of EmoTwitter is not to classify a whole tweet into an emotion category or polarity category but to identify the emotions in the tweet, traditional metrics such as precision and recall are not applicable. However, to get a sense of the coverage of the lexicon, we compared our measurements to a hand-annotated tweet dataset that aims to serve as the gold standard for model evaluation, the STS-Gold dataset [24]. In addition, as EmoTwitter is an exploratory system, in order for us to get feedback on the usefulness of its capability in accomplishing a variety of analytical tasks, we carried out a user-based evaluation. We evaluate the possible use of the visualization: viewing tweets, and polarity distribution, topics, emotion flow, and like and dislike patterns.

4.1 Lexicon Evaluation

As currently no lexicon exists that has been annotated with Plutchik's eight emotions, we evaluated the coverage of the lexicon based on polarity since there exist publicly available tweet datasets that have been annotated for polarity. For this we compared our system measurements to the hand annotated STS-Gold dataset. The dataset was developed to allow for the evaluation of sentiment classification models at tweet level.

From the STS-Gold dataset, we randomly selected a sample of 50 positive and negative tweets in order to compare the polarity output categorizations by EmoTwiter. In EmoTwitter, if a tweet had a higher proportion of positive words than negative, we counted it as a positive tweet and negative if it had a higher proportion of negative words. Using a chi-square test, we found that the categorizations from EmoTwitter were related with the hand annotations for the whole sample set ($p=0.146$, $df=2$), with an actual agreement of 74% between the hand-annotated and the EmoTwitter results. The agreement number is not itself impressive; however, the lexicon was built independently from the data to which it was applied. These scores however provide an indication that the lexicon we used correlates with the hand annotations from the STS-Gold dataset.

4.2 User-Based Evaluation

To further assess the potential use of EmoTwitter, we performed a formative user-based evaluation with seven participants, one PhD holder, four PhD candidates and two master degree students, all in the computer science field. Three participants had previously interacted with other SA tools while for four participants it was their first time. We first briefed the participants on the purpose of the session and then explained how the system works. We then asked them to freely experiment with the system and fill in a questionnaire after using the system. The average time spent with the experiment was 22 minutes. One participant took about 50 minutes as they were very interested in reading the tweets and seeing how the visualization changes with each tweet. The questionnaire covered questions on display, understanding of all the visual components (on a scale of five from very easy to understand, to not so easy), whether the users found the like and dislike patterns informative, suggestions for improvements, and possible applications for EmoTwitter.

The most liked aspect in interacting with EmoTwitter was that it allowed the participant to follow a person's emotions over time. Many participants became very interested in the emotion flow that they started searching online for information of what could have happened in the Twitter user's life to create spikes and dips in the emotions.

As we are interested in the emotion and sentiment representation, the users' answers to these questions were of interest. The emotion distribution representation was found easy or rather easy to understand by all the participants. The emotion flow was found very easy or easy to understand by all but one participant who found it not easy to understand. This particular participant felt that a user manual was needed in order to interact with EmoTwitter. For the like and dislike patterns, all participants found it either very easy or easy to understand.

When asked whether the like and dislike information in the text area was informative, four out of the seven participants asserted that it was informative. One of the participants stated that the representation would be useful as a way for further analysis.

Improvements suggested by the participants included a better organization of each of the visual display components, and more guiding information for new users. In addition, one participant mentioned that it would be nice to have the ability to analyze more than one Twitter user's emotions at a time. One participant also suggested that EmoTwitter would work better as a web application, which is of the things we plan to implement in the future.

When asked in which situations the participants found EmoTwitter useful; one participant mentioned that it would be nice to analyze a friend's, husband's, competitors', boss's or famous people's tweets. With famous people, one participant mentioned that for example they would like to see what their favorite NBA player says on Twitter and whether the player deserves to be respected.

Another participant pointed out that EmoTwitter can be helpful to law enforcement agencies whereby the agencies can monitor a person's messages and see if anything unpleasant is taking place, like cyberbullying, or if a Twitter user is too negative as stated by another participant. Other participants pointed out that EmoTwitter would be useful for news, marketing and advertising agencies.

Content filtering was another practical usage mentioned for EmoTwitter. Additionally, EmoTwitter was said to be useful as an augmented tool to other systems.

5 Conclusion and Future Work

We have presented EmoTwitter, a multifaceted visualization system. In this paper we have explored the automatic analysis and tracking of emotions within tweets. The developed system presented here aimed to function as an improvement into the way sentiments are currently analyzed and reported. In order to move beyond just the analysis of sentiment polarity we have made an attempt to identify regular occurring patterns of like and dislike over a period of time.

The preliminary evaluation showed that the system successfully presented information in an easy-to-understand manner and that the emotional flow of tweets can be meaningfully extracted.

Future work involves analyzing the like and dislike patterns for a longer period of time so as to observe the enduring sentiments a Twitter user has formed towards topics. This will involve constantly following certain Twitter users and storing their tweets as they appear.

In addition, we plan to conduct a deeper linguistic analysis to better understand the expressed emotions. As the current version of the system makes use of a lexicon-based approach for detecting emotions, we intend to extend the capabilities of the system by incorporating approaches such as aspect-level emotion analysis and common-sense analysis for broader emotion detection.

From the user-based evaluation, we learned that users want to know more about the events causing spikes in the emotion flow, thus we also aim to include event analysis in the emotion flow chart.

We conclude that EmoTwitter is potentially valuable for marketing, advertisement, security, and we plan to develop it further into a full system available online.

Acknowledgement. This work was supported by Detecting and Visualizing Emotions and their Changes in Text, grant No. 14166, Academy of Finland.

References

1. Broad, C.D.: Emotion and sentiment. Journal of Aesthetics and Art Criticism 13(2), 203–214 (1971)
2. Cattell, R.B.: Sentiment or attitude? the core of a terminology problem in personality research. Journal of Personality 9, 6–17 (2006)
3. Duan, L., Xu, D., Tsang, I.W.: Learning with augmented features for heterogeneous domain adaptation. In: Proc. of the 29th International Conference on Machine Learning, pp. 711–718. Omnipress, Edinburgh (2012)
4. Esuli, A., Sebastiani, F.: Sentiwordnet: A publicly available lexical resource for opinion mining. In: Proc. of the 5th Conference on Language Resources and Evaluation (LREC), Genoa, Italy, pp. 512–422 (2006)
5. French, V.V.: The structure of sentiments: I. A restatement of the theory of sentiments. Journal of Personality 15(4), 247–287 (1947)
6. Fukuhara, T., Nakagawa, H., Nishida, T.: Understanding sentiment of people from news articles: Temporal sentiment analysis of social events. In: Proc. of the International Conference on Weblogs and Social Media (ICWSM), Boulder, Colorado, USA (2007)

7. Go, A., Bhayani, R., Huang, L.: Twitter sentiment classification using distant supervision. Tech. rep., Technical report, Stanford Digital Library Technologies Project (2009)
8. Gordon, S.L.: The sociology of sentiments and emotion. In: Rosenberg, M., Turner, R.H. (eds.) Social Psychology: Sociological Perspectives, pp. 562–592. Basic Books, New York (1981)
9. Havre, S., Hetzler, B., Nowell, L.: Themeriver: Visualizing thematic changes in large document collections. IEEE Transactions on Visualizition and Computer Graphics 8(1), 9–20 (2002)
10. Hogenboom, A., Bal, D., Frasincar, F., Bal, M., de Jong, F., Kaymak, U.: Exploiting emoticons in sentiment analysis. In: Proc. of the 28th Annual ACM Symposium on Applied Computing, Coimbra, Portugal, pp. 703–710 (2013)
11. Kempter, R., Sintsova, V., Musat, C., Pu, P.: Emotionwatch: Visualizing fine-grained emotion in event-related tweets. In: Proc. of the 8th International AAAI Conference on Weblogs and Social Media (2014)
12. Kim, S., Hovy, E.: Determining the sentiment of opinions. In: Proc. 20th International Conference on Computational Linguistics (COLING 2004). ACL, Stroudsburg (2004)
13. Marcus, A., Bernstein, M.S., Badar, O., Karger, D.R., Madden, S., Miller, R.C.: Tweets as data: Demonstration of tweeql and twitinfo. In: Proc. of the 2011 ACM SIGMOD International Conference on Management of data (SIGMOD 2011), pp. 1259–1262 (2011)
14. Martínez-Cámara, E., Martín-Valdiva, T.M., Ureña-López, A.L., Montejo-Ráez, A.: Sentiment analysis in twitter. Journal of Natural Language Engineering 20(1), 1–28 (2014)
15. Mishne, G., de Rijke, M.: Moodviews: Tools for blog mood analysis. In: AAAI Spring Symposium on Computational Approaches to Analysing Weblogs, pp. 153–154 (2006)
16. Mohammad, S.M., Turney, P.D.: Emotions evoked by common words and phrases: Using mechanical turk to create an emotion lexicon. In: Proc. of the NAACL-HLT 2010 Workshop on Computational Approaches to Analysis and Generation of Emotion in Text, LA, California, pp. 26–34 (2010)
17. Munezero, M., Montero, S.C., Mozgovoy, M., Sutinen, E., Pajunen, J.: Are they different? affect, feeling, emotion, sentiment, and opinion detection in text. IEEE Transaction on Affective Computing 5(2), 101–111 (2014)
18. Murray, H.A., Morgan, C.D.: A clinical study of sentiments i and ii. Genetic Psychological Monographs 32(1-2),3–149, 153–311 (1945)
19. OxfordDictionaries: Enduring. Oxford University Press (2014)
20. Pang, B., Lee, L.: Opinion mining and sentiment analysis. Foundations and Trends in Information Retrieval 2(1-2), 1–135 (2008)
21. Plutchik, R.: A general psychoevolutionary theory of emotion. Emotion 1(3), 3–33 (1980)
22. Porter, M.F.: An Algorithm for Suffix Stripping. Cambridge University Press, Cambridge (1980)
23. Robinson, D.T., Smith-Lovin, L., Wisecu, A.K.: Symbolic interactionist roots of affect control theory. In: Stets, J.E., Turner, J.H. (eds.) Handbook of the Sociology of Emotions, pp. 179–199. Springer, New York (2007)
24. Saif, Y.H., Fernandez, M., Alani, H.: Evaluation datasets for twitter sentiment analysis: A survey and a new dataset, the sts-gold. In: CEUR Workshop Proceedings of ESSEM Workshop, pp. 9–21 (2013)

25. Scherer, K.S.: What are emotions? and how can they be measured? Social Science Information 44(4), 693–727 (2005)
26. Shelly, R.K.: Emotions, sentiments and performance expectations. In: Turner, J. (ed.) Theory and Research on Human Emotions: Advances in Group Processes, vol. 21, pp. 141–165. Emerald Group Publishing Limited (2004)
27. Strapparava, C., Valitutti, A.: Wordnet-affect: An affective extension of wordnet. In: Proc. of the 4th International Conference on Language Resources and Evaluation, pp. 1413–1416 (2004)
28. Thelwall, M.: Sentiment analysis and time series with twitter. In: Weller, K., Bruns, A., Burgess, J., Mahrt, M., Puschmann, C. (eds.) Twitter and Society, pp. 83–96. Peter Lang, New York (2014)
29. Wiebe, J., Wilson, T., Cardie, C.: Annotating expressions of opinions and emotions in language. Language Resources and Evaluation 39(2-3), 165–210 (2005)
30. Wilson, T., Wiebe, J.: Annotating attributions and private states. In: Proc. of the ACL Workshop on Frontiers in Corpus Annotation II, Pie in the Sky, pp. 53–60 (2005)
31. Yu, H., Hatzivassiloglou, V.: Towards answering opinion questions: Separating facts from opinions and identifying the polarity of opinion sentences, pp. 129–136 (2003)

Feature Selection for Twitter Sentiment Analysis: An Experimental Study

Riham Mansour[1], Mohamed Farouk Abdel Hady[2], Eman Hosam[1], Hani Amr[1], and Ahmed Ashour[1]

[1] Microsoft Research, Advanced Technology Lab, Cairo, Egypt
{rihamma,t-emadel,v-haniam,aash}@microsoft.com
[2] Microsoft, Redmond, WA, USA
Mohamed.Abdel-Hady@microsoft.com

Abstract. Feature selection is an important problem for any pattern classification task. In this paper, we developed an ensemble of two Maximum Entropy classifiers for Twitter sentiment analysis: one for subjectivity and the other for polarity classification. Our ensemble employs surface-form, semantic and sentiment features. The classification complexity of this ensemble of linear models is linear with respect to the number of features. Our goal is to select a compact feature subset from the exhaustive list of extracted features in order to reduce the computational complexity without scarifying the classification accuracy. We evaluate the performance on two benchmark datasets, CrowdScale and SemEval. Our selected 20K features have shown very similar results in subjectivity classification to the NRC state-of-the-art system with 4 million features that has ranked first in 2013 SemEval competition. Also, our selected features have shown a relative performance gain in the ensemble classification over the baseline of uni-gram and bi-gram features of 9.9% on CrowdScale and 11.9% on SemEval.

1 Introduction

Twitter is a popular micro-blogging service where users post status messages (called tweets). The users use tweets to share their personal feelings and some- times express opinions in the form of user-defined hashtags, emoticons or normal words about different topics such as events, movies, products or celebrities. Sentiment Analysis (SA) can be formulated as a text classification task where the categories are polarities such as positive and negative. There has been a large amount of NLP research on this user-generated content in the area of sentiment classification. Traditionally most of the research work has focused on large pieces of text, such as product and movie reviews that represent summarized thoughts of authors. Although tweets became publicly available for the research community, they are different from reviews primarily because they are limited to 140 characters and have a more colloquial linguistic style. The frequency of misspellings and slang in tweets is much higher than in reviews.

The sentiment classification task can be handled either by a lexicon-based approach that requires an extensive set of manually supplied sentiment-bearing words or a supervised machine learning approach that requires a large amount of hand-labeled training tweets. The sentiment analysis research has shown that a two-stage approach is more effective [4]: the first stage is subjectivity classification in which subjective instances are

A. Gelbukh (Ed.): CICLing 2015, Part II, LNCS 9042, pp. 92–103, 2015.
DOI: 10.1007/978-3-319-18117-2_7

distinguished from objective ones, then whether the subjective instances has "positive" or "negative" polarity is detected.

In this paper, we aim to investigate all types of features introduced in the literature for sentiment analysis. Then evaluate their discrimination ability on a number of benchmark datasets. By the end of this study, we find out the compact feature subset of the exhaustive list of that features that can maintain the classification accuracy while reduce the computational complexity as it is linear with respect of feature set size.

The remainder of the paper is organized as follows. In Section 2, related work on machine learning based on sentiment classification and feature selection for sentiment analysis are reviewed. Section 3 defines the feature types investigated in this paper. In Section 4, presents the performance evaluation of our approach for feature selection. Finally, some conclusive remarks are given in Section 5.

2 Related Work

Machine learning is used extensively to automatically extract sentiment from text. While traditional work [17] focused on movie reviews, more recent research has explored social networks for sentiment analysis. The methods involved differ somewhat since texts like tweets have a different purpose and a more colloquial linguistic style [9]. Go et al. [8] have trained a sentiment classifier to label tweets' sentiment polarities as "positive" or "negative". Pak et al. [16] trained classifiers to also detect "neutral" tweets that do not contain sentiment. Sentiment classifier training requires a large amount of labeled training data, but the manual annotation of tweets is expensive and time-consuming. To collect the training data, Go, Pak and others used a heuristic method introduced by Read [19] to assign sentiment labels to tweets based on emoticons.

The n-grams representation and specifically Bag-of-Words are commonly used for sentiment classification, resulting in high-dimensional feature space. Agarwal and Mittal [2] has extracted uni-grams and bi-grams from product and movie review text then they have used Information Gain (IG) and Minimum Redundancy maximum Relevancy [18] feature selection criteria to select prominent features.

Barbosa and Feng [4] followed the two-stage approach but instead of n-gram features they have used polarity lexicons, part-of-speech tags, lexical and special micro-blogging features such as emoticons, hashtags, punctuation and character repetitions and words in capital letters to build SVM classifiers. Because the language used on Twitter is often informal and differs from traditional text types [9], most approaches include a preprocessing step. Usually emoticons are detected, URLs removed, abbreviations expanded and twitter markup is removed or replaced by markers. Zhang et al. [23] combined a lexicon-based classifier with an SVM to increase the recall of the classification.

3 Feature Selection

Our goal is to select a compact feature subset from the exhaustive list of extracted features in order to reduce the computational complexity without scarifying the classification accuracy. An ensemble of two binary classifiers is composed of two members: a subjectivity classifier indicating whether the tweet carries sentiment or not and the other

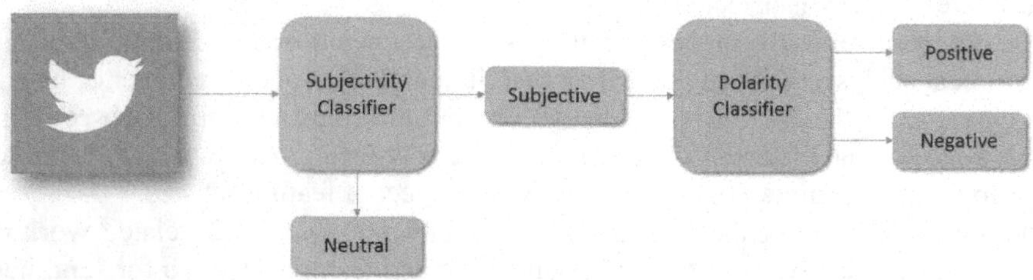

Fig. 1. Ensemble of Subjectivity and Polarity Classifiers

is a polarity classifier indicating whether the subjective tweet is positive or negative. Figure 1 depicts the ensemble of classifiers. We trained Maximum Entropy classifiers on the Azure Machine Learning (AzureML) platform for both subjectivity and polarity on two training data sets.

Because the language used on Twitter is often informal and differs from traditional text types [9]. We pre-process the tweets by removing stop words, using the Natural Language Toolkit (NLTK) library[1] and non-alphabetic characters. We detect emoticons using a regular expression adopted from Christopher Potts' tokenizing script [2]. We normalized URLs to http://someurl and userids to @someuser. We tokenized and part-of-speech tagged the tweets with the Carnegie Mellon University (CMU) tool [7]. All negations (e.g. not, no, never, n't, cannot) are replaced by ""NOT"". Repeated character sequences in words like "coooooool" are replaced with three characters, so it becomes "coool". In this paper, we adopt multiple features from the literature[13,1,21] in an attempt to find the best combinations for sentiment subjectivity and polarity classification. We use surface-form, semantic and sentiment features. The following subsections describe the three feature sets along with the baseline features.

3.1 Baseline Features

The baseline features of both our subjectivity and polarity classifiers are the unigrams and bigrams of the tweets. We apply feature reduction using Log Likelihood Ratio (LLR) to select the top 20K features that highly co-relate with the training data. All feature types are combined into a single feature vector. Pang et al. [17] have shown that feature presence (binary value) is more useful than feature frequency. Therefore, we use binary feature presence instead of feature frequency.

3.2 Senti-Features

These features refer to the subset of features we adopted from the 100 senti-features presented by Agarwal et al. [1]. Additional pre-processing steps have been performed

[1] http://nltk.org/
[2] http://sentiment.christopherpotts.net/tokenizing.html

on the tweets for the extraction of the senti-features as follows: All emoticons are replaced with their sentiment polarity specified in the sentiment lexicon presented in [1]. Social acronyms are expanded using an online resource [3] that has 5,184 expansions.

A number of the senti-features are based on prior polarity of words and emoticons. We adopted the same emoticon dictionary in [1] that is composed of 170 manually-annotated emoticons listed on Wikipedia [4] with their emotional state. We adopted the subjectivity lexicon [5] from Opinion Finder [22], an English subjectivity analysis system which classifies sentences as subjective or objective. The lexicon was compiled from manually developed resources augmented with entries learned from corpora. It contains 6,856 entries including 99 multi-word expressions. The entries in the lexicon have been labeled (1) for polarity either positive, negative, or neutral, (2) for part of speech, and (3) for reliability those that appear most often in subjective contexts are strong cludes of subjectivity, while those appearing less are often labeled weak. In order to increase the coverage of the lexicon, we adopt an approach similar to [3]. We used synonym relations from English WordNet[6]to expand the initial seed MPQA English polarity lexicon. The assumption is that synonyms carry same sentiment/polarity as compared to the root words. We make a hypothesis of traversing WordNet like a graph where words are connected to each other based on synonym or antonym relations. Consider each word in this list as a node of the graph. Each node has many in-links and many out-links. This is an undirected graph which is not fully connected i.e. not all the nodes are connected to every other node. For every word in the seed lexicon, we identify its synonym and append with appropriate polarity label in the seed lexicon. Unlike [3], we performed one iteration of traversal, we did not identify the synonyms of the seed lexicon words synonyms. As a post-processing step, we exclude any term that appears more than once with different polarity.

Unlike Agarwal et al. in [1], the features are calculated for the whole tweet only instead of calculating them for the last one-third as well. The senti-features are either counts of other features, lexicon-based features, and Boolean features including bag of words, presence of exclamation marks and capitalized text. We only re-implemented the counts and lexicon-based features as our unigram/bigram baseline covers the rest of the senti-features. Features whose prior polarity are calculated from the emoticon dictionary or the lexicon are classified as polar while all other features are non-polar. The following list depicts the polar and non-polar features we adopted.

Polar features:

- # of (+/-) POS (JJ, RB, VB, NN)
- # of negation words, positive words, negative words
- # of positive and negative emoticons
- # of (+/-) hashtags and capitalized words
- For POS JJ, RB, VB, NN, sum of the prior polarity scores of words of that POS
- Sum of prior polarity scores of all words

[3] `http://www.noslang.com/`

[4] `http://www.wikipedia.org/wiki/List_of_emoticons`

[5] `mpqa.cs.pitt.edu/lexicons/subj_lexicon/`

[6] `http://wordnet.princeton.edu/`

Non-polar features:

- # of JJ, RB, VB, NN
- # of slangs, latin alphabets, dictionary words, words
- # of hashtags, URLs, targets
- Percentage of capitalized text
- Exclamation, capitalized text

3.3 Sentiment-Specific Word Embedding Features (SSWE)

Sentiment-specific word embedding features (SSWE) are introduced in [21] where a model for learning SSWE is introduced. Unlike continuous word representations that typically models the syntactic context of words only, SSWE encodes sentiment information in the continuous word representation. SSWE addresses the problem of mapping words with similar syntactic context but opposite sentiment polarity such as "'good"' and "'bad"' to neighboring word vectors. The SSWEs are obtained in [21] from large-scale training corpora of distant-supervised tweets collected by positive and negative emoticons. The training data are passed to three neural networks whose loss functions incorporate the supervision from sentiment polarity of text.

The work in [21] extends the existing word embedding learning algorithm in [6] where the C&W model is introduced to learn word embeddings based on the syntactic contexts of words. Given an ngram, C&W replaces the center word with a random word to derive a corrupted ngram. The training objective is that the original ngram is expected to obtain a higher language model score than the corrupted ngram by a margin of 1. Following the C&W model, SSWE incorporate the sentiment information into the neural network through predicting the sentiment distribution of text based on input ngram. A sliding window on the ngram input is used to predict the sentiment polarity based on each ngram with a shared neural network. The higher layers of the neural network are interpreted as features describing the input. Instead of hand-crafted features for Twitter sentiment classification under a supervised framework like in [17], SSWE framework incorporates the continuous representations of words and phrases as the features of a tweet. The sentiment classifier is built from the tweets with manually annotated sentiment polarity. We have obtained the SSWEs from the authors of [21]. They trained their model on 10M tweets labeled automatically using emoticons in [11].

3.4 NRC Features

Mohammad et al. [13] presented the top-performed system in SemEval 2013 Twitter sentiment classification track. The work incorporates diverse sentiment lexicons and many hand-crafted features. We re-implemented this system as the codes are not available publicly. [13] presented two lexicons namely the hashtag sentiment lexicon and the sentiment140 lexicon. The former contains 308,808 entries of terms and their associated sentiment score calculated from tweets selected with seed positive and negative hashtags. The sentiment140 lexicon has been generated in the same way as the hashtag sentiment lexicon, but from the sentiment140 corpus [8] that is labeled automatically annotated based on emoticons. The sentiment140 lexicon has entries for 62,468 unigrams, 677,698 bigrams, and 480,010 non-contiguous pairs. The authors also use three other lexicons in their features from that were previously presented in [10,14] along with MPQA. Mohammad et al.[13] proposed the following features:

- Word ngrams for the presence or absence of contiguous sequences of 1, 2, 3, and 4 tokens. We have not used this feature as it is very similar to our unigram and bigram features. 3 and 4 grams are costly to calculate especially when training on large text corpus.
- Character ngrams for the presence or absence of contiguous sequences of 3, 4 and 5 characters. We have not used these features as we used unigram and bigram features as a replacement.
- all-caps for the number of words with all characters in upper case.
- POS capturing the number of occurences of each part-of-speech tag.
- hashtags for the number of hashtags.
- lexicon features are based on the 5 lexicons mentioned above. The lexicon features are created for all tokens in the tweet, each POS, hashtags, and all-caps tokens. For each token w, the polarity score p in the lexicon is used to determine
  - total count of tokens in the tweet with score(w,p)¿0.
  - total score adding all the score(w,p) for all tokens.
  - the maximal score among all tokens.
  - the score of the last token in the tweet with score(w,p)¿0.
- Punctuation for the number of contiguous sequences of exclamation marks, question marks, and both exclamation and question marks. Another feature checks whether the last token contains an exclamation or question mark.
- Emoticons where the polarity of an emoticon is determined with a regular expression adopted from Christopher Potts' tokenizing script as described earlier in this section. We did not implement this feature as it is already part of our pre-processing steps.
- Elongated words refers to the number of words with repeated character sequences. This feature is a little different from our pre-processing steps as it counts the number of such elongated words.
- Clusters refer the presence or absence of tokens from 1000 clusters provided by the CMU pos-tagger and produced with the Brown clustering algorithm on 56 million English tweets.
- Negation counts the number of negated contexts. A negated context is defined as a segment of a tweet that starts with a negation word and ends with one of the punctuation marks"' ',', '.', ':', ';', '!', '?'. Throughout the rest of the paper, we refer to the whole set of NRC features as the full feature set of NRC while the set that we implemented as NRC features. The subset of features that we employed from NRC are all the features except for the word ngram and character ngram features. The reason we excluded these features from NRC is mainly their size. For example, the full NRC are around 4 million features for the subjectivity classifier when trained on the SemEval dataset. Similar results in this example could be retained with 20,200 features with some of our feature combinations as shown in Section 4.

4 Experiments

In this section, we describe the set of experiments conducted in order to select a compact feature subset from the different feature types described in section 3. The aim is to reduce the dimensionality of the feature space used for sentiment classification without scarifying the classification accuracy.

4.1 Datasets

We have employed the following two benchmark datasets for our experiments as summarized in Table 1:

SemEval. This dataset was constructed for the Twitter sentiment analysis task (Task 2) [15] in the Semantic Evaluation of Systems challenge (SemEval-2013) [7]. All the tweets were manually annotated by 5 Amazon Mechanical Turk workers with negative, positive and neutral labels. The turkers were also asked to annotate expressions within the tweets as subjective or objective. The statistics of each sentiment label is shown in Table 1. Participants in the SemEval-2013 Task 2 used this dataset to evaluate their systems for expression-level subjectivity detection [5,13] as well as tweet-level subjectivity detection [12,20].

CrowdScale Dataset. The CrowdScale dataset [8] is the sentiment analysis judgment dataset in CrowdScale 2013. The tweets in the dataset is from the weather domain. Each tweet was evaluated by at least 5 raters. The possible answers are: "Negative", "Neutral"; the author is just sharing information, "Positive", "Tweet not related to weather condition" and "I can't tell". The total number of tweets in addition to the number of tweets of each sentiment for training and testing is shown in Table 1.

Table 1. Description of datasets used in our experiments

Dataset	Training Set				Development Set				Testset			
	Positive	Negative	Neutral	Total	Positive	Negative	Neutral	Total	Positive	Negative	Neutral	Total
SemEval	3,168	1,380	4,111	8,659	500	340	1160	2000	1,570	601	1,638	3,809
CrowdScale	14,253	15,513	20,234	50,000	3,237	3,496	4,510	11,243	3,237	3,496	4,510	11,243

4.2 Experimental Setup

We have conducted extensive experiments to explore the various combinations of features among the four most significant feature sets in the literature. Our experiments target exploring the best features for the subjectivity, polarity and ensemble classifier. All the experiments have been performed on the development sets to select the features while the test sets have been used to compare our system to the baseline system

[7] http://www.cs.york.ac.uk/semeval-2013/task2/

[8] http://www.crowdscale.org/shared-task/sentiment-analysis-judgment-data

using the selected features. The four feature sets discussed in Sections 3.1, 3.2, 3.3, and 3.4 resulted in 16 combinations on each of the three data sets discussed in Section refsec:allDatasets. The baseline is the model trained with the unigram and bigram featuers only. We then add one or more feature set(s) to the baseline to measure the improvement both in the subjectivity, polarity and ensemble classifiers independently. We used macro-F1 as an evaluation metric for all our experiment results. Macro-F1 is the average F1 across the positive, negative and neutral classes. For each feature set combination, we trained 2 types of models. subjectivity model trained using the neutral tweets as class 0 and the subjective tweets (positive or negative) as class 1 and polarity model trained using the positive tweets as class 1 and the negative tweets as class 0.

4.3 Results

This section depicts the results of the subjectivity and polarity classifiers independently followed by the results of the ensemble. This is motivated by our research goals in finding the best combination of features for each classifier alone as well as the ensemble.

Subjectivity Classification. The macro-F1 results of the subjectivity classifier for each feature set combination on the three datasets are shown in Figure 2 For the SemEval dataset, the maximum macro-F1 0.74 is from the NRC and SSWE combination while the maximum macro-F1 for CrowdScale is 0.83 from the full NRC features. For deciding the best feature set combination for the subjectivity classifier, we took the average macro-F1 across the two datasets and found that the best feature set combination for subjectivity classifier is the full NRC features for an average macro-F1 of 0.796. However, the feature vector spans multiple millions. For instance, the SemEval dataset, the feature vector had about 4 million features. The combination of features that retained the second best average macro-F1 of 0.79 is the NRC, the unigram-bigragm and the senti-features with a difference in the macro-F1 of 0.006 less than the full NRC features. We chose the second best combination of (NRC + unigram-bigram + senti-features) combination to serve as the feature set for the subjectivity classifier when integrated in the ensemble.

Polarity Classification. Figure 3 shows the macro-F1 scores of the polarity classifier for each dataset and for each feature set combination. The NRC and SSWE combination is the best combination for SemEval while unigram-bigram and SSWE is the best for CrowdScale. After averaging the macro-F1 of all the datasets, we found that the best feature set combination is NRC, unigram-bigram and SSWE with macro-F1 of 0.89.

Ensemble Results. The ensemble results shown in Table 2 are retained from the best combination of features for the subjectivity classifier (NRC + unigram-bigram + senti-features) and the polarity classifier (NRC + unigram-bigram + SSWE). The results are obtained from applying the best combinations on the test sets. Our baseline system is the ensemble with unigram-bigram features for both the subjectivity and polarity classifiers. The baseline results are shown for each dataset in Table 2. We can see from the results

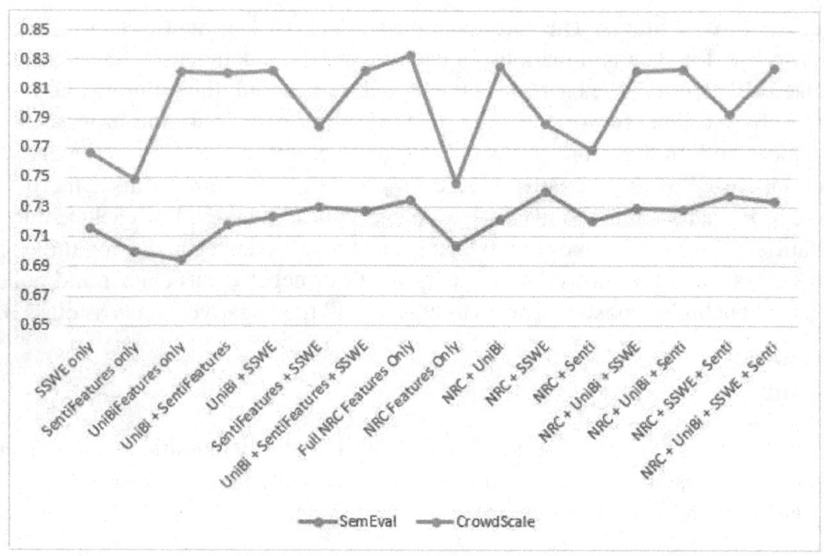

Fig. 2. Macro-F1 of subjectivity classifier for each feature set on each dataset

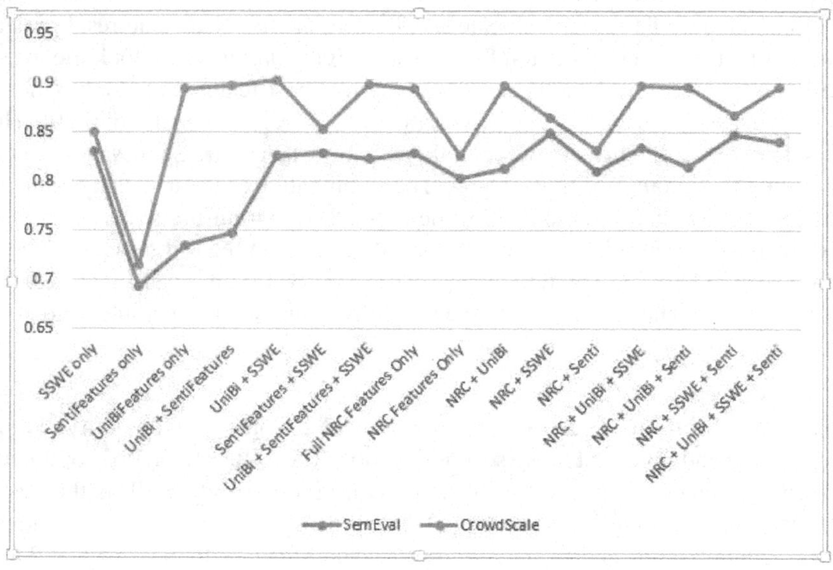

Fig. 3. Macro-F1 of polarity classifier for each feature set on each dataset

that our proposed combination of features have achieved 9.9% and 11.0% relative gain in the macro-F1 of both CrowdScale and SemEval over the unigram-bigram baseline. The size of the feature vector in each data set is few hundreds above the 20,000 unigram-bigram selected features, which maps to much reduced classification complexity.

Table 2. Macro-F1 Results for the Sentiment Ensemble

Dataset	Features	Positive	Negative	Neutral	Macro-F1	Relative Gain
CrowdScale	Baseline	*0.71*	*0.63*	*0.79*	*0.71*	
	Best	*0.77*	*0.76*	*0.82*	*0.78*	**9.9%**
SemEval	Baseline	*0.66*	*0.43*	*0.67*	*0.59*	
	Best	*0.70*	*0.57*	*0.70*	*0.66*	**11.9%**

4.4 Discussion

Figure 4 shows the results of McNemar significance test for subjectivity classifier trained on different feature subsets. The significance test has shown that there is not significant difference in terms of classification accuracy between the full NRC feature set and the selected feature subset. That means that we have reduced the computational complexity of the subjectivity classifier without sacrificing its prediction ability. Our explanation to this result is that the unigram and bigrams features with LLR feature selection capture the highly correlated ngram features for subjectivity classification. The selected features in the significance test we performed are the NRC (the statistical features that do not include the character and word ngrams), unigram-bigram and senti-features. The combination of the unigram-bigram and senti-features are the main difference between the two models we compared in the significance test. This demonstrates that the selected ngram features are equally effective compared to the large size of character and word ngrams in the full NRC feature set. Our ongoing work involves more automatic feature selection techniques to further reduce the computational complexity of both the subjectivity and polarity classifiers.

	[Full NRC] vs. [UniBi + NRC + Senti]		[Full NRC] vs. [Unigram]	
	McNemar Test	p-value	McNemar Test	p-value
SemEval	2.538	Not Significant Difference p = 0.1	13.26	Significant Difference p is less than 0.001
CrowdScale	4.232	Not Significant Difference p = 0.05	6.064	Significant Difference p = 0.01

Fig. 4. McNemar significance test for subjectivity classifier for different pairs of feature set combinations on each dataset

5 Conclusions

In this paper, we explore multiple sets of features used in the literature for the task of sentiment classification including surface-form, semantic and sentiment features. An ensemble of a subjectivity and polarity classifiers is used for sentiment classification. The classification complexity of this ensemble of linear models is linear with respect to the number of features. Our approach aims to select a compact feature subset from the exhaustive list of extracted features in order to reduce the computational complexity without scarifying the classification accuracy. We evaluate the performance on two benchmark datasets. Our selected 20K features have shown very similar results in subjectivity classification to the NRC state-of-the-art system with 4 million features that has ranked first in 2013 SemEval competition. Also, our selected features have shown a relative performance gain in the ensemble classification over the ngram baseline of 9.9% on CrowdScale and 11.9% on SemEval.

References

1. Agarwal, A., Xie, B., Vovsha, I., Rambow, O., Passonneau, R.: Sentiment analysis of twitter data. In: Proceedings of the Workshop on Languages in Social Media, pp. 30–38 (2011)
2. Agarwal, B., Mittal, N.: Optimal feature selection for sentiment analysis. In: Gelbukh, A. (ed.) CICLing 2013, Part II. LNCS, vol. 7817, pp. 13–24. Springer, Heidelberg (2013)
3. Bakliwal, A., Arora, P., Varma, V.: Hindi subjective lexicon: A lexical resource for hindi adjective polarity classification. In: Proceedings of the Eight International Conference on Language Resources and Evaluation (LREC 2012) (2012)
4. Barbosa, L., Feng, J.: Robust sentiment detection on twitter from biased and noisy data. In: Proceedings of the 23rd International Conference on Computational Linguistics (COLING 2010), pp. 36–44 (2010)
5. Chalothorn, T., Ellman, J.: Tjp: Using twitter to analyze the polarity of contexts, Atlanta, Georgia, USA, p. 375 (2013)
6. Collobert, R., Weston, J., Bottou, L., Karlen, M., Kavukcuoglu, K., Kuksa, P.: Natural language processing (almost) from scratch. The Journal of Machine Learning Research 12, 2493–2537 (2011)
7. Gimpel, K., Schneider, N., O'Connor, B., Das, D., Mills, D., Eisenstein, J., Heilman, M., Yogatama, D., Flanigan, J., Smith, N.A.: Part-of-speech tagging for twitter: Annotation, features, and experiments. In: Proceedings of the 49th Annual Meeting of the Association for Computational Linguistics: Human Language Technologies: Short papers, vol. 2, pp. 42–47. Association for Computational Linguistics (2011)
8. Go, A., Bhayani, R., Huang, L.: Twitter sentiment classification using distant supervision, vol. 150, pp. 1–6. Ainsworth Press Ltd (2009)
9. Han, B., Baldwin, T.: Lexical normalisation of short text messages: Makn sens a# twitter. In: Proceedings of the 49th Annual Meeting of the Association for Computational Linguistics: Human Language Technologies (2011)
10. Hu, M., Liu, B.: Mining and summarizing customer reviews. In: Proceedings of the Tenth ACM SIGKDD International Conference on Knowledge Discovery and Data Mining, pp. 168–177. ACM (2004)
11. Hu, X., Tang, J., Gao, H., Liu, H.: Unsupervised sentiment analysis with emotional signals. In: Proceedings of the 22nd International Conference on World Wide Web, pp. 607–618. International World Wide Web Conferences Steering Committee (2013)
12. Martınez-Cámara, E., Montejo-Ráez, A., Martın-Valdivia, M., Urena-López, L.: Sinai: Machine learning and emotion of the crowd for sentiment analysis in microblogs, Atlanta, Georgia, USA, p. 402 (2013)
13. Mohammad, S.M., Kiritchenko, S., Zhu, X.: Nrc-canada: Building the state-of-the-art in sentiment analysis of tweets. arXiv preprint arXiv:1308.6242 (2013)
14. Mohammad, S.M., Turney, P.D.: Emotions evoked by common words and phrases: Using mechanical turk to create an emotion lexicon. In: Proceedings of the NAACL HLT 2010 Workshop on Computational Approaches to Analysis and Generation of Emotion in Text, pp. 26–34. Association for Computational Linguistics (2010)
15. Nakov, P., Kozareva, Z., Ritter, A., Rosenthal, S., Stoyanov, V., Wilson, T.: Semeval-2013 task 2: Sentiment analysis in twitter (2013)

16. Pak, A., Paroubek, P.: Twitter as a corpus for sentiment analysis and opinion mining. In: Proceedings of the Seventh Conference on International Language Resources and Evaluation (LREC 2010) (2010)
17. Pang, B., Lee, L., Vaithyanathan, S.: Thumbs up?: sentiment classification using machine learning techniques, pp. 79–86 (2002)
18. Peng, H., Long, F., Ding, C.: Feature selection based on mutual information: Criteria of max-dependency, max-relevance, and min-redundancy. IEEE Transactions on Pattern Analysis and Machine Intelligence 27(8), 1226–1238 (2005)
19. Read, J.: Using emoticons to reduce dependency in machine learning techniques for sentiment classification. In: Proceedings of the ACL Student Research Workshop, ACLstudent 2005, pp. 43–48 (2005)
20. Remus, R.: Asvuniofleipzig: Sentiment analysis in twitter using data-driven machine learning techniques. Atlanta, Georgia, USA 3(1,278), 450 (2013)
21. Tang, D., Wei, F., Yang, N., Zhou, M., Liu, T., Qin, B.: Learning sentiment-specific word embedding for twitter sentiment classification. In: Proceedings of the 52nd Annual Meeting of the Association for Computational Linguistics, pp. 1555–1565 (2014)
22. Wilson, T., Wiebe, J., Hoffmann, P.: Recognizing contextual polarity in phrase-level sentiment analysis. In: Proceedings of HLT/EMNLP 2005 (2005)
23. Zhang, L., Ghosh, R., Dekhil, M., Hsu, M., Liu, B.: Combining lexicon-based and learning-based methods for twitter sentiment analysis, vol. 89 (2011)

An Iterative Emotion Classification Approach for Microblogs

Ruifeng Xu[1], Zhaoyu Wang[1], Jun Xu[1,*], Junwen Chen[3], Qin Lu[2], and Kam-Fai Wong[3]

[1] Shenzhen Engineering Laboratory of Performance Robots at Digital Stage, Harbin Institute of Technology Shenzhen Graduate School, Shenzhen, China
[2] Dept. of Computing, The Hong Kong Polytechnic University, Hong Kong
[3] Dept. of Systems Engineering & Engineering Management, The Chinese University of Hong Kong, Hong Kong
xuruifeng@hitsz.edu.cn, hit.xujun@gmail.com, kfwong@se.cuhk.edu.hk

Abstract. The typical emotion classification approach adopts one-step single-label classification using intra-sentence features such as unigrams, bigrams and emotion words. However, single-label classifier with intra-sentence features cannot ensure good performance for short microblogs text which has flexible expressions. Target to this problem, this paper proposes an iterative multi-label emotion classification approach for microblogs by incorporating intra-sentence features, as well as sentence and document contextual information. Based on the prediction of the base classifier with intra-sentence features, the iterative approach updates the prediction by further incorporating both sentence and document contextual information until the classification results converge. Experimental results obtained by three different multi-label classifiers on NLP & CC2013 Chinese microblog emotion classification bakeoff dataset demonstrates the effectiveness of our iterative emotion classification approach.

Keywords: Emotion Classification, Iterative Classification, Microblogs.

1 Introduction

Microblogging website has become an important channel for producing and propagating information. It also provides a platform for users to share their ideas and emotions. Identifying and understanding the microblog emotions has great social and commercial value. For example, sociologists can utilize public emotion of relevant microblogs to detect important moment of breaking events. However, the short and flexible expression in microblog can render current emotion analyzer ineffective.

Text emotion analysis aims to automatically identify personalized emotions in the text. The emotion here is referring to one person's intra psychological reactions and feelings, such as *like*, *anger*, *sadness* and *happiness,* etc. The basic issue in the emotion

---

* Corresponding author.

A. Gelbukh (Ed.): CICLing 2015, Part II, LNCS 9042, pp. 104–113, 2015.
DOI: 10.1007/978-3-319-18117-2_8

analysis is emotion classification which aims to classify the text into emotion categories. Currently, most text emotion classification is normally regarded as a single-label classification problem. Actually, many texts contain more than one emotional category. Even the smallest linguistic unit in Chinese, namely a word, may contain different emotional tendencies, such as "喜忧参半(*Mingled hope and fear*) and 悲喜交加(*grief with joy*)". Quan's study showed that about 15.1% of the Chinese words in vocabulary have more than one emotional attributes [1]. Therefore, in this study, the text emotion classification problem is regarded as a multi-label classification problem.

The current text emotion classification techniques can be camped into two major approaches, namely the lexicon and rule based approach and the machine learning based approach. The former approach mainly uses emotion dictionary and rules to identify the emotional expression keywords for emotion classification. Some works use syntax parser to refine the emotion classification [2-5]. This approach is limited by the coverage and description capability of the emotion dictionary resources. The machine learning based approach mainly uses words, emotional words, topic words, etc. as features. The annotated text is used to train the classifier for emotion classification [6-7]. Currently, the intra-sentence features, such as unigram features, bigram features, etc. are adopted in machine learning based approaches [8-12]. Few works utilize the transformation features between sentences and features of the overall document emotional tendency. However, intra-sentence features cannot fully capture the emotion patterns of short and informal microblog messages. For example, in the following microblog,

<microblog id="1">
<sentence_1>尼玛！！*(Damn!!)*</sentence>
<sentence_2>好好听！！！！！*(Sounds good!!!!!)* </sentence>
<sentence_3>我太喜欢这首歌了！！*(I like this song very much!)*</sentence>
</microblog>

There are three sentences while all of them expressed "like" emotions. In sentence_1, "尼玛!!(damn!)" is a common word in microblog to expresses negative emotions such as "anger" and "disgust". The Classifier only with intra-sentence features is likely to misclassify sentence_1 as "anger" or "disgust". However, the relationship between neighboring sentences and emotion trend of the whole microblog indicates the chances that sentence_1 should be classified as positive emotion "like". Similar problems also occur in the emotion classification for Twitter text.

Motivated by this observation, this paper proposes an iterative multi-label microblog emotion classification approach using intra-sentence, sentence and document contextual information. Firstly, a multi-label classifier with intra-sentence features acts as a base classifier for classifying emotions of sentences in microblog. Based on the initial classification results, the contextual information, namely the emotion transformation between neighboring sentences and the emotion tendency of the whole microblog are further incorporated to update the prediction for each sentence. Such iterative update terminates until the classification results converge. Experimental results on NLP&CC 2013 Chinese microblog emotion classification bakeoff dataset show that the emotion classification accuracy of the proposed approach achieved 83.26% which is 22.97% higher than the base classifier. Meanwhile, such result outperforms the best results in NLP&CC

2013 bakeoff for 18.98%. It is the highest achieved performance on this dataset, based on our knowledge. Furthermore, this approach improves the classification performances for three different base classifiers. These results demonstrate the effectiveness of our iterative emotion classification approach with intra-sentence, sentence and document contextual features.

The rest of this paper is organized as follows: Section 2 briefly reviews the related works on emotion classification; Section 3 describes the proposed iterative multi-label microblog emotion classification approach; Section 4 gives the experimental results and discussions. Finally, Section 5 gives the conclusion.

2 Related Works

The majority of current text emotion classification techniques can be categorized as dictionary and rule based approach and machine learning based approach.

The former approach mainly uses emotions lexicon resource to identify the keyword of emotional expression for emotion classification. Shen Yang et al. [2] calculated the emotion index by using attitude lexicon, weight lexicon, denial lexicon, degree lexicon, and conjunctions lexicon. Chunling Ma et al. [3] used keyword recognition to identify the emotional contents in the text messages, and then used the syntax feature to detect the emotional significance. Carlo Strapparava et al. [4] established the mechanism to calculate the semantic similarity between universal semantic words and emotions vocabulary based on Latent Semantic Analysis for emotion classification. Aman et al. [5] developed the knowledge-based classification method using emotional dictionary, knowledge base and other resources. The approach based on dictionary and rules has the difficulty to solve the problem of unknown words. Meanwhile, the study of the impact of grammatical structure and semantics for emotional expression are insufficient.

Machine learning based approach uses emotion labeled training set to train machine learning based classifier using the words, emotional words, topics and other features for future emotion classification. Pang Lei et al. [6] constructed a pseudo-labeled corpus from large unlabeled corpus automatically, and used pseudo-labeled data as training set to training the classifier. The disadvantage of this method is that the existence of negative sentences. Pak and Paroubek et al. [7] established a Twitter text emotional polarity dataset, and implemented the emotion classifiers based on Naïve Bayes (NB), Support Vector Machine (SVMs) and Conditional Random Fields (CRF) models, respectively. The adopted features are normally emotional keywords and other words features within sentences which neither takes into account of the order of features and the correlation of semantic nor the use of contextual features.

Multi-label classification learning algorithms have two major categories. One is problem conversion method which uses single-label classifiers to solve multi-label classification. Another one is an adaptive algorithm, that is, to improve some single-label classification algorithms for enabling multi-label classification. Nadia Ghamrawi et al. [14] proposed a multi-label CRFs model in which the co-occurrence labels are used as parameters directly. Hullermeier et al. [15] proposed a classifier based on a comparison of the labels (Pairwise Comparison). It generates all possible binary classifiers through any relationship between the two labels collections, and then votes the classification results between the two labels. Finally, it makes a combination of voting results as the final output. Konstantinos Trohidis et al. [16] implemented and compared several classification algorithms in multi-label classification of music, including BR, LP, random k-tag set (Random k-Label sets, RAKEL) and Multi-Label k-Nearest Neighbor (MLkNN).

3 An Iterative Multi-label Emotion Classification Approach

Emotion classification of microblogs is hard due to most of them are short and informal sentences. Meanwhile, sometimes, the same word and phrase even express different emotions within different context. For example, "我靠!(Damn!)", "呵呵(Hehe.)". Such ambiguities are hard to resolve by using the intra-sentence features. The observations on more cases show that the exploring of the context information is a potential way to capture the emotions expressed in these sentences. Thus, we propose an iterative approach which combines the intra-sentence features and context features. Firstly, we build the base classifier to classify each sentence in the microblog by using intra-sentence features. Secondly, with the initial classification results, we estimate the emotion transfer probabilities between neighboring sentences, and the emotion transfer probabilities between the sentence and the overall emotions of the microblog. Then all of these probabilities are incorporated with base classifier to update the classified categories of each sentence, iteratively, until the classification results converge. During this procedure, the overall emotions of the microblog are determined based on the emotion classification categories of all of the sentences in the previous round.

Let L be the set of emotion labels. For an emotion label $l \in L$, let Y_l^0 denotes the event that sentence s does not contain label l and Y_l^1 denotes the event that sentence s contains label l. Let w be the microblog which contains sentence s, s_P be the previous sentence of s and s_N be the following sentence. Let W_ϵ^1 denotes the event that the microblog has emotion label ϵ, and W_ϵ^0 denotes the event that the microblog does not have emotion label ϵ. Similarly, P_ϵ^1 and N_ϵ^1 denotes the event that s_P and s_N with emotion label ϵ, and P_ϵ^0 and N_ϵ^0 denotes that s_P and s_N does not have emotion label ϵ, respectively. Assuming that: 1) the emotion transformation between the adjacent sentences is a sequence of events and each transfer event is independent, and 2) the emotion transformation between the whole microblog and each sentence in the microblog is also independent, $\vec{y_s}(l)$ can be denoted as follows.

$$\vec{y_s}(l) = \arg\max_{e \in \{0,1\}} p\left(Y_l^e \middle| F, W_\epsilon^{\vec{y}(\epsilon)} \dots, P_\epsilon^{\vec{y}(\epsilon)} \dots, N_\epsilon^{\vec{y}(\epsilon)} \dots\right)$$

$$= \arg\max_{e \in \{0,1\}} p(Y_l^e | E) \times \prod_{\epsilon \in L} p\left(W_\epsilon^{\vec{y}(\epsilon)} \middle| Y_l^e\right) p\left(P_\epsilon^{\vec{y}(\epsilon)} \middle| Y_l^e\right) p\left(N_\epsilon^{\vec{y}(\epsilon)} \middle| Y_l^e\right) \quad (1)$$

$p(Y_l^e | E)$ is the posteriori probability where the evidence E are the intra-sentence features in this case. The transfer probability from emotion ϵ to emotion l can be calculated as equation (2).

$$p(\epsilon \to l) = \frac{count[\overrightarrow{y_{SP}}(\epsilon)=1,\ \overrightarrow{y_S}(l)=1]}{count[\overrightarrow{y_S}(l)=1]} \tag{2}$$

Likewise, the emotion transfer probability between the overall microblog and each sentence can be estimated by using the same equation. The conditional probability in equation (1) such as $p\left(W_\epsilon^{\vec{y}(\epsilon)}\middle|Y_l^e\right)$ and $p\left(P_\epsilon^{\vec{y}(\epsilon)}\middle|Y_l^e\right)$ can be calculated from the transfer probability in training dataset. For example, $p(P_\epsilon^1|Y_l^1) = p(\epsilon \to l)$.

The algorithm 1 gives the pseudo code of the proposed iterative approach. The approach outputs the initial classification results, namely $p(Y_l^e|E)$ by base classifier using only intra-sentence features (line 2-6). Then, sentence and document contextual information are converted to emotion transfer probabilities, and are incorporated to update the classification result iteratively (line 7-14). In the iteration process, the sentence level classification results of $(t\text{-}1)^{th}$ iteration are used to predict the microblog's emotion set by majority voting over all sentences. After that, equation (1) is applied to obtain the sentence level classification results of the tth iteration, that is $\overrightarrow{y_s}\,(l,t)$. Such iteration terminates until the classification results converge or the iteration number reaches the specified iteration number *I*.

Algorithm 1. The Iterative Multi-label Emotion Classification Approach

Input : Training set D_{train}, Test set D_{test}, Iteration number *I*.
1. Estimate the transfer probability between adjacent sentences and the transition probability between the overall microblog and each sentence in the microblog from the training dataset according to equation (2);
2. **For** each sentence s $\in D_{test}$
3. **For** each label $l \in L$
4. Calculate the $p(Y_l^e|F)$ with base classifier.
5. **End for**
6. **End for**
7. **For** each t $\in [1, \mathrm{I}]$
8. Find out the majority emotion category in the sentences as the category of the microblog by majority voting;
9. **For** each sentence s$\in D_{test}$
10. **For** each label $l \in L$
11. Iteratively update $\overrightarrow{y_s}\,(l,\mathrm{t})$ using equation (2);
12. **End for**
13. **End for**
14. **End for**

4 Experiments and Discussion

4.1 Experiment Setting

In this paper, we used the dataset from NLP&CC 2013 Chinese microblog emotion classification bakeoff [8-10]. The dataset consists of 14,000 microblogs and 45,431 sentences. Each sentence is annotated with up to two emotional category labels. The training set contains 4,000 microblogs and 13,246 sentences while the test set contains 10,000 microblogs and 32,185 sentences. Table 1 gives the emotion category distribution in the NLP&CC 2013 dataset.

Table 1 Distribution of Emotions in the Training and Test sets in the NLP&CC2013 Microblog Emotion Classification Bakeoff

	Emotion category	Primary Emotion		Secondary Emotion	
Training Set	happiness	729	14.70%	95	14.80%
	like	1,226	24.80%	138	21.60%
	anger	716	14.50%	129	20.20%
	sadness	847	17.10%	45	7.00%
	fear	114	2.30%	14	2.20%
	disgust	1,008	20.40%	187	29.20%
	surprise	309	6.20%	32	5.00%
Test Set	happiness	2,145	20.50%	138	17.60%
	like	2,888	27.60%	204	26.10%
	anger	1,147	10.90%	82	10.50%
	sadness	1,565	14.90%	84	10.70%
	fear	186	1.80%	20	2.60%
	disgust	2,073	19.80%	212	27.10%
	surprise	473	4.50%	43	5.50%

To evaluate the effectiveness of the proposed approach, three multi-label classification algorithms, namely Multi-label k-Nearest Neighbors (ML-kNN), the Binary Relevance method (BR), the random k-labelsets (RAKEL). They are chosen because they are three different kind of multi-label classification methods. They are commonly used for multi-label classification problem and perform well. Based on the kNN algorithm which for single-label classification, MLkNN algorithm counts the category labels of the k number of nearest neighbors and calculates the maximum a posteriori probability of the samples to be classified. BR and RAkEL belong to problem conversion method, namely decomposing the multi-label problems into multiple single-label problems. BR algorithm labels each category separately through binomial classification, and thereby transfers to multi-label classification. The Label Powerset (LP) transformation creates one binary classifier for every label combination attested in the training set. The random k-labelsets (RAKEL) algorithm uses multiple LP classifiers. Each classifier is trained on a random subset of the actual labels. The classifiers are ensemble to generate the final multi-labels by following a voting scheme.

We used unigram and bigram of Chinese words as intra-sentence features, and $\chi 2$ for feature selection. Different from the common used metric in single-label classification, namely precision, recall and F, the evaluation metric in multi-label classification consider the classification accuracy for more than one labeled categories and especially the ranking of labeled categories. Here, we adopted the commonly used metric in multi-label classification, namely, Average Precision (AVP). Formally, we define AVP for a ranking H by system f to be :

$$\text{AVP}(H) = \frac{1}{m}\sum_{i=1}^{m}\frac{1}{|Y_i|}\sum_{1\in Y_i}\frac{\left|\{1' \in Y_i \mid \text{rank}_f(x_i,1') \le \text{rank}_f(x_i,1)\}\right|}{\text{rank}_f(x,1)} \tag{3}$$

AVP estimates the average fraction of labels ranked above a particular label $\ell \in Y_i$ which actually are in Y_i. Note that AVP(f)=1for a system f which ranks perfectly the labels for all documents so that there is no document x_i for which a label not in Y_i is ranked higher than a label in Y_i.

4.2 Experimental Results and Discussions

Table 2 shows the performances of the emotion classification by different approaches. In the experiment, $k = 8$ is selected as the number of neighbors in ML-kNN algorithm. For the BR and RAkEL algorithm, Naïve Bayes (NB) and linear Support Vector Machines (SVMs) are adopted as the internal single-label classification algorithm, respectively. To evaluate the contributions of iterative incorporation of the sentence and document contextual information, different combinations are performed. The baseline methods only use the unigram and bigram features without considering the context and document information. C denotes that this iterative approach considers sentence context information, while D considers document information. The iteration number is set to 1 and 50, respectively.

Table 2. Emotion Classification Performance of Different Iterative Approaches (AVP)

Base Classifier	Baseline	Our Iterative Approach					
		D (I=1)	D (I=50)	C (I=1)	C (I=50)	C+D (I=1)	C+D (I=50)
ML-kNN	0.5818	0.5951	0.6074	0.6844	0.7851	0.6818	**0.8326**
BR_NB	0.6469	0.6482	0.6501	0.7210	0.7928	0.7214	0.7943
BR_SVM	**0.6651**	**0.6671**	**0.6685**	**0.7317**	**0.7968**	**0.7321**	0.7975
RAkEL_NB	0.3722	0.3803	0.3914	0.5039	0.6089	0.5035	0.6365
RAkEL_SVM	0.4592	0.4629	0.4715	0.5420	0.5922	0.5405	0.6253

As shown in Table 2, it is easy to observe that the proposed iterative approach, which combines the sentence and document contextual information, improves the emotion classification performance significantly. But the contributions of the sentence and document context features are varied. The combination of sentence context information, namely the inter-sentences emotion transformation probabilities improved the performance largely for all base classifiers. For example, ML-kNN classifier achieved the initial AVP of 58.18%. While the iterative approaches incorporation with sentence and document context information improves the performance by 20.33% and 2.57%, respectively. The performance of iterative approaches with only document context information is lower than those with sentence context information.

It is observed that the classification performances of all of the five base classifiers are improved. The best performance is achieved by the iterative approach based on ML-kNN with 50 iterations. Compared to the baseline, the AVP increases 43.10%.

Fig 1. shows the iteration curves of iterative approach with MLkNN classifier over different iterations and different intra-sentence features, from 1 to 50. It is observed that the incorporation of contextual sentence and document features achieves the top performance. Meanwhile, the contextual sentence features are shown more contributions compare to document features.

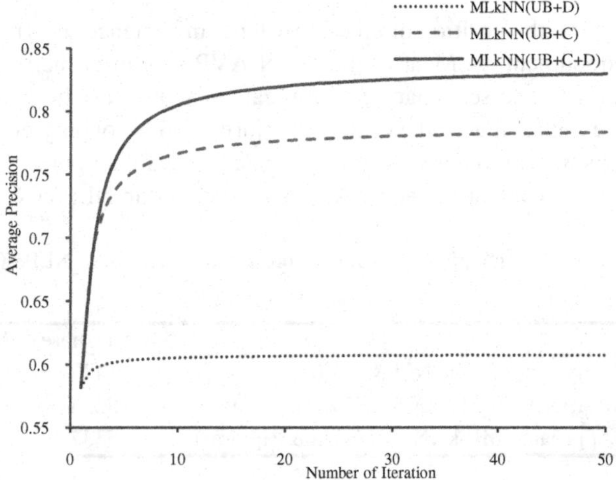

Fig. 1. MLkNN Classification Performance (AVP) vs. Iterations

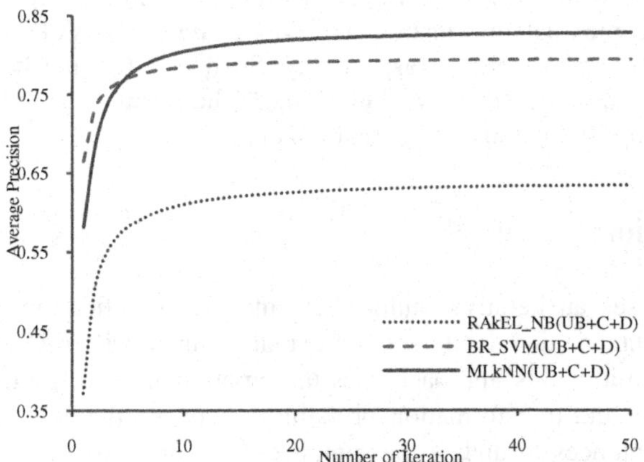

Fig. 2. Classification Performance of MLkNN, BR and RAkEL vs. Iterations

Fig. 2 shows the comparisons between different base classifiers by varying the iterations. It can be observed that, the iterative approaches always converge well, within 15 iterations. Furthermore, iterations from 0 to 10 give the largest improvement.

Table 3 compares the best performance we achieved and the performance achieved by the top ranked system in NLP&CC2013 Chinese microblog emotion bakeoff. The bakeoff adopted the average accuracy of an emotional sentence (labeled as N_AVP) as the evaluation metric, which is different from AVP. For the top-2 emotion classification, N_AVP take average precision in multi-label classification with the consideration of ranking order of emotion categories as metric:

$$N_{AVP} = \frac{1}{n}\sum_{i=1}^{n}\frac{1}{|Y_i|}\sum_{k=1}^{2}\frac{|(emotion_k \in Y_i | rank(x_i, emotion_k) \leq rank(x_i, y))|}{rank(x_i, emotion_k)} \quad (4)$$

Here $rank(x_i, y)$ is the position of emotion *y* label in sentence x_i. $|Y_i|$ is the number of emotions in document x_i. Meanwhile, the N-AVP sets strict metric which requires the order of majority and secondary emotion category must be the same as in the answer while loose metric ignore the order of majority and secondary emotion category. For fair comparison with other reported systems, we evaluate our approach by using the same dataset and evaluation metric which are adopted in NLP&CC 2013 Bakeoff.

Table 3. Performance Comparison of Our Approach and The best NLP&CC2013 Bakeoff System

	N-AVP(loose)	N-AVP(strict)
Top 1 System of the NLP&CC2013	0.3439	0.3305
Our Iterative Approach (ML-kNN, I=50, Unigrams)	0.5264	0.5140
Our Iterative Approach (ML-kNN, I=50, Uni&Bigrams)	**0.5337**	**0.5216**

In NLP&CC 2013 bakeoff, the top ranked system of the 19 participation systems achieved N-AVP of 0.3439 (loose metric) and 0.3305 (strict metric), respectively. Our iterative approach with MLkNN, which incorporates unigram, bigram, and sentence/document contextual information achieves a higher N-AVP of 0.5537 (loose metric) and 0.5216 (strict metric), respectively. Such result is the best performance achieved on this dataset based on our knowledge. The results also indicate the effectiveness of our proposed iterative classification approach.

5 Conclusion

This paper proposes an iterative multi-label emotion classification approach using intra-sentence, sentence and document contextual features. Different from the regular one-step classification, this approach uses the emotion transfer probability between adjacent sentences and transformation probability between the whole microblog and its component sentences to update the sentence emotion classification results iteratively. The evaluations on the NLP&CC 2013 bakeoff dataset demonstrates the best performance of the proposed iterative classification approach and the effectiveness to integrate intra-sentence features, context features and document features on this dataset. In particular, the proposed approach improves all of the MLkNN, BR (NB and SVM) and RAkEL (NB and SVM) classifiers which show the expansibility of our proposed approach.

Acknowledgement. This work is partially supported by the National Natural Science Foundation of China (No. 61370165, 61203378), the Natural Science Foundation of Guangdong Province (No. S2013010014475), MOE Specialized Research Fund for the Doctoral Program of Higher Education 20122302120070, Shenzhen Development and Reform Commission Grant No.[2014]1507, Shenzhen Peacock Plan Research Grant KQCX20140521144507925, Baidu Collaborate Research Funding, General Research Fund of Hong Kong (417112, 417613).

References

1. Quan, C., Ren, F.: Construction of a Blog Emotion Corpus for Chinese Emotional Expression Analysis. In: Proceedings of the 2009 Conference on Empirical Methods in Natural Language Processing, pp. 1446–1454 (2009)
2. Shen, Y., Li, S.: Emotion Mining Research on Micro-blog. In: Proceedings of 1st IEEE Symposium on Web Society, pp. 71–75 (2009)
3. Ma, C., Osherenko, A., et al.: A Chat System Based on Emotion Estimation from Text and Embodied Conversational Messengers. In: Proceedings of IEEE International Conference on Active Media Technology (2005)
4. Strapparava, C., Mihalcea, R.: Learning to Identify Emotions in Text. In: Proceedings of 2008 ACM Symposium on Applied Computing, pp. 1556–1560 (2008)
5. Aman, S., Szpakowicz, S.: Identifying Expressions of Emotion in Text. In: Matoušek, V., Mautner, P. (eds.) TSD 2007. LNCS (LNAI), vol. 4629, pp. 196–205. Springer, Heidelberg (2007)
6. Pang, L., Li, S., Zhou, G.: Emotion Classification Method of Chinese Micro-blog Based on Emotional Knowledge. Computer Engineering 38(13), 156–158 (2012)
7. Pak, A., Paroubek, P.: Twitter as a Corpus for Emotion Analysis and Opinion Mining. In: Proceedings of Language Resources and Evaluation Conference, pp. 1320–1326 (2010)
8. Zhang, J., Zhu, B., et al.: Recognition and Classification of Emotions in the Chinese Microblog based on Emotional Factor. Journal of Peking University 50(1), 79–84 (2014)
9. He, F., He, Y., et al.: A Microblog Short Text Oriented Multi-class Feature Extraction Method of Fine-Grained Emotion Analysis. Journal of Peking University 50(1), 48–54 (2014)
10. Ouyang, C., Yang, X., Lei, L., et al.: Multi-strategy Approach for Fine-grained Emotion Analysis of Chinese Micro-blog. Journal of Peking University 50(1), 67–72 (2014)
11. Liu, Z., Liu, L.: Empirical Study of Emotion Classification for Chinese Microblog based on Machine Learning. Computer Engineering and Applications 48(1), 1–4 (2012)
12. Lin, J., Yang, A., Zhou, Y.: Classification of Microblog Emotion Based on Naïve Bayesian. Computer Engineering and Science 34(9), 160–165 (2012)
13. Tsoumakas, G., Katakis, I., Vlahavas, I.: Mining Multi-label Data. In: Data Mining and Knowledge Discovery Handbook, pp. 667–685. Springer (2010)
14. Ghamrawi, N., McCallum, A.: Collective Multi-label Classification. In: Proceedings of the 2005 ACM Conference on Information and Knowledge Management, pp. 195–200 (2005)
15. Hullermeier, E., Furnkranz, J., Cheng, W., et al.: Label Ranking by Learning Pairwise Preferences. Artificial Intelligence 172(16), 1897–1916 (2008)
16. Trohidis, K., Tsoumakas, G., Kalliris, G., et al.: Multilabel Classification of Music into Emotions. In: Proceedings of 2008 International Conference on Music Information Retrieval, pp. 325–330 (2008)

Aspect-Based Sentiment Analysis Using Tree Kernel Based Relation Extraction

Thien Hai Nguyen and Kiyoaki Shirai

School of Information Science,
Japan Advanced Institute of Science and Technology
1-1 Asahidai, Nomi, Ishikawa 923-1292, Japan
{nhthien,kshirai}@jaist.ac.jp

Abstract. We present an application of kernel methods for extracting relation between an aspect of an entity and an opinion word from text. Two tree kernels based on the constituent tree and dependency tree were applied for aspect-opinion relation extraction. In addition, we developed a new kernel by combining these two tree kernels. We also proposed a new model for sentiment analysis on aspects. Our model can identify polarity of a given aspect based on the aspect-opinion relation extraction. It outperformed the model without relation extraction by 5.8% on accuracy and 4.6% on F-measure.

Keywords: Sentiment Analysis, Relation Extraction, Tree Kernel, Support Vector Machine.

1 Introduction

Relation extraction is a task of finding relations between pairs of entities in texts. Many approaches have been proposed to learn the relations from texts. Among these approaches, kernel methods have been used increasingly for the relation extraction [4,15,17,18]. The main benefit of kernel methods is that they can exploit a huge amount of features without an explicit feature representation. In the relation extraction task, many kinds of relations, from general to specific ones, are considered. This paper focuses on aspect-opinion relation, which is a relation between an aspect of an entity (eg. a price of a PC) and an opinion word or phrase that expresses evaluation on that aspect. It is still an open question if the kernel methods also work well for aspect-opinion relation extraction.

On the other hand, sentiment analysis is considered as an important task in an academic as well as commercial point of view. Many researches attempted to identify polarity of a sentence or paragraph regardless of the entities such as restaurants and their aspects such as food or service. While, this research focuses on aspect-based sentiment analysis, which is a task to identify the sentiment expressed towards aspects of entities.

The goal of our research is to develop a model to predict the sentiment categories (positive, neutral or negative) of the given aspect in the sentence. Intuitively, the opinion words related to the given aspect will have more influence

A. Gelbukh (Ed.): CICLing 2015, Part II, LNCS 9042, pp. 114–125, 2015.
DOI: 10.1007/978-3-319-18117-2_9

on the sentiment of that aspect. Our method firstly identifies the aspect-opinion relations in the sentence by tree kernel method. Then, it calculates the sentiment score for each aspect in the sentence by using these extracted relations.

Our contributions are summarized as follows:

1. We applied two existing tree kernels for aspect-opinion relation extraction.
2. We proposed a new tree kernel based on the combination of two tree kernels for aspect-opinion relation extraction.
3. We proposed a new method for aspect-based sentiment analysis enhanced by the automatically identified aspect-opinion relations.

The rest of this paper is organized as follows. Section 2 introduces some previous approaches on relation extraction and aspect-based sentiment analysis. Section 3 discusses methods for aspect-opinion relation extraction. Section 4 examines how to apply aspect-opinion relation extraction for aspect-based sentiment analysis. Finally, Section 5 concludes our research.

2 Previous Work

2.1 Relation Extraction

Some previous work used the dependency tree kernels for general relation extraction [4,15,18]. In these researches, they tried to extract all of the predefined relations in a given sentence. The predefined relations are *person-affiliation*, *organization-location* and so on. Nguyen et al. used tree kernel based on the constituent, dependency and sequential structures for relation extraction [13]. They focused on seven relation types such as *person-affiliation* in the ACE corpus, which was well-known as a dataset for general relation extraction. However, aspect-opinion relation was not considered in these researches. For the aspect-based sentiment analysis, it is very important to know whether there is a relation between an aspect and opinion word. To the best of our knowledge, there is a lack of researches trying to use tree kernel for aspect-opinion relation extraction.

Wu et al. proposed a phrase dependency parsing for extracting relations between product features and expression of opinions [17]. Their tree kernel is based on a phrase dependency tree converted from an ordinary dependency tree. However, they did not apply this model for calculating a sentiment score for a given aspect.

Bunescu and Mooney extracted the shortest path between two entities in a dependency tree to extract the relation between them [2]. The dependency kernel was calculated based on this shortest path. They suggested that the shortest path encodes sufficient information for relation extraction.

Kobayashi et al. combined contextual and statistical clues for extracting aspect-evaluation and aspect-of relations [7]. Since the contextual information is domain-specific, their model cannot be easily used in other domains.

2.2 Aspect-Based Sentiment Analysis

Aspect-based sentiment analysis has been found to play a significant role in many applications such as opinion mining on product reviews or restaurant reviews. The popular approach is to define a sentiment score of a given aspect by the weighted sum of opinion scores of all words in the sentence, where the weight is defined by the distance from the aspect [10,14]. Because this approach is simple and popular, it will be a baseline model in our experiment in Section 4. To the best of our knowledge, there is no research trying to apply aspect-opinion relation extraction for calculating the sentiment score of the given aspect in the sentence.

Other researches have attempted to use unsupervised topic modeling methods for aspect-based sentiment analysis. To identify the sentiment category of the aspect, topic models which can simultaneously exploit aspect and sentiment have been proposed, such as ASUM [5], JST [9] and FACTS model [8]. However, it is not obvious to map latent (inferred) aspects/sentiments to aspects/sentiments in the text.

3 Aspect-Opinion Relation Extraction

For a given sentence where an aspect phrase and opinion phrase have been already identified, we will determine whether there is a relationship between the aspect and opinion phrase. To achieve this goal, four supervised machine learning methods will be presented in the following subsections. One is Support Vector Machine (SVM) with a linear kernel and the others are SVM with tree kernels.

3.1 SVM-B: A Baseline Model

SVM has long been recognized as a method that can efficiently handle high dimensional data and has been shown to perform well on many applications such as text classification [6,12]. A set of features used for training SVM is shown in Table 1. Because this model was also used in previous work [7,17] for relation extraction, we chose it as a baseline model to compare with other methods.

Table 1. Features used in SVM-B

Feature	Values
Position of opinion word in sentence	{start, end, other}
Position of aspect word in sentence	{start, end, other}
The distance between opinion and aspect	{1, 2, 3, 4, other}
Whether opinion and aspect have direct dependency relation	{True, False}
Whether opinion precedes aspect	{True, False}
Part of Speech (POS) of opinion	Penn Treebank Tagset
POS of aspect	Penn Treebank Tagset

3.2 CTK: Constituent Tree Based Tree Kernel

Tree kernel for the constituent tree has been used successfully in many applications. Various tree kernels have been proposed such as subtree kernel [16] and subset tree kernel [3]. We applied the subtree kernel for this research. Figure 1 shows an example of a constituent tree for the sentence "It has excellent picture quality and color."

Given a constituent tree of a sentence, we represented each $r(e_1, e_2)$, aspect-opinion relation between the aspect entity e_1 and opinion entity e_2, as a subtree T rooted as the lowest common parent of e_1 and e_2. Notice that the aspect and opinion entity can be phrases in general. The subtree T must contain all of the words in these phrases. For example, the relation between the aspect "picture quality" and opinion "excellent" in Figure 1 is represented by the subtree rooted at "NP" node [1], which is the lowest common parent of "picture", "quality" and "excellent" node. The main idea of this tree kernel is to compute the number of the common substructures between two tree T_1 and T_2 which represent two relation instances. The kernel between two trees T_1 and T_2 is defined as in Equation (1).

$$K(T_1, T_2) = \sum_{n_1 \in N_1} \sum_{n_2 \in N_2} C(n_1, n_2) \tag{1}$$

N_1 and N_2 are the set of the nodes in T_1 and T_2. $C(n_1, n_2)$ is the number of common subtrees of two trees rooted at node n_1 and n_2. It is calculated as follows:

1. If n_1 and n_2 are pre-terminals with the same POS tag: $C(n_1, n_2) = \lambda$
2. If the production rules at n_1 and n_2 are different: $C(n_1, n_2) = 0$
3. If the production rules at n_1 and n_2 are the same:

$$C(n_1, n_2) = \lambda \prod_{j=1}^{nc(n_1)} (1 + C(ch(n_1, j), ch(n_2, j)))$$

where $nc(n_1)$ is the number of the children of n_1 in the tree. $ch(n_i, j)$ is the jth child-node of n_i. Since the production rules at n_1 and n_2 are the same, $nc(n_1) = nc(n_2)$. We set $\lambda = 0.5$ in our experiment.

Finally, since the value of $K(T_1, T_2)$ will depend greatly on the size of the trees T_1 and T_2, we normalize the kernel as in Equation (2).

$$K'(T_1, T_2) = \frac{K(T_1, T_2)}{\sqrt{K(T_1, T_1)K(T_2, T_2)}} \tag{2}$$

3.3 DTK: Dependency Tree Based Tree Kernel

A dependency tree kernel has been proposed by Culotta and Sorensen for general relation extraction [4]. This paper applies it for aspect-opinion relation extraction. Given a dependency tree of a sentence, we represent each relation $r(e_1, e_2)$

[1] It is denoted by the circle in Figure 1.

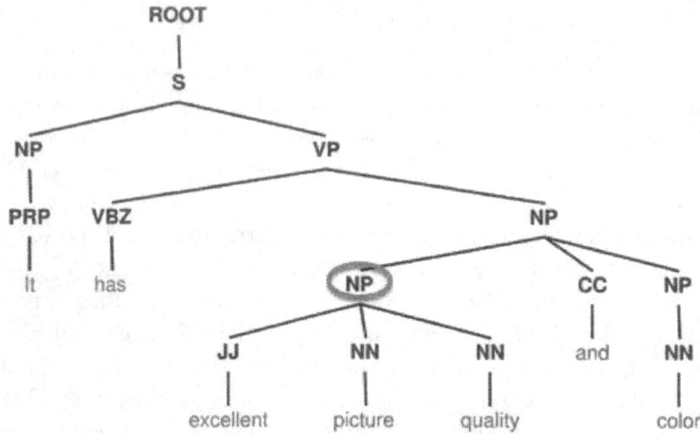

Fig. 1. An Example of Constituent Parsing Tree

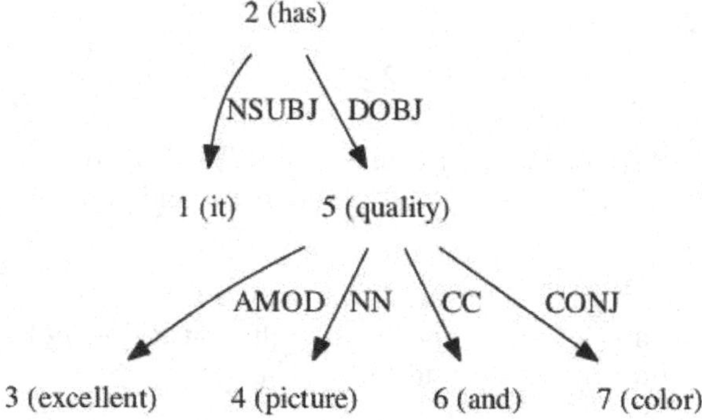

Fig. 2. An Example of Dependency Tree

as a subtree T rooted as the lowest common parent of the aspect e_1 and opinion e_2. For example, the relation between the aspect "picture quality" and opinion "excellent" in Figure 2 is the subtree rooted at "quality" node, which is the lowest common parent of "picture", "quality" and "excellent" node.

A subtree T of a relation instance can be represented as a set of nodes $\{n_0, \cdots, n_t\}$. Each node n_i is augmented with a set of features $f(n_i)$=$\{v_1, \cdots, v_d\}$. They are subdivided into two subsets $f_m(n_i)$ (features used for matching function) and $f_s(n_i)$ (for similarity function). A matching function $m(n_i, n_j) \in \{0, 1\}$ in Equation (3) checks if $f_m(n_i)$ and $f_m(n_j)$ are the same. A similarity function $s(n_i, n_j)$ in $(0, \infty]$ in Equation (4) evaluates the similarity between $f_s(n_i)$ and $f_s(n_j)$.

$$m(n_i, n_j) = \begin{cases} 1 \text{ if } f_m(n_i) = f_m(n_j) \\ 0 \text{ otherwise} \end{cases} \tag{3}$$

$$s(n_i, n_j) = \sum_{v_q \in f_s(n_i)} \sum_{v_r \in f_s(n_j)} C(v_q, v_r) \tag{4}$$

In Equation (4), $C(v_q, v_r)$ is a compatibility function between two feature values as:

$$C(v_q, v_r) = \begin{cases} 1 \text{ if } v_q = v_r \\ 0 \text{ otherwise} \end{cases} \tag{5}$$

For two given two subtrees T_1 and T_2 which represent for two relation instances with root nodes r_1 and r_2, the tree kernel $K(T_1, T_2)$ is defined as in Equation (6):

$$K(T_1, T_2) = \begin{cases} 0 & \text{if } m(r_1, r_2) = 0 \\ s(r_1, r_2) + K_c(r_1[c], r_2[c]) & \text{otherwise} \end{cases} \tag{6}$$

where K_c is a kernel function over children. Let **a** and **b** be sequences of children nodes' indices of node n_i and n_j, respectively. We denote the length of **a** by $l(\mathbf{a})$. K_c is defined as Equation (7). $n_i[\mathbf{a}]$ stands for the subtree consisting of children indicated by **a**, while $n_i[a_h]$ is h^{th} child of n_i. In this equation, we consider the contiguous kernel enumerating children subsequences that are not interrupted by not matching nodes. In our experiment, λ is set to 0.5.

$$K_c(n_i[c], n_j[c]) = \sum_{\mathbf{a},\mathbf{b},l(\mathbf{a})=l(\mathbf{b})} \lambda^{l(\mathbf{a})} K(n_i[\mathbf{a}], n_j[\mathbf{b}]) \prod_{h=1}^{l(\mathbf{a})} m(n_i[a_h], n_j[b_h]) \tag{7}$$

Finally, we also normalize the kernel as in Equation (2).

The augmented features are shown in Table 2. Note that *POS*, *isAspectNode* and *isOpinionNode* are used for matching between two nodes, while the rest is used for measuring the similarity of them.

Table 2. Features for Each Node in the Dependency Tree

	Feature	**Values**
f_m	*POS*	Penn Treebank POS Tagset
	isAspectNode	{0,1}
	isOpinionNode	{0,1}
f_s	*NER*	StanfordCoreNLP Name Entity Tagset
	relationToParentNode	StanfordCoreNLP Dependency Relation
	POSofParentNode	Penn Treebank POS Tagset
	NERofParentNode	StanfordCoreNLP Name Entity Tagset

3.4 CTK + DTK: Combination of Two Kernels

We proposed a new tree kernel based on the combination of two kernels CTK and DTK for aspect-opinion relation extraction. That is, we try to utilize the information from both the constituent and dependency tree. Equation (8) defines the combined kernel function.

$$K_{CTK+DTK}(T_1, T_2) = K_{CTK}(T_1, T_2) + K_{DTK}(T_1, T_2) \tag{8}$$

$K_{CTK}(T_1, T_2)$ and $K_{DTK}(T_1, T_2)$ are the CTK and DTK tree kernels described in Subsection 3.2 and 3.3, respectively. Since the summation of two kernels is valid, $K_{CTK+DTK}$ is obviously a valid kernel.

3.5 Evaluation of Tree Kernel Based Relation Extraction

Dataset: We conducted experiments with labeled dataset developed by Wu et al. [17]. We also corrected some errors such as typing errors, aspect and opinion marking errors, and removed redundant relations. There are 5 domains (DVD Player, Cell Phone, Digital Camera, Diaper, MP3 Player) in this dataset. Stanford CoreNLP [11] was used to parse constituent and dependency tree for each sentence.

Two experiments were designed. The first one is in-domain evaluation. This experiment tries to answer the question how well the models classify the data in the test set which is the same domain of the training data. We divided each domain into 80% for training and 20% for testing. The second experiment is cross-domain evaluation. This evaluates the models on the test set which is different domain to the training data. We used the sentences in "Digital Camera" and "Cell Phone" domains for training, and evaluated the models on "DVD Player", "Diaper" and "MP3 Player" domains. Accuracy, Precision, Recall and F-measure are used as the evaluation metrics. F-measure is the main metric to compare among four models SVM-B, CTK, DTK and CTK + DTK.

In-Domain Results: Table 3 summarizes the results of each domain in four metrics for each method. SVM-B performed worst in all of the domains in F-measure. Our method CTK + DTK improves F-measure of SVM-B method by 6.2%, 8.2%, 3.7%, 1.1% and 3.3% in "DVD Player", "Cell Phone", "Digital Camera", "Diaper" and "MP3 Player", respectively. Therefore, our CTK + DTK method is better than SVM-B in in-domain evaluation. In addition, CTK + DTK method beats CTK in "Cell Phone", "Digital Camera" and "MP3 Player" domain and achieves competitive performance in "DVD Player" and "Diaper" domain. Thus, we can concluded that CTK + DTK is better than CTK method. Finally, CTK + DTK method is better than DTK in "DVD Player" and "MP3 Player" domain, comparable in "Cell Phone" and "Diaper" domain. To sum, DTK and CTK + DTK are the best methods for aspect-opinion relation extraction in in-domain evaluation.

Cross-Domain Results: The results of four methods in cross-domain are shown in Table 4. Our method CTK + DTK outperformed the baseline SVM-B in all domains in F-measure. Improvements of 5.4%, 2.5% and 3.9% of F-measure are found in "DVD Player", "Diaper" and "MP3 Player" domain, respectively. Therefore, our CTK + DTK method is better than SVM-B. In addition, CTK + DTK is better than CTK by 1.3%, 1.6% and 4.5% F-measure in each domain. Finally, compared with DTK, CTK + DTK shows 2.1% and 1.9% F-measure improvement in "DVD Player" and "Diaper" domain, and achieves competitive performance in "MP3 Player" domain. Therefore, we can conclude that CTK + DTK is the best method for extraction of aspect-opinion relations in the cross-domain evaluation.

Table 3. In-domain Results of Aspect-Opinion Relation Extraction

Domain	Metric	SVM-B	CTK	DTK	CTK + DTK
DVD Player	A	0.804	**0.902**	0.863	**0.902**
	P	**0.905**	0.898	0.878	0.898
	R	0.864	**1.00**	0.977	**1.00**
	F	0.884	**0.946**	0.925	**0.946**
Cell Phone	A	0.728	0.712	**0.837**	0.815
	P	0.817	0.704	**0.884**	0.811
	R	0.764	**0.984**	0.870	0.943
	F	0.790	0.820	**0.877**	0.872
Digital Camera	A	0.721	0.652	**0.756**	0.709
	P	**0.798**	0.648	0.741	0.690
	R	0.746	**0.980**	0.940	0.975
	F	0.771	0.780	**0.829**	0.808
Diaper	A	**0.783**	0.739	0.739	0.739
	P	**0.929**	0.739	0.739	0.739
	R	0.765	**1.00**	**1.00**	**1.00**
	F	0.839	**0.850**	**0.850**	**0.850**
MP3 Player	A	0.800	0.705	0.800	**0.813**
	P	**0.923**	0.718	0.832	0.818
	R	0.769	**0.932**	0.885	**0.932**
	F	0.839	0.811	0.857	**0.872**

Table 4. Cross-domain Results of Aspect-Opinion Relation Extraction

Domain	Metric	SVM-B	CTK	DTK	CTK + DTK
DVD Player	A	0.749	0.778	0.787	**0.808**
	P	**0.863**	0.793	0.859	0.834
	R	0.787	**0.952**	0.855	0.928
	F	0.824	0.865	0.857	**0.878**
Diaper	A	0.804	0.780	0.794	**0.812**
	P	**0.910**	0.786	0.846	0.823
	R	0.810	**0.964**	0.881	0.949
	F	0.857	0.866	0.863	**0.882**
MP3 Player	A	0.765	0.686	**0.792**	0.772
	P	**0.833**	0.683	0.805	0.760
	R	0.774	**0.954**	0.894	0.942
	F	0.802	0.796	**0.847**	0.841

4 Aspect-Based Sentiment Classification Based on Relation Extraction

As mentioned in Section 1, aspect-based sentiment analysis is a task to identify the sentiment categories for a given aspect in a sentence. In this section, we tried to integrate the relation extraction model to aspect-based sentiment analysis. Intuitively, not all opinion words in the sentence represent emotion on the given aspect. Therefore, CTK + DTK described in Subsection 3.4 will be used to identify the strong relations between aspect and opinion entities.

4.1 Aspect-Based Sentiment Analysis Without Relation Extraction: A Baseline Model

We used a popular algorithm for calculating a score of a given aspect [10,14]. Even though this algorithm is simple, it can perform well in many cases. Given a sentence, which contains a set of aspects $A = \{a_1, \cdots, a_m\}$ and a set of opinion words $OW = \{ow_1, \cdots, ow_n\}$, the sentiment score for each aspect a_i is calculated as in Equation (9). The closer between the aspect phrase and the opinion word, the higher affection of that opinion on the aspect. Therefore, the sentiment value of the aspect is defined as the summation over all opinion values divided by their distances to that aspect. The aspect is categorized as positive, negative and neutral if $sentimentValue(a_i)$ is greater than 0.25, less than -0.25 and other.

$$sentimentValue(a_i) = \sum_{j=1}^{|OW|} \frac{opinionValue(ow_j)}{distance(a_i, ow_j)} \tag{9}$$

Opinion words were identified based on SentiWordNet [1] that is a lexical resource for opinion mining. Three sentiment scores (positivity, objectivity and negativity) are assigned to each word in SentiWordNet. $opinionValue(ow)$ is defined as Equation (10).

$$opinionValue(ow) = \frac{positivityScore - negativityScore}{positivityScore + negativityScore} \tag{10}$$

4.2 Aspect-Based Sentiment Analysis with Relation Extraction

We proposed a new method for identifying the sentiment category of a given aspect based on the aspect-opinion relations. The method supposes that the opinion words having relation with the aspect will more influence the polarity of it. Identification of the aspect-opinion relations in the sentence can help to improve the prediction of sentiment categories of the given aspect. In other words, aspect-opinion relation extraction enables us to distinguish opinion words of the target aspect and other aspects.

For a given sentence, the aspect-opinion relations were extracted by using the tree kernel method CTK + DTK. Then, we put more weight on the important opinion words in the sentiment score of the aspect as shown in Equation (11).

$$sentimentValue(a_i) = \sum_{j=1}^{|OW|} weight(a_i, ow_j) \cdot \frac{opinionValue(ow_j)}{distance(a_i, ow_j)} \tag{11}$$

The weight of opinion is calculated as:

$$weight(a, ow) = \begin{cases} 2 \text{ if } r(a, ow) = 1 \\ 1 \text{ otherwise} \end{cases} \tag{12}$$

4.3 Evaluation of Aspect-Based Sentiment Analysis

Dataset: Because the data used in Subsection 3.5 is not annotated with the sentiment categories of the aspects, we used the restaurant reviews dataset in SemEval2014 Task 4 [2]. It consists of over 3000 English sentences of the restaurant reviews. For each sentence, the aspect terms and their polarity are annotated. The possible values of the polarity field are "positive", "negative", "neutral" and "conflict". Since, we do not deal with "conflict" category in our model, 84 sentences including the aspects with "conflict" polarity are removed from the dataset. CTK + DTK was trained from the sentences in "Digital Camera" and "Cell Phone" domains in Wu et al.'s dataset.

To investigate the effectiveness of integrating aspect-opinion relation extraction to aspect-based sentiment analysis, we compared the model with and without relation extraction (we call "ASA with RE" and "ASA w/o RE", respectively). Table 5 shows Accuracy, Precision, Recall and F-measure for all aspect phrases in the dataset. Precision, Recall and F-measure are the average for three polarity categories weighted by the number of true instances. Accuracy of "ASA with RE" was 0.523 [3]. It outperformed the baseline by 5.8%. Furthermore, Recall and F-measure of "ASA with RE" were greatly improved. Table 6 shows the results of the sentence-based evaluation. Exact Match Ratio (EMR) is defined as a ratio of correctly classified sentences where the polarity of all aspects in the sentence are successfully identified. Partial Match Ratio (PMR) is the average of the partial matching scores of individual sentences, that is the proportion of the number of correctly classified aspects to all aspects in the sentence. "ASA with RE" was better 3.8% EMR and 3.6% PMR than "ASA w/o RE". From these results, we can conclude that using the aspect-opinion relation extraction is useful for sentiment analysis of aspects.

Table 5. Results of Aspect-based Sentiment Identification

Metric	ASA w/o RE	ASA with RE
A	0.465	**0.523**
P	**0.532**	**0.532**
R	0.465	**0.523**
F	0.477	**0.523**

Table 6. Results of Sentence-based Sentiment Identification

Metric	ASA w/o RE	ASA with RE
EMR	0.596	**0.634**
PMR	0.666	**0.702**

[2] `http://alt.qcri.org/semeval2014/task4/`

[3] Accuracies of participating systems in SemEval 2014 Task 4 were between 0.42 and 0.81. However, these results cannot be simply compared to Table 5. Our method was evaluated on a training data of the task, while the participating systems were trained on it and evaluated on a separate test data. Furthermore, our system was not developed by supervised machine learning, unlike the top participating system.

5 Conclusion

We applied two kernels of constituent and dependency trees and proposed the new tree kernel for aspect-opinion relation extraction. The results showed that the models using tree kernels outperformed the baseline SVM-B. Furthermore, we proposed the new method for identifying the sentiment categories of the aspects in the sentences with the relation extraction module. Our method achieved better performance in almost all metrics compared to the method without relation extraction.

Our tree kernel based model for aspect-opinion relation extraction can be further improved by using semantic information from semantic trees. In addition, combining the syntactic tree and semantic tree for calculating tree kernel will be explored in our future work.

References

1. Baccianella, S., Esuli, A., Sebastiani, F.: Sentiwordnet 3.0: An enhanced lexical resource for sentiment analysis and opinion mining. In: Proceedings of the Seventh Conference on International Language Resources and Evaluation (LREC), vol. 10, pp. 2200–2204 (2010)
2. Bunescu, R.C., Mooney, R.J.: A shortest path dependency kernel for relation extraction. In: Human Language Technology Conference and Conference on Empirical Methods in Natural Language Processing (HLT/EMNLP), pp. 724–731. ACL (2005)
3. Collins, M., Duffy, N.: New ranking algorithms for parsing and tagging: Kernels over discrete structures, and the voted perceptron. In: Proceedings of the 40th Annual Meeting on Association for Computational Linguistics (ACL), pp. 263–270. ACL (2002)
4. Culotta, A., Sorensen, J.S.: Dependency tree kernels for relation extraction. In: Scott, D., Daelemans, W., Walker, M.A. (eds.) Proceedings of the 42nd Annual Meeting of the Association for Computational Linguistics (ACL), pp. 423–429. ACL (2004)
5. Jo, Y., Oh, A.H.: Aspect and sentiment unification model for online review analysis. In: Proceedings of the Fourth ACM International Conference on Web Search and Data Mining, pp. 815–824. ACM (2011)
6. Joachims, T.: Text categorization with support vector machines. In: Nédellec, C., Rouveirol, C. (eds.) ECML 1998. LNCS, vol. 1398, pp. 137–142. Springer, Heidelberg (1998)
7. Kobayashi, N., Inui, K., Matsumoto, Y.: Extracting aspect-evaluation and aspect-of relations in opinion mining. In: Proceedings of the 2007 Joint Conference on Empirical Methods in Natural Language Processing and Computational Natural Language Learning (EMNLP-CoNLL), pp. 1065–1074. ACL (2007)
8. Lakkaraju, H., Bhattacharyya, C., Bhattacharya, I., Merugu, S.: Exploiting coherence for the simultaneous discovery of latent facets and associated sentiments. In: Proceedings of the Eleventh SIAM International Conference on Data Mining (SDM), pp. 498–509. SIAM/Omnipress (2011)
9. Lin, C., He, Y.: Joint sentiment/topic model for sentiment analysis. In: Proceedings of the 18th ACM Conference on Information and Knowledge Management, pp. 375–384. ACM (2009)

10. Liu, B., Zhang, L.: A survey of opinion mining and sentiment analysis. In: Mining Text Data, pp. 415–463. Springer (2012)
11. Manning, C.D., Surdeanu, M., Bauer, J., Finkel, J., Bethard, S.J., McClosky, D.: The Stanford CoreNLP natural language processing toolkit. In: Proceedings of 52nd Annual Meeting of the Association for Computational Linguistics: System Demonstrations (ACL), pp. 55–60 (2014)
12. Nguyen, T.H., Shirai, K.: Text classification of technical papers based on text segmentation. In: Gulliver, T.A., Secord, N.P. (eds.) Information Theory 1993. LNCS, vol. 7934, pp. 278–284. Springer, Heidelberg (2003)
13. Nguyen, T.T., Moschitti, A., Riccardi, G.: Convolution kernels on constituent, dependency and sequential structures for relation extraction. In: Proceedings of the 2009 Conference on Empirical Methods in Natural Language Processing (EMNLP), pp. 1378–1387. ACL (2009)
14. Pang, B., Lee, L.: Opinion mining and sentiment analysis. Foundations and Trends in Information Retrieval 2(1-2), 1–135 (2008)
15. Reichartz, F., Korte, H., Paass, G.: Dependency tree kernels for relation extraction from natural language text. In: Buntine, W., Grobelnik, M., Mladenić, D., Shawe-Taylor, J. (eds.) ECML PKDD 2009, Part II. LNCS, vol. 5782, pp. 270–285. Springer, Heidelberg (2009)
16. Smola, A.J., Vishwanathan, S.: Fast kernels for string and tree matching. In: Becker, S., Thrun, S., Obermayer, K. (eds.) Advances in Neural Information Processing Systems 15 (NIPS), pp. 585–592. MIT Press (2003)
17. Wu, Y., Zhang, Q., Huang, X., Wu, L.: Phrase dependency parsing for opinion mining. In: Proceedings of the 2009 Conference on Empirical Methods in Natural Language Processing (EMNLP), pp. 1533–1541. ACL (2009)
18. Zelenko, D., Aone, C., Richardella, A.: Kernel methods for relation extraction. Journal of Machine Learning Research (JMLR) 3, 1083–1106 (2003)

Text Integrity Assessment: Sentiment Profile vs Rhetoric Structure

Boris Galitsky[1], Dmitry Ilvovsky[2], and Sergey O. Kuznetsov[2]

[1] Knowledge Trail Inc., San Francisco, USA
bgalitsky@hotmail.com
[2] Higher School of Economics, Moscow, Russia
{dilvovsky,skuznetsov}@hse.ru

Abstract. We formulate the problem of text integrity assessment as learning the discourse structure of text given the dataset of texts with high integrity and low integrity. We use two approaches to formalizing the discourse structures, sentiment profile and rhetoric structures, relying on sentence-level sentiment classifier and rhetoric structure parsers respectively. To learn discourse structures, we use the graph-based nearest neighbor approach which allows for explicit feature engineering, and also SVM tree kernel–based learning. Both learning approaches operate on the graphs (parse thickets) which are sets of parse trees with nodes with either additional labels for sentiments, or additional arcs for rhetoric relations between different sentences. Evaluation in the domain of valid vs invalid customer complains (those with argumentation flow, non-cohesive, indicating a bad mood of a complainant) shows the stronger contribution of rhetoric structure information in comparison with the sentiment profile information. Both above learning approaches demonstrated that discourse structure as obtained by RST parser is sufficient to conduct the text integrity assessment. At the same time, sentiment profile-based approach shows much weaker results and also does not complement strongly the rhetoric structure ones.

Keywords: sentiment profile, rhetoric structure, text integrity, text cohesiveness.

1 Introduction

Integrity is an important property of text in terms of style, communication quality, trust and overall reader impression. Text integrity assessment is an important NLP task for customer relationship management, automated email answering, text quality analysis, spam detection, disinformation and low quality content, as well as other domains. Text integrity assessments helps in recognizing a mood of an author, the implicit intent of his message, trustworthiness of the subject being communicated, and can assist in a decision on how to react to this message.

Text integrity is high when the author provides an acceptable argumentation for his statements, sufficient details are provided to substitute the claims. The text looks truthful and cohesive: entities are first defined when introduced, and then related to each other. Text is authoritative: it sounds valid and can be trusted even by a reader unfamiliar with given knowledge domain.

A. Gelbukh (Ed.): CICLing 2015, Part II, LNCS 9042, pp. 126–139, 2015.
DOI: 10.1007/978-3-319-18117-2_10

Text integrity is low when flaws in argumentation and communication can be detected. A reader can identify missing pieces of information, and claims are not substituted. There are problems in text cohesiveness, it is hard to believe in what is being communicated. There is noticeable inconsistency in writer's logic and also in writer's discourse representing complaint scenario.

In this study we focus on such area of text integrity assessment as validity of a customer complaint. Complaint processing is a field of customer relationship management where an automated system needs to "comprehend" textual complaints and to make a decision on how to react to it. Complaints fall into two classes:

- Complaints with proper text integrity. These complaints need to be trusted, and the customer needs to be compensated in one or another way. We refer to this class as valid complaints.
- Complaints with issues in text integrity. These complaints cannot be trusted, they do not read as genuine description of how a complainant was communicating his case with his opponents.

For the domain of customer complaints, these are the text features we want the validity assessment to be independent: quality of writing, language grammar, professionalism using language, writing creativity, educational background, familiarity with topic, emotional state. The task is to identify the cases of "artificial" complainants with plausible facts but faulty discourse (invalid class), and also the cases of genuine dissatisfaction with a product written poorly in grammar and style (valid class).

In [3] authors represented complaints as graphs and learned these graphs to classify complaints into the classes of valid and invalid. Complainants were inputting complaints via a form with the focus on the structure of communicative actions-based dialogue, to avoid NLP analysis. Since then, performance of sentiment analysis and rhetoric parsers has dramatically increased, and a discourse structure of text to be learned can be formed from text automatically. Taking into account the discourse structure of conflicting dialogs, one can judge on the validity of these dialogs. In this work we will evaluate the combined system, discourse structure extraction and its learning.

Text integrity is tightly connected with how the author estimates attitudes and actions of himself and his opponents (what we call a sentiment profile) on one hand, and how proper the author composes discourse structure, on the other hand. In the hostile/contradictory/controversial environments, it is hard for an author to objectively reflect the state of affairs, adequately describe opinions of his opponents. In this case, a consistent presentation of sentiments associated with proponents and opponents is an important component of text integrity.

It is also hard to make an assessment for a given text in a stand-alone mode, and we built a training set of complaints manually tagged as valid or invalid.

To represent the linguistic features of text, we use the following data:

- Syntactic parse trees
- Coreferences
- Sentiments, attached to phrases
- Rhetoric relations between the parts of the sentences

To combine the above data for the purpose of learning, we take parse trees as a basis and add the other data in the form of additional arcs and node labels. We refer to this representation as Parse Thickets (PT) [31].

For a complaint to be classified, it has to be similar to the elements of a given class to be assigned to this class. To evaluate the contribution of sentiment profile and discourse structure, we use two types of learning:

1. Nearest Neighbor (kNN) learning with explicit feature engineering. We measure similarity as an overlap between the pair of graphs for a given text and for a given element of training set
2. Statistical learning of structures with implicit feature engineering. We apply kernel learning to parse trees [36], extended by discourse relations [32]. Similarity is measured as a distance in the space of features (subtrees of extended parse trees).

The goal of this research is to estimate the contribution of each data type and the above learning methods to the problem of text integrity assessment.

2 Related Work

Text readability is a characteristic fairly close to text integrity as we formulate it. As syntactic and lexico-semantic features are broadly used in literature, the authors focus on discourse-level analysis as cohesion and coherence. Coherence and cohesion are two important properties of texts. Text coherence is considered as a "semantic property of discourses, based on the interpretation of each individual sentence relative to the interpretation of other sentences" [12].

Successful writing is expected to follow three main rules:

1. Writer has a clear communicative intent and goals
2. Writer should properly select descriptive words and phrases
3. Thoughts need to be organized in a logical, readable way

A skilled writer needs to address text coherence, sentence-level cohesion and word-level cohesion. Text integrity, the characteristic we focus on in this work, is associated with the last rule.

Cohesion refers to the presence or absence of explicit cues in the text that allow the reader to make connections between the ideas in the text. For example, overlapping words and concepts between sentences indicate that the same ideas are being referred to across sentences. Likewise, connectives such as because, therefore, and consequently, inform the reader that there are relationships between ideas and the nature of those relationships. Whereas cohesion refers to the explicit cues in the text, coherence refers to the understanding that the reader derives from the text, which may be more or less coherent depending on a number of factors, such as prior knowledge and reading skill [6], [14, 15].

A text is represented as a sequence of related utterances. Some theories describe coherence relations by the existence of explicit linguistic markers reinforcing cohesion [23, 24]. However, cohesive markers are not mandatory elements to obtain coherent texts, although they contribute to the overall text interpretation [23]. In [20] authors identified several cohesive devices helpful for the semantic interpretation of the whole text: coreference relations (various expressions referring to the same entity), discourse connectives, lexical relations such as synonymy, hypernymy, hyponymy, meronymy, and thematic progressions. Among these cohesive devices, coreference relations are expressed via anaphoric chains [27] or reference chains [25, 26].

Another approach of coherence in readability is based on the latent semantic analysis (LSA) [13]. This method projects sentences in a semantic space in which each dimension roughly corresponds to a semantic field. Therefore, it better allows assessing the semantic similarity between sentences, since it can capture lexical repetitions, even though synonyms or hyponyms. However, this method is not sensitive to cohesive clues such as ellipsis, pronominal anaphora, substitution, causal conjunction, etc. An alternative approach to LSA was suggested in [21]. Text is considered as a matrix of the discourse entities presented in each sentence. The cohesive level of a text is then computed based on the transitions between those entities.

Essay quality is usually related to its cohesion and coherence of the essay. This is reflected in the literature about writing [18], as well as textbooks that teach students how to write [19].

The interplay of coherence and cohesion is an intensely studied, but still not fully understood issue in discourse organization. Both are known to vary with genre [5]. In expository prose, for instance, the coherence structure is strongly determined by content-oriented relations, while instructive, argumentative, or persuasive texts are structured according to the writer's discursive strategy, involving relations between speech acts and thus what in [28] was called the intentional structure of the discourse. This difference corresponds to the distinction between semantic (or 'ideational') and pragmatic coherence relations [22]. Similarly, expository texts have shorter cohesive chains than for instance narratives [16] and generally can be expected to have more lexical cohesive (thematic) links than other text types.

In [1] authors investigate the hypothesis that lexical cohesion is closely aligned with coherence structure in thematically organized (expository) texts, but less so in texts with a predominantly intentional structure (e.g., persuasive texts). The validity of this hunch has been confirmed w.r.t. local relations in a small pilot study comparing texts from the Wall Street Journal (WSJ) corpus [29] to a sample of fundraising letters [4]. The number of cohesive links between elementary (clause-level) discourse units was greater for units that were directly connected in the discourse structure than for units that had no direct coherence link, and this difference was much larger for the expository (WSJ) texts than for the fundraising letters.

Rhetorical Structure Theory (RST) [2], [35] has been used to describe or understand the structure of texts [11], and to link rhetorical structure to other phenomena, such as anaphora or cohesion. Authors in [7] compare written and spoken discourse, and examine the relationship between rhetorical structure and anaphoric relations. Many studies use RST to analyze second language writing, and determine the coherence of the text, as a measure of the proficiency of the learner [8, 9]. Authors in [10] use it to investigate the process of text creation by naive writers, from planning phase to final product.

Unlike the above studies, we attempt to evolve text quality assessment towards trustworthiness and make it less dependent on author's writing skills. The domain of customer complaints demonstrates that less skillful writers can be more honest describing their experience with a company. On the contrary, skillful writers can be at advantage describing something which has never happened to take advantage of what they think of customer support policies.

3 Text Integrity and Sentiments

As most of the studies of text coherence indicate, the determining features are those of discourse. Intuitively, text integrity is high if sentiments are neutral, consistently positive or consistently negative. Communicative actions need to form a plausible sequence [33]. Argumentative patterns, which are reflected via rhetoric structure of text, need to be acceptable as well.

This is our first example of a complaint which is very emotional. However it is hard to see integrity flaws here.

Ex. 1. Valid complaint

I placed an order on your system. I used a $50 gift card. My total was $50 card and a promo code for 1 cent shipping bringing my total to 5.67.

I later checked my bank account and noticed a charge for 25.99. I called your customer service department and got the most unusual explanation. I was told that they had to charge me for the cost of the basket that was covered by the gift card. They kept saying it was pre authorized. I only authorized 5.67 to be charged from my card. Your explanation makes no sense. You have committed bank fraud! I will tell everyone I know not to order from you. You people are thieves! I don't even know where to start. How do you make it up to someone when you steal from them?

In our second example, the complaint author has an issue with defeating his own claims.

Ex. 2. Invalid complaint

I explained that I made a deposit, and then wrote a check, which bounced due to a bank error. A customer service representative confirmed that it usually takes a day to process the deposit. I reminded that I was unfairly charged an overdraft fee a month ago in a similar situation.

They explained that the overdraft fee was due to insufficient funds as disclosed in my account information. I disagreed with their fee because I made a deposit well in advance and wanted this fee back. They denied responsibility saying that nothing can be done at this point. They also confirmed that I needed to look into the account rules closer.

Most complaints are very emotional. By the way people express their anger and dissatisfaction we can judge whether a given complaint follows the common sentence and valid arguments, or is driven just by emotions. Applying only manual rules complaint may be considered invalid if:

– all actions associated with an opponent have negative polarity;
– all actions of both opponent and proponent have negative polarity;
– a positive (from the complainant's standpoint) action of an opponent is followed by a negative action of the proponent: *They thoroughly explained, but I would not follow the explanation because I knew.*

Actually the rules similar to the ones listed above are not able to handle all of the features of invalid/valid complaints properly. Looking through real texts it's easy to find the counter examples for all of them because they are too "rude" and can follow to false generalizations. So we need to pick discourse or/and sentiment features from the text to take into account linguistic knowledge instead of common sense.

We will learn a sentiment profile as a set of parse trees with some nodes having additional labels for sentiment polarity. If a sentiment profile is similar to the one of positive dataset of valid complaints and dissimilar to all of the elements of the dataset of invalid complaints, we classify the respective complaint as valid.

Fig. 1 depicts the sentiment profile for sentences 6..14 of the above text. For the sentences 1..5 the polarity is zero. We can see that each sentence has a negative sentiment

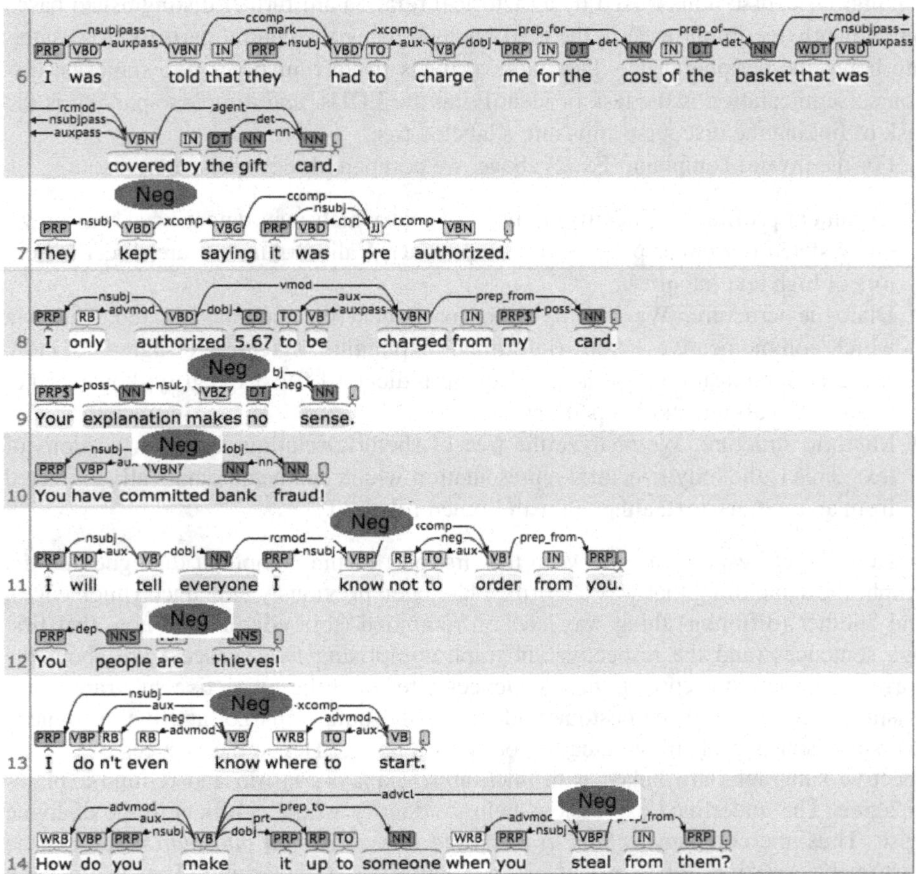

Fig. 1. Sentiment profile of a valid complaint

attached to both the proponent's own mental attitudes, and also proponent actions. This looks like a genuine monotonic negative profile which is a result of dissatisfaction with the service.

For the invalid complaint we have the negative sentiment associated with the action of an opponent, and a neutral sentiment associated with the action of the proponent. On its own, the sentiment profile does not indicate a flaw in text integrity here: one needs to look further into the argument structure of how the proponent argues that he is right and his opponent is wrong.

4 Rhetorical Structure and Integrity

RST is one of the most popular approach to model extra-sentence as well as intra-sentence discourse. RST represents texts by labeled hierarchical structures. Their leaves correspond to contiguous Elementary Discourse Units; adjacent ones are connected by rhetorical relations (e.g., Elaboration, Contrast), forming larger discourse units (represented by internal nodes), which in turn are also subject to this relation linking. Discourse units linked by a rhetorical relation are further distinguished based on their relative importance in the text: nucleus being the central part, whereas satellite being the peripheral one. Discourse analysis in RST involves two subtasks: discourse segmentation is the task of identifying the EDUs, and discourse parsing is the task of linking the discourse units into a labeled tree.

For the invalid complaint (Ex. 2) above, we compare three representations:

1. Argument profile. We identify portions of text which may defeat what was previously stated (by either proponent or opponent). Valid arguments are direct indicators of high text integrity.
2. Dialogue structure. We identify who (proponent or opponent) stated what via which communicative actions (informing, explaining, agreeing, disagreeing, etc.). Valid communication style is an adequate indicator of text integrity, however not as strong as argumentation patterns.
3. Rhetoric structure. We analyze the tree of rhetoric relations between portions of text. This is the only structural representation which can be automatically extracted from an arbitrary text, although with limited reliability.

Note the correspondence between the first part of the complaint dialogue and the graph: the same thing that was confirmed had been previously explained (thick edge), and another (different) thing was later on reminded (thin edge). Also note that first two sentences (and the respective subgraph comprising two vertices) are about the current transaction (deposit), three sentences after (and the respective subgraph comprising three vertices) the customer addresses the unfair charge, and the customer's last statement is probably related to both issues above. Hence the vertices of two respective subgraphs are linked with thick arcs: explain-confirm and remind-explain-disagree. The underlined expressions help to identify where conflicts in the dialogue arise. Thus, the company's *claim as disclosed in my account information* attacks the customer's assertion *due to a bank error*. Similarly, the expression "I made a deposit

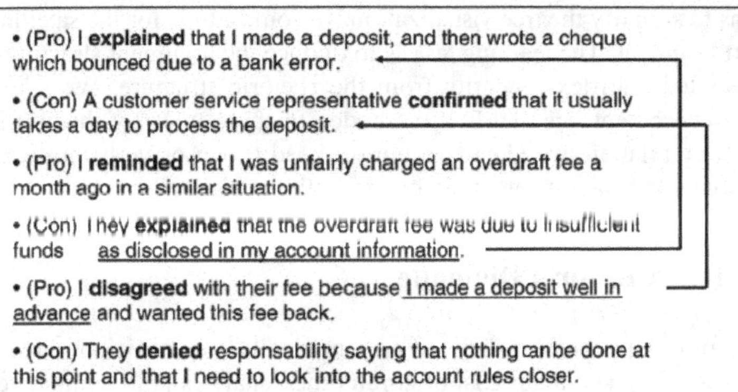

- (Pro) I **explained** that I made a deposit, and then wrote a cheque which bounced due to a bank error.
- (Con) A customer service representative **confirmed** that it usually takes a day to process the deposit.
- (Pro) I **reminded** that I was unfairly charged an overdraft fee a month ago in a similar situation.
- (Con) They **explained** that the overdraft fee was due to insufficient funds as disclosed in my account information.
- (Pro) I **disagreed** with their fee because I made a deposit well in advance and wanted this fee back.
- (Con) They **denied** responsability saying that nothing can be done at this point and that I need to look into the account rules closer.

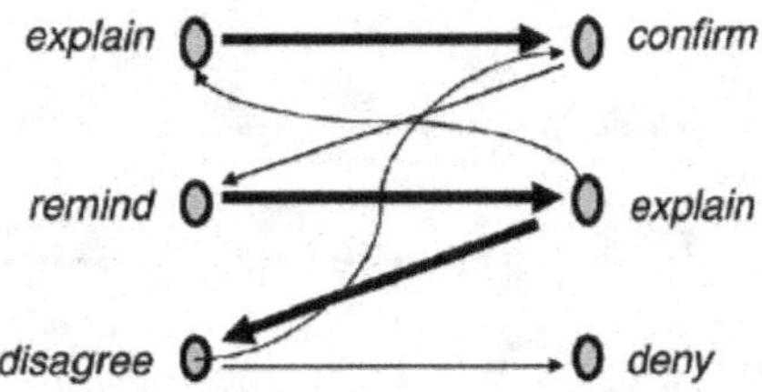

Fig. 2. The structure of communicative actions and arguments for invalid complaint

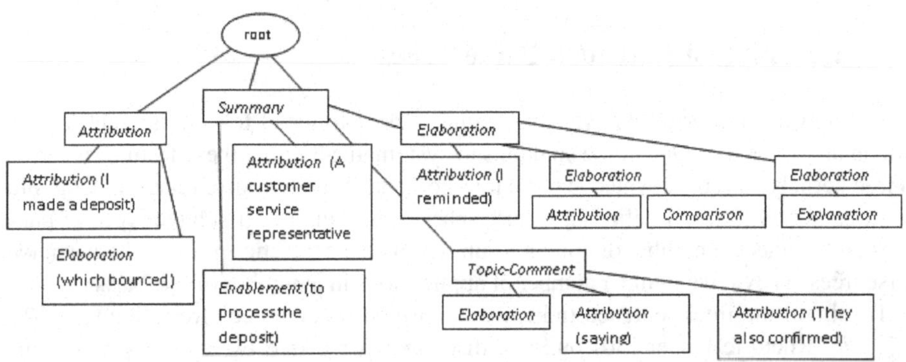

Fig. 3. Simplified rhetoric tree for the invalid complaint

well in advance" attacks the statement "it usually takes a day to process the deposit" (makes it non-applicable). The former attack has the intuitive meaning "existence of a rule or criterion of procedure attacks an associated claim of an error", whereas the latter would have the meaning "the rule of procedure is not applicable to this particular case".

Although we cannot extract argumentation structure from the dialogue directly, we can attempt to learn it from the rhetoric relations. Fig. 3 depicts the reduced rhetoric tree for the whole complaint. We only show the presentational and subject matter

RTS relations to simplify the tree visualization. In some cases, for the satellite part we show the text fragment. The learning task is to deduce the argument structure which is directly associated with text integrity from the rhetoric structure. We take a set of parse trees for each sentence and link the nodes of these trees by rhetoric relations. For a given RST relation, we identify a phrase head (noun or verb) in the text fragments for nucleus and satellite and link the respective nodes.

5 Coreferences for a Dialogue

The structure of coreferences, if they are grammatically correct, is not directly correlated to text integrity. The main goal of coreference analysis is to identify references to the proponent and references to opponents (she, them, customer support, company) to properly attach sentiments and RST relations (Fig. 4).

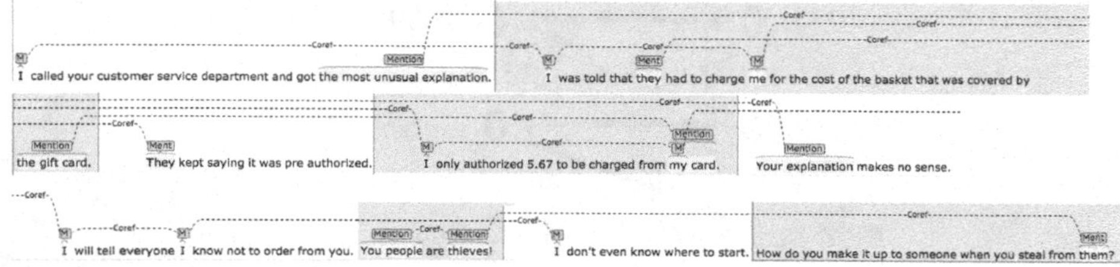

Fig. 4. Coreferences for the portion of the valid complaint

6 Learning of Extended Parse Trees

The design of discourse and syntactic features for automated text assessment tasks is still an art nowadays. One of the solutions to systematically treat these features is the set of tree kernels built over syntactic parse trees, extended by discourse relations. Convolution tree kernel [39, 40] defines a feature space consisting of all subtree types of parse trees and counts the number of common subtrees as the syntactic similarity between two parse trees. They have found a number of applications in a number of NLP tasks.

To obtain the inter-sentence links, we employed coreferences from Stanford NLP [37, 38]. Rhetoric parser [30] builds a discourse parse tree by applying an optimal parsing algorithm to probabilities obtained from two conditional random fields, intra-sentence and multi-sentence parsing. We extend and confirm the extracted rhetoric relations by our own combined manual rules and automatically learned rules derived from the available discourse corpus by means of syntactic generalization. We rely on the extractor of communicative actions and our own web-mined ontology of entity-entity relations [31].

For every arc which connects two parse trees, we obtain the extension of these trees, extending branches according to the arc. For a given parse tree, we will obtain a set of its extension, so the elements of kernel will be computed for many extensions, instead of just a single tree [32]. The problem here is that we need to find common sub-trees for a much higher number of trees than the number of sentences in text, however by subsumption (sub-tree relation) the number of common sub-trees will be substantially reduced. The resultant trees are not the proper parse trees for a sentence, but nevertheless form an adequate feature space for tree kernel learning.

In addition to the statistical learning of extended parse trees, we use the kNN approach which allows explicit feature engineering. Instead of representing the linguistic feature space of the totality of subtrees, we compute the similarity between the trees as finding the maximum common sub-trees. For the discourse level analysis, instead of building multiple extended trees, we combine them in Parse Thickets and compute similarity as the cardinality of their intersections. The construction of parse thicket follows the algorithm combining coreferences, entity-entity relations, communicative actions and RST relations [31].

We perform the comparative evaluation of text integrity assessment in both learning settings to obtain more accurate results on which discourse or syntactic features contribute most.

7 Evaluation

We formed a dataset of 400 complaints downloaded from PlanetFeedback.com between 2003 and 2014. We selected texts describing interaction between a customer and a company, such that it included non-trivial sentiments and non-trivial discourse structure. The texts were manually split into 254 valid and 146 invalid complaints. The baseline for our evaluation is a set of pure syntactic features (a set of parse trees for the text).

We conclude that the most important source of data for text integrity assessment is the set of Presentational relations, which include Antithesis, Background, Concession, Enablement, Evidence, Justification, Motivation, Preparation, and Restatement. Proper order of these relations is a determining feature for text integrity. This source works when anaphora information is available (without it a sentiment or a communicative action is not properly attached to a proponent/opponent).

The other sources are enumerated in the order of importance (taking into account both learning approaches):

– RST-Subject Matter (such as Circumstance, Condition, Elaboration, Evaluation, Interpretation, Means, Non-volitional Cause, Non-volitional Result, Otherwise, Purpose, Solutionhood, Unconditional, Unless, Volitional Cause/Result)
– Sentiment profile
– Multinuclear RST (such as Conjunction, Contrast, Disjunction, Joint, List, Multinuclear Restatement, Sequence

We observe that SVM TK [36] insignificantly outperforms PT kNN for most sources. In this work we do not evaluate a manual improvement to PT kNN by focusing on selecting particular subgraphs critical for relating a text to a class. These subgraphs are formed as a result of a manual analysis of the most frequent maximal common class-determining subgraphs. After manual feature engineering, we expect PT kNN to outperform SVM TK at least in specific domains.

Table 1. Evaluation of text integrity assessment in two learning settings, and relying on various discourse features

Method / sources	Precision	Recall	F-measure	Improvement over baseline
SVM TK parse trees	59.2	63.8	61.4	1.00
SVM TK extended parse trees with anaphora only	64.5	65.7	65.1	1.06
SVM TK extended parse trees with anaphora and sentiment profiles	68.2	64.9	66.5	1.08
SVM TK extended parse trees with anaphora and RST Presentational	73.4	67.3	70.2	1.14
SVM TK extended parse trees with anaphora and RST Presentational+ Subject Matter	78.1	73.4	75.7	1.23
SVM TK extended parse trees with anaphora, RST (full) and sentiment profile	83.6	75.1	79.1	1.29
Unconnected parse trees	58.7	67.1	62.6	1.00
PT with anaphora only	63.4	66.2	64.8	1.03
PT with anaphora and sentiment profiles	76.3	70.3	73.2	1.17
PT with anaphora and RST	82.3	75.8	78.9	1.26

Sentiment profiles determine the text validity in a significantly lesser degree than presentational RST relation, at least in the domain of customer complaints. In spite of the fact that both rhetoric relations and sentence-level sentiments have low accuracy of detection (below 50%), the former turned out to be significantly stronger correlated with text integrity compared to the latter.

This tool is open source and available at:
https://code.google.com/p/relevance-based-on-parse-trees/

8 Conclusions

We proposed a text integrity assessment algorithm based on rich linguistic features, from syntax to sentiment to discourse structure. We applied two distinct learning mechanisms to our text representation structure: tree kernel learning and nearest neighbor learning, and demonstrated that rhetoric structures used by the text author are the strongest indicators of text integrity. To evaluate out approach, we select the domain of customer complaints with an extensive corpus of texts which vary from consistent, coherent and truthful to the ones with argumentation flaws, emotionally driven and possibly written in a bad mood. The estimate of contribution of various sources of discourse information shed the light into how these sources affect the overall reader's impression with a text.

References

1. Berzlánovich, I., Egg, M., Redeker, G.: Coherence structure and lexical cohesion in expository and persuasive texts. In: Proceedings of the Workshop on Constraints in Discourse III (2008)
2. Mann, W., Matthiessen, C., Thompson, S.: Rhetorical Structure Theory and Text Analysis. In: Mann, W.C., Thompson, S.A. (eds.) Discourse Description: Diverse Linguistic Analyses of a Fund-Raising Text, Amsterdam, pp. 39–78 (1992)
3. Galitsky, B., González, M., Chesñevar, C.: A novel approach for classifying customer complaints through graphs similarities in argumentative dialogues. Decision Support Systems (2009)
4. Egg, M., Redeker, G.: Underspecified discourse representation. In: Benz, A., Kühnlein, P. (eds.) Constraints in Discourse, pp. 117–138. Benjamins, Amsterdam (2008)
5. Taboada, M.: The Genre Structure of Bulletin Board Messages. Text Technology 13(2), 55–82 (2004)
6. Todirascu, A., François, T., Gala, N., Fairon, C., Ligozat, A., Bernhard, B.: Coherence and Cohesion for the Assessment of Text Readability. In: Proceedings of NLPCS 2013, Marseille, France (October 2013)
7. Fox, B.A.: Discourse Structure and Anaphora: Written and Conversational English. Cambridge University Press, Cambridge (1987)
8. Kong, K.C.C.: Are Simple Business Request Letters Really Simple? A Comparison of Chinese and English Business Request Letters. Text 18(1), 103–141 (1998)
9. Pelsmaekers, K., Braecke, C., Geluykens, R.: Rhetorical Relations and Subordination in L2 Writing. In: Sánchez-Macarro, A., Carter, R. (eds.) Linguistic Choice Across Genres: Variation in Spoken and Written English, pp. 191–213. John Benjamins, Amsterdam (1998)
10. Torrance, M., Bouayad-Agha, N.: Rhetorical Structure Analysis as a Method for Understanding Writing Processes. In: Degand, L., Bestgen, Y., Spooren, W., van Waes, L. (eds.) Multidisciplinary Approaches to Discourse, Nodus, Amsterdam (2001)
11. Taboada, M., Mann, W.: Rhetorical Structure Theory: Looking Back and Moving Ahead. Discourse Studies 8(3), 423–459 (2006)
12. Van Dijk, T.: Text and context. Explorations in the semantics and pragmatics of discourse. Longman, London (1977)
13. Foltz, P.W., Kintsch, W., Landauer, T.K.: The measurement of textual Coherence with Latent Semantic Analysis. Discourse Processes 25, 285–307 (1998)
14. McNamara, D., Kintsch, E., Songer, N., Kintsch, W.: Are good texts always better? Interactions of text coherence, background knowledge, and levels of understanding in learning from text. Cognition and Instruction (1996)

15. O'reilly, T., McNamara, D.: Reversing the reverse cohesion effect: Good texts can be better for strategic, high-knowledge readers. Discourse Processes (2007)
16. Goutsos, D.: Modeling Discourse Topic: Sequential Relations and Strategies in Expository Text. Ablex, Norwood (1997)
17. Grosz, B., Sidner, C.: Attention, intentions, and the structure of discourse. Comput. Linguist. 12, 175–204 (1986)
18. DeVillez, R.: Writing: Step by step. Kendall Hunt, Dubuque (2003)
19. Golightly, K.B., Sanders, G.: Writing and Reading in the Disciplines. Pearson Custom Publishing, New Jersey (2000)
20. Halliday, M.A.K., Hasan, R.: Cohesion in English. Longman, London (1976)
21. Barzilay, R., Lapata, M.: Modeling Local Coherence: An Entity-based Approach. Computational Linguistics 34(1), 1–34 (2008)
22. Redeker, G.: Coherence and structure in text and discourse. In: Black, W., Bunt, H. (eds.) Abduction, Belief and Context in Dialogue. Studies in Computational Pragmatics, pp. 233–263. Benjamins, Amsterdam (2000)
23. Charolles, M.: Cohesion, coherence et pertinence de discours. Travaux de Linguistique 29, 125–151 (1995)
24. Hobbs, J.: Coherence and Coreference. Cognitive Science 3(1), 67–90 (1979)
25. Schnedecker, C.: Nom propre et chaînes de reference. Recherches Linguistiques 21.Klincksieck, Paris (1997)
26. Schnedecker, C.: Les chaînes de reference dans les portraits journalistiques: éléments de description. Travaux de Linguistique 2, 85–133 (2005)
27. Kleiber, G.: Anaphores et pronoms. Duculot, Louvain-la-Neuve (1994)
28. Grosz, B., Sidner, C.: Attention, intentions, and the structure of discourse. Comput. Linguist. 12(3), 175–204 (1986)
29. Carlson, L., Marcu, D., Okurowski, M.E.: Building a discourse-tagged corpus in the framework of rhetorical structure theory. In: van Kuppevelt, J., Smith, R. (eds.) Current Directions in Discourse and Dialogue, pp. 85–112. Kluwer Academic Publishers, Dordrecht (2003)
30. Joty, S., Carenini, G., Ng, R., Mehdad, Y.: Combining Intra- and Multi-sentential Rhetorical Parsing for Document-level Discourse Analysis. In: Proceedings of the 51st Annual Meeting of the Association for Computational Linguistics (ACL 2013), Sofia, Bulgaria (2013)
31. Galitsky, B., Ilvovsky, D., Kuznetsov, S.O., Strok, F.: Matching sets of parse trees for answering multi-sentence questions. In: Proceedings of the Recent Advances in Natural Language Processing, RANLP 2013, pp. 285–294. INCOMA Ltd., Shoumen (2013)
32. Ilvovsky, D.: Going beyond sentences when applying tree kernels. In: Proceedings of the Student Research Workshop ACL 2014, pp. 56–63 (2014)
33. Galitsky, B., Kuznetsov, S.O.: Learning communicative actions of conflicting human agents. J. Exp. Theor. Artif. Intell. 20(4), 277–317 (2008)
34. Vapnik, V.: The Nature of Statistical Learning Theory. Springer (1995)
35. Marcu, D.: From Discourse Structures to Text Summaries. In: Mani, I., Maybury, M. (eds.) Proceedings of ACL Workshop on Intelligent Scalable Text Summarization, Madrid, pp. 82–88 (1997)
36. Severyn, A., Moschitti, A.: Fast Support Vector Machines for Convolution Tree Kernels. Data Mining Knowledge Discovery 25, 325–357 (1997, 2012)

37. Recasens, M., de Marneffe, M.-C., Potts, C.: The Life and Death of Discourse Entities: Identifying Singleton Mentions. In: Proceedings of NAACL (2013)
38. Lee, H., Chang, A., Peirsman, Y., Chambers, N., Surdeanu, M., Jurafsky, D.: Deterministic coreference resolution based on entity-centric, precision-ranked rules. Computational Linguistics 39(4) (2013)
39. Collins, M., Duffy, N.: Convolution kernels for natural language. In: Proceedings of NIPS, pp. 625–632 (2002)
40. Moschitti, A.: Efficient Convolution Kernels for Dependency and Constituent Syntactic Trees. In: Proceedings of the 17th European Conference on Machine Learning, Berlin, Germany (2006)

Sentiment Classification with Graph Sparsity Regularization*

Xin-Yu Dai, Chuan Cheng, Shujian Huang, and Jiajun Chen

National Key Laboratory for Novel Software Technology
Nanjing University, Nanjing 210023, China
{daixinyu,mtlab,huangsj,chenjj}@nju.edu.cn

Abstract. Text representation is a preprocessing step in building a classifier for sentiment analysis. But in vector space model (VSM) or bag-of -features (BOF) model, features are independent of each other when to learn a classifier model. In this paper, we firstly explore the text graph structure which can represent the structural features in natural language text. Different to the BOF model, by directly embedding the features into a graph, we propose a graph sparsity regularization method which can make use of the the graph embedded features. Our proposed method can encourage a sparse model with a small number of features connected by a set of paths. The experiments on sentiment classification demonstrate our proposed method can get better results comparing with other methods. Qualitative discussion also shows that our proposed method with graph-based representation is interpretable and effective in sentiment classification task.

Keywords: Text Graph Representation, Graph Regularization, Sentiment Classification.

1 Introduction

With more and more user-generated-content (UGC) appearing on the internet, sentiment classification is becoming a hotspot research topic in natural language processing, data mining, and information retrieval research areas[1,2,5,6,8]. It is a task of binary polarity (e.g., positive or negative) classification for natural language text. Following the text classification approaches, a sentiment text is firstly represented as a vector with bag of words (BOW), then machine learning classifiers such as logistical regression (LR) and support vector machine (SVM) are applied for classification.

The big disadvantage of BOW is that it ignores word order information, syntactic structures and semantic relationships between words, which are essential attributes for sentiment analysis. So, many linguistics features such as Part-of-Speech[8], wordnet[7] and sentiment lexicon[4], are proposed to represent the sentiment text. Some structure features, such as high-order n-grams, word pairs[18], text graph[18], and dependency relations[3,23] are also used to represent the text.

* This research was supported by NSFC (61472183, 61170181).

A. Gelbukh (Ed.): CICLing 2015, Part II, LNCS 9042, pp. 140–151, 2015.
DOI: 10.1007/978-3-319-18117-2_11

However, when we apply traditional classifiers like SVM or LR for sentiment classification, text are always converted to the feature vectors with the vector space model (VSM), also named as bag of features(BOF) model. [18] proposed a graph structure to represent the text, but when performing sentiment classification task, the graph structure is still converted to the BOF representation. There is a common problem of BOF model that features are considered independent of each other while learning a classifier model. As for a graph, when a graph structure is degenerated to a vector, the separated features are independent. That is to say, some path information will be discarded in BOF Model.

To overcome the problem, in this paper, inspired by sparsity models[12,16,19], we propose a sentiment classification method with graph sparsity regularization. Firstly, we explore the text graph representation which can keep enough structural information in text. In our method, the graph can be viewed as a set of paths of connectivity among features, we can name our text representation as bag of path (BOP) model. Our proposed graph sparsity method is following the sparse model property which performs feature selection and model learning simultaneously. Additionally, by embedding the features into a graph, our method can enforce sparsity in the connectivity of the graph, and obtain a sparse model with a set of paths containing a small number of connected features.

The rest of this paper is organized as follows. In next section, we explore the graph representation for sentiment text. In Section 3, we will discuss the graph sparsity methods. In Section 4, we propose the sentiment classification method with graph sparsity regularization. In Section 5, we present several experimental results and give some further interpretable qualitative analysis of the selected features. And Section 6 concludes.

2 Text Graph Representation

There are some inherent structures in natural language text. We would like to firstly demonstrate why text sentiment classification can benefit from text structural information with a case. Consider the following examples, which are extracted from real online reviews:

- **Sentence 1:** lens visible in optical viewfinder.
- **Sentence 2:** lens is visible in the viewfinder.
- **Sentence 3:** lens barrel in the viewfinder.
- **Sentence 4:** the lens barrel does obstruct part of the lower left corner in the lens viewfinder.

These four negative reviews have the same meaning: *the lens of the camera is visible in viewfinder.* Intuitively, $lens \rightarrow viewfinder$ is a critical feature to determine the polarity of these four reviews. But both bag-of-words and N-grams cannot represent this noncontinuous high-order word co-occurrence feature.

So, the graph structure is suitable to represent the natural language text where structured knowledge can be conserved. In the text graph, each node in the graph represents a feature. And each edge represents the co-occurrence or dependency relationship between two features. The cost on each edge will represent the connectivity strength between two nodes.

In this section, we explore two graph-based text representation methods. The first one is a method presented by [18] which use high-order word co-occurrence prior knowledge to construct the graph. The other one is to use dependency relation knowledge to construct the graph. We also introduce the cross entropy method to assign the weight for each edge in the constructed graph.

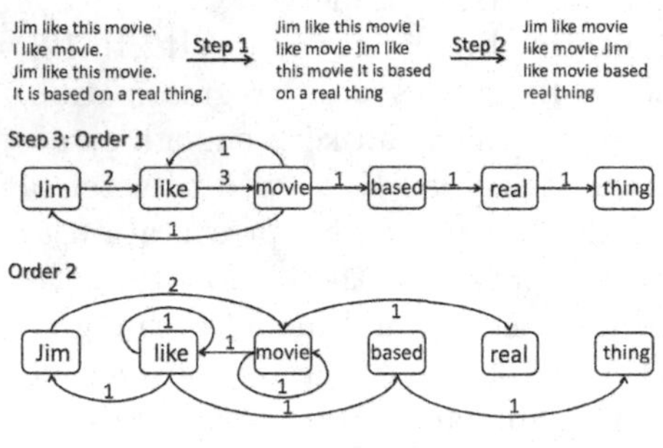

Fig. 1. An example of order and distance graph structure

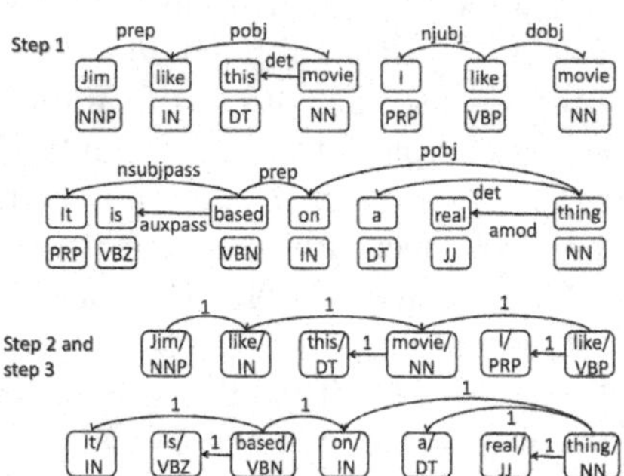

Fig. 2. An example of dependency tree based graph structure

2.1 Order and Distance Graph Structure

The key point of [18] 's graph representation is to conserve the sequence information in the text. The graph can represent the structure information about the ordering and distance between the words in the document.

Suppose we have a document with 4 sentences as follows:

- **Sentence 1:** Jim like this movie.
- **Sentence 2:** I like movie.
- **Sentence 3:** Jim like this movie.
- **Sentence 4:** It is based on a real thing.

Denote the graph structure as $G = (V, E)$ where V is denoted as the vertex set which refers to the feature set(all words in this case), and E means the edge set. We construct the graph structure as follows (shown in Fig. 1):

- **Step 1:** In each document, combine all sentences into one big sentence.
- **Step 2:** Remove the stopwords and then obtain a dictionary: *jim, like, movie, based, real, thing* . The dictionary formulates the feature space to represent the vertex set in the graph.
- **Step 3:** Construct a text graph on the feature set according to the order. With the order N, each word is linked to the Nth word after it. And the count of each edge is computed by the frequency. We have an edge $Jim \rightarrow movie$ with order 2.

In the Fig.1, the graph can be constructed with the above three steps. After the step 1 and step 2, we have a sentence of *Jim like movie like movie Jim like movie based real thing.* So, we have a edge $Jim \rightarrow Movie$ with order 2 . And this edge has the count of two because the word 'movie' appears as the second word after 'Jim' twice

Considering the simplicity and efficiency, in this paper, we use the order 1, order 2 and order 3 together to construct the text graph. In a word, [18] 's graph can represent the high-order non-continuous word co-occurrence information.

2.2 Dependency Tree Based Graph Structure

In order to explore more graph structure, we apply the dependency parsing to obtain the dependency tree as the graph structure. Dependency relation is another kind of high level structure information in the document, especially in modeling the long-range relations. Denote the graph structure as $G = (V, E)$ where V is denoted as vertex set which refers to the feature set, and E is the edge set. The dependency graph can be constructed as follows (shown in Fig. 2):

- **Step 1:** In each document, obtain the dependency tree structure of each sentence.
- **Step 2:** Use word together with its POS tag as a feature. The feature space is composed of all of these features from the documents set.
- **Step 3:** Two features are connected according to their dependency relation in the dependency tree. And the count of each edge is computed by how many times this dependency relation occurs in all of the dependency trees.

In this paper, we use the Stanford parser[10] to obtain the dependency trees for sentences.

In the Fig.2, we show an example graph constructed by the above three steps. With the dependency tree structure, we have an edge of $Jim/NNP \rightarrow like/IN$. And this edge has the count of 1 because this edge appears only once.

We have to note that in our graph, for simplicity, we don't use the dependency labels, such as "*prep*" and "*prbj*" in the Fig.2.

2.3 Compute the Cost of Each Edge

To represent the connectivity strength between two nodes, we compute the cost for each edge based on the original count in the graph. Inspired by the cross entropy measure, we design the following equation to compute the cost of each edge e.

$$c_e = -\left(p_e^{neg} \log p_e^{neg} + p_e^{pos} \log p_e^{pos}\right) \tag{1}$$

p_e^{neg} and p_e^{pos} stand for the probability of edge e belonging to negative class and positive class, respectively. They are computed as follows:

$$p_e^{neg} = \frac{count_e^{neg} + 1}{count_e^{neg} + count_e^{pos} + 2} \tag{2}$$

$$p_e^{pos} = \frac{count_e^{pos} + 1}{count_e^{neg} + count_e^{pos} + 2} \tag{3}$$

where the $count_e^{neg}$ stands for the frequency of edge e in negative class, and the $count_e^{pos}$ stands for the frequency of edge e in positive class. 1 and 2 are used for smoothing. Suppose one edge appears in the negative class many times and appears in the positive class rarely, according to Eq.(1), the cost of this edge is small which means this edge is important for classification.

3 Graph Sparsity Model

Estimation of linear models penalized by a regularization term has become a powerful method for model selection and has been used in many applications. Formally, the model selection can be formulated to minimize the following equation:

$$\min_{w \in R^p} L(w) + \lambda \Omega(w) \tag{4}$$

where $L(w)$ is the loss function and $\Omega(w)$ is the regularizer. The λ is a factor to control the trade-off between loss function and regularizer. With a good setup of λ, we can obtain estimators with good prediction performance, even in a high dimensional space.

Some kinds of regularizers, like $L1$-norm ($\Omega(w) = \|w\|_1$), can encourage sparsity, so that only a few features with non-zero coefficients are selected to improve the prediction performance and achieve interpretable results [15]. The typical sparsity models with L1-norm regularization include Lasso[12] and L1-LR(L1 regularization logistic regression)[26], and so on. The sparsity model has been used in text classification[11] and sentiment classification[14].

However, with $L1$-norm regularization, the sparsity model considers all features to be independent. There is some prior structural knowledge in some data, such as the data of bioinformatics and natural language. Considering the importance of structural priori knowledge for learning on structural data, some new regularization terms are proposed in recent years, which are usually called structured sparsity as the extension of sparsity[16,19].

Graph sparsity is a typical approach for structured sparsity. With the graph regularization term of $\Omega(w)$, features are directly embedded in a graph. Researchers try to automatically select a subgraph (a set of paths) with a small number of connected features[19,17,16] which are more interpretable in model selection. It has been used in many structural data learning problems , such as the gene sequence prediction in bioinformatics[20] and name entity recognition in natural language processing[9].

3.1 The Formulation

In graph sparsity, we have a directed acyclic graph (DAG) $G = (V, E)$ on the index set $I = \{1, 2, ..., p\}$ of the estimated parameter $w \in R^p$, where $V = I$

is the vertex set and $E = \{(i,j)\,|i,j \in V\}$ is the edge set. Denote g as the path on the graph of G, as $g = (v_1, v_2, ..., v_k)$, where $v_i \in V, i = 1, ..., k$ and $(v_i, v_{i+1}) \in E, i = 1, ..., k-1$. Let $\mathcal{G}$ as the set of all paths on graph G. Denote $\eta_g > 0$ as positive weight of each path $g \in \mathcal{G}$. The graph regularization term is defined as follows:

$$\Omega(w) = \min_{\mathcal{J} \subseteq \mathcal{G}} \left\{ \sum_{g \in \mathcal{J}} \eta_g \quad s.t. \quad \text{Supp}(w) \subseteq \bigcup_{g \in \mathcal{J}} g \right\} \tag{5}$$

where $Supp(w) = \{j\,|w_j \neq 0\}$. $\mathcal{J}$ is a subset of $\mathcal{G}$ whose union covers the support of w. This penalty encourages solutions whose set of non-zero coefficients is in the union of a small number of paths.

In sum, the graph G can be viewed as a set of all paths ($\mathcal{G}$). The graph regularizer defined in Eq.(5) is actually to select some paths with small weights which form a subgraph $\mathcal{J}$. The union of these paths will cover the nonzero features (That is to cover $Supp(w)$).

3.2 Optimization

We firstly describe how to compute the weight η_g of each path g in Eq.(5). Define a source node s and sink node t for graph G. G can be extended to $G' = (V', E')$, where $V' = V \cup \{s,t\}$ and $E' = E \cup \{su\,|u \in V\} \cup \{ut\,|u \in V\}$. Then, for a path $g = (u_1, u_2, ..., u_k)$, where $u_i \in V, i = 1, ..., k$ and $(u_i, u_{i+1}) \in E, i = 1, ..., k-1$ on G, the weight η_g can be defined as follows:

$$\eta_g = c_{su_1} + \sum_{i=1}^{k-1} c_{u_i u_{i+1}} + c_{u_k t} \tag{6}$$

In this paper, c_{su} and c_{ut} for $u \in V$ are defined as follows:

$$c_{su} = 0 \; \forall u \in V \tag{7}$$

$$c_{ut} = \frac{\sum_{(a,b) \in E} c_{ab}}{|E|} \; \forall u \in V \tag{8}$$

c_{ut} can be viewed as average cost of the whole graph. From Eq.(6), it is easy to see that the smaller the cost of an edge is, the more important it is.

With the path weight of η_g, we follow the path coding method with minimum cost network flow[19] to optimize the graph regularization of Ω in Eq.(5).

4 Sentiment Classification with Graph Sparsity Regularization

Based on the text graph representation and graph sparsity model, we now propose a framework of sentiment classification with the graph sparsity regularization method.

Algorithm 1. Sentiment classification with graph sparsity regularization

Input: Parameter λ for the logistic regression model with graph regularization. Graph scale constraint M. Dataset and the binary sentiment labels.
Output: Model Parameter: w
Step 1: Following the methods in section 2, construct the feature space F and the graph structure $G = (V, E)$ according to the dataset. Denote the index of features as I. And $V = I$ is composed of features in F. E is edge set.
Step 2: Use χ^2-statistic to compute a score for each feature in F and obtain a rank list of F.
Step 3: Select the top M features to construct a small feature space $\bar{F}$ (index set denoted as $\bar{I}$).
Step 4: Construct a subgraph $\bar{G} = (\bar{V}, \bar{E})$ from G, where $\bar{V} = \bar{I} \cup \{s, t\}$ and $\bar{E} = \{(i,j) \,|\, i, j \in \bar{V}, (i,j) \in E'\}$. Remove some edges along cycles in the graph $\bar{G} = (\bar{V}, \bar{E})$ randomly.
Step 5: According to the Graph $\bar{G} = (\bar{V}, \bar{E})$, optimize the logistic regression penalized by the graph sparsity regularization as follows:

$$w = \arg\min_{w} \sum_{i=1}^{N} \log\left(1 + e^{-y_i w^T x_i}\right) + \lambda \Omega(w)$$

where x_i corresponds to the text vectors with $tf * idf$ measure in the feature space of $\bar{F}$, and y_i corresponds the binary labels. $\Omega(w)$ is computed by Eq.(5).

For sentiment classification, we firstly construct the graphs $G = (V, E)$ according to the two ways described in the section 2. And following the graph sparsity regularization method, we can get a sparse model for sentiment classification with a set of paths containing a small number of connected features.

We want to note that the graph structure $G = (V, E)$ is assumed as a directed acyclic graph in this method. So, in our text graph, we remove some edges along cycles in the graph randomly.

For classification, we use the model of logistic regression penalized by the graph regularization. The detailed sentiment classification method with graph sparsity regularization is presented in the Algorithm 1.

The optimization is so inefficiency if the graph scale is large too much. The step 3 and step 4 is to limit the scale of the text graph which controls the

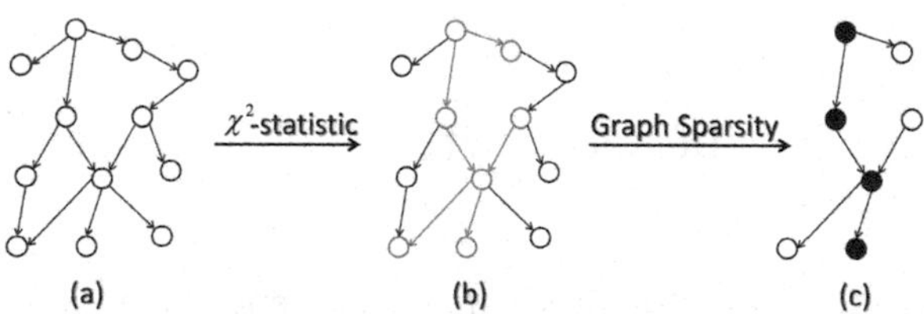

Fig. 3. A summary of our proposed framework

complexity of graph sparsity optimization. Since χ^2-statistic[13] is a popular preprocessing feature selection method, we use this method to firstly reduce the scale of feature space into a reasonable scale. For the discarded features, the corresponding edges on graph are also discarded.

The Fig.3 presents a intuitive processing of our proposed framework. After the step 1 in our algorithm, we have a original graph G shown as the Fig.3-(a). And after the step 2-4, with the limitation of the graph scale, we have a graph $\bar{G}$ with red color dots and edges as shown in the Fig.3-(b). And with our graph sparsity regularization, in the step 5 ,we will finally learn a model with a set of paths containing a small number of connected features which is represented as solid black dots and edges as shown in the Fig.3-(c).

5 Experiments

5.1 Setup

We assess our method in sentiment classification task. We use SPAMS from `http://spams-devel.gforge.inria.fr/` to train the graph sparsity model. We use 10-fold cross validation in our experiments. The parameter λ for logistic regression model with graph sparsity regularization is selected by cross validation. For reducing the graph scale, we set the graph scale constraint $M = 4000$ in the Algorithm 1.

We use a polarity dataset $V2.0$[25] as our experimental data. Dataset $V2.0$ is a full-length movie review dataset containing long reviews. It contains 1000 positive reviews and 1000 negative reviews. This dataset can be obtained from `http://www.cs.cornell.edu/people/pabo/movie-review-data/review_polarity.tar.gz`.

5.2 Comparing with Traditional Text Classification Methods

[18] proposed a text graph representation and applied it to sentiment classification. Following [18], the text graph structure is degenerated to a bag-of-features representation when performing sentiment classification. For each edge in the graph, a unique 'token' or 'pseudo-word' is created as features to represent the nodes and edges. From the linguistics view, [18]'s representations for sentiment classification include unigram, bi-gram and noncontinuous bi-grams features. With this kind of representation, classical classifiers, such as SVM, $L1$-norm regularized SVM ($L1$-SVM), $L1$-norm regularized Logistic Regression ($L1$-LR) are performed to sentiment classification for comparing. $L1$-LR and $L1$-SVM are trained by Liblinear [21]. SVM is trained by LibSVM [22]. χ^2-statistic is also used for feature selection before learning.

Fig. 4 shows that we can get the best results, which demonstrates the effectiveness of our method which directly learn a model on the graph structure. Directly with the graph structure and the graph sparsity regularization, our model can learn meaningful (statistical) structures such as the nodes, edges, and paths in the graph.

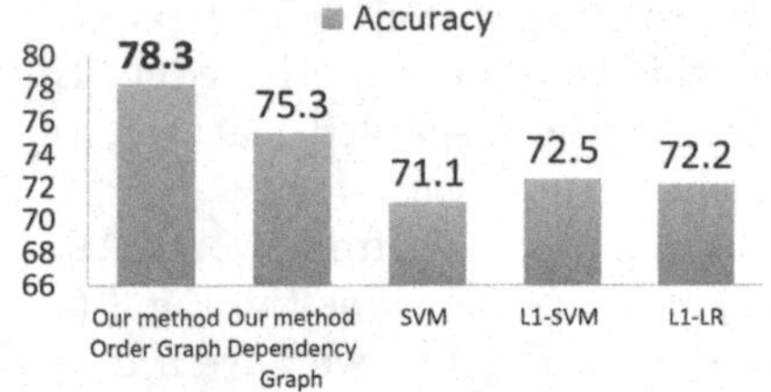

Fig. 4. Comparing with other classic classifiers

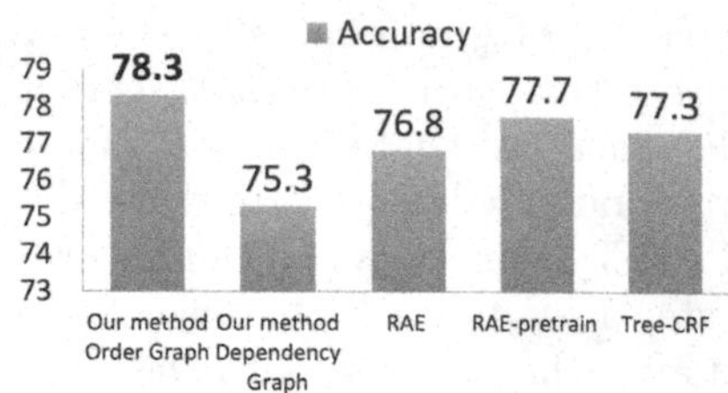

Fig. 5. Comparing with other text structure representations

In our paper, we explore two kinds of text graph structures. We also compare the results from our proposed methods with the two graph structures. In the Fig. 4 , the Order Word Co-occurrence Graph structure can get better result than the dependency-tree graph structure. The reason we think is that the quality of the dependency tree is not good enough considering that the reviews in the dataset are commonly informal sentences.

5.3 Comparing with Other Text Structure Representation

There are also some sentiment classification methods with structure text representation. Tree-CRF is a dependency tree-based sentiment classification method [23]. RAE and pre-trained RAE learn the vector space representation for multi-word phrases with a semi-supervised recursive autoencoders method [24]. And the embedded structure representation is applied for sentiment classification.

We also do experiments to compare our method with above methods on the polarity dataset $V2.0$. Fig. 5 shows our method with order graph outperform the other structure representation-based methods.

It is noteworthy that our method with dependency graph does not get good performance. From this interesting comparison, we can know that the performance of our methods may depend on the quality of the graph. With good prior graph structure knowledge, our method can get significant better performance. But with a low-quality graph like dependency tree, our method is not robust enough to get good result.

5.4 Qualitative Analysis from the Selected Subgraph

In order to show the good interpretability of our method, we extract the selected subgraph features from order and distance graph and dependency tree based graph, as shown in Fig. 6.

From the order and distance graph, we extract the top 20 features and the corresponding edges from the learned model as shown in Fig. 6(a) (for clear visualization, 3 isolated points are removed from the graph). We can find some meaningful sentiment units, such as, the edge *great* $\rightarrow$ spielberg reflects that some reviewers prefer spielberg's movies. And this edge should be an important

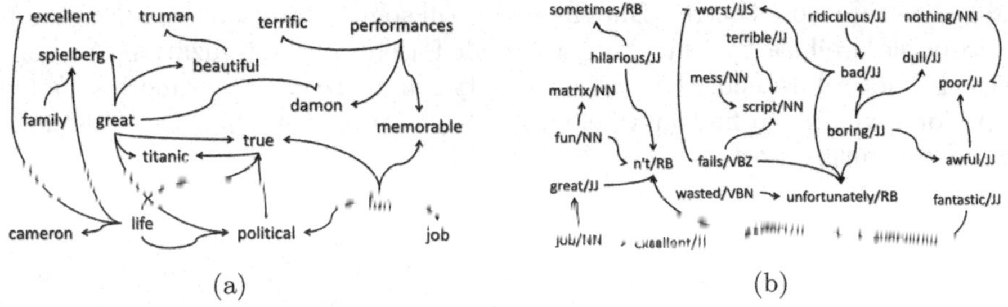

Fig. 6. Some selected features from graph representation. (a) from order and distance graph, and (b) from dependency tree-based graph.

Path and Node	#in Positive Doc	#in Negative Doc
boring/JJ→bad/JJ→unfortunately/RB	1	35
boring	54	175
bad	545	280
unfortunately	212	117

Fig. 7. Appearance count of path and nodes

discriminative feature for positive. Take life → cameron, life → *true*, *true* → titanic and *great* → titanic together as another example, this subgraph reflects that cameron's movie titanic is preferred by reviewers. In addition, the edge performances → terrific maybe means that some actor's performance impresses reviewers deeply.

From dependency tree based graph, we extract the top 30 features and the corresponding edges from the learned model shown in Fig. 6(b) (for clear visualization, 5 isolated points are removed from the graph). We find that the paths are useful in classification, such as the path boring/JJ → bad/JJ → unfortunately/RB. To validate that this selected path is really useful for classification, we explore the original reviews and dependency trees and count the appearance frequency of the paths and the three separated nodes in the positive class and negative class. From the Fig. 7, we can intuitively know that the path is more significantly useful for classification than each node respectively.

6 Conclusion

In this paper, we propose a framework that combines text graph representation with graph-regularized logistic regression on sentiment classification task. We make some experiments on a sentiment classification dataset to validate our method. Through quantitative and qualitative analysis, we present why the graph structure works better. But there are still some limitations of our method. For example, we have to limit the graph scale for efficient optimization. And our method is not robust enough with low-quality graph. To construct more valuable graph for text and to find more efficient graph sparsity optimization methods are our two future works.

References

1. Duric, A., Song, F.: Feature Selection for Sentiment Analysis Based on Content and Syntax Models. In: Proceedings of the 2nd Workshop on Computational Approaches to Subjectivity and Sentiment Analysis, ACL-HLT (2011)
2. Socher, R., Pennington, J., Huang, E.H., Andrew, Y.N., Manning, C.D.: Semi-supervised recursive autoencoders for predicting sentiment distributions. In: Proceedings of the 2011 Conference on Empirical Methods in Natural Language Processing (EMNLP) (2011)
3. Xia, R., Zong, C.Q., Li, S.S.: Ensemble of feature sets and classification algorithms for sentiment classification. Information Sciences 181(6), 1138–1152 (2011)
4. Das, S.R., Chen, M.Y.: Yahoo! for Amazon: Sentiment extraction from small talk on the Web. Management Science 53(9), 1375–1388 (2007)
5. Ponomareva, N., Thelwall, M.: Do neighbours help? an exploration of graphbased algorithms for cross-domain sentiment classification. In: Proceedings of the 2012 EMNLP-CoNLL, Jeju Island, Korea, pp. 655–665 (2012)
6. Wu, Y.B., Zhang, Q., Huang, X.J., Wu, L.D.: Structural Opinion Mining for Graph-based Sentiment Representation. In: Proceedings of the 2011 EMNLP, Edinburgh, Scotland, UK, pp. 1332–1341 (2011)
7. Kim, S.-M., Hovy, E.: Determining the Sentiment of Opinions. In: Proceedings of the 20th International Conference on Computational Linguistics. Association for Computational Linguistics (2004)
8. Pang, B., Lee, L., Vaithyanathan, S.: Thumbs up? Sentiment classification using machine learning techniques. In: Proceedings of EMNLP (2002)
9. Martins, A.F.T., Smith, N.A., Aguiar, P.M.Q., Figueiredo, M.A.T.: Structured Sparsity in Structured Prediction. In: Proceedings of EMNLP (2011)
10. Socher, R., Bauer, J., Manning, C.D., Andrew, Y.N.: Parsing With Compositional Vector Grammars. In: Proceedings of ACL (2013)
11. Fu, Q., Dai, X., Huang, S., Chen, J.: Forgetting word segmentation in Chinese text classification with L1-regularized logistic regression. In: Pei, J., Tseng, V.S., Cao, L., Motoda, H., Xu, G. (eds.) PAKDD 2013, Part II. LNCS, vol. 7819, pp. 245–255. Springer, Heidelberg (2013)
12. Tibshirani, R.: Regression Shrinkage and Selection Via the Lasso. Journal of the Royal Statistical Society, Series B 58(1), 267–288 (1994)
13. Yang, Y.M., Pedersen, J.O.: A comparative study on feature selection in text categorization. In: Proceedings ICML (1997)
14. Rafrafi, A., Guigue, V., Gallinari, P.: Coping with the Document Frequency Bias in Sentiment Classification. In: AAAI ICWSM (2012)
15. Buhlmann, P.L., van de Geer, S.A., Van de Geer, S.: Statistics for High-dimensional Data Methods, Theory and Applications. Springer, Heidelberg (2011)
16. Huang, J., Zhang, Z., Metaxas, D.: Learning with structured sparsity. Journal of Machine Learning Research 12, 3371–3412 (2011)

17. Bach, F., Jenatton, R., Mairal, J., Obozinski, G.: Optimization with sparsity-inducing penalties. Foundations and Trends in Machine Learning 4(1), 1–106 (2012)
18. Aggarwal, C.C., Zhao, P.X.: Towards graphical models for text processing. Knowledge and Information Systems (2013)
19. Mairal, J., Yu, B.: Supervised Feature Selection in Graphs with Path Coding Penalties and Network Flows. Journal of Machine Learning Research (2013)
20. Jacob, L., Obozinski, G., Vert, J.-P.: Group Lasso with overlap and graph Lasso. In: Proceedings of the International Conference on Machine Learning (ICML) (2009)
21. Fan, R.E., Chang, K.W., Hsieh, C.J., Wang, X.R., Lin, C.J.: LIBLINEAR: A library for large linear classification. Journal of Machine Learning Research 9, 1871–1874 (2008)
22. Chang, C.C., Lin, C.J.: LIBSVM: A library for support vector machines. ACM Transactions on Intelligent Systems and Technology 2 (2011)
23. Nakagawa, T., Inui, K., Kurohashi, S.: Dependency tree-based sentiment classification using CRFs with hidden variables. In: Proceedings of ACL:HLT (2010)
24. Socher, R., Pennington, J., Huang, E.H., Andrew, Y.N., Manning, C.D.: Semi-Supervised Recursive Autoencoders for Predicting Sentiment Distributions. In: Proceedings of EMNLP (2011)
25. Pang, B., Lee, L.: A Sentimental Education: Sentiment Analysis Using Subjectivity Summarization Based on Minimum Cuts. In: Proceedings of ACL (2004)
26. Goodman, J.: Exponential priors for maximum entropy models. In: Proceedings of the Annual Meeting of the Association for Computational Linguistics (2004)

Detecting Emotion Stimuli in Emotion-Bearing Sentences

Diman Ghazi, Diana Inkpen, and Stan Szpakowicz

School of Electrical Engineering and Computer Science
University of Ottawa, Ottawa, Ontario, Canada
diman.ghazi@gmail.com, Diana.Inkpen@uOttawa.ca,
szpak@eecs.uottawa.ca

Abstract. Emotion, a pervasive aspect of human experience, has long been of interest to social and behavioural sciences. It is now the subject of multi-disciplinary research also in computational linguistics. Emotion recognition, studied in the area of sentiment analysis, has focused on detecting the expressed emotion. A related challenging question, *why* the experiencer feels that emotion, has, to date, received very little attention. The task is difficult and there are no annotated English resources. FrameNet refers to the person, event or state of affairs which evokes the emotional response in the experiencer as emotion *stimulus*.[1] We automatically build a dataset annotated with both the emotion and the stimulus using FrameNet's *emotions-directed* frame. We address the problem as information extraction: we build a CRF learner, a sequential learning model to detect the emotion stimulus spans in emotion-bearing sentences. We show that our model significantly outperforms all the baselines.

1 Introduction

Causality is a semantic relation defined as "the relationship between cause and effect",[2] where the latter is understood as a consequence of the former. Causality detection used to depend on hand-coded domain-specific knowledge bases [17]. [16] defined semantic constraints to rank possible causality, and machine learning techniques now prevail.

[15] described automatic detection of lexico-syntactic patterns which express causation. There are two steps: discover lexico-syntactic patterns, and apply inductive learning to automatically detect syntactic and semantic constraints (rules) on the constituent components. [5] extracted possible causal relations between noun phrases, via a bootstrapping method of causality extraction using cue phrases and word-pair probabilities. A simple supervised method in [3] trained SVM models using features derived from WordNet and the Google N-gram corpus; providing temporal information to the causal relations classifier boosted the results significantly. All that work, however, addresses causality in general, by nature very different from detecting the cause of an emotion. The causal relation between two parts of the text are sought. We look for the emotion stimulus when the emotion can be conveyed explicitly in the text or be implicit.

---

[1] Most authors talk generically of *cause*.

[2] http://www.oxforddictionaries.com

A. Gelbukh (Ed.): CICLing 2015, Part II, LNCS 9042, pp. 152–165, 2015.
DOI: 10.1007/978-3-319-18117-2_12

Table 1. The elements of the emotion frame in FrameNet

Core	Event	The occasion or happening in which Experiencers in a certain emotional state participate.
	Experiencer	The person or sentient entity who experiences or feels the emotions.
	Expressor	It marks expressions that indicate a body part, gesture or other expression of the Experiencer that reflects his or her emotional state.
	State	The abstract noun that describes a more lasting experience by the Experiencer.
	Stimulus	The person, event, or state of affairs that evokes the emotional response in the Experiencer.
	Topic	The general area in which the emotion occurs. It indicates a range of possible Stimulus.
Non-Core	Circumstance	The condition(s) under which the Stimulus evokes its response.
	Degree	The extent to which the Experiencer's emotion deviates from the norm for the emotion.
	Empathy-target	The individual or individuals with which the Experiencer identifies emotionally and thus shares their emotional response.
	Reason/ Explanation	The explanation for why the Stimulus evokes a certain emotional response.
	Manner	It is any description of the way in which the Experiencer experiences the Stimulus which is not covered by more specific frame elements. Manner may also describe a state of the Experiencer that affects the details of the emotional experience.
	Parameter	A domain in which the Experiencer experiences the Stimulus.

A more specific task in causality analysis, most similar to our task, is to identify sources of opinions. [4] used machine-learning techniques to identify propositional opinions and their holders (sources). That pioneering work was limited in scope: only propositional opinions (which function as the sentential complement of a predicate), and only direct sources. [9], also among the pioneers in this field, viewed the problem as an information extraction task and tackled it using sequence tagging and pattern matching techniques simultaneously. They hypothesized that information extraction techniques would be well-suited to source identification because an opinion statement can be viewed as a kind of speech event with the source as the agent. [8] identify two types of opinion-related entities, expressions of opinions and sources of opinions, along with the linking relation between them. [19] analyzed judgment opinions, which they define as consisting of valence, a holder and a topic. All that work invokes interchangeably the terms *source of an opinion* and *opinion holder*. Although the source can be the reason that implies the opinion, it is mainly seen as the opinion holder – thus, it is a task very different from ours.

We focus on detecting the *emotion stimulus*. In FrameNet [13], an *experiencer* has a particular emotional *state*, which may be described in terms of a specific *stimulus* that invokes it, or a *topic* which categorizes the kind of stimulus. An explicitly named experiencer may be replaced by an *event* (with participants who are experiencers of the emotion) or an *expressor* (a body-part of gesture which would indicate the experiencer's state to an external observer). There can also be a *circumstance* in which the response occurs or a *reason* why the stimulus evokes the particular response in the experiencer.

Consider, for example, the sentence "In the Commons, Labour MPs unleashed their anger at the Liberal Democrats for promising to back the Government." "Labour MPs" is the *Experiencer*, "anger" is the expression of emotion, "the Liberal Democrats" is the *Emotion Stimulus*, and "for promising to back the Government" is the *Explanation*. We want our system to find the reason why Labour MPs were angry: to return the span *the Liberal Democrats* as the emotion stimulus.

[13] define six core frame elements for an emotion and six non-core elements (see Table 1). Of these, emotion stimulus seems to be the closest to saying *why* the experiencer feels that emotion. Therefore, here we focus on this aspect of emotion analysis.

We are particularly interested in the stimulus because determining *why* an emotion occurs has such intriguing applications as consumer behaviour analysis or mental-health care. It would be very useful for systems which answer question such as "how [x] feels about [y]" or "why [x] feels [y]". It also has practical importance to text summarization, because emotion expressions and emotion stimuli tend to be the most informative in an expressive sentence, so they can get higher weight in abstractive summarization.

We discuss emotion stimuli, a dataset we automatically built with emotion stimulus and emotion expression labels, and ways of detecting emotion stimuli. Section 2 covers the related work. Section 3 explains the process of collecting and annotating an emotion stimulus dataset using FrameNet data.[3] Section 4 discusses the features and the baselines for detecting emotion stimuli in emotion-bearing sentences as well as the experiments and the results. Section 5 concludes the paper and suggests future work.

2 Related Work

Researchers in the field of affective computing have investigated recognition, interpretation and representation of affect [30]. They consider a wide range of modalities such as affect in speech, facial display, posture and physiological activity. Due to the large volume of text data available on the Internet – blogs, email and chats – which are full of emotions, recently there has been a growing interest in automatic identification and extraction of sentiment, opinions and emotions in text. Besides, textual data on the Web take up little physical space and are easily transferred, so they have a high potential to be used for sharing ideas, opinion and emotions. It is also such an active area of research because its applications have spread to multiple domains, from consumer product reviews, health care and financial services to social events and political elections [14].

In order to recognize and analyze affect in written text – seldom explicitly marked for emotions – NLP researchers have come up with a variety of techniques, including the use of machine learning, rule-based methods and the lexical approach [28] [1] [2] [18] [37] [33] [6] [20]. Detecting emotion stimuli, however, is a very new concept in sentiment analysis. Emotion/stimulus interaction, an eventive relation, potentially yields crucial information in terms of information extraction. For example, we can predict future events or decide on the best reaction if we know the emotion cause [7].

[3] http://www.eecs.uottawa.ca/ diana/resources/emotion_stimulus_data/

Event-based emotion detection has been addressed in some previous research [38] [35] [24] but, to the best of our knowledge, only [7], [22] and [23] have worked on emotion cause detection. They explore emotions and their causes, focusing on five primary emotions – happiness, sadness, fear, anger and surprise – in Chinese texts. They have constructed a Chinese emotion corpus, but they focus on explicit emotions presented by emotion keywords, so each emotion keyword is annotated with its corresponding causes, if existing. In their dataset, they observe that most causes appear within the same clause of the representation of the emotion, so a clause might be the most appropriate unit to detect a cause. We find such granularity too large to be considered an emotion stimulus in English. Also, clauses were distinguished by punctuation: comma, period, question mark and exclamation mark. Just four punctuation marks are not enough to capture English clauses adequately.

Using linguistic cues, including causative verbs and perception verbs, the authors create patterns to extract general cause expressions or specific constructions for emotion causes. They formalize emotion cause detection as multi-label classification. Each instance may contain more than one label, such as "left-1, left-0", to represent the location of the clauses which are part of the cause. We have no evidence that their findings can be valid for English data. Their work is more of a solution for an over-simplified version of a complicated problem. Also, keyword spotting is used to detect the emotion word in the sentence and try to find the cause of that emotion, but not all emotions are expressed explicitly by emotion keywords.

In the end, what stands out is the fact that, as far as we know, there has been no significant previous work on emotion cause detection in *English* data. This may be due to the relative complexity of English in expressing emotions in text, or to the limitation in existing resources and datasets either for supervised machine learning or for evaluation purposes.[4] Therefore, we have collected data and built a new dataset for detecting emotion stimuli. Using the dataset, we also explore different baselines to establish the difficulty of the task. Finally, we train a supervised information extraction model which detects the emotion stimulus spans in emotion-bearing sentences.

3 Data Collection and Annotation

[13] define 173 emotion-directed lexical units which correspond to different emotions. A lexical unit is a word/meaning pair (essentially a lexeme).[5] Typically, each sense of a polysemous word belongs to a different semantic frame. "Happy", "angry" and "furious" are examples of LUs in the Emotion-directed frame. For each of them, FrameNet annotators have labelled some sentences. We built a set of sentences marked with the emotion stimulus (cause), as well as the emotion itself. To collect a larger set of data, we used synonyms of emotion LUs to group the data into fewer basic emotions. In the manual synonym annotation task, we suggested Ekman's six emotions (*happiness*, *sadness*, *suprise*, *disgust*, *anger*, *fear*) [11], as well as *shame*, *guilt* and *hope*, posited in literature [29], to consider if the emotion LUs did not fit an Ekman emotion. We also

[4] [27] recently built a small dataset with 523 sentences for 22 emotions. We found it too scarce for machine learning methods of cause detection.

[5] The term will be henceforth abbreviated as LU.

allowed the annotators to propose a more appropriate emotion not on the list. In the end, we chose Ekman's emotions and *shame*.[6]

Some emotions fit a basic one; for example, *fury* clearly belongs to the *anger* category. On the other hand, *affront* is not so obvious. We will now discuss the dictionaries and thesauri we used to get emotions and their synonyms. Then we will describe the manual annotation. Next, we will apply the first round of tagging the emotions with one of the seven classes we chose. We separate the emotion LUs with a strong agreement between sources and annotations from those with weaker agreement. For the latter set of LUs, we relax one of the conditions. We use two more emotion lexicons to break the ties and classify more LUs. Finally we build two datasets, one with the group of emotions with strong agreement and the other with the result of the second phase of annotation added.

3.1 Annotating Emotion Lexical Units

We collected the synonyms from two trustworthy online sources.[7] Of the LUs collected from FrameNet, 14 are not covered at all. The two sources also do not always agree. To get a tie breaker, we resorted to manual annotation. We grouped the list of 173 emotion LUs from FrameNet into Ekman's six emotion classes by using the synonym list from the Oxford Dictionary and from thesaurus.com, joined the results and asked human annotators (fluent English speakers) to verify those annotations. We gave them an Excel table with each LU displayed in a row and each emotion in a column. For each word, the corresponding emotions were marked. A word could be classified into:

I one of Ekman's six emotions (110 words);
II two classes at the same time (14 words);
III none of the six emotion classes (49 words).

For group I, the annotators indicated if they disagreed with the existing classification by crossing the existing mark and indicating the emotion they think is more appropriate (if there is one). For group II, they chose only one of the two emotions they thought was closer to the LU and crossed out the other one. For group III, they chose one of the three suggested classes, *guilt*, *shame* and *hope*, and grouped the LU into one of them. Finally, there was a column for comments, where the annotators could write any other emotion they thought was more suitable as a synonym of the emotion LU.

The results of these annotations are used to build the dataset containing sentences with emotion stimulus tags. Section 3.2 presents the process of building the dataset.

3.2 Building the Dataset

Considering both thesauri and human annotators' tags, each LU was tagged with the emotion that had the highest number of votes. We combined the result of the Oxford

[6] *Guilt* and *hope* were not found useful, but *confusion* was suggested as another emotion with many synonym matches among the emotion LUs. This requires further study. *Confusion* is not considered in related research, and anyhow we need more evidence in the literature.

[7] `http://www.thesaurus.com`
`http://www.oxforddictionaries.com/thesaurus/`

Dictionary and thesaurus.com, so we had four annotations in total. As a result, 102 out of 173 LUs were labeled with high agreement (at least 3 out of 4 votes), which are shown as strong agreement in Table 2.

For the other LUs, we did not have enough information to group them into one of the seven emotions. We used two more sources – the NRC emotion lexicon [26] and the WordNet affect lexicon [34] – to break some of the ties. In this step, the LUs were labeled with an emotion for which the number of votes was more than half of the votes (three in our case) and the difference of the votes for the two top emotion classes was at least 2. This added 21 LUs to our set. While the NRC emotion lexicon was not very useful,[8] many ties were broken by WordNet Affect. The result of this extended list is shown as weaker agreement in Table 2.

Table 2. Distribution of LUs for each emotion class

Agreement	happiness	sadness	surprise	disgust
Strong	22	31	12	3
Weaker	22	32	12	8
	anger	fear	shame	total
Strong	19	13	2	102
Weaker	26	17	6	123

Using the list of grouped emotion synonyms, we collected FrameNet data manually labeled for each LU. Next, we selected the sentences which contain emotion stimuli and we assigned each sentence to its corresponding emotion class. The distribution of the instances in the dataset is shown in Table 3. Each instance is a complete sentence, 18 tokens on average, and contains one stimulus assigned to the emotion LU.

Table 3. Distribution of labels in the emotion stimulus datasets. Dataset 1 contains synonym groups with strong agreement, Dataset 2 – also those with weaker agreement.

	happiness	sadness	surprise	disgust
Dataset 1	211	98	53	6
Dataset 2	211	107	53	38
	anger	fear	shame	total
Dataset 1	168	129	34	699
Dataset 2	199	144	68	820

As a complementary dataset, we also collected the sentences with no stimulus tag, yet containing expressions of one of the seven emotions. This dataset is much larger than the dataset with stimulus. This makes us wonder whether it is due to the existence of many emotion causes implicit in the context; or if it is because of other possible frame elements in the sentence we disregard, such as circumstances and explanation, which can indicate emotion stimuli; or if a sentence is not enough to always contain

[8] It puts many words into multiple negative classes such as *sadness*, *anger*, *disgust*. For example, *Despair* is labeled as anger, disgust, fear, and sadness.

the emotion stimulus and we should consider larger text portions (maybe the current sentence and the previous and next sentence). Nonetheless, we believe that building this dataset is useful in choosing one of these three reasons. As a future work, we also would like to extend the emotion-stimulus dataset by considering other frame elements such as circumstances and explanation, which can indicate emotion stimuli. The distribution of the emotion instances in this dataset is presented in Table 4.

Table 4. Distribution of labels in the emotion datasets with no stimulus

happiness	sadness	surprise	disgust
268	468	160	57
anger	fear	shame	total
284	279	77	1594

The next section explains how we use the emotion stimulus dataset to build a supervised model to learn emotion stimulus spans in emotion-bearing sentences.[9]

4 Automatically Detecting Emotion Stimulus

To assess the difficulty of detecting emotion stimuli, we develop baseline systems which work with intuitive features . We also explore various features and their effect on stimulus detection results. We have built labeled data annotated with emotion stimuli, so we also have the privilege to explore supervised learning methods for information extraction. Of the datasets explained in Table 3, we use the second one, with 820 instances. That gives us more data to learn from.

Statistical learning models help avoid the biases and insufficiency of coverage of manual rule and pattern detection methods. One of the most common learning paradigms for performing such labelling tasks are Hidden Markov Models or probabilistic finite-state automata to identify the most likely sequence of labels for the words in any given sentence [36]. Such models, however, do not support tractable inference, and they represent the data by assuming their independence. One way of satisfying both these criteria is to use a model which defines a conditional probability over label sequences given a particular observation sequence rather than using a joint distribution over both label and observation sequences.

Conditional random fields (CRFs) [21] are a probabilistic framework for labelling and segmenting sequential data, based on a conditional model which labels a novel observation sequence x by selecting the label sequence y maximizing the conditional probability p(y|x). We use CRF from MinorThird [10], because it allows error analysis: comparing the predicted labels with the actual labels by highlighting them on the actual text. It also ranks the features based on the weight, so we can see which features have contributed the most.

[9] The dataset we built indicates both the emotion expressed and the emotion stimulus in each sentence. In this work, however, we only detect emotion stimulus, and assume that the emotion expression is present (in our data, it is). Our future work will address both emotion expression and emotion stimulus detection at the same time.

4.1 Baselines

The baselines we explain here set the ground for comparing our results and evaluating the performance of different models. One of the main properties of emotions is that they are generally elicited by stimulus events: something happens to the organism to stimulate or trigger a response after having been evaluated for its significance [32]. That is why events seem to be the most obvious indicators of emotions; we build our first two baselines upon events. We are aware that they are not the only emotion stimuli, but we believe them to be important enough.

One problem with using events as emotion stimuli is that event detection itself is a challenging task. The literature suggests verbal and nominal events; the former are much more numerous [7]. A verb conveys an action, an occurrence or a state of being. We use verbs as a textual signal of events; as our first baseline, we mark verbs in a sentence as the emotion stimuli. We retrieve verbs with the OpenNLP POS tagger.[10] The second baseline is Evita [31], a tool which detects both nominal and verbal events; as an emotion stimulus, we select an event in a sentence at random.

We noted earlier that not only events can be stimuli. FrameNet defines a stimulus as the person, event or state of affairs that evokes the emotional response in the Experiencer. For the third baseline, we recognize an emotion stimulus in a larger portion of the sentence, a phrase or a syntactically motivated group of words. We use the OpenNLP chunker,[11] and randomly select as an emotion stimulus a chunk which contains a verb.

[23] used as a baseline the clause with the first verb to the left of the emotion keyword. In English, however, there are single-clause sentences such as "My grandfather's death made me very sad." or "I was surprised to hear the ISIS news." with both the emotion state and the stimulus. The whole sentence would be returned as an emotion stimulus. Even so, we believe that it is worth exploring and investigating how useful a clause will be in detecting emotion stimuli in English. As the next baseline, we select a random clause as the stimulus. In OpenNLP parse trees,[12] we take the S, and SBAR tags as indicators of independent and dependent clauses in a sentence. Next, we randomly choose one of the clauses as the emotion stimulus.

Finally, we use Bag-of-Words as a typical baseline for all NLP tasks and for the sake of comparison. The previous baselines were rule-based systems with simple heuristics. For this baseline, we apply CRF sequence learner from MinorThird to all the unigrams in the text. In the *sequence annotator learner* we select *CRF Learner* as the classifier, 100 (a MinorThird default) as the number of iterations over the training set, and *5-fold cross validation* as the evaluation option.[13]

---

[10] http://opennlp.apache.org/documentation/manual/opennlp.html#tools.postagger

[11] http://opennlp.apache.org/documentation/1.5.2-incubating/manual/opennlp.html#tools.chunker

[12] http://opennlp.apache.org/documentation/manual/opennlp.html#tools.parser

[13] The dataset is too small for 10-fold cross validation. We only wanted to use unigrams as features in the baseline, but the CRF learner in MinorThird adds the previous labels of each token as a feature. The results, then, are higher than when using only unigram features.

Table 5. Baselines, the results. For random chunks, a quick experiment shows that verb phrase and noun phrase chunks are only 30% of the text. For Bag-of-Words, the span-level evaluation scores are 0.3293, 0.2132 and 0.2588.

	Precision	Recall	F-measure
Verb	0.212	0.059	0.093
Evita events	0.265	0.044	0.076
Random Chunk	0.292	0.0692	0.112
Random Clause	0.419	0.532	0.469
Bag-of-Words	0.5904	0.5267	0.5568

The baseline results are presented in Table 5. In span detection problems, the evaluation measures can either be based on the number of matching tokens or be more strict and consider the exact spans and the number of exact matches. Consider the sentence "His doctors were astounded that he survived the surgery." The emotion stimulus span ought to be "that he survived the surgery." If we return "that he survived" instead, token-based measures find three tokens matches, but span-based measures treat this as no match. Naturally, the value of token-level measures is higher than span-level measures. That is why, to build the higher-bound baseline, we report the token level precision, recall and F-measure.

The results indicate very low coverage in the first three baselines while the clause and Bag-of-Words baselines are much higher. The reason can be that the data in FrameNet are well-formed and carefully collected. Having a quick look at the instances shows that the emotion stimulus tends to be longer than just a verb or a phrase. Stimuli are long enough to say why a particular emotion was experienced. Therefore as a baseline the random clause and Bag-of-Words experiments have higher coverage. We believe that, although the first baselines' results are really low when used as the only feature in a simple rule-based system, they still are interesting features to study as features in our machine learning methods. In the next section we will discuss adding these features, and compare the results with the baseline.

4.2 Features and Results

Corpus-Based. We use a set of corpus-based features built in MinorThird's text analysis package. Among the features there are the lower-case version of each single word, and analogous features for tokens in a small window to either side of the word. Here we set the window size to three as suggested by the literature [9]. Additional token-level features also include information whether the token is a special character such as a comma, and orthographic information. For example, there is a feature, the character pattern "X+", which indicates tokens with all capital letters. The features are grouped into positive and negative. *Positive* refers to the group of features built and weighted based on the tokens within the stimulus span. *Negative* are the features related to all the tokens outside of the targeted span. This feature extraction process results in 23,896 features used in our learning process.

We applied *CRF Learner* with the same settings as the Bag-of-Words baseline in the previous section. The result of these experiments are shown in Table 6.

Table 6. Results of detecting emotion stimulus using different features

	Token Precision	Token Recall	Token F-measure	Span Precision	Span Recall	Span F-measure
Corpus-Based	0.7460	0.7017	0.7232	0.5658	0.5402	0.5527
Corpus-Based + Event	0.766	0.756	0.761	0.567	0.561	0.5644
Corpus-Based + Chunker	0.776	0.761	0.7688	0.564	0.556	0.5603
Corpus-Based + Clause	0.809	0.731	0.768	0.623	0.564	0.592
Corpus-Based + Event + Chunker + Clause	0.811	0.746	0.777	0.666	0.593	0.6280

An analysis of our learnt model and the feature weights shows that, for the positive tokens, the left-side token features have a higher weight than the right-side tokens. It is the opposite for the negative tokens. Also, the highest-weighted token features include "at", "with", "about", "that" and emotion words such as "delight", "concerned", "ashamed", "anger" for the left-side tokens.

Although the result of these experiment significantly outperform all the baselines, we notice that the span precision and recall are much lower than at the token level. The reason is that the syntactic structure of a sentence is not considered in this set of features. According to the ranked features, many function words are among the highest-weighted features. This means that this task is very structure-dependent.

A few examples showcase some of the shortcomings of this model by comparing what is learnt (blue) versus what is the actual stimulus (green).

- "Colette works at marshalling our feelings of revulsion {{at this} voracious creature who has almost killed the poor box thorn.}" This example shows that, although these features might be useful to detect the beginning of the emotion stimulus, detecting the end of the span seems more challenging for them.
- "He was petrified {of the clippers} {at first}." In this case the model has learned that many emotion stimuli start with the word "at", so it chooses "at first" regardless of its semantic and syntactic role in the sentence.
- "At a news conference {at the Royal Geographical Society in London} , they described the mental and physical anguish {of their 95-day trek}." Lacking semantic features, the model does not recognize that a location cannot be an emotion stimulus alone.

Looking at the predicted labels and comparing them with the actual labels shows that we need deeper semantic and syntactic features (explained in the next sections).

Events. FrameNet's definition of emotion stimulus treats events as one of the main factors in detecting stimuli. That is why we use a tool to automatically detect events and add them to the features. The following examples show how events can be the main part of emotion stimuli.

- "I am desolate that Anthony has died."
- "His last illness was the most violent , and his doctors were astounded that he survived it ."
- "I join the Gentleman in expressing our sorrow at that tragic loss."

Evita [31] is a tool which develops algorithms to tag mentions of events in text, tag time expressions, and temporally anchor and order the events. The EVENT tag is used to annotate those elements in a text which mark the semantic events described by it. Syntactically, events are typically verb phrases, although some nominals, such as "crash" in "killed by the crash", will also be annotated as events. Evita's event classes are *aspectual*, *I-action*, *I-state*, *occurrence*, *perception*, *reporting* and *state*. The result of adding the event tags to the previous features in presented in Table 6.

Chunker. Text chunking divides a text into syntactically coherent segments like noun groups or verb groups, but does not specify their internal structure, nor their role in the main sentence. We use the OpenNLP chunker to tag the data with the chunks, because we believe that the chance of an emotion stimulus starting or ending in the middle of a chunk is very low. A chunker should help improve the span precision and recall.

Here are examples with the actual and predicted emotion stimulus label, using the previous model. We believe that considering chunks should help reduce the kind of errors which these examples illustrate.

- "Their cheerfulness and delight {{at still being} alive} only made Charlie feel more guilty." "Being alive" should be placed in one chunk, therefore the span will not end in the middle of a chunk.
- "Feeling a little frightened {{of the dead body behind} him in the cart} , he stopped for some beer at a pub , where he met Jan Coggan and Laban Tall." Again, "behind him in the cart" should be in one chunk, so by using a chunker we would know the predicted span was incorrect.

Clause. In English grammar, a clause is the smallest grammatical unit which can express a complete proposition. There are two different types of clauses, independent and dependent. An independent clause can stand alone as a complete sentence. Dependent clauses can be nominal, adverbial or adjectival. Noun clauses answer questions like "who(m)?" or "what?" and adverb clauses answer questions like "when?", "where?", "why?".[14] Although there might not be many cases when the whole clause is the emotion stimulus, there are some cases, as mentioned below, which make it worthwhile to look into clauses and considering them among the features.

To mark the clauses in a sentence, we use the OpenNLP parser. As suggested in the literature [12], we use the *SBAR* tag, which represents subordinate clauses in the parse trees, to identify dependent clauses in a sentence. We use the *S* tag inside the sentence to indicate independent clauses. The output of the parser is shown in the following example which shows how the emotion stimulus tag exactly aligns with the SBAR tag which indicates the subordinate clause.

- "I am pleased that they have responded very positively."[15]
- "I was so pleased she lived until just after Sam was born."[16]

---

[14] `http://www.learnenglish.de/grammar/clausetext.html`

[15] The parse is "I am pleased [SBAR that [S they have responded very positively.]]"

[16] The parse is "I was so pleased [SBAR [S she lived until [SBAR just after [S Sam was born.]]]]"

The result of adding the clause tags to the previously discussed features in presented in Table 6. At the end, we show the result of combining all the discussed features.

These results show that each set of features improves our span-learning model, while clause-based features are most effective among events, chunks and clause feature sets. Also, the combination of all features significantly outperforms every baseline. More improvement could come from adding more fine-grained features to each feature group. For example, we can add the type, tense and aspect of an event provided by the Evita tool. We can also improve our chunk-based features by postprocessing the chunker's result: combining relevant chunks into longer chunks. For example, two noun phrases with a preposition between them can give a longer noun phrase; this could be more useful in our task. Finally, although the CRF results are promising, we ought to explore other sequential learning methods such as maximum-entropy Markov models (MEMM), or conditional Markov models (CMM).

5 Conclusion and Future Directions

We have framed the detection of emotion causes as finding a stimulus element as defined for the emotion frame in FrameNet. We have created the first ever dataset annotated with both emotion stimulus and emotion statement;[17] it can be used for evaluation or training purposes. We used FrameNet's annotated data for 173 emotion LUs, grouped the LUs into seven basic emotions using their synonyms and built a dataset annotated with both the emotion stimulus and the emotion. We applied sequential learning methods to the dataset. We also explored syntactic and semantic features in addition to corpus-based features. We built a model which outperforms all our carefully built baselines.

The set we built in this work is small and well-formed, and contains carefully built data annotated by humans. To show the robustness of our model and to study the problem thoroughly,we would like in the future to extend our dataset in two ways: first to study *Circumstances* and *Explanation* frame elements to investigate whether they can also indicate emotion stimuli to be added to our dataset. Secondly, we would like to use semi-supervised bootstrapping methods to add instances of other existing emotion datasets which do not have emotion cause labels.

Also as a preliminary step to emotion stimulus detection, we would like first to define whether the sentence contains an emotion stimulus and then detect the emotion stimulus span. In this work, we built a dataset with emotion statements with no stimulus tag which could be used for this purpose.

Last but not least, we believe that an emotion stimulus and the emotion itself are not mutually independent. Although in this work we did not take the emotion of the sentences into account, in the future we would like to detect both the emotion and the emotion stimulus at the same time and to investigate whether indicating emotion causes can improve emotion detection and vice versa.

[17] [25] see stimulus narrowly as one towards whom the emotion is directed.

References

1. Alm, C.O., Roth, D., Sproat, R.: Emotions from Text: Machine Learning for Text-based Emotion Prediction. In: HLT/EMNLP, pp. 347–354 (2005)
2. Aman, S., Szpakowicz, S.: Identifying expressions of emotion in text. In: Matoušek, V., Mautner, P. (eds.) TSD 2007. LNCS (LNAI), vol. 4629, pp. 196–205. Springer, Heidelberg (2007)
3. Bethard, S., Martin, J.H.: Learning Semantic Links from a Corpus of Parallel Temporal and Causal Relations. In: Proc. ACL 2008 HLT Short Papers, pp. 177–180 (2008)
4. Bethard, S., Yu, H., Thornton, A., Hatzivassiloglou, V., Jurafsky, D.: Automatic Extraction of Opinion Propositions and their Holders. In: 2004 AAAI Spring Symposium on Exploring Attitude and Effect in Text, pp. 22–24 (2004)
5. Chang, D.S., Choi, K.S.: Incremental cue phrase learning and bootstrapping method for causality extraction using cue phrase and word pair probabilities. Information Processing and Management 42(3), 662–678 (2006)
6. Chaumartin, F.R.: UPAR7: A knowledge-based system for headline sentiment tagging. In: Proc. 4th International Workshop on Semantic Evaluations, SemEval 2007, pp. 422–425 (2007)
7. Chen, Y., Lee, S.Y.M., Li, S., Huang, C.R.: Emotion cause detection with linguistic constructions. In: Proc. 23rd International Conference on Computational Linguistics, COLING 2010, pp. 179–187 (2010)
8. Choi, Y., Breck, E., Cardie, C.: Joint extraction of entities and relations for opinion recognition. In: Proc. 2006 Conference on Empirical Methods in Natural Language Processing, EMNLP 2006, pp. 431–439 (2006)
9. Choi, Y., Cardie, C., Riloff, E., Patwardhan, S.: Identifying sources of opinions with conditional random fields and extraction patterns. In: Proc. Human Language Technology and Empirical Methods in Natural Language Processing, HLT 2005, pp. 355–362 (2005)
10. Cohen, W.W.: Minorthird: Methods for Identifying Names and Ontological Relations in Text using Heuristics for Inducing Regularities from Data (2004), `http://minorthird.sourceforge.net`
11. Ekman, P.: An argument for basic emotions. Cognition & Emotion 6(3), 169–200 (1992)
12. Feng, S., Banerjee, R., Choi, Y.: Characterizing Stylistic Elements in Syntactic Structure. In: Proc. the 2012 Joint Conference on Empirical Methods in Natural Language Processing and Computational Natural Language Learning, EMNLP-CoNLL 2012, pp. 1522–1533 (2012)
13. Fillmore, C.J., Petruck, M.R., Ruppenhofer, J., Wright, A.: FrameNet in Action: The Case of Attaching. IJL 16(3), 297–332 (2003)
14. Ghazi, D., Inkpen, D., Szpakowicz, S.: Prior versus contextual emotion of a word in a sentence. In: Proc. 3rd Workshop in Computational Approaches to Subjectivity and Sentiment Analysis, WASSA 2012, pp. 70–78 (2012)
15. Girju, R.: Automatic detection of causal relations for Question Answering. In: Proc. ACL 2003 Workshop on Multilingual Summarization and Question Answering, MultiSumQA 2003, vol. 12, pp. 76–83 (2003)
16. Girju, R., Moldovan, D.: Mining Answers for Causation Questions. In: AAAI Symposium on Mining Answers from Texts and Knowledge Bases (2002)
17. Kaplan, R.M., Berry-Rogghe, G.: Knowledge-based acquisition of causal relationships in text. Knowledge Acquisition 3(3), 317–337 (1991)
18. Katz, P., Singleton, M., Wicentowski, R.: SWAT-MP: the SemEval-2007 systems for task 5 and task 14. In: Proc. 4th International Workshop on Semantic Evaluations, SemEval 2007, pp. 308–313 (2007)
19. Kim, S.M., Hovy, E.: Identifying and Analyzing Judgment Opinions. In: Proc. HLT/NAACL 2006, pp. 200–207 (2006)

20. Kozareva, Z., Navarro, B., Vázquez, S., Montoyo, A.: UA-ZBSA: A headline emotion classification through web information. In: Proc. 4th International Workshop on Semantic Evaluations, SemEval 2007, pp. 334–337 (2007)
21. Lafferty, J.D., McCallum, A., Pereira, F.C.N.: Conditional Random Fields: Probabilistic Models for Segmenting and Labeling Sequence Data. In: Proc. Eighteenth International Conference on Machine Learning, ICML 2001, pp. 282–289. Morgan Kaufmann Publishers Inc., San Francisco (2001)
22. Lee, S.Y.M., Chen, Y., Huang, C.R.: A text-driven rule-based system for emotion cause detection. In: Proc. NAACL HLT 2010 Workshop on Computational Approaches to Analysis and Generation of Emotion in Text, CAAGET 2010, pp. 45–53 (2010)
23. Lee, S.Y.M., Chen, Y., Li, S., Huang, C.R.: Emotion Cause Events: Corpus Construction and Analysis. In: Proc. Seventh International Conference on Language Resources and Evaluation (LREC 2010). European Language Resources Association (ELRA), Valletta (2010)
24. Lu, C.Y., Lin, S.H., Liu, J.C., Cruz-Lara, S., Hong, J.S.: Automatic event-level textual emotion sensing using mutual action histogram between entities. Expert Systems With Applications 37(2), 1643–1653 (2010)
25. Mohammad, S., Zhu, X., Martin, J.: Semantic Role Labeling of Emotions in Tweets. In: Proc. 5th, ACL Workshop on Computational Approaches to Subjectivity, Sentiment and Social Media Analysis, pp. 32–41 (2014)
26. Mohammad, S.M., Turney, P.D.: Emotions evoked by common words and phrases: using mechanical turk to create an emotion lexicon. In: Proc. NAACL HLT 2010 Workshop on Computational Approaches to Analysis and Generation of Emotion in Text, CAAGET 2010, pp. 26–34 (2010)
27. Neviarouskaya, A., Aono, M.: Extracting Causes of Emotions from Text. In: International Joint Conference on Natural Language Processing, pp. 932–936 (2013)
28. Neviarouskaya, A., Prendinger, H., Ishizuka, M.: Affect Analysis Model: novel rule-based approach to affect sensing from text. Natural Language Engineering 17(1), 95–135 (2011)
29. Ortony, A., Collins, A., Clore, G.L.: The cognitive structure of emotions. Cambridge University Press (1988)
30. Picard, R.W.: Affective Computing. The MIT Press (1997)
31. Pustejovsky, J., Lee, K., Bunt, H., Romary, L.: ISO-TimeML: An International Standard for Semantic Annotation. In: Proc. the Seventh International Conference n Language Resources and Evaluation (LREC 2010) (2010)
32. Scherer, K.R.: What are emotions? And how can they be measured? Social Science Information 44, 695–729 (2005)
33. Strapparava, C., Mihalcea, R.: Learning to identify emotions in text. In: Proc. 2008 ACM Symposium on Applied Computing, SAC 2008, pp. 1556–1560 (2008)
34. Strapparava, C., Valitutti, A.: WordNet-Affect: an Affective Extension of WordNet. In: Proc. 4th International Conference on Language Resources and Evaluation, pp. 1083–1086 (2004)
35. Tokuhisa, R., Inui, K., Matsumoto, Y.: Emotion classification using massive examples extracted from the web. In: Proc. 22nd International Conference on Computational Linguistics, COLING 2008, vol. 1, pp. 881–888 (2008)
36. Wallach, H.M.: Conditional random fields: An introduction. Tech. rep., University of Pennsylvania (2004)
37. Wilson, T., Wiebe, J., Hoffmann, P.: Recognizing Contextual Polarity: An Exploration of Features for Phrase-Level Sentiment Analysis. Computational Linguistics 35(3), 399–433 (2009)
38. Wu, C.H., Chuang, Z.J., Lin, Y.C.: Emotion recognition from text using semantic labels and separable mixture models. ACM Transactions on Asian Language Information Processing (TALIP) 5(2), 165–183 (2006)

Sentiment-Bearing New Words Mining: Exploiting Emoticons and Latent Polarities

Fei Wang and Yunfang Wu

Key Laboratory of Computational Linguistics (Peking University), Ministry of Education
sxjzwangfei@163.com, wuyf@pku.edu.cn

Abstract. New words and new senses are produced quickly and are used widely in micro blogs, so to automatically extract new words and predict their semantic orientations is vital to sentiment analysis in micro blogs. This paper proposes *Extractor* and *PolarityAssigner* to tackle this task in an unsupervised manner. *Extractor* is a pattern-based method which extracts sentiment-bearing words from large-scale raw micro blog corpus, where the main task is to eliminate the huge ambiguities in the un-segmented raw texts. *PolarityAssigner* predicts the semantic orientations of words by exploiting emoticons and latent polarities, using a LDA model which treats each sentiment-bearing word as a document and each co-occurring emoticon as a word in that document. The experimental results are promising: many new sentiment-bearing words are extracted and are given proper semantic orientations with a relatively high precision, and the automatically extracted sentiment lexicon improves the performance of sentiment analysis on an open opinion mining task in micro blog corpus.

Keywords: new words, new senses, semantic orientation, LDA model.

1 Introduction

Micro blogs have become an important medium for people to post their ideas and share new information. In micro blogs, new words and new senses are produced largely and quickly, so to automatically recognize these words and predict their semantic orientations (SO) is vital to sentiment analysis.

This paper aims to mine sentiment-bearing words and predict their SO from large-scale raw micro blogs in an unsupervised manner. Our approach includes two modules: ***Extractor*** and ***PolarityAssigner***.

Extractor: *Extractor* aims to extract sentiment-bearing words from micro blog corpus by using lexical patterns. The difficult point is that we're working on a raw corpus without word segmentation and part of speech (POS) tagging. Both word segmentation and POS tagging are not trivial tasks in Chinese processing, especially in micro blogs where new words and ungrammatical sentences are used widely. Our approach overcomes the difficulties of word segmentation and POS tagging in an unsupervised manner. The

A. Gelbukh (Ed.): CICLing 2015, Part II, LNCS 9042, pp. 166–179, 2015.
DOI: 10.1007/978-3-319-18117-2_13

experimental results show that *Extractor* works well, automatically extracting 2,056 sentiment-bearing new words.

***PolarityAssigner*:** PolarityAssigner aims to assign a positive/negative SO to each sentiment-bearing word by exploiting emoticons and latent polarities. In micro blogs, emoticons are frequently used by people to emphasize their emotions. As a result, large quantities of emoticons co-occur with sentiment-bearing words. We assume that each sentiment-bearing word is represented as a random mixture of latent polarities, and each polarity is characterized by a distribution over emoticons. We apply a Latent Dirichlet Allocation (LDA) model to implement our assumption by giving it a novel interpretation. The experimental results show that our PolarityAssigner performs significantly better than previous work.

This paper is organized as follows: Section 2 discusses related work; Section 3 describes *Extractor*; Section 4 presents *PolarityAssigner*; Section 5 gives experimental results; Section 6 proposes future work.

2 Related Work

Work on SO Detecting. Mostly, SO detecting is regarded as a classification problem, which aims to classify a document into positive or negative category. Great bulk of work has devoted to this task, and researches like Turney et al.(2003), Wilson et al.(2005), Pang et al.(2005), Li et al.(2009), Li et al.(2010), Tu et al. (2012), Mukherjee et al.(2012) are quite representative. Most of the existing approaches exploit supervised or semi-supervised learning methods by incorporating human-labeled corpora or existing lexicons. On the contrary, our method is unsupervised and needs no labeled data, alleviating greatly the labor intense work of annotation. Vegnaduzzo (2004) describes a bootstrapping method to acquire a lexicon of subjective adjectives based on a POS tagged corpus. Compared with this method, we do not need segmentation or POS tagging in the corpus. Baroni et al. (2004) rank a large list of adjectives according to their subjectivity scores acquired through mutual information from a web text. In this method, they need an adjective list in advance, which is not available as new words are quite active in Micro blog corpus. Velikovich et al. (2010) build a large polarity lexicon using graph propagation method. Different from their work, we focus on words rather than n-grams.

Work on Lexical Patterns. Lexical patterns are widely used to extract semantic relations (e.g. is-a, part-of) between terms, generate semantic lexicons (e.g. to generate list of words of actors), etc. Researches like Pantel et al. (2006), Kozareva et al. (2008), Talukdar et al. (2006) are some representative work. Most of these works focus on nouns (esp. named entities). Different from previous work, our Extractor focuses on adjectives, because adjectives are more likely to carry sentiment. The work of Riloff et al. (2003) is close to our work, which presents a bootstrapping process to learn lexical patterns to identify subjective sentences. Different from it, we use lexical patterns to find sentiment words.

Work on Topic Model for Sentiment Analysis. Much work has devoted to topic models. We would like to mention only two researches considering both topic and sentiment. One is the joint sentiment-topic (JST) model proposed by Lin et al. (2009), which simultaneously detects topic and sentiment. Another is a joint model of text and aspect ratings proposed by Titov et al. (2008), which extracts sentient aspects in automatic summarization. Both of them are extensions of the original LDA model proposed by Blei et al. (2003). Compared with their work, PolarityAssigner is simple but novel, giving LDA a new interpretation with emoticons.

Work on Emoticons. In sentiment analysis on micro blogs, emoticons are mainly used in two ways: act as indicators of class labels in corpus building or serve as features in classification tasks. In Read (2005), Pak et al. (2010), Davidov et al. (2010), Hannak et al. (2012) and Zhao et al. (2012), emoticons are used to label a document as positive or negative; thus avoiding the labor intense work of annotation. In Liu et al.(2012), the "noisy" data labeled by emoticons is used to smooth the language model trained from manually labeled "pure" data. In Rao et al. (2010), Barbosa et al. (2010) and Zhang et al. (2012), emoticons are used as features for user attribute detection and sentiment classification. The work of Yang et al. (2007) is close to our study, which mines the relationship between words and emotions based on weblog corpora. A word that gets higher PMI with pre-defined emoticons is regarded as an emotion word, and the emoticons that get higher PMI with the emotion word vote for the SO of that word. Finally the extracted emotion words serve as features of a polarity classifier. Different from this work, we do not need to label each emoticon as positive or negative, but the polarities of emoticons are hidden in our LDA model.

3 Extractor

3.1 Overview

Extractor is a pattern-based method which makes use of linguistic knowledge in Chinese. A pattern is defined by a two-character string, which specifies the left-context and right-context of target instances. A pattern is like a container where sentiment-bearing words fill in. Using these containers we can extract sentiment-bearing words from large and somewhat noisy micro blog corpus. An instance is defined as a character string that fills in a pattern. An ideal instance should be, in this paper, a word and carries subjective sentiment.

In most of pattern-based approaches, word segmentation and POS tagging eliminate much of the ambiguities. For instance, only those instances with a particular POS tag are extracted, and only those words with a particular POS tag act as context indicators. More challengeable than previous work, our study works on an un-segmented raw corpus, and thus the ambiguities have to be solved by the patterns themselves.

Extractor consists of three phases. (1) Pattern induction: Extractor infers a set of patterns P which contain as many seed instances as possible. (2) Pattern selection: Extractor ranks all candidate patterns according to the reliability and reserves only the top-*L* patterns; (3) Instance extraction: Extractor retrieves from the corpus to get the set of instances I that match the selected patterns, and heuristic rules are applied to guarantee the quality of the extracted instances.

3.2 Pattern Induction

Pattern induction starts from four seed words, which are sentiment-bearing and are popular new words in micro blogs.

Seeds={坑爹|the reverse of one's expectation, 给力|excellent, 有爱|lovely, 蛋疼|wholly shit}

Considering that the overwhelming part of sentiment-bearing words is adjectives, the patterns should target on adjectives rather than other kinds of POS tags. According to Chinese grammar, adjectives can be preceded by adverbs and followed by auxiliaries or modal particles. Further according to Guo (2002), 98% adjectives can be preceded by degree adverbs (e.g. 很|very, 非常|very), and 72% adjectives can be followed by auxiliaries like 着|zhe, 了|le,过|guo. What's more, when people use adverbs especially degree adverbs to modify adjectives, the adjectives are very likely to convey positive/negative feelings. A pattern is composed by a left-context indicator and a right-context indicator, so adverbs are good choices for left-context indicators, and auxiliaries and model particles are good choices for right-context indicators. Roughly, the candidate patterns can be described as: "(adverb)_(auxiliary/ modal particle)", where the underline "_" denotes the place that an instance can fill in.

Adverbs, auxiliaries and model particles are all functional words whose word set is fixed and members can be enumerable. According to The Grammatical Knowledge-base of Contemporary Chinese (GKB) (Yu et al., 2009), there are 1,562 adverbs, 370 auxiliaries and modal particles. Following pattern (2), we will get 1562×370 candidate patterns. For instance, (不|not)_(了|le), (很|very)_(啊|a), (太|very)_(啦|la), (非常|very)_(啊|a), (特别|very)_(的|de) are all candidate patterns.

We compute the score of each candidate pattern using the following formula:

$$score(p) = \sum_{s \in seeds} hit(p,s) \tag{1}$$

where *hit(p,s)* denotes the number of times that the seed s occurs in pattern *p* in the corpus. Only the top-*K* patterns are reserved for further selection, and K is set to 50 in our experiment. All the reserved patterns constitute a set *P*.

3.3 Pattern Selection

Pattern selection ranks all patterns in *P* according to the reliability r_π and picks out only the top-L patterns. *L* is set to 20 in our experiment.

We define the reliability r_π of a pattern p as:

$$r_\pi(p) = \underbrace{\frac{\sum_{s \in seeds}\left(\frac{PMI(s,p)}{\max_{PMI}}\right) * r(s)}{|\,seeds\,|}}_{Espresso} * \underbrace{[1-\alpha(p)]}_{\substack{WordFormation\\Penalty}} * \underbrace{[1-\beta(p)]}_{\substack{VerbExtraction\\Penalty}} \tag{2}$$

$r_\pi(p)$ consists of three parts: Espresso, word formation penalty and verb extraction penalty.

Espresso
This part is borrowed from Pantel and Pennacchiotti (2006), which computes p's average association strength with each s in seeds. $r(s)$ is the reliability of s, and we set $r(s)$ to 1 for each s in our experiment. $\max_{PMI}$ is the maximum pointwise mutual information between all patterns and all seeds. The pointwise mutual information between a seed s and a pattern p is estimated by:

$$PMI(s, p) = \log_2\left(\frac{hit(s, p)}{hit(s,*)hit(*, p)} N\right) \tag{3}$$

where $hit(s, p)$ denotes the number of times that s fills in pattern p in our corpus; the aster (*) represents a wildcard; N is the number of micro blogs in the collected corpus.

Word Formation Penalty
This part penalizes those patterns whose left-contexts are likely to be a part of another word, namely combinational ambiguity. For example, "太|too" is an adverb, and "了|le" is a modal particle, so "太_了" is a candidate pattern. Given a string "太阳出来了| here comes the sun", the string "阳出来" would be extracted as an instance, which makes no sense. Actually, in the above string, "太" is a part of word "太阳|the sun" rather than a single adverb.

Take this into consideration, the more the left-context is productive, the more the pattern should be penalized. The word formation capability $a(p)$ is estimated using formula (6):

$$\alpha(p) = \frac{\sum_{wpl} frequency(wpl)}{frequency(pl)} \tag{4}$$

where *pl* represents the left-context indicator of a pattern p (e.g. "太"), *wpl* represents those words with *pl* as their right character (e.g. "太阳"), and it is obtained by looking up GKB. $frequency(str)$ returns the number of times that *str* occurs in the corpus, and it is estimated by going through the collected micro blog corpus using string match. $a(p)$ is between [0,1]. Word formation penalty is defined as $1-a(p)$ in formula (2), ranging between [0,1].

Only left-contexts are considered here. Right-contexts, which are auxiliaries or modal particles, have little chance to be a part of another word.

Verb Extraction Penalty
According to Chinese grammar, adverbs can be followed by verbs and adjectives acting as modifiers. Verb extraction penalty penalizes those patterns that frequently extract verbs rather than adjectives.

For example, "不|not" is an adverb, and "了|le" is a modal particle, so "不_了" is a candidate pattern. Both adjectives and verbs can fill in this pattern. For instance, the adjective "美|beautiful" can fill in this pattern to form "不美了|not as beautiful as before", and the verb "睡|sleep" can also fill in this pattern to form "不睡了|do not sleep". As a result, both adjective "美|beautiful" and verb "睡|sleep" can be extracted by this pattern. Please keep in mind that our aim is to extract sentiment-bearing adjectives, so the more a pattern is capable of extracting verbs, the more the pattern should be punished.

Verb extraction capability $\beta(p)$ is defined using formula (5)

$$\beta(p) = \frac{\sum_{i \in I_p \text{ AND } i \in Verbs} hit(i,p)}{\sum_{i \in I_p} hit(i,p)} \tag{5}$$

where I_p denotes all instances extracted by pattern *p*, *Verbs* denotes the verbs specified in GKB. $hit(i,p)$ is the number of times that an instance *i* fills in pattern *p* in our corpus. $\beta(p)$ is between [0,1]. Verb extraction penalty is defined as $1-\beta(p)$ in formula (2), ranging between [0,1].

3.4 Instance Extraction

Instance extraction retrieves from the corpus to get instances *I* that can fill in any of the patterns. An instance i would be extracted if it obeys the following principles.

1. The character length of instance *i* is between [1,4].
2. An instance *i* can be extracted by at least two patterns with different left-contexts, aiming to filter errors triggered by individual adverbs.
3. An instance *i* co-occurs with more than ϕ emoticons, and here ϕ is set to 10. This is motivated by: 1) frequent co-occurrence with emoticons is also a clue that *i* is very likely to be sentiment-bearing; 2) enough co-occurrence is necessary for the next *PolarityAssigner*.

4 PolarityAssigner

PolarityAssigner tries to assign a positive/negative SO to each sentiment-bearing word by exploiting emoticons and latent polarities. Unlike Extractor, PolarityAssigner method is language independent and can be applied to many other languages, since emoticons are widely and frequently used in almost all languages.

Emoticons (e.g. |shy, |angry) are frequently used in micro blogs. According to our observation, every 22 out of 100 micro blogs contain emoticons. Intuitively, an emoticon conveys feelings more vividly and straightforwardly than a word. So in micro blogs, when expressing feelings, people often use emoticons to emphasize their emotions and strengthen the expressive effect of the texts. Consequently, large quantities of emoticons co-occur with sentiment-bearing words, leading to the innovation of our approach.

We regard each sentiment-bearing instance (word) as a document, and treat all the emoticons co-occurring with that instance as the words of the document. For example, providing that the sentiment-bearing word "萌|cute" co-occurs with 20 emoticons in a corpus, including seven, four, two, one, one, three and two, the document representing "萌|cute" can be organized as in Figure 1. Each emoticon is one word in this document, and the SO of this document is then assigned to "萌|cute".

We assume that a document (representing a sentiment-bearing word) is represented as a random mixture of latent polarities, where each polarity is characterized by a distribution over emoticons. This assumption is consistent with the assumption of LDA, which says that each document is represented as a random mixture of latent topics, and each topic is characterized by a distribution over words. This leads to our model, as shown in Figure 2.

Fig. 1. Document representing"萌|cute

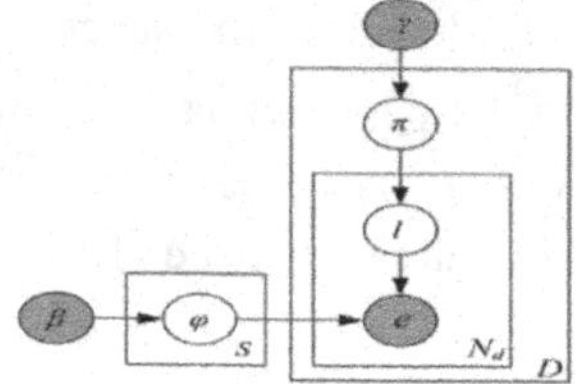

Fig. 2. LDA Model

Our model looks the same as LDA, but has a novel meaning. Assuming that we have a corpus with a collection of D documents that represent D sentiment-bearing words, each document d can be represented as a sequence of N_d emoticons denoted by $d = (e_1, e_2, ...e_{Nd})$, and each emoticon e_i in the document is an item derived from an emoticon vocabulary. Let S be the number of distinct polarities, which is set to 2 in our situation: positive and negative. The procedure of generating an emoticon e_i in a document d boils down to the following stages:

- For each document *d* choose a distribution $\pi_d \sim Dir(\gamma)$
- For each emoticon e_i in document *d*
  - Choose a polarity label $l_{di} \sim \pi_d$
  - Choose an emoticon e_i from the distribution over emoticons defined by polarity l_{di}.

The resolution of our model is the same as LDA. We can get $P(polarity_i \mid d)$ for each document *d*, and then we assign *d* with the polarity that gets higher probability.

LDA can automatically generate topics, but the model itself cannot name the topics. In our case, our model generates two polarities, but it cannot tell which one is positive and which one is negative. We propose a minimum-supervised method to solve this problem. Up to the present, all the documents have been split into two clusters with opposite polarities. In Chinese, "好|good" is unambiguously positive and is frequently used, so we assign the cluster containing "好|good" as positive, and the other cluster as negative. In this simple way, the polarity of each cluster is settled.

We define the SO of a document d as:

$$SO(d) = P(positive \mid d) - P(negative \mid d) \tag{6}$$

A document d is classified as positive when $SO(d)$ is positive, and negative when $SO(d)$ is negative. The magnitude (absolute value) of $SO(d)$ can be considered as the confidence of our judgment. Recall that each document represents one sentiment-bearing word, so the SO of the document is then assigned to the corresponding sentiment bearing word.

5 Experiment

5.1 Data

Our experiment is based on a three-day Microblog corpus (Sep 27, Sep 28, and Nov 4, 2011 year) provided by Sina Microblog, containing about 120 million texts. At the time of our experiment, Sina Microblog provided 82 emoticons and 200 more extended ones, and all of them constitute the emoticon vocabulary.

Totally 4,060 instances were extracted by Extractor. Each instance co-occurs with 336 emoticons on average. We apply Gibbs Sampling with 1,000 iterations to solve the model proposed in Figure 2.

Two human annotators familiar with micro blog language annotated these 4,060 instances independently. Each instance was annotated as one of the five categories: (1) P: positive; (2) N: negative; (3) P/N: it is ambiguous with both positive and negative meanings; (4) NE: neutral; (5) Error: it is not a meaningful word. Table 1 shows the labeled data. The inter-annotator agreement is 0.78 measured by the Kappa value. For those disagreed instances, two annotators explained their reasons and the final labeling were reached on the basis of fully discussion.

Table 1. Distribution of labeled instances

<table>
<tr><td>P</td><td>N</td><td>P/N</td><td>NE</td><td>Error</td></tr>
<tr><td>1526</td><td>1687</td><td>12</td><td>340</td><td>495</td></tr>
<tr><td colspan="3">3225, 79.4%</td><td colspan="2">20.6%</td></tr>
<tr><td colspan="4">3565, 87.8%</td><td>12.2%</td></tr>
</table>

5.2 Performance of Extractor

Precision

The distribution of extracted instances across each class is shown in Table 1. Among the 4,060 extracted instances, 3,565 instances extracted by Extractor are words, up to 87.8%; 3,225 instances are sentiment-bearing words, up to 79.4%. Please keep in mind that Extractor aims at extracting sentiment-bearing words, so the precision of Extractor is 79.4%. Considering the huge noises in Chinese raw micro blog corpus, the precision of 79.4% is promising.

Discovery of New Words

Among the extracted 3,225 sentiment-bearing words, 2,056 (63.8%) of them are new words. A word would be regarded as a new word if it is not recorded in sentiment

HowNet[1]. HowNet was manually constructed and contains 8,746 sentiment words, which is widely used in Chinese sentiment analysis. The high percentage of new words (63.8%) indicates that new words are very active in micro blogs, and identifying new words is vital to sentiment analysis in micro blogs.

Coverage of New Words

It is difficult to compute the recall of Extractor because it is difficult to build the new word set. We can only estimate the coverage of Extractor on new words. Kukuci[2] is a website collecting popular new words and net neologisms. We collected 18 sentiment-bearing new words from the homepage, as is listed in Table 2.

Table 2. New words from Kukuci

晕|dizzy, 牛B|terrific, 囧|embarrassed, 萌|cute, 雷|shocking, 闷骚|cold and dull, 顶|support, 郁闷|depressed, 靠谱|believable, 稀饭|like, 鸡动|excited, 碉堡|remarkable, 恶搞|spoof, 山寨|copycat, 杯具|miserable, 汗|perspire from embarrassment, 河蟹|irony usage of harmonious, FB| an enjoyable dinner

To our delight, all the listed words in Table 2 are extracted by Extractor from our micro blog corpus. It proves that Extractor works quite well in extracting new words.

5.3 Performance of *PolarityAssigner*

Baselines

To evaluate PolarityAssigner, we conducted two kinds of baselines: simple baseline using MFS (Most Frequent Sense) and previous Turney method.

MFS. On the basis of class distribution shown in Table 1, we naively guess that all words are negative.

Turney. Turney et al. (2003) define the SO of a word as in formula (7). PMI is estimated using the hit number returned by Baidu search engine.

$$SO(word) = \sum_{pword \in Pwords} PMI(word, pword) - \sum_{nword \in Nwords} PMI(word, nword) \qquad (7)$$

where *Pwords* is a list of positive words and *Nwords* is a list of negative words. *Pwords* and *Nwords* are defined as:

Pwords={漂亮|beautiful, 善良|kind, 高兴|happy, 大方|generous, 聪明|smart}
Nwords={难看|ugly, 邪恶|evil, 伤心|sad, 小气|stringy, 笨|stupid}

Evaluation Metrics

Evaluation is performed in terms of macro/micro accuracy (*Acc*) and precision (*Pre*). Precision is used to evaluate the performance of each $class_j$ (positive or negative in

[1] http://www.keenage.com/html/c_index.html.
[2] http://www.kukuci.com/

our case), while accuracy is used to evaluate the overall performance. Macro precision/accuracy is used to evaluate at the type level while micro at the token level.

Experimental Results

Table 3 and 4 show the performance of PolarityAssigner (PA) compared with two baselines. The difference between Table 3 and Table 4 is: all 4,060 instances extracted by Extractor are fed into PolarityAssigner and are evaluated in Table 3, while only the 3,225 true sentiment-bearing (P, N, P/N) words are fed into PolarityAssigner and are evaluated in Table 4. In other words, Table 3 lists the performance with accumulated errors from Extractor while Table 4 without accumulated errors.

Table 3 and Table 4 show that PolarityAssigner performs much better than both baselines. The trends presented in Table 3 and Table 4 are almost the same.

Table 3. Performance of *PolarityAssigner* with accumulated errors from *Extractor*

	Macro			Micro		
	Overall (Acc)	Each Class (Pre)		Overall(Acc)	Each Class (Pre)	
		Pos	Neg		Pos	Neg
MFS	41.6	NA	41.6	24.8	NA	24.8
Turney	42.4	40.1	60.9	61.9	62.2	**55.8**
PA	**63.5**	**66.6**	**62.4**	**75.3**	**90.1**	54.3

Table 4. Performance of PolarityAssigner without accumulated errors from Extractor

	Macro			Micro		
	Overall(acc)	Each Polarity (Pre)		Overall(acc)	Each Polarity (Pre)	
		Pos	Neg		Pos	Neg
MFS	52.3	NA	52.3	28.7	NA	28.7
Turney	53.4	50.4	77.3	71.7	72.1	62.8
PA	**79.5**	**85.0**	**77.5**	**87.1**	**95.2**	**72.7**

In the case with accumulated errors (Table 3), the macro (micro) accuracy of PolarityAssigner is 21.9% (50.5%) higher than MFS and 21.1% (13.4%) higher than Turney. In the case without accumulated errors (Table 4), the macro (micro) accuracy of *PolarityAssigner* is 27.2% (58.4%) higher than MFS and 26.1% (15.4%) higher than Turney.

With respect to precision, both tables show that PolarityAssigner performs better on positive class than negative class. We guess this is because some positive emoticons (e.g. laugh, titter) may co-occur with negative words when people are making fun of others or joking about themselves in a sarcastic manner.

The SO of Emoticons

In our assumption, each polarity is characterized by a distribution over emoticons, so $P(emoticon_j \mid polarity_i)$ is a by-product of our model. Table 5 lists some emoticons with the highest $P(emoticon_j \mid polarity_i)$.

Table 5. Emoticons with top $P(emoticon_j \mid polarity_i)$

emoticons with highest $P(emoticon_j \mid pos)$	|laugh |heart |love you |infatuation |titter | grin |applause |good job |so happy |pretty rabbit
emoticons with highest $P(emoticon_j \mid neg)$	|crying |drive me crazy |what |angry |weak |pathetic |embarrassed |faint |pick nose |sick

It is promising that emoticons with the highest $P(emoticon_j \mid pos)$ are all positive emoticons, and emoticons with the highest $P(emoticon_j \mid neg)$ are all negative ones.

Furthermore, if needed, the SO of each emoticon can be automatically computed using Bayes Formula:

$$P(polarity_i \mid emoticon_j) = \frac{P(emoticon_j \mid polarity_i)}{P(emoticon_j)} P(polarity_i) \quad (8)$$

Both $P(emoticon_j)$ and $P(polarity_j)$ can be easily estimated in the corpus. In future work, we would like to apply formula (8) to automatically detect the SO of emoticons.

5.4 Application of Sentiment-Bearing New Words

To prove the effectiveness of the extracted sentiment words and their predicted polarities, we incorporate the learned lexicon in a sentiment analysis task. We use an open microblog corpus for opinion mining[3], including 407 positive texts and 1,766 negative texts which are human annotated.

We use the simple lexicon-based method, and the polarity of one microblog is determined by the number of positive words minus the number of negative words. As we're working on an unsegmented corpus, one-character words are removed from the lexicon to reduce the ambiguities. We exploit five methods, by combing different lexicons. Table 6 shows the experimental results.

- **HN(HowNet).** Only use HowNet Lexicon.
- **A(Auto Lexicon).** Use the automatically extracted lexicon and their automatically predicted polarities.
- **HN+A.** Use the union of HowNet lexicon and Auto lexicon.
- **M(Manual Lexicon).** Use manual-annotated words and polarities on the results of Extractor, referred to Table 1.
- **HN+M.** Use the union of HowNet Lexicon and Manual Lexicon.

As shown in Table 6, M gets the highest F value on positive class and HN+A gets the highest F value on negative class. Compared with HowNet, the automatically

---

[3] `http://tcci.ccf.org.cn/conference/2012/pages/page04_evappt.html#`

Table 6. Experiment Results of the automatically learned lexicon

	Positive			Negative		
	pre	rec	F	pre	rec	F
HN	40.5	46.0	43.0	91.4	28.2	43.1
A	37.8	44.2	40.8	89.9	47.0	61.7
HN+A	37.6	49.1	42.6	91.3	**47.3**	**62.3**
M	**44.5**	56.3	**49.7**	94.0	39.5	55.6
HN+M	42.0	**60.4**	49.5	**94.4**	40.1	56.3

learned lexicon A gets comparable performance on positive class, and much higher performance on negative class, validating the effectiveness of the learned lexicon in microblog sentiment analysis.

Table 6 shows an interesting result: the automatic lexicon performs even better than the manually annotated lexicon on Negative class. This is because that PolarityAssigner is apt to assign some words with negative sentiment according to the co-occurring emoticons, while these words are neutral labeled by human out of context. For example, "什么|what" is a neutral word, however, it frequently appears in negative texts like "什么世道| what a fuck world", "凭什么|for what". It shows that PolarityAssigner captures more context meanings other than individual words.

6 Conclusion and Future Work

This paper proposes Extractor to extract sentiment-bearing words from raw micro blog corpus, and PolarityAssigner to predict the SO of words by exploiting emoticons and latent polarities. The precision of Extractor is 79.4%, out of which 63.8% (2.056 words) are new words. Considering the huge noises in Chinese raw micro blog corpus, the experimental results of Extractor are promising. The performance of PolarityAssigner is 79.5% in term of macro accuracy and 87.1% in term of micro accuracy, which outperforms both MFS and Turney baselines significantly. Finally, we apply the automatically learned lexicon in a sentiment analysis task on an open micro blog corpus, and the results show that our auto lexicon gets better performance than the traditional human-labeled HowNet lexicon.

Our method is easy to implement: Extractor needs no word segmentation or POS tagging, which are hard tasks for Chinese language processing in micro blog corpus. Our method is unsupervised: PolarityAssigner needs no polarity-labeled data, which is expensive and time-consuming to construct. Moreover, PolarityAssigner can be applied to other languages, since emoticons are widely and frequently used in almost all languages.

Acknowledgement. This work is supported by Humanity and Social Science foundation of Ministry of Education (13YJA740060), National Key Basic Research Program of China (2014CB340504), Key Program of Social Science foundation of China (12&ZD227), and the Opening Project of Beijing Key Laboratory of Internet Culture and Digital Dissemination Research (ICDD201302, ICDD201402).

References

1. Pak, A., Paroubek, P.: Twitter Based System: Using Twitter for Disambiguating Sentiment Ambiguous Adjectives. In: Proceedings of the 5th International Workshop on Sentiment Evaluation, pp. 436–439 (2010)
2. Hannnak, A., Anderson, E., Barrett, L.F., Lehmann, S., Mislove, A., Riedewald, M.: Tweetin' in the Rain: Exploring Societal-scale Effects of Weather on Mood. In: Proceedings of the Sixth International AAAI Conference on Weblog and Social Media, pp. 479–482 (2012)
3. Pang, B., Lee, L.: Seeing Stars: Exploiting Class Relationships for Semantic Categorization with Respect to Rating Scales. In: Proceedings of the 43rd Annual Meeting of the Association for Computational Linguistics, pp. 115–124 (2005)
4. Yang, C., Hsin-Yih Lin, K., Chen, H.-H.: Building Emotion Lexicon from Weblog Corpora. In: Proceedings of the 45th Annual Meeting of the ACL on Interactive Poster and Demonstration Sessions, pp. 133–136 (2007)
5. Lin, C., He, Y.: Joint Sentiment/Topic Model for Sentiment Analysis. In: Proceeding of the 18th ACM Conference on Information and Knowledge Management, pp. 375–384 (2009)
6. Blei, D.M., Ng, A.Y., Jordan, M.I.: Latent Dirichlet Allocation. The Journal of Machine Learning Research 3, 993–1022 (2003)
7. Rao, D., Yarowsky, D., Shreevats, A., Gupta, M.: Classifying Latent User Attributes in Twitter. In: Proceedings of the 2nd International Workshop on Search and Mining User-Generated Contents, pp. 37–44 (2010)
8. Davidov, D., Tsur, O., Rappoport, A.: Enhanced Sentiment Learning Using Twitter Hashtags and Smileys. In: Proceedings of the 23rd International Conference on Computational Linguistics, Posters, pp. 241–249 (2010)
9. Riloff, E., Wiebe, J.: Learning Extraction Patterns for Subjective Expressions. In: Proceedings of the 2003 Conference on Empirical Methods in Natural Language Processing, pp. 105–112 (2003)
10. Titov, I., McDonald, R.: A Joint Model of Text and Aspect Ratings for Sentiment Summarization. In: Proceedings of 46th Annual Meeting of the Association for Computational Linguistic, pp. 308–316 (2008)
11. Zhao, J., Dong, L., Wu, J., Xu, K.: MoodLens: an Emoticon-based Sentiment Analysis System for Chinese Tweets. In: Proceedings of the 18th ACM SIGKDD International Conference on Knowledge Discovery and Data Mining, pp. 1528–1131 (2012)
12. Read, J.: Using Emoticons to Reduce Dependency in Machine Learning Techniques for Sentiment Classification. In: Proceedings of the ACL 2005 Student Research Workshop, pp. 43–48 (2005)
13. Liu, K.-L., Li, W.-J., Guo, M.: Emoticon Smoothed Language Models for Twitter Sentiment Analysis. In: Proceedings of the 26th AAAI Conference on Atificial Intelligence, pp. 1678–1684 (2012)
14. Velikovich, L., Blair-Goldensohn, S., Hannanm, K., McDonald, R.: The Viability of Web-derived Polarity Lexicons. In: Proceedings of Human Language Technologies: The 2010 Annual Conference of the North American Chapter of the ACL, pp. 777–785 (2010)
15. Barbosa, L., Feng, J.: Robust Sentiment Detection on Twitter from Biased and Noisy Data. In: Proceedings of the 23rd International Conference on Computational Linguistics: Posters, pp. 36–44 (2010)

16. Zhang, L., Jia, Y., Zhou, B., Han, Y.: Microblogging Sentiment Analysis using Emotional Vector. In: Proceedings of the 2nd International Conference on Cloud and Green Computing, pp. 430–433 (2012)
17. Baroni, M., Vegnaduzzo, S.: Identifying Subjective Adjective through Web-based Mutual Information. In: Proceedings of KONVENS 2004, pp. 17-24 (2004)
18. Talukdar, P.P., Brants, T., Liberman, M., Pereira, F.: A Context Pattern Induction Method for Named Entity Extraction. In: Proceedings of the 10th Conference on Computational Natural Language Learning, pp. 141–148 (2006)
19. Pantel, P., Pennacchiotti, M.: Espresso: Leveraging Generic Patterns for Automatically Harvesting Semantic Relations. In: Proceedings of the 21st International Conference on Computational Linguistics and the 44th Annual Meeting of the Association for Computational Linguistics, pp. 113–120 (2006)
20. Turney, P.D., Littman, M.L.: Measuring Praise and Criticism: Inference of Semantic Orientation from Association. ACM Transactions on Information Systems (2003)
21. Guo, R.: Studies on Word Classes of Modern Chinese. The Commercial Press, Beijing (2002)
22. Yu, S., et al.: The Grammatical Knowledge-base of Contemporary Chinese. Tsinghua University Press, Beijing (2009)
23. Li, S., Huang, C.-R., Zhou, G., Lee, S.Y.M.: Employing Personal/Impersonal Views in Supervised and Semi-supervised Sentiment Classification. In: Proceedings of the 48th Annual Meeting of the Association for Computational Linguistics, pp. 414–423 (2010)
24. Vegnaduzzo, S.: Acquisition of Subjective Adjectives with Limited Resources. In: AAAI Spring Symposium Technical Report: Exploring Affect and Attitude in Text (2004)
25. Mukherjee, S., Bhattacharyya, P.: Sentiment Analysis in Twitter with Lightweight Discourse Analysis. In: Proceedings of the 24th International Conference on Computational Linguistics, pp. 1847–1864 (2012)
26. Li, T., Zhang, Y., Sindhwani, V.: A Non-negative Matrix Tri-factorization Approach to Sentiment Classification with Lexical Prior Knowledge. In: Proceedings of the 47th Annual Meeting of the ACL and the 4th IJCNLP of the AFNLP, pp. 244–252 (2009)
27. Wilson, T., Wiebe, J., Hoffmann, P.: Recognizing Contextual Polarity in Phrase-Level Sentiment Analysis. In: Proceedings of Human Language Technology Conference and Conference on Empirical Methods in Natural Language, pp. 347–354 (2005)
28. Tu, Z., He, Y., Foster, J., van Genabith, J., Liu, Q., Lin, S.: Identifying High-Impact Substructures for Convolution Kernels in Document-level Sentiment Classification. In: Proceedings of the 50th Annual Meeting of the Association for Computational Linguistics, pp. 338–343 (2012)
29. Kozareva, Z., Riloff, E., Hovy, E.: Semantic Class Learning from the Web with Hyponym Pattern Linkage Graphs. In: Proceedings of ACL 2008: HLT, pp. 1048–1056 (2008)

Identifying Temporal Information and Tracking Sentiment in Cancer Patients' Interviews

Braja Gopal Patra[1], Nilabjya Ghosh[2], Dipankar Das[1], and Sivaji Bandyopadhyay[1]

[1]Dept. of Computer Science & Engineering, Jadavpur University, Kolkata, India
[2]Yahoo, Bangalore, India
{brajagopal.cse,dipankar.dipnil2005}@gmail.com,
nilabjyaghosh@hotmail.com, sivaji_cse_ju@yahoo.com

Abstract. Time is an essential component for the analysis of medical data, and the sentiment beneath the temporal information is intrinsically connected with the medical reasoning tasks. The present paper introduces the problem of identifying temporal information as well as tracking of the sentiments/emotions according to the temporal situations from the interviews of cancer patients. A supervised method has been used to identify the medical events using a list of temporal words along with various syntactic and semantic features. We also analyzed the sentiments of the patients with respect to the time-bins with the help of dependency based sentiment analysis techniques and several Sentiment lexicons. We have achieved the maximum accuracy of 75.38% and 65.06% in identifying the temporal and sentiment information, respectively.

Keywords: temporal information, sentiment analysis, support vector machines, cancer patients' interviews.

1 Introduction

The extraction of temporal information and identification of the sentiment(s) associated with it from the clinical text is an interesting and popular research area. Several related work can be found on clinical temporal information, based on different admission and discharge summary reports of the patients [5, 13]. Various types of research have been carried out related to sentiment analysis from the textual data [4]. To the best of our knowledge, the extraction and analysis of clinical temporal data along with sentiment analysis has not been explored much till date.

Cancer is one of the most dangerous diseases in the world is associated with less chances of post-diagnosis survival as compared to other diseases. For all types of cancer, the latest available records are found during 2000 to 2001[1]. One of such reports from the Cancer Research in UK shows that 43% of men and 56% of women were alive for more than 5 years and might have been cured. It is also observed that

[1] http://www.cancerresearchuk.org/about-cancer/cancers-in- general/cancer-questions/what-perc entage-of-people-with-cancer-are-cured

A. Gelbukh (Ed.): CICLing 2015, Part II, LNCS 9042, pp. 180–188, 2015.
DOI: 10.1007/978-3-319-18117-2_14

so many cancer survivors face physical and mental challenges as resulted out of their disease and treatment[2] even after a decade of being cured. Another research in [3] shows that around 22%–29% newly diagnosed cancer patients suffered from major depressive disorder (MDD). It is observed that if the diagnosis is carried out in prior to the critical temporal stages, somehow, we could improve the possibilities of recovery by incorporating proper treatment. Thus, the proposed labeled corpora and a prototype system could be helpful and therefore explored, in the context of cancer diagnosis, treatment and recovery.

In the present paper, we have labeled the temporal information in each of the sentences of the cancer patients' interviews. We considered four time-bins to capture a medical event, such as *before*, *during*, *after* and *not-informative* where the *not-informative* sentences do not contain any temporal information. We also annotated the underlying sentiments (*positive, negative* and *neutral*) of the patients associated with the specific situations or temporal bins. However, the present research aims to help psychiatrists in order to resolve the psychological problems of cancer patients aroused for different emotions at different timestamps. Later on, we employed different machine learning algorithms for automatic identification of temporal and sentiment information. We used different syntactic and semantic features to extract the temporal and sentiment information from the clinical texts.

The rest of the paper is organized in the following manner. Section 2 provides related work in details. Section 3 provides an elaborative description of the data used in the task. Similarly, the features used in these experiments are described in Section 4. The detailed setup of experimentation and analysis of the results are described in Section 5. Finally, conclusions and future directions are presented in Section 6.

2 Related Work

Recently, several tasks have been carried out on identifying the temporal bins from different types of medical reports. Jindal and Roth [5] presented a joint inference model for the task of concept recognition in clinical domain. The major contribution of this paper was to identify the boundaries of medical concepts and to assign types to such concepts. Each concept was categorized to have 3 possible types e.g., Test, Treatment, and Problem. They used the datasets provided by i2b2/VA team as part of 2010 i2b2/VA shared task[3]. They achieved an overall accuracy of 86.1% in identifying the concept types.

On the other hand, Raghavan et al., [13] modeled the problem of clinical temporal extraction as a sequence tagging task using Conditional Random Fields (CRFs). They extracted a combination of lexical, section-based and temporal features from medical events in each of the clinical narratives. Learning temporal relations, for fine-grained temporal ordering of medical events in clinical text is challenging: the temporal cues typically found in clinical text may not always be sufficient for these kinds of

[2] http://as.wiley.com/WileyCDA/PressRelease/pressReleaseId-114982.html

[3] https://www.i2b2.org/NLP/Relations/

tasks [13]. The time-bins used in the above tasks were: *way before admission*, *before admission*, *on admission*, *after admission* and *after discharge*. The corpus used in this task consists of narratives specifically from MRSA[4] cases, admission notes, radiology and pathology reports, history and physical reports and discharge summaries.

It has been observed that the temporal expressions are sometimes appeared as fuzzy [18]. Zhou et al., [18] experimented on a corpus of 200 random discharge summaries from the Columbia University Medical Center data repository. The corpus covers the period from the year 1987 to 2004. They used features like the endpoint(s) of an event, anchor information, qualitative and metric temporal relations, and vagueness.

To determine the sentiment in a temporal bin, we need a prior sentiment annotated lexicon. Several works have been conducted on building emotional corpora in English language such as SentiWordNet [1], WordNet Affect [16] and Emotion Lexicon [10] etc. The SentiWordNet has the maximum level of disagreement with other Sentiment Lexicons like MPQA, Opinion Lexicon and Lexicon Inquirer [12], though it is widely used in research community, because of its large coverage.

Smith and Lee [14] examined the role of discourse functions in sentiment classification techniques. They used the National Health Service (NHS) feedback corpus and used the bag of words feature to achieve the maximum accuracy in Sentiment analysis. Niu et al., [7] applied the natural language processing and machine learning techniques to detect four possibilities in medical text: no outcome, positive outcome, negative outcome, and neutral outcome.

Some of the works can be found on identifying the events from the natural language texts and correlating them with the appropriate sentiment. Das et al., [4] identified the events and their sentiments from the News texts using CRFs. They have also analyzed the relationship in between several events (*after, before* and *overlap*) and their sentiments. In this paper, we tried to identify both temporal relations as well as the sentiment associated with it using several features stated in the literature.

3 Data Collection

Healthtalkonline[5] is an award winning website that shares more than 2000 patients' experiences over 60 health-related conditions and ailments. To accomplish our present task, we prepared a corpus of 727 interviews of cancer patients collected from the above mentioned web archive. The corpus contains data of 17 different types of cancer and an average of 30 to 45 documents in each of the cancer types. We developed an in-house web crawler which has been used to collect the data available on the Healthtalkonline web archive. We supplied the URL of the main cancer page to the crawler and it was able to hop all other pages containing the cancer-related interviews. As such, URLs of all the webpages containing cancer interviews were spotted and thereafter, data was extracted from these pages. A preliminary manual observation revealed that the webpages are different in format. Thus, three kinds of patterns were observed. All unnecessary information like XML tags and unrecognized characters were eliminated and the refined data along with the metadata were stored

---

[4] `https://uts.nlm.nih.gov/home.html`

[5] `www.healthtalkonline.org`

into XML format. The metadata such as disease type, sub type, age group, interview ID, Age at interview, sex, Age at diagnosis, background of the patient and brief outline were stored into XML format. A snapshot of a portion of such a XML document is shown in Figure 1.

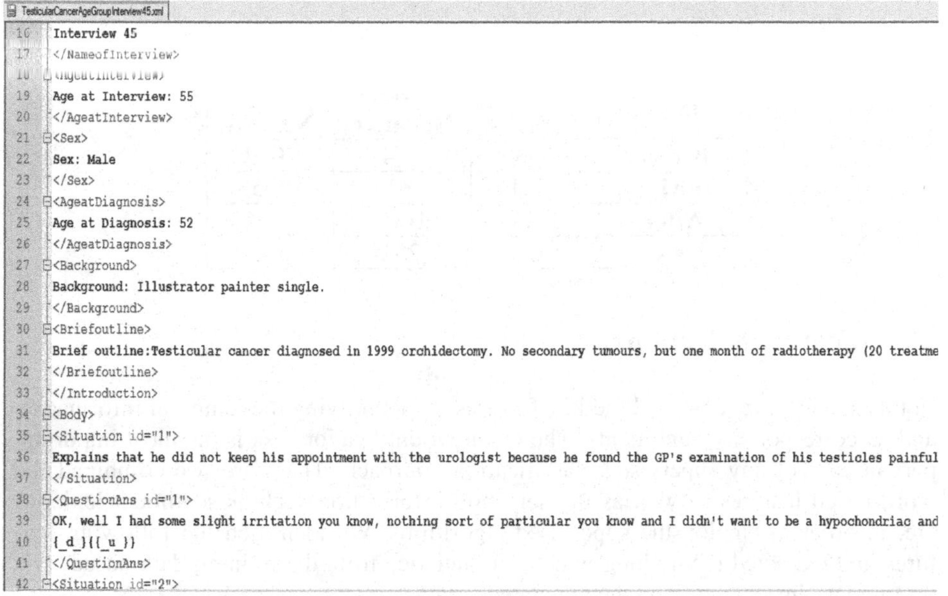

Fig. 1. Snapshot of the dataset

The corpus contains interviews only and is thus comprised of questions and the corresponding answers. Each line of an actual interview is either a narration/question indicative of the patient's conditions or is a response from the patient.

We tagged each of the sentences with sentiment information namely *positive*, *negative* and *neutral*. Similarly, the sentences were also tagged with temporal information such as *during, after, before* and *NA* classes. These classes were assigned relative to the cancer disease. For example, when a sentence contains any information related to the time before diagnosis of cancer, then we tagged the sentence as *before class*. The sentences were tagged with *during class*, if the sentences contain any information about the time during the cancer disease. We considered the time, after positive diagnosis of cancer to total cure as the class *during*. We tagged the sentences as *after* class, if the disease cancer is cured totally. We also tagged *NA* or *not informative*, if a sentence does not contain any information related to three temporal classes.

We observed the biasness in case of sentiments in the corpus. In case of the sentiment annotation, most of the sentences belong to *Neutral* class, whereas we also observed that in case of temporal annotation, mostly the sentences belong to *During* class. The statistics of the detail annotated data is provided in Table 1.

The whole dataset is annotated by three users. The inter-annotator agreement achieved is around 86% in case of Sentiment annotation whereas the inter-annotator agreement for Temporal information annotation is about 82%. The inter-annotator agreements for Temporal information annotation is low, because we observed some of the sentences that appear in between two sentences tagged with *During* class do not have any temporal information. In such cases, mostly one of the annotators assigned *During* class, whereas another two marked such instances with ***NA*** class. The statistics of the Temporal and Sentiment annotated sentences are shown in Table 1.

Table 1. Statistics of Corpus used in the Experiment

Temporal	Sentiment		
	Positive	Negative	Neutral
Before	668	774	2022
During	6201	6833	14654
After	1978	1005	2866
NA	206	251	567

4 Feature Selection

In this section, we identified the key features for identifying the temporal information and its corresponding sentiments. The feature identification task is the most important part in case of any supervised classification approach. Therefore, we combined the word-based features as well as the semantic information such as sentiment lexicons etc. in order to employ the supervised algorithms. We identified the following features and extracted them along with their statistics from the training data based on a preliminary investigation on the annotated data.

Positive and Negative Word Counts: We used the sentiment words as a feature for the sentiment identification tasks [8]. The words are identified as *positive*, *negative* or *neutral* using SentiWordNet (SWN) [1]. Again, we tagged each of the words in a sentence using WordNetAffect (WA) [16] in order to grasp more number of sentiment words. However, WA contains six classes of sentiments i.e. *angry, disgust, fear, joy, sad* and *surprise*. We grouped all the six classes into two classes i.e. *positive* (*joy* and *surprise*) and *negative* (*angry, disgust, fear* and *sad*) and each word is tagged with these two classes and if not found, then tagged as neutral [8].

We have removed the stop words while calculating the frequency. The words like '*not*', '*never*', '*no*' and '*neither*' were not removed as stop words as they used to play the roles of valence shifter and can convey the sentiment polarity of a sentence. The statistics of the sentiment words found are given in Table 3. It is noticed that the total of 117055 and 75436 numbers of *positive* and *negative* words are found in the corpus using SentiWordNet whereas 8251 and 20486 number of *positive* and *negative* words were found using the WordNetAffect. The number of *positive* words are less in the *positive* class of WA as it contains only the words form *happy* and *surprise* classes.

Phrase Level Polarity: We identified the adjective and adverbial phrases of a sentence and then assigned the polarity to each of the phrases using SentiWordNet. We parsed the sentences using the Stanford Parser [6]. We observed that there are at most five adjective or adverbial phrases that are present in a single sentence. We found 24052 and 10060 number of *positive* and *negative* phrases in our collected corpus as shown in Table 3.

Sentence Level Polarity: We split each of the compound sentences into multiple simple sentences using Stanford Parser [6] by the symbolic mark, "(S" as shown in Figure 2. Then, we identified the sentiment of each of the simple sentences using some handcrafted rules as given below. For example, in the following sentence, "*But they all said that I've handled it well and I never complained.*" three simple sentences are present and we identified the polarity of each small sentence using our handcrafted rules. Mainly, we have implemented four rules, i.e. 1. *Not*+NEG→POS, 2. *Not*+POS→NEG, 3. POS+POS→POS, 4. NEG+NEG→NEG.

(ROOT (S (CC But) (NP (PRP they)) (ADVP (DT all)) (VP (VBD said)
(SBAR (IN that)
(S (NP (PRP I)) (VP (VBP 've) (VP (VBN handled) (NP (PRP it))
(ADVP (RB well)))))) (CC and)
(S (NP (PRP I)) (ADVP (RB never)) (VP (VBD complained)))))) (. .)))

Fig. 2. Parsed tree of a sentence

Parts-of-Speech (POS): *POS* plays an important role in the *temporal* identification task. We used the *Stanford CoreNLP* [6] tool to parse each of the sentences to find out the *POS* tag for each of the words. We considered the frequency of the POS for different verbs, nouns, adjectives and adverbs.

Tense Information: This feature is useful to capture the standard distinctions among the grammatical categories of verbal phrases [4]. The tense attribute can have values, PRESENT, PAST, FUTURE, INFINITIVE, PRESPART, PASTPART, or NONE.

Temporal Words: We have identified the temporal words, which can relate two or more events together [4]. These can help to identify the temporal information from the text. For example, the sentence "***Before** the screening I was quiet happy.*" contains '***before***' temporal word and '*screening*' event, thus the sentence belongs to the *before* class. We collected the temporal words for the following three classes, *before*, *after* and *during* and the statistics of occurrence of these words in the corpus is given below in Table 2.

Medical Temporal Words: Similarly, we have identified some medical terms that can provide the temporal information. For example, the word *chemotherapy* is always used in *during* temporal class. We prepared a manual list of words that can occur during the diseases. There are 320 words belonging to the '*during*' class whereas there are only 14 words that are appeared in '*before*' class. It has to be mentioned that we have not found any medical word in the '*after*' class and '*before*' class containing the symptoms of cancer. Examples and the statistics of the occurrences of these words in the corpus are given in Table 2.

5 Experiments and Results

It is obvious that in order to achieve good results, we require a huge amount of annotated data for applying the supervised algorithms. But, to the best of our knowledge, no such annotated corpus is available on the web for cancer patients. We classified the annotated data using several machine learning algorithms and found that the Support

Vector Machines (SVMs) performs well. We used the API of Weka 3.7.7.5[6] to accomplish our classification experiments. Weka is an open source data mining tool. It presents a collection of machine learning algorithms for data mining tasks.

Table 2. Statistics of Temporal Words used in the Experiment

	Classes	**Examples**	**Total No. of words collected**	**Total No. of Instances in corpus**
Temporal Words	***Before***	Previously, back, before	15	458
	During	Today, now, during	12	1878
	After	Afterwards, after, thereafter	10	567
Medical Temporal Words	***Before***	Smoking, bleeding, screening	14	2594
	During	chemotherapy, biopsy, radiotherapy	320	25842

Table 3. Statistics of Different Features in Corpus

	Number of Instances
Positive Words using SentiWordNet	117055
Negative Words using SentiWordNet	75436
Positive Words using WordNetAffect	8251
Negative Words using WordNetAffect	20486
Adjective and Adverbial Phrases (positive)	24052
Adjective and Adverbial Phrases (Negative)	10060

Classification of Temporal Information: At first, we identified the temporal information in the sentence. We used the features like *POS, Tense Information, Temporal Words* and *Medical Temporal words* to identify the temporal information present in a sentence. It is observed that the number of the sentences in *during class* is much more compared to the other temporal classes.

We used LibSVM of Weka for identifying the temporal information in the text. We achieved the maximum accuracy of 75.38% for temporal classification after inclusion of all the features. The confusion matrix while achieving the maximum accuracy is given in Table 4.

Sentiment Classification: We used the *Positive* and *Negative* Word Counts, Phrase level polarity, Sentence level polarity, Parts-of-Speech (POS) and temporal information features for identifying sentiments of the sentences. We used Lib SVM of Weka for the above task. We achieved the maximum accuracy of 65.06% for sentiment classification after inclusion of all the features stated above. Confusion matrix for the above accuracy is given in Table 5.

---

[6] http://weka.wikispaces.com/

Table 4. Confusion Matrix for Temporal Classification

Classified as →	Before	During	After	NA
Before	**776**	2311	32	345
During	320	**27091**	71	206
After	304	5039	**292**	214
NA	111	358	50	**505**

Table 5. Confusion Matrix for Sentiment Classification

Classified as →	Positive	Negative	Neutral
Positive	**1814**	1377	5862
Negative	778	**2629**	5456
Neutral	785	1426	**17898**

Discussion: We observed that the accuracy for the temporal classification and sentiment classification is low. The main reason may be the biasness towards a specific category in the data set. In case of the temporal classification, the occurrence of the *during* class is much more as compared to other classes, whereas in sentiment classification, the number of instances from the *neutral* class is much more as compared to other two classes. The *Tense Information* feature does not work well in identifying the temporal information as the interviews were recorded after a long period of being cured. Most of the cases, the verbs are in the past tense only.

We also observed implicit sentiment in the corpus. For example, "*Yes, the report was positive.*" This sentence is identified as *positive* one, but the sentence is *negative*. Our system fails to identify such cases. Some of the sentences contain multiple sentiments and is thus difficult to annotate single sentiment to that sentence.

6 Conclusion

In this paper, we mainly annotated the interviews of the cancer patients with temporal and sentiment information. Then, we have classified the annotated data using SVM machine learning algorithm and achieved the maximum accuracy of 75.38% and 65.06% for temporal information and sentiment.

In future, we will include more suitable features to improve accuracy of our system. Our immediate goal is normalize the biasness of the dataset using some machine learning algorithm like SMOTE (Synthetic Minority Over-sampling Technique). Later on, we will try to perform the phrase based sentiment analysis on this corpus.

Acknowledgements. The work reported in this paper is supported by a grant from the project "CLIA System Phase II" funded by Department of Electronics and Information Technology (DeitY), Ministry of Communications and Information Technology (MCIT), Government of India.

References

1. Baccianella, S., Esuli, A., Sebastiani, F.: Sentiwordnet 3.0: An enhanced lexical resource for sentiment analysis and opinion mining. In: 7th Conference on International Language Resources and Evaluation (LREC 2010), Valletta, Malta (2010)
2. Bethard, S., Derczynski, L., Pustejovsky, J., Verhagen, M.: Clinical TempEval. arXiv preprint arXiv:1403.4928 (2014)
3. Brothers, B.M., Yang, H.C., Strunk, D.R., Andersen, B.L.: Cancer patients with major depressive disorder: Testing a bio-behavioral/cognitive behavior intervention. Journal of Consulting and Clinical Psychology 79(2), 253–260 (2011)
4. Das, D., Kolya, A.K., Ekbal, A., Bandyopadhyay, S.: Temporal Analysis of Sentiment Events–A Visual Realization and Tracking. In: Gelbukh, A.F. (ed.) CICLing 2011, Part I. LNCS, vol. 6608, pp. 417–428. Springer, Heidelberg (2011)
5. Jindal, P., Roth, D.: Using Soft Constraints in Joint Inference for Clinical Concept Recognition. In: Empirical Methods on Natural Language Processing, pp. 1808–1814 (2013)
6. Manning, C.D., Surdeanu, M., Bauer, J., Finkel, J., Bethard, S.J., McClosky, D.: The Stanford CoreNLP Natural Language Processing Toolkit. In: 52nd Annual Meeting of the Association for Computational Linguistics: System Demonstrations, pp. 55–60 (2014)
7. Niu, Y., Zhu, X., Li, J., Hirst, G.: Analysis of polarity information in medical text. In: Annual Symposium American Medical Informatics Association (2005)
8. Patra, B.G., Mukherjee, N., Das, A., Mandal, S., Das, D., Bandyopadhyay, S.: Identifying Aspects and Analyzing their Sentiments from Reviews. In: 13th Mexican International Conference on Artificial Intelligence (MICAI 2014) (2014)
9. Patra, B.G., Mandal, S., Das, D., Bandyopadhyay, S.: JU_CSE: A Conditional Random Field (CRF) Based approach to Aspect Based Sentiment Analysis. In: International Workshop on Semantic Evaluation (SemEval-2014) (COLING-2014), Dublin, Ireland, pp. 370–374 (2014)
10. Patra, B.G., Takamura, H., Das, D., Okumura, M., Bandyopadhyay, S.: Construction of Emotional Lexicon Using Potts Model. In: 6th International Joint Conference on Natural Language Processing (IJCNLP 2013), Nagoya, Japan, pp. 674–679 (2013)
11. Patra, B.G., Kundu, A., Das, D., Bandyopadhyay, S.: Classification of Interviews–A Case Study on Cancer Patients. In: 2nd Workshop on Sentiment Analysis Where AI Meets Psychology (COLING 2012), pp. 27–36. IIT Bombay, Mumbai (2012)
12. Potts, C.: Sentiment Symposium Tutorial (2011), `http://sentiment.christopherpotts.net/lexicons.html#relationships`
13. Raghavan, P., Fosler-Lussier, E., Lai, A.M.: Temporal classification of medical events. In: 2012 Workshop on Biomedical Natural Language Processing, pp. 29–37. ACL (2012)
14. Smith, P., & Lee, M.: Cross-discourse development of supervised sentiment analysis in the clinical domain. In: 3rd Workshop in Computational Approaches to Subjectivity and Sentiment Analysis, pp. 79-83 ACL (2012)
15. Stacey, M., McGregor, C.: Temporal abstraction in intelligent clinical data analysis: A survey. Artificial Intelligence in Medicine 39(1), 1–24 (2007)
16. Strapparava, C., Valitutti, A.: Wordnet-affect: An affective extension of wordnet. In: 4th International Conference on Language Resources and Evaluation (LRE), Lisbon, pp. 1083–1086 (2004)
17. Kar, R., Konar, A., Chakraborty, A., Nagar, A.K.: Detection of signaling pathways in human brain during arousal of specific emotion. In: International Joint Conference on Neural Networks (IJCNN), pp. 3950–3957. IEEE (2014)
18. Zhou, L., Melton, G.B., Parsons, S., Hripcsak, G.: A temporal constraint structure for extracting temporal information from clinical narrative. Journal of Biomedical Informatics 39(4), 424–439 (2006)

Using Stylometric Features for Sentiment Classification

Rafael T. Anchiêta[1], Francisco Assis Ricarte Neto[2], Rogério Figueiredo de Sousa[3], and Raimundo Santos Moura[3]

[1] Instituto Federal do Piauí
rta@ifpi.edu.br
[2] Universidade Federal de Pernambuco
farn@cin.ufpe.br
[3] Universidade Federal do Piauí
{rfigsousa,rsm}@ufpi.edu.br

Abstract. This paper is a comparative study about text feature extraction methods in statistical learning of sentiment classification. Feature extraction is one of the most important steps in classification systems. We use stylometry to compare with TF-IDF and Delta TF-IDF baseline methods in sentiment classification. Stylometry is a research area of Linguistics that uses statistical techniques to analyze literary style. In order to assess the viability of the stylometry, we create a corpus of product reviews from the most traditional online service in Portuguese, namely, Buscapé. We gathered 2000 review about Smartphones. We use three classifiers, Support Vector Machine (SVM), Naive Bayes, and J48 to evaluate whether the stylometry has higher accuracy than the TF-IDF and Delta TF-IDF methods in sentiment classification. We found the better result with the SVM classifier (82,75%) of accuracy with stylometry and (72,62%) with Delta TF-IDF and (56,25%) with TF-IDF. The results show that stylometry is quite feasible method for sentiment classification, outperforming the accuracy of the baseline methods. We may emphasize that approach used has promising results.

1 Introduction

Web 2.0 triggered an explosion of Web services, as well as a large increase in users on the network. Hence, has spread the creation of blogs, discussion forums, debate sites, social networks and shopping sites, favoring the interaction among individuals.

In this scenario, the search for reviews on products and/or services has become a quite common activity. In general, the vast majority of users visit opinion sites, discussion forums or even social networks, in search of the experience of other users before make a decision. However, despite being easy to find reviews on the Web, their manual analysis is not a trivial process due to the difficulty in analyzing the large amount of subjectivity in reviews (e.g. sarcasm, irony) as well as vagueness and ambiguity inherent in natural language.

A. Gelbukh (Ed.): CICLing 2015, Part II, LNCS 9042, pp. 189–200, 2015.
DOI: 10.1007/978-3-319-18117-2_15

Therefore, it is necessary automate the task of analyze and classify these opinions and/or sentiments. One of the Sentiment Analysis (SA) concerns, also known as Opinion Mining [1], it is to classify opinions expressed in texts concerning a particular object (product, service, institution or person) as positive, negative or neutral. In this context, various techniques based on supervised [2], [3], [4] and unsupervised [5], [6] learning may be used.

In general, the classification is based on adjectives found in the text and the text representation is based on the bag of words, using the Term Frequency Inverse Document Frequency (TF-IDF) as feature extraction. However, choose which features in a textual document is a crucial point for a good result in text classification.

Besides the adjectives, we may observe other features of the text for sentiment classification such as: total number of words, total different words, hapax legomena, syntactic features, frequency of synonyms of adjectives and so on.

In this context, this work presents the stylometric features, i.e., an alternative to traditional method of features extraction TF-IDF and its variants. Stylometry, the measure of style, is a burgeoning interdisciplinary research area that integrates literary stylistics, statistics and computer science in the study of the "style" or the "fell" [7].Therefore, we carried out a comparison of the TF-IDF and Delta TF-IDF with stylometry using three classifiers: Support Vector Machine (SVM), Naive Bayes, and J48 in order to evaluate whether the stylometry has higher accuracy in sentiment classification.

The main contribution in this paper is the evaluation of different feature categories highlighting the stylometric features for the sentiment classification. We propose three different feature categories (Word-based, Syntactic and Content specific) and compared with TF-IDF and Delta TF-IDF methods. We find out that the stylometric features improved the sentiment classification compared with the baseline methods.

The rest of the paper is organized as follows. In Sect. 2 we review the related works. Section 3 presents the created corpus to serve as basis to evaluate the methods of text feature extraction. Section 4 presents an overview about baseline methods of text feature extraction. In Sect. 5 we describe and we detail the stylometric features used in this work. Section 6 discusses the experiment and the comparative study. Finally, in Sect. 7 we show conclusion and future works.

2 Related Works

We can easily find in literature several different approaches related to Sentiment Analysis. These approaches are conducted either at the document, sentence or feature level, and can be distinguished between unsupervised or supervised methods. The focus in this work is evaluate feasibility of stylometric features in sentiment classification on an annotated corpus of product review, so, we used TF-IDF and Delta TF-IDF to develop a baseline for comparison with our stylometric approach.

Pang et al., (2002) [8] was the first paper that performs the classification movie reviews into two classes, positive and negative. The authors showed that using unigrams as features in classification performed quite well with either Naive Bayes or SVM, although the authors also tried a number of other feature opinions, such as: bigrams and unigrams+bigrams.

Another supervised approach for sentiment classification was developed by Yu and Hatzivassiloglou (2003) [9]. Unlike [8], the authors improved their task developing a Bayesian machine learning algorithm to classify the reviews on factual or opinion comments. Therefore, the authors identified the polarity of the sentence-level opinions as positive and negative in terms of the main perspective expressed in the opinion.

Turney (2002) [5] developed a non-supervised approach for the task of sentiment classification. The author introduces the use of Semantic Orientation to predict the polarity of the review. In order to perform this task, the author identified phrases that contains adjective and adverbs clues words and used a variant Pointwise Mutual Information-Information Retrieval (PMI-IR) to find if each phrase has more associativity with positive or negative words based on web queries. With this approach the author found a average classification precision of 74%.

Sharma and Dey (2012) [10] developed an Artificial Neural Network (ANN) to sentiment classification in movies and hotel reviews. The authors used three sentiment lexicons and Information Gain to extract features to train their model. This approach combine a ANN model with subjective knowledge available in sentiment lexicon and thus achieving 95% of accuracy in movies reviews. In another work [11], the same authors uses an Ensemble of Classifiers (Boosting) using SVM as base classifier for sentiment classification in on-line reviews. They indicated that the ensemble classifier has performed better than single SVM classifier.

Njolstad et at., (2014) [12] proposed and experimented four different feature categories (Textual, Categorical, Grammatical and Contextual), composed of 26 article features, for sentiment analysis. The authors used five different machine learning methods to train sentiment classifiers of Norwegian financial internet news articles, and achieved classification precisions up to ~71%. The experiment carried out with different feature subsets showed that the category relying on domain-specific sentiment lexicon ('contextual' category), able to grasp the jargon and lingo used in Norwegian financial news, it is of cardinal importance in classification - these features yielded a precision increase of ~21% when added to the other feature categories.

He and Rasheed (2004) [13] proposed recognize different authors based on their style of writing (without help from genre or period). The authors trained decision trees and neural networks to learn the writing style of live Victorian authors and distinguish between them based on certain features of their writing which define their style. The authors used three feature subsets (Tagged text, Parsed text and Interpreted text), composed of 22 features. Through this method, they achieved 82,4% accuracy on test set using decision trees and 88,2% accuracy on the test set using neural networks.

In this paper we use a set features used for authorship identification and we apply in the sentiment classification for Portuguese language. Next section presents the created corpus to evaluate our approach.

3 Corpus

In order to assess the feasibility of the stylometry in sentiment analysis, we firstly create a corpus[1] of product reviews. We have crawled database from the most traditional on-line service in Portuguese, namely, Buscapé[2], where users post their pros and cons comments about several products, services and companies. We gathered 2000 reviews about Smartphones from Buscapé database from a crawling in October 2013, being 1000 positive and 1000 negative reviews. They account 652KB, showing 209,250 characters, 35,594 tokens and 3,871 types for positive reviews and showing 440,958 characters, 76,201 tokens and 6,487 types for negative reviews. Hartmann et al., (2014)[14] shows more detail about the Buscapé database.

We have kept the preprocessing of the corpus to a minimum. Thus, we have lower-cased all words but we have not removed stop words or applied stemming.

4 Baseline Feature Extraction

The text classification study has become a project with great importance for Text Mining, sub-area of the Data Mining, and Artificial Intelligence. The feature extraction is the premise and basis to perform the text classification effective [15]. It is the first step of preprocessing which is used to presents the text into a clear format, i.e. set of features extraction is used to transform the input text into a feature set (feature vector). In this task, we may also remove stop words, words with little semantic information, and use stemming.

The following subsections present the TF-IDF and Delta TF-IDF methods that are popular techniques used for extracting text features to the task of the sentiment classification.

4.1 TF-IDF

According to Rajamaran (2011) [16] TF-IDF is a numerical statistic that is intended to reflect how important a word is to a document in a collection or corpus. It is often used as a weighting factor in Information Retrieval (IR) and Text Mining. The TF-IDF value increase proportionally to the number of times a word appears in the document, but is offset by the frequency of the word in the corpus, which helps to control for the fact that some words are generally more common than others. The TF-IDF value is defined by Equation 1.

1 `http://goo.gl/SG69B7`

2 `http://www.buscape.com.br`

$$TF - IDF = tf_{k,j} \times \log\left(\frac{N}{df_k}\right) \tag{1}$$

Where:

- $tf_{k,j}$ - number of times term k appears in a document j;
- N - total number of documents;
- df_k - number of documents with term k in it.

In other words, TF-IDF assigns to terms k a weight in document j that is:

1. Highest when k occurs many times within a small number of documents (thus lending high discriminating power to those documents);
2. Lower when the term k occurs fewer times in a document, or occurs in many documents (thus offering a less pronounced relevance signal);
3. Lowest when the term k occurs in virtually all documents.

4.2 Delta TF-IDF

Martineau and Finin (2009) [17] constructed vectors to classify a term based on term frequency vector as well as term presence vectors. Unlike TF-IDF which used single term presence vector, two vectors were separately constructed for presence in positively tagged documents and negatively tagged documents. Equation 2.

$$V_{t,d} = C_{t,d} \times \log_2\left(\frac{|N_t|}{|P_t|}\right) \tag{2}$$

Where,
$V_{t,d}$ = is the feature value for term t in document d.
$C_{t,d}$ = is the number of times term t occurs in document d.
$|N_t|$ = is the number of documents in negatively labeled training set with term t.
$|P_t|$ = is the number of documents in positively labeled training set with term t.

The results produced by the authors show that the approach produces better results than the simple tf or binary weighting scheme.

In this method, besides we have lower-cased all words, we removed all punctuation and we remove the words that occurs only once.

5 Stylometry

According to Zheng et al., (2006) [18] stylometry is a research area of Linguistics that uses statistical techniques to analyze literary style. They are used in several researches, such as: authorship identification of texts as emails [19], general online messages [18], journalistic texts [20], among others.

There are many stylometric features comprising of lexical, syntactic, word-based, structural and content-specific that are presented in various studies about authorship identification [21], [22] and [18]. A brief description of the relative discriminating capabilities of each five different types of stylometric features are given below.

Lexical-based features: are collected in terms of characters. For example, total number of characters, total number of alphabetic characters, total number of capital letters, total number of digit characters and total number of white-space characters are the most relevant metrics. These indicate the preference of an individual for certain special characters or symbols or the preferred choice of using certain units. We do not use lexical features in this work, because we do not consider characters or white-space as significant features.

Word-based features: includes word length distribution, average number of words per sentence, functional words and vocabulary richness. These features indicate the preference for specific words. Table 1 shows an example of word-based features.

In most reviews, the total number of words is greater in negative comments than positive comments, i.e., the total number of words is an important feature to detect the polarity of the review.

Syntactic features: including punctuation and Part of Speech (POS), can capture an author's writing style at the sentence level. The discriminating power of syntactic features is derived from people's different habits of organizing sentences. In this work, we use the TreeTagger [23], a tagger based on probabilistic decision trees obtained from annotated corpora trained by Pablo Gammalo [24], to extract the adjectives from the reviews.

Structural features: in general, represents the way as an author organizes the layout of a piece of writing. For example, average paragraph length, numbers of paragraphs per document, presence of greetings and their position within a document. These features are specifically for emails. We do not use structural features in this study.

Table 1. Word-based features in Portuguese language

Comment	Polarity	# words
O aparelho em si é bom, porém é extremamente lento, trava demais, independente da quantidade de aplicativos instalados ou abertos no aparelho. Até mesmo na agenda ele trava.	Negative	28
Excelente produto.	Positive	2

Content-specific features: are collections of certain key words commonly found in a specific domain and may vary from context even for the same author. In this work, we use the synonyms of the 56 adjectives present in the corpus. The synonyms were obtained by Electronic Thesaurus for Brazilian Portuguese (TEP) 2.0[3] [25]. The TEP 2.0 tool generates a set of synonyms from each adjective, thus the frequency of an adjective or its synonyms is a feature.

[3] `http://www.nilc.icmc.usp.br/tep2/index.htm`

In this work, we extracted 93 stylometric features: 34 word-based, 3 syntactic, and 56 content-specific. We use TF to compute the weight of each stylometric feature. Table 2 shows the stylometric features used. It is noteworthy that we do not use feature selection techniques, selection feature is a topic for future work.

Table 2. Stylometric Features

Feature set	Name
Word-based (1-34)	Total number of words (M)
	Average word length
	Total different words/M
	Hapax legomena on different words
	Hapax legomena *
	Hapax dislegomena *
	Guirad's R measure *
	Herdan's C measure *
	Herdan's V measure *
	Rubet's K measure *
	Maas' A measure *
	Dugast's U measure *
	Lukjanenkov and Nesitoj *
	Honore's R measure *
	Word length frequency distribution/M (20 features)
Syntactic (35-37)	Frequency of punctuations (P) (3 features) "." , ",", ":"
Content-specific (38-93)	Frequency of synonyms of adjectives (56 features)

Note. The definitions of measures with "*" can be found in Tweedie and Baayen (1998) [26].

Note that, most of these features are used for authorship identification of texts, however, the goal of our work is to use and analyze stylometric features in the sentiment classification, i.e., our approach differs from the other because we use a different method for text feature extraction in the sentiment classification and we compare with baseline methods for text feature extraction.

6 Experiment and Analysis

We have conducted experiments on the created corpus to evaluate the effectiveness of the stylometric features. We have gathered 2000 reviews about Smart phones from Buscapé database, being 1000 positive and 1000 negative reviews. We use the accuracy that is common in sentiment analysis to evaluate the effectiveness of the text feature extraction.

In order to show the feasibility of the stylometric features, we compared its use with TF-IDF and Delta TF-IDF in the sentiment classification. We used three different machine learning methods for training and testing: SVM, Naive

Bayes and J48 classifiers to perform the evaluation in the sentiment classification on the created corpus. The former three were chosen as they have both been widely used in sentiment classification. In this work, we used the Weka toolkit[4] to performs these classifications. SVM is, in particular, commonly regarded as the highest performing classifiers in sentiment analysis [27]. In this work, we used a polynomial kernel and all parameters are set to their default value. Naive Bayes and J48 were also included due to their recently revealed effectiveness in sentiment classification [28] and [29].

We evaluate the average of the accuracy based over the standard 10 fold cross-validation. The experiments were conducted according to the following sequence of steps:

1. TF-IDF and Delta TF-IDF
2. Only content-specific
3. Content-specific and word-based
4. Content-specific and syntactic
5. All stylometric features

Firstly, we execute the TF-IDF and Delta TF-IDF methods. Table 3 and 4 show the results for TF-IDF and Delta TF-IDF respectively of each classifier.

Table 3. Result for TF-IDF features

Classifiers	Feature	Accuracy
SVM		0.5625
Naive Bayes	TF-IDF	0.5605
J48		0.5534

Table 4. Result for Delta TF-IDF features

Classifiers	Feature	Accuracy
SVM		0.7262
Naive Bayes	Delta TF-IDF	0.7125
J48		0.7043

The results produced by the Delta TF-IDF were higher than the results produced by TF-IDF. Secondly, we separated the stylometric features into four feature sets which are: only content-specific, content-specific and word-based, content-specific and syntactic, and all stylometric features. The result for these feature sets are presented in Tables 5, 6, 7, and 8.

Table 5 shows the results only of the content-specific features. Using SVM classifier the accuracy is higher than TF-IDF method. We believe that higher accuracy is because we use a set of synonyms of adjectives belonging to the corpus, because adjectives are important features in the sentiment analysis.

[4] `http://www.cs.waikato.ac.nz/ml/weka/`

Table 5. Only content-specific

Classifiers	Feature	Accuracy
SVM		0.7035
Naive Bayes	Content-specific	0.6950
J48		0.6910

Table 6. Content-specific and word-based

Classifiers	Feature	Accuracy
SVM		0.7795
Naive Bayes	Content-specific and word-based	0.72
J48		0.7255

Table 7. Content-specific and syntactic

Classifiers	Feature	Accuracy
SVM		0.8090
Naive Bayes	Content-specific and syntactic	0.7895
J48		0.8005

Table 8. Stylometric features

Classifiers	Feature	Accuracy
SVM		0.8275
Naive Bayes	All stylometric	0.776
J48		0.7850

According to Tables 3 to 8 we can observe that the best results were obtained by the SVM classifier (56,25%), (72,62%), and (82,75%) TF-IDF, Delta TF-IDF, and stylometry respectively.

Analyzing the SVM classifier, the accuracy of the content-specific features is 70,35%, adding the word-based features the accuracy improved to 77,95%. On the other hand, adding the syntactic to content-specific features the accuracy improved to 80,90%, i.e., the use of the syntactic features had better improvement in the classification than the word-based features.

In this work, we use three syntactic features: dot, comma, and colon. These marks are widely used in Portuguese language. For instance, dot mark was used 7171 times in the corpus, comma mark was used 9139 times and colon mark was used 1262. Verifying the created corpus, colon mark is used 73 times in positive comments and 1191 times in negative comments.

We believe these marks are very important features because they are responsible for maintain the coherence of the text when one person is making your opinion. For instance, it is commonly possible to verify commas separating adversative sentences like concessions. Also, colon mark is used to enumerate features about a product. We verify on the corpus that colon mark is frequent in negative comment, and it occurs because the Brazilian customers likes to enumerate negative features to improve their arguments.

Finally, we analyzed all stylometric features in the classification. The accuracy obtained by stylometric features was 82,75% that is higher than the TF-IDF and Delta TF-IDF methods. The results show that stylometry is quite feasible method for sentiment classification, outperforming the accuracy of the baseline methods. We use TF to compute the wight of each stylometric feature.

It is important to say that we have not removed stop words or applied stemming. In a quick observation it is possible to notice positive comments in negative reviews and otherwise. This occurs for many reason already discussed by several works [3] and [1]. We believe that with a manual revision and/or preprocessing on the corpus, we can improve the classification, because the same terms positives or negatives can be used in both positive as in negative reviews, disturbing the final classification.

7 Conclusions and Future Works

In this work, we presented the feasibility study of stylometric features in the sentiment classification. We created a corpus of product reviews from the most traditional on-line service in Portuguese, namely, Buscapé, where users post their pros and cons comments about several products, services and companies.

In order to analyze the stylometric features we use three classifiers (SVM, Naive Bayes and J48) to compare and analyze with the baseline method of text feature extraction in the sentiment classification. We compared the stylometric features with TF-IDF and Delta TF-IDF methods and we measured the performance using the accuracy metric.

We found the better result with the SVM classifier using TF-IDF (56,25%), Delta TF-IDF (72,62%), and stylometry (82,75%) of accuracy. For the SVM classifier we use a polynomial kernel and all parameters are set to their default value.

The experiment carried out showed that only content-specific features accuracy is higher than TF-IDF method and the combination of content-specific and syntactic features present higher accuracy than content-specific and word-based features.

The results show that stylometry is quite feasible method for sentiment classification, outperforming the accuracy of the baseline methods.

With regard to future works, we highlight some efforts that are currently been development: (i) a feature selection technique to improve the classification (ii) a method to identify entities related to opinions; (iii) a study about on fuzzy logic to identifier which opinion is more relevant.

References

1. Liu, B.: Sentiment Analysis and Opinion Mining. In: Synthesis Digital Library of Engineering and Computer Science. Morgan & Claypool (2012)
2. Kotsiantis, S.B.: Supervised machine learning: A review of classification techniques. In: Proceedings of the 2007 Conference on Emerging Artificial Intelligence Applications in Computer Engineering: Real Word AI Systems with Applications in eHealth, HCI, Information Retrieval and Pervasive Technologies, pp. 3–24. IOS Press, Amsterdam (2007)

3. Pang, B., Lee, L.: Opinion mining and sentiment analysis. Found. Trends Inf. Retr. 2, 1–135 (2008)
4. Liu, B.: Web Data Mining: Exploring Hyperlinks, Contents, and Usage Data. Data-centric systems and applications. Springer (2007)
5. Turney, P.D.: Thumbs up or thumbs down?: Semantic orientation applied to unsupervised classification of reviews. In: Proceedings of the 40th Annual Meeting on Association for Computational Linguistics, ACL 2002, pp. 417–424. Association for Computational Linguistics, Stroudsburg (2002)
6. Liu, B.: Sentiment analysis and subjectivity. In: Handbook of Natural Language Processing, 2nd edn., Taylor and Francis Group, Boca (2010)
7. He, R.C., Rasheed, K.: Using machine learning techniques for stylometry. In: Arabnia, H.R., Mun, Y. (eds.) IC-AI, pp. 897–903. CSREA Press (2004)
8. Pang, B., Lee, L., Vaithyanathan, S.: Thumbs up?: Sentiment classification using machine learning techniques. In: Proceedings of the ACL 2002 Conference on Empirical Methods in Natural Language Processing, EMNLP 2002, vol. 10, pp. 79–86. Association for Computational Linguistics, Stroudsburg (2002)
9. Yu, H., Hatzivassiloglou, V.: Towards answering opinion questions: separating facts from opinions and identifying the polarity of opinion sentences. In: Proceedings of the 2003 Conference on Empirical Methods in Natural Language Processing, EMNLP 2003, pp. 129–136. Association for Computational Linguistics, Stroudsburg (2003)
10. Sharma, A., Dey, S.: A document-level sentiment analysis approach using artificial neural network and sentiment lexicons. SIGAPP Appl. Comput. Rev. 12, 67–75 (2012)
11. Sharma, A., Dey, S.: A boosted svm based ensemble classifier for sentiment analysis of online reviews. SIGAPP Appl. Comput. Rev. 13, 43–52 (2013)
12. Njolstad, P., Hoysaeter, L., Wei, W., Gulla, J.: Evaluating feature sets and classifiers for sentiment analysis of financial news. In: 2014 IEEE/WIC/ACM International Joint Conferences on Web Intelligence (WI) and Intelligent Agent Technologies (IAT), vol. 2, pp. 71–78 (2014)
13. He, R.C., Rasheed, K.: Using machine learning techniques for stylometry. In: Proceedings of the International Conference on Artificial Intelligence, IC-AI 2004, Proceedings of the International Conference on Machine Learning; Models, Technologies & Applications, MLMTA 2004, Las Vegas, Nevada, USA, June 21-24, vol. 2, pp. 897–903 (2004)
14. Hartmann, N., Avanço, L., Filho, P.P.B., Duran, M.S., das Graças Volpe Nunes, M., Pardo, T., Aluísio, S.M.: A large corpus of product reviews in portuguese: Tackling out-of-vocabulary words. In: LREC, pp. 3865–3871 (2014)
15. Aghdam, M.H., Ghasem-Aghaee, N., Basiri, M.E.: Text feature selection using ant colony optimization. Expert Syst. Appl. 36, 6843–6853 (2009)
16. Rajaraman, A., Ullman, J.D.: Mining of Massive Datasets. Cambridge University Press, New York (2011)
17. Martineau, J., Finin, T.: Delta tfidf: An improved feature space for sentiment analysis. In: ICWSM (2009)
18. Zheng, R., Li, J., Chen, H., Huang, Z.: A framework for authorship identification of online messages: Writing-style features and classification techniques. J. Am. Soc. Inf. Sci. Technol. 57, 378–393 (2006)
19. Iqbal, F., Khan, L.A., Fung, B.C.M., Debbabi, M.: e-mail authorship verification for forensic investigation. In: Proceedings of the 2010 ACM Symposium on Applied Computing 2010, pp. 1591–1598. ACM, New York (2010)

20. Pavelec, D., Justino, E., Oliveira, L.S.: Author identification using stylometric features. Inteligencia Artificial. Revista Iberoamericana de Inteligencia Artificial 11, 59–65 (2007)
21. Abbasi, A., Chen, H.: Writeprints: A stylometric approach to identity-level identification and similarity detection in cyberspace. ACM Trans. Inf. Syst. 26, 7:1–7:29 (2008)
22. Iqbal, F., Hadjidj, R., Fung, B.C., Debbabi, M.: A novel approach of mining write-prints for authorship attribution in e-mail forensics. Digital Investigation 5(suppl.), S42–S51 (2008), The Proceedings of the Eighth Annual {DFRWS} Conference
23. Schmid, H.: Probabilistic part-of-speech tagging using decision trees (1994)
24. Pablo Gamallo, M.G.: Freeling e treetagger: um estudo comparativo no âmbito do português. Technical report, Universidade de Santiago de Compostela (2013)
25. Maziero, E.G., Pardo, T.A.S., Di Felippo, A., Dias-da Silva, B.C.: A base de dados lexical e a interface web do tep 2.0: Thesaurus eletrnico para o portugus do brasil. In: Companion Proceedings of the XIV Brazilian Symposium on Multimedia and the Web, WebMedia 2008, pp. 390–392. ACM, New York (2008)
26. Tweedie, F.J., Baayen, R.H.: How variable a constant be? measures of lexical richness in perspective. Computers and the Humanities 32, 323–352 (1998)
27. Wang, S., Manning, C.D.: Baselines and bigrams: Simple, good sentiment and topic classification. In: Proceedings of the 50th Annual Meeting of the Association for Computational Linguistics: Short Papers, ACL 2012, vol. 2, pp. 90–94. Association for Computational Linguistics, Stroudsburg (2012)
28. Thelwall, M., Buckley, K., Paltoglou, G., Cai, D., Kappas, A.: Sentiment in short strength detection informal text. J. Am. Soc. Inf. Sci. Technol. 61, 2544–2558 (2010)
29. Castillo, C., Mendoza, M., Poblete, B.: Information credibility on twitter. In: Proceedings of the 20th International Conference on World Wide Web, WWW 2011, pp. 675–684. ACM, New York (2011)

Opinion Mining and Social Network Analysis

Automated Linguistic Personalization of Targeted Marketing Messages Mining User-Generated Text on Social Media

Rishiraj Saha Roy[1], Aishwarya Padmakumar[2], Guna Prasaad Jeganathan[3], and Ponnurangam Kumaraguru[4]

[1] Big Data Intelligence Lab, Adobe Research
rroy@adobe.com
[2] Computer Science and Engineering, IIT Madras, India
aishu@cse.iitm.ac.in
[3] Computer Science and Engineering, IIT Bombay, India
gunaprsd@cse.iitb.ac.in
[4] Precog, IIIT Delhi, India
pk@iiitd.ac.in

Abstract. Personalizing marketing messages for specific audience segments is vital for increasing user engagement with advertisements, but it becomes very resource-intensive when the marketer has to deal with multiple segments, products or campaigns. In this research, we take the first steps towards automating message personalization by algorithmically inserting adjectives and adverbs that have been found to evoke positive sentiment in specific audience segments, into basic versions of ad messages. First, we build language models representative of linguistic styles from user-generated textual content on social media for each segment. Next, we mine product-specific adjectives and adverbs from content associated with positive sentiment. Finally, we insert extracted words into the basic version using the language models to enrich the message for each target segment, after statistically checking in-context readability. Decreased cross-entropy values from the basic to the transformed messages show that we are able to approach the linguistic style of the target segments. Crowdsourced experiments verify that our personalized messages are almost indistinguishable from similar human compositions. Social network data processed for this research has been made publicly available for community use.

1 Introduction

Personalization is one of the key aspects of success in the present marketing landscape. Alongside aspects like the product advertised, offer presented and the ad layout, the linguistic style of the marketing message plays an important role in the success of the advertising campaign [1–4]. People from different demographics talk differently [5] and we hypothesize that communicating to specific audience segments in their own linguistic styles is expected to increase engagement with advertisements. This hypothesis assumes an even greater importance

A. Gelbukh (Ed.): CICLing 2015, Part II, LNCS 9042, pp. 203–224, 2015.
DOI: 10.1007/978-3-319-18117-2_16

in targeted marketing like email or social campaigns, where different versions of advertisements are communicated to different groups of people. In such targeted campaigns, the marketer has to produce multiple versions of the same ad such that it appeals to each audience segment. However, this requires additional resources like time, people and money for the hiring marketer, which may often be unavailable. Our proposed technology helps an individual copywriter to automatically create several variations of the same message, each containing words appealing to a specific target segment.

Approach. Adjectives and adverbs[1] make advertisement messages sound more urgent and exciting. However, different adjectives and adverbs are expected to evoke positive sentiment in different demographic segments. In this research, we take the first steps in automated message personalization by algorithmically inserting segment-wise preferred adjectives and adverbs into basic versions of marketing text (lacking or with minimal use of modifiers, usually the first versions created by ad copywriters). We use country and occupation as representative features defining linguistic style, and collect significant amounts of Tweets generated by 12 such segments. Next, we build language models (characterizing linguistic style) from each of these segment-specific corpora. We choose a product and collect Tweets that talk about the product. We extract Tweets with positive sentiment from this set and derive modifiers from the positive Tweets. Since it is difficult to have copywriters create fresh ad messages for us, we collect a set of public advertisements about the product, and manually remove modifiers from these ad messages to create ad skeletons (basic message versions). Subsequently, we use the set of product-specific modifiers and the language models to personalize these ad skeletons for each audience segment by suitably inserting modifiers for candidate keywords at appropriate locations. Finally, we evaluate our message transformation algorithms using cross-entropy and also ensure syntactic and semantic coherence with crowdsourced annotations.

Contributions. The primary contribution of this research is to take the first steps towards automating linguistic personalization of natural language text, with evidence of styles or word usage patterns mined from user-generated textual content. We demonstrate the effectiveness of our novel approach through a practical application in the marketing scenario, where we automatically enrich ad messages specific to several demographic segments. So far we have not come across previous research that algorithmically transforms a body of text to a form guided by a target linguistic model without altering the intent of the message. To facilitate this line of study, we are making the datasets (containing thousands of Tweets from 12 demographic segments) used in this research publicly available[2].

Organization. The rest of this paper is organized as follows. In the next section, we briefly survey literature relevant for this research. In Sec. 3, we describe our

---

[1] We refer to adjectives and adverbs as *keyword modifiers* in this work, and use *keywords* for nouns and verbs.

[2] `http://goo.gl/NRTLRA`, Accessed 31 January 2015.

message personalization algorithm in detail. We present details of our dataset in Sec. 4 and experiments on cross-entropy in Sec. 5. Evaluation of coherence for transformed messages using crowdsourcing is described in Sec. 6. We present a discussion in Sec. 7 and make concluding remarks with potential avenues for future research in Sec. 8.

2 Related Work

In this section, we present a brief survey of past literature that is relevant to our current research.

2.1 Document and Text Transformation

We first outline some of the works that have dealt with automatic transformation of document contents. One line of research includes changing contents at the *structural level*, for example, transforming linear text documents into hypertext[6], or transforming XML documents into OWL ontologies[7]. Such works focus on making use of the formatting of the input document to determine relations between its components. In contrast, we make an attempt to modify the actual text. In *text normalization* [8], words written in non-standard forms in communications (such as SMS) are converted to standard dictionary forms (for example, automatic spelling correction). Text normalization differs from our goal in that it involves word-level transformations for existing words in a text, and does not involve insertion of new words. Automatic *text summarization* [9] examines the textual content to determine important sentences but still uses original sentences from the text to compose a summary. More generally, the aim of *text adaptation* is to *enrich* a given text for "easier" use. Text adaptation [10] makes use of text summarization and other tools to create marginal notes like in-context meanings to enrich a piece of text, while *text simplification* [11, 12] aims at automatically simplifying a body of text for easier comprehension. They also identify low frequency words and make an attempt to obtain in-context meanings. Our enrichment is constrained so as to match the linguistic style of a target audience segment by inserting new words, and it is not for making the text easier or quicker to understand. Identifying language features for predicting *readability* or *reading levels* for text documents is another area allied to our research [13]. However, current work has not yet addressed the issue of automatically transforming the reading level of a given text based on relevant features. Template-based personalization is what is common in the industry today, where a copywriter has to manually populate message templates with different words for each segment. To a large extent, templates restrict the style and content of an ad message, and a method that enriches basic messages with free style is expected to be very helpful to copywriters.

2.2 Linguistic Style and Word Usage

Linguistic style involves *word usage patterns*, tendency of using different forms of part-of-speech (POS) like adjectives, levels of formalism, politeness, and sentence lengths [14]. Prior research has revolved around the *characterization* of these features. Tan et al. [15] study the effect of word usage in message propagation on Twitter and try to predict which of a pair of messages will be retweeted more. Interestingly, they find that making one's language align with both the community norms and with one's prior messages is useful in getting better propagation. Bryden et al. [16] find that social communities can be characterized by their most significantly used words. Consequently, they report that the words used by a specific user can be used to predict his/her community. Danescu-Niculescu-Mizil et al. [17] use word usage statistics to understand user lifecycles in online communities. They show that changes in word occurrence statistics can be used to model linguistic change and predict how long a user is going to be active in an online community. Hu et al. [18] measure features like word frequency, proportion of content words, personal pronouns and intensifiers, and try to characterize *formalism* in linguistic styles of several mediums like Tweets, SMS, chat, email, magazines, blogs and news. In this research, we focus only on word usage patterns, as the first step towards automatic generation of stylistic variations of the same content.

3 Method

In this section, we discuss the various steps in our algorithm for automatic message personalization.

3.1 Mining Dependencies from Corpora

As the first step, we identify segments in our target audience for whom we want to personalize our ad messages. Next, we extract textual content from the Web and social media that has been generated by members of each target segment. Once we have collected a significant amount of text for each segment (i.e., created a *segment-specific corpus*), we proceed with the following processing steps. First, we run a POS tagger [19] on each corpus to associate each word with a part-of-speech (like nouns, verbs and adjectives). Next, we perform a dependency parsing [20] of the POS-tagged text to identify long-range or non-adjacent dependencies or associations within the text (in addition to adjacent ones). For example, dependency parsing helps us extract noun-adjective associations like the following: the adjective `fast` is associated with the noun `software` in the sentence fragment `a fast and dependable software`, even though the pair does not appear adjacent to each other. After this, we build language models (LMs) from each corpus as described in the next subsection. Throughout this research, we first apply *lemmatization* on the words so that different forms of the same word are considered equivalent during relation extraction. Lemmatization normalizes words to their base forms or *lemmas* – for example, `radius` and `radii` are lemmatized to `radius` (singular/plural), and `bring`, `bringing`, `brought` and `brings` are all converted to `bring` (different verb forms).

3.2 Defining Language Models

A statistical language model is a probability distribution over all strings of a language [21]. In this research, we primarily use the 1-gram and 2-gram LMs, which measure the probabilities of occurrence of unigrams (single words) and bigrams (pairs of words). So we extract distinct unigrams and bigrams from each corpus, compute their occurrence probabilities and build the LM for the corpus. We elaborate on the computation of the probabilities in Sec. 5. Additionally, we store the probabilities of all distinct adjective-noun pairs (like `cheap-software`) and verb-adverb pairs (like `running-quickly`) in our LMs. The LM for each segment is used as a source to search for the most appropriate enrichment of words from the basic message.

3.3 Mining Positive Modifiers

Now that LMs have been built for each segment, we select the specific product that we wish to create ad messages for. Without loss of generality, our method can be extended to a set of products as well. We then extract textual content from the Web and social media content that concerns the selected product. This textual content is analyzed by a sentiment analysis tool, and we retain only the sentences that have positive sentiments associated with them. This step is very important as we will use words from this content to personalize our ad messages. We do not want our system to use words associated with negative sentiment for message transformation. Next, we run the POS-tagger on these positive sentiment sentences. Using the POS-tagged output, we extract adjectives and adverbs from these sentences. These adjectives and adverbs that are known to evoke positive sentiment in users, henceforth referred to as *positive modifiers*, will be used for the automatic transformation of ads. Our personalization involves insertion of adjectives for nouns and adverbs for verbs.

3.4 Identifying Transformation Points

We now have the resources necessary for performing an automatic message transformation. The copywriter now selects an audience segment and creates a basic version of an ad message, lacking or with minimal use of modifiers. We refer to this as the *ad skeleton*, which we wish to enrich. We run a POS tagger on the skeleton and identify the nouns and verbs in the message. Next, we compute term weights for nouns and verbs using the concept of inverse document frequency (IDF) as shown below:

$$IDF(\text{keyword}) = log_{10} \frac{\text{Product-specific messages}}{\text{Product-specific messages with keyword}} \tag{1}$$

In general, in text mining applications, the concept of IDF is used in combination with term frequencies (TF) to compute term weights. In our case however, since ad messages are short, each keyword generally appears only once in an ad. Hence, using IDF suffices. The intuition behind term weighting is to suggest enrichment for only those keywords that are discriminative in a context and not include words of daily usage like `have` and `been`. We choose term weight thresholds α_N and α_V (for nouns and verbs respectively) manually based on our ad corpus. Only the nouns and verbs that exceed α_N and α_V respectively are considered to be transformation points in the message.

3.5 Inserting Adjectives for Nouns

For each noun n in the ad message that has term weight more than the threshold α_N, we fetch the set of adjectives $ADJ(n)$ that appear in the content with positive sentiment and have a non-zero probability of co-occurrence with the corresponding noun n in the target LM. Adjectives in $ADJ(n)$ need to have appeared a minimum number of times, defined by a threshold β, in the segment-specific corpus to be considered for insertion (candidates with frequency $< \beta$ are removed). Next, we prune this list by retaining only those adjectives *adj* that have a pointwise mutual information (PMI) greater than a threshold γ_N on the right side with the noun n and the left side with the preceding word w (possibly null, in which case this condition is ignored) in the ad. PMI is a word association measure computed for a pair of words or a bigram $(a,\ b)$ (ordered in our case) that takes a high value when a and b occur more frequently than expected by random chance, and is defined as follows:

$$PMI(a\ b) = log_2 \frac{p(a,b)}{p(a)p(b)} \tag{2}$$

where $p(a)$ and $p(b)$ refer to the occurrence probabilities of a and b, and the joint probability $p(a,b)$ is given $p(a)p(b|a)$. Thus, $PMI(a,\ b) = log_2 \frac{p(b|a)}{p(b)}$. Hence, if the word sequence $< a\ b >$ has a high PMI, it is an indication that the sequence is syntactically coherent. Thus, choosing an adjective *adj* such that has $PMI(w\ adj) > \gamma_N$ (left bigram) and $PMI(adj\ n) > \gamma_N$ (right bigram) ensures that inserting *adj* before n will ensure a readable sequence of three words. For example, if the original text had `with systems`, and we identify `complex` as a candidate adjective for `systems`, we would expect the PMI scores of `with complex` and `complex systems` to be higher than γ_N, which ensures that the adjective `complex` fits in this context and `with complex systems` produces locally readable text. We now have a list of adjectives that satisfies the PMI constraints. We sort this list by $PMI(adj,\ n)$ and insert the highest ranking *adj* to the left of n. We now provide a formal description of the adjective insertion algorithm in Algorithm 1, that takes as input a *sentence*, the target *noun* in the sentence for which we wish to insert an adjective, the target language model LM, the list of positive adjectives *adj_list* and γ_N. Before that, we provide descriptions for functions used in the algorithms in this text in Tables 1, 2 and 3.

3.6 Inserting Adverbs for Verbs

The process of inserting adverbs for verbs is in general similar to the one for adjective insertion, but with some additional constraints imposed by verb-adverb

Table 1. General functions used in our algorithms

CREATE-EMPTY-LIST()	Returns an empty list
GET-SENTENCES (*text*)	Breaks the input *text* into sentences and converts each sentence into a sentence object
GET-TEXT (*sentences*)	Converts the input list of *sentence* objects to text

Table 2. Functions of a *sentence* object

GET-PREV-WORD (*word*)	Fetches the word that occurs before input *word*, in the sentence
GET-NEXT-WORD (*word*)	Fetches the word that occurs after input *word*, in the sentence
INSERT-BEFORE (*word_to_insert*, *word*)	Inserts input *word_to_insert* before input *word*, in the sentence
INSERT-AFTER (*word_to_insert*, *word*)	Inserts input *word_to_insert* after input *word*, in the sentence
TOKENIZE()	Returns a list of tokens
GET-NOUN-DEPENDENCIES()	Returns a list of *dependency* objects where *dependency.secondary_word* describes / modifies *dependency.primary_word* to form part of a noun phrase
PARSE()	Returns a list of *tagged_token* objects where *tagged_token.text* is the text of the token and *tagged_token.pos* is the POS. *tagged_token.pos* has a value of NOUN for nouns, VERB for verbs and PHRASE for phrases.

ordering principles. For each verb v in the ad message that has term weight $> \alpha_V$, we fetch the set of adverbs $ADV(v)$ that appear in the positive content and have a non-zero probability of co-occurrence with v in the target LM. In addition to the filtering on verbs imposed by γ_V, we remove *modal* and *auxiliary verbs* like `have`, `are`, `will` and `shall` that only add functional or grammatical meaning to the clauses in which they appear, and focus on main verbs only, that convey the main actions in a sentence. The candidate adverbs in $ADV(v)$ need to have appeared a minimum number of times β in the segment-specific corpus to be considered for insertion (candidates with frequency $< \beta$ are removed). Next, we prune $ADV(v)$ by retaining only those adverbs that either have $PMI(adv\ v) > \gamma_V$ or have $PMI(v\ adv) > \gamma_V$. The adverbs in $ADV(v)$ are ranked in descending order of their PMI scores (whichever of the two previous PMI is higher is considered for ranking) and the highest ranking adverb *adv* is selected for insertion. If $PMI(adv,\ v) > PMI(v,\ adv)$, and there is no word in the sentence that precedes v, then *adv* is inserted before v. If there is a word w preceding v, then *adv* is inserted only if $PMI(w,\ adv) > \gamma_V$. If $PMI(adv,\ v) < PMI(v,\ adv)$, and there is no word in the sentence that succeeds v, then *adv* is inserted after v. If there is some w succeeding the v, then *adv* is inserted only if $PMI(adv,\ w) > \gamma_V$. If the two PMIs are equal, then an

Table 3. Functions of an *LM* object

GET-TERM-WEIGHT (*word*)	Returns the term weight of the input *word*
GET-NOUN-ADJ-PROB (*noun*, *adjective*)	Returns the probability of a sentence containing the input *adjective* describing the input *noun*
GET-VERB-ADV-PROB (*verb*, *adverb*)	Returns the probability of a sentence containing the input *adverb* modifying the input *verb*
GET-PMI (*first_word*, *second_word*)	Calculates the PMI of the two input words, in that order
GET-MAX-PMI-BEFORE (*word_list*, *word_after*)	Returns word in input *word_list* which has maximum PMI(*word*, *word_after*)
GET-MAX-PMI-AFTER (*word_list*, *word_before*)	Returns word in input *word_list* which has maximum PMI(*word_before*, *word*)

Algorithm 1. Insert adjective for a noun in a sentence

```
function INSERT-ADJ-FOR-NOUN(sentence, noun, LM, adj_list, γ_N)
    prev_word ← sentence.GET-PREV-WORD(noun)
    adjs ← CREATE-EMPTY-LIST()
    for all adj ∈ adj_list do
        if LM.GET-NOUN-ADJ-PROB(noun, adj) > 0 then
            if LM.GET-PMI(adj, noun) > γ_N then
                if LM.GET-PMI(prev_word, adj) > γ_N then
                    adjs.INSERT(adj)
                end if
            end if
        end if
    end for
    best_adj = LM.GET-MAX-PMI-BEFORE(adjs, noun)
    sentence.INSERT-BEFORE(best_adj, noun)
return sentence
end function
```

arbitrary decision is made with respect to insertion to the left or the right side of v. If the highest-ranking adverb *adv* is found unsuitable for insertion with respect to any of the constraints mentioned earlier, then the next ranked adverb is considered in its place. This process is repeated until an insertion is made or the set $ADV(v)$ is exhausted. We now provide a formal description of the adverb insertion algorithm in Algorithm 2, that takes as input a *sentence*, the target *verb* in the sentence for which we wish to insert an adverb, the target LM, the list of positive adverbs *adv_list* and γ_V.

3.7 Enhancement with Noun Phrase Chunking

A noun phrase is a phrase which has a noun (or an indefinite pronoun) as its head word. Nouns embedded inside noun phrases constitute a special case when the usual steps for inserting adjectives for nouns produce unusual results. For example, for the noun phrase `license management tools`, we may get usual adjective insertions for `license`, `management`, and `tools`, resulting in strange possibilities like `general license easy management handy tools`. To avoid such situations, we perform noun phrase chunking [22] on the original text to detect noun phrases in the ad message. We do not insert adjectives within the

Algorithm 2. Insert adverb for a verb in a sentence

```
function INSERT-ADV-FOR-VERB(sentence, verb, LM, adv_list, γ_V)
    prev_word ← sentence.GET-PREV-WORD(noun)
    next_word ← sentence.GET-NEXT-WORD(noun)
    advs ← CREATE-EMPTY-LIST()
    for all adv ∈ adv_list do
        if LM.GET-VERB-ADV-PROB(verb, adv) > 0 then
            if LM.GET-PMI(verb, adv) > γ_V then
                if LM.GET-PMI(adv, next_word) > γ_V then
                    advs.INSERT(adv)
                end if
            else if LM.GET-PMI(adv, verb) > γ_V then
                if LM.GET-PMI(prev_word, adv) > γ_V then
                    advs.INSERT(adv)
                end if
            end if
        end if
    end for
    adv_b = LM.GET-MAX-PMI-BEFORE(advs, verb)
    adv_a = LM.GET-MAX-PMI-AFTER(advs, verb)
    pmi_b ← LM.GET-PMI(adv_b, verb)
    pmi_a ← LM.GET-PMI(verb, adv_a)
    if pmi_b > pmi_a then
        sentence.INSERT-BEFORE(adv_b, verb)
    else
        sentence.INSERT-AFTER(adv_a, verb)
    end if
return sentence
end function
```

noun phrase. Next, it is apparent from the example that inserting an adjective for the first word in a chunk is not always the best choice, where we would require an adjective for `tools` and not `license` for the phrase to make sense. The *chunk head* is the word in a chunk on which other words depend, and we wish to insert an adjective for the chunk head noun. Dependency parsing helps us to identify the chunk head, using the *dependency tree* of the sentence. We now follow the process of adjective insertion for the noun phrase head and insert it before the first word of the chunk. For checking the PMI for compatibility in context, we use the word immediately preceding the chunk. Also, we do not insert adjectives which are already part of the noun phrase. We now provide a formal description of the adjective insertion algorithm for a noun inside a noun phrase in Algorithm 3, that takes as input a *sentence*, the target noun *phrase* in the sentence for which we wish to insert an adjective for an embedded noun, the target language model LM, the list of positive adjectives adj_list and γ_N.

3.8 Message Personalization

Our message personalization technique incorporates the various steps discussed above. The algorithm takes as input an ad message *text*, the target language model LM, the list of positive adjectives and adverbs adj_list and adv_list respectively, and the term weight thresholds α_N and α_V, and produces the enriched segment-specific message as output. Finally, we present a formal version of our message personzalization technique in Algorithm 4, that incorporates the various steps discussed above. The algorithm takes as input an ad message *text*, the

Algorithm 3. Insert adjective for a noun phrase in a sentence

```
function INSERT-ADJ-FOR-PHRASE(sentence, phrase, LM, adj_list, γ_N)
    words_in_phrase ← phrase.TOKENIZE()
    first_word ← words_in_phrase[0]
    prev_word ← sentence.GET-PREV-WORD(first_word)
    noun_deps ← sentence.GET-NOUN-DEPS()
    head_candidates ← CREATE-EMPTY-LIST()
    for all noun_dep ∈ noun_deps do
        head_candidates.INSERT(noun_dep.primary_word)
    end for
    if head_candidates.length = 1 then
        head ← head_candidates[0]
        adjs ← CREATE-EMPTY-LIST()
        for all adj ∈ adj_list do
            if LM.GET-NOUN-ADJ-PROB(head, adj) > 0 then
                if LM.GET-PMI(adj, first_word) > γ_N then
                    if LM.GET-PMI(prev_word, adj) > γ_N then
                        adjs.INSERT(adj)
                    end if
                end if
            end if
        end for
        best_adj = LM.GET-MAX-PMI-BEFORE(adjs, head)
        sentence.INSERT-BEFORE(best_adj, first_word)
    end if
return sentence
end function
```

target language model LM, the list of positive adjectives and adverbs *adj_list* and *adv_list* respectively, and the term weight thresholds α_N and α_V.

4 Dataset

In this section, we describe the various datasets that were used in this research. The entire dataset is available for public use at `http://goo.gl/NRTLRA`.

4.1 Segment-Specific Corpus

Twitter has evidence of diverse linguistic styles and is not restricted to any particular type of communication [18]. We chose location (country) and occupation to be the demographic features on which we define customer segments, which we hypothesized to have effect on a person's word usage styles. (access to other possible factors like age or gender mostly restricted). We also considered analyzing gender and age as factors determining linguistic style, but we were unable to collect significant amount of textual data constrained by such personal information. To be specific, we consider three countries: USA (US), United Kingdom (UK), Australia (AU), and four occupations: students, designers, developers and managers, resulting in twelve distinct segments, producing $3 \times 4 = 12$ demographic segments in total.

For each city in AU, UK and US (as obtained through Wikipedia), a manual Google search was done for people who reside in that city, restricting the search results to Google+ pages (using Google Custom Search) and the number of result pages to about 15 to 25. These result pages were downloaded and stripped

Algorithm 4. Personalize a message

```
function PERSONALIZE-TEXT(text, LM, adj_list, adv_list, α_N, α_V)
    sentences ← GET-SENTENCES(text)
    for all sentence ∈ sentences do
        tagged_tokens = sentence.PARSE()
        for all token ∈ tagged_tokens do
            if token.pos = NOUN then
                noun ← token.text
                if LM.GET-TERM-WEIGHT(noun) > α_N then
                    sentence = INSERT-ADJ-FOR-NOUN(sentence, noun, LM, adj_list)
                end if
            else if token.pos = VERB then
                verb ← token.text
                if LM.GET-TERM-WEIGHT(verb) > α_V then
                    sentence = INSERT-ADV-FOR-VERB(sentence, verb, LM, adv_list)
                end if
            else if token.pos = PHRASE then
                phrase ← token.text
                sentence = INSERT-ADJ-FOR-PHRASE(sentence, phrase, LM, adj_list)
            end if
        end for
    end for
    personalized_text ← GET-TEXT(sentences)
return personalized_text
end function
```

to obtain Google+ identifiers inside them. Then, using the Google+ API, the public profiles of these identifiers were obtained. Some of these public profiles contained information about their occupation and Twitter handles, giving us a set of Twitter handles for each demographic segment defined by location and occupation. We found that data for some segments were sparse. In an attempt to increase the number of users in these segments, we listened to the public stream of Twitter for three days from these three countries (using bounding boxes with latitudes and longitudes) and collected new Twitter handles. Using the Twitter API, we searched their Twitter descriptions for the selected four occupations. This significantly increased the number of Twitter handles for the twelve segments. Finally, we mined Tweets for the mined Twitter handles using the Twitter streaming API. For all accesses to the Twitter API, we directly used the library Tweepy[3], which internally uses the Twitter stream API and the Twitter search API. Table 4 reports the details of corpora used for each segment (where k = thousand and M = million). The minimum number of sentences for a segment varied between $183k$ (AU-developer) and $777k$ (UK-manager). Thus, we had a reasonable amount of text for each segment. Number of sentences per Tweet varied between 1.22 (UK-student) and 1.43 (AU-manager). Number of words per sentence was observed to range from 7.44 (US-designer) to 8.29 (UK-student). Thus, even at aggregate levels, we observed noticeable distinctions in linguistic preferences.

[3] `http://www.tweepy.org/`, Accessed 31 January 2015.

Table 4. Data collected for each audience segment

Segment	#Tweets	#Sentences	#Words
AU-designer	180k	247k	2.0M
AU-developer	140k	183k	1.5M
AU-manager	240k	343k	2.7M
AU-student	400k	530k	4.3M
UK-designer	520k	678k	5.2M
UK-developer	480k	632k	5.2M
UK-manager	580k	777k	6.2M
UK-student	500k	610k	5.1M
US-designer	310k	414k	3.1M
US-developer	160k	209k	1.7M
US-manager	260k	356k	2.8M
US-student	500k	648k	5.1M

4.2 Product-Specific Corpus

Since choice of appropriate words vary from product to product, we have to select a product to run our experiments on. Out of different types of products, software is a category where the text of the marketing message often plays a very important role to initiate the first level of user engagement, unlike smartphones and other gadgets where the image often plays a deciding role. We chose the popular graphics suite *Adobe Creative Cloud* as our product. We collected a total of $1,621$ Tweets about Creative Cloud through the Twitter API. We performed a sentiment analysis on these $1,621$ Tweets using the `pattern.en` Python library which scores each message (Tweet) on a scale of -1 to $+1$. We retained only the Tweets with positive sentiment (sentiment score > 0), which were found to be $1,364$ in number. We performed a POS tagging on these Tweets using the Stanford NLP POS tagger and extracted modifiers from them. We were able to obtain 370 adjectives and 192 adverbs for Creative Cloud as a result of this extraction from positive Tweets. These modifiers will be used for the linguistic personalization of our ad messages.

4.3 Advertisements

We collected 60 advertisement fragments for Adobe Creative Cloud. 56 of these were obtained from different Adobe websites, and 4 from marketing emails received by the authors. These 60 ad messages contained 765 nouns and 350 verbs, which are our potential transformation points. We performed POS tagging on these ad messages and manually removed adjectives and adverbs from these messages to convert them to ad skeletons, which are basic versions of the messages that can be quickly created by copywriters. We performed our personalization experiments on these ad skeletons. The manual removal of modifiers ensured that the syntactic constraints of the messages were not violated, i.e. in some cases the modifiers were retained if deleting them made the sentence ungrammatical.

Table 5. Sample message transformations (Adjectives: bold, *adverbs*: bold + italics)

Ad Skeleton	**Transformation 1**	**Transformation 2**
Even as the landscape continues to change, MAX will remain the place to learn about generating graphics content for devices, and discovering about tools, development approaches, and formats. Learn about license management tools and all the things you wanted to know!	Even as the landscape continues to change ***dramatically***, MAX will ***always*** remain the **first** place to learn about generating **original** graphics content for **mobile** devices, and discovering about tools, **unique** development approaches, and formats. Learn about **valuable** license management tools and all the **greatest** things you wanted to know! (Segment: US-student)	Even as the landscape continues to change ***daily***, MAX will remain the **first** place to ***quickly*** learn about generating **adaptive** graphics content for devices, and discovering about tools, **critical** development approaches, and formats. Learn about **handy** license management tools and all the **best** things you wanted to know ***right***! (Segment: AU-designer)

5 Experimental Results

We perform our personalization experiments using the datasets described earlier. Concretely, we personalize the 60 Adobe Creative Cloud ad messages for each of the 12 demographic segments defined by location and occupation. We used Stanford NLP resources[4] for POS tagging [19] and dependency parsing [20], the NLTK python library[5] [23] for tokenization, the `pattern.en` Python library for lemmatization and sentiment analysis[6] [24], and the TextBlob Python library[7] for extracting noun phrases. We choose the following values for our thresholds: $\alpha_N = 0, \alpha_V = 6, \beta = 10, \gamma_N = \gamma_V = 0$. These thresholds must be tuned by a copywriter empirically for a given context. To give readers a feel of the segment-specific personalizations that our algorithm performs, we present two representative examples in Table 5. As we can see, the set of words that are inserted vary noticeably from segment to segment. While the set of adjectives inserted for US-designer is {`first, mobile, unique, valuable, greatest`}, the set for AU-designer is {`first, adaptive, critical, handy, best`}. Also, a decision is always made in context, and it is not necessary that corresponding locations in ads for different segments will always contain adjectives or adverbs. The set of adverbs is also observed to vary – being {`dramatically, always`} for US-student and {`daily, quickly, right`} for AU-designer. The 60 ad messages contained 4,264 words in total (about 71 words per ad). An average of 266 adjectives were inserted for these messages for each segment (varying between 312 (UK-designer) and 234 (UK-student)). Corresponding counts for adverbs was 91 (varying between 123 (UK-designer) and 63 (AU-student)).

[4] `http://goo.gl/dKF1ch`, Accessed 31 January 2015.
[5] `http://www.nltk.org/`, Accessed 31 January 2015.
[6] `http://goo.gl/bgiyxq`, Accessed 31 January 2015.
[7] `http://goo.gl/J0OE5P`, Accessed 31 January 2015.

Table 6. Summary of results for experiments with cross-entropy

Model	Unigram (Sentence)		Unigram (Frequency)		Bigram (Sentence)		Bigram (Frequency)	
Segment	**Ads with drop**	**% Drop in CE**	**Ads with drop**	**% Drop in CE**	**Ads with drop**	**% Drop in CE**	**Ads with drop**	**% Drop in CE**
AU-designer	60/60	11.65	60/60	10.82	60/60	10.17	60/60	46.08
AU-developer	60/60	11.58	60/60	10.74	60/60	10.09	60/60	46.04
AU-manager	60/60	11.10	60/60	10.34	60/60	9.64	60/60	45.81
AU-student	60/60	11.02	60/60	10.28	60/60	9.68	60/60	45.76
UK-designer	60/60	13.71	60/60	12.89	60/60	12.00	60/60	47.34
UK-developer	60/60	12.65	60/60	11.77	60/60	11.10	60/60	46.66
UK-manager	60/60	12.39	60/60	11.64	60/60	10.90	60/60	46.59
UK-student	60/60	10.63	60/60	9.86	60/60	9.27	60/60	45.51
US-designer	60/60	11.98	60/60	11.21	60/60	10.57	60/60	46.32
US-developer	60/60	12.27	60/60	11.43	60/60	10.71	60/60	46.44
US-manager	60/60	12.18	60/60	11.27	60/60	10.58	60/60	46.36
US-student	60/60	11.05	60/60	10.23	60/60	9.61	60/60	45.73

For an intrinsic evaluation of our message transformation algorithm, we need to find out that whether the changes we make to the basic message take it closer to the target LM. *Cross-entropy* (CE) is an information-theoretic measure that computes the closeness between LMs (or equivalent probability distributions) by estimating the amount of extra information needed to predict a probability distribution given a reference probability distribution. The CE between two LMs p and q is defined as:

$$CE(p, q) = -\sum_{i=1}^{n} -p_i log_2 q_i \tag{3}$$

where p_i and q_i refer to corresponding points in the two probability distributions (LMs). In our experiments, we treat the LM derived from the ad message as p and the LM for the target audience segment as q. Cross entropy is computed for both the original and transformed ad messages with respect to the target LM. Decreased CE values from original to transformed messages show that we are able to approach the target LM. We perform experiments using both unigram and bigram LMs, where points in the probability distributions refer to unigram and bigram probabilities, respectively. For computing probabilities, we considered both sentence-level and frequency-level probabilities. In sentence-level probabilities, we define the probability of an n-gram $\mathcal{N}$ as:

$$P_s(\mathcal{N}) = \frac{No.\ of\ sentences\ in\ corpus\ with\ \mathcal{N}}{No.\ of\ sentences\ in\ corpus} \tag{4}$$

while the frequency-level probabilities are computed as shown below:

$$P_f(\mathcal{N}) = \frac{Frequency\ of\ \mathcal{N}\ in\ corpus}{Total\ no.\ of\ n-grams\ in\ corpus} \tag{5}$$

where $n = 1$ or 2 according as the probability distribution followed is unigram or bigram. Since a word usually appears once in a sentence (except some function

Table 7. Ads (/60) with drop in cross-entropy by only adjective (adverb) insertions

Segment	Unigram (Sentence)	Unigram (Frequency)	Bigram (Sentence)	Bigram (Frequency)
AU-designer	24(60)	55(60)	37(59)	60(60)
AU-developer	18(58)	54(58)	34(58)	60(60)
AU-manager	24(58)	53(59)	35(58)	60(60)
AU-student	22(58)	50(58)	35(59)	60(60)
UK-designer	26(60)	55(60)	39(60)	60(60)
UK-developer	23(60)	55(60)	40(60)	60(60)
UK-manager	23(60)	57(59)	37(60)	60(60)
UK-student	23(58)	55(59)	32(59)	60(60)
US-designer	23(60)	54(60)	34(60)	60(60)
US-developer	24(59)	54(60)	37(60)	60(60)
US-manager	23(59)	54(60)	32(60)	60(60)
US-student	18(58)	55(58)	32(60)	60(60)
Mean CE drop %	4.98(4.55)	6.76(3.56)	4.50(3.74)	43.22(41.67)

words), the numerators are generally the same in both cases, and the normalization is different. Sentence-level probabilities, by themselves, do not add to one for a particular LM (1-gram or 2-gram) and hence need to be *appropriately normalized* before computing entropies. In making decisions for inserting adjectives and adverbs, we have used sentence-level probabilities because normalizing by the number of sentences makes probabilities of unigrams and bigrams comparable (in the same probability space). This is essential for making PMI computations or comparisons meaningful. The event space for the CE computations is taken to be union of the space (as defined by unigrams or bigrams) of the segment-specific corpus and the ad message. Add-one Laplace smoothing [21] is used to smooth the zero probability (unseen) points in both distributions p and q. We present the results of the CE computations below in Table 6.

From the CE results in Table 6, we observe that our method of personalization is successful in making the transformed ads approach the target LMs in 100% of the cases, i.e. for all 60 ad messages for all segments. This shows that our principles are working well. We also report the average percentage drops in CE values for each segment. A higher magnitude of the average drop represents a bigger jump towards the target model. For most of the models, we observe decreases in a similar range, which is $9-12\%$, and this is consistent across the segments. For the bigram LM based on frequencies, we observe much larger drops in CE, being in the range of $45-47\%$, again without much variance across segments. These results show the robustness of our method with respect to segments, and in turn, with regard to the size of corpus used. The magnitudes of the drops in CE were found to be much higher for the frequency-based bigram LM than the other three models.

Exclusive Effects of Adjectives and Adverbs. To observe the difference in effects of insertions of adjectives and adverbs on our personalization, we transformed messages with only one of the steps being allowed. Table 7 shows results for the adjective-only and adverb-only experiments. We report the number of ads (out of 60) showing decrease in CE. Values within parentheses show corresponding numbers for adverbs. We observed that even though the total number of adverbs inserted is low (91 on an average for each segment), they have a more pronounced effect on approaching the target LMs. This is reflected in the adjective-only experiments, where the number of "successful" transformations (showing decrease in CE) is noticeably lower than 60 for some of the models (e.g., between 18 and 26 for the normalized sentence LM for unigrams). The corresponding numbers are higher for adverb-only experiments, mostly between 58 and 60 for all metrics. Only mean values for the magnitudes of the drops in CE are reported due to shortage of space. The standard deviations of these drops were found to be quite low.

6 Evaluation of Coherence

While an intrinsic evaluation using LMs and cross-entropy ensures consistency with respect to word usage, human judgment is the only check for syntactic and semantic coherence of an algorithmically transformed ad message. In general, an experimental setup presenting annotators with a standalone ad message and asking whether it could be completely generated by a human writer is questionable. This is because the annotator cannot infer the criteria for evaluating a single message on its likelihood of being human generated. To get around this problem, we present annotators with a triplet of human generated and machine transformed messages, with exactly one human generated message hidden in the triplet with two machine transformed messages. The test would require identification of the human generated message in the triplet and score it as 5, and score the remaining two messages on a scale of $1 - 4$ on their likeliness of being completely generated by human. Such a setup makes it more intuitive for an annotator to give messages comparative ratings, and the promise of exactly one human generated message in a triplet makes him/her look for the abstract features that define a human generated message himself/herself from the set of provided messages.

We used crowdsourcing performed through Amazon Mechanical Turk (AMT) for collecting the human judgments. Each unit task on AMT is referred to as a Human Intelligence Task (HIT) and each worker or annotator is referred to as a Turker. For us, rating the three messages in a triplet constituted one HIT. In generating triplets, we considered types kinds of comparisons: one where three versions of the same ad message were shown to annotators, and one where the three messages in a triplet come from different advertisements. Since we wish to judge the semantic sense of the ad messages, it is not a requirement to have messages generated from the same advertisement in one triplet. We constructed 300 triplets for both kinds of tasks by randomly mixing messages for different

Table 8. Details about task posted on AMT

Feature	Details
Task description	Given a set of three ad messages, pick the one which is most likely to be written by a human, and score the other two relative to this one.
Keywords	Ad messages, ratings, comparisons, Human, Machine, Computer
Qualification	Task approval rate >= 50%
Annotations per HIT	Three
Payment per HIT	$0.05
Time allotted per HIT	5 minutes
Avg. time required per HIT	35 seconds

segments with original human generated messages. We requested three annotations for each triplet, so that we can have a measure of general inter-annotator agreement (even though the same Turker may not solve all the HITs). For each ad message in a triplet, we considered its final rating to be an average of the ratings provided by the three annotators. Details of our crowdsourcing task are presented in Table 8. We provide a screenshot of the task below accompanied by a solved example in Fig. 1.

A total of 105 Turkers participated in our task. We rejected annotations that were inconsistent (triplets having no rating of 5 or multiple ratings of 5), which had obvious answering patterns for a Turker or which took a negligible time to complete, and re-posted the tasks on AMT. We found all the triplets in

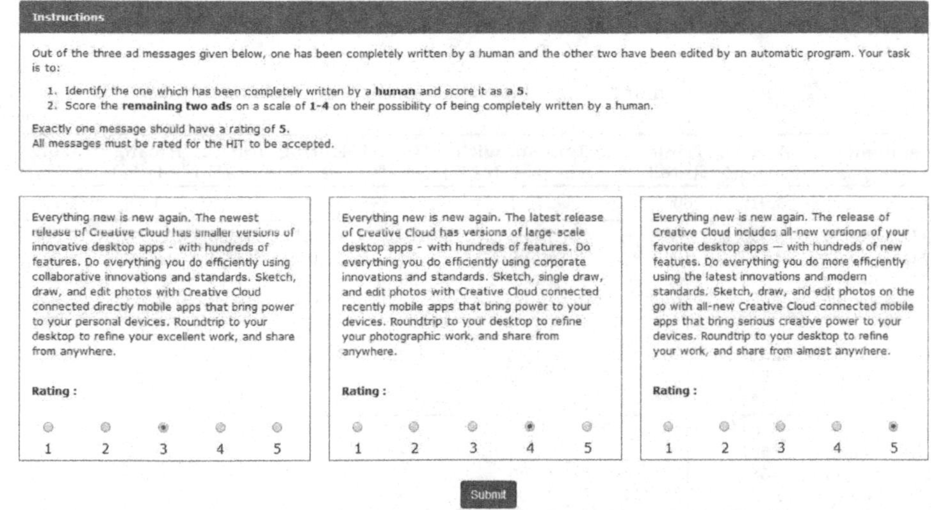

Fig. 1. Guidelines and a solved example for our crowdsourcing experiment

the results of our second round to have consistent ratings. Our success would be measured by the fraction of times machine transformed messages are able to confuse the user, and on the average score received by such messages. We present our results in Table 9. From Table 9, we observe a very exciting result: while the average rating received by human-generated messages is 3.90, those received by our transformations was 3.59, falling below the human average by less than a "point" (5-point scale) at only 0.35. Human messages do get the highest average rating, implying that annotators have a cognitive model in place for evaluating the syntactic and semantic structure of real ad messages. But human messages getting a mean rating noticeably below 5 implies that the machine transformed messages are able to confuse the annotators a significant number of times. Also, the variability of the average rating is not very high among segments, ranging from 3.49 (UK-student) to 3.73 (AU-developer). This shows the robustness of our transformation with respect to semantics across segments, that have varying amounts of data. The distributions of ratings (1 to 5) for each segment are also shown in Table 9. While ratings of 1, 3, 4, and 5 seem to have reasonable shares, peculiarly we did not obtain a single rating of 2. We used an additional 60 triplets without a human message (without the knowledge of the annotators) to observe if there was unusual annotator behavior in such cases. We did not find any interesting behavior to report for such cases.

Next, we note that human generated messages in triplets receive a rating greater than 4.5 (implying that at least one annotator rated it as 5) in 25.67% of the triplets. This is still the highest, but it is much less than 100%, implying that the machine transformed messages frequently obtained high ratings in triplets from multiple annotators. For messages generated for specific segments, the percentage of triplets where they received ratings greater than 4.5 varied from 9.57% to 16.52%. Thus, no particular segment dominated the results, and the good results for the machine messages can be attributed to good ratings received in more or less uniform shares by messages personalized for all segments.

Table 9. Summary of results obtained through crowdsourcing

Segment	Avg. Rating	#Triplets Rated	#Triplets with Score >= 4.5 (%)	Rating =1 (%)	Rating =2 (%)	Rating =3 (%)	Rating =4 (%)	Rating =5 (%)
Human	3.91	600	25.67	12.28	0	14.23	31.78	41.73
AU-designer	3.54	100	11.00	19.67	0	15.34	37.00	28.00
AU-developer	3.71	100	14.00	14.67	0	16.34	38.00	31.00
AU-manager	3.50	100	14.00	20.67	0	17.67	31.67	30.00
AU-student	3.57	100	12.00	17.00	0	19.00	36.67	27.34
UK-designer	3.50	100	13.00	18.34	0	16.00	33.00	32.67
UK-developer	3.69	100	16.00	14.34	0	18.00	37.34	30.34
UK-manager	3.55	100	11.00	19.00	0	20.00	29.34	31.67
UK-student	3.44	100	10.00	20.67	0	19.67	34.34	25.34
US-designer	3.50	100	10.00	17.67	0	22.00	35.00	25.34
US-developer	3.56	100	14.00	18.67	0	19.34	31.00	31.00
US-manager	3.58	100	10.00	16.00	0	21.67	34.00	28.34
US-student	3.54	100	13.00	19.00	0	18.00	34.34	28.67

Table 10. Results for triplets with variants of the same message (Type 1). Values in parentheses correspond to triplets with different messages (Type 2).

Segment	Avg. Rating	#Triplets Rated	#Triplets with Score >= 4.5 (%)
Human	4.07(3.80)	240(360)	29.17(23.33)
AU-designer	3.55([illegible])	40([illegible])	15.00([illegible])
AU-developer	3.76(3.67)	40(60)	12.50(15.00)
AU-manager	3.62(3.43)	40(60)	22.50(8.33)
AU-student	3.53(3.61)	40(60)	7.50(15.00)
UK-designer	3.67(3.58)	40(60)	10.00(15.00)
UK-developer	3.66(3.72)	40(60)	10.00(20.00)
UK-manager	3.40(3.64)	40(60)	10.00(11.67)
UK-student	3.60(3.33)	40(60)	10.00(10.00)
US-designer	3.63(3.42)	40(60)	7.50(11.67)
US-developer	3.47(3.62)	40(60)	5.00(20.00)
US-manager	3.64(3.56)	40(60)	2.50(15.00)
US-student	3.55(3.53)	40(60)	15.00(11.67)

We now present results separated by the types of annotations requested – *Type 1*, where each triplet contains variants of the same message, and *Type 2*, where each triplet can contain different messages. While humans get an average rating of 4.07 and highest scoring triplets at 29.17% for Type 1, the numbers decrease to 3.80 and 23.33% for Type 2. This implies that while annotators can make out certain small human traits when versions of the same message are provided, their task becomes harder when different messages come into the fray. This generally reflects the good quality of the transformations, but also highlights the room for improvement for the transformation method. The fall in human scores is absorbed in a fair way by transformations for the various segments, most of them showing small increase or decrease in rating points.

Inter-Annotator Agreement (IAA). Traditional measures for computing IAA when there are multiple annotators, like Fleiss' Kappa [25], are not applicable in a typical crowdsourced setup where one annotator need not complete all the tasks. Thus, to get an idea of IAA in our context, we computed the average standard deviation over the three annotator ratings for a particular ad message within a triplet, across all rated triplets. We found this value to be 0.919, which, on a 5-point scale, reflects fairly good agreement among Turkers for this task.

7 Discussion and Error Analysis

Result Correlation with Corpus Size. To check if the quality of our personalized messages is strongly correlated with the amount of data that we gathered for each segment, we computed the Kendall-Tau Rank correlation coefficients

(τ) between the vectors obtained from the cross-entropy values (all *four* LMs) in Table 6 and the dataset sizes for each segment (as measured by the *three* factors Tweets, sentences, and words) in Table 4. We also computed τ between the data sizes and the *two* AMT measures (average ratings and percentage of Tweets with high ratings in Table 9). The first set of computations resulted in 12 ($= 4 \times 3$) values of τ ranging between 0.05 and 0.12, and the second set in 6 values between -0.27 and -0.10. Since τ varies between -1 (all *discordant* pairs) and $+1$ (all *concordant* pairs), these values can be interpretated as implying very little correlation between dataset size and result quality. Since our results are in general satisfactory, we can conclude that the quantity of data collected by us is substantial.

Imperative Verbs and Indefinite Pronouns. Verbs in their imperative mood, when appearing at the beginning of a sentence, like ***Sketch** with our pencil tool.*, triggered an error in the Stanford POS tagger. They were labeled as nouns instead, and even though we had adverbs in our repository for the verbs, we were unable to make insertions in such cases. Pronouns like `everything` and `someone`, called indefinite pronouns, were incorrectly labeled as nouns by the Stanford tagger. Hence, adjective insertions were performed on such words as well. We observed one notable difference in these cases: while adjectives generally precede nouns in sentences, adjectives for indefinite pronouns usually make more sense if the adjective succeeds the pronoun (*everything* ***useful***, *someone* ***good***).

Personalization and Privacy. There is concern among users that too much personalization may result in a breach of privacy. While this may be true in several cases like recommending highly specific products to individual users, the technology that we propose in this research is safe in this perspective. This is because we are operating on the general linguistic style of a segment of users, and do not personalize for a particular user. Also, communicating in a style, i.e., with a choice of words that is known to evoke positive sentiment and is common in the audience segment, only tries to ensure that the target users understand meanings of words in context and is expected to elicit higher levels of engagement, and not raise concerns about privacy violation.

Direct Evaluation of Message Personalization. An interesting future work would be to validate our hypothesis of message personalization increasing user engagement, with a more direct evaluation using clickthrough rates (CTR). Alternative approaches may include direct evaluation like pushing hypothetical ads or reaching out to real Twitter users/bloggers who follow desired demographic patterns. However, challenges include eliciting significant response rates from participants, and exclude the effects of other factors in the final clickthrough as far as possible. Nevertheless, the focus of our research in this paper is on text transformation, and we evaluated how well we were able to do it.

8 Conclusions and Future Work

In this research, we have proposed and evaluated approaches on how we can automatically enrich a body of text using a target linguistic style. As an application scenario, we have transformed basic ad messages created by ad copywriters for demographic segments, based on linguistic styles of the corresponding segments. To this end, we have used well-established techniques, models and measures in NLP like POS tagging, dependency parsing, chunking, lemmatization, term weighting, language models, mutual information and cross-entropy. Decreased cross entropy values from the original messages to the transformed messages with respect to the target language models show that our algorithm does take ads closer to specific linguistic styles computationally. In addition, we have shown that automatically transformed messages are semantically coherent as they have been rated highly by users on their likelihood of being completely composed by humans. With our approach, while creation of the original ad message still remains in the hands of the human copywriter, it helps cut down on the additional resources required for hand-crafting personalized messages for a large set of products or demographic clusters. Finally, we are making our demographic-specific Tweets public for use by the research community.

Automatically transforming text with respect to deeper linguistic features like more general word usage patterns, formal or informal usage, sentence lengths, and aspects of sentiments are potential avenues for future research, with most of the current work restricted to measurement and reporting of these aspects. Through our research, we have tried to lay down the stepping stones in the area of guided text transformation, a field that we believe has immense potential.

Acknowledgments. We wish to acknowledge Prof. Atanu Sinha from University of Colorado, Boulder, USA, for providing useful suggestions at various stages of this work. We also thank Priyank Shrivastava and his team from Adobe Systems India for insights on advertisement campaigns.

References

1. Gunsch, M.A., Brownlow, S., Haynes, S.E., Mabe, Z.: Differential forms linguistic content of various of political advertising. Journal of Broadcasting and Electronic Media 44, 27–42 (2000)
2. Kitis, E.: Ads - Part of our lives: Linguistic awareness of powerful advertising. Word and Image 13, 304–313 (1997)
3. Kover, A.J.: Copywriters' Implicit Theories of Communication: An Exploration. Journal of Consumer Research 21, 596–611 (1995)
4. Lowrey, T.M.: The effects of syntactic complexity on advertising persuasiveness. Journal of Consumer Psychology 7, 187–206 (1998)
5. Schwartz, H.A., Eichstaedt, J.C., Kern, M.L., Dziurzynski, L., Ramones, S.M., Agrawal, M., Shah, A., Kosinski, M., Stillwell, D., Seligman, M.E.P., Ungar, L.H.: Personality, gender, and age in the language of social media: The open-vocabulary approach. PLoS ONE 8, e73791 (2013)

6. Furuta, R., Plaisant, C., Shneiderman, B.: Automatically transforming regularly structured linear documents into hypertext. Electron. Publ. Origin. Dissem. Des. 2, 211–229 (1989)
7. Thuy, P.T.T., Lee, Y.K., Lee, S.Y.: DTD2OWL: Automatic Transforming XML Documents into OWL Ontology. In: ICIS 2009, pp. 125–131 (2009)
8. Liu, F., Weng, F., Wang, B., Liu, Y.: Insertion, deletion, or substitution?: Normalizing text messages without pre-categorization nor supervision. In: HLT 2011, pp. 71–76 (2011)
9. Barzilay, R., Elhadad, M.: Using lexical chains for text summarization. In: Advances in Automatic Text Summarization, pp. 111–121 (1999)
10. Burstein, J., Shore, J., Sabatini, J., Lee, Y.W., Ventura, M.: The automated text adaptation tool. In: NAACL Demonstrations 2007, pp. 3–4 (2007)
11. Chandrasekar, R., Doran, C., Srinivas, B.: Motivations and methods for text simplification. In: Proceedings of the 16th Conference on Computational Linguistics, COLING 1996, vol. 2, pp. 1041–1044. Association for Computational Linguistics, Stroudsburg (1996)
12. De Belder, J., Moens, M.F.: Text simplification for children. In: Proceedings of the SIGIR Workshop on Accessible Search Systems, pp. 19–26 (2010)
13. Feng, L., Jansche, M., Huenerfauth, M., Elhadad, N.: A comparison of features for automatic readability assessment. In: COLING 2010, pp. 276–284 (2010)
14. Tausczik, Y.R., Pennebaker, J.W.: The psychological meaning of words: LIWC and computerized text analysis methods. Journal of Language and Social Psychology 29, 24–54 (2010)
15. Tan, C., Lee, L., Pang, B.: The effect of wording on message propagation: Topic- and author-controlled natural experiments on Twitter. In: ACL 2014, pp. 175–185 (2014)
16. Bryden, J., Funk, S., Jansen, V.: Word usage mirrors community structure in the online social network twitter. EPJ Data Science 2 (2013)
17. Danescu-Niculescu-Mizil, C., West, R., Jurafsky, D., Leskovec, J., Potts, C.: No country for old members: User lifecycle and linguistic change in online communities. In: WWW 2013, pp. 307–318 (2013)
18. Hu, Y., Talamadupula, K., Kambhampati, S.: Dude, srsly?: The Surprisingly Formal Nature of Twitter's Language. In: ICWSM 2013 (2013)
19. Toutanova, K., Klein, D., Manning, C.D., Singer, Y.: Feature-rich part-of-speech tagging with a cyclic dependency network. In: NAACL 2003, pp. 173–180 (2003)
20. Socher, R., Bauer, J., Manning, C.D., Ng, A.Y.: Parsing with compositional vector grammars. In: ACL 2013, pp. 455–465 (2013)
21. Zhai, C., Lafferty, J.: A study of smoothing methods for language models applied to ad hoc information retrieval. In: SIGIR 2001, pp. 334–342 (2001)
22. Abney, S.P.: Parsing by chunks. In: Principle-Based Parsing, pp. 257–278. Kluwer Academic Publishers (1991)
23. Bird, S., Klein, E., Loper, E.: Natural Language Processing with Python: Analyzing Text with the Natural Language Toolkit. O'Reilly, Beijing (2009)
24. De Smedt, T., Daelemans, W.: Pattern for Python. JMLR 13, 2063–2067 (2012)
25. Fleiss, J.L.: Measuring nominal scale agreement among many raters. Psychological Bulletin 76, 378–382 (1971)

Inferring Aspect-Specific Opinion Structure in Product Reviews Using Co-training

Dave Carter[1,2] and Diana Inkpen[1]

[1] University of Ottawa, School of Electrical Engineering and Computer Science
david.carter@cnrc-nrc.gc.ca
[2] National Research Council Canada
diana.inkpen@uottawa.ca

Abstract. Opinions expressed about a particular subject are often nuanced: a person may have both negative and positive opinions about different aspects of the subject of interest, and these aspect-specific opinions can be independent of the overall opinion. Being able to identify, collect, and count these nuanced opinions in a large set of data offers more insight into the strengths and weaknesses of competing products and services than does aggregating overall ratings. We contribute a new confidence-based co-training algorithm that can identify product aspects and sentiments expressed about such aspects. Our algorithm offers better precision than existing methods, and handles previously unseen language well. We show competitive results on a set of opinionated sentences about laptops and restaurants from a SemEval-2014 Task 4 challenge.

1 Introduction

Humans are opinionated beings. Some opinions may be arbitrary, but many are nuanced and explicitly supported. People share their opinions online in great numbers. The deluge of available text makes these opinions accessible but, paradoxically, due to their sheer number, it becomes increasingly difficult to synthesize and generalize these opinions. The goal of this work is to develop usable and useful software that, given a set of casually written product reviews, identifies products' aspects (features) and infers writers' opinions about these aspects. Such aspect-specific sentiments can be aggregated to support decision making.

For example, consider the sentence: *I love my new iPhone because of its amazing screen but the battery is barely sufficient to get me through the day.*

There are three sentiments expressed in this sentence:

- a positive sentiment about the iPhone itself;
- a positive sentiment about the screen; and
- a negative sentiment about the battery or battery life.

The *screen* and the *battery [life]* are two aspects of the product *iPhone*. We seek to automatically annotate these two aspects in such a sentence and correctly infer that the writer has a positive sentiment about the screen and a

A. Gelbukh (Ed.): CICLing 2015, Part II, LNCS 9042, pp. 225–240, 2015.
DOI: 10.1007/978-3-319-18117-2_17

negative sentiment about the battery life, without being sidetracked by the positive sentiment about the phone itself. (Perhaps a very simple natural language processing system might see that *battery* and *love* appear in the same sentence and infer that the writer has a positive opinion of the battery life; avoiding such incorrect inferences is a challenge of doing aspect-based sentiment analysis well.)

This work uses co-training to try to take advantage of unlabelled data to find these aspects, rather than using only human-annotated data; the former is much cheaper and easier to procure, and is more readily available. Co-training has been used for various tasks since 1998, but has never, to our knowledge, been applied to the task of aspect-specific sentiment analysis.

The co-training algorithm we developed for aspect-specific sentiment analysis offers high precision: it is likely to get its predictions correct, at the expense of making fewer predictions (or, put more archaically: it makes sins of omission, but few sins of commission). High precision matches a naïve intuition of "correctness" fairly well; and high-precision, lower-recall systems can be combined in ensemble learning to create powerful voting systems like IBM's Watson [1].

While sentiment classification of text has been attempted computationally for roughly twenty years now, aspect-specific sentiment identification is a newer task in natural language processing that is undergoing active research at present (e.g., as part of the SemEval-2014 competition, wherein there was a shared task called Aspect Based Sentiment Analysis that attracted submissions from 31 teams).

This paper unfolds as follows. Related work is presented in the following section. An overview of our experimental methods ensues, and includes a description of the algorithm developed. The algorithm is evaluated on data from the SemEval-2014 Task 4 challenge and results are compared to the results of the teams that participated in the challenge. Finally, conclusions follow.

2 Related Work

There have been attempts at inferring the sentiment of sentences using computers for twenty years, with some approaches based on manually coded rules derived from observed linguistic phenomena, and some using machine learning and other forms of artificial intelligence. Our work draws on both approaches.

The broad field of sentiment analysis is well-established. Commendable works that survey the state-of-the-art in sentiment analysis include [2], [3], and [4].

2.1 Sentiment Analysis Using Product Reviews

Product reviews are useful data to work on for aspect-based sentiment analysis. One conclusion of [4] is that reviews of restaurants, hotels, and the like have a significant influence on consumers' purchasing decisions, and that, depending on the type of product or service, 20% to 99% of consumers will pay more for an item that is rated five stars out of five than a competing item ranked four stars out of five. In a similar vein, [5] discusses abstractly why product reviews are useful, while [6] describes how reviews impact the pricing power of an item.

Various experiments have been performed on Amazon reviews. One of the first sets of Amazon data used for sentiment analysis was annotated in [7] and then used in an experiment that predicted aspect-specific sentiments expressed in the data. A similar experiment was performed on the same data in [8], using rules based on parse trees. A system working on a subset of the same data set is described in [9]; it tries to classify (using linguistic rules) the polarity of sentiment-bearing words in context (one person might like his or her phone to be *small* and their car *big*, while another might prefer a *big* phablet and a *small* sporty car). Opinions about aspects of products as stated in Amazon reviews were analyzed by [10], using reviews of books, DVDs, electronics, and kitchen appliances; impressive domain adaptation results were achieved.

Further efforts to identify aspect-specific sentiments expressed in text have used other data sets. An effort to identify aspect-opinion pairs at the sentence level is described in [11], using a mix of web pages, camera reviews, and news articles. Experiments in using latent discourse analysis (LDA) are described in [12] and [13], identifing product aspects in reviews and then matching tokens from the reviews' sentences that correspond to each product aspect. A similar LDA-based experiment is described in [14]; while it does not perform as well as supervised models, it also doesn't need sentiment-bearing words to be labelled in the input data. Finding topics and associated opinions, a task not unlike that of aspect-specific sentiment extraction, is pursued by [15] using opinionated text about laptops, movies, universities, airlines, and cities. Product aspect-sentiment pairs are identified in online message board postings in [16] and are used to generate marketing intelligence summaries. Unsupervised methods are used to mine aspects and sentiments for restaurants and netbooks in [17]. Interestingly, they found that the extracted aspects were more representative than a manually-constructed list on the same data, avoiding problems of over-generalization or over-representation (being too granular or too fine-grained in combining similar aspects). The work of [18] tries to be aspect-specific, first mining the product aspects and then the opinion polarities of each aspect using CNet and Amazon reviews; in practice, the only experiment in which they have reasonable results is classifying the polarity of the review (i.e., at the document level). Topics and sentiment orientations are identified in car reviews in [19], using clustering techniques to mine unigrams mentioned in positive and negative contexts for different makes and models of cars; some aspect-specific sentiments are found in this manner, though results are noisy.

2.2 Co-training

Co-training is a semi-supervised learning approach that uses both labelled and unlabelled data. Two classifiers try to classify the same data into the same classes using different and uncorrelated sets of features ("views", in co-training parlance). The algorithm iteratively builds larger and larger sets of training data.

Co-training was introduced by Blum and Mitchell [20]. They present an approach for using a small set of labelled data and a large set of unlabelled data to iteratively build a more complete classifier model. Classification features are divided into two views. The main example they provided was a task to classify web pages by topic, where one view was the textual content of the pages, and the other view was composed of the text in links used to access the pages. Two assumptions are made: that each view is sufficient to classify the data, and that the views are conditionally independent given the class label.

Many saw promise in Blum and Mitchell's proposed co-training algorithm but sought to alleviate some concern about the two assumptions it made about the co-training views. Evidence supporting that the conditional independence requirement can be relaxed is offered in several works ([21], [22], [23], [24], [25]). An alternative to the conditional independence assumption is offered in [26]; an *expansion* assumption of the data is proposed and a proof is offered that data meeting this assumption will derive benefit from a co-training approach (also assuming that the underlying machine learning classifiers are never confident in their classifications in cases when they are incorrect). The authors assume that the views need be at most "weakly dependent", rather than assuming conditional independence; and are, in fact, quite explicit in stating that this assumption is the "right" assumption compared to the earlier assumption. (It is worth noting that Blum is a co-author of this paper). Finally, a very practical analysis of the assumptions underlying co-training is offered by [27].

Confidence-based co-training (where a classifier estimates class probabilities, and only those with high probabilities are added to training data in subsequent iterations) worked well in [28]. Confidence-based co-training was also used in [29]; they sample the data where the two views' classifiers agree the most.

Limitations of co-training have been posited. Co-training improves classifier performance to a certain threshold (as high-confidence data are added to the training models in early iterations), and then as more examples are added, performance declines slightly [30]. The usefulness of co-training depends largely on (and is roughly proportional to) the difference between the two views [31].

Co-training has been used in a few sentiment analysis tasks. Blum and Mitchell's algorithm is used in [32] to do sentiment classification on reviews, using Chinese data as one view and English data as a second view. Sentiment classification of tweets is offered in [33] and [34] using co-training, while [35] uses co-training to identify sentiment in an online healthcare-related community. Co-training has been applied to other natural language processing tasks including email classification [36], sentence parsing [37], word sense disambiguation [38], co-reference resolution [39], and part-of-speech tagging [40].

3 Methods

We work with text from product reviews collected in [41]. This text is written by the general public, and each sentence contains one or more specific aspects of products along with, in most cases, a stated opinion about each aspect.

We aim to identify the aspect-specific opinions expressed in a sentence. We construe this as a classification task, where each word in a sentence can be classified as one of three mutually exclusive cases:

- The word is inside (is part of) a product aspect
- The word is inside (is part of) an expression of a sentiment
- The word is outside of a product aspect or expression of a sentiment

The software we developed takes a sentence as input and returns a list of zero or more tagged stated opinions about aspects of a product. The system is based on machine learning plus a handful of heuristics. Co-training underlies our method, as it offers two advantages over fully supervised learning: the (optional) ability to use unlabelled data to build a larger set of training data, and its use of independent and sufficient views.

Co-training is a semi-supervised classification algorithm that augments a small set of labelled data with a large set of unlabelled data to reduce the error rate in a classification task [20]. A main motivation of such an approach is that labelled data is "expensive" (as it is usually hand-labelled by humans, which incurs time and/or monetary costs), and so any improvement in results that can be gleaned from unlabelled data is essentially "free" [20]. Co-training uses two conditionally independent "views" of the data being classified. Each such view must be (at least theoretically) sufficient to classify the data. Co-training iteratively builds up each classifier's knowledge by adding high-confidence classified cases to the training set; the expertise of one classifier is used to train the other.

We use [surface-level] lexemes (including predicted part-of-speech) and syntactic features as the two views. In English, as in many languages, there is not a one-to-one relation between lexemes and their part of syntax; they may be independent for all but the most basic of functional words (conjunctions, particles, and the simplest adverbs and personal pronouns, for example).

The lexical view is inspired by collocations. For example, a word following the fragment "I like my phone's ..." is fairly likely to be a product aspect, and is unlikely to be a sentiment-bearing word (unless, perhaps, it is a superlative adjective followed by the aspect). Features in the lexical view include the surface form of the token, its lemma, and its (predicted) part-of-speech. In addition, these same features are recorded for the preceding and following three tokens for each given token; this is somewhat inspired by work on extraction patterns, where a pattern like "I like my *some_product_name* despite its rather poor *some_product_attribute*" can be used to extract product names and product attributes with fairly high confidence.

The syntactic view is inspired by the observation that both product aspects and sentiment-bearing words appear in a limited number of grammatical structures. For example, a noun that is the direct object of a verb may be more likely to be a product aspect than the verb itself; in contrast, a verb is more likely to express sentiment than it is to be a product aspect. Features in the syntactic view include the node in the parse tree immediately above the token; the chain of nodes above the token in the parse tree up to the nearest sentential unit; the chain of nodes above the token in the full parse tree; whether the token is referred to by a pronoun elsewhere in the sentence; a list of dependency relations in which the token participates (e.g., whether it is a direct object of another word in the sentence); its predicted semantic role (e.g., whether it is a subject in a sentence); and whether it participates in a negation clause. The intent is to work in a manner similar to extraction patterns but to do so in a way that reflects the complexity of language, particularly long-distance dependencies that might not be accounted for in an n-gram model.

We begin with the Blum & Mitchell algorithm (Algorithm 1, left column) [20], and modify it to classify using a confidence score (Algorithm 2, right column).

Both algorithms begin with set (L) of labelled training examples and a set (U) of unlabelled examples. The Blum and Mitchell algorithm then selects a random pool of unlabelled examples (U') that is much smaller than the full unlabelled set, and enlarges this pool every iteration; whereas our algorithm considers all remaining unlabelled examples (U') at each iteration. The Blum and Mitchell algorithm iterates a fixed number of times (k); whereas our algorithm keeps running as long as the number of unlabelled examples that could be classified in the previous iteration (n_{i-1}) is greater than zero. Each iteration, both methods have a classifier train itself on a single view of all labelled data (including data that have been labelled successfully in previous iterations), then classify the data in (U'). At this point, Blum and Mitchell randomly pick one positive and three negative examples to add to the set of labelled data; whereas our algorithm adds to the set of labelled data those data about which it was most confident. This confidence metric is defined as the confidence of the most confident classifier; a more complex scoring function could be used, such as the amount of agreement or the amount of disagreement between the two classifiers, as suggested by [29].

The most notable divergence from the Blum and Mitchell algorithm is the decision to add a large number of examples at each iteration, so long as the classifiers are confident in their classification of such unlabelled examples. We have chosen to implement an upper limit in the algorithm, so that, rather than accepting all new unlabelled examples that can be classified with a confidence of, say, 55%, it will only accept the top m-most cases. The intuition is that it may be desirable, especially in the early iterations, to add only the most confident examples and retrain so as to be able to more confidently label the next set; the expertise added by only accepting the highly confidently-labelled examples may be sufficient to more confidently classify the merely marginal unlabelled examples that may have been classified with confidence at or just above the threshold. In practice, the algorithm tends to use this upper limit in only the first several iterations; after roughly the fifth iteration, the confidence threshold determines the number of unlabelled examples added at each iteration, as the most obvious examples have already been added to the labelled set.

While Blum and Mitchell's algorithm takes as an input the maximum number of iterations k (which would also presumably have to scale proportionally to the size of the data set), our algorithm requires the maximum number of new examples to label in each iteration m, which roughly determines the number of iterations ($i_{\max}$) for a given confidence threshold. In practice, there tend to be roughly $i_{\max} = 8$ iterations required to process the SemEval-2014 task 4 data.

Given:

- a set L of labelled training examples with features x
- a set U of unlabelled examples with features x

Create a pool *U'* of examples by choosing u examples at random from U

Result: An enlarged pool *L'*

initialization;

for $i \leftarrow 1$ **to** k **do**

  Use L to train a classifier h_1 that considers only the x_1 portion of x;

  Use L to train a classifier h_2 that considers only the x_2 portion of x;

  Allow h_1 to label p positive & n negative examples from *U'*;

  Allow h_2 to label p positive & n negative examples from *U'*;

  Add these self-labelled examples to L;

  Randomly choose $2p + 2n$ examples from U to replenish *U'*;

end

with typical $p = 1$, $n = 3$, $k = 30$, $u = 75$;

Algorithm 1. Blum and Mitchell's co-training algorithm [20] (largely verbatim)

Given:

- a set L of labelled training examples with features x
- a set U of unlabelled examples with features x

Create a pool *U'* of all examples from U

Result: An enlarged pool *L'*

initialization;
$i = 1$;

while $i = 1$ **or** $n_{i-1} > 0$ **do**

  Use L to train a classifier h_1 that considers only the x_1 portion of x;

  Use L to train a classifier h_2 that considers only the x_2 portion of x;

  Allow h_1 to label all examples from *U'*;

  Allow h_2 to label all examples from *U'*;

  Sort these self-labelled examples in descending order of *max(confidence of* h_1*, confidence of* h_2*)*;

  Add the top n most confidently labelled examples to L where $n \leq m$ and the confidence of the prediction of every such example is greater than c;

  $i \leftarrow i + 1$;

end

with typical $m = 2500$, $c = 0.55$, $i_{\max} \approx 8$;

Algorithm 2. Our co-training algorithm using confidence-based classification

The confidence threshold c in our algorithm is tuneable. This parameter serves as classification confidence floor; the algorithm will not include any labelled examples when the confidence in that example's classification is less than this floor. The support vector machine classifier we selected offers fairly good classification performance, so we set this threshold to a relatively low 0.55 for all experiments described herein. (A grid search classifying the development data with confidence thresholds $c \in \{0.00, 0.45, 0.50, 0.55, 0.65, 0.75, 0.85, 0.95\}$ revealed that 0.55 was close to optimal.)

LibSVM [42], a support vector machine classifier, was used for classification; a radial basis function (RBF) kernel was used. SVM tuning parameters are provided in our source code, which we make available.[1] The SVM classifiers were tuned on the first 20% of the sentences in each of the five data sets in [7].

4 Evaluation

4.1 Data

The data set we selected for experimentation was originally developed in [43], containing restaurant review text from Citysearch New York, and was modified and enlarged for an aspect-specific sentiment analysis task at SemEval-2014 [41]. This SemEval-2014 data set contains sentences extracted from reviews of restaurants (3041 training sentences and 800 test sentences) and reviews of laptops of different brands (3045 training sentences and 800 test sentences). Aspects that appear in the sentence are tagged and assigned a sentiment polarity of positive, neutral, negative, or "conflict", the latter referring to cases where both positive and negative sentiments about the aspect appear in the same sentence (along the lines of "the service was friendly but slow"). The data are written by casual writers, but, subjectively, the quality of the writing appears to be rather good; spelling errors and instances of odd formatting (like informally-bulleted lists) that plague some other data sets seem to be rare.

This particular data set offers a good basis for comparison for our approach to sentiment analysis. The competition drew 57 submissions for the first phase of evaluation and 69 for the second phase of evaluation.

The aspects are carefully tagged in the data, including character positions. The sentiment-bearing words themselves are not tagged, so it is up to the software to determine in some other manner how and where the sentiment is expressed in the sentence; we used the lexicon from [7] for this. An example is:

```
<sentence id="337">
    <text>However, the multi-touch gestures and large tracking area make having an
        external mouse unnecessary (unless you're gaming).</text>
    <aspectTerms>
        <aspectTerm term="multi-touch gestures" polarity="positive" from="13" to="33"/>
        <aspectTerm term="tracking area" polarity="positive" from="44" to="57"/>
        <aspectTerm term="external mouse" polarity="neutral" from="73" to="87"/>
        <aspectTerm term="gaming" polarity="neutral" from="115" to="121"/>
    </aspectTerms>
</sentence>
```

[1] https://github.com/davecart/cotraining

We tokenized and processed the sentences using the Stanford CoreNLP tools, which labelled each token with its predicted part-of-speech, its lemma, whether it is believed to be a named entity (e.g., a brand name); and built a parse tree for the sentence, with coreferences labelled.

4.2 Classifying Product Aspects

We compared our system's ability to label product aspects in sentences (independent of any effort to glean associated sentiments) to the results of those who participated in the SemEval-2014 task 4 subtask 1 challenge.

Our system, operating in a supervised manner and allowing only an exact match in cases where aspects were composed of multiple words, offered higher precision than all other systems on the laptop reviews, though perhaps not significantly so. Our system – whether running in a supervised manner or using co-training with only half of the training data being labelled – offered precision on the restaurant reviews that was roughly tied with the top competitor, and much higher than the mean and median.

Table 1. Comparing aspect classification results on SemEval-2014 task 4 (subtask 1) data

Data set	Laptop reviews				Restaurant reviews			
	P	R	F_1	A	P	R	F_1	A
SemEval-2014 task 4 subtask 1 (aspect term extraction)								
mean performance	0.690	0.504	0.562	-	0.767	0.672	0.708	-
lowest performance	0.231	0.148	0.239	-	0.371	0.340	0.383	-
median performance	0.756	0.551	0.605	-	0.818	0.720	0.727	-
highest performance	0.848	0.671	0.746	-	0.909	0.827	0.840	-
Our results								
- fully supervised, using all training data	0.863	0.401	0.547	0.632	0.915	0.681	0.781	0.647
- training with first half, co-training with second half	0.822	0.292	0.430	0.581	0.909	0.587	0.713	0.589
- training with second half, co-training with first half	0.829	0.224	0.353	0.559	0.910	0.616	0.734	0.606

Our system offered weaker performance in recall: somewhat below the mean and median when processing the laptop reviews in a supervised manner, and roughly tied with the mean and well below the median when examining restaurant reviews. Co-training offered much worse recall, though still better than the weakest of the SemEval-2014 task 4 competitors.

With high precision and relatively weak recall, our system achieved F_1 scores that placed mid-pack among SemEval-2014 competitors when considering all test data in a supervised manner. When co-training with the laptop data, our

F_1 was below average; whereas when co-training with the restaurant reviews, our F_1 scores were slightly above the mean and tied with the median.

The product aspect classification performance of our system on the SemEval-2014 data can be described as being roughly average among the 31 teams, and is characterized by very high precision and rather low recall.

Accuracy was not reported in the competition results, but we offer our system's accuracy performance for future comparison and to illustrate that, even in cases where recall is low, accuracy remains at reasonable levels.

4.3 Classifying the Sentiments of Aspects

The second subtask in the SemEval-2014 task 4 challenge was to predict the stated aspect-specific sentiment of sentences where the product aspect(s) were already labelled. We compared our system's performance on this task to that of those who entered the challenge (Table 2).

Table 2. Comparing sentiment orientation classification results (given tagged aspects) on SemEval-2014 task 4 (subtask 2) data

Data set	Laptop reviews	Restaurant reviews
	Accuracy	Accuracy
SemEval-2014 task 4 subtask 2 (determine polarity, given aspects)		
mean performance	0.590	0.691
lowest performance	0.365	0.417
median performance	0.586	0.708
highest performance	0.705	0.810
Our results		
- fully supervised, using all training data)	0.719	0.690
- training with first half, co-training with second half)	0.668	0.643
- training with second half, co-training with first half)	0.662	0.631

Accuracy was the only metric reported by the challenge organizers; accordingly, this is the only metric that we report.

When our system was used in an entirely supervised manner, it (just barely, and probably not significantly) bested all competitors in the laptops portion of the SemEval-2014 task 4 challenge. Even the co-training results are well above both mean and median on the laptop reviews. On the other hand, when trying to classify sentiments in the restaurant reviews, performance of the supervised system was tied with the mean and very slightly lower than the median competitor; and the accuracy when co-training was almost 10% worse than the mean and median competitor, though still much better than the least successful teams that participated in the challenge.

Our system thus appears to offer fairly compelling performance in classifying the sentiments expressed about known product aspects in these data sets, even when co-training with only half of the training data being labelled.

4.4 Performance Finding All Aspect-Sentiment Pairs in a Sentence

It is a more challenging and more interesting task to classify both product aspects and the associated sentiments in a sentence than is classifying aspects in isolation. Sadly, this was not a part of the SemEval-2014 task 4 challenge, although it would be a natural extension thereof. Our system's sentence-level results are listed in Table 3.

Table 3. Aspect-specific opinion inference results on SemEval-2014 task 4 competition data (classifying aspects and sentiments simultaneously, given unlabelled sentences)

Data set	Laptop reviews				Restaurant reviews			
	P	R	F_1	A	P	R	F_1	A
SemEval-2014 Task 4 sentences								
- fully supervised, using all training data, all test data	0.890	0.268	0.412	0.596	0.936	0.507	0.658	0.600
- training with first half, co-training with second half	0.880	0.121	0.213	0.528	0.933	0.321	0.477	0.468
- training with second half, co-training with first half	0.935	0.103	0.185	0.523	0.923	0.354	0.512	0.488
- mean performance loss when using co-training with only half of the labelled training data	-2%	58%	52%	12%	1%	34%	25%	20%

The performance of co-training in a real-world and suitably difficult task can be analyzed here. The co-trained models were trained using only half as much labelled data as the supervised model. Precision remained sufficiently high to conclude that it was tied with the supervised model. Recall dropped quite a bit. In the laptop reviews, the F_1 score roughly halved, whereas in the restaurant reviews it dropped an average of 25%. Accuracy suffered 22% in the worst of the trials. These results are somewhat comforting: using only half as much training data seems to reduce the F_1 by half, at worst, while maintaining high precision. This could be an acceptable trade-off in a particular application domain, since labelled data is both difficult and expensive to produce.

By comparison, [44] offers insight into humans' classification performance. Humans seem to be able to classify polarity at the sentence level with roughly 88% precision and 70% recall. Our system, performing a more nuanced task of classifying aspect-specific sentiments at the sentence level, meets this level of precision, if not exceeding it; though it does nowhere near as well at recall. (Human brains, viewed as a natural language processing machine, are trained on much larger language models than our system, so one could intuitively expect that humans might have better recall than NLP software trained on a mere 3000 sentences.)

5 Conclusions and Further Work

Useful and useable software was developed that can label sentiments expressed about specific aspects of a product. The software developed is characterized by its very high precision and somewhat weak (or, in some cases, very weak) recall. It is better at classifying the sentiments expressed about known attributes in laptop reviews than any of the 31 teams who performed the same task in a recent international NLP challenge (SemEval-2014 task 4).

The software can be trained with only labelled data (supervised learning), or can be trained with fewer labelled data and a set of unlabelled data (co-training); unlabelled data are more readily available and much cheaper to procure or produce. When using co-training to perform this aspect-specific sentiment analysis, precision remains high or improves very slightly, at the expense of some recall. This appears to be the first application of co-training to aspect-based sentiment analysis. The algorithm implemented differs from the commonly accepted co-training algorithm of [20], offering better scalability and taking advantage of the ability of newer machine learning classifiers to estimate the confidence in their own predictions. We believe that the tuneable parameters of the algorithm herein are more intuitive than those in [20].

The co-training algorithm developed in our work could be applied to other tasks, both within the natural processing domain and outside of it. (By comparison, Blum and Mitchell's co-training algorithm has found diverse applications).

In the future, it could be interesting to incorporate work on opinion strength. At present, we lump together all positive and all negative opinions, whereas in natural language, opinions are more nuanced. If a consumer is using comparative ratings of an aspect-specific sentiment classification system to make informed choices, it is probably advantageous that the strength of the opinions be known and aggregated (e.g., a cell phone with many weakly negative opinions about the battery life might be preferable to one with a similar number of very strong negative opinions about its battery life). There is some existing academic work on strength-based sentiment classification, e.g., [45] and [46], so that would seem a natural pairing.

One necessary compromise in trying to learn only aspect-specific sentiments herein was a willful ignorance of sentiments expressed about the products (atomically) or the products' brands. A step forward might be incorporating classifiers designed to label such expressions at the same time as labelling aspects and sentiments; the sentiment-bearing word classifier could likely be used as-is. The architecture of the system developed herein can be extended to any n lexically mutually exclusive classes; this could include named entities, competing brands, or retailers. With additional learning models for products (and synonyms thereof) and brands (perhaps by using a named entity tagger), a better picture of both the broader and more specific opinions expressed in text might be gleaned, for a better overall understanding of the text.

Some semi-supervised algorithms (e.g., that in [47]) run a prediction on all *training* data at each iteration to see if, for example, a borderline example that was added in a previous iteration should now be rejected from the training data because it now falls below a particular threshold due to the new knowledge gained by the classifier in the meantime (termed "escaping from initial misclassifications" in the Yarowsky paper). That could be a compelling addition to our approach.

References

1. Ferrucci, D., Brown, E., Chu-Carroll, J., Fan, J., Gondek, D., Kalyanpur, A.A., Lally, A., Murdock, J.W., Nyberg, E., Prager, J., et al.: Building Watson: An overview of the DeepQA project. AI Magazine 31, 59–79 (2010)
2. Liu, B., Zhang, L.: A survey of opinion mining and sentiment analysis. In: Mining Text Data, pp. 415–463. Springer (2012)
3. Liu, B.: Sentiment analysis and opinion mining. Synthesis Lectures on Human Language Technologies 5, 1–167 (2012)
4. Pang, B., Lee, L.: Opinion mining and sentiment analysis. Foundations and Trends in Information Retrieval 2, 1–135 (2008)
5. Ghose, A., Ipeirotis, P.G.: Estimating the helpfulness and economic impact of product reviews: Mining text and reviewer characteristics. IEEE Transactions on Knowledge and Data Engineering 23, 1498–1512 (2011)
6. Archak, N., Ghose, A., Ipeirotis, P.G.: Show me the money!: Deriving the pricing power of product features by mining consumer reviews. In: Proceedings of the 13th ACM SIGKDD International Conference on Knowledge Discovery and Data Mining, KDD 2007, pp. 56–65. ACM, New York (2007)
7. Hu, M., Liu, B.: Mining and summarizing customer reviews. In: Proceedings of the Tenth ACM SIGKDD International Conference on Knowledge Discovery and Data Mining, KDD 2004, pp. 168–177. ACM, New York (2004)
8. Popescu, A.M., Etzioni, O.: Extracting product features and opinions from reviews. In: Natural Language Processing and Text Mining, pp. 9–28. Springer (2007)
9. Ding, X., Liu, B., Yu, P.S.: A holistic lexicon-based approach to opinion mining. In: Proceedings of the 2008 International Conference on Web Search and Data Mining, WSDM 2008, pp. 231–240. ACM, New York (2008)
10. Blitzer, J., Dredze, M., Pereira, F.: Biographies, bollywood, boom-boxes and blenders: Domain adaptation for sentiment classification. In: ACL, vol. 7, pp. 440–447 (2007)
11. Nasukawa, T., Yi, J.: Sentiment analysis: Capturing favorability using natural language processing. In: Proceedings of the 2nd International Conference on Knowledge Capture, K-CAP 2003, pp. 70–77. ACM, New York (2003)
12. Titov, I., McDonald, R.: A joint model of text and aspect ratings for sentiment summarization. In: Proc. ACL 2008: HLT, pp. 308–316 (2008)
13. Titov, I., McDonald, R.: Modeling online reviews with multi-grain topic models. In: Proceedings of the 17th International Conference on World Wide Web, WWW 2008, pp. 111–120. ACM, New York (2008)

14. Jo, Y., Oh, A.H.: Aspect and sentiment unification model for online review analysis. In: Proceedings of the Fourth ACM International Conference on Web Search and Data Mining, WSDM 2011, pp. 815–824. ACM, New York (2011)
15. Mei, Q., Ling, X., Wondra, M., Su, H., Zhai, C.: Topic sentiment mixture: modeling facets and opinions in weblogs. In: Proceedings of the 16th International Conference on World Wide Web, pp. 171–180. ACM (2007)
16. Glance, N., Hurst, M., Nigam, K., Siegler, M., Stockton, R., Tomokiyo, T.: Deriving marketing intelligence from online discussion. In: Proceedings of the Eleventh ACM SIGKDD International Conference on Knowledge Discovery in Data Mining, KDD 2005, pp. 419–428. ACM, New York (2005)
17. Brody, S., Elhadad, N.: An unsupervised aspect-sentiment model for online reviews. In: Human Language Technologies: The 2010 Annual Conference of the North American Chapter of the Association for Computational Linguistics, HLT 2010, pp. 804–812. Association for Computational Linguistics, Stroudsburg (2010)
18. Dave, K., Lawrence, S., Pennock, D.M.: Mining the peanut gallery: Opinion extraction and semantic classification of product reviews. In: Proceedings of the 12th International Conference on World Wide Web, WWW 2003, pp. 519–528. ACM, New York (2003)
19. Gamon, M., Aue, A., Corston-oliver, S., Ringger, E.: Pulse: Mining customer opinions from free text. In: Proc. of the 6th International Symposium on Intelligent Data Analysis, pp. 121–132 (2005)
20. Blum, A., Mitchell, T.: Combining labeled and unlabeled data with co-training. In: Proceedings of the Eleventh Annual Conference on Computational Learning Theory, COLT 1998, pp. 92–100. ACM, New York (1998)
21. Collins, M., Singer, Y.: Unsupervised models for named entity classification. In: Proceedings of the Joint SIGDAT Conference on Empirical Methods in Natural Language Processing and Very Large Corpora, pp. 100–110 (1999)
22. Goldman, S., Zhou, Y.: Enhancing supervised learning with unlabeled data. In: Proceedings of the 17th International Conference on Machine Learning, pp. 327–334. Morgan Kaufmann (2000)
23. Dasgupta, S., Littman, M.L., McAllester, D.: Pac generalization bounds for co-training. Advances in Neural Information Processing Systems 1, 375–382 (2002)
24. Abney, S.: Bootstrapping. In: Proceedings of the 40th Annual Meeting on Association for Computational Linguistics, ACL 2002, pp. 360–367. Association for Computational Linguistics, Stroudsburg (2002)
25. Wang, W., Zhou, Z.H.: Co-training with insufficient views. In: Asian Conference on Machine Learning, pp. 467–482 (2013)
26. Balcan, M.F., Blum, A., Yang, K.: Co-training and expansion: Towards bridging theory and practice. In: Advances in Neural Information Processing Systems, pp. 89–96 (2004)
27. Du, J., Ling, C.X., Zhou, Z.H.: When does cotraining work in real data? IEEE Trans. on Knowl. and Data Eng. 23, 788–799 (2011)
28. Nigam, K., Ghani, R.: Analyzing the effectiveness and applicability of co-training. In: Proceedings of the Ninth International Conference on Information and Knowledge Management, CIKM 2000, pp. 86–93. ACM, New York (2000)
29. Huang, J., Sayyad-Shirabad, J., Matwin, S., Su, J.: Improving multi-view semi-supervised learning with agreement-based sampling. Intell. Data Anal., 745–761 (2012)
30. Pierce, D., Cardie, C.: Limitations of co-training for natural language learning from large datasets. In: Proceedings of the 2001 Conference on Empirical Methods in Natural Language Processing, pp. 1–9 (2001)

31. Wang, W., Zhou, Z.-H.: Analyzing co-training style algorithms. In: Kok, J.N., Koronacki, J., Lopez de Mantaras, R., Matwin, S., Mladenič, D., Skowron, A. (eds.) ECML 2007. LNCS (LNAI), vol. 4701, pp. 454–465. Springer, Heidelberg (2007)
32. Wan, X.: Bilingual co-training for sentiment classification of chinese product reviews. Computational Linguistics 37, 587–616 (2011)
33. Liu, S., Li, F., Li, F., Cheng, X., Shen, H.: Adaptive co-training svm for sentiment classification on tweets. In: Proceedings of the 22nd ACM International Conference on Conference on Information & Knowledge Management, CIKM 2013, pp. 2079–2088. ACM, New York (2013)
34. Liu, S., Zhu, W., Xu, N., Li, F., Cheng, X.Q., Liu, Y., Wang, Y.: Co-training and visualizing sentiment evolvement for tweet events. In: Proceedings of the 22nd International Conference on World Wide Web Companion, WWW 2013 Companion, pp. 105–106. International World Wide Web Conferences Steering Committee, Republic and Canton of Geneva (2013)
35. Biyani, P., Caragea, C., Mitra, P., Zhou, C., Yen, J., Greer, G.E., Portier, K.: Co-training over domain-independent and domain-dependent features for sentiment analysis of an online cancer support community. In: Proceedings of the 2013 IEEE/ACM International Conference on Advances in Social Networks Analysis and Mining, ASONAM 2013, pp. 413–417. ACM, New York (2013)
36. Kiritchenko, S., Matwin, S.: Email classification with co-training. In: Proceedings of the 2001 Conference of the Centre for Advanced Studies on Collaborative Research, CASCON 2001, p. 8. IBM Press (2001)
37. Sarkar, A.: Applying co-training methods to statistical parsing. In: Proceedings of the Second Meeting of the North American Chapter of the Association for Computational Linguistics on Language Technologies, NAACL 2001, pp. 1–8. Association for Computational Linguistics, Stroudsburg (2001)
38. Mihalcea, R.: Co-training and self-training for word sense disambiguation. In: Proceedings of the Conference on Computational Natural Language Learning, CoNLL 2004 (2004)
39. Ng, V., Cardie, C.: Bootstrapping coreference classifiers with multiple machine learning algorithms. In: Proceedings of the 2003 Conference on Empirical Methods in Natural Language Processing, EMNLP 2003, pp. 113–120. Association for Computational Linguistics, Stroudsburg (2003)
40. Clark, S., Curran, J.R., Osborne, M.: Bootstrapping pos taggers using unlabelled data. In: Proceedings of the Seventh Conference on Natural Language Learning at HLT-NAACL 2003, CONLL 2003, vol. 4, pp. 49–55. Association for Computational Linguistics, Stroudsburg (2003)
41. Pontiki, M., Galanis, D., Pavlopoulos, J., Papageorgiou, H., Androutsopoulos, I., Manandhar, S.: Semeval-2014 task 4: Aspect based sentiment analysis. In: Proceedings of the International Workshop on Semantic Evaluation (SemEval) (2014)
42. Chang, C.C., Lin, C.J.: LIBSVM: A library for support vector machines. ACM Transactions on Intelligent Systems and Technology 2, 27:1–27:27 (2011), `http://www.csie.ntu.edu.tw/~cjlin/libsvm`
43. Ganu, G., Elhadad, N., Marian, A.: Beyond the stars: Improving rating predictions using review text content. In: Proceedings of the 12th International Workshop on the Web and Databases, WebDB 2009 (2009)
44. Nigam, K., Hurst, M.: Towards a robust metric of opinion. In: AAAI Spring Symposium on Exploring Attitude and Affect in Text, pp. 598–603 (2004)
45. Wilson, T., Wiebe, J., Hwa, R.: Just how mad are you? finding strong and weak opinion clauses. In: Proceedings of AAAI, pp. 761–769 (2004)

46. Turney, P.D., Littman, M.L.: Measuring praise and criticism: Inference of semantic orientation from association. ACM Transactions on Information Systems 21, 315–346 (2003)
47. Yarowsky, D.: Unsupervised word sense disambiguation rivaling supervised methods. In: Proceedings of the 33rd Annual Meeting on Association for Computational Linguistics, ACL 1995, pp. 189–196. Association for Computational Linguistics, Stroudsburg (1995)

Summarizing Customer Reviews through Aspects and Contexts

Prakhar Gupta[1], Sandeep Kumar[1], and Kokil Jaidka[2]

[1] Department of Computer Science and Engineering, IIT Roorkee, Roorkee, India
prakharguptage@gmail.com,
sgargfec@iitr.ac.in
[2] Adobe Research India Labs, Bangalore, Karnataka, India
koki0001@e.ntu.edu.sg

Abstract. This study leverages the syntactic, semantic and contextual features of online hotel and restaurant reviews to extract information aspects and summarize them into meaningful feature groups. We have designed a set of syntactic rules to extract aspects and their descriptors. Further, we test the precision of a modified algorithm for clustering aspects into closely related feature groups, on a dataset provided by Yelp.com. Our method uses a combination of semantic similarity methods- distributional similarity, co-occurrence and knowledge base based similarity, and performs better than two state-of-the-art approaches. It is shown that opinion words and the context provided by them can prove to be good features for measuring the semantic similarity and relationship of their product features. Our approach successfully generates thematic aspect groups about food quality, décor and service quality.

Keywords: Aspect Detection, Text Classification, Clustering, Text Analysis, Information Retrieval, Opinion Mining, Online Reviews.

1 Introduction

Online reviews are an important resource for people, looking to make buying decisions, or searching for information and recommendations about a product or business. Online review websites like Yelp provide a way for information seekers to browse user reviews, ratings and opinions about the different aspects of service at restaurants and hotels. However, sifting through a large number of reviews to understand the general opinion about a single aspect, is a tedious task. This is the research problem addressed in approaches for aspect mining and analysis, where the aim is to automatically analyze user reviews and generate a summary around the various aspects of a product.

The approach followed in aspect mining studies is to extract parts of speech or aspect-sentiment pairs [1]. In the current work, we extract aspects-descriptor pairs through the syntactic, contextual and semantic features of text, and cluster them into meaningful, related feature groups. People also tend to mention their thoughts about related aspects in the same sentence, which can be leveraged to provide context for

A. Gelbukh (Ed.): CICLing 2015, Part II, LNCS 9042, pp. 241–256, 2015.
DOI: 10.1007/978-3-319-18117-2_18

aspects. The context provided by words such as "delicious" and "uncooked" can prove to be a good indicator of the category of the aspect (in this case, "food") they are used with. Using sources such as WordNet [2] can further help to relate similar aspects – for example, "water" is related to "drink", and "pasta" and "italian pasta" should belong to the same group. Together, these features comprise the heart of our aspect clustering method.

The application of this work is in summarizing a large set of reviews around the aspects they comprise. There are two major contributions of this work-

1. A set of syntactic rules to find aspects, and their opinion carrying descriptors, within sentences of reviews.
2. A clustering algorithm for identifying and clustering similar aspects, using similarity features based on context and thesauri.

This paper is organized as follows: Section 2 describes the related work. Section 3 discusses our problem statement in detail. Section 4 presents the methodology. Section 5 gives the experiments and results. Section 6 discusses results and Section 7 contain conclusions and future works.

2 Related Work

Work in aspect detection is diverse, and syntactic approaches [3] are as popular as knowledge-rich approaches [1]. Several studies have focused on extracting aspects along with their opinions by using dependency parsing [3] or relationships between noun and verb phrases [6]. Hu and Liu [4] and Yi and Niblack in [5] extracted aspect as noun phrases, by using association rule mining and a set of aspect extraction rules and selection algorithms respectively; however, these methods did not perform well with low frequency aspects, such as specific dishes in a restaurant. Several studies have focused on extracting aspects along with their opinions - [3] used dependency parsing to find relationships between opinion words and target expressions, and [6] identified noun and verb phrases as aspect and opinion expressions. These ideas motivated our approach for developing syntactic rules for extracting aspects and their describing adjectives and other parts of speech.

Clustering similarity measures may rely on pre-existing knowledge resources like WordNet [1][7]. Popular similarity metrics include Cosine function, Jaccard Index and PMI (Pointwise Mutual Information) to calculate similarity between words. The method proposed by [8] mapped feature expressions to a given domain product feature taxonomy, using lexical similarity metrics. In [9], a latent semantic association model is used to group words into a set of concepts according to their context documents and then they categorize product features according to their latent semantic structures. The authors in [10] grouped words using a graph-based algorithm based on PMI or Chi-squared test. Knowledge-based approaches have usually showed increased precision but lower recall compared to previous work; furthermore, they are also not able to handle cases where the knowledge bases do not contain domain specific knowledge, or do not use word distribution information. In this work, we have tested our own clustering algorithm against the state-of-the-art, MCL clustering algorithm, and compared the results.

Probabilistic approaches for summarizing reviews include applied topic modeling to identify major themes. However, according to Blei et al. [11], topic models like LDA are not suitable for aspect detection in reviews, as they capture global topics, rather than aspects mentioned in the review. Nevertheless, several significant works have aimed to overcome this problem, notably the multi-grain topic model MG-LDA [12][13] was constructed, which attempts to capture global and local topics, where the local topics correspond to aspects. More recent approaches include creation of hierarchies of aspects [14], extraction of aspects using word frequency and syntactic patterns [15], and semi-supervised methods [16]. We have also used one variant of LDA [11] described in section 5 as a baseline for comparing topic models and clustered aspects.

3 Problem Description

Our first research objective is to extract aspects and the words used to describe them.

Task 1- *Extract the aspects and their descriptors.*

Definition (Descriptor) - A word, especially an adjective or any other modifier used attributively, which restricts or adds to the sense of a head noun. Descriptors express opinions and sentiments about an aspect, which can be further used in generation of summaries for the aspects.

Definition (Aspect-Descriptor Pair) - An aspect-descriptor pair consists of an aspect and the descriptor of that aspect. e.g. (sandwich, tasty) in "This is a tasty sandwich".

Sometimes there may be more than one aspect-descriptor pair in a sentence for the same aspect if more than one descriptor is present. In some cases, the descriptor may not modify the aspect directly but may modify the verb, any adjective or any other descriptor of the aspect. In such cases, a separate pair is created for that descriptor and the aspect (e.g. waiter, angry) in "The *waiter* looked at me *angrily*".

Task 2- *Clustering of aspects into natural groups.*

The next tasks is to group aspects which fall into natural groups; for example, in restaurant reviews, natural groups of aspects may be about food, some particular type of food like Chinese, décor of restaurant etc. This is done by aggregating aspects based on their term similarity, and then using the following features for clustering the aspects and their descriptors-

- Context or co-occurrence of aspects
- External knowledge base based similarity
- Semantic similarity based on aspects' descriptors

4 The Methodology

The proposed framework for our method is shown in Figure 1. It comprises the detailed workflow for the two main research objectives - Discovery of aspect-descriptor pairs and clustering of discovered aspects.

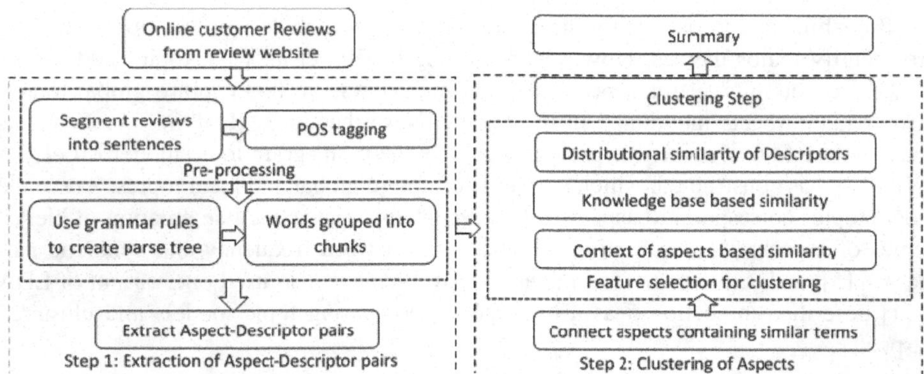

Fig. 1. Framework for aspect-descriptor extraction and clustering

4.1 Extraction of Aspect-Descriptor Pairs

A set of syntactic rules for identifying aspects and descriptors was developed, based on the following observations of the English reviews on Yelp:

- Adjectives, Participles, Articles, Possessive Pronouns and Prepositional Phrases can describe, modify or pointed to a noun.
- Words which describe an aspect are mostly modifiers like adjectives and participles. Adjectives modify or describe a noun. Sometimes Adjectives precede the noun they modify (e.g. "I like the *spicy* burger"), or they may follow a linking verb (e.g. "The burger was *spicy*.").
- Participles are verb forms that can be used as adjectives. An example of Past participles as descriptor is "*bored*" in "I was *bored* at the theatre." Present participle as descriptors can occur in front of a noun (e.g. "I like the *sizzling* dish.") or otherwise (e.g. "I like the dish *sizzling*.").
- Sometimes adverbs also carry information regarding an aspect (e.g. "The restaurant is open *daily*.").

Table 1 shows the custom syntactic rules we have created to identify and extract aspects and their descriptors. The first column provides the "chunk labels" we use to identify the type of aspect-descriptor pair created. In the second column, words in '<>' represent a tag, '<tag.*>' represent a tag can be followed by other letters like VBZ for VB, '<>*' represents zero or more occurrence, '<>+' represents one or more occurrence '?' represents zero or one occurrence. The rules are applied in the sequence given in the table so that a rule with higher priority is detected before a lower priority rule.

Before the extraction of aspect-descriptors pairs, reviews undergo certain preprocessing steps. First, each is review document is segmented into sentences. Next, the reviews are tokenized and lemmatized and parts of speech tagged are obtained, using the Stanford POS tagger, and further transformed into parse trees with the help of Stanford chunker. The result is a tree with words grouped into syntactic, labelled chunks, as in the first column in Table 1. Finally, these are passed to a function which then extracts Aspect-Descriptor pairs from the structures. There may be more than one chunk present in a sentence. A processed chunk can be a used in processing of another rule and become a sub-part of another chunk. If no rule is recognized and still descriptor are present in the sentence (e.g. "It was elegant."), then the aspect-descriptor pair ("forbusiness", "elegant") is generated. Such descriptors are assumed to describe the whole business entity as the aspect.

Table 1. Grammar Rules with examples

<table>
<tr><th>Chunk labels</th><th>Rule*</th><th>Pair extracted</th><th>Example</th></tr>
<tr><td>A</td><td>{<JJ>*<VJ><JJ>*<RB>*<IN|DT>*<NP>}</td><td>(NP,JJ/VJ)</td><td>They have broken <u>windows</u>.</td></tr>
<tr><td><b>B</b></td><td>{<b><JJ>+<CC>?<JJ>?</b><RB>*<IN|DT>*<NP> <IN|CC|DT>?<<b>NP</b>>?}/{<JJ><VJ>}</td><td>(NP.JJ)</td><td><i>Dirty</i> and <i>wet</i> <u>bedsheets</u> were found in the room</td></tr>
<tr><td><b>C</b></td><td>{<<b>NP</b>>+<W.*|PRP>*<VB.*|VJ><b><RB></b>+ <DT>?<<b>JJ|VJ</b>>}</td><td>(NP,JJ/RB)</td><td><u>Opening</u> is <i>always hectic</i>.</td></tr>
<tr><td><b>D</b></td><td>{<<b>B|A</b>>+<VB*|VJ><DT>?<<b>JJ|VJ|RB</b>>*}</td><td>(B|A,JJ/VJ)</td><td><i>Hot</i> <u>sizzler</u> is <i>amazing</i>.</td></tr>
<tr><td><b>E</b></td><td>{<<b>NP</b>>+<W.*|PRP>*<VB.*|VJ><DT>? <<b>JJ|VJ</b>>+}</td><td>(NP,JJ/VJ)</td><td><u>Rooms</u> are <i>clean</i>.</td></tr>
<tr><td><b>F</b></td><td>{<<b>NP</b>.*><W.*>*<VB.*><DT>?<RB>? <<b>B|A</b>>+}</td><td>(NP,B|A)</td><td><u>Weekends</u> are <i>great for people</i></td></tr>
<tr><td><b>G</b></td><td>{<<b>NP</b>>+<W.*>*<RBR|RBS>?<<b>JJ|VJ</b>></td><td>(NP,JJ/VJ)</td><td>I liked the <u>fish</u> <i>fried</i>.</td></tr>
</table>

**Where JJ are* Adjectives; *VJ ->{<VBG|VBN>*}* are Participle verbs; *RB* are Adverbs; *VB* are Verbs; *NBAR ->{<NN.*|JJ>*} }<JJ*> are nouns and nouns with adjectives*; *NP->{<NBAR><CC|IN><NBAR>}/{<NBAR>}* are noun phrases and noun phrases with conjunctions.

4.2 Clustering of Aspects

For generating aspect-based summaries, aspects which contain the similar terms are aggregated; then, feature values are calculated for every pair of aspects. After calculating the features, the aspects are clustered based on the calculated values. These steps are described in detail below.

Step 1- Connect Aspects Containing Similar Terms

In this step, aspects which are exactly similar or are almost exactly similar in case of multigrams are aggregated into a list of similar aspects, or an aspect-set. It is based on the fact that aspects sharing some words are likely to belong to the same cluster, for example "pool table" and "wooden pool table" most likely refer to the same aspect. Unigrams aspects, which are already lemmatized, are added only to a list of aspect which contain the exact same unigram. For multigrams, an approximate string matching is used. To paraphrase, for every new aspect from list of aspect descriptor pairs, if the incoming aspect is a multigram, say x, its term similarity is first measured against a list of multigrams aspects and if the similarity with another multigram aspect, say y, comes out to be greater than a threshold value, then the multigram aspect x is added to the list of the multigram aspect y, otherwise a new list is initialized with x. The similarity metric used is Jaccard similarity coefficient, eq. 1.

$$Similarity = s(C_i, C_j) = \frac{|s(C_i) \cap s(C_j)|}{|s(C_i) \cup s(C_j)|} \tag{1}$$

Here, $S(C_i)$ represents set of words in string C_i.

Multigrams are not grouped with unigrams because in some cases, if a multigram aspect approximately matching the unigram aspect is added to the set of unigram, then all new incoming unigrams which match with any word of the multigram will be added to the set. For example, if the multigram "poker face" is initialized as an aspect list, then both unrelated unigram aspects "poker" and "face" will be added to the same set.

Step 2- Feature Value Calculation for Clustering

Context or Co-occurrence of Aspects

In this step, aspects which are used in the same contexts are aggregated together based on their co-occurrence patterns, from the following observations of data:

- In general, a sentence is a collection of related aspects with a single focus, which creates a rough semantic boundary. Thus, word distribution information of aspects can be leveraged to cluster aspects based on their context similarity.
- People often express aspects and opinions about them in the same or repetitive syntactic structure, within a single sentence. For example, people often express their opinions about various dishes they ate, their experience with staff, etc. in the same sentence. For example "The fish was tasty but the chicken was overcooked." and "The waiter were friendly and the manager was understanding".
- Unrelated sentences can be a part of same review; but in the sentences itself, related aspects are usually mentioned together.

To gather context information, for every sentence in every review, a context vector is created which comprises all the aspects in the sentence. PMI (or Pointwise Mutual Information) measures the strength of association between two words by comparing the pair of words' bigram frequency to the unigram frequencies of the individual words. It is an indicator of collocation between the terms.

$$PMI(x,y) = \log\left(\frac{p(x,y)}{p(x).p(y)}\right) \tag{2}$$

It has been noticed that bigrams with low frequency constituents may gain high PMI value even when their occurrence probability is low. This problem can be addressed by multiplying the PMI value by an additional term of log of bigram frequency of *x* and *y*, *bigramfreq(x,y),* as suggested in [18]. The final PMI value is given in equation 3.

$$PMI(x,y) = \log(bigramfreq(x,y)) * \log\left(\frac{p(x,y)}{p(x).p(y)}\right) \tag{3}$$

PMI scores are calculated for each pair of aspect-set and saved to a record. The PMI values of multi-gram aspects is taken as the average value of PMI values of every combination of unigram aspect terms present in the multi-gram aspects. We also incorporate co-occurrence pattern information from this step to group together pairs of co-occurring unigram sets into multigrams. Furthermore, at this stage, pairs which have either low individual probabilities or low PMI (below a threshold) are removed from the record and from the clustering procedure.

External or Knowledge Base Based Similarity

In the next step, given two aspect terms, a1 and a2, we need to find their semantic similarity. We followed a WordNet based similarity approach similar to [19], to take into account both the minimum path length and the depth of the hierarchy path, so that specialized words with concrete semantics are grouped closer together than words in the upper levels of the hierarchy [19] as described in equation 4.

$$\mathrm{sw}(a_1, a_2) = e^{-\alpha l}.\frac{e^{\beta h} - e^{-\beta h}}{e^{\beta h} + e^{-\beta h}} \tag{4}$$

Here, α and β are parameters which scale the contribution of shortest path length and depth respectively and their optimal values depend on the knowledge base used, which in case of WordNet is proposed as 0.2 and 0.45 [19]. For the first term in the equation which is a function of path length, the value of *l* is 0 if both aspects are in the same synsets and are synonyms. If they are not in the same synset but their synsets contains one or more common words, value of *l* is 1. In all other cases, the value of *l* is the actual path length between the aspects. For the function of depth, due to the reason explained above, the function scales down the similarity for subsuming words at upper layers and scales it up for subsuming words at the lower layers.

The final similarity score is a value between 0 and 1. In case aspect is not present in WordNet, this value of the feature is taken as 0. If a similarity score is found, then if the pair has a minimum support in the corpus, the value is used in clustering. The similarity values of multi-gram aspects is taken as the average value of similarity values of every combination of unigram aspect terms present in the multi-gram aspects.

Distributional Similarity of Descriptors

Descriptors of an aspect contain semantic information that reflects the relationship of the aspects. Using the similarity of the virtual contexts provided by the descriptors, we can find similar aspects which may not themselves co-occur in the same contexts. Such information can capture implicit aspects which are not evident in the reviews by other features.

We model the semantic similarity of aspects as a function of semantic similarity of their descriptor words, by using a metric of word to word similarity of descriptors which indicates the semantic similarity of the two input aspects. Suppose we have a set for both aspects consisting of their descriptors of the form $S_1 = \{d_{11}, d_{12},.., d_{1m}\}$ and $S_2=\{d_{21}, d_{22},..., d_{2n}\}$, where d_{ij} are descriptor words. We first calculate semantic similarity of every pair of descriptors in both set. For semantic similarity of descriptor words, we use the metric normalized PMI as it indicates the degree of statistical dependence between two words. NPMI in equation 5 gives the semantic similarity of two descriptors x and y based of their occurrence and co-occurrence probabilities in the dataset.

$$NPMI(x, y) = \log\left(\frac{p(x, y)}{p(x).p(y)}\right)/p(x, y) \tag{5}$$

We have used another corpus based metric called inverse document frequency which was first introduced by Jones [20]. It is calculated as the log of total number of aspects divided by the number of aspects for which the descriptor is used, which is log(N/n_s). It is based on the fact that descriptors which occur with few aspects contain a greater amount of discriminatory ability than the descriptors that occur with many aspects in the data with a high frequency. This is because such aspects have meaning which relate to particular type of aspects like "delicious" relate to food related aspects.

This metric works well for similar aspects like "decor", "ambiance", "decoration", "furnishing" etc., their descriptors often share the same words like "colorful", "elegant", "modern", "sophisticated" etc. However, descriptors like "great" and "awesome" can be used with a large variety of aspects. Although such descriptors will get low *idf* values, we have manually created a list of very common descriptors which are not included in the calculation of similarity values of aspects. Once we have pairwise similarity values of descriptors from equation 5, they are used to calculate the similarity value of the aspects using the equation 6.

$$sim(A_1, A_2) = \frac{1}{2}\left(\frac{\sum_{d\in A_1}(maxSim(d, A_2) * log(N/n_d))}{\sum_{d\in A_1} log(N/n_d)} + \frac{\sum_{d\in A_2}(maxSim(d, A_1) * log(N/n_d))}{\sum_{d\in A_2} log(N/n_d)}\right) \quad (6)$$

Here, A_1 and A_2 are aspects, d is a descriptor, N is the total number of aspects in the corpus and n_d is the number of aspects d appears with. For each descriptor d in the aspect A1, we identify the descriptor in the aspect A2 with which it gets the maximum similarity value using equation 5. The equation is inspired by work in [21]. The descriptor similarities are weighted with the corresponding inverse document frequencies, then summed up, and an average is taken with the value we get by repeating the same procedure with descriptors of aspect A2. The final value $sim(A_1, A_2)$ is an estimate of the similarity between the aspects.

Step 3- Clustering step

Once the above similarity metrics are calculated, aspects are clustered together. We have used two approaches for clustering, one is our algorithm, which we call **Simset clustering** and other is a graph based algorithm called Markov Clustering (MCL) [22]. For both algorithms, first, the values of the 3 features described in section 4.2 are calculated; then, a graph is created with aspects as nodes and the value of the features as weight of edges between them. Both algorithms do not require a predefined number of clusters.

In Simset algorithm, first, the aspects are sorted by the total sum of their PMI values with other aspects. Then, for every aspect, a set called simset is initialized which will contain aspects similar to it. Then similarity with every other aspect is measured and if the similarity with aspect a_j is greater than a pre-fixed threshold, for any of the 3 features, then the a_j is either added to the simset of a_i if a_i has not been clustered earlier, otherwise the a_i is added to the simset of a_j and every element a_k in simset of a_i which has similarity value greater than any threshold with a_j, is moved to simset of a_j.

The major difference between Simset and MCL is that in MCL, the edges between each pair of aspects is weighted by their similarity values while in Simset, an edge is present between a pair of aspects only if the similarity value between them is greater than threshold values and the weight is same for each edge.

The MCL algorithm takes the graph as input, and simulates random walk through the graph. It is based on the principle that transition from one node to another within a cluster is much more likely than those in different clusters, and takes into account the weight of their links. The clustering process proceeds in an iterative manner and consists of two steps, one called *expansion,* which corresponds to computing random walks of higher length, which means random walks with many steps and the other called *inflation*, which boosts the probabilities of intra-cluster walks and demotes inter-cluster walks. Increasing the inflation parameter produces more fine-grained clustering. The algorithm converges after some iterations and results in the separation of the graph into different clusters.

Algorithm 1. Simset Clustering of Aspects

for a_i in A={a_1, a_2,… ,a_n} **do**
  initialize simset(a_i)
  for a_j in {a_{i+1}, a_{i+2},… ,a_n} **do**
   if similarity(a_i,a_j) > any(threshold$_h$) (h={1,2,3}) **then**
     if a_j not already clustered with another aspect **then**
       add a_j to simset(a_i), remove a_j from A
     else add a_i to simset(a_j)
       for all a_k ϵsimset(a_i) **do**
         if similarity(a_k,a_j) > any(threshold$_h$) **then**
           move a_k to simset(a_j)
   else : create simset(a_i) as a new cluster

5 Experiments

This section evaluates the proposed algorithm using dataset obtained from yelp.com. We analyze the performance of Aspect-Descriptor extraction and clustering in detail.

5.1 Dataset Description

Our dataset consists of online reviews of businesses provided by Yelp for "Yelp Dataset Challenge 2014[1]". The dataset consisted of 1,125,458 reviews of different businesses. The reviews were aggregated for every business id and filtered for a minimum number of reviews per business id. Then among the remaining businesses, the reviews of one hotel related business was taken as the final dataset, as it contained a lot of diverse aspects which could be detected and clustered by our approach. The reviews were segmented into sentences giving a total of 6,784 sentences.

1 Yelp (Dataset Challenge) `http://www.yelp.com/dataset_challenge`.

5.2 Qualitative Comparison

We show the top 10 clusters (according to size of clusters) detected from our model, in Table 2. The aspects which do not belong to the cluster are struck through. We have also identified a list of descriptor words for every aspect in every cluster in the output. It can be seen that the clusters extracted by Simset are descriptive and informative. Clusters have been manually assigned a label, provided in the first column of the table, and can easily be distinguished as related to service, art, parking etc. For example for the aspect “bed”, the list of descriptor words is ["comfort", "super comfi", etc.].

Table 2. Top aspect clusters detected by Simset clustering

Cluster label	Aspects terms	Common Descriptors
Parking related	parking, spot, level, parking garage, park space, fixture parking, light space, spot strip, slot, parking tip, self parking system, plenty of parking, light spot, live space, foot of space, light parking lot, garage, light bulb, ~~reaction, bond~~, flight, garage with number, ~~slot machine, lot of thought , lot of people, lot of restaurant, classic atmosphere lot~~, light alert, support, lot of celeb	Easy, good, easy to find, underground, plenty, open, free, biggest, available
Hotel related	hotel, casino resort, attractive hotel , hotel group, type of hotel, technology hotel, cosmopolitan hotel, detailed hotel, time hotel, thing hotel, genre of hotel, end hotel, boutique hotel, part hotel	Beautiful, amazing, sophisticated
Casino related	casino lounge, casino floor property, floor, ~~2nd floor~~, 3rd floor, ~~ground floor, floor balcony~~, three floor, casino/hotel, casino strip, lobby and casino, local casino, elevator, shelf liquor, casino tour, casino tour, background for picture, ~~entrance,~~ hotel/casino , level casino, casino area, control decade, casino/resort, store sell	Hippest, favorite, modern, unique, new, amazing, open
Night-club marque related	identity club, marque club, ~~card time, card key,~~ credit card, marque management,~~Identity, identity reward program, henry,~~ one word club, marque nightclub, ~~gold card status~~, glitch on check, pocket marque, line at marque, douchebag club, club downstair, sign marque, 2 night	hot, new, good music, popular, breathtaking
Bar related	chandelier, bartender, bar, casino bar, bar option, time bartend, bartender service, chandelier bar, mini-bar, mini bar space, bar in paris, casino bar restaurant, care minibar, bar sip, ~~pizzeria~~, lobby bar, bar and food event, bond bar, bar stool, chandelier middle, buffet, ~~fridge,~~ bond, vesper, spoon buffet, ~~kink buffet,~~ crystal chandelier, ~~minifridge~~	Crystal, sparkling, marvelous, multi- level, massive
Bathroom related	bathtub knob, deck shower, shower and tub, soak tub, shower area, shower pool, tub, bathroom, sink tub, tub outside, roll of toilet paper, toiletries, hair, bath, ~~bottle of water,~~ bathroom toiletries, bathroom 2, water pressure,~~ check in~~, whirlpool tub, ~~hallway with north~~	Huge, chic, cool, spacious, open
Art related	art, piece, piece of art, artwork, art from artist world, art work, art book, column of art, ~~pong and sport~~, graffiti artist, art display, ~~sport book~~, ~~foosball~~, ~~ping pong~~	Interesting, modern,original, great

Table 2. *(Continued)*

Art related	art, piece, piece of art, artwork, art from artist world, art work, art book, column of art, ~~pong and sport~~, graffiti artist, art display, ~~sport book~~, ~~foosball~~, ~~ping pong~~	Interesting, mod-ern,original, great
Pool related	place, boulevard, feel place, ~~table~~, issue, feel, pool table, pool deck, pool experience, ~~end place~~, ~~room area~~, dining place, ~~bedside table~~, area level, ~~town~~, pool with cabana, ~~pizza place~~, pool day, pool ideal	Beautiful, amazing, mo-dern, edgeless
Cus-tomer service related	cocktail service, week customer service line, customer service, cock-tail waitress, beverage service, room service, service waiting, food from room service, desk service, notch service, player club service, ~~gambler~~, factor and service, ~~internet service~~, food and service, service before	Good, quick, friendly, horrible, poor, great, terrible
Room related	suit, terrace, one bedroom, room terrace studio, terrace suit, bed room studio, ~~one person~~, one bedroomsuite, ~~one thing, one complaint~~, com-fort tower suit, city suit, conference center	Expensive, impressive, special, view

The top 10 clusters from DLDA are provided in Table 3. For DLDA, data was clustered by keeping n=10, and α and β to their proposed values. It is evident from the comparison of two tables that our system gives a better understanding of the aspects. The most reviewed aspects like the nightclub, bar and Casino are also detected explicitly, unlike the DLDA model. In the results of the DLDA model, since all words are unigrams, some of the aspects do not make sense like "lot" in "parking lot".

Table 3. Clusters detected from DLDA

Topic 1	room, service, time, desk, check, hour, customer, minute, glass, anything, money, coffee, security, employee, food, identity, call, cosmopolitan, gambling, plenty
Topic 2	bar, chandelier, drink, cosmo, friend, crystal, review, elevator, music, ~~design, story, end, top,~~ vibe, ~~desk, art, work, bag, touch~~, ceiling
Topic 3	night, day, club, marquee, property, ~~line, hour~~, party, crowd, weekend, ~~guest, front, way,~~ middle, ~~window, shop, name,~~ woman, ~~course, anyone~~
Topic 4	~~casino~~, lot, parking, lobby, spot, level, space, light, door, shower, garage, ma-chine, ~~thing, art~~, someone, slot, valet, wall~~, three, part~~
Topic 5	place, ~~everything~~, food, stay, card, star, year, guest, book, ~~eye~~, ~~pizza,~~ morning, couple, access, trip, ~~guy, mini, player~~, choice, ok
Topic 6	pool, ~~floor, area~~, restaurant, casino, table, lounge, game, chair, wow, bartender, ~~boulevard, living,~~ atmosphere, box, seat, nightclub, movie, week
Topic 7	room, ~~one,~~ tv, bathroom, bed,~~ two,~~ bedroom, ~~friend,~~ screen, ~~side, star, tub~~, fridge, kitchen, ~~time, city, size, idea, part, rest~~
Topic 8	strip, view, suite, balcony, terrace, ~~one, bellagio~~, hotel, ~~something~~, fountain, tow-er, ~~point, phone,~~ people, show, ~~kind~~, building, ~~wrap, word~~
Topic 9	~~hotel,~~ people, ~~cosmo, thing, decor,~~ staff,~~ buffet, fun, way, bit, experience, time, spoon, weekend,~~ person, conference, ~~water, system, entrance~~
Topic 10	hotel, vega, cosmopolitan, la, hip, ~~nothing,~~ reason, ~~spa, detail, service,~~ aria, re-view, ~~select, thing,~~ center, ~~note, everyone, cocktail, charge,~~ event

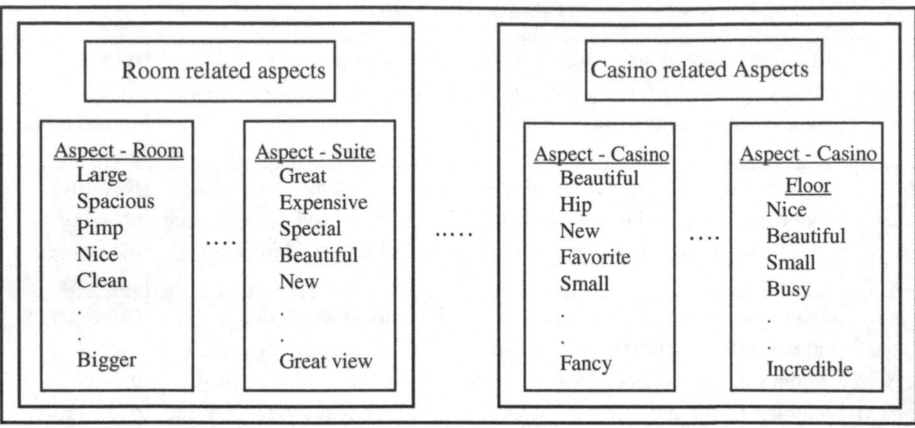

Fig. 2. Example of summary generated for aspects

An example of summary generated for aspects by our method is shown in figure 2. Different aspects in a group along with the descriptor for each aspects form a summary for the reviews which is easy to understand. One advantage of our method is that the aspect-descriptor pairs extracted from reviews can be grouped with any algorithm to produce summaries for aspects and their descriptors.

5.3 Results

Evaluation of Aspect-Descriptor Pair Discovery

The results for aspect-descriptor extraction obtained from manually labelling first 750 sentences in the dataset are presented in Table 4. It is notable that in our results, along with explicit Aspect-Descriptor pairs, our system could identify most descriptors that were used for expressing opinions about the business entity as the aspect like for "Stylish and modern" gives pairs ("forbusiness", "stylish") and ("forbusiness", "modern").

Table 4. Performance metrics

Precision	Recall	F-score
0.871	0.894	0.882

Evaluation of Aspect Clustering

For evaluation of our clustering method, we used the reviews of a particular business from the collection, from which we selected first 500 reviews, which after segmentation consisted of 6784 review sentences. We compared Simset and MCL algorithms with a modified version of LDA [12]. It takes as input a set of documents, and outputs groups of terms, and each group belongs to a topic. For input to this version of LDA (denoted as ***DLDA***), all words except the aspects in the documents are removed and only distributional information of aspects is used for grouping of aspects. The topic modeling parameters were set to their default values. The number of iteration of Gibbs sampling is set to 2000. Number of topics is set to 20.

MCL algorithm takes only inflation parameter I as input. Low value of inflation parameter produces coarser clustering, and high value fine-grained clustering. In the experiments, we adjusted I for each experiment which gives the best performance. We experimented with MCL by giving different features and their combinations as input to the algorithm. The different settings are- Only value of co-occurrence based similarity set as edge weights in graph (experiment denoted by ***MCL-pmi***), only value of Wordnet based similarity set as edge weights in graph (experiment denoted by ***MCL-word***), only value of distributional similarity of descriptors) set as edge weights in graph (experiment denoted by ***MCL-desc***) and maximum value of a features set as edge weights in graph after normalization (experiment denoted by ***MCL-mix***). In MCL-word and MCL-desc, only those pairs of aspects are considered for calculation which have a minimum co-occurrence support in the dataset, to avoid grouping unrelated aspects together.

Only needed parameters in Simset are the different threshold values we have used which we set as 7.0 for feature 1 (co-occurrence based similarity), 0.75 for feature 2 (WordNet based similarity) and 0.75 for feature 3 (descriptor based similarity) after a series of manual evaluations.

In this section we will show some objective results of our clustering. Measures like perplexity and topic coherence are often used for evaluation for clustering, however perplexity does not reflect the semantic coherence of aspects and can sometimes be contrary to human judgments and topic coherence cannot take into account the knowledge base feature used by us as it relies upon only co-occurrence statistics. So for evaluation of our system, we have used Purity and Entropy as metrics as in [23]. Since we did not have any form of ground truth for the dataset, we had to evaluate the performance of clustering manually. Purity P_j of each cluster j and the total purity of clustering solution are defined as follows

$$P_j = \frac{1}{n_j} Max(n_j^i)$$

$$Purity_{total} = \sum_{j=1}^{m} \frac{n_j}{n} P_j$$

where P_j is the purity and n_j is size of the jth cluster. n is the sum of sizes of all clusters.

Entropy measures the purity of clusters class labels. The smaller the entropy values, better the clustering is. The entropy and the total entropy are defined as follows.

$$E(S_r) = -\sum_{i=0}^{q} \frac{n_r^i}{n_r} log \frac{n_r^i}{n_r}$$

$$Entropy_{total} = \sum_{j=1}^{m} \frac{n_j}{n} E(S_r)$$

where q is the number of classes, and n_r^i is the number of documents of the ith class that were assigned to the rth cluster. The entropy of the whole clustering is defined as the sum of the individual cluster entropies weighted according to the cluster size.

The results for evaluation of clustering algorithms are summarized in Table 5. We did our evaluation of top 20 clusters according to the number of aspects contained by them for every method. It clearly show that the algorithms MCL-mix and Simset outperform all other baseline methods by a large margin.

Table 5. Experimental results for clustering algorithms

Clustering algorithm	Average no. of aspects in a cluster	Total Purity	Total Entropy
LDA	20.0	0.29	3.24
MCL-word (I=4.0)	15.1	0.65	1.50
MCL-desc (I=2.0)	8.8	0.57	1.68
MCL-pmi (I=5.0)	17.9	0.70	1.32
MCL-mix (I=5.0)	13.6	**0.71**	**1.16**
Simset	18.0	**0.75**	**0.84**

6 Discussion

The following paragraphs discuss the performance and analyze the errors observed, in increasing order of performance.

DLDA performs worst of all methods. Since DLDA depends heavily on distributional information and only considers unigrams. Aspects such as "design" and "decor" are very less likely to be put in the same groups. The results of MCL-desc have been confounded by generic descriptors for aspects, like "good" and "amazing", which are often used for a lot of aspects (e.g., food), alongside specific descriptors such as "hot", "tasty" and so on. If only common descriptors were used for aspects, then even unrelated aspects like "bar", "tub", etc. got high similarity scores resulting in poor results.

MCL-word was able to give put aspects like in "design" and "decor" in similar groups. However, if aspects were unavailable in Wordnet, it was not able to group similar aspects together. Aspect terms which were not related in the dataset, but were related in meaning and in Wordnet, got high Wordnet similarity scores and will be grouped together. Although we have set a condition for minimum support required for consideration of Wordnet similarity, still aspects like "staff" and "crowd" which although occur with minimum support, but yet are unrelated get good similarity scores in Wordnet. MCL-pmi performed slightly better.

MCL-mix considered both context and Wordnet relation into account. It proved that combining multiple criteria for aspect similarity results in better clustering. Finally, Simset clustering gives best result overall with 5% increase in purity and 40 % lower entropy than MCL-mix. The major difference between both was that in MCL, edges were given weights and in Simset, all edges had same weights but they were removed if similarity values were less than a threshold. The increase in performance can be attributed to removal of noise due to removal of edges. For example, in MCL, aspects with moderate similarity values like "game" and "pool party" were considered in clustering and if there were not enough connections of these aspects with other aspects, they would be clustered together, but in Simset such aspects with moderate value similarities are not considered for clustering.

7 Conclusion and Future Work

In this paper, we study the problem of aspect discovery and clustering. We first discovered aspect-descriptor pairs from reviews. Then we proposed three features and metrics for aspect similarity and an unsupervised clustering method. The aspect similarity features proposed performed worked well with both clustering algorithms and have proven to be better than baseline method. The experiments are preliminary, and our method has yet to be tested on different datasets and domains. More parts of speech like verbs can be considered as descriptor words. Sometimes review about an aspect may extend to more than one sentence of a review, for such possibilities, a co-reference method is needed. In future work, we plan to generate natural language summaries of aspect clusters, to highlight the constituent aspects and their descriptors in a meaningful manner.

References

1. Popescu, A.M., Etzioni, O.: Extracting product features and opinions from reviews. In: Proceedings of Human Language Technology Conference/Conference on Empirical Methods in Natural Language Processing, HLT-EMNLP 2005, Vancouver, CA, pp. 339–346 (2005)
2. Miller, G.A.: WordNet: A Lexical Database for English Comm. ACM 38(11), 39–41 (1995)
3. Qiu, G., Liu, B., Bu, J., Chen, C.: Opinion word expansion and target extraction through double propagation. Computational Linguistics 37, 9–27 (2011)
4. Hu, M., Liu, B.: Mining and summarizing customer reviews. In: Proceedings of ACM-KDD, pp.168–177 (2004)
5. Yi, J., Niblack, W.: Sentiment mining in WebFountain. In: Proceedings of the 21st International Conference on Data Engineering, ICDE 2005, pp. 1073–1083. IEEE Computer Society, Washington, DC (2005)
6. Wu, Y., Zhang, Q., Huang, X., Wu, L.: Phrase dependency parsing for opinion mining. In: Proceedings of the 2009 Conference on Empirical Methods in Natural Language Processing, vol. 3, pp. 1533–1541 (2009)
7. Liu, B., Hu, M., Cheng, J.: Opinion Observer: Analyzing and Comparing Opinions on the Web. In: Proceedings of WWW, pp. 342–351 (2005)
8. Carenini, G., Ng, R., Zwart, E.: Extracting knowledge from evaluative text. In: Proceedings of International Conference on Knowledge Capture, pp. 11–18 (2005)
9. Guo, H., Zhu, H., Guo, Z., Zhang, X.X., Su, Z.: Product Feature Categorization with Multilevel Latent Semantic Association. In: Proceedings of International Conference on Information and Knowledge Management, pp.1087–1096 (2009)

10. Matsuo, Y., Sakaki, T., Uchiyama, K., Ishizuka, M.: Graph based word clustering using a web search engine. In: Proceedings of the 2006 Conference on Empirical Methods in Natural Language Processing, EMNLP 2006, pp. 542–550. Association for Computational Linguistics, Stroudsburg (2006)
11. Blei, D.M., Ng, A.Y., Michael, J.: Latent dirichlet allocation. Journal of Machine Learning Research 3, 993–1022 (2003)
12. Titov, I., McDonald, R.: Modeling online reviews with multi-grain topic models. In: Proceedings of the 17th International Conference on World Wide Web, pp. 111–120 (2008)
13. Titov, I., McDonald, R.: A joint model of text and aspect ratings for sentiment summarization. In: Proceedings of ACL, HLT, pp. 308–316 (2008)
14. Kim, S., Zhang, J., Chen, Z., Oh, A., Liu, S.: A Hierarchical Aspect-Sentiment Model for Online Reviews. In: Proceedings of AAAI, pp. 526–533 (2013)
15. Liu, K., Xu, L., Zhao, J.: Syntactic Patterns versus Word Alignment: Extracting Opinion Targets from Online Reviews. In: Proceedings of ACL 2013, pp. 1754–1763 (2013)
16. Mukherjee, A., Liu, B.: Aspect Extraction through Semi-Supervised Modeling. In: Proceedings of ACL, pp. 339–348 (2012)
17. Liu, Q., Gao, Z., Liu, B., Zhang, Y.: A Logic Programming Approach to Aspect Extraction in Opinion Mining. In: Proceedings of the 2013 IEEE/WIC/ACM International Joint Conferences on Web Intelligence (WI) and Intelligent Agent Technologies (IAT), WI-IAT 2013, vol. 01, pp. 276–283. IEEE Computer Society, Washington, DC (2013)
18. Hoang, H.H., Kim, S.N., Kan, M.Y.: A re-examination of lexical association measures. In: Proceedings of Workshop on Multiword Expressions (ACL-IJCNLP), pp. 31–39 (2009)
19. Li, Y.H., Bandar, Z., McLean, D.: An Approach for Measuring Semantic Similarity Using Multiple Information Sources. IEEE Trans. Knowledge and Data Eng. 15(4), 871–882 (2003)
20. Jones, K.S.: A statistical interpretation of term specificity and its application in retrieval. Journal of Documentation 28(1), 11–21 (1972)
21. Rada, M., Courtney, C., Carlo, S.: Corpus-based and knowledge-based measures of text semantic similarity. In: Proceedings of the 21st National Conference on Artificial Intelligence, vol. 1, pp. 775–780 (2006)
22. Dongen, S.V.: A cluster algorithm for graphs, Ph.D. dissertation, University of Utrecht (2000)
23. Zhai, Z., Liu, B., Xu, H., Jia, P.: Clustering product features for opinion mining. In: Proceedings of the 4th ACM International Conference on Web Search and Data Mining, pp. 347–354 (2011)

An Approach for Intention Mining of Complex Comparative Opinion Why Type Questions Asked on Product Review Sites

Amit Mishra and Sanjay Kumar Jain

Computer Engineering Department, NIT Kurukshetra, Haryana, India-136119

Abstract. Opinion why-questions require answers to include reasons, elaborations, explanations for the users' sentiments expressed in the questions. Sentiment analysis has been recently used for answering why type opinion questions. Existing research addresses simple why-type questions having description of single product in the questions. In real life, there could be complex why type questions having description of multiple products (as observed in comparative sentences) given in multiple sentences. For example, the question, "I need mobile with good camera and nice sound quality. Why should I go for buying Nokia over Samsung?" Nokia is the main focus for the questioner who shows positive intention for buying mobile. This calls for natural requirement for systems to identify the product which is centre of attention for the questioners and the intention of the questioner towards the same. We address such complex questions and propose an approach to perform intention mining of the questioner by determining the sentiment polarity of the questioner towards the main focused product. We conduct experiments which obtain better results as compared to existing baseline systems.

Keywords: Question Answering, Information retrieval, Natural language processing, Natural language understanding and reasoning.

1 Introduction

Question Answering Systems (QASs) provide specific answers to questioners' questions. Why Questions type are the questions that require answers to include reasons, elaborations, explanations for the questioners questions. Most of the research dealing with 'why' type questions in QASs consults information source based on facts i.e., newspaper, technical documents etc. [1] Research on why type questions consulting opinion data has been very limited. With the emergence of Web 2.0, there are massive user generated data on the web such as social networking sites, blogs, and review sites etc. [7,8, 9]. These opinionated data sources contain public opinions which could help the questioners in making judgment about the products. Hence they could contribute in answering 'why type questions' such as why should I look for product x?

From literature [1, 2, 4, 6, 12, 17], we find that current state of art addresses simple why-type questions having description of single product expressed in single sentence.

A. Gelbukh (Ed.): CICLing 2015, Part II, LNCS 9042, pp. 257–271, 2015.
DOI: 10.1007/978-3-319-18117-2_19

In real life, there could be complex why questions having multiple product descriptions (as in comparative sentences) expressed in multiple sentences. For example, "I need mobile with good camera. Why it is better for me to have Nokia over Micromax?" here 'Nokia' is center of attention as a product for the questioner. Systems need to determine the intention of the questioner along with main focused product for which he needs information (Nokia in this example). The task is Intention mining. Murthy Ganapathibhotla, Bing Liu [12] performs Opinion Mining in Comparative Sentences and assumes that objects generally appear on both sides of a comparative word. The techniques fall flat when applied on complex sentences like why it is better to have Nokia over Micromax? Here the opinion bearing word is 'better' and objects are 'Nokia' and 'micromax'. In the question, objects are on the same side of opinion bearing word. The authors do not deal with opinion mining from multiple sentences. Such technique does not address such complex scenario

Another prominent system, SenticNet [15, 17] detects sentiment polarity of single sentence by using machine-learning and knowledge-based techniques. It performs tasks of Aspect mining. The SenticNet captures the conceptual and affective information in the sentence by using the bag-of-concepts model. The system assumes that input text is opinionated. It does not address problem of opinion mining from multiple sentences and comparative opinion mining from text. Another systems, OpinionFinder[14] performs sentiment analysis at document level and identify subjective sentences and sentiment expressions in the text.

Such systems could fall flat when applied to multi sentences and multiple products based Why type questions.

We present an approach for determining the sentiment polarity of complex opinion why-questions that could be expressed in multiple sentences or could have multiple product descriptions. To the best of our knowledge, we do not find any work which could address such problems.

We assume that products are already extracted from questions through the work reported in literature [Jindal Liu 2006]. For instance, "why it is better to have Nokia over Micromax in England than usa?" Here the products like Nokia and Micromax are already identified. The questioner sentiment on main focused product is to be determined. This is a key piece of information while searching for answers of the questions. As in the example, questioner is looking with positive intent for Nokia as compared to Micromax.

Rest of the paper is organized as follows Section 2 discuss related literature work. Section 3 presents our Approach for determining sentiment polarity of Why-questions. We have results and discussion in Section 3. Finally, we have conclusions and scope for future research in Section 5.

2 Related Work

Based on works on opinion question answering [1, 2, 4, 5, 6], we find that question analysis, document analysis, retrieval method and answer processing are the steps in drawing answers to opinion 'why' questions. Output of the question analysis phase has cascade effects on other phases in generating correct answers. Further, we find that question analysis comprises of several sub processes i.e., recognizing entity in question, identifying its aspects, detecting sentiment polarity of question and question form. Determining polarity of why-questions is a significant phase for generating correct answers as it searches for intention of users expressed in questions related to products.

Sentiment polarity of opinion questions is determined through identification of opinion bearing words and computation of their polarity score through opinion lexical resources [1, 2, 4, 5, 6].

S. Moghaddam et al develop a opinion question answering system in which they consider only adjectives as opinion bearing words for the task of determining sentiment polarity of questions [4,8]. They use a subset of seed words containing 1,336 adjectives. These words are manually classified into 657 positives and 679 negatives by Hat Zivassiloglov et al. [14]. In another work, Farah Benamara found that adjectives and adverbs work better than adjectives alone for the task of sentiment polarity detection [16]. Muhammad Abuliash et al. use adjectives or adjectives headed by adverbs as opinion bearing words in text documents to produce summary of review documents on the basis of features through semantic and linguistic analysis using SentiWordNet[13]. These researchers ignore nouns and verbs which could also behave as opinion words. Turney found that adjectives, verbs, nouns and adverbs play significant role as opinion bearing words for the task of opinion mining [17].

Jong hu et al. consult a Japanese polarity dictionary distributed via Alagin forum in their question answering [2].The dictionary is not available in English. Jianxing Yu et al. present a opinion question answering system for products review sites by exploiting hierarchical organization of the product reviews [5]. They use SVM sentiment classifier to determine sentiment polarity of questions. For doing this, they consult the MPQA project sentiment lexicon. Most of the words in MPQA project are objective words such as buy; purchase, choose etc. hence we consider the corpus as not a good choice.

SenticNet detect sentiment polarity of single sentence by using machine-learning and knowledge-based techniques [15, 17]. The SenticNet capture the conceptual and affective information in the sentence by using the bag-of-concepts model. The system assumes that input text is opinionated. It does not deal with multiple sentences and sentences with comparative opinions. OpinionFinder [7] perform document level analysis and identify subjective sentences and sentiment expressions in the text.

Murthy Ganapathibhotla, Bing Liu perform Opinion Mining in Comparative Sentences [12] and assume that objects generally appear on both sides of a comparative word which is not true in complex cases like why it is better to have Nokia over Micromax?

Stanford Sentiment [14] determine the sentiment of movie reviews by considering the order of words through representation of whole sentences based on the sentence structure through new deep learning model.

From the work done in area of opinion question answering, we don't find any work which focus on intention mining from comparative multiple sentences.

Based on [1, 2, 3, 4, 5, 6] Research issues related to opinion polarity detection in 'why' questions are as follows:

1. Identification of question sentiment polarity (positive or negative)

   1.1 Question could be single or multiple sentences.
   1.2 Question could have single opinion or mixed opinion.
   1.3 Question could be comparative.

2. Identification of opinion expressing words as Verb, noun, adjective, all play important role in opinion mining.

3 Our Approach for Intention Mining

In this section, we determine the sentiment polarity of the questioner towards main focused product from multiple sentences why questions. The task can be categorized into three steps –

1. We first identify more important text span from multiple sentence why questions which is important in view of opinion mining.
2. We compute sentiment polarity of such text span.
3. We finally extract the main focused product from the questions.

3.1 Extraction of the Opinionated Segment from Why Questions for Intention Mining

The objective is to extract the text span from a why type question which is significant for intention mining. For example - I went to a good shop X. I saw two category of mobiles- Micromax and Nokia category. Why should it is better for me to buy Nokia over micromax?

Here, 'Why should it is better for me to buy Nokia over Micromax?" is the text span in comparison to others for opinion mining.

We analyze intra-sentential and inter-sentential discourse structure of why-questions to determine more important spans. The algorithm is as follows:

1. The question is parsed through A PDTB-Styled End-to-End Discourse Parser developed by Ziheng Lin et al. [9]. We first analyze inter-sentential discourse structure then intra-sentential discourse structure of the question.
2. The analysis of inter-sentential discourse structure:
   2.1 If discourse relation is either Cause, or Conjunction or Contrast in between 'why' keyword containing sentence and other sentence then, we choose Argument Arg(2) span as first priority.
   2.2 Else If discourse relation is Condition or others, in between 'why' keyword containing sentence and other sentence, then, we choose Argument Arg(1) span as first priority.
3. The analysis of intra-sentential discourse structure:

3.1 If discourse relation is Cause, or Conjunction or Contrast in between pair of arguments in the sentence having 'why' keyword, then we choose argument Arg(2) span as first priority.
3.2 Else If discourse relation is Condition or others, in between pair of arguments in the sentence having 'why' keyword, then we choose Arg(1) span as first priority.

For Example:

I need a mobile with good sound quality and nice looks. I went to market. I found three good shops. I went to shop number 3. Why should it is better to have nokia over micromax?

We see the output file as shown below:

{NonExp_0_Arg1 {NonExp_0_Arg1 I need a mobile with good sound quality and nice looks . NonExp_0_Arg1} NonExp_0_Arg1}

{NonExp_1_Arg1 {NonExp_0_Arg2_EntRel {NonExp_1_Arg1 {NonExp_0_Arg2_EntRel I went to market . NonExp_0_Arg2} NonExp_1_Arg1} NonExp_0_Arg2} NonExp_1_Arg1}

{NonExp_2_Arg1 {NonExp_1_Arg2_EntRel {NonExp_2_Arg1 {NonExp_1_Arg2_EntRel I found three good shops . NonExp_1_Arg2} NonExp_2_Arg1} NonExp_1_Arg2} NonExp_2_Arg1}

{NonExp_3_Arg1 {NonExp_2_Arg2_EntRel {NonExp_3_Arg1 {NonExp_2_Arg2_EntRel I went to shop number 3 . NonExp_2_Arg2} NonExp_3_Arg1} NonExp_2_Arg2} NonExp_3_Arg1}

{NonExp_3_Arg2_Cause {NonExp_3_Arg2_Cause Why should it is better to have nokia over micromax ? NonExp_3_Arg2} NonExp_3_Arg2}

We first analyze inter sentential discourse structure. We find that there is cause relations in between 4th sentence and 5th sentence with later being Arg2 i.e., for a relation- Non Exp 3 cause, we see Arg 1 as 'I went to shop number 3', and Arg 2 as 'Why it is better to have nokia over micromax?'. Hence we select Arg 2 as more important text span.

Hence the overall intention of user with which he is looking for product is expressed in Arg 2 text span 'Why it is better to have nokia over micromax'.

3.2 Computing the Sentiment Polarity of Why-Questions for Intention Mining

Method 1 Consulting MPQA [7]: We perform subjectivity analysis of why questions using OpinionFinder System. Opinion Finder recognizes subjective sentences as well as different aspects of subjectivity within sentences.

Method 2 Consulting SenticNet [15]: we performed the sentiment analysis through work reported in literature [15, 17].

Method 3 Consulting Stanford Opinion Mining: Stanford Sentiment determine the sentiment of reviews by considering the order of words through representation of whole sentences based on the sentence structure through recursive neural network model.

Method 4: Our Method
We follow average scoring methods in computing sentiment scores of the target opinionated text span of questions as their sentiment polarity is same as of that of questions. We compute scores on the basis of sentiment scores of opinion words of the text spans [1, 2, 4, 6]. From literature survey, we find that adjectives, nouns, adverb, verb could behave as opinion bearing words. In this regard, we parse the question through the Stanford Parser [10] to determine the part of speech of each word. We remove Pre compiled Stopwords from the question words to get opinion words. We change opinion words to their root form through morphological analysis.

We classify the sentiment polarity (i.e. Positive, or negative or neutral) of Question text span through following steps as discussed below:

3.1.1 Computing Score of Opinion Word

We compute the score of each opinion word of question text span through methods described in literature using different popular sentiment lexicons ie, SentiWord Net, and MPQA. We propose a method which performs better in comparison to the discussed methods.

3.1.2 Computing Score of Question Text Span (Question Polarity Scoring (QPS))

We take average of scores of words to determine overall sentiment polarity of question text span [5, 7, 8, 10, 12, and 28].

3.1.3 Computing Score of Opinion Word

In this section, we compute the score of each opinion word of question text span by using different popular opinion lexicons i.e, SentiWordNet, MPQA. We recompute the score of words through our approach.

In our approach, we search for synonymous words to improve the sentiment polarity of why-questions. From our experiments, we find that MPQA and SentiWordnet is the effective dictionary for the purpose. Our approach is as follows:

1. Calculate score of each argument.
2. We compute the score of opinion word extracted in section 3.2. We calculate the score of the word through following rules. As there are two values for subjective score (strong or weak), and two values of positive score (strong or weak) and two values of negative scores (strong or weak) hence there are (2*2*2=8) combinations. And there is one combination of words not found in corpus. Each word score in each argument is calculated from MPQA dictionary
3. If the polarity of word is positive or negative regardless of its score and strength is strongsubj or weaksubj. Then final score of word will be made same.

Strong positive with strong subj of word has score equivalent to 1.00
Strong positive with strong subj of word has score equivalent to .75
Weak positive with strong subj of word has score equivalent to .50

Weak positive with weak subj of word has score equivalent to .25
The word which is not found in the corpus is assigned score 0.00.
Weak negative with weak subj of word has score equivalent to -0.25
Weak negative with strong subj of word has score equivalent to -0.50
Strong negative with strong subj of word has score equivalent to -0.75
Strong negative with strong subj of word has score equivalent to -1.00

4. Else the score of the word is calculated with the help of SentiWordNet

We update Scoring Method 1 consulting SentiWordNet by doing modification. We do some extra computation on WordScore(w) if it equals to zero. We compute WordScore(w). If WordScore(w) equals to zero, then we search for other synonymous words falling in same synonymous set. We compute WordScore(w)

For example: if I need average mobile, why should I choose the product X? , Choose is synonymous with take#10, select#1, pick_out#1 , prefer#2 opt#1 in sentiWord Net. Hence the updated positive score of the 'choose' word is average sum of all positive scores of synonymous words. Same is done for negative score computation.

Computing score of Question text span (Question Polarity Scoring (QPS)): QPS is computed by averaging the score (both positive and negative) of the opinion words present in the question text span related to the feature M:

$$QScore(q) = \frac{1}{n}\sum_{i=1}^{n} WordScore(i)$$

QScore(q) score of question text span Q which is related to product feature M. WordScore(i) is score found of ith word (w) in question text span S .
n = Total Number of words in Question text span.
Based on value of QScore(q), we determine polarity of question span text q.
IF QScore(q) is positive, hence question span text q have positive polarity.
QScore(q) is negative, hence question span text q have negative polarity.
QScore(q) is neutral, hence question span text q is neutral.

3.3 Determine Centre of Attention of the Questioner

We propose an algorithm to determine the main focus (product) for which questioners require information in answers. We parse the questions through a semantic role labeler [18]. It detects semantic arguments related with predicate or verb in the sentence.

Following rules are followed to extract entities.

1. Extract semantic arguments connected with each predicate or verb in the questions and group them separately.
2. If there is a product labeled as either Arg(0) or Arg(1), select the product as main focus
3. Else if a product is Subject(subj), select product as main focus
4. Else if the product is labeled as 'pmod' i.e., object of preposition and the preposition is Subject, select product as main focus.

We analyze 78 manually constructed opinion why Questions prepared by our colleagues which consist of different ways by which why-questions could be asked on product review sites [list of questions and their analysis presented after reference section]. There is no standard data set for opinion 'why' questions to the best of our knowledge. We find accuracy of Question Fragmentation module for opinion mining in Table 1. The details are given after reference section. We do the analysis of the questions and determine their sentiment polarity. In Table 2, we present the accuracy observed in different methods.

Table 1. Accuracy of Question Fragmentation module

Method	Our Method
Accuracy	76.9%

Table 2. Accuracy of different Methods [23]

Method	Method 1	Method 2	Method 3	Our Method
Accuracy	0.50	0.70	0.24	0.79

We compared our results with work [15,17] and [12] in Table 3. In the work, researchers have extracted products or service features (aspects) from text using aspect parser of Senticnet. We performed extra processing by deriving more focused product from questions. For the purpose of comparison, we derived the aspects from the questions using aspect parser, and considered the output as successful even if the parser list of aspects matches with focused product or its aspect.

Table 3. Accuracy of different Methods in Finding Main Focused Product

Method	Method by [24,20]	Method by [12]	Our Method
Accuracy	0.6	0.56	0.78

4 Results and Discussion

We analyze the results and get following observations. We find that our proposed Method gives maximum accuracy of 79% in sentiment polarity detection of opinion why questions. For the task of main focus detection we find that our method excels the current state of art by 18%.

We observe that there are various factors which affect our accuracy. The main reason for outperforming Stanford RNTN is the absence of proper training sample to train the RNTN model. Because of this RNTN can't learn the context of most of the why type questions and predict them as neutral or if any why type contains polar words then, RNTN labelled that sentence according to its deep learning framework. On the other hand though Sentic Patterns obtained comparable results to our system but it fails to achieve as high result we got. The linguistic patterns for “should”,”could” help to increase the accuracy of the Sentic Patterns system on the dataset. However, due to the lack of “why” question type specific patterns it does not perform very well. Both RNTN and Sentic Paterns can’t detect the main focus of an opinionated why type question e.g. the opinion target in an opinionated why type question. This also means both of the system are not capable to detect the sentiment of comparative why type question. Below, we give other reasons why the proposed method is novel and effective:

4.1 WSD (Word Sense Disambiguation)

We calculate the average sum of all scores of the word related to a given part of speech in SentiWord Net. Words behave differently in terms of polarity in different context. Hence identification of the word sense and allotting the score of the sense directly could improve the performance of the systems. Such as *Why I need camera x?* here average sum of need word leads to negative polarity.

4.2 Opinion Bearing Words

Identification of opinion bearing words in the sentence could increase the performance of the proposed system. Our system calculates the scores of all words of the sentences.

4.3 Discourse Analysis

We use PDTB-Styled End-to-End Discourse Parser developed by Ziheng Lin et al [45] as the accuracy of discourse parser in today’s era is not very promising hence it affect our performance.

4.4 Domain Specific Lexicon

SentiWord Net, MPQA, Bing Liu lexicon are open domain dictionary. Some domain specific lexicons behave differently in polarity than general domain lexicons. E.g. long. If the *camera coverage* is *long* then it is good. But the *movie is long* it expresses negative sentiments.

4.5 Informal Language

Use of informal language effect the method.

5 Conclusions and Future Works

In this paper, we determine the polarity of the questions that could be single or multiple sentence(s) ‘why’ type questions that could have multiple products description in the questions. We perform discourse-based analysis of why type questions which is dependent on performance of automatic discourse parser. Instead of calculating score for all words, we observe that detecting opinion bearing words and computing their sentiment scores could improve the performance of 'why' QAS. We know SentiWord Net, MPQA is general domain dictionary hence there should be domain specific learning to use same. In the future, we will use different discourse parsers to evaluate and compare our methods on different parameters. We will use machine-learning methods for the task of sentiment polarity detection of questions as it could be effective in different domains. We will use machine-learning methods for the task of main focused determination from questions. We will review the semantic role labelers and identify their usefulness for our work. Future work [17-28, 33-40] will also explore the use of commonsense knowledge to better understand the contextual polarity as well as the opinion target of a why type question. We will also explore the use of contextual information carried by syntactic n-grams [29-32] and complex representations of text [41, 42].

References

1. Fu, H., Niu, Z., Zhang, C., Wang, L., Jiang, P., Zhang, J.: Classification of opinion questions. In: Serdyukov, P., Braslavski, P., Kuznetsov, S.O., Kamps, J., Rüger, S., Agichtein, E., Segalovich, I., Yilmaz, E. (eds.) ECIR 2013. LNCS, vol. 7814, pp. 714–717. Springer, Heidelberg (2013)
2. Oh, J.-H., et al.: Why-question answering using sentiment analysis and word classes. In: Proceedings of EMNLP-CoNLL 2012 (2012)
3. Hung, C., Lin, H.-K.: Using Objective Words in SentiWordNetto Improve Sentiment Classification for Word of Mouth. IEEE Intelligent Systems (January 08, 2013)
4. Moghaddam, S., Ester, M.: AQA: Aspect-based Opinion Question Answering. IEEE-ICDMW (2011)
5. Yu, J., Zha, Z.-J., Wang, M., Chua, T.-S.: Answering opinion questions on products by exploiting hierarchical organization of consumer reviews. In: Proceedings of the Conference on Empirical Methods on Natural Language Processing (EMNLP), Jeju, Korea, pp. 391–401 (2012)
6. Ku, L.W., Liang, Y.T., Chen, H.H.: Question Analysis and Answer Passage Retrieval for Opinion Question Answering Systems. International Journal of Computational Linguistics & Chinese Language Processing (2007)
7. Wilson, T., Wiebe, J., Hoffmann, P.: Recognizing Contextual Polarity in Phrase-level Sentiment Analysis. HLT/EMNLP (2005)
8. Moghaddam, S., Popowich, F.: Opinion polarity identification through adjectives. CoRR, abs/1011.4623 (2010)
9. A PDTB-Styled End-to-End Discourse Parser developed by Lin, Z., et al. http://wing.comp.nus.edu.sg/~linzihen/parser/
10. Bu, F.: Function-based question classification for general QA. In: Proceedings of the 2010 Conference on Empirical Methods in Natural Language Processing, October 9-11, pp. 1119–1128, Massachusetts, USA (2010)
11. Padmaja, S., et al.: Opinion Mining and Sentiment Analysis - An Assessment of Peoples' Belief: A Survey. International Journal of Adhoc, Sensor & Uboquitos Computing 4(1), 21 (2013)

12. Ganapathibhotla, M., Liu, B.: Mining Opinions in Comparative Sentences. In: Proc. of the 22nd International Conference on Computational Linguistics, Manchester (2008)
13. Heerschop, B., et al.: Polarity Analysis of Texts Using Discourse Structure. In: Proc. 20th ACM Int'l Conf. Information and Knowledge Management, pp. 1061–1070. ACM (2011)
14. Stanford sentiment analysis, http://nlp.stanford.edu:8080/sentiment/rntnDemo.html
15. Cambria, E., Hussain, A.: Sentic Computing. Techniques, Tools, and Applications. Springer, Dordrecht (2012)
16. Jindal, N., Liu, B.: Mining comparative sentences and relations. In: Proceedings of National Conf. on Artificial Intelligence, AAAI 2006 (2006)
17. Poria, S., Agarwal, B., Gelbukh, A., Hussain, A., Howard, N.: Dependency-Based Semantic Parsing for Concept-Level Text Analysis. In: Gelbukh, A. (ed.) CICLing 2014, Part I. LNCS, vol. 8403, pp. 113–127. Springer, Heidelberg (2014)
18. Björkelund, A., Hafdell, L., Nugues, P.: Multilingual semantic role labeling. In: Proceedings of the Thirteenth Conference on Computational Natural Language Learning, CoNLL 2009, June 4-5, pp. 43–48, Boulder (2009)
19. Poria, S., Gelbukh, A., Cambria, E., Yang, P., Hussain, A., Durrani, T.: Merging SenticNet and WordNet-Affect emotion lists for sentiment analysis. In: 2012 IEEE 11th International Conference on Signal Processing (ICSP), October 21-25, vol. 2, pp. 1251–1255 (2012)
20. Poria, S., Cambria, E., Winterstein, G., Huang, G.-B.: Sentic patterns: Dependency-based rules for concept-level sentiment analysis. Knowledge-Based Systems 69, 45–63 (2014), http://dx.doi.org/10.1016/j.knosys.2014.05.005. ISSN 0950-7051
21. Poria, S., Gelbukh, A., Das, D., Bandyopadhyay, S.: Fuzzy Clustering for Semi-supervised Learning–Case Study: Construction of an Emotion Lexicon. In: Batyrshin, I., González Mendoza, M. (eds.) MICAI 2012, Part I. LNCS, vol. 7629, pp. 73–86. Springer, Heidelberg (2013)
22. Cambria, E., Fu, J., Bisio, F., Poria, S.: AffectiveSpace 2: Enabling Affective Intuition for Concept-Level Sentiment Analysis. In: Twenty-Ninth AAAI Conference on Artificial Intelligence (2015)
23. Poria, S., Cambria, E., Hussain, A., Huang, G.-B.: Towards an intelligent framework for multimodal affective data analysis. Neural Networks 63, 104–116 (2015), http://dx.doi.org/10.1016/j.neunet.2014.10.005, ISSN 0893-6080
24. Poria, S., Cambria, E., Ku, L.-W., Gui, C., Gelbukh, A.: A rule-based approach to aspect extraction from product reviews. SocialNLP 2014, 28 (2014)
25. Poria, S., Gelbukh, A., Cambria, E., Das, D., Bandyopadhyay, S.: Enriching SenticNet polarity scores through semi-supervised fuzzy clustering. In: 2012 IEEE 12th International Conference on Data Mining Workshops (ICDMW), pp. 709–716. IEEE (2012)
26. Poria, S., Gelbukh, A., Hussain, A., Bandyopadhyay, S., Howard, N.: Music genre classification: A semi-supervised approach. In: Carrasco-Ochoa, J.A., Martínez-Trinidad, J.F., Rodríguez, J.S., di Baja, G.S. (eds.) MCPR 2012. LNCS, vol. 7914, pp. 254–263. Springer, Heidelberg (2013)
27. Poria, S., Gelbukh, A., Cambria, E., Hussain, A., Huang, G.-B.: EmoSenticSpace: A novel framework for affective common-sense reasoning. Knowledge-Based Systems 69, 108–123 (2014)
28. Poria, S., Gelbukh, A., Hussain, A., Howard, N., Das, D., Bandyopadhyay, S.: Enhanced SenticNet with Affective Labels for Concept-Based Opinion Mining. IEEE Intelligent Systems 28(2), 31,38 (2013), doi:10.1109/MIS.2013.4
29. Sidorov, G.: Should syntactic n-grams contain names of syntactic relations. International Journal of Computational Linguistics and Applications 5(1), 139–158 (2014)

30. Sidorov, G., Gelbukh, A., Gómez-Adorno, H., Pinto, D.: Soft Similarity and Soft Cosine Measure: Similarity of Features in Vector Space Model. Computación y Sistemas 18(3) (2014)
31. Sidorov, G., Kobozeva, I., Zimmerling, A., Chanona-Hernández, L., Kolesnikova, O.: Modelo computacional del diálogo basado en reglas aplicado a un robot guía móvil. Polibits 50, 35–42 (2014)
32. Ben-Ami, Z., Feldman, R., Rosenfeld, B.: Using Multi-View Learning to Improve Detection of Investor Sentiments on Twitter. Computación y Sistemas 18(3) (2014)
33. Poria, S., Gelbukh, A., Agarwal, B., Cambria, E., Howard, N.: Common sense knowledge based personality recognition from text. In: Castro, F., Gelbukh, A., González, M. (eds.) MICAI 2013, Part II. LNCS, vol. 8266, pp. 484–496. Springer, Heidelberg (2013)
34. Cambria, E., Poria, S., Gelbukh, A., Kwok, K.: Sentic API: A common-sense based API for concept-level sentiment analysis. In: Proceedings of the 4th Workshop on Making Sense of Microposts (# Microposts2014), co-located with the 23rd International World Wide Web Conference (WWW 2014), vol. 1141, pp. 19–24. CEUR Workshop Proceedings, Seoul (2014)
35. Agarwal, B., Poria, S., Mittal, N., Gelbukh, A., Hussain, A.: Concept-Level Sentiment Analysis with Dependency-Based Semantic Parsing: A Novel Approach. In: Cognitive Computation, pp. 1–13 (2015)
36. Poria, S., Cambria, E., Howard, N., Huang, G.-B., Hussain, A.: Fusing Audio, Visual and Textual Clues for Sentiment Analysis from Multimodal Content. Neurocomputing (2015)
37. Chikersal, P., Poria, S., Cambria, E.: SeNTU: Sentiment analysis of tweets by combining a rule-based classifier with supervised learning. In: Proceedings of the International Workshop on Semantic Evaluation, SemEval 2015 (2015)
38. Minhas, S., Poria, S., Hussain, A., Hussainey, K.: A review of artificial intelligence and biologically inspired computational approaches to solving issues in narrative financial disclosure. In: Liu, D., Alippi, C., Zhao, D., Hussain, A. (eds.) BICS 2013. LNCS, vol. 7888, pp. 317–327. Springer, Heidelberg (2013)
39. Pakray, P., Poria, S., Bandyopadhyay, S., Gelbukh, A.: Semantic textual entailment recognition using UNL. Polibits 43, 23–27 (2011)
40. Das, D., Poria, S., Bandyopadhyay, S.: A classifier based approach to emotion lexicon construction. In: Bouma, G., Ittoo, A., Métais, E., Wortmann, H. (eds.) NLDB 2012. LNCS, vol. 7337, pp. 320–326. Springer, Heidelberg (2012)
41. Das, N., Ghosh, S., Gonçalves, T., Quaresma, P.: Comparison of Different Graph Distance Metrics for Semantic Text Based Classification. Polibits 49, 51–57 (2014)
42. Alonso-Roris, V.M., Gago, J.M.S., Rodríguez, R.P., Costa, C.R., Carballa, M.A.G., Rifón, L.A.: Information Extraction in Semantic, Highly-Structured, and Semi-Structured Web Sources. Polibits 49, 69–75 (2014)

List of Questions

1. Why people say that it is better to have micromax over Nokia in england than usa
2. Why people say that Nokia is better than micromax in england than usa
3. Why people say that it is better to neglect Nokia over micromax in england than usa
4. Why people say that Nokia is not as good as micromax in england than usa
5. Why people say that Nokia is more valuable than micromax in england than usa
6. Why people say that micromax is good but Nokia is better in england than usa
7. Why people say that in market Nokia is more popular in england than usa
8. Why people say that Nokia is much better in england than usa
9. Why people say that Nokia is more efficient to buy in england than usa
10. Why people say that people prefer Nokia over micromax in england than usa
11. Why it is better to have micromax over Nokia?
12. Why Nokia is better than micromax
13. Why it is better to neglect Nokia over micromax?
14. Why Nokia is not as good as micromax?
15. Why Nokia is more valuable than micromax
16. Why micromax is good but Nokia is better
17. Why in market Nokia is more popular
18. Why Nokia is much better
19. Why Nokia is more efficient to buy
20. Why people prefer Nokia over micromax
21. Why should I not buy Nokia if I need mobile with good looks and nice sound quality?
22. Why should I buy Nokia if I need mobile with good looks and nice sound quality?
23. Why should one feel sad after buying X?
24. I need mobile with good sound quality and nice looks. Why should one feel sad after buying x?
25. If I need mobile with good looks and nice sound quality, Why should I buy Nokia?
26. Why Nokia should be good option as a mobile?
27. Why should one regret for long time after buying Nokia?
28. I went to market because I need mobile with good camera. Why Should I go for Nokia?
29. Why I bought Nokia at cheaper price but feel cheated?
30. Why should one consider buying Nokia as an alternative to x?
31. I went to market and bought Nokia. Why should I feel satisfied finally?
32. Why I went to market for buying Nokia?

33. I went to shop. I heard good things about Nokia. Hence I bought it. Why Should I be not happy?
34. If I need Nokia then why should I purchase ni?
35. Why one feel cheated in the end after spending money on Nokia?
36. Why one gets sick but need Nokia for daily purpose?
37. Why should one buy Nokia next after getting salary?
38. I went to shop. I took money from atm. I want good mobile. Why should I buy Nokia?
39. Why should one buy Nokia instead of looking for its bad reviews?
40. Why should I buy Nokia?
41. Why should I like Nokia?
42. Why should I go for Nokia?
43. Why should I look for Nokia?
44. Why should I accept Nokia?
45. Why should I choose Nokia?
46. Why should I forget Nokia?
47. Why should I get fond of Nokia?
48. Why should I overlook Nokia?
49. Why should I suggest Nokia?
50. Why should I recommend Nokia?
51. Why should I propose Nokia?
52. Why should I advise for Nokia?
53. Why should I need Nokia?
54. Why should I feel sad?
55. Why should I demand for Nokia?
56. Why should I call for Nokia?
57. Why should I require Nokia?
58. Why should I want Nokia?
59. Why should I prefer Nokia?
60. Why should I desire for Nokia?
61. Why should I opt for Nokia?
62. Why should I pick Nokia?
63. Why should I select Nokia?
64. Why should I wish for Nokia?
65. Why should I aspire for Nokia?
66. Why Nokia is first choice?
67. Why I am inclined towards Nokia?
68. Why should I favor Nokia?
69. Why should I order Nokia?
70. Why should I insist for Nokia?
71. Why should I neglect Nokia?

72. Why should I stop thinking about Nokia?
73. Why should I put Nokia out of his mind?
74. Why should I feel cheated in the end?
75. Why should I be happy?
76. Why should I feel satisfied finally?
77. Why should one leave Nokia?
78. Why should one love Nokia?

TRUPI: Twitter Recommendation Based on Users' Personal Interests

Hicham G. Elmongui[1,2], Riham Mansour[3], Hader Morsy[4], Shaymaa Khater[5], Ahmed El-Sharkasy[4], and Rania Ibrahim[4]

[1] Alexandria University, Computer and Systems Engineering, Alexandria 21544, Egypt
elmongui@alexu.edu.eg

[2] GIS Technology Innovation Center, Umm Al-Qura University, Makkah, Saudi Arabia
elmongui@gistic.org

[3] Microsoft Research Advanced Technology Lab, Cairo, Egypt
rihamma@microsoft.com

[4] Alexandria University, SmartCI Research Center, Alexandria 21544, Egypt
{hader,sharkasy,ribrahim}@mena.vt.edu

[5] Virginia Tech, Computer Science Department, Blacksburg, VA 24061, USA
skhater@vt.edu

Abstract. Twitter has emerged as one of the most powerful micro-blogging services for real-time sharing of information on the web. The large volume of posts in several topics is overwhelming to twitter users who might be interested in only few topics. To this end, we propose TRUPI, a personalized recommendation system for the timelines of twitter users where tweets are ranked by the user's personal interests. The proposed system combines the user social features and interactions as well as the history of her tweets content to attain her interests. The system captures the users interests dynamically by modeling them as a time variant in different topics to accommodate the change of these interests over time. More specifically, we combine a set of machine learning and natural language processing techniques to analyze the topics of the various tweets posted on the user's timeline and rank them based on her dynamically detected interests. Our extensive performance evaluation on a publicly available dataset demonstrates the effectiveness of TRUPI and shows that it outperforms the competitive state of the art by 25% on nDCG@25, and 14% on MAP.

Keywords: Twitter, personalized recommendation, dynamic interests.

1 Introduction

Twitter has emerged as one of the most powerful micro-blogging services for real-time sharing of information on the web. Twitter has more than 500 million

A. Gelbukh (Ed.): CICLing 2015, Part II, LNCS 9042, pp. 272–284, 2015.
DOI: 10.1007/978-3-319-18117-2_20

users [1], and the volume of tweets one receives is persistently increasing especially that 78% of Twitter users are on the ubiquitous mobile devices [34].

The high volume of tweets is getting overwhelming and reduces one's productivity to the point that more than half US companies do not allow employees to visit social networking sites for any reason while at work [28]. In addition, a large base of twitter users tend to post short messages of 140 characters reflecting a variety of topics. Individual users' interests vary over the time, which is evidenced by the dynamic *Trends* feature of Twitter, which suggests hashtags, metadata tags prefixed by the symbol #, to the users based on her followees [32].

For the above reasons, Twitter users share a need to digest the overwhelming volume. Recommending interesting content to the user is harder in the case of Twitter, and microblogs in general, as the tweet is limited in size and thus leaks the context in which it was posted. Existing systems that recommend tweets to the users either 1) provide ranking models that are not personalized to the users' interests; 2) do not capture the dynamic change in the user's interest over the time; 3) use Latent Drichlet Allocation (LDA) [5] to represent user's interests and hence is not scalable to large datasets [25] ; or, 4) assume special user marking to the tweets of interests [7, 9, 10, 13, 19, 27, 36, 39]

In this paper, we propose TRUPI, a Twitter Recommendation based on User's Personal Interests. TRUPI aims at presenting the tweets on the user timeline in an order such that tweets that are more interesting to her appears first. In order to do so, TRUPI learns the changing interests of the users over the time, and then ranks the received tweets accordingly. More specifically, TRUPI employs an ensemble of interest classifiers that indicate the most probable interest label of each tweet on the user's timeline. Tweets are then fed into a ranking model to order the tweets based on the current user's interests. The user's interests are modeled as a time variant level of interests in different topics.

The rest of the paper is organized as follows. Section 2 highlights related work. Section 3 presents an overview of TRUPI. Section 4 gives details of the interest detection and the tweet ranking. In Section 5, TRUPI's extensive performance evaluation is presented before we conclude by a summary in Section 6.

2 Related Work

Many recommendation systems have been proposed in the literature for Twitter. Research efforts go into many directions: from recommending hashtags [12, 21, 41] and recommending URLs [6] to providing news recommendations [26] and suggesting followees [15, 16, 20].

New approaches have been proposed to deal with recommending tweets on the user's timeline. Duan et al. use a learning-to rank algorithm using content relevance, account authority, and tweet-specific features to rank the tweets [9]. Uysal and Croft construct a tweet ranking model making use of the user's re-tweet behavior. They rank both the tweets and the users based on their likelihood of getting a tweet re-tweeted [36]. Similarly, Feng and Wang propose personalized

[1] `http://www.statisticbrain.com/twitter-statistics/` visited March 2014

recommendation for tweets [10]. In their evaluation, the metric of measuring the interest in a tweet is whether the user would re-tweet or not. Our proposed recommender is different from these approaches as their ranking models are not personalized, and they do not capture the dynamic change in the user's interest.

Nevertheless, Pennacchiotti et al. introduce the problem of recommending tweets that match a user's interests and likes [27]. Also, Chen et al. recommend tweets based on collaborative ranking to capture personal interests [7]. These two propositions neither account for the change in the user interest over time nor work on the semantic level.

Guo et al. propose *Tweet Rank*, a personalized tweet ranking mechanism that enables the user to mark tweets as interesting by defining some interest labels [13]. Our proposed approach does not assume any special user marking.

Yan et al. propose a graph-theoretic model for tweet recommendation [39]. Their model ranks tweets and their authors simultaneously using several networks: the social network connecting the users, the network connecting the tweets, and the network that ties the two together. They represent user interest using LDA, which is not scalable for large datasets [25].

Little work has been done in the dynamic personalized tweet recommendation that accounts for the change in the user interests. Abel et al. explore the temporal dynamics of users' profiles benefiting from semantic enrichment [2]. They recommend news articles, *and not tweets*, for topic-based profiles.

Our previous work proposes an approach for dynamic personalized tweet recommendation using LDA [19]. In that work, a model is defined to classify the tweet into important or not important tweet. In TRUPI, tweets are ranked based on the user's dynamic level of interest in the tweet topic. TRUPI explores the tweet content (and semantic) along with numerous additional social features.

3 Overview of TRUPI

TRUPI recommender consists of two phases. The first phase is to create the user profiles, which contains the topics of interests to the user. The user profile is dynamic; i.e., it changes over time. The second phase occurs in an online fashion to give a ranking score to the incoming tweet. This ranking score would provide for presenting the tweet to the user by its importance.

User profiles contain the different topics of interest to the users. Each profile is characterized by a weighted set of interests or topics (e.g., sports, politics, movies, etc.) The weights represent the probability that a user is interested in a certain topic. Such probabilities are learned from the user history as follows.

First, the tweets are classified into different topics. The topics are learned from a large chunk of tweets. Those tweets are clustered so that each cluster contains tweets corresponding to the same topic of interest. Next, each cluster is labeled with the topic with the help of a series of topic classifiers. This process results into the capability of placing an incoming tweet to its matching cluster, and hence, the tweet would be labeled with the same topic as the cluster.

The dynamic user profiles are created from the history of each individual user. Her last history tweets are used to compute the probability distribution of her topics of interests. The history tweets may either be time-based or count-based. A *time-based history* of length t contains the tweets that appeared on her timeline in the last t time units. A *count-based history* of size k contains the last k tweets that appeared on her timeline. While the former might reflect the recent history, the latter might have to be used in the case of low activity on the timeline.

When the system receives an incoming tweet, it consults the user profile in order to give a ranking score to this tweet. Many features are used in this scoring technique. Those features reflect how much interest the user may have in the topic of the tweet or in the sender (e.g., a close friend or celebrity). Such features would be extracted from the history of the user. They would reflect how she interacted with the recent tweets of the same tweet topic or the past tweets coming from this sender.

Tweets are preprocessed before entering our system. The aim of this preprocessing is to prepare the tweet for the subsequent algorithms. The text of the tweet is normalized as follows:

- Ignore tweets that do not convey much information. This includes tweets with number of tokens less than 2.
- Replace slang words with their lexical meaning using a slang dictionary. We use the Internet Slang Dictionary & Translator [1].
- Lexically normalize extended words into their canonical form. We replace any successive repetition of more than two characters with only two occurrences. For instance, *coooool* will be transformed into *cool*.
- Normalize Out-of-Vocabulary (OOV) words. We detect OOV words in a tweet using GNU Aspell [11]. For each OOV word, we get candidate replacements based on lexical and phonemic distance. We then replace the OOV word with the correct candidate based on edit distance and context.
- Identify named entities in the tweet and associate them with it for later usage. We use the Twitter NLP tools [3].
- Extract hashtags from the tweet and associate them with it.
- Run a Twitter spam filter [30] on any tweet containing a URL. If a tweet turns out to be spam, it is ignored. Otherwise, extract the URLs from the tweet and associate them with it.
- Represent the tweet as a feature vector using TF-IDF representation [29]. To emphasize on the importance of the tweet, we doubled the weights for the hashtags and named entities. This is in line with the fact that tweets with hashtags get two times more engagement as stated in [17].

4 Interest Detection and Tweet Ranking

This section provides details for how the user's interests are detected and how his tweets would be ranked in TRUPI, the proposed Twitter recommendation system.

4.1 Tweet Topic Classification

Since our main target is to capture the user's dynamic interests on the semantic level, we will need first to understand the tweet by classifying the tweets into topics (i.e., sports, music, politics, etc.)

Tweet Clustering and Classification. It is worth mentioning that since the tweet is short in nature, its context is leaked and it is not easy to capture it. Hence, we enrich the tweets' text by grouping tweets that talk about the same topic in one cluster using the Online Incremental Clustering approach [4].

Online incremental clustering is applied by finding the most similar cluster to a given tweet, which is the one whose centroid has the maximum cosine similarity with the tweet [24]. If the cosine similarity is above a certain threshold, the tweet is inserted into that cluster. Otherwise, the tweet forms a new cluster by itself.

Next, topic-based binary SVM classifiers are applied to classify the cluster into one of the topics [8]. The cluster is classified into a topic if the confidence score exceed a certain threshold. The binary SVM classifiers were trained by labeling the tweets using a predefined list of keyword-topic pairs that is obtained by crawling DMOZ – the Open Directory Project [31]. The pairs are constructed as follows. From `dmoz.org/Recreation/Food/`, for instance, we get <drink, cheese, meat, ...>. We create the list of *keyword-topic* pairs as <drink, Food>, <cheese, Food>, <meat, Food> ...

Two white lists are automatically constructed for each topic; one for hashtags, and one for named entities. The white lists consist of hashtags and named entities that belong to a certain topic. The construction is described in Section 4.1.

Upon the arrival of a tweet to the user, it is clustered using the used online incremental clustering algorithm. The tweet will be labeled with the same label of the cluster it belongs to. If the tweet does not belong to any cluster, or if the tweet belongs to an unlabeled cluster, we check whether the tweet contains hashtags or named entities that belong to our white lists. In this case, the tweet is labeled with the corresponding topic. Otherwise, we try URL labeling, which labels the tweet using the content and the slugs of the URL.

White Lists Construction. Two white lists are automatically constructed for each topic. One list for named entities and one for hashtags contained in the tweets. These white lists would be looked up for incoming tweets. The rational behind these white lists is that some named entities or hashtags are associated with specific topics. For instance, *Ronaldo* and *#WorldCup* would be associated with *sports*, whereas *Madonna* and *#Beatles* are associated with the topic *music*.

Constructing Named Entities White Lists: We used DBpedia [22], a large-scale multilingual knowledge base extracted from the well-known online encyclopedia, Wikipedia [37]. First, we retrieved the different resources; i.e., named entities, along with their types. Then we grouped the types that belong to the same topic together. This is done by projecting the types on the formed clusters

Table 1. Examples of Wikipedia's Named Entities Retrieved from DBpedia

Sports	Music	Politics	Food
Jim Hutto	Nelson Bragg	Dante Fascell	Pisto
Tiger Jones	Maurice Purtill	James Kennedy	Teacake
Stephon Heyer	Michael Card	Riffith Evans	Apple pie
Allen Patrick	Faizal Tahir	Barack Obama	Pancake
Fujie Eguchi	Claude King	Daniel Gault	Potato bread

and assigning the topic of the cluster to the named entity type. For instance, the types *Musical Artist* and *Music Recording* would fall, among others, in clusters that are labeled with the topic *music*. This grouping would results into having white lists for the topics of interests (e.g., *music*). Each white list contains the named entities associated with the corresponding topic. Table 1 contains examples of the named entities associated with some topics of interests.

Since we get the canonical form of the named entities from DBpedia, we extend the white lists with the synonyms of the named entities as well. For instance, for *Albert Einstein*, we also add *Einstein, Albert Eienstein, Albert LaFache Einstein, Alber Einstein*, etc. We use WikiSynonyms to retrieve the synonyms of the named entities given the canonical form [38].

Constructing Hashtags White Lists: The construction of the white lists for the hashtags is more involved than that for the named entities. Different hashtags are to be associated with their corresponding topics. This association needs to be learned from the historical tweets as follows.

The procedure starts by looping on each tweet w in the labeled clusters. Each tweet is assigned a confidence score, denoted by $C(w)$, that refers to how confident we are with respect to the learned topic of this tweet. This confidence score is the same as the classification confidence of the binary SVM topic classifier used to label the cluster in the aforementioned topic classification step.

Each hashtag, denoted by h, is assigned a score, that is a measure of its relatedness to each topic p. The score function $score(h|p)$ is defined as

$$score(h|p) = \frac{\sum\limits_{\substack{w \sim p \\ h \in w}} C(w)}{|P| + \sum\limits_{h \in w} C(w)} \quad (1)$$

where P is the set of the adopted topics of interests and $w \sim p$ means that the tweet w is labeled as topic p. The rational behind this scoring function is: the more tweets belonging to a certain topic, tagged with a certain hashtag, the closer relation is between this hashtag and the topic. Also, $|P|$ is added in the denominator to prevent the score to be 1 and to discriminate between the heavily-used and lightly-used hashtags in the cases where they are not used in more than one topic.

Hashtags with score above 0.7 were chosen to be added to our hashtags white lists. Table 2 gives the top 5 related hashtags to each topic obtained using our approach.

Table 2. Top-5 Topic Related Hashtags

Sports	Music	Politics	Food
#running	#dance	#radio	#organic
#baseball	#jazz	#liberty	#wine
#cycling	#opera	#libertarian	#beer
#lpga	#christian	#environmental	#coffee
#basketball	#rock	#politicalhumor	#chocolate

URL Labeling. We found that 56% of the extracted URLs have slugs in their path. A slug is the part of a URL which identifies a page using human-readable keywords (e.g., heavy metal in `http:www.example.org/category/heavy-metal`). Slug words are extracted from the URL using regex and are added to the tweet text. The expanded tweet is labeled again using the binary SVM classifier.

In addition, 6% of the extracted URLs are for videos from [40]. For such URLs, the *category* field on the video page is used as the label of the tweet.

4.2 Dynamic Interests Capturing

In this subsection, we describe how TRUPI captures the dynamic level of interest of the user in a certain topic.

For a certain user u, the dynamic level of interest is computed for each topic from the recent history of the user interaction on the microblog. This interaction includes her tweets, re-tweets, replies, and favorites. As described before, the tweets of the user u are either unlabeled or are successfully labeled as described in Section 4.1. For each tweet that is labeled when it fell in one of the clusters, it gets assigned the same confidence score of the binary SVM topic classifier. For tweets that are labeled using a white list, their confidence score is treated as 1. For unlabeled tweets, the confidence score is 0.

Without loss of generality, we assume a time-based history. On any day d, a user u is active if he interacts with a tweet. The set of tweets with which she interacted is denoted by W_d. Her level of interest at a topic p is computed as

$$L_{u,d}(p) = \sum_{\substack{w \sim p \\ w \in W_d}} C(w) \tag{2}$$

where $C(w)$ is the confidence score in the topic label of a tweet w and $w \sim p$ means that the tweet w is labeled as topic p.

Upon the arrival of a new tweet w', the user's level of interest in this tweet would be a function of two things: 1) the topic of the tweet, which is labeled as p' with the same method as in Section 4.1, and 2) the dynamic level of interest in the different topics of the tweets earlier this day and in the past week as in [19]. The user's level of interest in the tweet w', which arrived on day d', is computed as

$$I_u(w') = C(w') \sum_{d=d'-7}^{d'} L_{u,d}(p') \tag{3}$$

where $C(w')$ is the confidence score in the topic label of a newly arrived tweet w'.

The user's profile contains the topics of interests to the user. Specifically, for each topic p', it contains the dynamic level of interest of the user in it as

$$\text{DynLOI}_u(p') = \sum_{d=d'-7}^{d'} L_{u,d}(p') \tag{4}$$

The dynamic level of interest is computed over a sliding window. At the beginning of each day, the oldest day in the window is flushed away and incoming tweets enters the window. This sliding window allows for the incremental evaluation of the dynamic level of interest. In other words, the dynamic level of interest does not have to be recomputed with every incoming tweet, which allows for TRUPI to work in a timely manner.

4.3 User Tweets Ranking

A machine-learned ranking model [18] is used to assign a ranking score to the incoming tweets to a certain user. The tweets are posted in a descending order of the assigned ranking scores.

RankSVM is the data-driven support vector machine used to rank the tweets [9]. It learns both the ranking function and the weights of the input features. For a user u_1 receiving a tweet w authored by user u_2, TRUPI uses five ranking feature categories, which are:

- The dynamic level of interest in the topic of w, as computed in Section 4.2.
- The popularity features of w. They include the number of favorites and retweets of w.
- The authoritative features of u_2. They include the number of u_2's followers, followees, and tweets.
- The importance features of w to u_1. They are divided into two groups: 1) globally importance features, which include whether w contains a URL or a hashtag, and 2) locally importance features, which include whether w mentions u_1 or whether it contains a hashtag that was mentioned by u_1 during last week.
- The interaction features. They include the number of times u_1 mentioned, favorited, or replied to a tweet authored by u_2. They also include the similarity between the friendship of u_1 and u_2 (the number of common users both of them follow). They also include the number of days since the last time they interacted together.

5 Experimental Evaluation

We perform extensive experiments to evaluate the quality performance of the proposed recommender. In our experiments, we compare with the state-of-the-art techniques. All used machine-learning algorithms were executed from the WEKA suite [14].

Table 3. 5-fold Cross Validation for Topic Classification

Topic	Precision	Recall	F1
Sports	94.76%	94.68%	94.72%
Music	94.29%	94.29%	94.29%
Politics	98.16%	98.19%	98.17%
Food	98.11%	98.09%	98.10%
Games	97.65%	97.62%	97.64%

We adopted the dataset used in [23], which is publicly available at [35]. This dataset contains 284 million following relationships, 3 million user profiles and 50 million tweets. We sampled this dataset to retrieve 20,000 users, 9.1 million following relationships and 10 million tweets. The sampling consisted of the first 20K users from a breadth first search that started from a random user (id = 25582718). We complemented the dataset by crawling Twitter using the Twitter REST API [33] to retrieve all the actions which have been done on the sampled tweets including Favorite, Re-tweet and Reply-to.

We consider the positive examples during periods which we call Active Periods. An active period is a time frame during which the user have performed an action. Only tweets within this active period are considered in our system in order to avoid the negative examples caused by users being inactive during long periods of time. We scan the tweets in a window of size 10 tweets before and after the re-tweeted, replied to, or favorited tweet.

For the topic classification, we use the micro-averaged F-measure, which considers predictions from all instances [24]. For the ranking module, we use the normalized discounted cumulative gain (nDCG), which measures the performance based on a graded relevance of the recommended entities [24]. We also use the Mean Average Precision (MAP), which has been shown to have especially good discrimination and stability [24].

5.1 Performance of Topic Extraction

In our experiments, we adopted 5 topics of interests, namely, sports, music, politics, food, and games. A binary SVM classifier is created for each topic. The tweets were labeled using the list of keyword-topic pairs obtained from DMOZ - the Open Directory Project [31]. The classifiers were evaluated by measuring the 5-fold cross validation accuracy on our labeled 10M tweets dataset. The results are shown in Table 3.

5.2 Performance of Personalized Binary Recommendation

The personalized binary recommendation model does not rank the tweets. It just tells whether an incoming tweet is important to the user or not if she acts upon it [19]. The ground truth adopted was that a tweet is important to a user if she replied to, re-tweeted, or favorited it. The feature used in TRUPI were

Table 4. 10-fold Cross Validation for Binary Recommendation Classifiers

Approach	Precision	Recall	F1
DynLDALOI(J48)	74.22%	88.61%	80.78%
TRUPI	85.70%	82.76%	84.20%

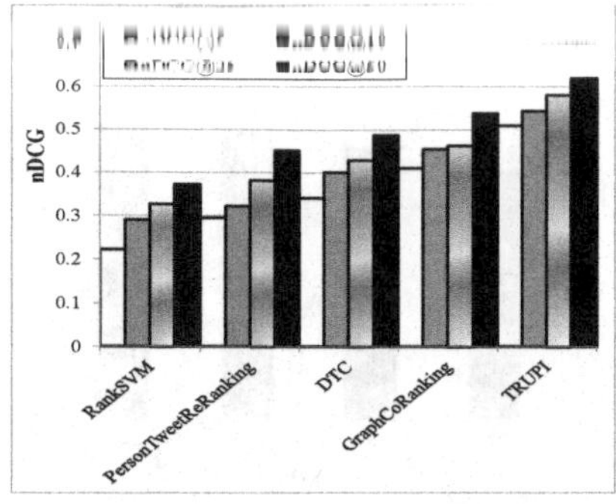

Fig. 1. Personalized Ranking Recommendation using (nDCG)

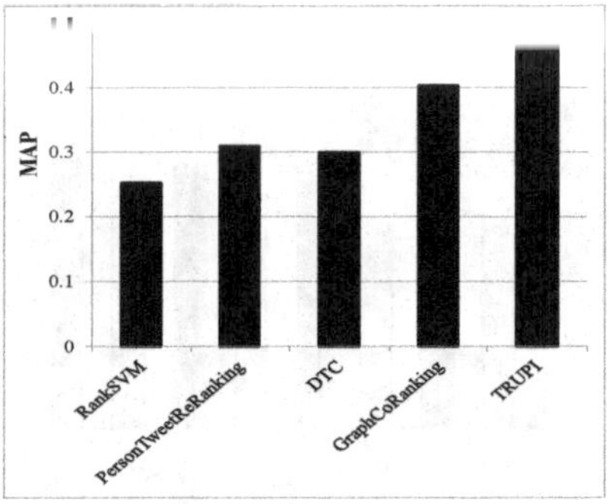

Fig. 2. Personalized Ranking Recommendation (MAP)

the same features used in the ranking model in Section 4.3. We compared TRUPI with the state-of-the-art binary recommendation technique, DynLDALOI, which introduced the notion of dynamic level of interest in the topics using LDA [19]. Table 4 shows the 10-fold cross validation of the binary recommendation. The F1 measure of TRUPI outperforms DynLDALOI by 4.23%.

5.3 Performance of Personalized Ranking Recommendation

We evaluated the TRUPI personalized ranking recommendation model through extensive experiments. TRUPI was compared with four state-of-the-art techniques: 1) RankSVM [9], which learns the ranking function and weight of the input features; 2) DTC [36], where a decision tree based classifier is build and a tweet ranking model is constructed to make use of the user's re-tweet behavior; 3) PersonTweetReRanking [10], which consider whether a tweet is important if the user re-tweets it; and 4) GraphCoRanking [39], where the tweets are ranked based on the intuition that there is a mutually reinforcing relation between tweets and their authors. Our ground truth was whether the user was interested, i.e., acted upon a top ranked tweet.

Figures 1 and 2 show the performance of the ranking recommendation techniques using nDCG@5, nDCG@10, nDCG@25, nDCG@50, and MAP. The figures show that TRUPI outperforms all the other competitors. TRUPI outperforms RankSVM by 130%, 87%, 78%, 67%, and 83% on nDCG@5, nDCG@10, nDCG@25, nDCG@50, and MAP respectively. It also outperforms PersonTweetReRanking by 73%, 69%, 52%, 37% and 49% on them respectively. Similarly,

Table 5. TRUPI's versions

Version	Features
TRUPI v1	A base version (without all the added features)
TRUPI v2	Adding the hashtag white lists to v1
TRUPI v3	Adding the named entities white lists to v2
TRUPI v4	Adding the URL labeling to v3

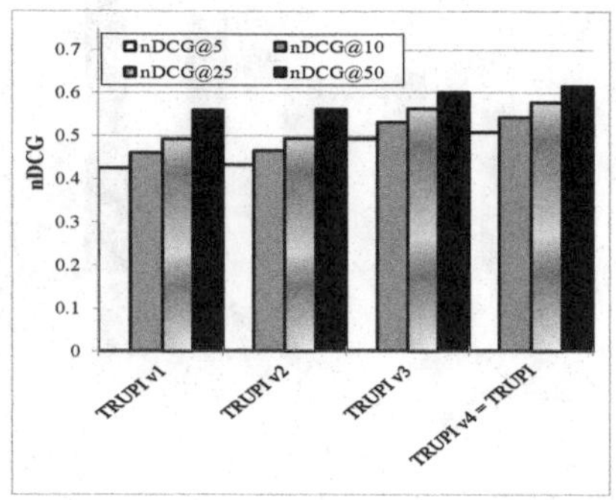

Fig. 3. Effect of TRUPI's Components (nDCG)

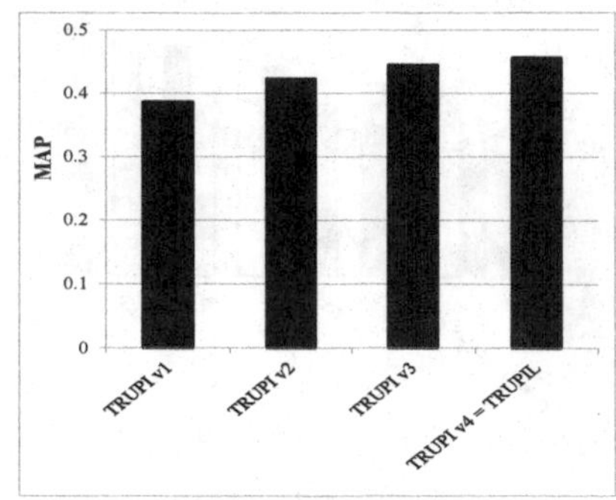

Fig. 4. Effect of TRUPI's Components (MAP)

it outperforms DTC by 49%, 35%, 35%, 27%, and 54% on the same metrics respecitively. Finally, TRUPI outperforms GraphCoRanking by 24%, 19%, 25%, 15%, and 14% on them respectively.

5.4 Analyzing TRUPI's Components

We scrutinized the different components of TRUPI. We illustrate the effect of the different components by looking at 4 versions of TRUPI. The versions are shown in Table 5. Note that TRUPI v4 is the full fledge proposed TRUPI.

Figures 3 and 4 show the performance of the different versions of TRUPI using nDCG@5, nDCG@10, nDCG@25, nDCG@50, and MAP. Adding the hashtags white lists gave a relative gain for nDCG@10 by 2% on the base case. The named entities white lists gave 14% on the same metric, whereas the URL labeling added 3%. On MAP, they added 10%, 5%, 2% respectively.

6 Conclusion

In this paper, we proposed TRUPI, a personalized recommendation system based on user's personal interests. The proposed system combines the user social features and interactions as well as the history of her tweets content to attain her interests. It also captures the dynamic level of users' interests in different topics to accommodate the change of interests over time. We thoroughly evaluated the performance of TRUPI on a publicly available dataset and have found that TRUPI outperforms the competitive state-of-the-art Twitter recommendation systems by 25% on nDCG@25, and 14% on MAP.

Acknowledgement. This material is based on work supported in part by Research Sponsorship from Microsoft Research.

References

[1] `http://www.noslang.com/`. Internet Slang Dictionary & Translator (last accessed January 06, 2014)
[2] Abel, F., Gao, Q., Houben, G.-J., Tao, K.: Analyzing User Modeling on Twitter for Personalized News Recommendations. In: Konstan, J.A., Conejo, R., Marzo, J.L., Oliver, N. (eds.) UMAP 2011. LNCS, vol. 6787, pp. 1–12. Springer, Heidelberg (2011)
[3] Ritter, A., Clark, S.: Twitter NLP Tools (2011), `https://github.com/aritter/twitter_nlp` (last accessed January 06, 2014)
[4] Becker, H., Naaman, M., Gravano, L.: Beyond Trending Topics: Real-World Event Identification on Twitter. In: Procs. ICWSM 2011 (2011)
[5] Blei, D.M., Ng, A.Y., Jordan, M.I.: Latent Dirichlet Allocation. The Journal of Machine Learnnig Research 3, 993–1022 (2003)
[6] Chen, J., Nairn, R., Nelson, L., Bernstein, M., Chi, E.: Short and Tweet: Experiments on Recommending Content from Information Streams. In: CHI (2010)
[7] Chen, K., Chen, T., Zheng, G., Jin, O., Yao, E., Yu, Y.: Collaborative Personalized Tweet Recommendation. In: Procs. of SIGIR 2012 (2012)
[8] Cortes, C., Vapnik, V.: Support-Vector Networks. Machine Learning 20(3), 273–297 (1995)
[9] Duan, Y., Jiang, L., Qin, T., Zhou, M., Shum, H.-Y.: An empirical study on learning to rank of tweets. In: COLING 2010 (2010)
[10] Feng, W., Wang, J.: Retweet or Not?: Personalized Tweet Re-ranking. In: Procs. of WSDM 2013, pp. 577–586 (2013)
[11] GNU Aspell (2011), `http://aspell.net/` (last accessed January 06, 2014)
[12] Godin, F., Slavkovikj, V., De Neve, W., Schrauwen, B., Van de Walle, R.: Using Topic Models for Twitter Hashtag Recommendation. In: Procs. of WWW 2013 Companion (2013)
[13] Guo, Y., Kang, L., Shi, T.: Personalized Tweet Ranking Based on AHP: A Case Study of Micro-blogging Message Ranking in T.Sina. In: WI-IAT 2012 (2012)
[14] Hall, M., Frank, E., Holmes, G., Pfahringer, B., Reutemann, P., Witten, I.H.: The WEKA data mining software: an update. SIGKDD Explorations 11(1), 10–18 (2009)
[15] Hannon, J., Bennett, M., Smyth, B.: Recommending twitter users to follow using content and collaborative filtering approaches. In: RecSys 2010 (2010)
[16] Hannon, J., McCarthy, K., Smyth, B.: Finding Useful Users on Twitter: Twittomender the Followee Recommender. In: Clough, P., Foley, C., Gurrin, C., Jones, G.J.F., Kraaij, W., Lee, H., Mudoch, V. (eds.) ECIR 2011. LNCS, vol. 6611, pp. 784–787. Springer, Heidelberg (2011)
[17] Huffington Post's Twitter Statistics, `http://www.huffingtonpost.com/belle-beth-cooper/10-surprising-new-twitter_b_4387476.html` (last accessed January 06, 2014)
[18] Joachims, T.: Optimizing Search Engines Using Clickthrough Data. In: Procs. of KDD 2002, pp. 133–142 (2002)

[19] Khater, S., Elmongui, H.G., Gracanin, D.: Tweets You Like: Personalized Tweets Recommendation based on Dynamic Users Interests. In: SocialInformatics 2014 (2014)
[20] Kwak, H., Lee, C., Park, H., Moon, S.: What is Twitter, a Social Network or a News Media. In: Procs. of WWW 2010, pp. 591–600 (2010)
[21] Kywe, S.M., Hoang, T.-A., Lim, E.-P., Zhu, F.: On Recommending Hashtags in Twitter Networks. In: Procs. of SocInfo 2012, pp. 337–350 (2012)
[22] Lehmann, J., Isele, R., Jakob, M., Jentzsch, A., Kontokostas, D., Mendes, P.N., Hellmann, S., Morsey, M., van Kleef, P., Auer, S., Bizer, C.: DBpedia - A Large-scale, Multilingual Knowledge Base Extracted from Wikipedia. Semantic Web Journal (2014)
[23] Li, R., Wang, S., Deng, H., Wang, R., Chang, K.C.-C.: Towards Social User Profiling: Unified and Discriminative Influence Model for Inferring Home Locations. In: Procs. of KDD 2012, pp. 1023–1031 (2012)
[24] Manning, C.D., Raghavan, P., Schütze, H.: Introduction to Information Retrieval. Cambridge University Press (2008)
[25] Mikolov, T., Chen, K., Corrado, G., Dean, J.: Efficient Estimation of Word Representations in Vector Space. In: ICLR 2013 Workshops (2013)
[26] De Francisci Morales, G., Gionis, A., Lucchese, C.: From Chatter to Headlines: Harnessing the Real-time Web for Personalized News Recommendation. In: Procs. of WSDM 2012, pp. 153–162 (2012)
[27] Pennacchiotti, M., Silvestri, F., Vahabi, H., Venturini, R.: Making Your Interests Follow You on Twitter. In: Procs. of CIKM 2012 (2012)
[28] Robert Half Technology. Whistle - But Don't tweet - While You Work (2009), http://rht.mediaroom.com/index.php?s=131&item=790 (last accessed January 06, 2014)
[29] Salton, G., Buckley, C.: Term-weighting Approaches in Automatic Text Retrieval. Information Processing & Management 24(5), 513–523 (1988)
[30] Santos, I., Miñambres-Marcos, I., Laorden, C., Galán-García, P., Santamaría-Ibirika, A., Bringas, P.G.: Twitter Content-Based Spam Filtering. In: Procs. of CISIS 2013, pp. 449–458 (2013)
[31] The Open Directory Project, http://www.dmoz.org/ (last accessed January 06, 2014)
[32] Twitter (2006). http://www.twitter.com/ (last accessed January 06, 2014)
[33] Twitter REST API, https://dev.twitter.com/docs (last accessed January 06, 2014)
[34] Twitter Usage, http://about.twitter.com/company (last accessed January 06, 2014)
[35] UDI-TwitterCrawl-Aug2012 (2012), https://wiki.cites.illinois.edu/wiki/display/forward/Dataset-UDI-TwitterCrawl-Aug2012 (last accessed January 06, 2014)
[36] Uysal, I., Croft, W.B.: User oriented tweet ranking: a filtering approach to microblogs. In: Procs. of CIKM 2011, pp. 2261–2264 (2011)
[37] Wikipedia (2001), http://www.wikipedia.org/ (last accessed January 06, 2014)
[38] WikiSynonyms, http://wikisynonyms.ipeirotis.com/ (last accessed January 06, 2014)
[39] Yan, R., Lapata, M., Li, X.: Tweet Recommendation with Graph Co-ranking. In: Procs. of ACL 2012, pp. 516–525 (2012)
[40] YouTube (2005), http://www.youtube.com/ (last accessed January 06, 2014)
[41] Zangerle, E., Gassler, W., Specht, G.: Recommending#-Tags in Twitter. In: Procs. of SASWeb 2011, pp. 67–78 (2011)

Detection of Opinion Spam with Character n-grams

Donato Hernández Fusilier[1,2], Manuel Montes-y-Gómez[3], Paolo Rosso[1], and Rafael Guzmán Cabrera[2]

[1] Natural Language Engineering Lab., Universitat Politècnica de València, Spain
[2] División de Ingenierías, Campus Irapuato-Salamanca, Universidad de Guanajuato, Mexico
[3] Laboratorio de Tecnologías del Lenguaje, Instituto Nacional de Astrofísica, Óptica y Electrónica, Mexico
{donato,guzmanc}@ugto.mx, mmontesg@ccc.inaoep.mx
prosso@dsic.upv.es

Abstract. In this paper we consider the detection of opinion spam as a stylistic classification task because, given a particular domain, the deceptive and truthful opinions are similar in content but differ in the way opinions are written (style). Particularly, we propose using character n-grams as features since they have shown to capture lexical content as well as stylistic information. We evaluated our approach on a standard corpus composed of 1600 hotel reviews, considering positive and negative reviews. We compared the results obtained with character n-grams against the ones with word n-grams. Moreover, we evaluated the effectiveness of character n-grams decreasing the training set size in order to simulate real training conditions. The results obtained show that character n-grams are good features for the detection of opinion spam; they seem to be able to capture better than word n-grams the content of deceptive opinions and the writing style of the deceiver. In particular, results show an improvement of 2.3% and 2.1% over the word-based representations in the detection of positive and negative deceptive opinions respectively. Furthermore, character n-grams allow to obtain a good performance also with a very small training corpus. Using only 25% of the training set, a Naïve Bayes classifier showed F_1 values up to 0.80 for both opinion polarities.

Keywords: Opinion spam, deceptive detection, character n-grams, word n-grams.

1 Introduction

With the increasing availability of review sites people rely more than ever on online opinions about products and services for their decision making. These reviews may be positive or negative, that is, in favour or against them. A recent survey found that 87% of people have reinforced their purchase decisions by

A. Gelbukh (Ed.): CICLing 2015, Part II, LNCS 9042, pp. 285–294, 2015.
DOI: 10.1007/978-3-319-18117-2_21

positive online reviews. At the same time, 80% of consumers have also changed their minds about purchases based on negative information they found online[1]. Additionally, there is a special class of reviews, the *deceptive opinions*, which are fictitious opinions that have been deliberately written to sound authentic in order to deceive the consumers. Due to their growing number and potential influence, the automatic detection of opinion spam has emerged as a highly relevant research topic [3,17,8].

Detecting opinion spam is a very challenging problem since opinions expressed on the Web are typically short texts, written by unknown people for very different purposes. Initially, opinion spam was detected by methods that seek for duplicate reviews [9]. It was only after the release of the gold-standard datasets by [16,17], which contain examples of positive and negative deceptive opinion spam, that it was possible to conduct supervised learning and a reliable evaluation of the task. The main conclusion from recent works is that standard text categorization techniques are effective at detecting deception in text. Particularly, best results have been approached using word n-grams together with other stylometric features [4,17].

We consider the detection of opinion spam as a stylistic classification task because, given a particular domain, the deceptive and truthful opinions are similar in content but differ in the way opinions are written (style). Furthermore, based on the fact that character n-grams are able to capture information from content and style, and motivated by their good performance in other tasks such as authorship attribution and polarity classification, we propose in this paper the use of character n-grams for the detection of opinion spam. Concretely, we aim to investigate in depth whether character n-grams are: (i) more appropriate than word n-grams, and (ii) more robust than the word n-grams in scenarios where only few data for training are available. Two are the main experiments we carried out. In the first experiment we considered 1600 hotel reviews. We analysed the classification of positive and negative opinions employing as features character n-grams and word n-grams. The best results were obtained using character n-grams with values for n of 5 and 4 respectively. The second experiment was varying the size of the training corpus in order to demonstrate the robustness of character n-grams as features. The obtained results show that with few samples in the training corpus, it is possible to obtain a a very classification performance, comparable to that obtained by word n-grams when using the complete training set.

The rest of the paper is organized as follows. Section 2 presents the related works on opinion spam detection and the use of character n-grams in other text classification tasks. Section 3 describes the corpus used for experiments as well as their configuration. Section 4 discusses the obtained results. Finally, Section 5 indicates the main contributions of the paper and provides some directions for future work.

[1] How Online Reviews Affect Your Business.
`http://mwpartners.com/positive-online-reviews`. Visited: April 2, 2014.

2 Related Work

The detection of spam on the Web has been mainly approached as a binary classification problem (spam vs. non-spam). It has been traditionally studied in the context of e-mail [2], and Web pages [5,15]. The detection of opinion spam, i.e., the identification of fake reviews that try to deliberately mislead human readers, is just another face of the same problem [18].

Due to the lack of reliable labeled data, most initial works regarding the detection of opinion spam considered unsupervised approaches which relied on meta-information from reviews and reviewers. For example, in [9], the authors proposed detecting opinion spam by identifying duplicate content. In a subsequent paper [10], they focussed on searching for unusual review patterns. In [14], the authors proposed an unsupervised approach for detecting groups of opinion spammers based on criteria such as the number of products that have been target of opinion spam and a high content similarity of their reviews. Similarly, in [20] it is presented a method to detect hotels which are more likely to be involved in spamming.

It was only after the release of the gold-standard datasets by [16,17], which contain examples of positive and negative deceptive opinion spam, that it was possible to conduct supervised learning and a reliable evaluation of the task. [16,13,3,17,7,8] are some examples of works that have approached the detection of opinion spam as a text classification task. In all of them word n-grams (unigrams, uni+bigrams and uni+bi+trigrams) have been employed as features. However, best results have been obtained combining word n-grams with style information. For example, [16] considered information from LIWC (linguistic inquiry and word count)[2], and [4] incorporated syntactic stylometry information in the form of deep syntax features.

In this work, we propose the use of character n-grams for detecting opinion spam. By using this representation our aim is to focus more on the writing style of the deceptive opinions than in their content. That is, our hypothesis is that somehow the writing style of a deceiver is different if compared to the one of honest users. This was also corroborated by Ott in [16].

Character n-grams have been used for email spam detection [11] and sentiment classification [1] with higher effectiveness than using word n-grams. They are also considered the state-of-the-art for authorship attribution [19]. To the best of our knowledge, this work is the first where character n-grams are used for the detection of opinion spam. The results that we will present in Section 4 show that they allow to address the problem more effectively than with word n-grams.

3 Experimental Setup

To test whether character n-grams allow to address the detection of opinion spam more effectively than word n-grams, we used the corpus of 1600 hotel

---

[2] `www.liwc.net/`

reviews that was facilitated by Ott[3]. These reviews are about 20 hotels of the downtown area of Chicago, where each hotel has 80 reviews, half of them are positive and the other half are negative. Each positive and negative subset is composed of 20 deceptive reviews and 20 truthful reviews. Deceptive opinions were generated using the Amazon Mechanical Turk, whereas (likely) truthful opinions were mined from reviews on TripAdvisor, Expedia, Hotels.com, Orbitz, Priceline, and Yelp.

The following paragraphs show four opinions for the same hotel. These examples are interesting since they show the great complexity of the automatically, and even manually, detection of deceptive opinions. The opinions are similar and just minor details can help distinguishing one from the other. For example, in [16] authors describe that there is a relationship between deceptive language and imaginative writing, and that deceptive reviews tend to use the words "experience", "my husband", "I", "feel", "business", and "vacation" more than genuine ones.

Example of a positive *deceptive* opinion

My husband and I stayed for two nights at the Hilton Chicago, and enjoyed every minute of it! The bedrooms are immaculate, and the linens are very soft. We also appreciated the free WiFi, as we could stay in touch with friends while staying in Chicago. The bathroom was quite spacious, and I loved the smell of the shampoo they provided-not like most hotel shampoos. Their service was amazing, and we absolutely loved the beautiful indoor pool. I would recommend staying here to anyone.

Example of a positive *truthful* opinion

We stay at Hilton for 4 nights last march. It was a pleasant stay. We got a large room with 2 double beds and 2 bathrooms, The TV was Ok, a 27' CRT Flat Screen. The concierge was very friendly when we need. The room was very cleaned when we arrived, we ordered some pizzas from room service and the pizza was ok also. The main Hall is beautiful. The breakfast is charged, 20 dollars, kinda expensive. The internet access (WiFi) is charged, 13 dollars/day. Pros: Low rate price, huge rooms, close to attractions at Loop, close to metro station. Cons: Expensive breakfast, Internet access charged. Tip: When leaving the building, always use the Michigan Ave exit. It's a great view.

Example of a negative *deceptive* opinion

I stayed two nights at the Hilton Chicago. That was the last time I will be staying there. When I arrived, I could not believe that the hotel did not offer free parking. They wanted at least $10. What am I paying for when I stay there for the night? The website also touted the clean linens. The room was clean and I believe the linens were clean. The problem was with all of the down pillows etc. Don't they know that people have allergies? I also later found out that this hotel allows pets. I think that this was another part of my symptoms. If you like a clean hotel without having allergy attacks I suggest you opt for somewhere else to stay. I did not like how they nickel and dimed me in the end for parking. Beware hidden costs. I will try somewhere else in the future. Not worth the money or the sneezing all night.

Example of a negative *truthful* opinion

My $200 Gucci sunglasses were stolen out of my bag on the 16th. I filed a report with the hotel security and am anxious to hear back from them. This was such a disappointment, as we liked the hotel and were having a great time in Chicago. Our room was really nice, with 2 bathrooms. We had 2 double beds and a comfortable hideaway bed. We had a great view of the lake and park. The hotel charged us $25 to check in early (10am).

[3] http://myleott.com/op_spam

For representing the opinion reviews we used a bag of character n-grams (BOC) and a bag of word n-grams (BOW); in both cases we applied a binary weighting scheme. Particularly, for building the BOW representation we pre-processed texts removing all punctuation marks and numerical symbols, i.e., we only considered alphabetic tokens. We maintained stop words, and converted all words to lowercase characters.

For classification we used the Naïve Bayes (NB) classifier, employing the implementation given by Weka [6], and considering as features those n-grams that occurred more than once in the training corpus. It is important to comment that we performed experiments using several classification algorithms (e.g., SVM, KNN and multinomial NB), and from all of them NB consistently showed the best results.

The evaluation of the classification effectiveness was carried out by means of the macro average F_1-measure of the deceptive and truthful opinion spam categories. We performed a 10 fold cross-validation procedure to assess the effectiveness of each approach, and we used the the Wilcoxon Signed Rank Test for comparing the results of BOC and BOW representations in all the evaluation scenarios. For these comparisons we considered a 95% level of significance (i.e., $\alpha = 0.05$) and a null hypothesis that both approaches perform equally well.

4 Experiments

In this section we describe the two experiments we carried out in order to see whether character n-grams allow to obtain a better performance than word n-grams (first experiment), and also to evaluate the robustness of character-based representation when only few examples of deceptive opinion spam are available for training (second experiment).

4.1 Experiment 1: Character vs. Word n-grams

In this first experiment, we aim to demonstrate that character n-grams are more appropriate than word n-grams to represent the content and writing style of opinion spam. We analysed the performance of both representations on positive as well as on negative reviews.

Table 1 shows the results obtained with word n-grams. These results indicate that the combination of unigrams and bigrams obtained the best results in both polarities; however, the difference in F_1 with unigrams was not statistically significant in any case. In contrast, the representation's dimensionality was increased 7.5 for the positive opinions and 8 times for the negative reviews, suggesting that word unigrams are a good representation for this task.

Another interesting observation from Table 1 is that classifying negative opinions is more difficult than classifying positive reviews; the highest F_1 measure obtained for negative opinions was 0.848, whereas for positive opinions the best configuration obtained 0.882. We figure out that this behaviour could be caused by the differences in the vocabularies' sizes; the vocabulary employed in negative opinions was 37% larger than the vocabulary from positives, indicating that

Table 1. Results using *word n-grams* as features, in positive and negative opinions. In each case, the reported results correspond to the macro average F_1 of the deceptive and truthful opinion categories.

FEATURES	POSITIVE		NEGATIVE	
	size	*macro* F_1	*size*	*macro* F_1
unigrams	5920	0.880	8131	0.850
uni+bigrams	44268	0.882	65188	0.854
uni+big+trigrams	115784	0.881	174016	0.840

their content is in general more detailed and diverse, and, therefore, that larger training sets are needed for their adequate modelling.

Figure 1 shows the results obtained with character n-grams for different values of n. It also compares these results against the best result using word n-grams as features. These results indicate that character n-grams allow to obtain better results than word n-grams on the positive opinions. The best result was obtained with 5-grams ($F_1 = 0.902$), indicating an improvement of 2.3% over the result using as features word unigrams and bigrams ($F_1 = 0.882$).

Regarding the negative opinions, results were very similar; character n-grams showed to be better than word n-grams. However, in this case the best results were obtained with character 4-grams. We presume, as before, that this behaviour could be related to the larger vocabulary used in the negative opinions, which make difficult the modelling of large n-grams from the given training set. The best result for character n-grams was $F_1 = 0.872$, indicating an improvement of 2.1% over the result using unigrams and bigrams as features $F_1 = 0.854$.

Using the Wilcoxon test as explained in Section 3, we found that the best results from character n-grams are significantly better that the best results from the word-based representations with $p < 0.05$ in the two polarities.

To have a deep understanding of the effectiveness of character n-grams as features, we analysed the 500 n-grams with the highest information gain for both polarities. From this analysis, we have observed that n-grams describing the location of the hotel (e.g. *block, locat, an ave*) or giving some general information about the rooms (e.g. *hroom, bath, large*) are among the most discriminative for positive spam. In contrast, some of the most discriminative n-grams for negative opinions consider general characteristics (e.g. *luxu, smel, xpen*) or they are related to negative expressions (e.g. *_don, (non, nt_b*). This analysis also showed us that the presence of n-grams containing personal pronouns in first person of singular and plural such as *I, my, we* are 20% more abundant in the list of n-grams from negative opinions than in the list from the positive reviews.

4.2 Experiment 2: Character n-grams Robustness

The second experiment aims to demonstrate the robustness of the character n-grams with respect to the size of the training corpus. To carry out this experiment, for each one of the ten folds used for evaluation, we considered 25%, 50%

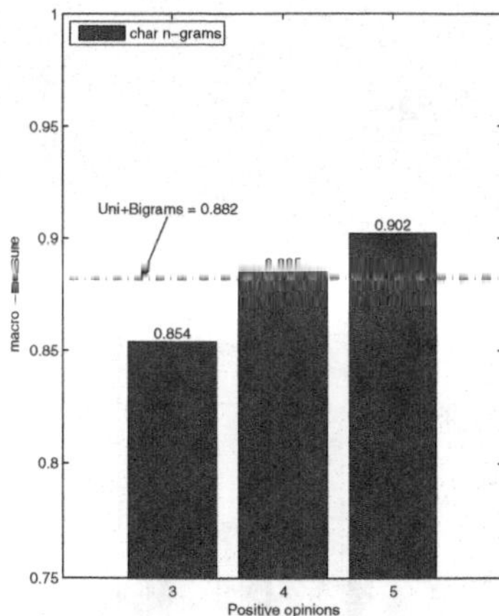

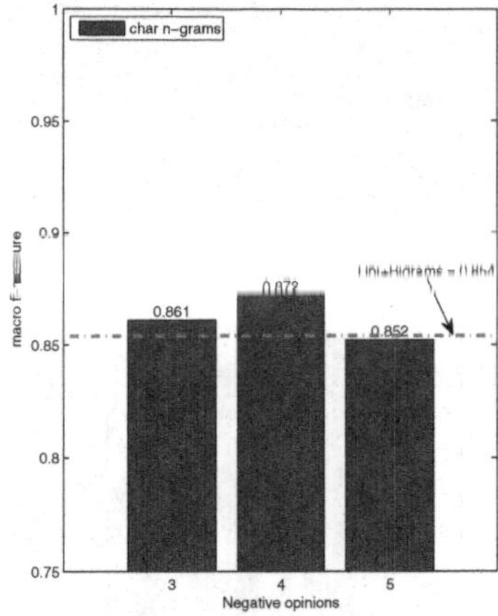

Fig. 1. Results using *character n-grams* as features, in positive and negative opinions. In each case, the reported results correspond to the macro average F_1 of the deceptive and truthful opinion categories. The dotted line indicates the results using word unigrams and bigrams as features.

and 100% of the training instances to train the classifier, while mantaining fixed the test set partition.

Figure 2 shows the results obtained with the Naïve Bayes classifier for both polarities, positive and negative opinions, as well as using both kinds of features, character n-grams and word n-grams. These results indicate that the performance obtained with character n-grams is consistently better that the performance of word n-grams. In particular, it is important to notice that using only 25% of the original training set, which consists of 180 opinions reviews, half of them deceptive and the other half truthful, the representation based on character n-grams shows F_1 values up to 0.80 for both polarities. Using the Wilcoxon test as explained in Section 3, we found that the results from character n-grams are significantly better that the results from the word-based representations with $p < 0.05$ in both polarities.

As an additional experiment we evaluated the variation in performance of the proposed representation using other classifiers. Particularly, Figure 3 compares the results obtained by the Naïve Bayes classifier with those obtained with SVM as well as with a multinomial Naïve Bayes classifier. These results indicate an important variation in F_1 measure caused by the selection of the classifier. On the one hand, the Naïve Bayes classifier shows the best results for the positive opinions; they are significatively better than those from SVM according to the Wilcoxon test with $p < 0.05$. On the other hand, SVM obtained the best results in the classification of deceptive and truthful negative reviews, significantly improving the results of the Naïve Bayes classifier only when using the complete (100%) training set. Somehow this results were not completely unexpected since previous works have showed that Naïve Bayes models tend to surpass the SVM classifiers when there is a shortage of positives or negatives [8].

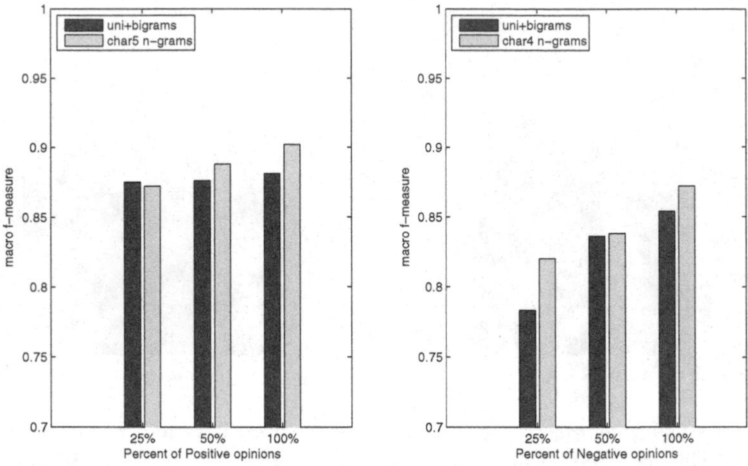

Fig. 2. Results of *character n-grams* and *word n-grams* varying the size of the training sets. The reported results correspond to the macro average F_1 of the deceptive and truthful opinion categories.

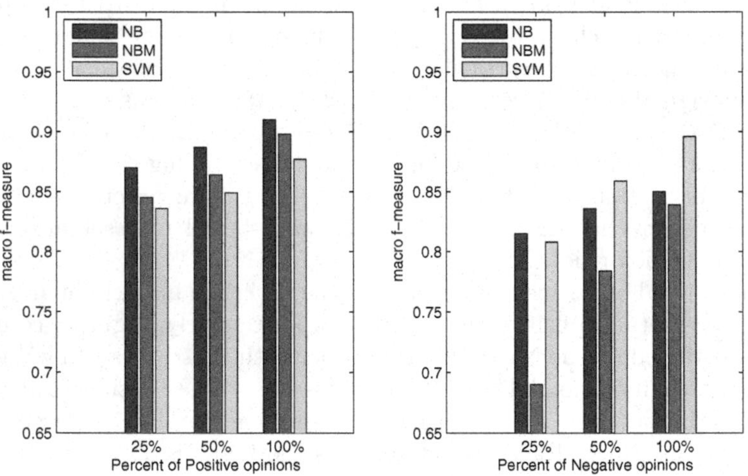

Fig. 3. Results of *character n-grams* using different classifiers and varying the size of the training sets. The reported results correspond to the macro average F_1 of the deceptive and truthful opinion categories.

5 Conclusions and Future Work

In this paper we proposed a novel approach for detecting deceptive opinion spam. We considered the detection of opinion spam as a stylistic classification task, and, accordingly, we proposed using *character n-grams* as features. Although character n-grams have been used in similar tasks showing higher effectiveness than word n-grams, to the best of our knowledge, this work is the first where character n-grams are used for the detection of opinion spam. Experiments were carried out employing Ott's corpus of 1600 hotel reviews, 800 deceptive and 800 truthful. Based on the experimental results it is possible to formulate the following two conclusions: (i) character n-grams showed to capture better than word n-grams the content of deceptive opinions as well as the writing style of deceivers, obtaining better results in both polarities. (ii) Character n-grams showed a better robustness than word-grams obtaining good performance with small training sets; using only 25% of the training data, character n-grams were able to obtained F_1 values up to 0.80 in both polarities.

As future work, we plan to investigate the possibility of combining character n-grams with word n-grams. Going a step forward, we also aim to evaluate other approaches from authorship attribution in the detection of opinion spam.

Acknowledgments. This work is the result of the collaboration in the framework of the WIQEI IRSES project (Grant No. 269180) within the FP 7 Marie Curie. The second author was partially supported by the LACCIR programme under project ID R1212LAC006. Accordingly, the work of the third author was in the framework the DIANA-APPLICATIONS-Finding Hidden Knowledge in Texts: Applications (TIN2012-38603-C02-01) project, and the VLC/CAMPUS Microcluster on Multimodal Interaction in Intelligent Systems.

References

1. Blamey, B., Crick, T., Oatley, G.: RU:-) or:-(? character-vs. word-gram feature selection for sentiment classification of OSN corpora. Research and Development in Intelligent Systems XXIX, 207–212 (2012)
2. Drucker, H., Wu, D., Vapnik, V.N.: Support Vector Machines for Spam Categorization. IEEE Transactions on Neural Networks 10(5), 1048–1054 (2002)
3. Feng, S., Banerjee, R., Choi, Y.: Syntactic Stylometry for Deception Detection. Association for Computational Linguistics, short paper. ACL (2012)
4. Feng, S., Xing, L., Gogar, A., Choi, Y.: Distributional Footprints of Deceptive Product Reviews. In: Proceedings of the 2012 International AAAI Conference on WebBlogs and Social Media (June 2012)
5. Gyongyi, Z., Garcia-Molina, H., Pedersen, J.: Combating Web Spam with Trust Rank. In: Proceedings of the Thirtieth International Conference on Very Large Data Bases, vol. 30, pp. 576–587. VLDB Endowment (2004)
6. Hall, M., Eibe, F., Holmes, G., Pfahringer, B., Reutemann, P., Witten, I.: The WEKA Data Mining Software: an Update. SIGKDD Explor. Newsl. 10–18 (2009)

7. Hernández-Fusilier, D., Guzmán-Cabrera, R., Montes-y-Gómez, M., Rosso, P.: Using PU-learning to Detect Deceptive Opinion Spam. In: Proceedings of the 4th Workshop on Computational Approaches to Subjectivity, Sentiment and Social Media Analysis for Computational Linguistics: Human Language Technologies, Atlanta, Georgia, USA, pp. 38–45 (2013)
8. Hernández-Fusilier, D., Montes-y-Gómez, M., Rosso, P., Guzmán-Cabrera, R.: Detecting Positive and Negative Deceptive Opinions using PU-learning. Information Processing & Management (2014), doi:10.1016/j.ipm.2014.11.001
9. Jindal, N., Liu, B.: Opinion Spam and Analysis. In: Proceedings of the International Conference on Web Search and Web Data Mining, pp. 219–230 (2008)
10. Jindal, N., Liu, B., Lim, E.: Finding Unusual Review Patterns Using Unexpected Rules. In: Proceedings of the 19th ACM International Conference on Information and Knowledge Management, CIKM 2010, pp. 210–220(October 2010)
11. Kanaris, I., Kanaris, K., Houvardas, I., Stamatatos, E.: Word versus character n-grams for anti-spam filtering. International Journal on Artificial Intelligence Tools 16(6), 1047–1067 (2007)
12. Lim, E.P., Nguyen, V.A., Jindal, N., Liu, B., Lauw, H.W.: Detecting Product Review Spammers Using Rating Behaviours. In: CIKM, pp. 939–948 (2010)
13. Liu, B.: Sentiment Analysis and Opinion Mining. Synthesis Lecture on Human Language Technologies. Morgan & Claypool Publishers (2012)
14. Mukherjee, A., Liu, B., Wang, J., Glance, N., Jindal, N.: Detecting Group Review Spam. In: Proceedings of the 20th International Conference Companion on World Wide Web, pp. 93–94 (2011)
15. Ntoulas, A., Najork, M., Manasse, M., Fetterly, D.: Detecting Spam Web Pages through Content Analysis. Transactions on Management Information Systems (TMIS), 83–92 (2006)
16. Ott, M., Choi, Y., Cardie, C., Hancock, J.T.: Finding Deceptive Opinion Spam by any Stretch of the Imagination. In: Proceedings of the 49th Annual Meeting of the Association for Computational Linguistics: Human Language Technologies, Portland, Oregon, USA, pp. 309–319 (2011)
17. Ott, M., Cardie, C., Hancock, J.T.: Negative Deceptive Opinion Spam. In: Proceedings of the 2013 Conference of the North American Chapter of the Association for Computational Linguistics: Human Language Technologies, Atlanta, Georgia, USA, pp. 309–319 (2013)
18. Raymond, Y.K., Lau, S.Y., Liao, R., Chi-Wai, K., Kaiquan, X., Yunqing, X., Yuefeng, L.: Text Mining and Probabilistic Modeling for Online Review Spam Detection. ACM Transactions on Management Information Systems 2(4), Article: 25, 1–30 (2011)
19. Stamatatos, E.: On the robustness of authorship attribution based on character n-gram features. Journal of Law & Policy 21(2) (2013)
20. Wu, G., Greene, D., Cunningham, P.: Merging Multiple Criteria to Identify Suspicious Reviews. In: RecSys 2010, pp. 241–244 (2010)
21. Xie, S., Wang, G., Lin, S., Yu, P.S.: Review Spam Detection via Time Series Pattern Discovery. In: Proceedings of the 21st International Conference Companion on World Wide Web, pp. 635–636 (2012)
22. Zhou, L., Sh, Y., Zhang, D.: A Statistical Language Modeling Approach to Online Deception Detection. IEEE Transactions on Knowledge and Data Engineering 20(8), 1077–1081 (2008)

Content-Based Recommender System Enriched with Wordnet Synsets

Haifa Alharthi and Diana Inkpen

University of Ottawa, ON, Canada
{halha060,Diana.Inkpen}@uottawa.ca

Abstract. Content-based recommender systems can overcome many problems related to collaborative filtering systems, such as the new-item issue. However, to make accurate recommendations, content-based recommenders require an adequate amount of content, and external knowledge sources are used to augment the content. In this paper, we use Wordnet synsets to enrich a content-based joke recommender system. Experiments have shown that content-based recommenders using K-nearest neighbors perform better than collaborative filtering, particularly when synsets are used.

1 Introduction

Recommender systems (RSs) have been an active research area in recent years, and most university computer science departments now offer RS-related courses [1]. Basically, what recommender systems do is to predict which items (e.g., movies, cameras, books, etc.) suit a user who has not seen them, and suggest those items to the user. RSs effectively solve the problem of having too many products on the internet to choose from [1] [2].

The process of making recommendations requires three components: items, users and user feedback on items. Things that RSs recommend to users are called items, regardless of if they are a service, a trip or any other product. The second component of a recommendation process is users. Users are the center of the system, as some researchers define RSs as software that develops and uses customers' profiles [1] . The information needed to build a user profile varies from system to system; RSs might exploit users' demographic information, ratings or personality. Feedback, the third component, is how users interact with items, and it can be collected explicitly or implicitly. Rating items can have many classes, such as 1 to 5 stars, two classes (like/dislike), or only one class (like). After receiving suggestions, a user may provide feedback to indicate whether the user likes or dislikes the item; the system stores the user's opinion in a database, and uses it for future recommendations [1] [2].

1.1 Collaborative Filtering (CF)

Collaborative filtering is the most widely used approach of recommender systems [3]. CF systems exploit the available ratings of some users to predict and recommend

A. Gelbukh (Ed.): CICLing 2015, Part II, LNCS 9042, pp. 295–308, 2015.
DOI: 10.1007/978-3-319-18117-2_22

items to another group of users. In general, CF predicts the preferences of users by exploiting a user-item matrix that has the ratings of m users $\{u_1, u_2, \ldots u_m\}$ on n items $\{i_1, i_2, \ldots, i_n\}$. A user u_i has rated a list of items I_{ui}. If the user has not rated the item, a missing value is shown in the matrix. There are two approaches of CF algorithms: user-based and item-based [4]. In the former, it is assumed that two users with similar tastes or rating histories will rate items similarly in the future. Each user is represented as a vector of items' ratings, which is called the user profile. The system finds users with similar profiles to the target user and exploits their ratings to predict the likeliness that the user likes a particular item [5]. The item-based CF system computes the similarity between two co-rated items, and the most similar items to those the target user has already preferred are recommended [6] [2].

1.2 Content Based Approach (CB)

In the content-based approach, a model for each user is developed based on the patterns and similarities in the content of previously rated items. After a user un gives a rating R_k on an item I_k, a training dataset TR_n that consists of pairs (I_k, R_k) is developed. Using this dataset, supervised machine learning algorithms are applied to create the user model which is used to predict the user's preferred items. The model can assess whether or not the user will be interested in an item that is not yet rated. Thus, the user profile has structural information about the user's preferences, and the system matches the profile and the items' descriptions in order to make recommendations [7] [2].

1.3 Collaborative Filtering vs. Content-Based RSs

CF does not require additional information about the content of items; it only needs the rating matrix in order to make suggestions. Another advantage is that CF is simple to implement. However, CF systems have the following issues. When there are no previous ratings, or not enough ratings related to a user, the system cannot find similar users and this is called the cold start or new-user problem. Also, the system cannot suggest a new item if no user has rated the item yet. CF also suffers from having users, called 'gray sheep', with completely different tastes from the rest of the other users. However, humans also fail to predict gray sheep preferences, so it is not considered a serious problem in RSs. In addition, because CF systems rely on the ratings of users, they are vulnerable to shilling attacks, which occur when some fake users rate specific products highly as a form of promotion, and give competing products low ratings. This may lead the system to recommend the promoted products to many users [4] [8].

By not relying on users' ratings, CB system avoids having the gray sheep and shilling attack problems. In addition, the reasons why a CB system makes recommendations are known, since it makes suggestions based on item attributes. Thus, the system can justify why it recommended an item. For example, it could suggest an action movie with a specific actor to a user because the user tends to favor action movies with that actor. Moreover, unlike CF, content-based systems do not suffer from the new-item problem, because as soon as an item is introduced, the system finds users who favored items with similar content, and recommends the new item to them [1] [7]. Nonetheless, CB does face some challenges. First, knowledge about items' content is required, and an inadequate item description could result in poor recommendations. Another issue is overspecialization; that is, when the system only recommends items with similar content to previously-favored items. In addition, CB has the same new-user problem as CF, in that a CB system can only build a representative model for a user when they have rated enough items. Otherwise, it has difficulty making quality recommendations [7].

1.4 Problem Statement

Insufficient item descriptions can be a problem with CB systems. To overcome this, knowledge sources such as thesauri and ontologies are used to obtain precise recommendations [7]. To build our CB system we employed the Jester dataset, which is typically used to evaluate CF systems; to our knowledge, this is the first work that uses Jester for CB recommendations. We chose this dataset because it is made up of jokes, which are text-based items. Jokes are smaller than news articles, and thus can be good examples of items that require enrichment from external resources such as Wordnet.

1.5 WordNet

Wordnet is a lexical database widely used in natural language processing applications. It groups English nouns, verbs, adjectives and adverbs into synsets. A synset is a set of synonyms which refer to one concept and can be substituted for another, such as car and automobile. If words have dissimilar senses, they belong to different synsets. At least, 117,000 synsets are connected with many semantic relationships [9].

The most important semantic relation linking two synsets is that of hyperonymy (and its inverse relation, hyponymy). A hyponym (e.g., Chihuahua) is a kind of hypernym (e.g., dog). This relationship connects general synsets to more specific synsets in a hierarchal structure. Though the highest level of noun hypernyms is the word entity, there are two kinds of nouns that Wordnet can deal with: type (e.g.,dog) which can have higher type (e.g., animal), and instance (e.g., Barack Obama) of another word (e.g., President). Synsets of verbs can be connected with 'is kind of' relationships, such as to perceive which is a superclass of to listen [9] [10].

The hypothesis behind this research is that enriching the content of items with hypernyms, hyponyms and synonyms will allow detecting more similarities between the items, which could result in better recommendations. The rest of the paper is organized as follows: Section 2 discusses work that exploited Wordnet's synsets in recommender systems, Section 3 examines how items are represented and the algorithms used to build user profiles, Section 4 describes information about the dataset, experimental settings and evaluation metrics, Section 5 presents the results, Section 6 discusses the results, and future work is considered in Section 7.

2 Related Work

To support RSs with linguistic knowledge and feed them additional information for better analysis, some researchers exploited external sources such as Wordnet. In news recommendations, content-based RSs often represent articles as TF-IDF[1] vectors, and apply the cosine similarity between items to find the articles most similar to the previously preferred ones. More advances were introduced to news recommendation systems when synsets were used to represent items. In [11], word sense disambiguation was applied on terms, and they were then replaced by synsets. Items were represented by SF-IDF, where SF refers to synset frequency; this representation allowed better recommendation performance than the traditional TF-IDF approach.

SiteIF, an RS that recommends news in English and Italian, was proposed in [12], and is the first CB system to represent items based on their senses. It uses MultiWordNet, an English-Italian database with matched senses. News articles are automatically annotated by applying word domain disambiguation. When a user reads a news article, the synsets related to that article are added to the user profile, then the system uses the semantic network and the user profile to recommend new articles. However, no results of this work have been reported.

ITR, another system that uses the textual description of items from multiple domains to make recommendations (e.g., movies vs. books) is presented in [13] [14]. This system takes advantage of WordNet's lexical ontology to build item representation based on senses, by using word sense disambiguation. Every item is represented as vector of synsets, and the vector space model is called bag-of-synsets (BOS). After a user rates many items, a Naïve Bayes model is developed that classifies the items as positive or negative, and the classifier learns the synsets that are most likely to be associated to preferred items. The experiments show that sense-based profiles achieve higher accuracy than regular keyword-based user profiles.

A content-based RS that recommends items in the multimedia domain was proposed in [15]. One approach was to find the similarity between images by using their metadata, including their genres. However, in the experiments, there were few items available and the annotators tagged them with numerous distinct genres, thus hypernyms of genres were used to determine more commonalities between items. For example, in one case an image was annotated with tiger and another image with cat, and both were tagged with animal. The level of introduced hypernyms was set relatively low, otherwise the number of genres would be too small and the genre attributes not as helpful in the classification task. However, in [15] the focus was to compare the use of metadata and affect information in content-based RSs, and the effect of hypernyms was not reported.

---

[1] Text frequency (TF) gives higher weight to frequent words, and inverse document frequency (IDF) gives higher weight to words that appear in fewer documents.

3 Content-Based Recommender System Enriched with Wordnet's Synsets

In the following sections, the process of representing our items (the jokes) is shown alongside the development of user profiles using two algorithms: nearest neighbors and support vector machines

3.1 Item Representation

Since the content is unstructured text with no fixed features, such as actors and directors, the preprocessing task involves complicated natural language processing. The text was tokenized, then stemmed, meaning that after the text is converted to separate words, the root of the words is found. An English language tokenizer was used on the text of jokes. Stop words were eliminated and words were transformed into lower case, so two identical words were not considered by the classifier as different terms due to the difference in their letters case. After this step, synonyms, hypernyms or hyponyms were added, followed by a stemmer that uses a database of a dictionary and replaces words using matching rules. If there are multiple stems for one term, the first stem is used. In addition, if the stem for the token is not found in the database, the term is kept. After preprocessing the textual data, the term-frequency matrix that represents items is created. Term frequency freq(d; t) means the number of times term *t* occurs in document *d*, as opposed to the number of times all terms occur in *d* [16].

The following is an example of how the original text of a joke is transformed when synonyms of nouns, verbs, adjectives and adverbs, as well as hyponyms and hypernyms of nouns and verbs, are added.

Joke: Employer to applicant: "In this job we need someone who is responsible." Applicant: "I'm the one you want. On my last job, every time anything went wrong, they said I was responsible."

Joke with synonyms: employer applicant/applier: " occupation/business/job/line_of_work/line responsible . " applicant/applier : " iodine/iodin/I/atomic_number_53 'm privation/want/deprivation/neediness . occupation/business/job/line_of_work/line , time/clip went wrong/wrongfulness , aforesaid/aforementioned/said iodine/iodin/I/atomic_number_53 responsible . "

Joke with hyponyms: boss/hirer aspirant/aspirer/hopeful/wannabe/wannabee : " confectionery responsible . " aspirant/aspirer/hopeful/wannabe/wannabee : " iodine-131 'm absence . confectionery , day went aggrieve , said iodine-131 responsible . "

Joke with hypernyms: leader person/individual/someone/somebody/mortal/soul : " activity responsible . " person/individual/someone/somebody/mortal/soul : " chemical_element/element 'm poverty/poorness/impoverishment . activity , case/instance/example went injustice/unjustness , said chemical_element/element responsible . "

3.2 User Profile

A user model contains information about a user's interests. The process of developing the model is considered a text classification task, where previously rated items are training examples, and their ratings are labels. Machine learning algorithms are used to develop a function that learns from the previous behavior of the user to predict her subsequent favored items. The algorithms could predict the numeric rating of the user (e.g., 5 stars) or might only predict if the user will like or dislike the item. The following algorithms are used in the experiments:

Nearest Neighbors Algorithm (KNN)

kNN is a lazy algorithm since it puts the training dataset in memory and when it needs to make a prediction in an item-based CF, it searches for the items which have similar vectors to the item in the test dataset. Based on the target class (e.g., like/dislike) or the numeric ratings (e.g., 1-5 stars) of the similar items (neighbors), the prediction of the unrated items is made. The process of searching for the similar items involves using a similarity function [17]. We have used the cosine similarity which is usually applied with items represented in a vector space model. Eq. 1 shows how cosine similarity between two items' vectors A and B is calculated [18].

$$similarity = cos(\theta) = \frac{A.B}{||A||||B||} \quad (1)$$

Support Vector Machines

SVM is a classification algorithm which in a linearly separable space discovers the maximum hyperplane that separates two target classes; e.g., like/dislike. Let L be the training set in the form of (x_i; y_i). Each instance x_i contains *D* dimensions and is associated with a target class y_i= 1 or -1 where:

i = 1 ... L; $x \in R^D$; $y_i \in$ (-1; 1). In this case an assumption is made that data is linearly separable so that a line, which is called hyperplane, can be drawn which is described by w.x+b=0 where:

w is normal vector to the hyperplane [19]; and

$\frac{b}{w}$ is the distance between the hyperplane and the origin. The closest instances to separating hyperplane are called Support Vectors [19].

4 Evaluation

In the following subsections, we describe the adopted dataset, evaluation metrics and experimental settings.

4.1 Dataset

Jester is an online system which recommends jokes. Users rate jokes in the range of [-10.00 , +10.00], and the rating is continuous. User ratings from 1999 to 2003 were collected. The dataset has 4.1 million ratings of 100 jokes, from 73,421 users [20]. Jokes in their textual format were mapped to users' ratings, so they could be used for CB recommendations. The first ten users who rated all 100 jokes are selected to be the target users, and their ratings comprise the target class or the label for the classification task. The dataset was decreased to save processing time and effort, so only the first 200 users were included in the training dataset of CF experiments.

The length of the jokes varies; some are only one sentence and others are a paragraph. The jokes in Jester are written in formal English, as spelling, grammar and punctuation are taken into consideration. The following example is of one of the jokes: "Q. What is orange and sounds like a parrot? A. A carrot." All questions and answers in the dataset are in the Q. and A. format.

4.2 Evaluation Metrics

The accuracy of recommendations is the most important aspect when evaluating RSs [8]. The metrics to evaluate the accuracy were categorized by [8], into ratings predictions, classification predictions and rank accuracy metrics.

Ratings Predictions Metrics

Many RSs attempt to predict a user's numeric rating. For example, in the movie domain (e.g., Netflix) the RS provides the users with the likelihood that they will favor a movie, in a range of one to five stars. In these cases, predictive accuracy metrics are applied, which take the distance from the predicted numeric rating to the actual rating into account. Two metrics are widely used to measure the ratings predictions: Root Mean Squared Error (RMSE) as in Eq. 2, and Mean Absolute Error (MAE) as in Eq. 3. In these equations, (u, i) refers to a user-item pair, and T is the test dataset with a predicted rating of ˆrui, and real rating of rui. Unlike MAE, large errors are not tolerated in RMSE [8]. MAE would prefer a system that makes few errors even if they are big errors (e.g., to predict a rating as 1 when in fact it is 5) whereas RMSE would prefer a system that makes many small errors rather than few big errors. The recommendation algorithms in the first group try to predict the exact ratings of a user.

$$\text{RMSE}=\sqrt{\frac{1}{|T|}\sum_{(u,i)\epsilon T}(\hat{}r_{ui}-r_{ui})^2} \tag{2}$$

$$\text{MAE}=\sqrt{\frac{1}{|T|}\sum_{(u,i)\epsilon T}|\hat{}r_{ui}-r_{ui}|} \tag{3}$$

Classification Accuracy Metric

The metrics of classification prediction are suitable for evaluating RSs when the aim is to classify the items as interesting or non-interesting to the user. The RSs goal here is not to show the user a numeric rating, but to recommend good items. To evaluate our approach, we used the popular information retrieval system metrics of 'precision' as in Eq. 4 and 'recall' as in Eq. 5. Many research papers have employed these to evaluate recommender systems, including [21], [22], and [23]. In the second group of experiments, the purpose is to only find whether or not a user will like the recommended item. To achieve this, the ratings of the target user has to be converted to binary classes. All the jokes with non-negative ratings are labeled with like while negative ratings are considered as dislike.

$$\text{Precision}=\frac{\textit{true positives}}{\textit{true positives}+\textit{false positives}}=\frac{\textit{relevant items recommended}}{\textit{all recommended items}} \quad (4)$$

$$\text{Recall}=\frac{\textit{true positives}}{\textit{true positives}+\textit{false negatives}}=\frac{\textit{relevant items recommended}}{\textit{all relevant items}} \quad (5)$$

4.3 Experiments Settings

The preprocessing and classification experiments were conducted in Rapidminer[2], a widely used machine learning and text mining tool. The number of neighbors for KNN in all experiments was ten, which showed the best results after many trails. KNN uses the cosine similarity measure, and the default settings of linear SVM in Rapidminer were not changed. A classifier was built for each user, and the reported MAE, RMSE, precision and recall values are the average values for the ten selected target users. The statistical test, Wilcoxon, was applied, and it is reported when there was a statistically significant increase. The test option in all experiments was set to ten-fold cross-validation, as this is widely applied in machine learning tasks.

Since there are no hyponyms or hypernyms for adjectives and adverbs, these experiments involved the use of nouns and verbs only, while synonyms for all four part-of-speech types were found. Different levels of recursion depth were applied for hyponyms and hypernyms. For example, if the recursion depth was set to two when searching for hyponyms, the hyponym of the found hyponym of a word was also returned. As well, if more than one meaning per word was found only the first meaning is used, and if there was more than one synset per meaning the first synset was used. After many trails, it was determined that using only the first word in a synset does not significantly affect the final results, so all the words in the selected synset were returned. Any words with no matching hyponym, hypernym or synonym were also included in the bag of words.

2 https://rapidminer.com/

5 Results

In this section, the ratings prediction metrics are used to evaluate the three approaches in the first group of experiments (which is about ratings on the 5 star scale): CF, KNN-based CB and SVM-based CB. The classification accuracy metrics are used to evaluate the same approaches in the second group of experiments (which is about classification into like/dislike).

5.1 First Group

First, the collaborative filtering KNN is compared to two content-based algorithms: KNN and Linear SVM in Graph 1. At this stage, only the text of jokes is used to make the CB recommendations, thus it is called the base form. As expected, SVM-based CB makes fewer prediction errors than KNN-based CB, and it is statistically better than CF in both MAE and RMSE. In this section, synsets were applied to enrich the bag of words, and then the highest performing algorithm, linear SVM was used. The effect of adding synonyms of words from different POS types was found in five experiments, as illustrated in Graph 2. The error rates produced by the base form of SVM are less than in the five experiments, and increased error rates were also observed when hyponyms and hypernyms were added, as illustrated in Graphs 3 and 4.

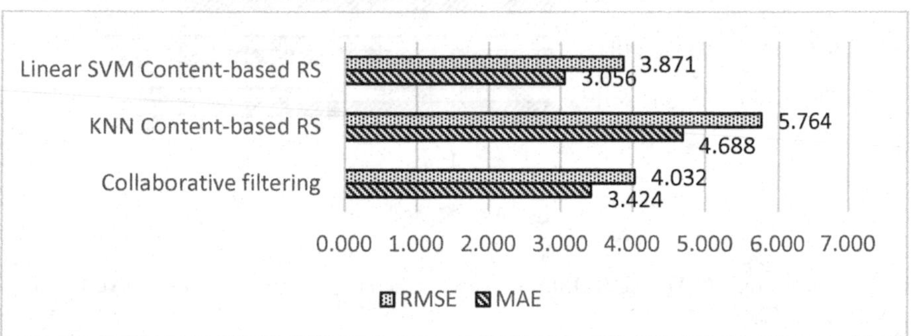

Graph 1. The average value of MAE and RMSE of 10 users when CF, KNN CB and SVM CB is used

Since the addition of Wordnet synsets did not improve the prediction of CB using SVM, determining whether this will have the same effect when working with KNN could be important. Thus, the same experiments that were performed with SVM CB were also conducted with KNN CB (only the highest performing experiments are reported here). Graph 5 indicates that when Wordnet synsets were used the MAE and RMSE decreased by approximately one point, and there was a significant difference at alpha= .005 when statistical tests were applied. Graph 5 also shows that the use of hyponyms of nouns achieves the highest performance, though it was less accurate than CF.

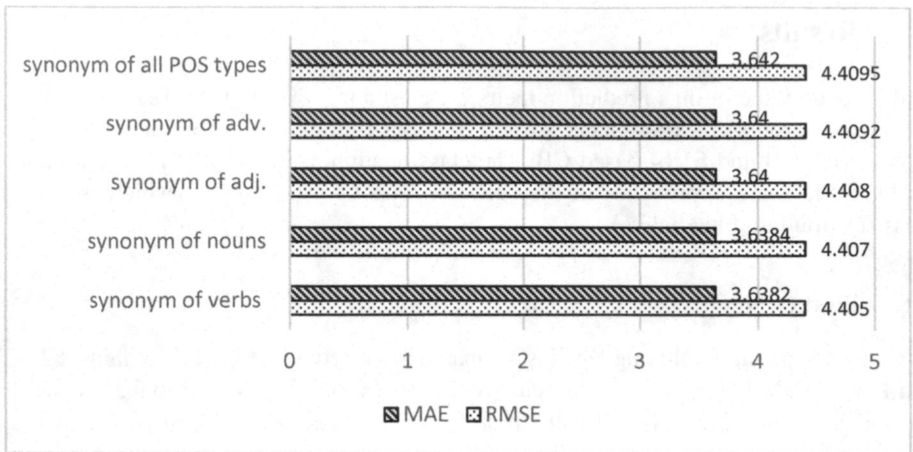

Graph 2. The average MAE and RMSE results of 10 users when synonyms are used with SVM CB

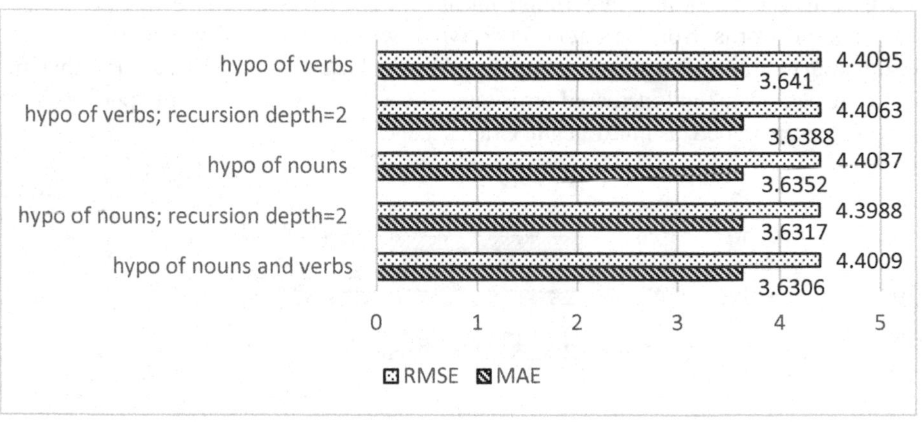

Graph 3. The average MAE and RMSE of 10 users when hyponyms of nouns and verbs are used with SVM CB

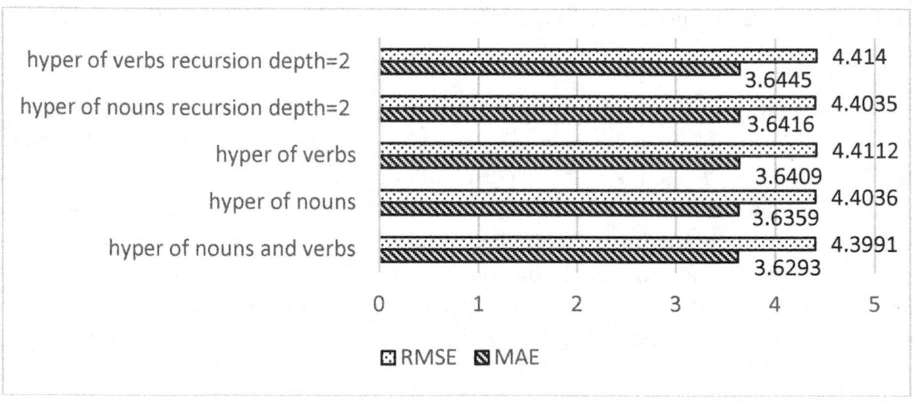

Graph 4. The average MAE and RMSE of 10 users when hypernyms are used with SVM CB

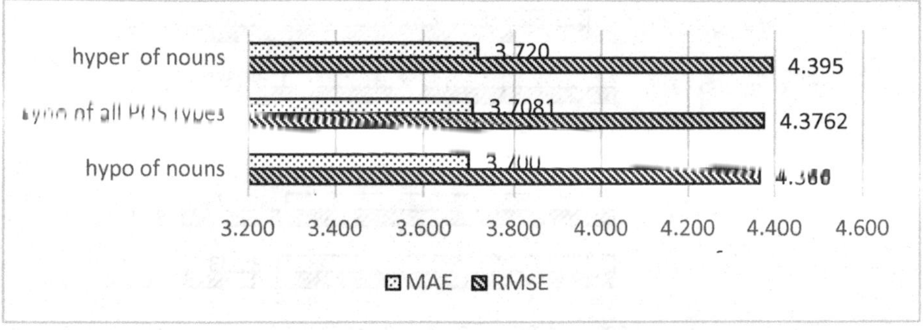

Graph 5. The average of best MAE and RMSE results for ten users when synonyms, hyponyms, and hypernyms are added to KNN CB

5.2 Second Group

Graph 6 compares collaborative filtering to the base form of two content-based algorithms, and shows that the precision and recall of KNN CB is highest. While the average recall of all the algorithms is similar, the precision of CF and KNN CB surpasses SVM CB by a minimum of 9%. The combinations of hyponyms, hypernyms and synonyms that gave the best results in group one were also tested, and, curiously, the precision scores for all experiments are higher than the base form of SVM, as shown in Graph 7. In addition, there are small increases in the precision and recall when Wordnet's synsets are used with KNN CB, as shown in Graph 8. The combinations that gave the highest recall and precision in all experiments were when using hypernyms of nouns.

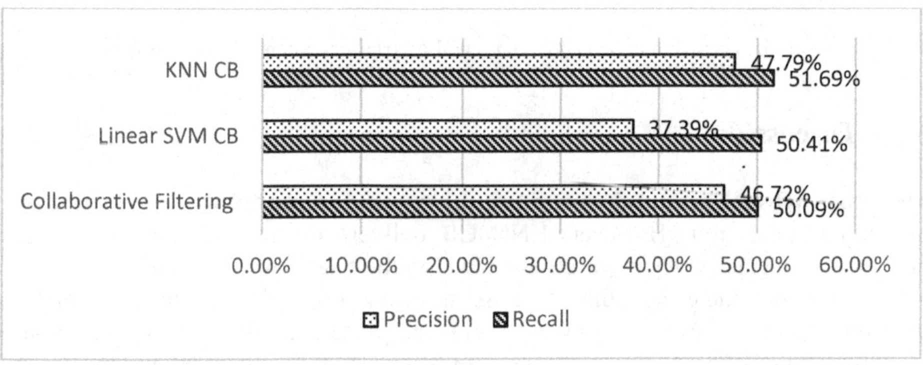

Graph 6. The average precision and recall of 10 users in CF, SVM CB and KNN CB

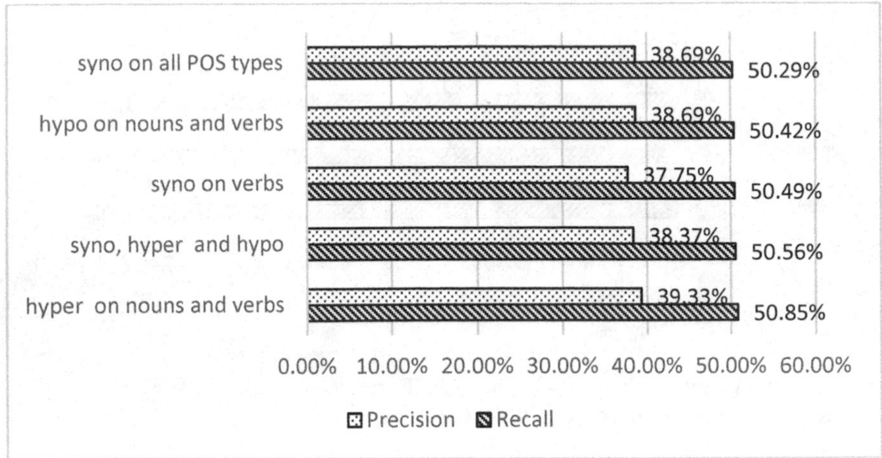

Graph 7. The average precision and recall of 10 users when SVM CB is used

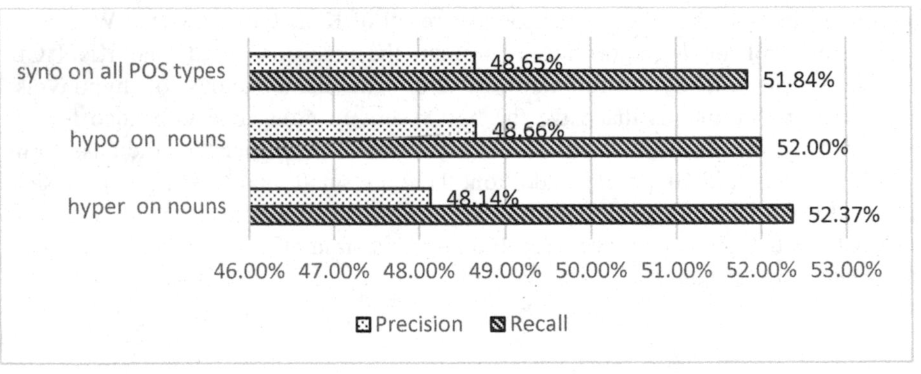

Graph 8. The average precision and recall of 10 users when KNN CB is used

6 Discussion

The Jester website uses collaborative filtering, which is powerful when predicting the real rating of a user. However, KNN CB delivers recommendations with high precision, particularly when synsets are added to the bag of words. Though KNN CB could not predict the exact ratings of a user as accurately as SVM, it had 10% higher precision than SVM when judging the interesting jokes to a user in the second group of experiments. Precision is significant in RSs, as it is important to give users recommendations they are interested in. Synsets helped KNN make predictions with far lower error rates, and a slight increase in precision and recall. Actually, all combinations of wordnet synsets could not improve the precision and recall of SVM and KNN CB more than two percentage points in group two.

As shown in Graph 2, synonyms of verbs gave the highest accuracy, followed by adjectives. This indicates that verbs and adjectives play an important role when classifying jokes. Synonyms of all POS types were also high when used with SVM and KNN. In all experiments, hyponyms and hypernyms of nouns performed better than verbs, though when hypernyms and hyponyms of nouns and verbs were combined, they also achieve high accuracy. As expected, adding hypernyms and hyponyms of nouns in all experiments resulted in higher accuracy than using synonyms. This was also expected, as hyponyms introduce specific information and hypernyms add more generic information.

7 Conclusion and Future Work

In this work, the effect of using hyponyms, hypernyms and synonyms in content-based RSs is examined in two groups of experiments. The first group of experiments evaluated how the prediction of ten users' ratings is affected by the introduction of Wordset's synsets. The second group evaluated the binary predictions. It was found that using the pure form of content based that employs linear SVM without any additions results in the least prediction errors for the ratings. SVM CB accuracy is less when Wordnet synsets are added, but it still outperforms collaborative filtering. However, KNN CB with or without the enrichment of hyponyms, hypernyms and synonyms performed less accurately than CF, though it outperformed the others in the second group of experiments, particularly when synsets are added. In future work, we plan to apply word sense disambiguation techniques, to study the effects of using other Wordnet relations such as meronym, holonym and troponym, and to test more content-based algorithms.

References

[1] Ricci, F., Rokach, L., Shapira, B.: Introduction to Recommender Systems Handbook. In: Recommender Systems Handbook, pp. 1–35. Springer, New York (2011)

[2] Alharthi, H., Tran, T.: Item-Based Collaborative Filtering Using the Big Five Personality Traits. In: Proceedings of the ACM RecSys 2014 Workshop on Recommender Systems for Television and Online Video, Silicon Valley (2014)

[3] Burke, R.: Hybrid Recommender Systems: Survey and Experiments. User Modeling and User-Adapted Interaction 12(4), 331–370 (2002)

[4] Su, X., Khoshgoftaar, T.M.: A Survey of Collaborative Filtering Techniques. Advances in Artificial Intelligence (4), 2 (2009)

[5] Wang, J., de Vries, A.P., Reinders, M.J.T.: Unifying Userbased and Itembased Collaborative Filtering Approaches by Similarity Fusion. In: Proceedings of the 29th Annual International ACM SIGIR Conference on Research and Development in Information Retrieval, Seattle, WA, USA (2006)

[6] Sarwar, B., Karypis, G., Konstan, J., Riedl, J.: Item-based Collaborative Filtering Recommendation Algorithms. In: Proceedings of the 10th International Conference on World Wide Web, Hong Kong (2001)

[7] Lops, P., de Gemmis, M., Semeraro, G.: Content-based Recommender Systems: State of the Art and Trends. In: Recommender Systems Handbook, pp. 73–105. Springer, Heidelberg (2011)
[8] Shani, G., Gunawardana, A.: Evaluating recommendation systems. In: Recommender Systems Handbook, pp. 257–298. Springer, Heidelberg (2011)
[9] Fellbaum, C.: WordNet. In: Theory and Applications of Ontology: Computer Applications, pp. 231–243. Springer, Heidelberg (2010)
[10] Bai, R., Wang, X., Liao, J.: Extract Semantic Information from WordNet to Improve Text Classification Performance. In: Kim, T.-h., Adeli, H. (eds.) AST/UCMA/ISA/ACN 2010. LNCS, vol. 6059, pp. 409–420. Springer, Heidelberg (2010)
[11] Capelle, M., Moerland, M.: Semantics-Based News Recommendation. In: WIMS 2012 Proceedings of the 2nd International Conference on Web Intelligence, Mining and Semantics, New York (2012)
[12] Stefani, A., Strapparava, C.: Personalizing Access to Web Sites: The SiteIF Project. In: Proceedings of the 2nd Workshop on Adaptive Hypertext and Hypermedia HYPERTEXT 1998, Pittsburgh (1998)
[13] Degemmis, M., Lops, P., Semeraro, G.: A content-collaborative recommender that exploits WordNet-based user profiles for neighborhood formation. User Modeling and User-Adapted Interaction 17(3), 217–255 (2007)
[14] Semeraro, G., Basile, P., de Gemmis, M., Lops, P.: User Profiles for Personalizing Digital Libraries. In: Handbook of Research on Digital Libraries: Design, Development and Impact, pp. 149–159. IGI Global (2009)
[15] Tkalčič, M., Burnik, U., Košir, A.: Using affective parameters in a content-based recommender system for images. User Modeling and User-Adapted Interaction 20(4), 279–311 (2010)
[16] Han, J., Kamber, M., Pei, J.: Data Mining: Concepts and Techniques, 2nd edn. The Morgan Kaufmann Series in Data Management Systems (2006)
[17] Pazzani, M.J., Billsus, D.: Content-Based Recommendation Systems. In: Brusilovsky, P., Kobsa, A., Nejdl, W. (eds.) Adaptive Web 2007. LNCS, vol. 4321, pp. 325–341. Springer, Heidelberg (2007)
[18] Pazzani, M.J., Billsus, D.: Content-Based Recommendation Systems. In: Brusilovsky, P., Kobsa, A., Nejdl, W. (eds.) Adaptive Web 2007. LNCS, vol. 4321, pp. 325–341. Springer, Heidelberg (2007)
[19] Fletcher, T.: Support Vector Machines Explained (2009), http://www.cs.ucl.ac.uk/staff/T.Fletcher/ (accessed August 2014)
[20] Goldberg, K.: Anonymous Ratings Data from the Jester Online Joke Recommender System, http://goldberg.berkeley.edu/jester-data/ (accessed February 15, 2014)
[21] Billsus, D., Pazzani, M.J.: Learning collaborative information filters. In: Proceedings of the Fifteenth International Conference on Machine Learning (1998)
[22] Basu, C., Hirsh, H., Cohen, W.: Recommendation As Classification: Using Social and Content-based Information in Recommendation. In: Proceedings of the Fifteenth National/Tenth Conference on Artificial Intelligence/Innovative Applications of Artificial Intelligence, Madison (1998)
[23] Sarwar, B.M., Karypis, G., Konstan, J.A., Riedl, J.: Analysis of Recommendation Algorithms for e-Commerce. In: Proceedings of the 2nd ACM Conference on Electronic Commerce, Minneapolis (2000)

Active Learning Based Weak Supervision for Textual Survey Response Classification*

Sangameshwar Patil[1,2] and B. Ravindran[1]

[1] Dept. of Computer Science and Engineering,
Indian Institute of Technology Madras, India
{sangam,ravi}@cse.iitm.ac.in
[2] TRDDC, Tata Consultancy Services, India
sangameshwar.patil@tcs.com

Abstract. Analysing textual responses to open-ended survey questions has been one of the challenging applications for NLP. Such unstructured text data is a rich data source of subjective opinions about a specific topic or entity; but it is not amenable to quick and comprehensive analysis. Survey coding is the process of categorizing such text responses using a pre-specified hierarchy of classes (often called a *code-frame*). In this paper, we identify the factors constraining the automation approaches to this problem and observe that a completely supervised learning approach is not feasible in practice. We then present details of our approach which uses multi-label text classification as a first step without requiring labeled training data. This is followed by the second step of active learning based verification of survey response categorization done in first step. This weak supervision using active learning helps us to optimize the human involvement as well as to adapt the process for different domains. Efficacy of our method is established using the high agreement with real-life, manually annotated benchmark data.

Keywords: Survey Text Mining, Active Learning, Noisy Text, Text Mining Application.

1 Introduction

Surveys typically consist of two major types of questions: questions with pre-determined choices, and questions with free form answers. In the literature [14] the former are referred to as closed-ended questions and the latter as open-ended questions. Closed-ended questions are typically multiple-choice objective questions where the respondents are expected to select the closest applicable answer(s) among pre-defined choices. While the close-ended questions provide a mechanism for structured feedback and enable quick analysis, the scenario is significantly different for the open-ended questions.

* A preliminary (work-in-progress) version of this paper was presented as a poster [11] at NLDB 2013.

A. Gelbukh (Ed.): CICLing 2015, Part II, LNCS 9042, pp. 309–320, 2015.
DOI: 10.1007/978-3-319-18117-2_23

In case of open-ended questions, the respondents are not constrained to choose from a set of options pre-conceived by the survey designer. Such questions enable the respondents to express their opinion and feelings freely using language as the medium. Predictably, data available from responses to open-ended questions have been found to be a rich source for variety of purposes such as:

- To identify specific as well as general suggestions for improvements
- To identify topics / issues which were not covered by the closed-ended questions
- To provide additional evidence to reason about and support the findings from quantitative analysis of the closed-ended questions.

To derive broad-based insights from the subjective answers to the open-ended questions, it is necessary to convert the unstructured textual responses to quantitative form.

1.1 Survey Coding Process

Survey coding is the process of converting the qualitative input available from the responses to open-ended questions to a quantitative format that helps in summarization, quick analysis and understanding of such responses. The set of customer responses in electronic text format (also known as verbatims in the survey analysis parlance) and a pre-specified set of codes, called as *code-frame*, constitute the input data to the survey-coding process. A code-frame consists of a set of tuples (called code or label) of the form <code-id, code-description>. Each code-id is a unique identifier (usually numeric) assigned to a code and the code-description usually consists of a short description that "explains" the code. Table 1 shows a small, representative code-frame. Figure 1 shows the responses to an open-ended question in a survey seeking students' feedback after an aptitude test at a college. The question asked to the students who had undertaken the test was: "What do you like about the test?".

Table 1. A sample code-frame

Code Id	Code Description
04	Verbal ability questions
05	Quantitative ability questions
25	Liked technical domain questions
27	Liked the puzzles
62	Support staff was prompt and courteous

Survey coding (also called tagging or labeling) is the task of assigning one or more codes from the given code-frame to each customer response. As per the current practice in market research industry, it is carried out by specially trained human annotators (also known as coders). The code description helps the human

Response R1: "Testing the language skill as well as the branch-wise engineering knowledge"
```
Codes- 04: Verbal ability questions
25: Liked Technical domain questions
```

Response R2: "Sudoku and synonym questions, helpful people."
```
Codes- 27: Liked the puzzles
04: Verbal ability questions
62: Support staff was prompt and courteous
```

Response R3: " puzzles and quant was cool."
```
Codes- 27: Liked the puzzles
05: Quantitative ability questions
```

Fig. 1. Sample output of survey coding process: Examples of survey responses and code-IDs assigned to them. (These are responses to the open-ended question "What do you like about the test?" by the students who had just undertaken an aptitude test.)

coder in identifying the responses to which a particular code can be assigned. Sample output of the survey-coding process is shown in Figure 1. Below each survey response, the applicable codes selected from the sample code-frame in Table 1 are given.

1.2 Challenges in the Survey Coding Process

Majority of the survey responses being extempore do not follow the orthographic rules and grammatical conventions of language. The typical "noise" observed in the survey responses may be categorized as:

- **Syntactic Noise:** They are typically incomplete sentences (e.g., see Figure 1). Spelling and grammatical errors are commonplace, so are other violations such as incorrect punctuation, incorrect capitalization etc.
- **Semantic Noise:** The meaning of a word or a phrase may not be apparent due to inherent ambiguity in natural language. This could be due to multiple reasons:
  - Informal or colloquial usage of words is common. For instance, we have found many real-life examples in which verbs `cover`, `remove`, `control`, `treat`, `combat`, `clean`, `eliminate`, `wipe off` and even `help with` have been used in place of the verb `kill` to describe the notion *`kill germs`*.
  - Label noise: Sometimes, the descriptions of two or more codes in the code-frame could be semantically overlapping and will lead to ambiguity, e.g., `protect against germs` and `neutralizes germs` are semantically equivalent and have occurred together in a code-frame[1].

[1] Such scenarios are likely because code-frames typically contain more than 100 codes and are updated by human coders for coding a given set of survey responses.

Analyzing such noisy text has proved to be challenging for existing Natural Language Processing (NLP) tools and techniques. Further, the task of survey coding becomes cumbersome for a human annotator due to business demands for quick turn-around time and the large amount of textual data that has to be read, understood and categorized. A more subtle and even more challenging problem with manual annotation is that it is prone to errors due to subjectivity in human interpretation. As elaborated in Section 2.2, survey coding task presents unique challenges and does not easily lend itself to automation.

In this paper, we outline existing approaches for classifying textual survey responses and their limitations in Section 2. In particular, we observe that a completely supervised learning approach is not feasible in practice and the unsupervised learning approach has severe limitations. Then, in Section 3, we present a weakly supervised approach for classifying responses to open-ended survey questions. Our approach makes use of active learning techniques to minimize the amount of supervision required. We present experimental validation of the proposed approach on real-life datasets in Section 4 and then conclude with our observations.

2 Related Work

The hardness of automating the coding problem is underscored by the advisory nature of available software that seeks human intervention as well as the apparent lack of fully automated solutions. Most of existing commercial technologies aid the human annotator to find responses that match with regular expression based pattern. These methods cannot handle responses which express same concept/feeling in other words; e.g. synonyms, hypernyms, etc. Such pattern matching methods also fall short in handling spelling errors, grammatical errors and ambiguity.

2.1 Research Literature

Academic and industry research community is well aware of the problems and challenges faced in the survey coding process for more than two decades [18], [12]. The gravity of the problem has increased exponentially with the Internet boom as well as ease and the lower cost of conducting online surveys compared to the traditional paper-pencil surveys. Over the past decade the problem has been attracting increasing attention both in the research community as well as the industry.

Research community has approached the problem of automatic code assignment from multiple perspectives. The most active research group in this area is led by Sebastiani and Esuli et al. [15,5,8,3]. They approach the survey coding problem primarily using supervised learning. They formulate the problem of coding as a classification problem and try to learn the model using pre-coded open answers from the survey responses. They have compared the results of three approaches [5]: dictionary based approach, naïve Bayesian, and SVM.

According to their observations, supervised learning methods provide more accurate and stable results than the dictionary based approach. At the same time, naïve Bayesian technique outperforms SVM by small margin. Esuli and Sebastini [2] have also used active learning to get positive and negative samples of each code and use it as training data to develop supervised learning techniques for automated survey coding.

Li and Yamanishi [7] apply classification rules and association rules to categorize survey responses for a car brand image survey. They use stochastic complexity to learn the classification rule and association rules. Classification rules are in the form of IF THEN ELSE and association rules are in the form of IF THEN OR. Given a phrase or a word in the phrase from the open answer, the decision rule assigns the target to the textual response.

Xu et al. [19] use weighted ridge regression for automatic coding in medical domain and show that it outperforms conventional ridge regression as well as linear SVM. Essentially, their approach assumes that sufficiently large amount of labeled dataset, i.e., training dataset is available. However, in real-life, especially in non-medical domain (for instance, the market research industry), often such training data is either not be available or generating the training data is an expensive, time-consuming proposition.

2.2 Limitations of Existing Approaches

Almost all the supervised learning techniques in research papers and commercial products need training data which is specific to each survey. This training data is not available with the survey and has to be created by the human annotators to begin with. In most of the cases, the cost and effort required to create necessary training data outweighs the benefits of using supervised learning techniques. Thus use of supervised learning techniques is not the best possible solution and it has been found to be a non-viable option in practice.

One may attempt to apply unsupervised techniques such as text clustering to the problem of survey coding. In text clustering, a set of given documents are grouped into one or more clusters, such that documents within a cluster are very similar and documents belonging to different clusters are quite dissimilar. The task of clustering is critically dependent on the notion of text similarity used. It may appear that all documents that have high similarity to a particular code description belong to the same cluster, i.e., each code description defines a cluster of documents. However, in survey coding, more than one code-ID may be assigned to a document; in (non-fuzzy) clustering, a document belongs to one and only one cluster. Even in the case of fuzzy clustering, the clusters formed may not correspond to the pre-specified codes in the code-frame. A much more serious problem is that there is no obvious and fool-proof way to compute the similarity of a document with a given code description. This is because code descriptions as well as the survey responses are typically very short. Also, the similarity of a document with a code is often quite indirect and requires background knowledge.

As a result, current standard practice is to do survey coding using specially trained work-force of human coders and use limited, but viable automation such as regular expression based pattern matching.

3 Our Approach

We now present our two stage iterative method in which first we use a new unsupervised multi-label text categorization algorithm. The output of this stage is then passed through a weakly supervised learning stage that uses active learning paradigm. Details of the solution including the individual components, algorithms are given below. A user needs to provide a code-frame F and the set of documents D, i.e., survey responses to which appropriate codes from F are to be assigned.

Feature Extraction: For the first stage of unsupervised multi-label classification, we use a new feature representation called semantic units (SemUnit) for each code. SemUnit tries to capture the concept expressed in the code description. It represents a word in terms of its semantics using its WordNet [4] synset ids and also attaches a weight to measure the relative importance of this word in that code's description. We use the unsupervised word sense disambiguation utilities [9] to estimate the likely word senses. For a given word, this enables us to find out synonyms, antonyms as well as other related words (hypernyms, hyponyms, holonyms, meronyms among many others). This concept-based representation is vital for the next phase of the algorithm. As an illustrative example of SemUnit, consider two sample codes in the sample code-frame shown in Table 1:

- code 04: Verbal ability questions
- code 27: Liked the puzzles

At the end of feature extraction phase, these codes are represented as:

- code 04: Verbal#j#4#i2 ability#n#1#i2 questions#n#1,3#i3
- code 27: Liked#v#2#i3 the#stopword#i0 puzzles#n#1#i1

The above representation denotes that out of all possible meanings of the word *"verbal"*, we consider Wordnet sense number 4 with its part-of-speech tag as adjective. Further, we use one of four pre-determined weights (fractional numbers) to capture the *relative importance* of each word in a particular code's description. Semantics associated with these weights is denoted using following labels:

- **i0** = 0.0 : a word with the importance **i0** is not important at all.
- **i1** = 0.64 : a word with the importance **i1** is the most important word for that particular code and will cause the code to be assigned to a survey response containing this word in the first round of code assignment. (Note that in subsequent rounds of assignment, this weight may get modified.)
- **i2** = 0.4 : a word with the importance **i2** is not sufficient alone to cause the code assignment, but it must be combined with another word from code description which has importance of **i2** or higher.

- **i3** $= 0.3$: a word with the importance **i3** is not sufficient alone to cause the code assignment, but it must be combined with at least two other words from code description which have importance of **i3** or higher.

Code-Assignment Stage: In the code-assignment stage, we propose **Code Assignment under Weak Supervision** (CAWS) algorithm (Figure 2) for multi-label text classification. We make use of the semantic unit based representation of each code to find out overlap between that code's textual description and the words in each sentence for each document. We group this lexical overlap along five major word categories, viz., nouns, verbs, adjectives, adverbs and cardinals. Each overlapping word is weighted with the importance of the word in the code description and quantifies our partial belief about whether the corresponding code can be assigned to the given document D_i.

To decide whether a code is applicable to a document, we need to combine the evidence presented by multiple such overlapping words. For this purpose, we

Algorithm: Code_Assignment_under_Weak_Supervision (CAWS)

Input: code frame F = $\{(a_1, C_1), (a_2, C_2), \ldots, (a_M, C_M)\}$ // a_i = code-ID, C_i = textual code description

Input: set of documents D = $\{D_1, D_2, \ldots, D_N\}$ // each document D_i is an ordered list of sentences

Output: $\{(D_1, L_1), (D_2, L_2), \ldots, (D_N, L_N)\}$ // a subset $L_i \subseteq \{a_1, a_2, \ldots, a_M\}$ of code-IDs assigned to each document $D_i \in$ D

1. Unsupervised multi-label classification
   a. Find out overlap between the semantic unit based representation of each code and the words in each sentence for each document. Each overlapping word is weighted with the importance of the word in the code description and represents a belief that this particular code is applicable to the given document D_i.
   b. Multiple such beliefs are combined using **certainty factor algebra** to find a single value for the belief that a particular code is applicable to a document D_i
   c. Assign a subset of labels $L_i \subseteq \{a_1, a_2, \ldots, a_M\}$ of code-IDs to each document $D_i \in$ D for which the belief is above the threshold **θ**.
2. Weak Supervision using Active Learning:
   a. "CORRECT or EXTRA CODE" feedback: Select a subset of documents for each of the assigned code-ID using **active learning (pool-based sampling and clustering)**. This subset of representative code-assignment decisions is reviewed by the human coder. Feedback regarding whether each assignment decision is correct or extra (i.e. incorrect) is sought from the human.
   b. "MISSED CODE" feedback: We also seek feedback from the human coder (i.e. "oracle" in active learning parlance) about whether a particular code should have been assigned to a document. For this purpose we cluster the set of documents D using K-means and silhouette coefficient. Cluster exemplars are chosen as query instances.
   c. **If** there are no corrections, **then stop. If** human coder wants to stop **then stop.**
3. Refine the importance of word senses for the codes for which the corrections were offered by the human coder.
4. If the number of iterations (or corrections) is more than a pre-set limit **then stop else** go to step 2.

Fig. 2. Code Assignment under Weak Supervision (CAWS) algorithm

use the certainty factor algebra (CFA)[2] to get a single, consolidated value for this belief.

CFA [1] is a simple mechanism to encode a measure of belief (or disbelief) for a hypothesis, rule or event given relevant evidence as well as combining multiple such partial beliefs into a single belief regarding the conclusion.

If the final belief regarding the conclusion is above certain threshold (denoted by θ), we assign the code to the document. Based on the given values for the importance factors (**i0, i1, i2, i3**) as described in previous subsection, the value of this threshold θ is chosen to be 0.6. One can easily note that the specific values of **i1, i2, i3** and θ do not matter much. The threshold value θ is actually a function of **i1, i2, i3**. Any choice of values which preserve the CFA semantics associated with **i1, i2, i3** would work for us.

Active Learning Based Weak Supervision: We exploit Quality checking (QC) step in survey coding process to improve the baseline classification done by the unsupervised multi-label classifier described in Section 3. Quality checking (QC) step is a necessary and well established part of the industry standard process to minimize the problem of inter-coder disagreement. QC step essentially consists of verification of the code-assignments by another human coder.

We use active learning [16] techniques to optimize feedback solicitation. We query a human coder, i.e., *"oracle"* in active learning parlance, regarding whether a subset of code-assignments to survey responses are correct. In particular, we use cluster based active learning [10]. For every code a_i in the code-frame, let S_i be the set of responses to which code a_i has been assigned. We cluster S_i using K-means algorithm and query a representative code-assignment instance for each cluster. We use silhouette coefficient [17,6,13] to decide number of clusters at run-time. Silhouette Coefficient (ShC) provides a quantified measure of the trade-off between intra-cluster cohesiveness and inter-cluster separation. Silhouette coefficient for i^{th} data point is given by $ShC_i = \frac{b_i - a_i}{max(a_i, b_i)}$, where a_i is the average distance between i^{th} data point and other points in the same cluster; and b_i is average distance between i^{th} data point and all other points in the next nearest cluster. Silhouette Coefficient for a given clustering of data-points is average of individual ShC_i values. We try out different clusterings and pick the one with maximum silhouette coefficient. Medoids of individual clusters (and potentially a few more data-points within each cluster which have maximum distance from the given medoid) are selected as exemplars for which feedback is sought using active learning.

For the query instance, the *oracle*, i.e., human can give feedback regarding whether the code-assignment was correct or extra, i.e., incorrect. If the feedback is correct, our belief regarding the word-senses and their importance is reinforced. If the code-assignment is incorrect, we seek corrective feedback from the *oracle* to know the correct word senses/meaning as well as the relative importance of words within the code-description. The *oracle* can also give feedback to identify

[2] A brief summary of CFA is also available at `http://www.cs.fsu.edu/~lacher/courses/CAP5605/documents/scfa.html`

"missed" code-assignments, i.e., code(s) which should have been assigned, but the multi-label classifier missed it. We update the knowledge base with this feedback so that it can be used to improve the baseline code-assignment in the multi-label classification step as well as future survey coding of surveys of similar category.

If there is corrective feedback provided by the *oracle*, the multi-label classification step is repeated with the additional input available from the feedback step. Thus the code assignments are further refined based on the input available from the QC step. The final set of codes assigned to each document, i.e., survey response are output after the validation as per the quality checking step.

Survey Domain	# of codes	# of responses	CAWS algorithm						Baseline_1 (SubString)			Baseline_2 (Bag of Words)		
			Without Feedback (Unsupervised)			Accuracy (%) after feedback								
			R	P	F1	R	P	F1	R	P	F1	R	P	F1
Medical_1	83	256	86.0	81.8	83.9	88.1	84.3	86.2	2.8	91.7	5.5	31.8	23.9	27.3
Medical_2	112	1075	69.4	72.3	70.8	69.3	79.7	74.1	0.4	87.5	0.8	33.3	13.4	19.1
Hygiene	140	763	82.6	70.2	75.9	83.4	70.1	76.2	7.0	42.0	12.0	49.4	14.3	22.2
Pet food	107	1153	74.5	68.1	71.2	74.4	72.6	73.5	4.2	82.1	7.9	40.9	13.6	20.4
Cosmetics	137	558	62.9	58.7	60.7	69.8	72.5	71.1	0.9	100.0	1.8	22.2	6.5	10.1
Detergent	167	1124	67.7	53.9	60.0	72.3	60.3	65.8	0.3	10.0	0.6	26.7	7.3	11.5

Fig. 3. Sample results for surveys in diverse domains. The accuracy (in %) is reported using the standard measures of Precision (P), Recall (R) and F1.

4 Experimental Results

We have evaluated our approach using a benchmark of multiple survey datasets from diverse domains such as health/medical, household consumer goods (e.g. detergents, fabric softners, etc.), food and snack items, customer satisfaction surveys, campus recruitment test surveys etc. Each dataset was annotated by a human expert. A sample set of responses from each dataset was independently verified by another domain expert. We have chosen datasets for which the sample annotation verification by experts had average agreement of 95%.

We did not come across any public-domain tools for automated survey coding against which we could compare our approach. To show the effectiveness of our method and to highlight the difficulty of survey coding task, we compare with two baseline approaches. In the first baseline approach (Baseline_1), we assign a code to a response if the code description appears as a substring of that response text. In the second baseline approach (Baseline_2), we relax the stringent requirement of exact substring match and use the bag of words (BoW) approach. We compute the word overlap between a code description and a response, after removing the stop words from both. Note that the code-frames for these surveys are organized in a hierarchy of two levels. In Baseline_2, for each parent-level category in a code-frame, we score each code with the fraction of its words overlapping with the response. Within each parent-level category, we assign the code with maximum, non-zero overlap with the response.

Figure 3 summarizes some of our results of unsupervised classification of survey responses (without using any feedback) as well as the improvement in the accuracy after feedback. In Figure 3, we see that Baseline_1 has excellent average precision; however, it performs very poorly in the recall. Baseline_2 does not demand exact match of code description with response. It looks for non-contiguous overlap between code description and response text. Expectedly, Baseline_2 improves the recall. However, it suffers in the precision due to inherent limitation of the bag of words approach which ignores the associated semantics. We contrast this with the high accuracy achieved by our approach even without any feedback and underscore its effectiveness.

Figure 4 shows that the amount of feedback required to achieve improvement in accuracy is quite less compared to the total number of responses and codes. This indicates that active learning is effective for minimizing the feedback required to improve the accuracy.

Survey Domain	# of responses for which feedback is given	# of codes for which feedback is given	# of MISSED CODE feedback given	# of EXTRA CODE feedback given
Medical_1	10 (4 %)	8 (1 %)	3	5
Medical_2	11 (1 %)	12 (10.7 %)	6	6
Hygiene	19 (2.49 %)	19 (13.6 %)	13	6
Pet food	9 (0.7 %)	8 (7.5 %)	4	4
Cosmetics	13 (2.3%)	13 (9.5%)	7	6
Detergent	9 (0.8 %)	10 (6 %)	7	3

Fig. 4. Details of feedback given for the exemplars selected using active learning for the output shown in Figure 3

5 Conclusion

Survey coding application has non-trivial challenges and does not lend itself easily to automation. In this paper, we suggested that standard machine learning approaches for text classification or clustering are not viable in practice for survey coding task. We presented a two step, iterative algorithm - **Code Assignment under Weak Supervision** (CAWS). In the first step, multi-label categorization is achieved in an unsupervised manner aided by a knowledge base. Then, we exploit the quality checking(QC) phase, which is an important part of survey coding process, to improve the accuracy further. We use active learning technique to optimize the weak supervision available in the form of human feedback in QC. We observed that our approach achieves good accuracy on human annotated benchmark data and works well for surveys from diverse domains.

References

1. Buchanan, B., Shortliffe, E.: Rule Based Expert Systems: The MYCIN Experiments of the Stanford Heuristic Programming Project. Addison-Wesley, Reading, MA (1984), iSBN 978-0-201-10172-0
2. Esuli, A., Sebastiani, F.: Active learning strategies for multi-label text classification. In: Boughanem, M., Berrut, C., Mothe, J., Soule-Dupuy, C. (eds.) ECIR 2009. LNCS, vol. 5478, pp. 102–113. Springer, Heidelberg (2009)
3. Esuli, A., Sebastiani, F.: Machines that learn how to code open-ended survey data. International Journal of Market Research 52(6) (2010), doi:10.2501/S147078531020165X
4. Fellbaum, C.: WordNet: An Electronic Lexical Database. MIT Press (1998)
5. Giorgetti, D., Prodanof, I., Sebastiani, F.: Automatic coding of open-ended surveys using text categorization techniques. In: Proceedings of Fourth International Conference of the Association for Survey Computing, pp. 173–184 (2003)
6. Kaufman, L., Rousseeuw, P.J.: Finding groups in data: An introduction to cluster analysis. Wiley series in Probability and Statistics. John Wiley and Sons, New York (1990)
7. Li, H., Yamanishi, K.: Mining from open answers in questionnaire data. In: Proceedings of Seventh ACM SIGKDD (2001)
8. Macer, T., Pearson, M., Sebastiani, F.: Cracking the code: What customers say in their own words. In: Proceedings of MRS Golden Jubilee Conference (2007)
9. Navigli, R.: Word sense disambiguation: A survey. ACM Computing Surveys (CSUR) 41(2), 10 (2009)
10. Nguyen, H., Smeulders, A.: Active learning using pre-clustering. In: Proceedings of the International Conference on Machine Learning, ICML, pp. 79–86. ACM (2004)
11. Patil, S., Palshikar, G.K.: Surveycoder: A system for classification of survey responses. In: Métais, E., Meziane, F., Saraee, M., Sugumaran, V., Vadera, S. (eds.) NLDB 2013. LNCS, vol. 7934, pp. 417–420. Springer, Heidelberg (2013)
12. Pratt, D., Mays, J.: Automatic coding of transcript data for a survey of recent college graduates. In: Proceedings of the Section on Survey Methods of the American Statistical Association Annual Meeting, pp. 796–801 (1989)
13. Rousseeuw, P.J.: Silhouettes: a graphical aid to the interpretation and validation of cluster analysis. Journal of Computational and Applied Mathematics 20, 53–65 (1987)

14. Schuman, H., Presser, S.: The open and closed question. American Sociological Review 44(5), 692–712 (1979)
15. Sebastiani, F.: Machine learning in automated text categorization. ACM Computing Surveys 34(1), 1–47 (2002)
16. Settles, B.: Active Learning. Morgan Claypool, Synthesis Lectures on AI and ML (2012)
17. Tan, P.N., Steinbach, M., Kumar, V.: Introduction to Data Mining. Addison-Wesley, Upper Saddle River (2005)
18. Viechnicki, P.: A performance evaluation of automatic survey classifiers. In: Honavar, V.G., Slutzki, G. (eds.) ICGI 1998. LNCS (LNAI), vol. 1433, pp. 244–256. Springer, Heidelberg (1998)
19. Xu, J.W., Yu, S., Bi, J., Lita, L.V., Niculescu, R.S., Rao, R.B.: Automatic medical coding of patient records via weighted ridge regression. In: Proceedings of Sixth International Conference on Machine Learning and Applications, ICMLA (2007)

Detecting and Disambiguating Locations Mentioned in Twitter Messages

Diana Inkpen[1], Ji Liu[1], Atefeh Farzindar[2], Farzaneh Kazemi[2], and Diman Ghazi[2]

[1] School of Electrical Engineering and Computer Science
University of Ottawa, Ottawa, ON, Canada
Diana.Inkpen@uottawa.ca
[2] NLP Technologies Inc., Montreal, QC, Canada
farzindar@nlptechnologies.ca

Abstract. Detecting the location entities mentioned in Twitter messages is useful in text mining for business, marketing or defence applications. Therefore, techniques for extracting the location entities from the Twitter textual content are needed. In this work, we approach this task in a similar manner to the Named Entity Recognition (NER) task focused only on locations, but we address a deeper task: classifying the detected locations into names of cities, provinces/states, and countries. We approach the task in a novel way, consisting in two stages. In the first stage, we train Conditional Random Fields (CRF) models with various sets of features; we collected and annotated our own dataset or training and testing. In the second stage, we resolve cases when there exist more than one place with the same name. We propose a set of heuristics for choosing the correct physical location in these cases. We report good evaluation results for both tasks.

1 Introduction

A system that automatically detects location entities from tweets can enable downstream commercial or not-for-profit applications. For example, automatic detection of event locations for individuals or group of individuals with common interests is important for marketing purposes, and also for detecting potential threats to public safety.

The extraction of location entities is not a trivial task; we cannot simply apply keyword matching due to two levels of ambiguities defined by [1]: *geo/non-geo ambiguity* and *geo/geo ambiguity*. Geo/non-geo ambiguities happen when a location entity is also a proper name (e.g., *Roberta* is a given name and the name of a city in *Georgia, United States*) or has a non-geographic meaning (e.g., *None* is a city in Italy in addition to the word *none* when lower case is ignored or when it appears at the beginning of a sentence). A geo/geo ambiguity occurs when several distinct places have the same name, as in *London, UK; London, ON, Canada; London, OH, USA; London, TX, USA; London, CA, USA*, and a few more in the USA and other countries. Another example is the country name *China* being the name of cities in the United States and in Mexico.

As a consequence of the ambiguities, an intelligent system smarter than simple keyword matching is required. Specifically, we propose to address the geo/non-geo ambiguities by defining a named entity recognition task which focuses on locations and

A. Gelbukh (Ed.): CICLing 2015, Part II, LNCS 9042, pp. 321–332, 2015.
DOI: 10.1007/978-3-319-18117-2_24

ignores other types of named entities. We train CRF classifiers for specific types of locations, and we experiment with several types of features, in order to choose the most appropriate ones. To deal with geo/geo ambiguities, we implement several heuristic disambiguation rules, which are shown to perform reasonably well. The consequent hybrid model is novel in the social media location extraction domain. Our contribution consists in the specific way of framing the problem in the two stages: the extraction of expressions composed of one or more words that denote locations, followed by the disambiguation to a specific physical location. Another contribution is an annotated dataset that we made available to other researchers. The fully-annotated dataset and the source code can be obtained through this link[1].

2 Related Work

Before the social media era, researchers focused on extracting locations from online contents such as news and blogs. [6] named this type of work *location normalization*. Their approach used a Maximum-entropy Markov model (MEMM) to find locations and a set of rules to disambiguate them. Their system is reported to have an overall precision of 93.8% on several news report datasets. [1] tried to associate each location mention in web pages with the place it refers to; they implemented a score-based approach to address both geo/non-geo and geo/geo ambiguities. Specifically, lexical evidences supporting the likelihood of a candidate location increases its score. When applied to Internet contents, their algorithm had an accuracy of 81.7%. [14] also focused on web pages; they assigned a weighted probability to each candidate of a location mentioned in a web page; they took into account the other locations in the same web page and the structural relations between them. [12] assumed that the true reference of a location is decided by its location prior (e.g., Paris is more likely the capital of France) and context prior (e.g., Washington is more likely the capital of USA if it has "Wizards" in its context); they developed a ranking algorithm to find the most likely location reference based on the two priors, which achieved a precision of 61.34%.

Social media text (especially tweets), is very different from traditional text, since it usually contains misspellings, slangs and is short in terms of length. Consequently, detecting locations from social media texts is more challenging. [2] looked at how to exploit information about location from French tweets related to medical issues. The locations were detected by gazetteer lookup and pattern matching to map them to physical locations using a hierarchy of countries, states/provinces and cities. In case of ambiguous names, they did not fully disambiguate, but relied on users' time zones. They focused on the locations in user's profile, rather than the locations in the text of tweets. [11] detected place names in texts in a multi-lingual setting, and disambiguated them in order to visualize them on the map.

Statistical techniques were used to resolve ambiguities. For example, [10] identified the locations referenced in tweets by training a simple log-near model with just 2 features for geo/non-geo ambiguity and geo/geo ambiguity; the model achieved a precision of 15.8%. [5] identified location mentions in tweets about disasters for GIS applications; they applied off-the-shelf software, namely, the Stanford NER software to this task and

[1] https://github.com/rex911/locdet

compared the results to gold standards. [7] also showed that off-the-shelf NER systems achieve poor results on detecting location expressions.

3 Dataset

Annotated data are required in order to train our supervised learning system. Our work is a special case of the Named Entity Recognition task, with text being tweets and target Named Entities being specific kinds of locations. To our knowledge, a corresponding corpus does not yet exist.[2]

3.1 Data Collection

We used the Twitter API[3] to collect our own dataset. Our search queries were limited to six major cell phone brands, namely iPhone, Android, Blackberry, Windows Phone, HTC and Samsung. Twitter API allows its users to filter tweets based on their languages, geographic origins, the time they were posted, etc. We utilized such functionality to collect only tweets written in English. Their origins, however, were not constrained, i.e., we collected tweets from all over the world. We ran the crawler from June 2013 to November 2013, and eventually collected a total of over 20 million tweets.

3.2 Manual Annotation

The amount of data we collected is overwhelming for manual annotation, but having annotated training data is essential for any supervised learning task for location detection. We therefore randomly selected 1000 tweets from each subset (corresponding to each cellphone brand) of the data, and obtained 6000 tweets for the manual annotation (more data would have taken too long to annotate).

We have defined annotation guidelines to facilitate the manual annotation task. [8] defined spatialML: an annotation schema for marking up references to places in natural language. Our annotation model is a sub-model of spatialML. The process of manual annotation is described next.

Gazetteer Matching. A gazetteer is a list of proper names such as people, organizations, and locations. Since we are interested only in locations, we only require a gazetteer of locations. We obtained such a gazetteer from GeoNames[4], which includes additional information such as populations and higher level administrative districts of each location. We also made several modifications, such as the removal of cities with populations smaller than 1000 (because otherwise the size of the gazetteer would be

[2] [7] recently released a dataset of various kinds of social media data annotated with generic location expressions, but not with cities, states/provinces, and countries).

[3] `https://dev.twitter.com`

[4] `http://www.geonames.org`

Table 1. The sizes of the gazetteers

Gazetteer	Number of countries	Number of states and provinces	Number of cities
GATE	465	1215	1989
GeoNames	756	129	163285

very large, and there are usually very few tweets in the low-populated areas) and removal of states and provinces outside the U.S. and Canada; we also allowed the matching of alternative names for locations. For instance, ATL, which is an alternative name for Atlanta, will be matched as a city.

We then used GATE's gazetteer matching module [4] to associate each entry in our data with all potential locations it refers to, if any. Note that, in this step, the only information we need from the gazetteer is the name and the type of each location. GATE has its own gazetteer, but we replaced it with the GeoNames gazetteer which serves our purpose better. The sizes of both gazetteers are listed in Table 1 [5]. In addition to a larger size, the GeoNames contains information such as population, administrative division, latitude and longitude, which will be useful later in Section 5.

Manual Filtering. The first step is merely a coarse matching mechanism without any effort made to disambiguate candidate locations. E.g., the word *Georgia* would be matched to both the state of Georgia and the country in Europe.

In the next phase, we arranged for two annotators, who are graduate students with adequate knowledge of geography, to go through every entry matched to at least one of locations in the gazetteer list. The annotators are required to identify, first, whether this entry is a location; and second, what type of location this entry is. In addition, they are also asked to mark all entities that are location entities, but not detected by GATE due to misspelling, all capital letters, all small letters, or other causes. Ultimately, from the 6000 tweets, we obtained 1270 countries, 772 states or provinces, and 2327 cities.

We split the dataset so that each annotator was assigned one fraction. In addition, both annotators annotated one subset of the data containing 1000 tweets, corresponding to the search query of Android phone, in order to compute an inter-annotator agreement, which turned out to be 88%. The agreement by chance is very low, since any span of text could be marked, therefore the kappa coefficient that compensates for chance agreement is close to 0.88. The agreement between the manual annotations and those of the initial GATE gazetteer matcher in the previous step was 0.56 and 0.47, respectively for each annotator.

Annotation of True Locations. Up to this point, we have identified locations and their types, i.e., geo/non-geo ambiguities are resolved, but geo/geo ambiguities still exist. For example, we have annotated the token *Toronto* as a city, but it is not clear whether it refers to *Toronto, Ontario, Canada* or *Toronto, Ohio, USA*. Therefore we randomly choose 300 tweets from the dataset of 6000 tweets and further manually annotated the locations detected in these 300 tweets with their actual location. The actual location is

[5] The number of countries is larger than 200 because alternative names are counted; the same for states/provinces and cities.

Table 2. An example of annotation with the true location

Mon Jun 24 23:52:31 +0000 2013
<location locType='city', trueLoc='22321'>Seguin **</location>**
<location locType='SP', trueLoc='12'>Tx **</location>**
RT @himawari01271, #RETWEET#TEAMFAIRYROSE #TMW #TFBJP #500aday #ANDROID #JP #FF #Yes #No #RT #ipadgames #TAF #NEW #TRU #TLA #THF 51

denoted by a numerical ID as the value of an attribute named *trueLoc* within the XML tag. An example of annotated tweet is displayed in Table 2.

4 Location Entity Detection

We looked into methods designed for sequential data, because the nature of our problem is sequential. The different parts of a location such as country, state/province and city in a tweet are related and often given in a sequential order, so it seems appropriate to use sequential learning methods to automatically learn the relations between these parts of locations. We decided to use CRF as our main machine learning algorithm, because it achieved good results in similar information extraction tasks.

4.1 Designing Features

Features that are good representations of the data are important to the performance of a machine learning task. The features that we design for detecting locations are listed below:

- Bag-of-Words: To start with, we defined a sparse binary feature vector to represent each training case, i.e., each token in a sequence of tokens; all values of the feature vector are equal to 0 except one value corresponding to this token is set to 1. This feature representation is often referred to as *Bag-of-Words* or unigram features. We will use *Bag-of-Words Features* or *BOW features* to denote them, and the performance of the classifier that uses these features can bc considered as the baseline in this work.
- Part-of-Speech: The intuition for incorporating Part-of-Speech tags in a location detection task is straightforward: a location can only be a noun or a proper noun. Similarly, we define a binary feature vector, where the value of each element indicates the activation of the corresponding POS tag. We later on denote these features by *POS features*.
- Left/right: Another possible indicator of whether a token is a location is its adjacent tokens and POS tags. The intuitive justification for this features is that locations in text tend to have other locations as neighbours, i.e., *Los Angeles, California, USA*; and that locations in text tend to follow prepositions, as in the phrases *live in Chicago*, *University of Toronto*. To make use of information like that, we defined another set of features that represent the tokens on the left and right side of the target token

and their corresponding POS tags. These features are similar to Bag-of-Words and POS features, but instead of representing the token itself they represent the adjacent tokens. These features are later on denoted by *Window features* or *WIN features*.
- Gazetteer: Finally, a token that appears in the gazetteer is not necessarily a location; by comparison, a token that is truly a location must match one of the entries in the gazetteer. Thus, we define another binary feature which indicates whether a token is in the gazetteer. This feature is denoted by Gazetteer feature or GAZ feature in the succeeding sections.

In order to obtain BOW features and POS features, we preprocessed the dataset by tokenizing and POS tagging all the tweets. This step was done using the Twitter NLP and Part-of-Speech Tagging tool [9].

For experimental purposes, we would like to find out the impact each set of features has on the performance of the model. Therefore, we test different combinations of features and compare the accuracies of resulting models.

4.2 Experiments

Evaluation Metrics. We report precision, recall and F-measure for the extracted location mentions, at both the token and the span level, to evaluate the overall performance of the trained classifiers. A token is a unit of tokenized text, usually a word; a span is a sequence of consecutive tokens. The evaluation at the span level is stricter.

Experimental Configurations. In our experiments, one classifier is trained and tested for each of the location labels city, SP, and country. For the learning process, we need to separate training and testing sets. We report results for 10-fold cross-validation, because a conventional choice for n is 10. In addition, we report results for separate training and test data (we chose 70% for training and 30% for testing). Because the data collection took several months, it is likely that we have both new and old tweets in the dataset; therefore we performed a random permutation before splitting the dataset for training and testing.

We would like to find out the contribution of each set of features in Section 4.1 to the performance of the model. To achieve a comprehensive comparison, we tested all possible combinations of features plus the BOW features. In addition, a baseline model which simply predicts a token or a span as a location if it matches one of the entries in the gazetteer mentioned in Section 3.2.

We implemented the models using an NLP package named MinorThird [3] that provides a CRF module [13] easy to use; the loss function is the log-likelihood and the learning algorithm is the gradient ascent. The loss function is convex and the learning algorithm converges fast.

4.3 Results

The results are listed in the following tables. Table 3 shows the results for countries, Table 4 for states/provinces and Table 5 for cities. To our knowledge, there is no previous work that extracts locations at these three levels, thus comparisons with other models are not feasible.

Table 3. Performance of the classifiers trained on different features for countries. Column 2 to column 7 show the results from 10-fold cross validation on the dataset of 6000 tweets; the last two columns show the results from random split of the dataset where 70% are the train set and 30% are the test set. (Same in Table 4 and Table 5)

Features	Token			Span			Separate train-test sets	
	P	R	F	P	R	F	Token F	Span F
Baseline-Gazetteer Matching	0.26	0.64	0.37	0.26	[illegible]	[illegible]	–	
Baseline-BOW	0.93	0.83	0.88	0.92	0.82	0.87	0.86	0.84
BOW+POS	0.93	0.84	0.88	0.91	0.83	0.87	0.84	0.85
BOW+GAZ	0.93	0.84	0.88	0.92	0.83	0.87	0.85	0.86
BOW+WIN	0.96	0.82	0.88	0.95	0.82	0.88	0.87	0.88
BOW+POS+ GAZ	0.93	0.84	0.88	0.92	0.83	0.87	0.85	0.86
BOW+WIN+ GAZ	0.95	0.85	0.90	0.95	0.85	0.89	0.90	0.90
BOW+POS+ WIN	0.95	0.82	0.88	0.95	0.82	0.88	0.90	0.90
BOW+POS+ WIN+GAZ	0.95	0.86	0.90	0.95	0.85	0.90	0.92	0.92

Table 4. Performance of the classifiers trained on different features for SP

Features	Token			Span			Separate train-test sets	
	P	R	F	P	R	F	Token F	Span F
Baseline-Gazetteer Matching	0.65	0.74	0.69	0.64	0.73	0.68	—	—
Baseline-BOW	0.90	0.78	0.84	0.89	0.80	0.84	0.80	0.84
BOW+POS	0.90	0.79	0.84	0.89	0.81	0.85	0.82	0.84
BOW+GAZ	0.88	0.81	0.84	0.89	0.82	0.85	0.79	0.80
BOW+WIN	0.93	0.77	0.84	0.93	0.78	0.85	0.80	0.81
BOW+POS+GAZ	0.90	0.80	0.85	0.90	0.82	0.86	0.78	0.82
BOW+WIN+GAZ	0.91	0.79	0.84	0.91	0.79	0.85	0.83	0.84
BOW+POS+WIN	0.92	0.78	0.85	0.92	0.79	0.85	0.80	0.81
BOW+POS+WIN+GAZ	0.91	0.79	0.85	0.91	0.80	0.85	0.84	0.83

4.4 Discussion

The results from Table 3, 4 and 5 show that the task of identifying cities is the most difficult, since the number of countries or states/provinces is by far smaller. In our gazetteer, there are over 160,000 cities, but only 756 countries and 129 states/provinces, as detailed in Table 1. A lager number of possible classes generally indicates a larger search space, and consequently a more difficult task. We also observe that the token level F-measure and the span level F-measure are quite similar, likely due to the fact that most location names contain only one word.

We also include the results when one part of the dataset (70%) is used as training data and the rest (30%) as test data. The results are slightly different to that of 10-fold cross validation and tend to be lower in terms of f-measures, likely because less data are used for training. However, similar trends are observed across feature sets.

The baseline model not surprisingly produces the lowest precision, recall and f-measure; it suffers specifically from a dramatically low precision, since it will predict

Table 5. Performance of the classifiers trained on different features for cities

Features	Token			Span			Separate train-test sets	
	P	R	F	P	R	F	Token F	Span F
Baseline-Gazetteer Matching	0.14	0.71	0.23	0.13	0.68	0.22	—	—
Baseline-BOW	0.91	0.59	0.71	0.87	0.56	0.68	0.70	0.68
BOW+POS	0.87	0.60	0.71	0.84	0.55	0.66	0.71	0.68
BOW+GAZ	0.84	0.77	0.80	0.81	0.75	0.78	0.78	0.75
BOW+WIN	0.87	0.71	0.78	0.85	0.69	0.76	0.77	0.77
BOW+POS+GAZ	0.85	0.78	0.81	0.82	0.75	0.78	0.79	0.77
BOW+WIN+GAZ	0.91	0.76	0.82	0.89	0.74	0.81	0.82	0.81
BOW+POS+WIN	0.82	0.76	0.79	0.80	0.75	0.77	0.80	0.79
BOW+POS+WIN+GAZ	0.89	0.77	0.83	0.87	0.75	0.81	0.81	0.82

everything contained in the gazetteer to be a location. By comparing the performance of different combinations of features, we find out that the differences are most significant for the classification of cities, and least significant for the classification of states/provinces, which is consistent with the number of classes for these two types of locations. We also observe that the simplest features, namely BOW features, always produce the worst performance at both token level and span level in all three tasks; on the other hand, the combination of all features produces the best performance in every task, except for the prediction of states/provinces at span level. These results are not surprising.

We conducted t-tests on the results of models trained on all combinations of features listed in Table 3, 4 and 5. We found that in *SP* classification, no pair of feature combinations yields statistically significant difference. In *city* classification, using only BOW features produces significantly worse results than any other feature combinations at a 99.9% level of confidence, except BOW+POS features, while using all features produces significantly better results than any other feature combinations at a 99% level of confidence, except BOW+GAZ+WIN features. In *country* classification, the differences are less significant; where using all features and using BOW+GAZ+WIN features both yield significantly better results than 4 of 6 other feature combinations at a 95% level of confidence, while the difference between them is not significant; unlike in *city* classification, the results obtained by using only BOW features is significantly worse merely than the two best feature combinations mentioned above.

We further looked at the t-tests results of *city* classification to analyze what impact each feature set has on the final results. When adding POS features to a feature combination, the results might improve, but never statistically significantly; by contrast, they always significantly improve when GAZ features or WIN features are added. These are consistent with our previous observations.

4.5 Error Analysis

We went through the predictions made by the location entity detection model, picked some typical errors made by it, and looked into the possible causes of these errors.

Example 1:

> Mon Jul 01 14:46:09 +0000 2013
> Seoul
> yellow cell phones family in South Korea #phone #mobile #yellow #samsung `http://t.co/lpsLgepcCW`

Example 2:

> Sun Sep 08 06:28:50 +0000 2013
> minnesnowta .
> So I think Steve Jobs' ghost saw me admiring the Samsung Galaxy 4 and now is messing with my phone. Stupid Steve Jobs. #iphone

In Example 1, the model predicted "Korea" as a country, instead of "South Korea". A possible explanation is that in the training data there are several cases containing "Korea" alone, which leads the model to favour "Korea" over "South Korea".

In Example 2, the token "minnesnowta" is quite clearly a reference to "Minnesota", which the model failed to predict. Despite the fact that we allow the model to recognize nicknames of locations, these nicknames come from the GeoNames gazetteer; any other nicknames will not be known to the model. On the other hand, if we treat "minnesnowta" as a misspelled "Minnesota", it shows that we can resolve the issue of unknown nicknames by handling misspellings in a better way.

5 Location Disambiguation

5.1 Methods

In the previous section, we have identified the locations in Twitter messages and their types; however, the information about these locations is still ambiguous. In this section, we describe the heuristics that we use to identify the unique actual location referred to by an ambiguous location name. These heuristics rely on information about the type, geographic hierarchy, latitude and longitude, and population of a certain location, which we obtained from the GeoNames Gazetteer. The disambiguation process is divided into 5 steps, as follows:

1. **Retrieving candidates.** A list of locations whose names are matched by the location name we intend to disambiguate are selected from the gazetteer. We call these locations candidates. After step 1, if no candidates are found, disambiguation is terminated; otherwise we continue to step 2.
2. **Type filtering.** The actual location's type must agree with the type that is tagged in the previous step where we apply the location detection model; therefore,

we remove any candidates whose types differ from the tagged type from the list of candidates. E.g., if the location we wish to disambiguate is *Ontario* tagged as a city, then *Ontario* as a province of Canada is removed from the list of candidates, because its type *SP* differs from our target type. After step 2, if no candidates remain in the list, disambiguation is terminated; if there is only one candidate left, this location is returned as the actual location; otherwise we continue to step 3.

3. **Checking adjacent locations.** It is common for users to put related locations together in a hierarchical way, e.g., Los Angeles, California, USA. We check adjacent tokens of the target location name; if a candidate's geographic hierarchy matches any adjacent tokens, this candidate is added to a temporary list. After step 3, if the temporary list contains only one candidate, this candidate is returned as the actual location. Otherwise we continue to step 4 with the list of candidates reset.
4. **Checking global context.** Locations mentioned in a document are geographically correlated [6]. In this step, we first look for other tokens tagged as a location in the Twitter message; if none is found, we continue to step 5; otherwise, we disambiguate these context locations. After we obtain a list of locations from the context, we calculate the sum of their distances to a candidate location and return the candidate with minimal sum of distances.
5. **Default sense.** If none of the previous steps can decide a unique location, we return the candidate with largest population (based on the assumption that most tweets talk about large urban areas).

5.2 Experiments and Results

We ran the location disambiguation algorithm described above. In order to evaluate how each step (more specifically, step 3 and 4, since other steps are mandatory) contributes to the disambiguation accuracy, we also deactivated optional steps and compared the results.

The results of different location disambiguation configurations are displayed in Table 6, where we evaluate the performance of the model by accuracy, which is defined as the proportion of correctly disambiguated locations. By analyzing them, we can see that when going through all steps, we get an accuracy of 95.5%, while by simply making sure the type of the candidate is correct and choosing the default location with the largest population, we achieve a better accuracy. The best result is obtained by using the adjacent locations, which turns out to be 98.2% accurate. Thus we conclude that adjacent locations help disambiguation, while locations in the global context do not. Therefore the assumption made by [6] that the locations in the global context help the inference of a target location does not hold for Twitter messages, mainly due to their short nature.

Table 6. Results on the subset of 300 tweets annotated with disambiguated locations

Deactivated steps	Accuracy
None	95.5 %
Adjacent locations	93.7 %
Global context	**98.2** %
Adjacent locations + context locations	96.4 %

5.3 Error Analysis

Similar to Section 4.5, this section presents an example of errors made by the location disambiguation model in Example 3. In this example, the disambiguation rules correctly predicted "NYC" as "New York City, New York, United States"; however, "San Francisco" was predicted as "San Francisco, Atlantida, Honduras", which differs from the annotated ground truth. The error is caused by step 4 of the disambiguation rules that uses contextual locations for prediction; San Francisco of Honduras is 3055 kilometres away from the contextual location New York City, while San Francisco of California, which is the true location, is 4129 kilometres away. This indicates the fact that a more sophisticated way of dealing with the context in tweets is required to decide how it impacts the true locations of the detected entities.

6 Conclusion and Future Work

In this paper, we looked for location entities in tweets. We extracted different types of features for this task and did experiments to measure their usefulness. We trained CRF classifiers that were able to achieve a very good performance. We also defined disambiguation rules based on a few heuristics which turned out to work well. In addition, the data we collected and annotated is made available to other researchers to test their models and to compare with ours.

We identify two main directions of future work. First, the simple rule-based disambiguation approach does not handle issues like misspellings well, and can be replaced by a machine learning approach, although this requires more annotated training data. Second, since in the current model, we consider only states and provinces in the United States and Canada, we need to extend the model to include states, provinces, or regions in other countries as well. Lastly, deep learning models were shown to be able to learn helpful document level as well as word level representations, which can be fed into a sequential tagging model; we plan to experiment with this approach in the future.

References

1. Amitay, E., Har'El, N., Sivan, R., Soffer, A.: Web-a-Where: Geotagging Web Content. In: Proceedings of the 27th Annual International Conference on Research and Development in Information Retrieval, SIGIR 2004, pp. 273–280. ACM Press, New York (2004), http://dl.acm.org/citation.cfm?id=1008992.1009040
2. Bouillot, F., Poncelet, P., Roche, M.: How and why exploit tweet ' s location information? In: Jérôme Gensel, D.J., Vandenbroucke, D. (eds.) AGILE 2012 International Conference on Geographic Information Science, pp. 24–27. Avignon (2012)
3. Cohen, W.W.: Minorthird: Methods for identifying names and ontological relations in text using heuristics for inducing regularities from data (2004)
4. Cunningham, H.: GATE, a general architecture for text engineering. Computers and the Humanities 36(2), 223–254 (2002)
5. Gelernter, J., Mushegian, N.: Geo-parsing messages from microtext. Transactions in GIS 15(6), 753–773 (2011)

6. Li, H., Srihari, R.K., Niu, C., Li, W.: Location normalization for information extraction. In: Proceedings of the 19th International Conference on Computational Linguistics, vol. 1, pp. 1–7. Association for Computational Linguistics, Morristown (2002), http://dl.acm.org/citation.cfm?id=1072228.1072355
7. Liu, F., Vasardani, M., Baldwin, T.: Automatic identification of locative expressions from social media text: A comparative analysis. In: Proceedings of the 4th International Workshop on Location and the Web, LocWeb 2014, pp. 9–16. ACM, New York (2014), http://doi.acm.org/10.1145/2663713.2664426
8. Mani, I., Hitzeman, J., Richer, J., Harris, D., Quimby, R., Wellner, B.: SpatialML: Annotation Scheme, Corpora, and Tools. In: Proceedings of the 6th International Conference on Language Resources and Evaluation, p. 11 (2008), http://www.lrec-conf.org/proceedings/lrec2008/summaries/106.html
9. Owoputi, O., OConnor, B., Dyer, C., Gimpel, K., Schneider, N., Smith, N.A.: Improved part-of-speech tagging for online conversational text with word clusters. In: Proceedings of NAACL-HLT, pp. 380–390 (2013)
10. Paradesi, S.: Geotagging tweets using their content. In: Proceedings of the Twenty-Fourth International Florida, pp. 355–356 (2011), http://www.aaai.org/ocs/index.php/FLAIRS/FLAIRS11/paper/viewFile/2617/3058
11. Pouliquen, B., Kimler, M., Steinberger, R., Ignat, C., Oellinger, T., Blackler, K., Fluart, F., Zaghouani, W., Widiger, A., Forslund, A., Best, C.: Geocoding multilingual texts: Recognition, disambiguation and visualisation. In: Proceedings of the Fifth International Conference on Language Resources and Evaluation, LREC 2006. European Language Resources Association, ELRA (2006), http://aclweb.org/anthology/L06-1349
12. Qin, T., Xiao, R., Fang, L., Xie, X., Zhang, L.: An efficient location extraction algorithm by leveraging web contextual information. In: proceedings of the 18th SIGSPATIAL International Conference on Advances in Geographic Information Systems, pp. 53–60. ACM (2010)
13. Sarawagi, S., Cohen, W.W.: Semi-markov conditional random fields for information extraction. In: NIPS, vol. 17, pp. 1185–1192 (2004)
14. Wang, C., Xie, X., Wang, L., Lu, Y., Ma, W.Y.: Detecting geographic locations from web resources. In: Proceedings of the 2005 Workshop on Geographic Information Retrieval, GIR 2005, p. 17. ACM Press, New York (2005), http://dl.acm.org/citation.cfm?id=1096985.1096991

Natural Language Generation and Text Summarization

Satisfying Poetry Properties Using Constraint Handling Rules

Alia El Bolock and Slim Abdennadher

German University in Cairo,
Computer Science Department
{alia.elbolock,slim.adbennadher}@guc.edu.eg

Abstract. Poetry is one of the most interesting and complex natural language generation (NLG) systems because a text needs to simultaneously satisfy three properties to be considered a poem; namely grammaticality (grammatical structure and syntax), poeticness (poetic structure) and meaningfulness (semantic content). In this paper we show how the declarative approach enabled by the high-level constraint programming language Constraint Handling Rules (CHR) can be applied to satisfy the three properties while generating poems. The developed automatic poetry generation system generates a poem by incrementally selecting its words through a step-wise pruning of a customised lexicon according to the grammaticality, poeticness and meaningfulness constraints.

1 Introduction

He trusts a tear to sing along
when he dances with the hearts
He hates to trust in a romance where
he loves to miss to dance with a care

If we consider the given text, we will probably notice that it is an English poem. But why so? First of all because it satisfies grammaticality which is a property that should be realized by any text. It means that a text should adhere to the linguistic rules of a language defined by a grammar and a lexicon. If we inspect the text we will see that it follows the rules of the English language. The second property which should also hold for any text is meaningfulness. It means that a text should convey a certain message that has a meaning under some interpretation given a specific knowledge base. The meaning of the presented text can be interpreted in many ways and thus it satisfies meaningfulness. The final property which actually distinguishes a poem from any other text is the poeticness which is the existence of poetic features. This includes both figurative as well as poetic and form-dependent ones. If we take a closer look at the text we will notice the presence of poeticness features. In [1], Manurung postulates that a text needs to simultaneously satisfy these three properties to be considered a poem. The numerous works presented by Manurung et al. [1] show why these

A. Gelbukh (Ed.): CICLing 2015, Part II, LNCS 9042, pp. 335–347, 2015.
DOI: 10.1007/978-3-319-18117-2_25

properties are sufficient to characterize poems. Thus, following this claim which defines poetry in a tangible way, the presented text could be considered a poem.

This paper maps out how the three properties can be ensured while generating poems using the constraint logic programming language Constraint Handling Rules (CHR). This is done in light of a rapid prototype for a CHR poetry generation system available under `http://www.CICLing.org/2015/data/18`. CHR has been successfully used for many applications e.g. automatic music generation [2]. Automatic poetry generation has never been tackled using CHR. CHR is a high-level programming language originally developed for writing constraint solvers. Since the poetry writing process is governed by constraints, CHR is will suited for such application. All of CHRs properties allow for the intuitive definition of the poetry generation process in a compact way (the system consists of only four main rules). Throughout this paper, we will show how the developed system is capable of generating poems like the one introduced at the beginning.

The paper will start by a brief overview of the state of the art and an overview of CHR. Then, the realization of each the three properties will be discussed in detail before moving on to a preliminary evaluation of the resulting poems and the conclusions.

2 Constraint Handling Rules

CHR [3] is a declarative rule-based constraint logic programming language. It allows for multi-headed rules and conditional rule application. CHR is a committed-choice single assignment language. CHR is best suited for rapid prototyping but it has also evolved into a general-purpose programming language. A CHR program consists of rules that add and remove constraints to and from a global constraint store. There are two types of constraints: user-defined constraints and built-in constraints that are defined by the host language of CHR (usually Prolog).

The most general rule format is the simpagation rule:

$$r_{sp} @ H^k \setminus H^r \Longleftrightarrow G \mid B \tag{1}$$

The rule is fired if the constraints in the constraint store match with the atoms in the head of the rule and the guard conditions hold. The rule removes the H^r constraints and adds the body constraints B. Consider the example of the one-ruled algorithm for finding the minimum number:

```
min(N) \ min(M) <=> N=<M | true.
```

Given the initial query `min(3)`, `min(5)` and `min(8)`. The first firing of the rule will be triggered by matching the `min(3)` and `min(5)` constraints with the head of the rule. The guard condition will hold as 3 is less than the 5 and the `min(5)` constraint will be removed. Similarly, the rule will now fire with `min(3)` and `min(8)` which will result in the removal of `min(8)`. Now the only constraint remaining in the constraint store is `min(3)` and thus no more rule head matching can occur. The program will terminate and return 3 as the minimum number.

3 Related Work

Automatic poetry generation started developing as a research field in the late nineties when the first promising systems started to emerge. Since then various systems using a large range of approaches started to appear [4]. Before describing how our system achieves the three poetry properties, the other approaches used in this growing field needs to be reviewed. Word salad, grammar- and template-based as well as form-aware text generators will not be considered here as they do not aim at simultaneously fulfilling the three properties of grammaticality, poeticness and meaningfulness [1].

Evolutionary Approaches. Evolutionary approaches utilizes techniques based on concepts of biological evolution e.g. natural selection and genetic inheritance.

McGonnagall [1] represents poetry generation as a state space search problem. Using evolutionary methods, it generates poems that are metrically constrained and following certain target semantics. McGonnagall is capable of separately finding optimal poems for moderately-sized target semantics and metre patterns but it has difficulties with simultaneously considering both evaluation functions [5]. The produced poems are very constrained and grammatically correct with the disadvantage of relying on a knowledge-intensive approach. The high adherence to constraints causes the poems to sound too repetitive.

POEVOLVE [6] takes the actual process of human poetry writing as a reference for creating an evolutionary computational model of poetry generation. The system generates limericks that fulfill rhyme and rhythm. The main drawback of the system is its lack of consideration for syntax and semantics.

Case-Based Approaches. Another popular approach for poetry generation is case-based reasoning (CBR), where existing poems are retrieved and then adapted based on the required content and a target message input by the user.

The most recent version of WASP [7] uses CBR to produce Spanish poetry. A poem is generated by replacing some words in a number of existing poems according to certain constraints. The lines of the original poems are then split. Words are selected based on their lexical category and relatedness to a user-defined meaning and set of existing case. Phonetic information is not considered in the word choice and thus the system lacks strong poeticness. This version of WASP can produce poems that are not always grammatically correct or coherent.

COLIBRI [8] is another Spanish poetry generation system that uses CBR. This approach ensures the conformity with the phonetic constraints while trying to satisfy a certain user-defined message. The system ensures the syntactic well-formedness of the generated poems.

Constraint-Based Approaches. Most poetry generation approaches define poetry through constraints but to the best of our knowledge only a few utilize constraint logic programming to generate poems. The poetry generation system presented in [9,10] consists of two sub-components: The first is a conceptual space specifier that fulfills grammaticality and coherence through a corpus-based approach. It produces the poem skeleton and a set of candidate words for each position in the poem. The second one is a constraint-based conceptual space explorer responsible for achieving poeticness features and generating the output poems. This system only uses constraint programming for the final part of the poetry generation process after all the information has been gathered by the first sub-component.

The generated poems usually are similar to a certain topic and to the poems on the corpus. The system can produce poems that are not always grammatically correct or meaningful.

4 Poetry Properties

As mentioned before, a text T needs to simultaneously satisfy grammaticality G, poeticness $P$and meaningfulness M to be considered a poem.

$$T \in G \cap P \cap M \tag{2}$$

The first two properties need to be fulfilled by any natural language text as they mainly refer to the presence of syntax and semantics in the text. The final property is specific to poem as it means the presence of poetic features including figurative, phonetic and form-dependent ones. In the following we will describe how our constraint-based approach is used to realize each of the properties while generating poems. In the introduced system the poetry generation process consists of pruning rules that are applied on a lexicon, to generate each word in the poem. The pruning rules narrow down the lexicon words according to the gramamticality, meaningfulness and poeticness constraints.

4.1 Grammaticality

Any natural language text needs to adhere to the linguistic rules of the language it belongs to. These rules are defined by a grammar and a lexicon. Naturally, the same holds for poems. A customized lexicon was complied for the described system. The grammar for representing the English language is a list-based sequence of lexical categories. We will go into the details of the lexicon and grammar implementation, separately.

Lexicon. For any system to be capable to generate natural language texts it needs a lexicon containing the words used in the language. The lexicon is designed specifically to simplify the word selection and pruning process by providing the system with the needed information in the format best suited for the implementation using CHR. The chosen lexicon design enables us to best make use of the declarative power or CHR and renders the generation process more compact. The lexicon design follows the approach described in [11].

The lexicon is a compilation of the "The Carnegie Mellon Pronouncing Dictionary" and "The Unofficial Alternate 12 Dicts Package".

Each entry in the lexicon has the following format:

```
word type;number of syllables;stress pattern list;pronunciation list;word
```

The word type represents the lexical category of the current word i.e. its Part of Speech (POS) tag. The number of syllables corresponds to the number of vowel phonemes in the word and the stress pattern corresponds to the stress of each syllable (0 for unstressed and 1 for stressed). The pronunciation list contains the phonemes for the pronunciation of the word. The final element in the entry contains the actual word. For example, the entry of the word 'trust' would be `verb;1;1;t,r,ah1,s,t;trust;`. There are 7 different main lexical categories in the lexicon: noun, verb, adjective, adverb, pronoun, preposition and conjunction. Additionally, each one can have some sub-categories, like the verb tense or noun types. Having the type of the word at the beginning of each lexicon entry, enables us to only import those words from the lexicon that belong to the required POS-tag.

The lexicon is restricted by removing the words that are not very common as well as those that are not likely to appear in poems (e.g. 'tipster'). Also, instead of always considering the whole lexicon whenever generating a poem, smaller theme-based sub-lexicons are generated and only one of them is considered for the creation of one poem. This would improve the coherence because only words related to each other with respect to the theme will be chosen for a poem. For the time being this is done manually as proof of concept on the sample theme of love. Using the love-themed lexicon instead of the main one, would enable us to generate poems such as 1. However, this would require the addition of some final information to the lexicon. So far the lexicon has the necessary information needed for phonetic features and initial meaningfulness. But it lacks information that ensures the generation of grammatically correct sentences because it does not give any insight about the selectional restrictions enforced by verbs. For example, sentences like *'He trusts about dearest lust'* could be generated. The use of the preposition 'about' after the verb 'trusts' is not an intuitive one. Thus, the lexicon is extended by adding the necessary information needed to construct the rest of the sentence after the appearance of any verb. The entry of the verb 'trust' is now extended to give all the possibilities:

1. `verb;1;1;t,r,ah1,s,t;trust;obj;`
2. `verb;1;1;t,r,ah1,s,t;trust;obj,inf;`
3. `verb;1;1;t,r,ah1,s,t;trust;obj,prep,obj;with,to`

In all the cases the 'trust' should be followed by an object. Then either the sentence can end or it can be continued by an infinitive verb or a preposition and another object. We have to state exactly which prepositions are allowed with each verb. In this case we can either use 'with' or 'to'. The system regards the two entries as two different ones of the same POS-tag. Now the system is capable of generating sentences like *'He trusts a tear to sing along'*.

Grammar. The grammar of the whole poem is represented by a list containing the lexical categories of all the words that should appear in the poem. This list is referred to as the grammar pattern list. The system narrows down the lexicon to only those words that match the word type at the head of the grammar pattern list. As long as the grammar represented in the grammar pattern list is a correct one the grammatical structure of the generated sentences should be correct. The system goes through the list and selects all the lexicon words that match the current POS-tag to then prune them further to match the remaining constraints. The grammaticality is only ensured through the POS-tag matching with the word types in the grammar pattern list and without any application of further constraints.

The grammar pattern list, which is the core grammaticality measure, can be generated in many ways. The first possibility is the obvious one, of manually deciding on the required grammar pattern list. The user could specify the exact grammar of the required poem. However, this possibility is trivial and tedious for the user to fully specify the whole grammatical pattern of the poem. The second option is to extract the grammar of the poem, from a corpus of existing poems. This approach has been investigated in numerous works so far and thus is was decided to pursue a different approach. The final option is the automatic generation of the grammar pattern list, which is the actual approach pursued in this work. The grammar pattern list is initially empty and the system just fills it with a certain number of nouns, to specify how many main sentences should appear in the final poem. Any noun is automatically preceded with an article, initially 'a'. Later, the article will be removed or modified as needed. After any noun a verb should appear. Optionally, two nouns can first be combined with a conjunction before following them by a verb. Another optional feature, is the conjunction of two sentences; at this point two nouns and two verbs. Also, some noun are randomly preceded by an adjective. The grammar pattern list of a poem at this point could be `article,noun,verb,conjunction,article, noun,verb`. All additions to the grammar pattern list are handled automatically, as will be discussed.

Grammar Correction. After ensuring that the sentence structure itself is correct, some further grammar rules need to be enforced. These additional corrections are selectional restrictions and thus need to be handled on their own because of the word dependencies.

For instance, the article can only be correctly set after the noun itself has been chosen. In case of a plural noun, the article is changed to 'the'. In case of a noun starting with a vowel the article becomes 'an'. And finally, if the chosen noun is a pronoun, then the article is entirely removed. For example if the first chosen noun is the pronoun 'he', the article will be removed and the grammar pattern list will become `noun,verb,conjunction,article, noun,verb`.

Another case were the grammar correction needs to be handled explicitly, is the s-form verbs. Whenever the noun of the sentence is a singular third person the verb types in the whole grammar pattern list have to be modified to become s-form verbs instead of regular ones. If we go back to the example of the pronoun 'he', then the corrected grammar pattern list would be `noun,s-form,conjunction,article,noun,s-form` In case of a past verb, this restriction is naturally ignored.

Sentences that consist of only a noun and verb would produce text of very crude nature. This is where the extra information added to the lexicon comes in. The information about the selectional restrictions after a verb is used to correctly expand the grammar pattern list. Depending on the choice of the verb, the list is updated with the necessary POS-tags, in the correct location, for example if the s-form verb chosen after the pronoun 'he' is 'trusts' then a possible expansion of the grammar pattern list could be `noun,s-form,object,prep:to,verb, conjunction,article,noun,s-form`. The expansion of the grammar pattern list proceeds accordingly until no more possibilities remain.

4.2 Poeticness

Next, the realization of the poeticness feature will be discussed; i.e. how we ensure that the generated texts are poems. The interesting thing about poetic features, is that they are many and that not all of them have to appear in a certain poem at once. Also, they are not very strictly defined and restricted like other linguistic features and are highly dependent on subjective taste and opinion.

Basic Features. One feature that most linguists agree upon, for a text to be considered a poem, is the rhythm. The rhythm represents the measured flow of words in the poem. It is determined by the alternating between stressed and unstressed syllables. The rhythm of the poem is realized similarly to the grammar of the poem: through a metre list. However, instead of defining the target metre list of the whole poem at once, the list is defined for each verse separately. This way, depending on the poem type, each verse can have a different rhythm, form and length. The target metre list, is defined based on the required rhythm pattern of the poem. It consists of zeros and ones to represent stressed and unstressed syllables. This enables, the pattern matching with the stress pattern lists of each word in the lexicon.

The pruning of the lexicon according to the rhythm, is performed after the selection of the POS-tag matching words. Only those words whose stress pattern list is a prefix of the current metre list of the poem are chosen. For example, if the current target metre list is of the format `[1,0,1,0]`, then words such as 'trusts' (with stress list `[1]`) and dances (with the stress list `[1,0]`) can be chosen from the lexicon.

The finishing of a single rhythm pattern list signals the termination of a poem's verse, which denotes the final chosen word as the last word of said verse. This brings us to the another popular feature of poetry; namely rhyme. If a user decided he wants to enforce rhyme while generating poetry, an extra constraint is enforced on the last word of each verse. Depending on the chosen rhyme scheme the last words of certain verses have to rhyme. In these cases, the lexicon is narrowed down further to allow only suitably rhyming last words. To decide

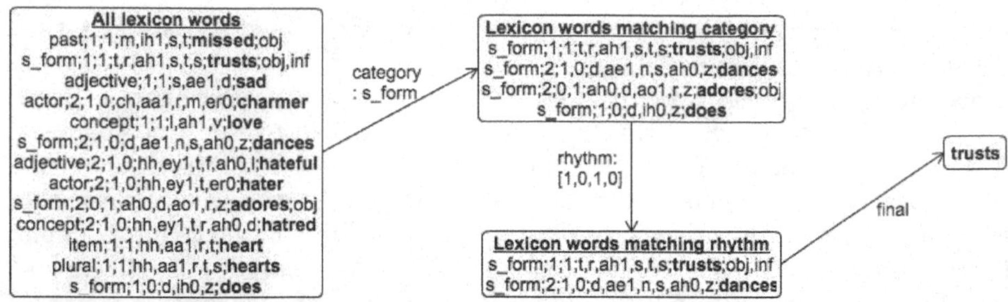

Fig. 1. Narrowing down the lexicon to satisfy the grammaticality and poeticness constraints

whether two words rhyme, the phonemes following their last stressed vowels are checked. If they match and the consonants preceding the vowel are different, then the words are considered to rhyme. For example, the words 'where' (with pronunciation list `[w,eh1,r]`) and 'care' (with pronunciation list `[k,eh1,r]`) are clearly considered to rhyme.

Rhythm and rhyme alone are enough to transform a text into a poem. Figure 1 shows an example of narrowing down the lexicon according to the constraints discussed so far. The lexicon is narrowed down until all the constraints have been satisfied. Finally the final poem word is chosen randomly from this list.

Figures of Speech. But the actual beauty of poetry comes from the added figures of speech that make for for the individual interpretation and imagination of each human. While it is possible to investigate the deliberate incorporation of figures of speech, in particular metaphors, the chosen approach focused on examining how metaphors could appear through underspecification.

This was achieved because the poem words were chosen randomly as long as they satisfied the harder constraints of grammar, rhythm and semantics. The metaphors that appear in the generated poems are different from the accustomed human generated metaphors which allows for more creativity and interpretation. An example of two metaphors generated in the poem 1 displayed above: `He trusts a tear to sing along` and `he dances with the hearts`. Rhyming and rhythm matching words that have no other restriction than the grammatical structure are chosen. Yet, the system produces two metaphors that can be interpreted by the reader because of the random word choice.

Enjambments are also implicitly achieved in the generated poems because the grammar pattern list is globally defined for the whole poem, while the rhythm pattern list is only defined locally for each verse. So again, through underspecification, enjambments are generated because a sentence could span two verses.

Form. The final important trait of poems is that they must have a unique form. Poems can have various different forms, which can be specified by the format of the rhythm pattern list. One clear example of poetry forms, is concrete or shape poems. For example, if the user requires the generation of a diamante, the rhythm pattern lists would have the following form:

```
[_,_],[_,_,_,_],[_,_,_,_,_],[_,_,_,_,_,_],[_,_,_,_,_],[_,_,_,_],[_,_]
```

Another example is if the user wants to generate a blank verse poem, in which case the rhythm pattern list would be: `[[0,1,0,1,0,1,0,1,0,1],..]`, as each verse should follows the iambic pentameter rhythm. Also, each poetry form can have a different required rhyme scheme, which can be specified to generate a poem in the required form. In the example of the blank verse poem, the rhyming should be ignored, as blank verse poems do not usually rhyme.

4.3 Meaningfulness

The final and most ambiguous feature of poetry, and any literature text in general, is the meaningfulness. Meaningfullness is difficult in that it cannot be specifically defined. For the purposes of this work we only consider meaningfulness in terms of textual coherence. Coherence has been ensured because of three measures: the generation of theme-based lexicons, the restriction of the actors in the poem and replacing the random choice at the final pruning step with a word relations function.

Theme-Based Lexicons. The main notion of coherence is achieved because of the extraction of smaller theme-based sub-lexicons from the general lexicon. By enforcing a poem to be generated solely from a lexicon that contains words belonging to a specific theme, it is ensured that the poem will have this theme. Also, all the words in the poem will have a certain unity in terms of the topic and the text will gain a meaning. This also allows for freedom in the meaning of the generated poems so that they do not sound too constrained but allow the reader to interpret the actual message of the text as he pleases.

Poem Actors' Restriction. Another measure taken to improve the coherence of the produced poems, is the restriction of the actors of the poem. In other words, to limit the number of subject and objects that act in one poem. This is achieved by starting out with one subject. Then, whenever choosing another subject or object there are two choices: either choose from the list of existing subjects and objects or choose a new one and add it to the list. The decision between the two options is done randomly with higher probability for the existing nouns. In the first verse of our poem 1 for example the existing chosen actor is the pronoun 'he'. When choosing the subject of the second verse the pronoun 'he' is chosen before going through the whole word selection process. This allows the poem to have more coherence, as it specifies the individuals the poem is revolving around.

Some extra restrictions can be enforced when choosing the subjects and objects. There is the soft constraint of the uniqueness subject and object of the a certain sentence. This constraint is thus not always enforced, to avoid excluding sentences that would provide a poetic meaning where the subject of the sentence acts upon himself. Also, the exact number of subject and objects of the poem can be specified. For example, a certain poem can have one specific subject and one object. Additionally, the user could specify the actual actors (subjects and objects) of the poem. So, one could generate a poem whose only actors are a 'king' and a 'queen', for example.

Relations Function. The coherence of the generated poems is improved by incorporating a basic word relations function to narrow down the words matching all the other constraints. A network of related words is externally generated for each lexicon word to make the poetry generation process more efficient. The networks are generated using the Prolog version of Wordnet [12].

Whenever a new word is added to the poem its list of word relations is imported to the system and added to a global list of the poem's related words. The final pruning step of the lexicon is thus updated to choose a word from the constraint matching words that is included in the poem's relations list. If no such word is present, solution finding is ensued by returning to the random choice instead.

5 Preliminary Evaluation

As in any creative work of art, the evaluation of poetry is very subjective. It is hard to find a measure to define, what is good poetry or even what is acceptable poetry. An initial evaluation of the quality of the generated poems was performed in form of an online survey. The survey consisted of 6 questions. It was posted online on `https://www.surveymonkey.com/s/8TNQMP2`. A total of about a hundred opinions have been gathered.

The first part of the survey consisted of Turing-like tests to determine whether the generated poems could be compared to human-written ones. The results shown in 2 (a) are an example of a question where the user had to rank 3 poems according to preferability. In total around 60% of the readers ranked the computer generated poem higher than atleast one of the two other poems. Figure 2 (b) shows that 47% of the raters mistakenly thought the computer generated poem was written by a human, when asked to decide which of two poems was human written.

The second part of the survey, is designed to give insight about the presence and quality of the three poetry properties.

The reader is given two computer generated poems to rate in terms of various different features:

- Rhythm: 'yes' and 'no'
  57% agreed that the first poem has rhythm, while 82% could hear the rhythm in the second one. This shows that the rhythm feature has been achieved. It should be noted that most readers do not read the poems out loud, which is usually required to hear the rhythm.

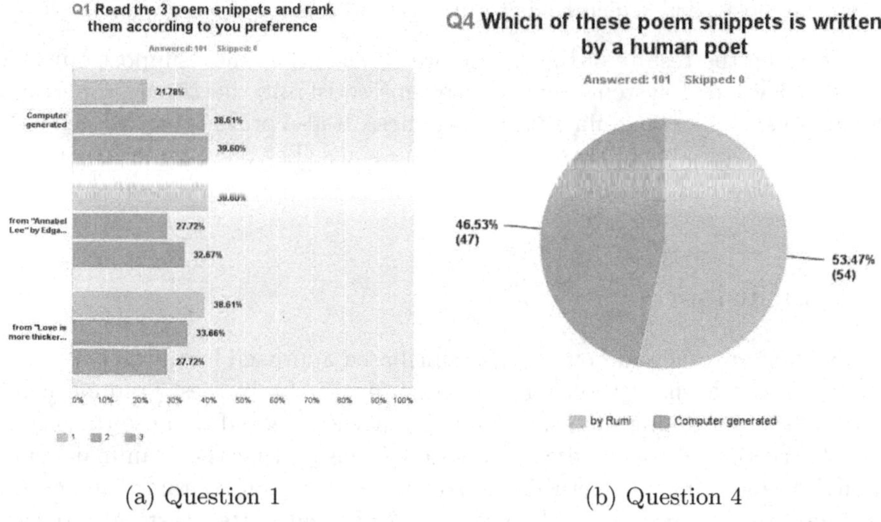

(a) Question 1

(b) Question 4

Fig. 2. Survey results

- Rhyme: 'yes' and 'no'
  The first poem has a loose rhyme and thus 70% reviewers said it contained no rhyme. The second poem has a strict rhyme scheme, which almost all the readers agreed with.
- Figurative Language: 'great', 'good', 'average', 'poor', 'no'
  Only 7% considered the poem to be lacking any figures of speech, while the remaining 93% all found the poem's language figurative. Additionally, around 51% of the readers considered the resulting figures of speech to be of good and even great quality (20 %). From these results we can extrapolate that the claim of achievement of figurative language and metaphors through underspecification has not only been satisfied, but that it is of acceptable good quality.
- Poetic style: 'very poetic', 'somewhat poetic' and 'not poetic'
  The style of the poems was considered somewhat poetic by a total average of 54%, and very poetic by 33%. This shows that 87% thought the poems satisfy the poeticness feature, and more than a quarter of the reviewers found it of high poetic. This shows that to the most part the poeticness feature has also been satisfied.
- Message: choice between 'clear' and 'unclear'
  64% and 46% of the readers found the message of the first and second poem clear, respectively. These results show that the meaningfulness property has been achieved, to the most part.
- Language: 'understandable', 'odd' and 'unclear'
  The language of both poems was understandable to an average of 69% of the reviewers, and unclear to around 9%; while an average of 22% found the

language odd. This shows that the grammaticality feature was also achieved to the most part, which leads to understandable text.

To sum up the results of the survey, we can say that they support our claim that the developed system, generates poems satisfying the three properties of grammaticality, meaningfulness and poeticness. It also proved that the generated poems are comparable to those written by humans. The quality of the generated poems can however be improved. A more comprehensive evaluation should also be performed.

6 Conclusions

In this paper, we showed how a constraint-based approach for automatic poetry generation can be used to ensure the three poetry properties: grammaticality, poeticness and meaningfulness. Grammaticality is achieved because the correctness of structure of the generated sentence is ensured by the grammar defined through the grammar pattern list. Also, the information contained in the lexicon, improves the grammatical correctness of the generated text. Any discrepancies in the grammar that arise throughout the generation process are handled through the grammar correction. Poeticness is achieved because the generated poems must have rhythm, rhyme and a certain form Rhythm has to be present in any generated poem, while rhyme is optional. The form of the poem can be manipulated using the rhythm pattern list, as well as the rhyme and rhythm scheme. Finally, figurative language in the form of metaphors and enjambments is achieved through under-specification for the time being. Meaningfulness is achieved in terms of text coherence. The coherence of the generated poems, is ensured because of the thematic unity of any generated poem. The generated poems are also coherent because they have clearly defined acting persons and some related words.

The quality of the produced poems with respect to the three properties can be enhanced in various ways. The grammaticality can be improved by handling more complex sentence structures and referencing referencing. A grammar different than the chosen one could be used and the results could be compared. Also, the grammar structure and style of the poem could be learned from existing authors, to reproduce similar poems according to the request of the users. The system could incorporate a more sophisticated approach for handling figurative language and generating metaphors to improve the poeticness of the generated poems. More phonetic features such as alliterations can easily be incorporated as extra constraints while narrowing down the lexicon. The meaningfulness can be improved by enforcing some semantic constraints and including a more advanced word relations function while narrowing down the lexicon. A more extensive evaluation that includes the lingual background of the reviewers should be performed. Another option would be the anonymous, online posting of the generated poetry in random poetry websites, to get the feedback of a poetry-familiar audience. Because the system is lexicon-based, it is easily portable to other languages, besides English given a similar lexicon of the target language. Another option, is to extend the approach developed in this work, to the generation of other types of natural language, by adapting the enforced constraints.

Acknowledgments. We thank Prof. Thom Frühwirth, Prof. Veronica Dahl, Assoc. Prof. Haythem Ismail and Amira Zaki for their valuable input and suggestions.

References

1. Manurung, H.: An evolutionary algorithm approach to poetry generation (2004)
2. Sneyers, J., De Schreye, D.: Apopcaleaps: Automatic music generation with chrism. In: 11th International Society for Music Information Retrieval Conference, ISMIR 2010, Utrecht, The Netherlands (August 2010) (Submitted)
3. Frühwirth, T.: Theory and practice of constraint handling rules. The Journal of Logic Programming 37(1-3), 95–138 (1998)
4. Oliveira, H.: Automatic generation of poetry: an overview. Universidade de Coimbra (2009)
5. Manurung, R., Ritchie, G., Thompson, H.: Using genetic algorithms to create meaningful poetic text. J. Exp. Theor. Artif. Intell. 24(1), 43–64 (2012)
6. Levy, R.P.: A computational model of poetic creativity with neural network as measure of adaptive fitness. In: Proceedings of the ICCBR 2001 Workshop on Creative Systems. Citeseer (2001)
7. Gervás, P.: Computational modelling of poetry generation. In: Artificial Intelligence and Poetry Symposium, AISB Convention 2013. University of Exeter, United Kingdom (2013)
8. Díaz-Agudo, B., Gervás, P., González-Calero, P.A., Craw, S., Preece, A.: Poetry generation in colibri. In: Proceedings of the 6th European Conference on Case Based Reasoning, Aberdeen, Scotland (2002)
9. Toivanen, J., et al.: Corpus-based generation of content and form in poetry. In: Proceedings of the Third International Conference on Computational Creativity (2012)
10. Toivanen, J.M., Järvisalo, M., Toivonen, H.: Harnessing constraint programming for poetry composition. In: Proceedings of the Fourth International Conference on Computational Creativity, p. 160 (2013)
11. Bolock, A.E., Abdennadher, S.: Towards automatic poetry generation using constraint handling rules. In: Proceedings of the 30th ACM Symposium on Applied Computing (2015)
12. Princeton, U.: About wordnet (2010)

A Multi-strategy Approach for Lexicalizing Linked Open Data

Rivindu Perera and Parma Nand

School of Computer and Mathematical Sciences,
Auckland University of Technology
Auckland 1010, New Zealand
{rivindu.perera,parma.nand}@aut.ac.nz

Abstract. This paper aims at exploiting Linked Data for generating natural text, often referred to as lexicalization. We propose a framework that can generate patterns which can be used to lexicalize Linked Data triples. Linked Data is structured knowledge organized in the form of triples consisting of a subject, a predicate and an object. We use DBpedia as the Linked Data source which is not only free but is currently the fastest growing data source organized as Linked Data. The proposed framework utilizes the Open Information Extraction (OpenIE) to extract relations from natural text and these relations are then aligned with triples to identify lexicalization patterns. We also exploit lexical semantic resources which encode knowledge on lexical, semantic and syntactic information about entities. Our framework uses VerbNet and WordNet as semantic resources. The extracted patterns are ranked and categorized based on the DBpedia ontology class hierarchy. The pattern collection is then sorted based on the score assigned and stored in an index embedded database for use in the framework as well as for future lexical resource. The framework was evaluated for syntactic accuracy and validity by measuring the Mean Reciprocal Rank (MRR) of the first correct pattern. The results indicated that framework can achieve 70.36% accuracy and a MRR value of 0.72 for five DBpedia ontology classes generating 101 accurate lexicalization patterns.

Keywords: Lexicalization, Linked Data, DBpedia, Natural Language Generation.

1 Introduction

The concept of Linked Data introduces the process of publishing structured data which can be interlinked. This structured data is represented in the form of triples, a data structure composed of a subject, a predicate, and an object (e.g., ⟨Brooklyn Bridge, location, New York City⟩). A Linked Data resource contains millions of such triples representing diverse and generic knowledge.

Recently, there has been a growing need to employ knowledge encoded in Linked Data in Natural Language Processing (NLP) applications [1]. However, Linked Data must be represented in natural text to support NLP applications. The process

A. Gelbukh (Ed.): CICLing 2015, Part II, LNCS 9042, pp. 348–363, 2015.
DOI: 10.1007/978-3-319-18117-2_26

of creating Linked Data does not support to associate such information because the creation process is a template based Information Extraction (IE).

This paper explores the appropriateness of employing lexicalization to convert Linked Data triples back to the natural text form. Lexicalization is the process of converting an abstract representation into natural text. This is a widely studies area in Natural Language Generation (NLG), but early approaches cannot be used with Linked Data because of the rapidly growing nature and domain diversity of Linked Data resources. The proposed framework is a pipeline of processes to generate patterns which can be used to lexicalize a given triple into its natural text form. The patterns are evaluated based on two factors: syntactic correctness and re-usability. The syntactic correctness for a lexicalization patterns is not only the adherence to specific syntactic representation, but also the coherence of the pattern (no unimportant additional text is included in the pattern). The re-usability factor measures whether the pattern can be generalized to a range of triples with same knowledge.

A lexicalization pattern is defined in this context as a generic pattern that can be used to convert a Linked Data triple into natural language sentence. The resulting sentence will essentially contain the subject and the object of the triple being considered. In some cases it is also possible to include the predicate of a triple in a sentence to represent the triple in a sentence. Table 1 shows three example triples and expected patterns to lexicalize them. However, there can be multiple patterns to lexicalize a given triple, but among them only one will be the most suitable pattern that can generate a coherent and syntactically correct sentence.

We hypothesise that employing Information Extraction (IE) on sentence collection related to Linked Data resources together with lexical semantic resources can generate syntactically correct lexicalization patterns. Further, such a model can cover most of the patterns required to lexicalize a triple. The proposed model uses Open IE based relation extraction model to extract relations present in text which are then converted into lexicalization patterns. In parallel, the framework mines lexicalization patterns using verb frames utilizing two lexical semantic resources: VerbNet [2] and WordNet [3]. The identified patterns are then ranked and categorized based on the ontology class hierarchy specified by DBpedia. We have released the version 1.0 of this pattern database which can be found in project web site[1] [4,5].

Table 1. Example lexicalization patterns for three triples. The symbol `s?` and `o?` represent the subject and the object of the triple respectively.

Triple			Lexicalization pattern
Subject (s)	Predicate (p)	Object (o)	
Steve Jobs	birth date	1955-02-24	*s?* was born on *o?*
Steve Jobs	birth place	San Francisco	*s?* was born in *o?*
Berlin	country	Germany	*s?* is situated in *o?*

[1] `http://staff.elena.aut.ac.nz/Parma-Nand/projects/realtext.html`

The remainder of the paper is structured as follows. In Section 2 we describe the proposed framework in detail. Section 2 presents the experiments used to validate the framework. Section 4 compares the proposed model with other similar works. Section 5 concludes the paper with an outlook on future work.

2 RealText$_{\text{lex}}$ Framework

Fig. 1 depicts the complete process of generating lexicalization patterns in the proposed framework. The process starts with a given DBpedia ontology class (e.g., person, organization, etc.) and then selects two DBpedia resources for the entity (e.g., person ⇒ Steve Jobs, Bill Gates). In next step, the related Wikipedia text for the DBpedia resource together with text from other websites related to the DBpedia resource are extracted and prepares the text by applying co-reference resolution. The framework also reads the triples related to the DBpedia resource in parallel. Using both text collection and triples, the framework then extracts sentences which contain the triple elements. The selected sentences are then subjected to an Open Information Extraction (Open IE) based relation extraction. The output of this process is triples and relations related to each triple.

The triples and related relations extracted in aforementioned process are then used to determine lexicalization patterns. First, each relation is aligned with the related triple and relation patterns are extracted. At the same time the model retrieves the verb frames from VerbNet [2] and WordNet [3]. Both verb frames and relation patterns are then processed to identify basic patterns. The leftover triples without a pattern are associated with a predetermined pattern based on predicate of the triple. Next, a combined list of patterns are sent to the pattern enrichment process. This process adds additional features to the pattern such as multiplicity and grammatical gender. Finally, based on the enriched patterns,

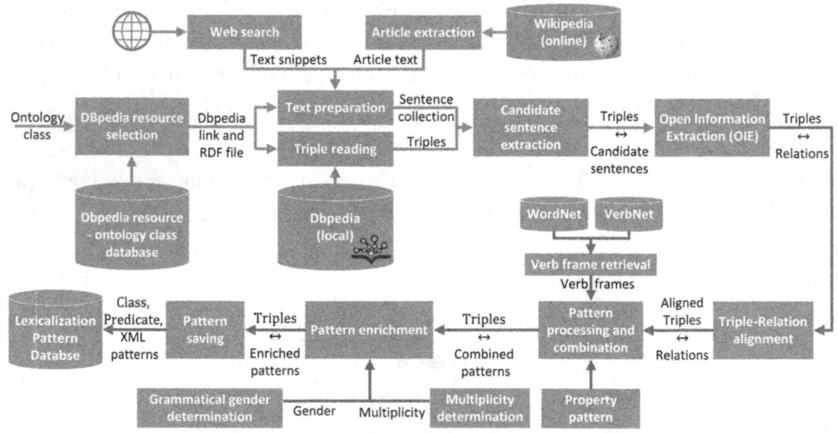

Fig. 1. Schematic representation of the complete framework

Extensible Markup Language (XML) patterns are built and stored in an indexed database.

Following sections describe the individual components of the framework in detail.

2.1 DBpedia Resource Selection

The function of this unit is to select two Dbpedia resources for the given ontology class. For an example, if the input is given as *Person* ontology class then the model should select two Dbpedia resources such as *http://dbpedia.org/resource/Walt_Disney* and *http://dbpedia.org/resource/Bill_Gates*.

Furthermore, to facilitate latter steps, this unit must also output RDF file names of these resources (e.g., *Walt Disney.rdf* and *Bill Gates.rdf*). This can be accomplished by executing a query in DBpedia online SPARQL endpoint[2]. However, executing SPARQL queries is a time consuming task and furthermore, if the endpoint is not functioning the whole lexicalization process is interrupted. To mitigate this, we have created a database containing DBpedia resource link, ontology class of the resource, and RDF file name. This database has 9010273 records and released with our framework for public usage. Since DBpedia does not offer dump which contains separate RDF file for each resource, we have downloaded partial set of RDF files related to records present in the database. If the required RDF file is not present in the repository, the model automatically downloads the resource for use as well as stores it in the repository for possible future usage.

2.2 Text Preparation

The objective of the text preparation unit is to output co-reference resolved set of sentences. To achieve this objective, it first retrieves text from web and Wikipedia and then performs co-reference resolution. The resulting text is split to sentence level and passed as a text collection to the candidate sentence extraction unit. Following sections describe the process in detail.

Text Retrieval. Text related to triples are retrieved from web as well as Wikipedia. Since, DBpedia is based on Wikipedia, the Wikipedia text contains natural language sentences corresponding to the DBpedia triples. However, this assumption is not always true. For example, in Wikipedia birth date of a person is mentioned in the format of *person name (birth date)* (e.g., "Walt Disney (December 5, 1901 - December 15, 1966) was an American business magnate..."). Such sentences cannot be used to identify a pattern to lexicalize the birth date of a person, because property birth date is not represented in natural text. Due to this we use text snippet retrieved from other web sites in conjunction with the Wikipedia text. For example, a sentence like "Walt Disney was born on

[2] http://dbpedia.org/sparql

December 5, 1901,..." appeared in *biography.com* has a natural language representation for the birth date property. Therefore, text retrieval is accomplished using both text extracted from Wikipedia text and snippets of text acquired through a web search.

Co-reference Resolution. The retrieved text in Section 2.2 can contain co-references to entities. For an example, consider following sentences extracted from Wikipedia page for *Walt Disney*.

- "*Walt Disney* (December 5, 1901 - December 15, 1966) was an American business magnate. *He* left behind a vast legacy, including numerous animated shorts and feature films produced during his lifetime. *Disney* also won seven Emmy Awards and gave his name to the Disneyland."

In above text, *He* and *Disney* both refer to the entity *Walt Disney*. Existence of such mentions which does not reflect the original entity when taken as individual sentences, can have a negative effect when finding sentences to extract lexicalization patterns. Therefore, as a preprocessing step all such mentions are replaced by actual entity that it referenced. The framework uses Stanford CoreNLP [6] to identify co-reference mentions and replaces them with the actual entity. The resulting co-reference resolved sentence will look like follows:

- "*Walt Disney* (December 5, 1901 - December 15, 1966) was an American business magnate. *Walt Disney* left behind a vast legacy, including numerous animated shorts and feature films produced during his lifetime. *Walt Disney* also won seven Emmy Awards and gave his name to the Disneyland."

The resulting co-reference resolved text is then passed to the sentence splitter.

Sentence Splitting. The sentence splitting is responsible for splitting the co-reference resolved text into sentences. This is accomplished using Stanford CoreNLP tokenizer [6]. The tokenizer is based on Penn Tree Bank guidelines and uses deterministic process to split sentences using set of heuristics.

2.3 Triple Reading

Parallel to text preparation, the framework retrieves triples related to the DBpedia resources selected in Section 2.1. These triples are read from a local repository of DBpedia RDF files. The repository currently contains 163336 RDF files and can be automatically download from DBpedia on demand.

Currently, DBpedia contains two types of triples, DBpedia properties and DBpedia ontology mapped properties. The DBpedia properties are those which are extracted using the DBpedia Information Extraction framework. Recently, the DBpedia team added an ontology property list to organize DBpedia properties by mapping them with human involvement. However, the mapping process is still in progress and many DBpedia properties still remain unmapped. Therefore, the proposed framework used DBpedia properties to read triples employing Jena RDF parser[3].

[3] `https://jena.apache.org/`

- ⟨Walt Disney, birth date, 1901-12-05⟩
- ⟨Walt Disney, birth place, Hermosa⟩

Fig. 2. Sample set of triples retrieved from triple reader

Fig. 2 depicts a sample set of triples that can be retrieved by the triple reader from DBpedia RDF file for the resource *Walt Disney.*

2.4 Candidate Sentence Extraction

The objective of candidate sentence extractor is to identify potential sentences that can lexicalize a given triple. The input is taken as a collection of co-reference resolved sentences (from text preparation unit described in Section 2.2) and a set of triples (from triple reader described in Section 2.3). This unit firstly verbalizes the triples using a set of rules. Following list shows the set of rules used to verbalize triples.

- Dates are converted to normal date format. (e.g., 1901-12-05 ⇒ December 05 1901)
- Object values pointing to another DBpedia resource are replaced with actual resource names. (e.g., dbpedia:Chicago ⇒ Chicago)

Then each sentence is analyzed to check either complete subject (s), the object (o) or the predicate (p) are mentioned in the sentence (S). The only exception is that the object is analysed as a complete phrase as well as set of terms (t_o). For example, the object value like "Walt Disney Company" will be split to three terms (e.g, Walt Disney Company ⇒ Walt, Disney, Company) and will check whether a subset of the terms appears in the sentence. This sentence analysis uses a set of cases which assigns a score to each sentence based on presence of a triple. The scores were determined based on preliminary experiments. The sentences which do not belong to any of the cases are not considered as candidate sentences. The cases with the respective score assigned by the candidate sentence extractor are shown in (1) below:

$$sc(T,S) = \begin{cases} 1, & \texttt{if } (\mathrm{s}\in\mathrm{S})\ \&\ (\mathrm{p}\in\mathrm{S})\ \&\ (\mathrm{o}\in\mathrm{S}) \\ 2/3, & \texttt{if } (\mathrm{s}\in\mathrm{S})\ \&\ (\mathrm{o}\in\mathrm{S}) \\ 2/3, & \texttt{if } (\mathrm{s}\in\mathrm{S})\ \&\ (\mathrm{p}\in\mathrm{S}) \\ 2/3, & \texttt{if } (\mathrm{p}\in\mathrm{S})\ \&\ (\mathrm{o}\in\mathrm{S}) \\ 1/3, & \texttt{if } (\mathrm{o}\in\mathrm{S}) \\ 1/3 \times cr_{(t_o,S)}, & \texttt{if } (\mathrm{cr}_{(t_o,S)} > 0.5) \end{cases} \tag{1}$$

where, $sc(T, S)$ represents candidate sentence score for triple (T) and sentence represented as set of terms (S). The ratio of terms contained in the sentence ($\mathrm{c}_{(t_o,S)}$) is calculated as:

$$cr_{(t_o,S)} = \frac{\#(t_o \cap S)}{\#t_o} \tag{2}$$

where, t_o represents set of terms appearing in the object.

The output of this module is a collection of triples and sentences where each sentence is associated with one or more triple. Example scenario for triples taken from *Walt Disney* is shown in Fig. 3.

- ⟨Walt Disney, birth date, 1901-12-05⟩
  - `sentence`: Walt Disney was born on December 5, 1901. `score:0.667`
- ⟨Walt Disney, birth place, Hermosa⟩
  - `sentence`: Walt Disney was born in Hermosa. `score:0.667`

Fig. 3. Example output from candidate sentence extraction

2.5 Open Information Extraction

Once the candidate sentences are selected for each triple, we then extract relations from these candidate sentences in order to identify lexicalization patterns.

Extracting relations is a well known and heavily researched area in the IE domain. The approaches to relation extraction can be broadly divided into two categories: traditional closed IE based models and Open IE based models. The closed IE based models rely on rule based methods [7], kernel methods [8], and sequence labelling methods [9]. The burden associated with hand-crafted rules, the need for hand-tagged data for training, and domain adaptability are three key drawbacks associated with these traditional relation extraction models.

In recent times a new concept is arisen to address the drawbacks. The Open IE [10] essentially focuses on domain independent relation extraction and predominantly targets the web as a corpus for deriving the relations.

The framework proposed in this paper uses textual content extracted from the web which works with a diverse set of domains. Specifically, the framework uses Ollie Open IE system [11] for relation extraction. Each sentence from the web associated with a triple is analyzed for relations and all possible relations are determined. This module then associates each relation with the triple and outputs a triple-relations collection. A relation is composed of first argument (arg1), relation (rel), and second argument (arg2). The module also assigns a score to each relation which is defined as the average of confidence score given by Ollie and a score assigned for the specific sentence in Section 2.4.

Relations extracted for the example scenario shown in Section 2.4 are shown in Fig. 4.

- ⟨Walt Disney, birth date, 1901-12-05⟩
  - **arg1**: Walt Disney, **rel**: was born on, **arg2**: December 5, 1901 score:0.934
- ⟨Walt Disney, birth place, Hermosa⟩
  - **arg1**: Walt Disney, **rel**: was born in, **arg2**: Hermosa score:0.786

Fig. 4. Example output from Open Information Extraction module

2.6 Triple-relation Alignment

The objective of this module is to determine how well each relation (arg1, rel, arg2) is aligned with the triple (s, p, o) and to generate a general pattern based on the alignment. This is accomplished by aligning verbalized triple (using triple verbalization rules described in Section 2.4) with the relation. Each relation is again scored in this step by assigning the final score which is average of score assigned by Open IE (in Section 2.5) module and score determined by the level of alignment. The alignment score is calculated as in (3):

$$sc(T, R_a) = \begin{cases} 1, & \texttt{if } (\text{s}\in \text{arg1} \cup \text{arg2}) \,\&\, (\text{p}\in\text{rel}) \,\&\, (\text{o}\in \text{arg1} \cup \text{arg2}) \\ 2/3, & \texttt{if } (\text{s}\in \text{arg1} \cup \text{arg2}) \,\&\, (\text{p}\in\text{rel}) \\ 2/3, & \texttt{if } (\text{s}\in \text{arg1} \cup \text{arg2}) \,\&\, (\text{o}\in \text{arg1} \cup \text{arg2}) \\ 2/3, & \texttt{if } (\text{p}\in\text{rel}) \,\&\, (\text{o}\in \text{arg1} \cup \text{arg2}) \\ 1/3, & \texttt{if } (\text{o}\in\text{arg2}) \\ 1/3 \times cr_{(t_o,arg2)}, & \texttt{if } (\text{cr}_{(t_o,arg2)} > 0.5) \end{cases} \tag{3}$$

where, $sc(T, R_a)$ represents triple-relation alignment score for triple (T) and aligned relation (R_a). The relation is represented in three components ($arg1, rel, arg2$). The ratio of object terms contained in the second argument of the aligned relation ($\text{cr}_{(t_o,arg2)}$) is calculated as:

$$cr_{(t_o,arg2)} = \frac{\#(t_o \cap arg2)}{\#t_o} \tag{4}$$

This module outputs a collection of triples with aligned relations including the associated score determined by the level of alignment. The relations that are not aligned (based on the cases in (3)) are not included in the collection.

2.7 Pattern Processing and Combination

This module generates patterns from aligned relations in Section 2.6. In addition to these patterns, verb frame based patterns are also determined and added to the pattern list. Any other pattern that is not associated with these patterns are assigned a generic pattern based on the predicate. The following sections describe these processes in detail.

Relation Based Patterns. Based on the aligned relations and triples, a string based pattern is generated. These string based patterns can get two forms as shown in Fig. 5 for two sample scenarios. The subject and object are denoted by symbols *s?* and *o?* respectively.

- ⟨Walt Disney, birth date, 1901-12-05⟩
  - `arg1`: Walt Disney, `rel`: was born on, `arg2`: December 5, 1901
    `pattern`: s? was born on o?
- ⟨Walt Disney, designer, Mickey Mouse⟩
  - `arg1`: Mickey Mouse, `rel`: is designed by, `arg2`: Walt Disney
    `pattern`: o? is designed by s?

Fig. 5. Basic patterns generated for two sample triples. *s?* and *o?* represent subject and object respectively.

Verb Frame Based Patterns. Verb frame based pattern generation process attempts to find patterns based on semantic frames. A semantic frame of a verb is a grammatical structure that shows the usage of the verb. For instance, the verb *create* has a semantic frame *Noun phrase, Verb, Noun phrase* which can be realized in an example as "Walt Disney created Mickey Mouse Clubhouse".

The framework utilizes two lexical semantic resources, VerbNet and WordNet to mine patterns. We have created a slightly moderated version of VerbNet which has all verbs and their associated frames merged into one indexed database and can be found in the RealText$_{lex}$ project website. Currently, the framework generate only one type of pattern (`s? Verb o?`), if the predicate is a verb and if that verb has the frame {*Noun phrase, Verb, Noun phrase*} in either VerbNet or WordNet.

Property Based Patterns. The predicates which cannot associate with a pattern in the above processes described in Section 2.7 and Section 2.7 are properties belonging to the DBpedia resources selected in Section 2.1. The left over predicates are assigned a generic pattern (`s? has` ⟨`predicate`⟩ `of o?`) based on the specific predicate. For example, a predicate like "`eye colour`" will be assigned with a pattern like "`s? has eye colour of o?`".

The patterns acquired from aforementioned modules are then passed to the enrichment module to add related information.

2.8 Pattern Enrichment

Pattern enrichment adds two types of additional information; grammatical gender related to the pattern and multiplicity level associated with the determined pattern. When searching a pattern in the lexicalization pattern database, these additional information is also mined in the lexicalization patterns for a given predicate of an ontology class.

Grammatical Gender Determination. The lexicalization patterns can be accurately reused later only if the grammatical gender is recorded with the pattern. For example, consider triple, ⟨`Walt Disney, spouse, Lillian Disney`⟩ and lexicalization pattern, "`s? is the husband of o?`". This pattern cannot be reused to lexicalize the triple ⟨`Lillian Disney, spouse, Walt Disney`⟩, because the grammatical gender of the subject is now different, even though the property (spouse) is same in both scenarios. The framework uses three types of grammatical gender types (male, female, neutral) based on the triple subject being considered in the lexicalization. The gender of the subject is determined by DBpedia grammatical gender dataset [1].

Multiplicity Determination. Some triples in DBpedia has same subject and predicate values for different object values. Fig. 6 shows set of triples taken from DBpedia resource for Nile river and for East river which have the same predicate (country).

- Nile River
  - ⟨Nile River, country, Burundi⟩
  - ⟨Nile River, country, Egypt⟩
  - ⟨Nile River, country, Kenya⟩
- East River
  - ⟨East River, country, United States⟩

Fig. 6. Object Multiplicity scenario for predicate type *country*

According to Fig 6, Nile River has three countries listed as it does not belong to one country, but flows through these countries. However, East River belongs only to United States. The lexicalization patterns generated for these two scenarios will also be different and cannot be shared. For example, lexicalization pattern for Nile river will in the form of "`s? flows through o?`" and for East River it will be like "`s? is in o?`". To address this variation, our framework checks whether there are multiple object values for the same subject and predicate, then it adds the appropriate property value (multiple/single) to the pattern.

These enriched patterns with grammatical gender and multiplicity are then passed to the pattern saving module to store these in a database.

2.9 Pattern Saving and Lexicalization Pattern Database

The pattern saving module first converts the patterns to an XML format and then it saves these to the lexicalization pattern database with the final score calculated as described in Section 2.6. A sample XML pattern generated using triple ⟨Walt Disney, birth date, 1901-12-05⟩ is shown in Fig. 7.

The XML patterns are written to the lexicalization database with three other attributes, ontology class, predicate and the final score. Sample set of records are shown in Table 2.

```
<lexpat>
<pat> s? was born on o? </pat>
<gender> male </gender>
<multiplicity> false </multiplicity>
</lextpat>
```

Fig. 7. Sample XML pattern

Table 2. Sample set of entries from lexicalization pattern database

Ontology class	Predicate	Pattern	Score
Person	birthDate	<lexpat>...</lexpat>	0.984
Person	birthPlace	<lexpat>...</lexpat>	0.791
River	country	<lexpat>...</lexpat>	0.454

3 Experimental Framework

The experimental framework analysed both lexicalization patterns as well as the underlying process to generate them. We used a subset of randomly selected DBpedia ontology classes to evaluate the accuracy of the framework. The following sections describe the results and evaluation details for the selected ontology classes.

3.1 DBpedia Ontology Class Selection

Table 3 shows the ontology classes selected for evaluation of the framework including various other related statistics. For the experiment, 5 ontology classes were randomly selected which gave us 254 triples in total. A predetermined rule set was used to filter out invalid triples such as "Wikipedia URL", "WordNet type", and "photo collection URL" which cannot be lexicalized as natural language representations. The further removed duplicate predicates, leaving a total of 132 predicates that were used in the experiment.

Table 3. Results of the ontology class selection

Ontology class	Valid triples	Invalid triples	Unique predicates to lexicalize
Bridge	61	10	39
Actor	60	12	37
Publisher	15	5	9
River	49	9	35
Radio Host	29	4	12

3.2 Experimental Settings and Results

This section presents results collected from each module contributing towards pattern mining as well as the overall evaluation results.

Table 4 shows the results for the individual modules in framework. The table gives the numbers for the sentences processed, the candidate sentences and relations extracted from the candidate sentences. Table 5 shows the summary of the breakdown of the results for pattern extraction. The last 5 columns of the table also shows the results for the pattern enrichment modules. To get a clear idea about on the accuracy of the framework, we checked how many syntactically correct lexicalization patterns appear as the highest ranked pattern for the given predicate. In this context syntactic correctness was considered as being both grammatically accurate and coherent. The results of this evaluation is shown in Fig. 8 for each of the ontology classes.

Table 4. Results of the sentence and relations processed

Ontology class	All sentences	Candidate sentences	Extracted relations
Bridge	202	146	445
Actor	129	126	422
Publisher	212	25	99
River	88	69	158
Radio Host	38	27	87

Table 5. Results of the pattern extraction module

Ontology class	Relational patterns	Verb frame patterns	Property based patterns	Pattern enrichment				
				Multiplicity		Grammatical gender		
				Multiple	Single	Male	Female	Neutral
Bridge	272	8	9	163	126	0	0	289
Actor	422	0	16	369	69	400	22	16
Publisher	39	1	4	32	12	0	0	44
River	157	2	10	158	11	0	0	169
Radio Host	30	1	1	14	18	0	0	32

Since, the framework ranks lexicalization patterns using a scoring system, we considered it as a method that provides a set of possible outputs. We decided to get a statistical measurement incorporating Mean Reciprocal Rank (MRR) as shown below to compute the rank of the first correct pattern of each predicate in each ontology class.

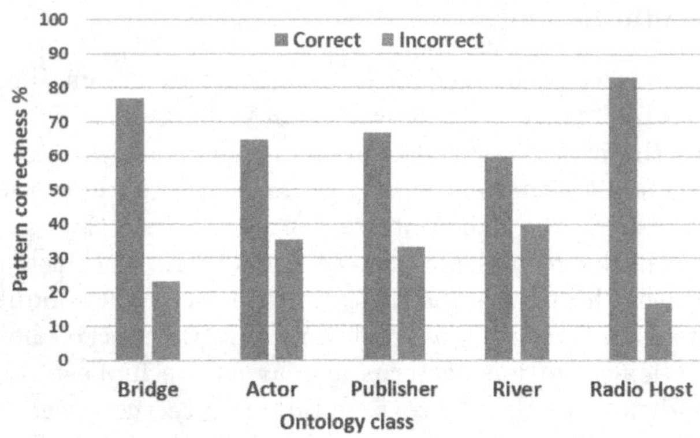

Fig. 8. Analysis of syntactic correctness of the extracted patterns

$$MRR = \frac{1}{|P|}\sum_{i=1}^{|P|}\frac{1}{rank_i} \tag{5}$$

where P and $rank_i$ represent predicates and the rank of the correct lexicalization for the ith predicate respectively. Table 6 depicts the MRR results for the 5 ontology classes being considered.

Table 6. Mean Reciprocal Rank analysis for ranked lexicalization patterns

Ontology class	MRR
Bridge	0.77
Actor	0.69
Publisher	0.72
River	0.61
Radio Host	0.83

Table 7 shows a statistical summary of proposed approach.

Table 7. Statistics of evaluation of proposed approach

	Candidate templates	Lexicalizations	Accuracy
Proposed approach	393	101	70.36%

3.3 Observations and Discussions

The following observations can be made based on the results of the experiment. Firstly, the triple filtering step was effective in being able to filter out all the invalid triples as shown in Table 3. The filtering process was able to filter out

40 triples from total of 254. All invalid filtered out triples were identified as true negatives.

Apart from the *Publisher* ontology class (Table 4) more than 50% of the sentences were able to be selected as candidates. This was mainly due to the variation of the sentence patterns appeared in the two DBpedia resources which were selected by resource selection module (Section 2.1). The two resources selected (*Marvel Comics* and *Elsevier*) had sentences which were hard to align with triples. This caused a dramatic decrease in selection score thus eliminating 88% of the sentences.

The number of extracted relations in Table 4 shows that 1211 relations were able to be extracted from a 393 candidate sentences. This emphasizes an interesting point that on average 3 relations are associated with each candidate sentence. This is one of the significant benefit that our framework received by incorporating Open IE model which can extract large number of accurate relations compared to traditional closed IE relation extraction approaches.

Fig. 8 shows that our framework has achieved 70.36% average accuracy for 5 ontology classes where the lowest accuracy was reported as 60%. This evaluation does not take into account the rank of the correct lexicalization patterns and measures the number of correct patterns present in the extracted set of patterns. On the other hand, MRR based evaluation provides an detailed look at ranking of the first correct lexicalization. Average MRR value of 0.724 achieved for 5 ontology classes. The level of alignment between triples under consideration and the relations had a positive effect on the accuracy. For example, the ontology class *Radio Host* has achieved good accuracy and a MRR value (accuracy: 83.34%, MRR: 0.83) with a smallest set of relations because of the alignment between relations and triples.

The verb frame and property patterns have not significant impact on pattern extraction, however, they are still crucial for identifying patterns that cannot be represented as semantic relations.

Finally, based on the comparison in Table 7, it is clear that proposed approach in this paper has advanced the way of deriving lexicalizations by generating reasonable number of valid patterns and with a higher accuracy.

4 Related Work

This section discusses the related works in which have used Linked Data for lexicalization and also compares our framework and concept with them.

Unsupervised template extraction model presented by Duma and Klein [12] is similar in goal to our framework. Duma and Klein [12] attempt to extract sentences from Wikipedia which have DBpedia triples mentioned and then extract templates based on a heuristic. This model differ from our framework in many ways. Firstly, it does not consider extracting more textual content to generate candidate sentences. Furthermore, Wikipedia data is taken in raw format without further processing.

Walter et al. [13] introduces the Lemon model which extracts lexicalizations for DBpedia using dependency patterns extracted from Wikipedia sentences.

They argue that dependency parsing can generate patterns out of sentences to lexicalize a given triple. However, the strategy of traversing through a dependency parse is not explained well. In our initial experiment the use of dependency parsing for Wikipedia sentences showed that determining a pattern is hard to accomplish by sole usage of a dependency parse. The framework proposed in this paper introduces Open IE to address this issue. Furthermore, our framework is equipped with additional functions to enrich the sentence collection (e.g, co-reference resolution). Gerber and Ngomo [14] discuss the string based model to extract patterns. This simple model limits it generating more coherent patterns for triples being under consideration. Furthermore, it is not possible to expect all different lexicalization patterns from text. In our model this is achieved by integrating two lexical semantic resources VerbNet and WordNet which in parallel contributing to the pattern extraction process.

Ell and Harth [15] present the language independent approach to extract RDF verbalization templates. The core of this model is the process of aligning a data graph with the extracted sentence. To extract patterns they utilize maximal sub-graph pattern extraction model. However, there are key differences in our framework compared to one introduced by Ell and Harth [15]. Our framework starts the process with a text preparation unit which provides a potential text to extract patterns. Furthermore, we have studied the effectiveness of Open IE in pattern extraction which is specifically targeting on extracting relational patterns using a web as the corpus.

5 Conclusion and Future Work

This paper presented a framework to generate lexicalization patterns for DBpedia triples using a pipeline of processes. The pipeline starts with ontology classes which is then used to mine patterns aligning triples with relations extracted from sentence collections from the web. Furthermore, two additional pattern identification modules, verb frame based and property based were used to add additional patterns. The framework also incorporates an enrichment process for the further alignment of the patterns to the human generated text. The framework generated patterns were human-evaluated and showed an accuracy of 70.36% and a MRR of 0.72 on test dataset.

In future, we aim to target on expanding the test collection to build a reasonable sized lexicalization pattern database for DBpedia. The results also showed that the verb frame based pattern mining can be further improved by incorporating additional lexical semantic resources. The individual composite modules can also be improved (especially the relation-triple alignment module) to generate high quality lexicalization patterns for DBpedia as well as to generalize the process for other Linked Data resources. This will allow us to introduce a generalizable lexicalization pattern generation framework for evolving Linked Data cloud.

References

1. Mendes, P.N., Jakob, M., Bizer, C.: DBpedia for NLP: A Multilingual Cross-domain Knowledge Base. In: International Conference on Language Resources and Evaluation, Istanbul, LREC (2012)
2. Kipper, K., Dang, H.T., Palmer, M.: Class-Based Construction of a Verb Lexicon. In: Seventeenth National Conference on Artificial Intelligence, pp. 691–696. AAAI Press, Austin (2000)
3. Miller, G.A.: WordNet: A Lexical Database for English. Communications of the ACM 38(11), 39–41 (1995)
4. Perera, R., Nand, P.: Real text-cs - corpus based domain independent content selection model. In: 2014 IEEE 26th International Conference on Tools with Artificial Intelligence (ICTAI), pp. 599–606 (November 2014)
5. Perera, R., Nand, P.: The role of linked data in content selection. In: The 13th Pacific Rim International Conference on Artificial Intelligence, pp. 573–586 (December 2014)
6. Manning, C., Bauer, J., Finkel, J., Bethard, S.J., McClosky, D.: The Stanford CoreNLP Natural Language Processing Toolkit. In: The 52nd Annual Meeting of the Association for Computational Linguistics, ACL (2014)
7. Brin, S.: Extracting Patterns and Relations from the World Wide Web. In: International Workshop on the World Wide Web and Databases, pp. 172–183. Springer-Verlag, Valencia (1998)
8. Zelenko, D., Aone, C., Richardella, A.: Kernel methods for relation extraction. In: Empirical Methods in Natural Language Processing, EMNLP 2002, vol. 10, pp. 71–78. Association for Computational Linguistics, Morristown (2002)
9. Kambhatla, N.: Combining lexical, syntactic, and semantic features with maximum entropy models for extracting relations. In: ACL 2004 on Interactive Poster and Demonstration Sessions. ACL, Morristown (2004)
10. Etzioni, O., Banko, M., Soderland, S., Weld, D.S.: Open information extraction from the web. Communications of the ACM 51(12), 68 (2008)
11. Mausam, S.M., Bart, R., Soderland, S., Etzioni, O.: Open language learning for information extraction. In: Joint Conference on Empirical Methods in Natural Language Processing and Computational Natural Language Learning, pp. 523–534. ACL, Jeju Island (2012)
12. Duma, D., Klein, E.: Generating Natural Language from Linked Data: Unsupervised template extraction. In: 10th International Conference on Computational Semantics, IWCS 2013. ACL, Potsdam (2013)
13. Walter, S., Unger, C., Cimiano, P.: A Corpus-Based Approach for the Induction of Ontology Lexica. In: 18th International Conference on Applications of Natural Language to Information Systems, pp. 102–113. Springer-Verlag, Salford (2013)
14. Gerber, D., Ngomo, A.C.N.: Bootstrapping the Linked Data Web. In: The 10th International Semantic Web Conference. Springer-Verlag, Bonn (2011)
15. Ell, B., Harth, A.: A language-independent method for the extraction of RDF verbalization templates. In: 8th International Natural Language Generation Conference. ACL, Philadelphia (2014)

A Dialogue System for Telugu, a Resource-Poor Language

Mullapudi Ch. Sravanthi, Kuncham Prathyusha, and Radhika Mamidi

IIIT-Hyderabad, India
{Mullapudi.sravanthi,prathyusha.k}@research.iiit.ac.in,
radhika.mamidi @iiit.ac.in

Abstract. A dialogue system is a computer system which is designed to converse with human beings in natural language (NL). A lot of work has been done to develop dialogue systems in regional languages. This paper presents an approach to build a dialogue system for resource poor languages. The approach comprises of two parts namely Data Management and Query Processing. Data Management deals with storing the data in a particular format which helps in easy and quick retrieval of requested information. Query Processing deals with producing a relevant system response for a user query. Our model can handle code-mixed queries which are very common in Indian languages and also handles context which is a major challenge in dialogue systems. It also handles spelling mistakes and a few grammatical errors. The model is domain and language independent. As there is no automated evaluation tool available for dialogue systems we went for human evaluation of our system, which was developed for Telugu language over 'Tourist places of Hyderabad' domain. 5 people evaluated our system and the results are reported in the paper.

1 Introduction

A dialogue system is a computer program that communicates with a human in a natural way. Many efforts are being done to make the conversations seem natural. Despite a lot of progress in computational linguistics and language processing techniques we do not see much usage of dialogue systems in real time. Some reasons for this may be the lack of domain expertise, linguistic experts and computational tools. Our approach to build a dialogue system is quick and does not require many language processing tools. Our approach can be described in two parts namely Data Management and Query Processing.

- ***Data Management:*** This component deals with categorization, text segmentation and storage of the data in a hierarchical manner which helps in fast retrieval of the output.
- ***Query Processing:*** This takes a natural language query from a user as input, process it to extract the keywords and update the context if necessary. Based on the extracted keywords and the context it either retrieves an answer from the database or asks the user an interactive question. In this way, it maintains the conversation with the user all the time.

A. Gelbukh (Ed.): CICLing 2015, Part II, LNCS 9042, pp. 364–374, 2015.
DOI: 10.1007/978-3-319-18117-2_27

This model can be implemented with minimal or no use of any language processing tool. Our approach is language independent and can be extended to any domain. It can also handle context and code-mixed queries to make the conversations look more natural.

Dialogue Example:
U1: hExarAbAxulo[*] mAnyumeVMts emi unnAyi?
(What are the monuments in Hyderabad?)
S1: cArminAr, golkoVMda, makkA masjix. mIku vItilo exi kAvAli?
(Charminar, Golconda, Mecca Masjid. What among these do you need?)

U2: cArminAr eVkkada uMxi?
(Where is Charminar?)
S2: cArminAr yoVkka cirunAmA - cArminAr, heVxarAbAx, weVlaMgANa.
(The address of Charminar is -Charminar, Hyderabad, Telangana.)

U2: eVppudu opeVn uMtuMxi?
(When is it open?)
S2: cArminar yoVkka samayaM anni rojulu 9:00 am - 5:30 pm
(Charminar open timings -All days 9:00 am - 5:30 pm)

U3: eMtrI PIju eMwa?
(What is the entry fee?)
S3: cArminAr yoVkka PIju - iMdiyans - 20, vixeSIyulu - 150
(The fee for Charminar - Indians - Rs.20, Foreigners -Rs.150)

For example, the above conversation shows a user-system interaction from 'Tourism' domain for Telugu language. User responses are represented as U1, U2, etc. and system responses are represented as S1, S2, etc. User and system response together make a discourse unit i.e. U1 and S1 is a discourse unit. We can observe that context is also handled up to many discourse units i.e. to answer U3, U4 we need context information from U2.

U3 is a code-mixed query as it contains 'opeVn' *(open),* an English word. We can see that U3 has been successfully processed and understood by the system. This shows that code-mixed queries are also handled by our system.

2 Related Work

There has been a lot of progress in the field of dialogue systems in last few years. In general dialogue systems are classified into three types. (a) Finite State (or graph) based systems, (b) Frame based systems, (c) Agent based systems.

(a)Finite State Based Systems: In this type of systems, conversation occurs according to the predefined states or steps. This is simple to construct but doesn't allow user

[*] Words are in wx format (sanskrit.inria.fr/DATA/wx.html). All the examples given in the paper are from Telugu language.

to ask questions and take initiative. [3] Proposed a method using weighted finite state transducer for dialogue management.

(b)Frame Based Systems: These systems have a set of templates which are filled based on the user responses. These templates are used to perform other tasks. [2] Proposed an approach to build natural language interface to databases (NLIDB) using semantic frames based on Computational Paninian Grammar. Context information in NLIDB is handled by [1]. In this paper different types of user-system interactions were identified and context was handled for one specific type of interaction. A dialogue based question answering system [6] which extracts keywords from user query to identify a query frame has been developed for Railway information in Telugu.

(c)Agent Based Systems: These systems allow more natural flow of communication between user and system than the other systems. The conversations can be viewed as interaction between two agents, each of which is capable of reasoning about its own actions. [4] Developed an agent based dialogue system called Smart Personal Assistant for email management. This has been further extended for calendar task domain in [5]. Our model can be categorized as an agent based system.

3 Our Approach

Fig.1 describes the flow chart of the internal working of our model.

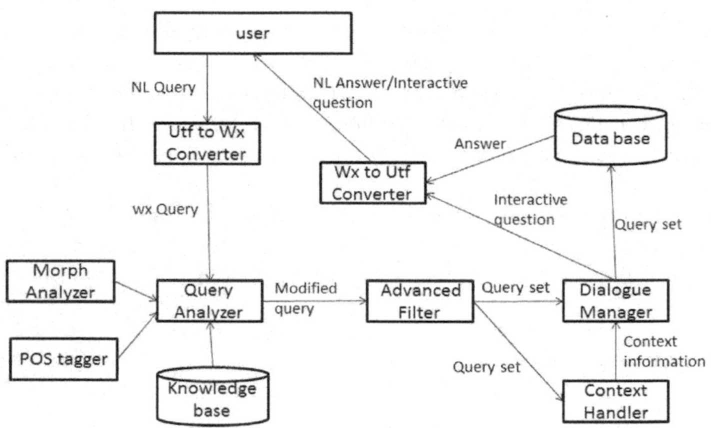

Fig. 1. System Architecture

The major components in our method are:

- Data Organization(Database)
- Query Processing

- Knowledge Base
- Query Analyzer
- Advanced Filter
- Context Handler
- Dialogue Manager

3.1 Data Organization

Every domain has an innate hierarchy in it. Study of the possible queries in a domain gives an insight about the hierarchical organization of the data in that domain. For example, consider 'Tourist places of Hyderabad' domain in Fig. 2.

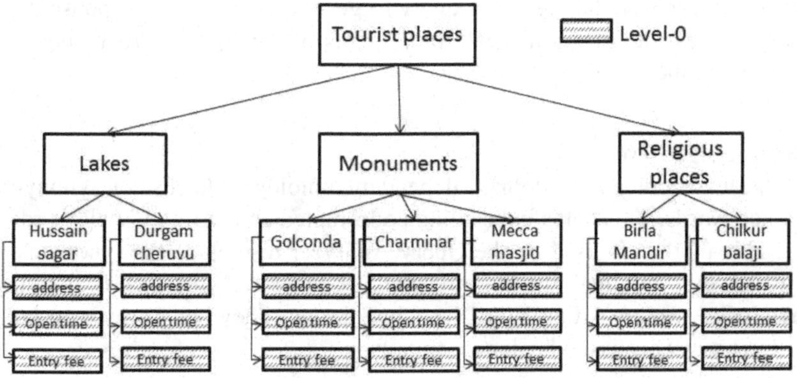

Fig. 2. Data organization of 'Tourist places of Hyderabad' domain

When we store data in this manner it becomes easy to add information and extend the domain. We can see the extension of 'Tourist places of Hyderabad' domain to 'Hyderabad Tourism' domain in Fig. 3.

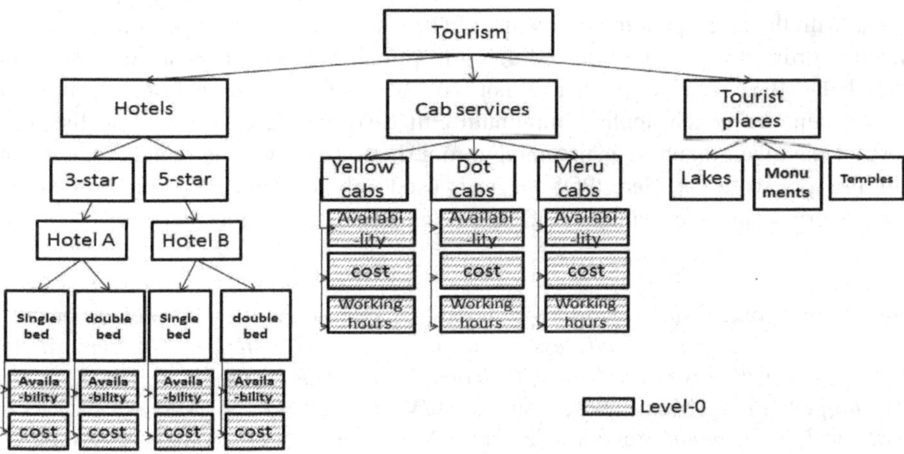

Fig. 3. Data organization of 'Tourism of Hyderabad' domain

In this hierarchical tree structure the leaf nodes are at level-0 and the level increases from a leaf node to the root node. The data at level-n (where n is number of levels) is segmented recursively until level-1. Then each segment at level-i (i=1... n) is given a level-(i-1) tag.

In physical memory all the layers above level-1 are stored as directories, level-1 as files and level-0 as tags in a file. The text in a file is stored in the form of segments and each segment is given a level-0 tag (address, open timings, entry fee etc.). The labels of all the files and directories along with the information in the files contribute to the data set.

3.2 Query Processing

The entire process from taking a user query to generating a system response is termed as 'Query Processing'. The different components of the 'Query Processing' module are described in the subsequent sections.

3.2.1 Knowledge Base

Knowledge base contains a domain dependent ontology like list of synonyms and code-mixed words. This helps in handling code-mixed and wrongly spelt words in the queries. This module is used by the Query Analyzer to replace the synonyms, code-mixed words etc., in a query with corresponding level-i (0…n) tags. If any language has knowledge resources like WordNet, dbpedia etc., they can be used to build the knowledge base. This has to be done manually.

3.2.2 Query Analyzer

The NL query given by the user is converted into wx query which is given as input to the Query Analyzer. The wx query is then tokenized and given as input to morphological analyzer and parts of speech (POS) tagger[11]. From the morphological analyzer's output, extract the root words of all the tokens in the query and replace these tokens with the corresponding root words. In this modified query the synonyms, code-mixed words etc., are replaced with corresponding level-i tags as discussed in Knowledge Base. Languages that do not have a morphological analyzer can build a simple stemmer which applies 'minimum edit distance algorithm' to find the root word of the given token by maintaining a root dictionary and then replaces the token with the root word [7]. Here, POS tagger is used only to identify the question words in the query. Languages with no POS tagger can have a list of question words.

Example:

User Query: golkoVMda addreVssu emiti, e tEMlo cUdavaccu?
Golconda address what what time can visit
(What is the address of Golconda, what time to visit?)

POS-tagger: golkoVMda addreVssu emiti/**WQ** e/**WQ** tEMlo cUda-vaccu

Root word replacement: golkoVMda addreVssu emiti/WQ e/WQ tEM cUdu

Query Analyzer's Output: golkoVMda cirunAmA emiti/WQ e/WQ samayaM cUdu

From the above output, we can see that English words like 'addreVssu' *(address)* and 'tEM' *(time)* are mapped to corresponding Telugu words i.e. 'cirunAmA' *(address)* and 'samayaM' *(time)* respectively by using knowledge base. If there is no corresponding Telugu word the English word remains as it is.

3.2.3 Advanced Filter

From the above modified query, question words and level-i words are extracted. From these words this module extracts only the words which play a role in the answer retrieval by applying heuristics like level-i words nearest to the question words, level-0 words etc. If these heuristics are not satisfied, all the level-i words are considered for further processing. These final sets of words are the keywords.

A user query can contain more than one question. In such cases, keywords belonging to a particular question are grouped together. All the groups of a query are collectively called a 'query set'.

Example:
U: nenu golkoVMda xaggara unnAnu, cArminAr PIju eVMwa iMkA cArminAr makkAmasjix eVkkada unnAyi ?
(I am near Golconda, what is the fee for Charminar and what is the address of Charminar and Mecca Masjid?)
Query analyzer output: nenu golkoVMda xaggara uMdu cArminAr PIju eVMwa/WQ iMkA cArminAr makkAmasjix cirunAmA uMdu
Extracted words: golkoVMda, cArminAr, PIju, eVMwa, cArminAr, makkAmasjix, cirunAmA
Keywords: cArminAr, PIju, eVMwa, cArminAr, makkAmasjix, cirunAmA
Query set: [cArminAr, PIju, eVMwa], [cArminAr, makkAmasjix, cirunAmA]

In 'U', '*I am near Golconda*' is unnecessary information. Therefore even if 'Golconda' is a word belonging to level-i, it will not be considered as a keyword.

3.2.4 Context Handler

Handling context is very important in any conversation to capture user's intention. This is a major challenge in present day dialogue systems. The query set given by Advanced Filter is used to update the context. If we identify any level-i (i=1, 2...n) word in the query set then there is a shift in the context. If there are no level-n(n<i) words in the query set we borrow level-n words relevant to level-i from the previous query set and add them to form the new query set. If the query set contains level-0 words and no words from level-i (i=1, 2...n), then we borrow level-i words from the previous query set.

Example:
U1: cArminAr eVkkada uMxi?
(Where is Charminar?)
S1: cArminAr yoVkka cirunAmA - cArminAr, heVxarAbAx, weVlaMgANa.
(The address of Charminar is -Charminar, Hyderabad, Telangana.)

U2: eVppudu opeVn uMtuMxi?
(When is it open?)
S2: cArminar yoVkka samayaM anni rojulu 9:00 am - 5:30 pm
(Charminar open timings -All days 9:00 am - 5:30 pm)

U3: eMtrI PIju eMwa?
(What is the entry fee?)
S3: cArminAr yoVkka PIju - iMdiyans - 20, vixeSIyulu - 150
(The fee for Charminar - Indians - Rs.20, Foreigners -Rs.150)

U4: golkoVMda eVkkada uMxi?
(Where is Golconda?)
S4: golkoVMda yoVkka cirunAmA - ibrahIM bAg, heVxarAbAx, weVlaMgANa 500008.
(The address of Golconda is Ibrahim Bagh, Hyderabad, Telangana 500008.)

In this example, to answer U2 and U3 we need contextual information *(*cArminAr [*Charminar*] *)* from U1. Context doesn't change for U2 and U3 as there is no level-i(i=1,2..n) word in them. We can observe the context switch from U3 to U4 i.e. switch from 'cArminAr' *(Charminar)* to 'golkoVMda' *(Golconda)* due to the occurrence of Golconda (a level-1 word in 'Tourism' domain) in U4.

3.2.5 Dialogue Manager

In any dialogue system, Dialogue Manager (DM) is the major component. It coordinates the activity of several subcomponents in a dialogue system and controls the flow of dialogue by giving relevant responses to user queries. The dialogue manager takes the query set and the context information as input. If the user query is ambiguous or no keywords are identified then the dialogue manager poses the user an interactive question from the set of canned questions. Otherwise it retrieves a relevant answer from the database.

Example:
U1: nenu golkoVMda cUdAli
(I have to visit Golconda)
S1: mIku memu e vidamugA sAyapadagalamu?
(How can we help you?)

U2: axi eVkkada uMxi?
(Where is it?)
S2: golkoVMda yoVkka cirunAmA - ibrahIM bAg, heVxarAbAx, weVlaMgANa 500008.
(The address of Golconda is Ibrahim Bagh, Hyderabad, Telangana 500008.)

In U1, the information provided is insufficient. Therefore the dialogue manager posed an interactive question to the user. Based on the query set and contextual information of U2, the Dialogue manager retrieves a relevant answer from the database.

4 Detailed Execution

User query: hExarAbaxlo makkAmasjix eVppudu opeVn uMtuMxi?
(*When is Mecca Masjid open in Hyderabad?*)

POS-tagger: hExarAbaxlo makkAmasjix eVppudu/WQ opeVn uMtuMxi

Replace with root word: hExarAbax makkAmasjix eVppudu/WQ opeVn uMdu

Replace with synonym: hExarAbax makkAmasjix eVppudu/WQ samayaM uMdu

Keywords: makkAmasjix, eVppudu, samayaM

Query set: [makkAmasjix, eVppudu, samayaM]

Answer: makkAmasjix yoVkka samayaM – annirojulu 4:00 am - 9:30 pm
(*Mecca Masjid open timings - All days 4:00 am -9:30 pm*).

5 Evaluation

Automatic evaluation is not available for dialogue systems. Human evaluation is only possible. There were 5 evaluators who used and evaluated the system based on the metrics given in Table 1. The mother tongue of all the evaluators is Telugu. Table 1 shows the criteria for the evaluation and the average of the rating given by human evaluators on the scale of 1-5 where 1 means poor and 5 means excellent.

Table 1. Human evaluation of our system

Metric		average rating
Speed	How fast are the responses?	4
Timeout	Does the system hang?	5
Recognition	Does the system understand your intention?	3.5
Reliability	Did you find all the information you were looking for?	4
Relevance	Are the responses appropriate?	4
Usability	Is the system easy to use?	4
Complexity	Does the system handle complex sentences?	3
Performance	Overall performance of the system	3.5

6 Error Analysis

Many efforts are being done to make the conversations seem natural i.e. closer to human conversation. For this, we need to handle many discourse issues such as Anaphora, ellipsis etc., and other issues like grammatical and spelling errors. Our system can handle these issues to some extent but as the complexity of the sentence increases, the performance of the system degrades. Some such issues are discussed in this section.

Anaphora Resolution: [8]
U1: cArminAr eVkkada uMxi ?
(Where is Charminar?)
S1: cArminAr - cirunAmA - cArminAr, hExarAbAx, weVlaMgANA
(The address of Charminar is - Charminar, Hyderabad, Telangana.)

U2: xAni PIju eMwa, golkoVMda eVkkada uMxi
(what is its fee and where is Golconda?)
S2: golkoVMda yoVkka cirunAmA - ibrahIM bAg, heVxarAbAx, weVlaMgANa 500008.
(The address of Golconda is Ibrahim Bagh, Hyderabad, Telangana 500008.)

golkoVMda yoVkka PIju - iMdiyans - 5, vixeSIyulu - 100
(The fee for Golconda - Indians - Rs.5, Foreigners -Rs.100)

Here, in U2, 'xAni' *(its)* may refer to 'cArminAr' *(Charminar)* or 'golkoVMda' *(Golconda).* Though the pronoun 'xAni' should map to 'cArminAr', it is mapped to 'golkoVMda' as golkoVMda' is nearer to the pronoun ('its').

U1: cArminAr eVkkada uMxi ?
(Where is Charminar?)
S1: cArminAr - cirunAmA - cArminAr, hExarAbAx, weVlaMgANA
(The address of Charminar is - Charminar, Hyderabad, Telangana.)
U2: xAni PIju eMwa ?
(What is the entry fee?)
S3: cArminAr yoVkka PIju - iMdiyans - 20, vixeSIyulu - 150
(The fee for Charminar - Indians - Rs.20, Foreigners -Rs.150)

Here, in U2, as there is no ambiguity in co-reference, 'xAni' *(its)* can only refer to 'cArminAr' *(Charminar)* in U1. The system successfully analyzed and gave the correct response.

Ellipsis: [10]
U1: cArminAr cirunAmA eMti, mari PIju ?
(What is the address of Charminar, (what about its) fees?)

S1: cArminAr - cirunAmA - cArminAr, hExarAbAx, weVlaMgANA
(The address of Charminar is - Charminar, Hyderabad, Telangana.)
cArminAr yoVkka PIju - iMdiyans - 20, vixeSIyulu - 150
(The fee for Charminar - Indians - Rs.20, Foreigners -Rs.150)

In this example, '*what about its*' in U1 is ellipsis. The system properly interprets the above query. In general if the sentence becomes complex the system may not be able to handle the ellipses like in the below example.

U1: cArminAr cirunAmA eMti, mari PIju, golkoVMda eVppudu opeVn uMtaxi, mari cArminAr
(What is the address of Charminar, (what about its) fees, When is Golconda open, (what about) Charminar)
S1: golkoVMda - samayaM - somavAraM wappa anni rojulu 5:00pm – 9:00 pm, AxivAraM 7:00pm - 8:00pm kAMwi praxarSana
(except on Monday Golconda is open on all other days from 5:00pm to 9:00pm and on Sundays there is alazer light show from 7:00pm to 8:00pm)

golkoVMda - PIju - iMdiyans - 5, vixeSIyulu – 100
(The fee for Golconda - Indians - Rs.5, Foreigners -Rs.100)

golkoVMda yoVkka cirunAmA - ibrahIM bAg, heVxarAbAx, weVlaMgANa 500008.
(The address of Golconda is Ibrahim Bagh, Hyderabad, Telangana 500008.)

cArminAr - cirunAmA - cArminAr, hExarAbAx, weVlaMgANA
(The address of Charminar is - Charminar, Hyderabad, Telangana.)

The output of the system has to be address, entry fee and open timings of Charminar along with the open timings of Golconda, but we can see that the system wrongly interpreted and gave output for open timings, entry fee and address of Golconda along with address of Charminar.

Sandhi: [9]
Sandhi is a common phenomenon in agglutinative languages. For example, consider the below dialogue.

U: cArminAreVkkaduMxi ?
(Where is Charminar?)
S: mIku memu e vidamugA sAyapadagalamu?
(How can I help you?)

In U1, we can see that a sentence is expressed as single word in Telugu language which cannot be analyzed by NLP applications. To handle these cases, there is a need for sandhi splitter which splits 'cArminAreVkkaduMxi ' to 'cArminAr ' *(Charminar)*, 'eVkkada' *(where)* and 'uMxi' *(present)*.

7 Conclusion

We have shown a new and quick approach to build a dialogue system. It can be readily adapted to other languages. In general, only language specific parts like Database and Knowledge base have to be replaced for this purpose. Our model is portable to any domain. It requires a stemmer and a set of question words which can be easily developed. This brings us one step closer to build dialogue systems for resource poor languages. Our system also maintains conversation by posing questions to the user.

In future, we intend to build a multi-lingual and multi-domain dialogue system by improving our current model which should be able to handle pragmatics and discourse. We also intend to handle sandhi, ellipses and anaphora resolution to make the conversations seem more natural. This system can also be integrated with speech input and output modules.

Acknowledgements. This work is supported by Information Technology Research Academy (ITRA), Government of India under, ITRA-Mobile grant ITRA/15(62)/Mobile/VAMD/01.

References

1. Akula, A.R., Sangal, R., Mamidi, R.: A novel approach towards incorporating context processing capabilities in nlidb system
2. Gupta, A., Akula, A., Malladi, D., Kukkadapu, P., Ainavolu, V., Sangal, R.: A novel approach towards building a portable nlidb system using the computational paninian grammar framework. In: 2012 International Conference on Asian Language Processing (IALP), pp. 93–96. IEEE (2012)
3. Hori, C., Ohtake, K., Misu, T., Kashioka, H., Nakamura, S.: Weighted finite state transducer based statistical dialog management. In: IEEE Workshop on Automatic Speech Recognition Understanding, ASRU 2009, pp. 490–495 (2009)
4. Nguyen, A., Wobcke, W.: An agent-based approach to dialogue management in personal assistants. In: Proceedings of the 10th International Conference on Intelligent User Interfaces, IUI 2005, pp. 137–144. ACM, New York (2005)
5. Nguyen, A., Wobcke, W.: Extensibility and reuse in an agent-based dialogue model. In: 2006 IEEE/WIC/ACM International Conference on Web Intelligence and Intelligent Agent Technology Workshops, WI-IAT 2006 Workshops, pp. 367–371. IEEE (2006)
6. Reddy, R.R.N., Bandyopadhyay, S.: Dialogue based question answering system in telugu. In: Proceedings of the Workshop on Multilingual Question Answering, MLQA 2006, pp. 53–60. Association for Computational Linguistics, Stroudsburg (2006)
7. Srirampur, S., Chandibhamar, R., Mamidi, R.: Statistical morph analyzer (sma++) for indian languages. In: COLING 2014, p. 103 (2014)
8. Mitkov, R.: Anaphora resolution: The state of the art. Technical report. University of Wolverhampton, Wolverhampton (1999)
9. Sandhi splitter and analyzer for Sanskrit(with special reference to aC sandhi), by Sachin kumar, thesis submitted to JNU special centre for Sanskrit (2007), `http://sanskrit.jnu.ac.in/rstudents/mphil/sachin.pdf`
10. Dalrymple, M., Shieber, S.M., Pereira, F.C.N.: Ellipsis and higher-order unification. Technical report, Computation and Language E-Print Archive (1991)
11. Brants, T.: TnT–A statistical part-of-speech tagger. In: Proceedings of the Sixth Applied Natural Language Processing Conference (ANLP 2000), pp. 224–231 (2000)

Anti-summaries: Enhancing Graph-Based Techniques for Summary Extraction with Sentiment Polarity

Fahmida Hamid and Paul Tarau

Department of Computer Science & Engineering,
University of North Texas,
Denton, TX-76201, USA
fahmidahamid@my.unt.edu, tarau@cs.unt.edu

Abstract. We propose an *unsupervised* model to extract two types of summaries *(positive, and negative)* per document based on sentiment polarity. Our model builds a *weighted polar digraph* from the text, then evolves recursively until some desired properties converge. It can be seen as an enhanced variant of *TextRank* type algorithms working with non-polar text graphs. Each positive, negative, and objective opinion has some impact on the other if they are semantically related or placed close in the document.

Our experiments cover several interesting scenarios. In case of a one author news article, we notice a significant overlap between the *anti-summary* (focusing on negatively polarized sentences) and the the summary. For a transcript of a debate or a talk-show, an anti-summary represents the disagreement of the participants on stated topic(s) whereas the summary becomes the collection of positive feedbacks. In this case, the anti-summary tends to be *disjoint* from the regular summary. Overall, our experiments show that our model can be used with TextRank to enhance the quality of the extractive summarization process.

Keywords: graph based text processing, sentiment polarity, extractive summarization, TextRank algorithm, positive-negative bias-based ranking algorithm.

1 Introduction

A document expresses a writer's opinions along with facts. Usually an article covering several issues will qualify some with positive feedback and some with negative. A high quality summary should reflect the most "important" ones amongst them. *Summarization* is thus closely related to *sentiment analysis*. There has been limited work done on the intersection of text summarization and sentiment analysis. Balahur et al. [1] showed a technique of sentiment based summarization on multiple documents. They used a supervised sentiment classifier to classify the blog sentences into three categories (positive, negative, and objective). The positive and the negative sentences are, then fed to the summarizer separately to produce one summary for the positive posts and another one for the negative posts. The success of their model mostly depends on the performance of the sentiment classifier. Besides, their summarizer does not consider the impact of positive (negative) sentiments while producing the summary of negative (positive) sentiments. It sounds unintuitive to totally separate the sentiment-flows before producing the

A. Gelbukh (Ed.): CICLing 2015, Part II, LNCS 9042, pp. 375–389, 2015.
DOI: 10.1007/978-3-319-18117-2_28

summaries. Manning et al. [2], in their *sentiment summary* paper used *Rotten Tomatoes* dataset (for training and testing) to extract the most important *paragraph* from the reviewer's article. They aimed at *capturing the key aspects of author's opinion* about the subject (movie). They worked with a *supervised* technique and articles with *single topic.*

In this work, we propose an unsupervised, mutually recursive model that can represent text as a graph labeled with polarity annotations. Our model builds a graph by collecting words, and their lexical relationships from the document. It handles two properties (*bias* and *rank*) for each of the important words. We consider *sentiment polarity* of words to define the bias. The *lexical definition* and *semantic interactions* of one word to others help defining edges of the text-graph. Thus we build a *weighted directed graph* and apply our model to get the top (positively and negatively ranked) words. Each word in our graph starts with the same *rank*, which eventually converges to some distinct values with the effect of *bias* of its neighbors and *weighted in-links*. Those words then specify the weight of each sentence and grant us a direction to choose important ones. The *bias* of a node gets updated from the *rank* of it's neighbors. The mutual dependency of the graph elements represents the impact of the author's sentiment. Our concept is analogous to TextRank algorithm, except -

- Our model works for a polar graph whereas TextRank works with non-polar one.
- The rank of a node in TextRank gets updated by the connectivity (weighted/ unweighted), whereas the rank in our model gets updated based on the weighted links and bias of its neighbors.

To the best of our knowledge, our concept of *anti-summary* is new. Hence it was hard to compare the results with a gold standard. We have chosen DUC2004 dataset and basic TextRank algorithm for comparative study. Through our experiments, we have found the following interesting facts -

- When the anti-summary and summary are mostly *disjoint*, the document is a collection of different sentiments on stated topics. It can be a transcript from some debate, political talk, controversial news, etc.
- When the anti-summary *overlaps* at a noticeable amount with the summary, the document is a news article stated from a neutral point of view.
- By blending anti-summary with TextRank generated one, we show another way of producing *opinion-oriented* summary which not only contains the flow of negative sentiment but also includes facts (or some positive sentiment).

2 Related Work

Automated text summarization dates back to the end of fifties [8]. A summarizer deals with the several challenges. To extract important information from a huge quantity of data while maintaining quality are two of them. A summarizer should be able to understand, interpret, abstract, and generate a new document. Majority of the works focus on *"summarization by text-span extraction"* which transforms the summarization task to a simpler one: ranking sentences from the original document according to their salience or their likelihood of being part of a summary [5].

Early research on extractive summarization was based on simple heuristic features of the sentences such as their position in the text, frequency of words they contain etc. More advanced techniques also consider the relation between sentences or the discourse structure by using synonyms of the words or anaphora resolution. To improve generic machine generated summaries, some researchers [5] converted some hand-written summaries (collected from the Reuters and the LosAngeles Times) to their corresponding extracted ones. Based on their experiments, they stated that *summary length is independent of document length*. Though Hovy and Lin [6] stated earlier, "A summary is a text that is produced out of one or more texts, that contains the same information of the original text, and that is *no longer than half* of the original text." For our experiments, we will generate summaries with at-most top ten sentences per document.

Graph based ranking algorithms have recently gained popularity in various natural language processing applications; specially in generating *extractive summaries*, selecting keywords, forming word clusters for sense disambiguation, and so on. They are essentially a way of deciding the importance of a vertex within a graph, based on global information recursively drawn from the entire graph [10]. The basic idea is of "voting" or "recommendation". When one vertex links to another one, it is basically casting a vote for the other vertex. The importance of the vertex casting the vote determines how important the vote itself is. Hence the score (usually called *"rank"*) associated with a vertex is determined by the votes cast for it, and the score of the vertices casting these votes. TextRank works well because it does not rely on the local context of a text unit and requires no training corpus, which makes it easily adaptable to other languages or domains. Erkan and Radev [3] in *LexRank* (another graph based ranking algorithm to produce multi-document summary) used *the centrality of each sentence in a cluster* to assess the importance of each sentence. To measure the similarity between two sentences, they used cosine similarity matrix (based on word overlap and *idf (inverse document frequency)* weighting). Being inspired by the success of textrank models, we had the idea to apply a *polar textrank* model in order to extract sentences from *negative (positive)* point of view.

It is important that we consider each sentiment of the author while producing the summary. In our work, we adopt a graph based ranking model which originally was proposed for *trust-based (social, peer-to-peer) networks* [13]. It intends to measure the prestige (rank) of nodes (participants in the event) present in the graph. Their hypothesis, *"the node which is prone to trust (mistrust) all its neighbors is less reliable than the node who provides unpredictable judgments,"* works also for producing summaries. Each node (word) has weighted (positive/negative/neutral) directed links to its neighbor nodes (other words, possibly collected from the same sentence or nearby sentences). The more weight it provides to its neighbors the more importance (either positive or negative) it indicates. The impact is higher when a node behaves differently (a positive biased node has a negative weighted outline or vice versa) towards its neighbors.

3 Anti-Summary

We propose an extractive summarization technique which produces anti-summaries as well as summaries for each document. We would discuss what anti-summary is and why

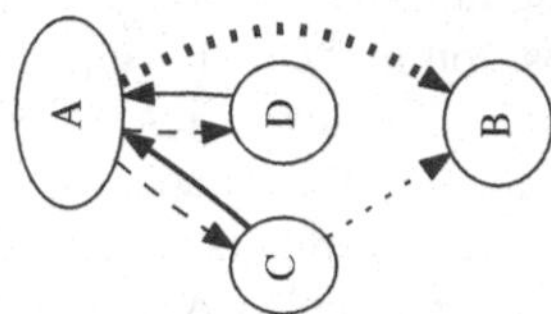

Fig. 1. A text-graph describing several topics

it is important. Sentences with upstream knowledge are the candidates of anti-summary. A sentence does not have to contain words like {*no, neither, never, not, ever, bother, yet,* ...}, to be the part of the anti-summary.

We can start with a generic example: A document is about topic A. It is comparing the qualities of A with related topics B, C, and D. Suppose, topic B is mostly receiving negative opinions in that document. Then a summary of the document should include positive feedbacks on A and the anti-summary should be more about the properties of B.

Anti-summaries are as important as summaries. They help us find relative materials on a specific topic. For example, from a news article, without any supervised topic detection, anti-summaries can indicate which parts of it represent negative/ suppressed opinion. In a scientific article, anti-summaries tell how system A is different/ better/ worse than system B where as summaries might only tell us the usefulness of system A.

Interestingly enough, some summary sentences are also present in anti-summary of the document. This means, anti-summaries are not exactly opposite to summaries, it is the reverse stream of the main news. Anti-summaries can help a search engine build comparative database. It is intuitive that two documents are related if there is a significant match between one's summaries and the other's anti-summaries.

4 Sentiment Analysis: Covering Minimal Issues

Sentiment Analysis has important aspect on fields which are affected by people's opinions, e.g., politics, economics, social science, management science and so on. It plays a vital role in every aspect of NLP; for example, co-reference resolution, negation handling, word sense disambiguation etc. *Sentiment words* are instrumental to sentiment analysis [7]. Words like *good, wonderful, amazing* convey positive connotation whereas *bad, poor, terrible* are used to flow negative sense. As an exception, some adjectives and adverbs (e.g. *super, highly*) are not oriented clearly to either one of the poles (positive, negative). A list of sentiment words are called *sentiment lexicon.* A sentiment lexicon is necessary but not sufficient for sentiment analysis.

A positive or negative sentiment word may have opposite orientations in different application domains, e.g., *"The vacuum cleaner sucks!"* vs. *"The camera sucks."* Sentiment words may be used objectively rather subjectively in some sentences, e.g., *"If I can find a good camera, I can buy that."* Sarcastic sentences are trickier to handle, as well as sentences having no sentiment word but with a sentiment expressed.

Based on the level of granularities (document level, sentence level, entity and aspect level) we choose the *entity level* analysis of sentiments. For example, the sentence, "The iPhone's call quality is good, but its battery life is short," evaluates two aspects: *call quality* and *battery life*. The two opinion targets for this sentence, *call quality* has positive sentiment and *battery life* has negative.

Our model is unsupervised, and we decided not to use statistical database to calculate the sentiment polarity for sentence/ paragraph/ document. Hence we have used only a *sentiment lexicon* to get the usual sentiment polarity at word level.

5 The Polarity Based TextRank Model

Jon Kleinberg's *HITS* or Google's *PageRank* are two most popular graph based ranking algorithms, successfully used for analyzing the link-structure of world wide web, citation graph, and social networks. A similar line of thinking is applied to semantic graphs from natural language documents, resulting in a graph based ranking model, TextRank [10]. The underlying hypothesis of TextRank is that in a cohesive text fragment, related text units tend to form a web of connections that approximates the model humans build about a given context in the process of discourse understanding. TextRank, with different forms (weighted, unweighted, directed, undirected) of graphs, was applied successfully for different applications, specifically for text summarization [11]. Based on the results so far, it performed well for summarizing general text documents. There are documents which present arguments, debates, competitive results and they are subjective reflections of the author(s). The limitation of TextRank (and other similar models) is that it does not handle negative recommendations different from positive ones. In this work, we present a different model [13] that can be adopted to have the impact of sentiments on the summary.

5.1 Trust Based Network

A network based on trust (e.g. facebook, youtube, twitter, blogs) is quite different from other networks; in each case, reputation of a peer as well as types of opinion (trust, mistrust, neutral) matters. An explicit link in a *trust-based* network indicates that two nodes are close (connected), but the link may show either *trust* or *mistrust*. A *neutral* opinion in a trust based network is different from *no-connection*. TextRank gives higher ranks with better connectivity. The situation changes dramatically in trust based networks as a highly disliked node may also be well connected. To take care of this situation, authors [13] correlated two attributes for each node: *Bias* and *Prestige*.

Bias & Prestige. The bias of a node is the propensity to trust/ mistrust other nodes. The prestige of a node is the ultimate rank (importance) of it in compared to other nodes. Formally, let $G = (V, E)$ be a graph, where an edge $e_{ij} \in E$ (directed from node i to node j) has weight $w_{ij} \in [-1, 1]$. $d^o(i)$ and $d^i(i)$ correspondingly denote the set of outgoing links from and incoming links to node i. Bias reflects the expected weight of an outgoing edge. Using bias, the inclination of a node towards trusting/ mistrusting is measured. The bias of node i can be determined by

$$bias(i) = \frac{1}{2|d^o(i)|} \sum_{j \in d^o(i)} (w_{ij} - rank(j)) \tag{1}$$

Prestige (rank) reflects the expected weight of an in-link from an un-biased node. Intuitively, when a highly biased node (either positive or negative) gives a rating, such score should be given less importance. When a node has a positive (negative) bias and has an edge with a negative (positive) weight, that opinion should weigh significantly. Hence, the prestige (rank) of node j could be determined as

$$rank(j) = \frac{1}{|d^i(j)|} \sum_{k \in d^i(j)} (w_{kj}(1 - X_{kj})) \tag{2}$$

where the auxiliary variable X_{kj} determines the change on weight w_{kj} based on the bias of node k.

$$X_{kj} = \begin{cases} 0 & \textit{if}\,(bias(k) \times w_{kj}) \leq 0, \\ |bias(k)| & \textit{otherwise}. \end{cases} \tag{3}$$

After each iteration of 1 and 2, edge weight w_{kj} is scaled from the old weight as follows:

$$w_{kj}^{new} = w_{kj}^{old}(1 - X_{kj}) \tag{4}$$

6 Text as Graph

In order to apply the graph based ranking algorithms, we convert the text document into a graph. We extract words (except stop-words) from each sentence and represent them as *nodes* of our graph. Each pair of related words (*lexically* or *semantically*) forms the *edges*. We use *SentiWordNet* (a publicly available lexical resource for opinion mining) to determine the sentiment polarity of each node (signature word). SentiWordNet [4] assigns to each synset of *WordNet* [12] three sentiment scores: positivity, negativity, and objectivity. We choose the highest (most common) sentiment polarity of a word as the *bias*. Edge weights are determined by the total outgoing edges from the node. If there is a {not, no, though, but,...} present between $word_a$ and $word_b$, the edge weight (w_{ab}) receives the opposite sign of $bias_a$. Our algorithm performs the following steps:-

- Phase I: Build the Text-Graph
  - Collect signature words; use them as *nodes* for the graph. Use their sentiment polarity as *bias*.
  - Add edges between nodes (words) that reside in the same sentence (within a chosen window size).
  - Assign edge-weights (w_{ab}) based on the total outgoing edges from each source node ($word_a$).
  - Update/ add edge-weights (w_{ab}) if they are semantically related (e.g., use a matching function on their definition/synonym list).

- Assign a random value as *rank* to all the nodes of the graph (initially all nodes are on the same level).
- Phase II: Apply Ranking Model
  - Apply formula [1,2,3,4] over the graph until the $rank$ value converges.
- Phase III: Find Ranked Word Vectors, & Extract Sentences
  - Create a positive word vector, W^{pos} of *keywords* by selecting all positively ranked words.
  - Create a negative word vector, W^{neg} of *keywords* by selecting all negatively ranked words.
  - Use W^{pos} and W^{neg} to determine the weight and orientation of the sentences.
  - Group top k (can be determined by the user) negatively (positively) oriented sentences as *anti-summary (summary)*.

The following subsections will discuss our process in detail. To demonstrate several intermediate outcomes of our process, we will use a sample article: `http://students.cse.unt.edu/~fh0054/SummaryAnti/Kennedy1961/kennedyPart1.txt` which is a small fragment (only 77 sentences are considered) of President Kennedy's speech in 1961.

6.1 Signature Words

Using a standard *parts of speech tagger*, we extract words that are labeled as either one from the set: {*noun, verb, adjective, adverb*}. These are our *signature words*. We also collect their *definition* and *sentiment polarity* for the next phase. Table 1, 2 show the intermediate data generated from example 01.

Example 01: The first and basic task confronting this nation this year was to turn recession into recovery.

6.2 Define Nodes & Edges: From a Single Sentence

Let, x and y are two words residing in the same sentence, and $|position_x - position_y| < window\ size$. We create distinct nodes (if not already exist) for x and y, and define their relations (edges) by either of the rules:

- If $parts\_of\_speech(x) = \{verb\}$, add $edge(x, y)$.
- If $parts\_of\_speech(x) \cup parts\_of\_speech(y)$ $\subset \{noun, adjective, adverb\}$, then add $edge(x, y)$ and $edge(y, x)$.
- Finally, we add edge-weight, $w_{xy} = sign(bias(x)) \times \frac{1}{|E|}$ to all the existing edges.

6.3 Connect Sentences through Words: Add More Edges/ Update Weights

Let x and y are two different words from two different sentences (or from the same sentence, $|position_x - position_y| \geq window\ size$). We use their *definition* (available in WordNet) to determine $similarity$ between them. If $def(x)$ stands for $definition$ of x,

$$similarity(x, y) = \frac{def(x) \cap def(y)}{def(x) \cup def(y)} \tag{5}$$

We add/ update edge-weight w_{xy} and w_{yx} using the following manners:

- We do not update the graph if the $similarity(x, y)$ is $zero$.
- For an *existing* edge between x and y, we adjust w_{xy} as $w_{xy} + similarity(x, y) \times sign(bias(x))$.
- For a *no edge* between x and y, we add two new edges ($edge(x, y)$ and $edge(y, x)$) where $similarity(x, y)$ acts as the weight for the new edges.

Table 1. Words & Their Entities

Word	PoS	Polarity	Definition
first	adj	0.0	preceding all others in time or space or degree
confront	v	−0.5	oppose, as in hostility or a competition
nation	n	0.0	a politically organized body of people under a single government
year	n	0.0	a period of time containing 365 (or 366) days
turn	v	0.0	change orientation or direction, also in the abstract sense
recession	n	0.0	the state of the economy declines; a widespread decline in the GDP and employment and trade lasting from six months to a year
recovery	n	0.0	return to an original state

Table 2. Degree of Similarity

Word	Definition	Similarity
recession	the state of the economy declines; a widespread decline in the GDP and employment and trade lasting from six months to a year	
recovery	return to an original state	0.035714
security	the state of being free from danger or injury	
progress	gradual improvement or growth or development	0.0
recovery	return to an original state	
security	the state of being free from danger or injury	0.071428

This phase helps relate semantically closer words in the document.

To demonstrate how the graph is formed, we randomly picked two sentences from the stated article: *'Our security and progress cannot be cheaply purchased; and their price must be found in what we all forego as well as what we all must pay'* and *'The first and basic task confronting this nation this year was to turn recession into recovery'*. The sentence graph with only these two sentences (with $window\ size = 4$) is shown in figure 2. We notice that word pairs {*(security, recession), (security, recovery), (progress, recovery)*}, for example, are connected to each other through the *similarity* relationship.

6.4 Keyword Extraction

Once the graph is built, we add a real value (can be chosen randomly) to every node as it's $rank$. This way, there is no discrimination beforehand. Then we apply set of equations [1,2,3,4] several times (until the rank value converges) over the graph. For real time output, one can control the repetition using a threshold. Table 3 shows a handful of positively ranked and negatively ranked keywords (out of 568 total words) from the same article.

6.5 Sentence Extraction

Our top (positive, and negative) ranked keywords define the weights of the sentences. Let W^{pos} (W^{neg}) be the list of words achieving positive (negative) rank values. Let W^{pos} is a list of size m and W^{neg} is a list of size n.Weight of a sentence, s_j is:

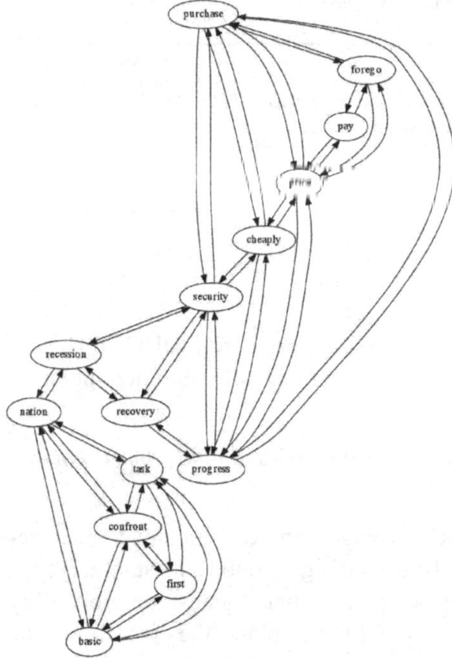

Fig. 2. Sentence Graph

Table 3. A subset of Keywords

keyword	Rank	
initiate	1.50503400134e-07	
wisely	2.37049201066e-19	
cheaply	2.03124423939e-19	
thailand	1.[illegible]e-28	
believe	2.53398571394e-38	positively
crucial	1.14205226912e-40	ranked
forego	7.26752110553e-46	
mind	3.3034186848e-46	
handicap	1.89292202562e-57	
progress	1.18505270306e-62	
cambodia	-0.0210985300569	
recovery	-0.126687287356	
recession	-0.0376108282812	
unwilling	-7.70602663247e-05	
congress	-0.064049452945	negatively
cooperate	-0.285579113282	ranked
building	-0.0387114701238	
havana	-0.0128506602917	
frontier	-0.0182778316133	
compete	-0.075853425471	

$$weight(s_j^{neg}) = (\sum_{v_i \in W^{neg} \wedge v_i \in s_j} |rank(v_i)|)/(n \times |s_j|), \tag{6a}$$

$$weight(s_j^{pos}) = (\sum_{v_i \in W^{pos} \wedge v_i \in s_j} rank(v_i))/(m \times |s_j|), \tag{6b}$$

Now each sentence has two weights associated with it; $weight(s_j^{pos})$ corresponds its weight on positively ranked words whereas $weight(s_j^{neg})$ corresponds its weight on negatively ranked words. Thus S^{neg} (S^{pos}) forms a weight vector of sentences on negatively (positively) ranked ones. One can select top k many sentences based on S^{neg} as the *anti-summary*. The similar line of thinking goes for generating regular summaries. To avoid promoting longer sentences, we are using length of the sentence as the *normalization factor*. Table 4 shows two top sentences(the first one is the 2nd top positively ranked and the second one is 5th top negatively ranked). The original file can be found at: `http://students.cse.unt.edu/~fh0054/SummaryAnti/Kennedy1961/kennedyPart1SA.txt`.

Our model uses a mutually recursive relation on *bias* and *rank* calculation. It incrementally updates the *edge-weights* as well. Hence, it helps get the final ranks (polarity and weights) of words on global context. Since a textrank model does not rely on the local context of a text unit, and requires no training corpus, it is easily adaptable to other languages or domains.

Table 4. Sample of top sentences

sentence	weight
Our security and progress cannot be cheaply purchased; and their price must be found in what we all forego as well as what we all must pay.	$1.98797587956e-14$
The first and basic task confronting this nation this year was to turn recession into recovery.	-0.00117486599011

7 Evaluation

We used TextRank (extracted) and Human (abstract) summaries from DUC 2004 (task 1) as the baseline. TextRank is unsupervised and it does not handle sentiment polarity. Hence, we started with hypothesis 1.

Hypothesis 1. *polarity based summaries and anti-summaries should almost equally intersect with TextRank generated ones.*

In order to verify the hypothesis, we calculated average number of sentence intersection between each pair of the three summaries (our model generated anti-summary(N), summary(P) and textrank summary(T)). Then we plotted them against the probability of intersection of two random generated summary. Table 5 explains the operations. The test cases are named as -

- case a: an average size of $(P \cap T)$,
- case b: an average size of $(N \cap T)$,
- case c: an average size of $(P \cap N)$,
- case d: an average size of $(S_1 \cap S_2)$, with any two randomly selected set S_1 and S_2 of the same size as P, N and T.

(Summaries of these set of articles are stored in link: `http://students.cse.unt.edu/~fh0054/cicling2015/`).

Quite interestingly, for shorter articles, *case a* and *case b* showed similar (and better) performance than *case c* and *case d* [table 5, & figure 3]. It also supports hypothesis 1. For larger articles, *case b* was the winner. The following section gives the mathematical background for case d.

Table 5. summary & their average size of intersection

Test Set	Total Files	no of sentences per file(n)	$avg(n)$	summary size(k sentences)	$avg(k)$	case a	case b	case c
1	163	$n > 30$	48	10	10.00	1.75	3.18	1.43
2	337	$n \leq 30$	16.5	$n/3$	5.10	1.63	1.88	1.08
				$n/2$	8.05	4.21	4.73	3.386
3	410	$n \leq 40$	20.02	$n/3$	6.34	2.626	2.304	1.44
				$n/2$	9.775	5.826	5.613	4.256

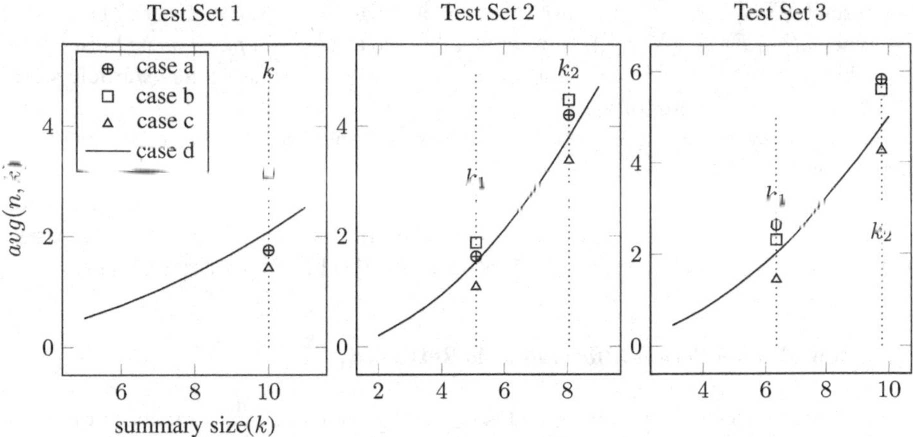

Fig. 3. Average of sentence intersection based on equation 7

7.1 Baseline: *Intersection* of Two Models vs. the *Random* Possibility

The average size of an intersection(avg) of subsets with k elements taken from a set with n elements can be determined by eq. 7.

$$avg(n,k) = \frac{\sum_{i=0}^{k} i\binom{k}{i}\binom{n-k}{k-i}}{\sum_{i=0}^{k}\binom{k}{i}\binom{n-k}{k-i}} \tag{7}$$

For two summaries of different sizes k and l this generalizes to:

$$avg(n,k,l) = \frac{\sum_{i=0}^{k} i\binom{k}{i}\binom{n-k}{l-i}}{\sum_{i=0}^{k}\binom{k}{i}\binom{n-k}{l-i}} \tag{8}$$

These formulas are justified as follows: Fix one of the subsets as $X = 0, 1, \ldots, k-1$. Then an intersection of size i is computed by taking a subset of X of size i (there are $(b = \binom{k}{i})$ such sets). We have $j = l - i$ elements in X that will be selected among the b' subsets of size j of the remaining $n-k$ elements in the complement of X counted as $b' = \binom{n-k}{l-i}$. So the numerator of the fraction, will be obtained by summing up for $i = 0$ to $k-1$ the product of i with the number of subsets b and and the number of subsets b', counting the total length of the subsets. The denominator of the fraction will count the total number of these subsets and the result of their division will give the average size of the intersections.

Knowing the average sizes of the overlap of two summaries of size k or sizes k and l when they are different (seen as sets of words), tells us whether our model-generated summaries, and anti-summaries have a better rate of intersecting with each other (and textrank) than *random summaries* would.

7.2 Does $(P \cap N)$ Indicate Something Interesting?

In each case, $(P \cap N)$ is minimal (fig. 3) which indicates that our model is successfully splitting the two flow of sentiments from documents. This raises a set of questions, e.g.,

- when $(N \cap T) \gg (P \cap T)$, should we label the article as a *negatively* biased one?
- when $(P \cap T) \gg (N \cap T)$, should we label the article as a *positively* biased one?
- when $(P \cap N) \gg (P \cap T)$ and $(P \cap N) \gg (N \cap T)$, is it a news/article stated from a *neutral* point of view?

We tried to answer these questions based on experimental results. We might need voluntary human judges to label the articles based on the the extractive summaries and compare our *summary based guesses*. We leave this phase as a future direction. Interested reader can get our result from the following link: `http://students.cse.unt.edu/~fh0054/cicling2015/summaryHalf/fileType.txt`

7.3 How Much *Relevant Information* Is Retrieved?

We needed to know whether our model is gathering some relevant sentences or not. We use abstractive summaries (provided with DUC2004 dataset) and TextRank extracted ones as base results; then use the *recall* measure to estimate the ratio of number of relevant information retrieved.

Recall & Precision. Recall (R) is the ratio of *number of relevant information received to the total number of relevant information in the system.* If D is the original document, A is the anti-summary, and S is the standard summary, then the recall(R) value can be calculated from eq. 9.

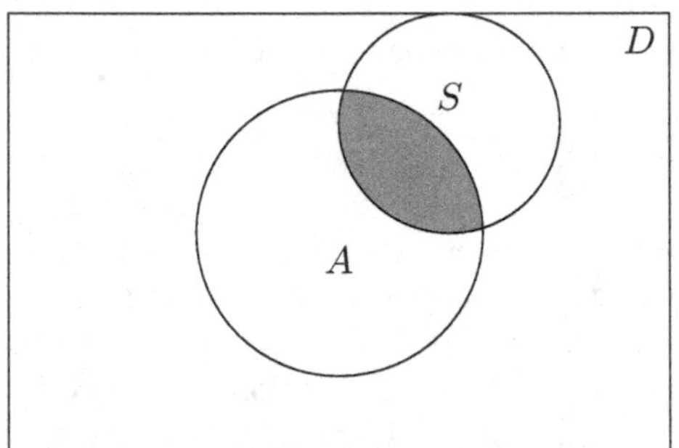

$$R = \frac{A \cap S}{S} \tag{9}$$

$$P = \frac{A \cap S}{A} \tag{10}$$

Another well-known measure is *Precision (P)* which is the ratio of number of relevant records retrieved to the total number (relevant and irrelevant) of records retrieved (eq. 10). As our test dataset had different file sizes, we can tune the anti-summary length as we want, and we cannot firmly state that $A \setminus (A \cap S)$ is irrelevant; we believe, interpretation of R is more relevant than P, in our case.

Table 6 shows the average recall value on the our model generated summary(P), anti-summary(N), and textrank(T) summary with respect to human (H) summary. We have used *uni-gram word matching* for computing recall rate. This result gives us an idea that -

Hypothesis 2. *Anti-summary can help extending TextRank summary in order to produce sentiment oriented summary.*

Table 6. Avg. Recall of P, N, T w.r.t. Human Summary

Test Set	Total Files	no of sentences per file(n)	summary size(k sentences)	recall(P) $= \frac{P \cap H}{H}$	recall(N) $= \frac{N \cap H}{H}$	recall(T) $= \frac{T \cap H}{H}$
1	163	$n > 30$	10	.469	.397	.417
2	337	$n \leq 30$	$n/3$	.408	.458	.457
			$n/2$	.545	.596	.583
3	410	$n \leq 40$	$n/3$	.449	.436	.531
			$n/2$	.588	.582	.607

7.4 Evaluation through Examples: $(P \cap N)$ Is Minimal

The next block is a summary and an anti-summary produced by our system, for the data file: `http://students.cse.unt.edu/~fh0054/SummaryAnti/googleCase.txt`. This example shows a clear distinction between the summary and the anti-summary sentences. The *summary* sentences represent the view of *European Union and other Companie*s questioning Googles privacy policy. On the other hand, the *anti-summary* sentences are about *Googles* steps and clarifications in the issue. So, anti-summary is a better approach to generate upstream information from a document, without knowing the topic(s) in the document.

summary:
"While there are two or three open minds on the company's advisory group that oversees the exercise, the process appears to be fundamentally skewed against privacy and in favor of publication rights". Its advisers include Wikipedia founder Jimmy Wales who has described the right as "deeply immoral," according to a report in the Daily Telegraph, as well as a former Spanish privacy regulator and an ex-justice minister of Germany. "It doesn't help to throw around big, loaded words like that when you're trying to find a convergence of views". "I hope Google take the opportunity to use these meetings to explain its procedures and be more transparent and receptive about how it could meet the requirements of this judgment," said Chris Pounder, director of Amberhawk, a company that trains data-protection officials. Anyone interested in attending can sign up online about two weeks before the events, Google said.
anti_summary:
Google chairman Eric Schmidt and Drummond are among the advisers who will draft a report on the ruling to discuss the implications of the court case for Internet users and news publishers and make recommendations for how the company should deal with requests to delete criminal convictions. Privacy regulators have criticized Mountain View, California-based Google's steps to tell Web publishers when it is removing links to their sites. Regulators are drafting guidelines on how they should handle any disputes by people who were unhappy at how Google handles their initial request for links to be removed. The first event takes place in Spain, the trigger for the EU court ruling that changed Google's business when the company fought an order by the country's data-protection regulator to remove a link with details to a state auction of houses to cover tax debt that popped up on a search for Mario Costeja Gonzalez. Al Verney, a spokesman for Google in Brussels, said the company will hear from invited experts and also from people in the audience at the events, who could sign up on the Internet to attend.

7.5 Evaluation through Examples: $(P \cap N)$ Is Maximal

We would like to show another sample summary and anti-summary which are generated from a news over the aid provided to the flood-damaged area in Honduras and Nikaragua. The news can be found at: `http://students.cse.unt.edu/`

~fh0054/SummaryAnti/testData02/APW19981106.0869. It shows two important features:

- One, out of pulled top three sentences is common between summary and the anti-summary.
- The summary sentences are mostly about *aid*, whereas anti-summary sentences are about *damage* and *delaying on delivering the foods to the sufferers.*
- Hence it is a good example of *non-polar* articles.

summary:
In the Aguan River Valley in northern Honduras, floodwaters have receded, leaving a carpet of mud over hundreds of acres (hectares). A score of cargo aircraft landed Thursday at the normally quiet Toncontin airport in the Honduran capital of Tegucigalpa, delivering aid from Mexico, the United States, Japan and Argentina. First lady Hillary Rodham Clinton added Nicaragua and Honduras to a trip she plans to the region beginning Nov. 16.
anti_summary:
Foreign aid and pledges of assistance poured into Central America, but damage to roads and bridges reduced the amount of supplies reaching hundreds of isolated communities to a trickle: only as much as could be dropped from a helicopter, when the aircraft can get through. A score of cargo aircraft landed Thursday at the normally quiet Toncontin airport in the Honduran capital of Tegucigalpa, delivering aid from Mexico, the United States, Japan and Argentina. "It's a coincidence that the ships are there but they've got men and equipment that can be put to work in an organized way," said International Development Secretary Clare Short.

From these two examples, we can state that:

Hypothesis 3. *Summary and Anti-summary* overlap *at a significant amount, if the article contains more* objective *sentences than* subjective *ones.*

Besides SentiWordNet, we search for more accurate sentence and word level sentiment analyzer. Mao and Lebanon[9]'s work focuses on a supervised model of *sentence level sentiment detection*. We can adopt their technique, apply sentence level sentiment as the bias and then rank the sentences. One important aspect of working with text is *noise reduction*. Not handling *anaphora resolution* is the weakest point for our experiments. But one can easily modify our *graph generation approach* to get rid of this issue.

8 Conclusion

Our approach is domain independent and *unsupervised*. From our experiments we can deduce that most of the important sentences contain a good mixture of positive and negative sentiments toward related topics; a noticeable amount of extracted sentences overlap between the summary and the anti-summary for a *non-biased* article; and a sentiment oriented summary can be produced by extending TextRank summary with anti-summary model generated one.

In future, we would like to apply this graph based technique as a *semi-supervised* approach. Using some sentiment training dataset, we can adjust the *bias* of each node in the graph, and then use a sentiment classifier or SentiWordNet to define the polarity-direction. Besides, we will be applying *anaphora resolution* techniques and *semantic parsing* while defining the graph. For shorter articles, we have applied anaphora resolution by hand. It performed better on defining sentence connectivity more accurately and ranked related words more precisely. We also plan to extend this work and build a model that can generate summary not only by extracting sentences but also by rephrasing some of them.

References

1. Balahur, A., Kabadjov, M.A., Steinberger, J., Steinberger, R., Montoyo, A.: Summarizing opinions in blog threads. In: Kwong, O. (ed.) PACLIC, pp. 606–613. City University of Hong Kong Press (2009), http://dblp.uni-trier.de/db/conf/paclic/paclic2009.html#BalahurKSSM09
2. Beineke, P., Hastie, T., Manning, C., Vaithyanathan, S.: Exploring sentiment summarization. In: Qu, Y., Shanahan, J., Wiebe, J. (eds.) Proceedings of the AAAI Spring Symposium on Exploring Attitude and Affect in Text: Theories and Applications. AAAI Press (2004), aAAI technical report SS-04-07
3. Erkan, G., Radev, D.R.: Lexrank: Graph-based lexical centrality as salience in text summarization. J. Artif. Int. Res. 22(1), 457–479 (2004), http://dl.acm.org/citation.cfm?id=1622487.1622501
4. Esuli, A., Sebastiani, F.: Sentiwordnet: A publicly available lexical resource for opinion mining. In: Proceedings of the 5th Conference on Language Resources and Evaluation (LREC 2006), pp. 417–422 (2006)
5. Goldstein, J., Kantrowitz, M., Mittal, V., Carbonell, J.: Summarizing text documents: Sentence selection and evaluation metrics. In: Proceedings of the 22nd Annual International ACM SIGIR Conference on Research and Development in Information Retrieval, SIGIR 1999, pp. 121–128. ACM, New York (1999), http://doi.acm.org/10.1145/312624.312665
6. Hovy, E., Lin, C.-Y.: Automated text summarization and the summarist system. In: Proceedings of a Workshop on TIPSTER 1998, Held at Baltimore, Maryland, October 13-15, pp. 97–214. Association for Computational Linguistics, Stroudsburg (1998), http://dx.doi.org/10.3115/1119089.1119121
7. Liu, B.: Sentiment analysis and opinion mining. Synthesis Lectures on Human Language Technologies 5(1), 1–167 (2012), http://dx.doi.org/10.2200/S00416ED1V01Y201204HLT016
8. Luhn, H.P.: The automatic creation of literature abstracts. IBM J. Res. Dev. 2(2), 159–165 (1958), http://dx.doi.org/10.1147/rd.22.0159
9. Mao, Y., Lebanon, G.: Isotonic conditional random fields and local sentiment flow. In: Advances in Neural Information Processing Systems (2007)
10. Mihalcea, R., Tarau, P.: TextRank: Bringing order into texts. In: Proceedings of EMNLP 2004 and the 2004 Conference on Empirical Methods in Natural Language Processing (July 2004)
11. Mihalcea, R.: Graph-based ranking algorithms for sentence extraction, applied to text summarization. In: Proceedings of the ACL 2004 on Interactive Poster and Demonstration Sessions, ACLdemo 2004 (2004)
12. Miller, G.A.: Wordnet: A lexical database for english. Communications of the ACM 38, 39–41 (1995)
13. Mishra, A., Bhattacharya, A.: Finding the bias and prestige of nodes in networks based on trust scores. In: Proceedings of the 20th International Conference on World Wide Web, WWW 2011, pp. 567–576. ACM, New York (2011), http://doi.acm.org/10.1145/1963405.1963485

A Two-Level Keyphrase Extraction Approach

Chedi Bechikh Ali[1], Rui Wang[2], and Hatem Haddad[3]

[1] ISG, Tunis University, Tunisia
LISI laboratory, INSAT, Carthage University, Tunisia
[2] School of Computer Science and Software Engineering,
The University of Western Australia
[3] Department of Computer Engineering,
Faculty of Engineering,
Mevlana University, Konya, Turkey
{chedi.bechikh,haddad.hatem}@gmail.com,
21224938@student.uwa.edu.au,
hhatem@mevlana.edu.tr

Abstract. In this paper, we present a new two-level approach to extract KeyPhrases from textual documents. Our approach relies on a linguistic analysis to extract candidate KeyPhrases and a statistical analysis to rank and filter the final KeyPhrases. We evaluated our approach on three publicly available corpora with documents of varying lengths, domains and languages including English and French. We obtained improvement of Precision, Recall and F-measure. Our results indicate that our approach is independent of the length, the domain and the language.

Keywords: Keyphrase extraction, linguistic analysis, statistical approach.

1 Introduction

The available growing amount of text data today is very large for using manual annotation. Finding automated approaches to text annotation is a fundamental and essential target for researchers in different domains. Authors usually tend to annotate their documents with a list of multi-word phrases, rather than single words. We call these KeyPhrases (KPs), instead of keywords, because they are often phrases (noun phrases in most cases). We define that KPs are multi-word phrases that describe and capture the main topics discussed in a given document [1,2].

More than single keywords, KeyPhrases can be very expressive, precise in their meaning, and have a high cognitive plausibility. This is because humans tend to think in terms of KPs more than single keywords. For these reasons, KPs are used in different fields, such as text summarisation [3], clustering [4], indexing and browsing [5], highlighting [6] and searching [7].

The automatic selection of important topical phrases from documents is automatic keyphrase extraction. We distinguish Keyphrase extraction from Keyphrase generation and Keyphrase inference. Indeed, Keyphrase generation associates a

A. Gelbukh (Ed.): CICLing 2015, Part II, LNCS 9042, pp. 390–401, 2015.
DOI: 10.1007/978-3-319-18117-2_29

set of KPs to a given text regardless to their origin, while Keyphrase inference associates a set of KPs to the text that may not be found inside it [8].

Generally, automatic Keyphrase extraction approaches can be classified into supervised machine learning and unsupervised approaches. In this paper, we only focus on unsupervised approaches since supervised machine learning approaches need extensive training data which is not always available.

The idea of unsupervised approaches is to rank the importance of each phrase occurring in a document, and then only a number of top ranked phrases are extracted as KPs. However, many previous studies only focus on ranking algorithms, but underestimate the importance of correctly identifying candidate phrases that are the inputs to ranking algorithms. Wang et al. [9] show that an efficient phrase chunking technique can make a significant improvement on the same ranking algorithm.

Many unsupervised KP extraction approaches were only evaluated on single datasets with one chosen language, which makes them highly length, domain and language specific. The capabilities and efficiencies of these approaches applying on different datasets, domains and languages remain unknown.

In this paper, we proposed a two-level approach that employs a pipeline architecture consisting of two levels: a deep linguistic analyser that identifies candidate phrases, and a statistical ranking algorithm that assigns scores for each of the candidate phrases identified by the linguistic analyser. We evaluated our work on three publicly available datasets, including two English datasets and one French dataset, with documents of varying length and domains.

This paper is organised as follows: in Section 2, we briefly review some linguistic approaches and statistical approaches for KP extraction. In Section 3, we describe our proposed approach. The details of the datasets for our evaluation is described in Section 4, and the experimental results and discussions are presented in Section 5. We conclude our study in Section 6 with an outlook to some future work.

2 Related Work

2.1 Linguistic Approaches for Keyphrase Extraction

Linguistic appproaches for keyphrase extraction are based on POS patterns and they use two steps processing: a set of POS tag sequences is defined; then, relying on a POS tagger, all potential KPs that match any POS pattern is extracted. Barker and Cornacchia [10] restrict candidates to noun phrases, and then rank them using heuristics based on: length, term frequency and head noun frequency. Bracewell et al. [6] also restrict candidates to noun phrases, and cluster them if they share a term. The clusters are ranked according to the noun phrase and words' frequencies in the document. Finally, the centroids of the top-N ranked clusters are selected as KPs. The *Exact* keywords matching strategy is only partially indicative of the performance of a keyphrase extraction method, because it fails to match when there is lexical semantic variations (automobile sales, car sales) or morphological variation. Zesch and Gurevych [11] introduce a keyphrases approximate matching strategy, *Approx*, that considers the morphological variation and the two cases where there is an overlap between the extracted and the proposed keyphrases. The first case is when the extracted keyphrase includes the standard keyphrase (Includes). The second case is when the extracted keyphrase is a part of the gold standard keyphrase (PartOf).

We note that KPs extracted statistically to represent the meaning of a document may contain noise or incoherent n-grams. In the following example:

> *The student will probably be attending a special reading on software engineering on Monday*

statistical approaches extract KPs, such as *will probably*, *student will*, *be attending*. However, these KPs have neither important significance in this sentence nor represent the meaning of the sentence. On the other hand, linguistic approaches based noun phrase identifier will extract keyphrases such as *special reading*, and *software engineering*. We note that these KPs are more likely to represent the content or the sentence topic than those extracted by statistical approaches.

2.2 Statistical Approaches for Keyphrase Extraction

Statistical approaches for keyphrase extraction is to rank phrases in a given document according to their statistical information that indicates the likelihood of being keyphrases .The most commonly used information are word frequency and co-occurrence frequency. Intuitively, *content words* with higher frequencies tend to carry more information than the ones with lower frequencies. Statistical approaches for keyphrase extraction can be seen as a two-step task: representing statistical information, and scoring phrases. Two common approaches are 1) matrix representation coupled with statistical techniques [12], and 2) graph representation coupled with graph ranking algorithms [13]. In the first approach, statistical information of each words or phrases are organised in matrices, such as *term-document matrix* and *co-occurrence matrix*, which enable applying statistical techniques to the corpora, such as Term Frequency Inverse Document Frequency (TF-IDF) and *Chi Squared Test*. In the second approach, the statistical information are organised in a graph, where words or phrases in a document are represented as vertices, and two vertices are connected if any lexical or semantic relation exist. For example, two vertices are connected if their corresponding words are co-occurred in a given window size, and then graph ranking algorithms, such as *PageRank* [14], are applied to the graph to assign the degree distribution of each vertex.

3 Our Approach for Extract Keyphrases

We propose a two-level approach to extract KPs. The approach is a combination of a linguistic analyser allocating linguistically correct candidate KPs as the first level, and a statistical approach to rank and select the KPs as the second level.

A set of candidate KPs is the output of the linguistic level where no ranking is conducted. The resulting list of candidate KPs is long since all the potential KPs are extracted. On the second level, the input is the set of candidate KPs and the output is weight score assigned to each candidate KP.

3.1 Our Linguistic Approach for Keyphrase Extraction

Our Linguistic approach for KPs extraction is based on two parts:

1. A linguistic analysis with a tagger which generates a tagged corpus. Each word is associated to a tag corresponding to the syntactic category of the word, example: noun, adjective, preposition, gerundive, proper noun, determiner, ect.
2. The tagged corpus is used to extract a set of KPs. These candidate KPs are extracted by the identification of syntactic patterns.

We adopt the definition of syntactic patterns in [15], where a pattern is a syntactic rule on the order of concatenation of grammatical categories which form a noun phrase:

- V: the vocabulary extracted from the corpus
- C: a set of lexical categories
- L: the lexicon $\subset V \times C$

A pattern is a syntactic rule of the form:

$$X := Y_1 Y_2 Y_k ... Y_{k+1} Y_n$$

where $Y_i \in$ C and X is a candidate Keyphrase.

For the English language, 12 syntactic patterns are defined: 4 syntactic patterns of size two (for example: Noun Noun, Adjective Noun, etc.), 6 syntactic patterns of size three (for example: Adjective Noun Noun, Adjective Noun Gerundive, etc.) and 2 syntactic patterns of size 4. For the French language, 31 syntactic patterns are defined: 8 syntactic patterns of size two (for example: Proper-Noun Proper-Noun), 16 syntactic patterns of size three (for example: Noun Noun Adjective, Noun Preposition Noun, etc.) and 7 syntactic patterns of size 4.

For the exemple of Section 3.1, the linguistic analysis with a tagger will generate the following tagged sentence:

> *The/Determiner student/Noun will/Modal probably/Adverb be/Verb attending/Gerundive a/Determiner special/Adjective reading/Noun on/ Preposition software/Noun engineering/Noun on/Preposition Monday/ Proper Noun*

Then candidate KPs are extracted by the identification of syntactic patterns. For example The candidate KP *software engineering* is extracted based on the syntactic pattern *Noun Noun.*

It worths to note that our proposed linguistic approach should not be seen as a simple preprocessing technique which usually relies on simple *n-gram* filters or arbitrary POS patterns to identify candidate phrases. Our proposed linguistic approach, on the other hand, is based on deep linguistic analysis. For example, a simple noun phrase chunker, such as *basic chunker* described by Wang et. al [9] is not able to identify any KP that contains a preposition, whereas our approach can identify them correctly.

3.2 Our Statistical Approach for Keyphrase Extraction

Authors in [2] conducted an evaluation and an analysis of five keyphrase extraction algorithms (TF-IDF, TextRank, SingleRank, ExpandRank and a Clustering-based approach) on four corpora. They concluded that TF-IDF [12] and TextRank [13] algorithms are offering the best performances across different datasets. For this reason, we re-implemented and used these two algorithms.

TF-IDF is a weighting scheme that statistically analyses how important a term (a single word or a multi-word phrase) is to an individual document in a corpus. The intuition is that a frequent term distributed evenly in a corpus is not a good discriminator. So, less weight should be assigned. In contrast, a term occurs frequently in a few particular documents should be assigned more weight.

TF-IDF provides a heuristic weighting scheme for scoring weights of words in a document against the corpus. The input to TF-IDF is a term-document matrix which describes frequency of terms that occur in each document. Fig. 1, left, shows an example where rows correspond to all the words in the corpus (documents 1 and 2), and each column represents a document in the corpus. TF-IDF is calculated as the product of two statistics; a term's TF score and its IDF score. The TF scheme analyzes the importance of a term against a document, thus a term with higher frequency is assigned a higher TF weight. While the IDF weighting scheme analyses the importance of a term against the entire corpus, a term occurring frequently in a large number of documents gains a lower IDF score. Let t denotes a term, d denotes a document, and D denotes a corpus, where $t \in D$, and $d \in D$, the *tfidf* is calculated as:

$$tfidf(t, d, D) = tf(t, d) \times idf(t, D) \tag{1}$$

Based on Equation 1, the *tfidf* weight can be calculated by any combination of tf and idf functions. We re-implemented the classic TF-IDF weighting scheme [12] that assigns weight to a term t_i in document d as:

$$tfidf(t_i) = tf_i \times idf_i = tf_i \times log \frac{|D|}{|\{d \in D : t_i \in d\}|} \tag{2}$$

where tf_i is the number of times that the term t_i occurs in d, $|D|$ is the total number of documents in corpus D, and $|\{d \in D : t_i \in d\}|$ is the number of documents in which term t_i occurs.

TextRank introduced by Mihalcea and Tarau [13], represents a text in an undirected graph, where words or phrases correspond to vertices, and edges represent co-occurrence relations between two vertices. Two vertices are connected if any co-occurrence relations are found within a defined window-size. For example, in Figure 1, co-occurrence relations are defined in sentence boundaries.

TextRank implements the idea of 'voting'. Indeed, if a vertex v_i links to another vertex v_j as v_i casting a vote for v_j, then the higher the number of votes v_j receives, the more important v_j is. Moreover, the importance of the vote itself is, also, considered by the algorithm; the more important the voter v_i is, the more important the vote itself is. The score of a vertex is, therefore, calculated based on the votes it received and the importance of the voters. The votes a vertex received can be calculated directly that is so called *local vertex-specific information*. The importance of a voter which is recursively drawn from the entire graph is the *global information*. Therefore, TextRank computes the importance of a vertex within a graph is not only determined based on *local vertex-specifc information* but also taking *global information* into account.

TextRank uses a derived PageRank [14] algorithm to rank the words (vertices). The original PageRank is for ranking a directed unweighted graph. In a directed graph $G = (V, E)$, let $in(v_i)$ be the set of vertices that point to a vertex v_i, and $out(v_i)$ be the set of vertices to which v_i point. The score of v_i is calculated by PageRank as:

$$S(v_i) = (1 - d) + d \times \sum_{j \in in(v_i)} \frac{1}{|out(v_j)|} S(v_j) \tag{3}$$

In TextRank, the in-degree of a vertex equals to its out-degree, since the graph is undirected. Formally, let D denotes a document, and w denotes a word, then $D = \{w_1, w_2, ..., w_n\}$. The weight of a vertex is calculated by TextRank as:

$$WS(v_i) = (1 - d) + d \times \sum_{v_j \in in(v_i)} \frac{w_{ji}}{\sum_{v_k \in out(v_j)} w_{jk}} WS(v_j) \tag{4}$$

where w_{ij} is the strength of the connection between two vertices v_i and v_j, and d is the dumping factor, usually set to 0.85 [13,14].

4 Experimental Datasets and Tagging

To evaluate the extracted Keyphrases and their abilities to represent documents, we use three corpora from varying domains and different characteristics. Hulth2003 and SemEval2010 are two English corpora. Deft2012 is a French corpus. Table 1 provides an overview of each corpus.

Table 1. Datasets Statistics for the three datasets

	French Dataset	English Dataset	
	Deft2012	Hulth2003	SemEval2010
Size	6.4M	1.3M	9.4M
# Documents	467	1599	244
# Terms/Document	6318	120	9647
# Pre-assigned KPs	809	13387	2931
# Pre-assigned KPs/Document	2.1	8.4	12

4.1 English Datasets

We selected two English datasets[1]:

- Hulth2003; a short document dataset consists of 1,599 abstracts of journal papers in computer science field collected from *Inspec.*
- SemEval2010; a long document dataset consists 244 full-length conference or workshop papers in computer science and social and behavioral science from the *ACM Digital Library.*

Both Hulth2003 and SemEval2010 consist of training and test sets for supervised extraction evaluations. Since no training data is required for our approaches, we use the evaluation data from both training and test datasets.

Each dataset article pairs with keyphrases assigned by authors, readers, or both. We use the combination of both authors and readers as the ground truth for our evaluation. It worths to note that not all assigned keyphrases appear in the actual content of the document in both of the datasets, and some studies tend to exclude these keyphrases from the evaluation. In this study, we do not exclude them. This is also the case for Deft2012 French corpus.

4.2 French Datasets

Started in 2005, the aim of DEFT[2] (Defi fouilles de textes) campaign is to evaluate, using the same corpora, methods and systems of different research teams. DEFT 2012 campaign challenge was to evaluate the ability of research systems to extract keywords that can be used to index the content of scientific papers published in journals of Humanities and Social Sciences [16]. Each scientific paper pairs with keyphrases assigned manually by experts (mainly the papers authors). We use four corpora from DEFT 2012 campaign that we merge to build Deft2012 corpus.

4.3 Datasets Tagging

To tag the English corpora, we use TreeTagger[3], a tool for annotating text with part-of-speech and lemma information. It was developed by Helmut Schmid in

[1] `http://github.com/snkim/AutomaticKeyphraseExtraction`

[2] `www.deft2012.limsi.fr`

[3] `http://www.cis.uni-muenchen.de/~schmid/tools/TreeTagger/`

the TC project at the Institute for Computational Linguistics of the University of Stuttgart. The TreeTagger has been successfully used to tag different languages.

The French corpus Deft2012 is POS tagged using Cordial[4], a tagger that is known to outperform TreeTagger on French texts. Thus, each word of the corpus is associated with its form, its lemma and its POS tag [17].

5 Experiment Setup and Results

Fig. 1 represents two real documents taken from the Hulth2003 corpora[5]. Applying the linguistic level, 11 KPs are extracted from document 1 (information interaction, information architecture, etc) and 17 KPs are extracted from Document 2 (bilingual web site, multi-language sites, structural analysis, existing bilingual web designs, etc).

Document 1: *Information interaction provides a framework for information architecture. Information interaction is the process that people use in interacting with the content of an information system. Information architecture is a blueprint and navigational aid to the content of information-rich systems. As such information architecture performs an important supporting role in information interactivity.*

Document 2: *Creating an information architecture for a bilingual web site presents particular challenges beyond those that exist for single and multilanguage sites. The development of the information architecture is based on a combination of two aspects: an abstract structural analysis of existing bilingual web designs focusing on the presentation of bilingual material, and a bilingual card-sorting activity conducted with potential users.*

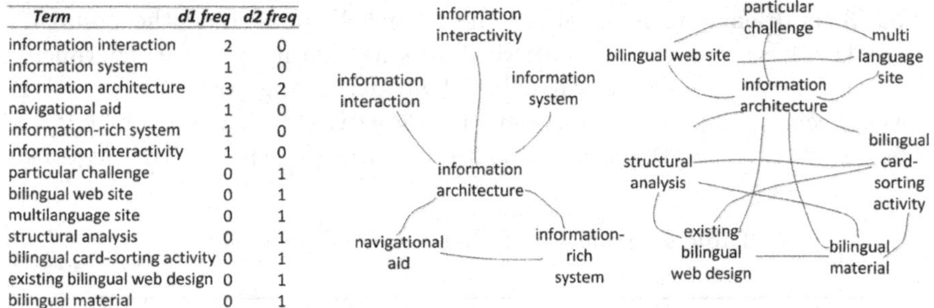

Term	*d1 freq*	*d2 freq*
information interaction	2	0
information system	1	0
information architecture	3	2
navigational aid	1	0
information-rich system	1	0
information interactivity	1	0
particular challenge	0	1
bilingual web site	0	1
multilanguage site	0	1
structural analysis	0	1
bilingual card-sorting activity	0	1
existing bilingual web design	0	1
bilingual material	0	1

Fig. 1. Statistical algorithm text representations: TF-IDF for Document 1 & 2 (*left*), TextRank Graph for Document 1 (*middle*) and Document 2 (*right*)

We use the outputs of the linguistic approach described in Section 3.1 as the inputs to the statistical approach described in Section 3.2. For TextRank, each phrase is considered as a node of the graph, two phrases are connected if they co-occur in the same sentence. The initial value of each node is set to $\frac{1}{n}$, where n is the total number of the nodes in the graph. The damping factor is set to 0.85, iterations number is 30, and threshold of breaking is 0.0001. TF-IDF does not require special setting.

[4] The Cordial tagger is developed by Synapse Developpement (www.synapse-fr. om).

[5] Document 1 refers to the document 207.abstr of Hulth2003 corpus and Document 2 refers to the document 209.abst of Hulth2003 corpus.

5.1 Evaluation Measures

In order to evaluate the extraction results, we use the *Precision*, *Recall*, and *F-measure* for evaluating the ranking algorithms. The *Precision* is defined as: $\frac{the\ number\ of\ correctly\ matched}{total\ number\ of\ extracted}$, *Recall* is defined as: $\frac{the\ number\ of\ correctly\ matched}{total\ number\ of\ assigned}$, and *F-measure* is defined as: $2 \times \frac{precision \times recall}{precision + recall}$.

5.2 Evaluation Results

There are 13,387 assigned KPs in Hulth2003 dataset, and 2,931 assigned keyphrases in SemEval2010 dataset. At the first level, our linguistic approach extracted 45,113 and 297,249 candidate keyphrases from Hulth2003 and SemEval2010 datasets respectively, and these candidates are inputs to the statistical approaches. After ranking, only 15,374 and 2,440 candidates are extracted (we extracted top 10 ranked ones for each document) as KPs for Hulth2003 and SemEval2010 datasets respectively.

We present Recall and F-score based on both the number of keyphrases that exist in the documents, and the number of all preassigned keyphrases (including the keyphrases that do not even appear in the document). For example, for the Deft2012 corpora, we have only 54% preassigned KPs appearing in the documents, which means the maximum recall we can obtain is 54%. Table 2 shows the evaluation results.

Table 3 represents statistics about the French Datasets and the number of retrieved keyphrases for the two statistical ranking algorithms. The experimental results for French dataset are done based on extracting top 10 ranked KPs. However, because the average number of KPs assigned to a document is 1.73 KPs per document, the Precision measures are 6.1% for the TF-IDF algorithm

Table 2. Evaluation Result for English Dataset

	Hulth2003		SemEval2010	
	TF-IDF	TextRank	TF-IDF	TextRank
# Extracted KPs	15374	15374	2440	2440
# Preassigned KPs	13387	13387	2931	2931
# KPs in the documents	10132	10132	2389	2389
# Matched KPs	3906	3854	504	536
Precision	25.4%	**25.1%**	**20.7%**	**22.0%**
Recall (KPs in the documents)	38.6%	**38.0%**	**21.1%**	**22.4%**
F-score (KPs in the documents)	30.6%	**30.2%**	**20.9%**	**22.2%**
Recall (All preassigned KPs)	29.2%	28.8%	17.2%	18.3%
F-score (All preassigned KPs)	27.2%	26.8%	18.8%	20.0%
Base line algorithm (shallow linguistic analysis)				
Precision	**28.7%**	24.1%	17.1%	13.6%
Recall (KPs in the documents)	**41.8%**	35.1%	11.8%	9.4%
F-score (KPs in the documents)	**34.0%**	28.6%	14.0%	11.2%

Table 3. Evaluation Result for French Dataset

	TF-IDF	TextRank
# Extracted KPs	3770	3770
# Preassigned KPs	809	809
# KPs in the documents	430	430
# Matched KPs	230	222
Precision	**6.1%**	**5.9%**
Recall (KPs in the documents)	**53.5%**	**51.6%**
F-score (KPs in the documents)	**11.0%**	**10.6%**
Recall (all preassigned KPs)	28.4%	27.4%
F-score (all preassigned KPs)	10.0%	9.7%
Base line algorithm (shallow linguistic analysis)		
Precision	**5.4%**	**4.8%**
Recall (KPs in the documents)	**47.7%**	**41.9%**
F-score (KPs in the documents)	**9.8%**	**8.6%**

and 5.9% for the TextRank algorithm. It is also worth noting that if we reduce the number of top ranked KPs to 5, we receive slightly better F-scores.

5.3 Results Discussion

Our approach, the deep linguistic analysis coupled with statistical ranking algorithms produces better performance than only using simple linguistic patterns. A commonly used POS pattern [9] is to search phrases that begin with a number of adjectives (optional) and end with a number of nouns. This POS pattern, when coupled with TF-IDF and TextRank, results F-scores of 14% and 11.2% on long English dataset (SemEval2010), and 34% and 28.6% on short English dataset (Hulth2003), respectively. Whereas, our approach made significant improvements of 6.9% with TF-IDF and 11% with TextRank on long dataset. On short dataset, our approach produced a better F-score of 30.2% with TextRank.

For the French dataset, we received much lower precision. This is due to the absence of a large number of preassigned KPs inside the documents. Indeed, for Deft2012 corpus, more than 46% of preassigned KPs are not in the documents. In the case of Hulth2003 and SemEval corpora, They are 24% and 18% respectively. Many reasons explain the non-appearence of these preassigned KPs inside the documents. In the case of the preassigned keyphrase *femme iranienne rurale*[6] for example, the extracted KPs are *femme iranienne*[7] and *femme rurale*[8]. The individual tokens *femme*, *iranienne*, and *rurale* do not occur in adjacency inside the documents. In the case of the English Datasets, the same phenomenon is observed. For example, the preassigned KP *non-linear distributed paramater model* does not exist as it is in the documents but the keyphrases *non-linear*

[6] Rural Iranian woman.

[7] Iranian woman.

[8] Rural woman.

and *distributed parameter model* do exist separately. Another example is related to the preassigned KPs *average-case identifiability* and *average-case controllability*. In the documents, the KP *average-case identifiability and controllability* exists, but it does not match with the preassigned KPs. To solve these issues, a more in-deep Natural Language Processing is planned to be used in our future research.

6 Conclusion

In this paper, we have introduced a two-level approach for automatic keyphrase extraction. The system employs pipeline architecture consisting of two levels: a deep linguistic analyser that identifies the candidate phrases, and a statistical ranking algorithm that uses the candidates from the linguistic analyser as inputs to score each phrase according to its importance towards the core topic of a given document. We evaluated our work on different datasets with documents of varying length, domains and languages. We demonstrated that our proposed approach outperformed the baseline algorithm. We conclude that our approach is corpus, domain and language independent in comparison with other approaches that are usually tuned specifically to a specific corpus and domain. Our future work includes investigating more in-deep linguistic analysis to identify the topical related phrases that do not appear in the actual document. The head-modifier structure, for example will be used to solve mismatching problems between the preassigned KPs and the extracted KPs to allow partial matching instead of exact matching. The use of KeyPhrase inference will be also combined with our approach to solve the problem of missing KPs. We will also use the document structure because KPs are most likely to appear in the titles and the abstracts.

Acknowledgment. This research was funded partially by the Australian Postgraduate Awards Scholarship, Safety Top-Up Scholarship by The University of Western Australia.

References

1. Turney, P.D.: Learning algorithms for keyphrase extraction. Inf. Retr. 2, 303–336 (2000)
2. Hasan, K.S., Ng, V.: Conundrums in unsupervised keyphrase extraction: Making sense of the state-of-the-art. In: 23rd International Conference on Computational Linguistics, Beijing, China, pp. 365–373 (2010)
3. Litvak, M., Last, M.: Graph-based keyword extraction for single-document summarization. In: Proceedings of the Workshop on Multi-source Multilingual Information Extraction and Summarization, pp. 17–24. Association for Computational Linguistics, Stroudsburg (2008)
4. Hammouda, K.M., Matute, D.N., Kamel, M.S.: CorePhrase: Keyphrase extraction for document clustering. In: Perner, P., Imiya, A. (eds.) MLDM 2005. LNCS (LNAI), vol. 3587, pp. 265–274. Springer, Heidelberg (2005)

5. Gutwin, C., Paynter, G., Witten, I., Nevill-Manning, C., Frank, E.: Improving browsing in digital libraries with keyphrase indexes. Decis. Support Syst. 27, 81–104 (1999)
6. Bracewell, D.B., Ren, F., Kuriowa, S.: Multilingual single document keyword extraction for information retrieval. In: IEEE Natural Language Processing and Knowledge Engineering, pp. 517–522 (2005)
7. Turney, P.D.: Coherent keyphrase extraction via web mining. In: Proceedings of the Eighteenth International Joint Conference on Artificial Intelligence, Acapulco, Mexico, pp. 434–442 (2003)
8. Nart, D.D., Tasso, C.: A keyphrase generation technique based upon keyphrase extraction and reasoning on loosely structured ontologies. In: Proceedings of the 7th International Workshop on Information Filtering and Retrieval, Turin, Italy, pp. 49–60 (2013)
9. Wang, R., Liu, W., McDonald, C.: How preprocessing affects unsupervised keyphrase extraction. In: Gelbukh, A. (ed.) CICLing 2014, Part I. LNCS, vol. 8403, pp. 163–176. Springer, Heidelberg (2014)
10. Barker, K., Cornacchia, N.: Using noun phrase heads to extract document keyphrases. In: Hamilton, H.J. (ed.) Canadian AI 2000. LNCS (LNAI), vol. 1822, pp. 40–52. Springer, Heidelberg (2000)
11. Zesch, T., Gurevych, I.: Approximate matching for evaluating keyphrase extraction. In: Recent Advances in Natural Language Processing, Borovets, Bulgaria, pp. 484–489 (2009)
12. Jones, K.S.: A statistical interpretation of term specificity and its application in retrieval. Journal of Documentation 28, 11–21 (1972)
13. Mihalcea, R., Tarau, P.: Textrank: Bringing order into texts. In: Lin, D., Wu, D. (eds.) Conference on Empirical Methods in Natural Language Processing, pp. 404–411. Association for Computational Linguistics, Barcelona (2004)
14. Brin, S., Page, L.: The anatomy of a large-scale hypertextual web search engine. Computer Networks and ISDN Systems 30, 107–117 (1998)
15. Haddad, H.: French noun phrase indexing and mining for an information retrieval system. In: Nascimento, M.A., de Moura, E.S., Oliveira, A.L. (eds.) SPIRE 2003. LNCS, vol. 2857, pp. 277–286. Springer, Heidelberg (2003)
16. Paroubek, P., Zweigenbaum, P., Forest, D., Grouin, C.: Indexation libre et contrôlée d'articles scientifiques. présentation et résultats du défi fouille de textes deft2012 (controlled and free indexing of scientific papers. presentation and results of the deft2012 text-mining challenge) (in french). In: JEP-TALN-RECITAL 2012, Atelier DEFT 2012: DÉfi Fouille de Textes, pp. 1–13. ATALA/AFCP, Grenoble (2012)
17. Quiniou, S., Cellier, P., Charnois, T., Legallois, D.: What about sequential data mining techniques to identify linguistic patterns for stylistics? In: Gelbukh, A. (ed.) CICLing 2012, Part I. LNCS, vol. 7181, pp. 166–177. Springer, Heidelberg (2012)

Information Retrieval, Question Answering, and Information Extraction

Conceptual Search for Arabic Web Content

Aya M. Al-Zoghby[1] and Khaled Shaalan[2]

[1] Faculty of Computers and Information Systems, Mansoura University, Egypt
aya_el_zoghby@mans.edu.eg
[2] The British University in Dubai, UAE
khaled.shaalan@buid.ac.ae

Abstract. The main reason of adopting Semantic Web technology in information retrieval is to improve the retrieval performance. A semantic search-based system is characterized by locating web contents that are semantically related to the query's concepts rather than relying on the exact matching with keywords in queries. There is a growing interest in Arabic web content worldwide due to its importance for culture, political aspect, strategic location, and economics. Arabic is linguistically rich across all levels which makes the effective search of Arabic text a challenge. In the literature, researches that address searching the Arabic web content using semantic web technology are still insufficient compared to Arabic's actual importance as a language. In this research, we propose an Arabic semantic search approach that is applied on Arabic web content. This approach is based on the Vector Space Model (VSM), which has proved its success and many researches have been focused on improving its traditional version. Our approach uses the Universal WordNet to build a rich concept-space index instead of the traditional term-space index. This index is used for enabling a Semantic VSM capabilities. Moreover, we introduced a new incidence measurement to calculate the semantic significance degree of the concept in a document which fits with our model rather than the traditional term frequency. Furthermore, for the purpose of determining the semantic similarity of two vectors, we introduced a new formula for calculating the semantic weight of the concept. Because documents are indexed by their topics and classified semantically, we were able to search Arabic documents effectively. The experimental results in terms of Precision, Recall and F-measure have showed improvement in performance from 77%, 56%, and 63% to 71%, 96%, and 81%, respectively.

Keywords: Semantic Web (SW), Arabic Language, Arabic web content, Semantic Search, Vector Space Model (VSM), Universal Word Net (UWN), Wikipedia, Concept indexing.

1 Introduction

Search engines are still the most effective tools for finding information on the Web. Ambiguities in users' queries make searching the web content a challenge. Searching becomes more sophisticated when dealing with linguistically rich natural language

A. Gelbukh (Ed.): CICLing 2015, Part II, LNCS 9042, pp. 405–416, 2015.
DOI: 10.1007/978-3-319-18117-2_30

such as Arabic, which has a number of properties that makes it particularly difficult to handle by a computational system [1]. The use of terminological variations for the same concept, i.e. Synonymous terms, creates a many-to-one ambiguity. Whereas, the use of the same terminology for different concepts, i.e. Polysemous terms, creates a one-to-many ambiguity [2, 3]. It is desirable that search engines have three main features: accurately interpret the user's intension, handle the relevant knowledge from different information sources, and deliver the authentic and relevant results to each user individually [4, 5, 6]. Our goal is to address all these features in our proposed semantic-based search approach.

From the performance perspective, the traditional search engines are characterized by trading off a high-recall for low-precision. The reasons are that the results of these systems are sensitive to the input keywords, and the consequences of misinterpreting the synonymous and polysemous terminologies [7]. In other words, not only all relevant pages are retrieved, but also some irrelevant pages are retrieved as well, which impact the Precision. On the other hand, if some relevant pages are missing, this obviously leads to low Recall. One suggested solution is to use the Semantic Search Engines (SSEs) which rely on ontological concepts for indexing rather than lexical entries of standard lexicons that are commonly used by traditional search engines. Thus, SSEs aim to retrieve pages referring to specific concepts indicated by the query, rather than pages mentioning the input keywords, which will resolve the semantic ambiguity [8, 9]. The ontology should resolve the issue of semantic similarity of terminologies that comprise the keyword since they can be interpreted via the ontological representation of their related concept. Moreover, the SSEs can benefit from the inherent Generalization/Specialization properties of the ontological hierarchy. When a semantic search engine fails to find any relevant documents, it might suggest generic answers. On the other hand, if too many answers are retrieved, the search engine might suggest more specialized answers [8, 9]. As a comparison with traditional search engines, the returned results will be more relevant, and those missing documents will also be retrieved, which means higher accuracy and better robustness [10].

The success and advances of Semantic Web technology with Latin languages can also be investigated in order to bridge the gap in other underdeveloped languages, such as Arabic. Statistics from WWW indicated that there are an increasing number of Arabic textual contents available on electronic media, such as Web pages, blogs, emails, and text messages, which make the task of searching Arabic text relevant. However, there are linguistic issues and characteristics facing the development of Semantic Web systems in order to be able to effectively search the Arabic web content. This is due to the richness of the Arabic morphology and the sophistication of its syntax. Moreover, the highly ambiguous nature of the language makes keyword-based approaches of the traditional search engines inappropriate [11, 12]. For example, the optional vowelization in modern written Arabic text gives different meaning to the same lexical form. Another example is the polysemous or multi-meaning of words which arise from terminologies that share the same orthography but differ in meaning [13, 14]. Nevertheless, there are specific issues that are related to the handling of Arabic text in computational systems. One of them is the differences in Arabic script encoding of web content [15]. Another issue is the availability of Arabic resources, such as corpora and gazetteers [1]. Existing Arabic resources are available but at significant expense. In many cases these resources are limited or not suitable for the desired task. So, researchers often develop their own resources, which require significant efforts in data collection, human annotation, and verification.

This research proposes enhancements to the semantic VSM-based search approach for Arabic Information Retrieval applications, and the like. VSM is a common information retrieval model for textual documents that has demonstrated its capability to represent documents into a computer interpretable form. Many researches have been successful in improving its traditional version [16]. In our proposed approach, we build a concept-space from which we construct a VSM index. This model has enabled us to represent documents by semantic vectors, in which the highest weights are assigned to the most representative concept. This representation permits a semantic classification within and across documents, and thus the semantic search abilities reflected in its Precision and Recall values can be obtained. The construction of the concept-space is derived from semantic relationships presented at the Universal WordNet (UWN). UWN is an automatically constructed cross-lingual lexical knowledge base from the multilingual WordNet. For more than 1,500,000 words in more than 200 languages, UWN provides a corresponding list of meanings and shows how they are semantically related [17]. The main reasons for choosing the UWN are: a) it is widely a standard, b) it is very powerful in supporting semantic analysis, and c) it has the ability to provide the meaning of missing Arabic terms from their corresponding translations from other languages. Moreover, the UWN would facilitate dealing with Arabic dialects, which is an active research topic. The proposed approach is used to develop a system that is applied on a full dump of the Arabic Wikipedia. The evaluation of the system's retrieval effectiveness using the constructed concept-space index resulted in noticeable improvements in the performance in terms of Precision and Recall as compared to the traditional syntactic term-space baseline. The experimental results showed an overall enhancement of the F-Measure score from 63% to 81% due to employing the semantic conceptual indexing.

The rest of this paper is organized as follows: Section 2 presents the architecture of the model and the implementation aspects. The experimental results are discussed in details at Section 3. Finally, the paper is concluded at Section 4.

2 System Architecture

This section describes the architecture of the proposed conceptual VSM-based search system and its components. The overall architecture is depicted in Fig. 1.

The *'Document Acquisition'* module acquires documents from the web which we use as the knowledge source.

The *'NE Extraction'* module extracts Arabic Named Entities from the acquired documents. The extracted "Named Entity" (NE) covers not only proper names but also temporal and numerical expressions, such as monetary amounts and other types of units [1]. ANEE [18], a popular rule-based named entity recognition tool that is integrated with the GATE development environment, is used for extracting the NEs. ANEE is capable of recognizing Arabic NEs of types: Person, Location, Organization,

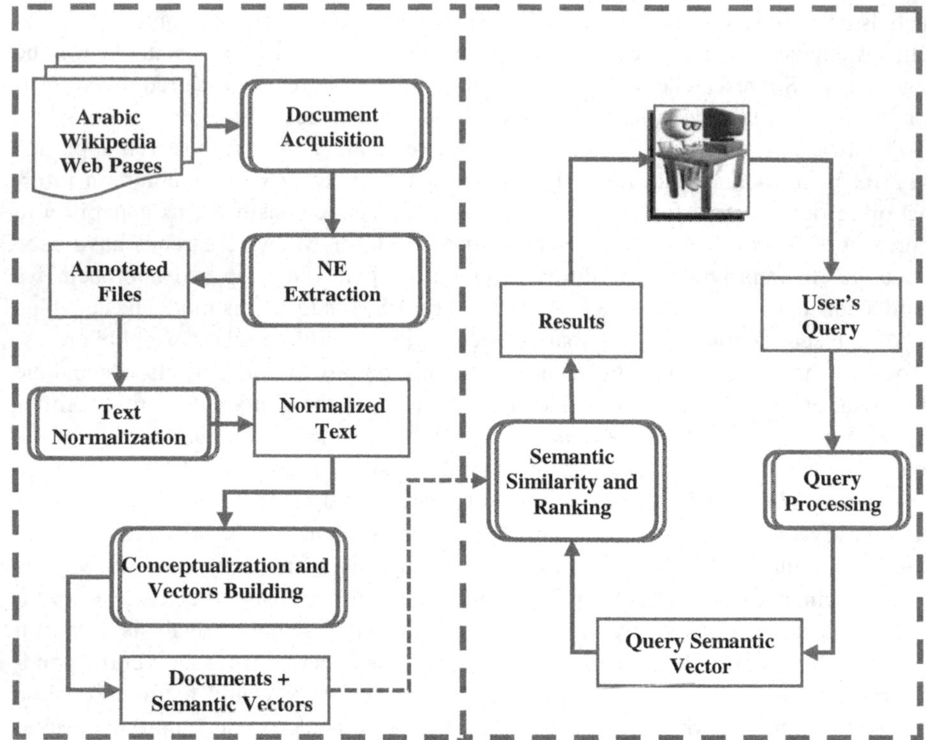

Fig. 1. The System Architecture

Measurement, Time, Number, Percent, and File name. A number of scholars have used the ANEE tool in their research studies on Arabic NER, including Maynard et al. (2002), Elsebai et al. (2009), Elsebai et al. (2011), and Abdallah et al. (2012) [1]. In our research, we also used ANEE tool to annotate the Arabic text with Person, Location, and Organization NEs. In the literature, these types are so called ENAMEX. Afterwards, these named entities are expanded, via the UWN, with their aliases[1].

The *'Text Normalization'* module performs four preprocessing steps. First, it splits the full dump into separate Wikipedia articles; each one has a designated title. Then, the text is normalized by filtering out non-Arabic letters[2]. Next, the RDI-Swift indexer[3] is applied on the cleaned texts to perform indexing. This indexing would facilitate eliminating all inflected forms of the Arabic stop-word[4] by searching for it within the text. Finally, stop-words are eliminated; excluding those that are part of NEs.

---

[1] For example: **Yasser Arafat** (Arabic: ياسر عرفات, *Yāsir `Arafāt*) and his aliases **Abu Ammar** (Arabic: أبو عمار, *'Abū `Ammār*).

[2] This is specific to the current version of the RDI-Swift indexer that is available to us which requires normalizing the text using the RDI morphological analyzer.

[3] `http://www.rdi-eg.com/technologies/arabic_nlp.htm`; `http://rdi-eg.com/Demo/SWiftSearchEngine/Default.aspx`; `http://www.rdi-eg.com/Downloads/Swift2WhitePaper.doc`

[4] For example: من, منه, منها, منهما, لمن...

The *'Conceptualization and Vectors Building'* module, which is the core of the proposed system, performs progressive development of the Arabic semantic vectors. The vectors are generated using three indexing methods that differ in their incidence measure, see Table 1. The three indices gradually developed: Morphological Term-Space (MTS), Semantic Term-Space (STS), and Concept-Space (CS) are used to evaluate the performance of the proposed model.

An entry of the MTS dictionary considers all inflected forms of the term. An entry of the STS dictionary is a semantic expansion of an MTS entry using UWN. In other words, each term in the space is represented by a set of related inflected forms of their semantic expansions. An entry of the CS dictionary consists of a main keyword that represents the concept and expansions of all encapsulated terms along with their morphological and semantic expansions. For the purpose of incidence measurement, we developed five different measurements that progress gradually from the *Term Frequency (tf)* to the *Concept Semantic Significance Degree (CSSD).* These measurements show the improvement we made in system's performance.

Query processing is handled by the *'Query Preprocessing'* module. Firstly, the query is normalized by removing stop words and non-Arabic words or letters. The query is then semantically expanded using the UWN. Finally, the *'Semantic Similarity and Ranking'* module calculates the similarity between user's query and documents-space. Then, if it finds similar results, the ranking is applied according to similarity.

Table 1. Indexing and Incidence Measurement methods

Indexing Method	Incidence Measurement	
	Measure	**Description**
Morphological Term-Space (MTS)	Term's Morphological Frequency (TMF)	It is the count of the morphological inflections occurrences of the term.
Semantically Expanded Term-Space (STS)	Term's Semantic Frequency (TSF)	It is the frequency of the morphological derivations of the term itself and its semantic expansions as well.
	Term's Semantic Significance Degree (TSSD)	In this measurement, the variation of the association degree of each expansion of the term is taken into consideration.
Concept-Space (CS)	Concept's Semantic Frequency (CSF)	This type is equivalent to that of TSF but in terms of Concepts instead of terms, so it will be computed as the count of matching occurrences of all concepts' ingredients.
	Concept's Semantic Significance Degree (CSSD)	It is the equivalent of the TSSD, but in terms of concepts.

3 Experimental Analysis

3.1 The Conceptualization Process

From the *AWDS*[5], a terms-space of 391686 terms is extracted, 31200 of which are NEs. Each extracted term is enclosed with its set of derivations that occurred in the document space, with an average of 16 root-based derivations each. The term-space is then expanded via UWN with 793069 in total; redundant expansions forms about 12% and takes the form of phrases. The shrinking algorithms are applied on the term-space, excluding the NEs set, to generate a concepts-space, which results in of 223502. As a result, the count of the concepts-space entries is shrunk to be just 62 % of that of the morphological-space. The ampler concept is defined by 66 merged terms while the narrower one is just of two. Some of the terms are never merged; most of them are NEs[6], out of vocabulary, or misspelled words. This, first level of, shrinking generated 299203 groups of terms from the 360486 terms. The second shrinking level then condensed them into just 223502 groups, each of which defines a distinct concept. This leads to the increment of the representation power of each item in the space, since the average of items weights is increased as shown by the last two columns in Table 2. Note that the weight average of V3 and V5 are lower than those of V2 and V4, respectively. However, it is noticeable that the weights of the semantic indexing (STS), are higher than those of the morphological indexing (MTS). Moreover, the weights of the conceptual indexing (CS) are higher than the weights of the

Table 2. The overall results of the Conceptualization Process

Exp.	Indexing	#Entries	Expanding Type	Average of Expansions / Entry #	Average of Expansions / Entry Classification		*Freq.* Avg	*df* Avg	Weight Average
V1	MTS	360486	Morpho-logically	4	Morphological Inflections		0.514	2.02 ≈ 2	1.02
V2	STS		Semantically	11	Inflections	4	0.704	2.6 ≈ 3	2.11
					Synonyms	2			
					Sub-Classes	1			
					Super-Classes	2			
V3					Has-Instances	1			1.65
					Instance-Of	1			
V4	CS	223502	Conceptually	33	Concept of 3 STS entries, each with its semantic expansion		0.7076	3.8 ≈ 4	2.83
V5									2.32

Morphological Based Weights Average

Semantic Based Weights Average

Conceptual Based Weights Average

---

5 AWDS stands for Arabic Wikipedia Documents Space.

6 Note that we excluded the already extracted NEs, however, other NEs are still impeded in the space and couldn't be recognized by the ANEE.

semantic indexing (STS). The important observation is that these results demonstrate the efficiency of the conceptual indexing in distinguishing documents by means of corresponding weights. The conceptualization process is fully described at [19].

3.1.1 Retrieval Capability.

The F-measure is very popular for calculating the performance of the retrieval system based on the harmonic mean of Precision and Recall scores, which we used to evaluate the retrieval accuracy of the proposed system. It is defined as follows:

$$F\_Measure = 2 \cdot \frac{Precision \cdot Recall}{Precision + Recall} \quad (1)$$

Where, *Precision* is the fraction of retrieved instances that are relevant, and

Recall is the fraction of relevant instances that are retrieved.

In order to conduct evaluation experiments, we have to determine beforehand the correct answer of the test query that will be applied on a dataset from *AWDS*. As an example, consider the query: مصادر الطاقة */msadr altaqh/ (Energy Resources).* Then, we need to determine the correct relevant documents, i.e. the actual Recall.

For the sake of evaluation, a gold standard test set of the first 200 documents from the AWDS is extracted, and classified according to relevancy to the test query: مصادر الطاقة */msadr altaqh/ (Energy Resources).* 77 documents, out of the 200 documents, are syntactically/semantically/conceptually relevant, whereas the rest are irrelevant. These documents are then used as the document-space of the search process. The query is preprocessed and semantically expanded using the UWN.

The performance of the system in terms of Precision, Recall, and F-Measure are shown in Table 3. As shown in the table, five experiments were conducted, i.e. V1 through V5 experiments, which uses MTS, STS, and CS indexing. Notice that both V2 and V3, have achieved the same values of the Precision, Recall, and F-Measure. Likewise, both V4 and V5 have achieved same scores. The reason is that these scores are not sensitive to the incidence measurement which differentiate between these experiments. However, the incidence measurement has an effective role in ranking results which we discuss in details in the subsequent section. From the results shown at Table 3, we observe the following:

First, the highest Precision is achieved by the V1 experiment, which uses the MTS index. They were just 12 irrelevant documents of a total of 123. It only contains the documents that have either the word مصادر*/msadr/ (Resources)* or the word طاقة*/taqh/ (Energy*) in an irrelevant context. However, despite its high precision, this experiment's *Recall* is the lowest, since some of the semantically and conceptually relevant documents are not considered; they are 32 out of the total 77 relevant documents.

Second, the *Precision* is degraded while using the STS indexing with experiments that use V2 and V3. This is mainly due to some issues in recognizing *multiword*

Table 3. The Retrieval Capapilities of the indecies: MTS, STS, and CS

Experiment	Indexing Method	Precision	Recall	F-Measure
V1	MTS	**76.92%**	55.50 %	63.04%
V2, V3	STS	68.57%	66.70%	67.62%
V4, V5	CS	71%	**96%**	81.62%

expressions, or phrases, by the RDI indexing system[7]. For example, the word طاقة */taqh/ (energy)* is expanded to the expression طاقة نووية */taqh nwwyh/ (Nuclear energy),* which is handled as two individual words. Consequently, documents related only to the word نووي */nwwy/ (nuclear or atomic)* should have been considered irrelevant. In this case, when V2 and V3 experiments are used, the count of the irrelevant documents becomes 22, which affects the *Precision*[8] to be lower than that obtained by V1. The *Recall,* on the other hand, is increased since the missing relevant documents are decreased from 32 achieved by V1 experiment to 24 achieved by V2 and V3 experiments. This improvement is due to the semantic expansions that included the words such as كهرباء */khrba'/ (Electricity),* and the phrases such as طاقة نووية */taqh nwwyh/ (Nuclear energy).*

Third, as for the V4 and V5 experiments, where the conceptual index CS is exploited, the *Precision* increased again, but not to the extent of the V1 experiment, since the issue of recognizing multiword expressions or phrases is still affecting it. This time, the individual words of an expression such as خلايا شمسية */ khlaya shmsyh/ (Solar Cells)* caused the retrieval of 6 additional irrelevant documents related to the term خلايا/khlaya/ (Cells). The Recall, however, is increased to the extent that makes the F-Measure of these experiments the highest. This means that the conceptual retrieval caught most of the relevant documents. Only three documents are escaped. They concern the terms: إشعاع / esh'ea'e/ (Radiation), and حركة المياه/ hrkh almyah/ (Water flow), since they are not covered in the concept's definition itself. The results showed an enhancement of the F-Measure value to be 81.62% using our proposed semantic conceptual indexing as compared to 63.04% achieved when using the standard indexing method.

3.2 Ranking Accuracy

In addition to measuring the proposed system's performance in terms of the retrieval capability, the ranking accuracy are also evaluated, see Table 4. The evaluation of

7 The set of all expansions of a term is collected as a set of individual words; separated by a space. The RDI indexing system is used to search for their inflected forms within the documents-space. This approach caused a problem in dealing with the multiword expansions, since the sequence of words forming a multiword expression is not recognized.

8 In order to resolve this problem, each expansion is indexed such that the multiword expression is discriminated. However, this solution is favorable if the indexing algorithm is not efficient in using the memory space.

Table 4. The Ranking accuracy of each Incidence Measurement

Exp.	Incidence Measurement	W Average	Ranking Distance Average[9]
V1	TMF	1.028692	4.625359
V2	TSF	2.113503	4.402002
V3	**TSSD**	1.653511	4.397337
V4	CSF	2.830778	4.165544
V5	**CSSD**	2.322311	**4.154426**

Enhancement caused by using STS indexing

Extra Error caused by using TSF Incidence Measurement

Extra Error caused by using CSF Incidence Measurement

Enhancement caused by CS Indexing

document ranking is aimed at measuring the accuracy of assigning the congruent weight that exactly represent the association degree of each entry of the index with a document in the space. This should indicate how the *Incidence Measurement* factor is directly affecting the capability of the accurate ranking.

Wherever the values of the weights averages of V2 and V4 experiments are greater than those of V3 and V5 experiments respectively, the ranking results show that V3 experiments is better than V2 experiment, and V5 experiments is the best. The reason is the extra error ratios caused by using the *frequency Incidence Measurements, TSF* and *CSF,* instead of the *Semantic Significance Degree Incidence Measurements, TSSD* and *CSSD.* However, these extra ratios are truncated via exploiting the TSSD and CSSD at the V3 and V5 experiments. This is directly reflected by the ranking efficiency of these experiments. Still, the V4 experiment gives better ranking that the V3 experiments since it is based on the *conceptual indexing CS*, even it suffers from the extra error ratio caused by using the *CSF Incidence Measurements.*

The *Ranking Accuracy* is measured by calculating the *Distance Average* between the experiment's ordering and that of a human specialist. *Distance Average* is defined as following:

Distance-Average(V_i)=

$$\frac{\sum_{j=1}^{n}\left|\frac{1}{Standard\_Rank_j}-\frac{1}{Experimental\_Rank_j}\right|}{n} \quad (2)$$

[9] The smallest the distance the closest to the correct ranking.

Where,

n = # retriever documents as a result to the user's query.
$Standard\_Rank_j$ = The standard rank of the document #j.
$Experimental\_Rank_j$ = The rank of document #j at V_i.

The closest ranking is obtained from the V5 experiment, which assures the capability of its features to rank the retrieved result more accurately.

4 Conclusion and Future Work

This study sheds the light on the inefficiency in the handling the Arabic language semantically, which in turn results in an unqualified Arabic Web content for being a suitable environment of the Semantic Web technology implementation. This inefficiency may be ascribed to the sophistication of the Arabic language that makes it complex to the extent that hinders its treatment electronically. Nevertheless, this should not stop more effective efforts for achieving the best possible solutions that enable the Arabic language and its users to take the advantages of the new electronic technologies generally, and the Semantic Web particularly. This might not be achieved unless the Arabic electronic infrastructure is adequately built. Briefly, we have to construct all of the necessary Arabic resources, such as the Ontologies, Gazetteers, WordNet, as well as the complementary utilities such as morphological analyzers and named entity extractors. Likewise, the support of the Arabic script in the Semantic Web tools such as OWL/RDF editors must be taken into consideration. If that infrastructure is achieved and is sufficiently strong, the development of wider applications will be better facilitated, and the obtained results will be more reliable and trustworthy.

In an attempt to take a step in that long pathway, we proposed an Arabic semantic search system that is based on the Vector Space Model. The Vector Space Model is one of the most common information retrieval models for textual documents due to its ability to represent documents into a computer interpretable form. However, as it is syntactically indexed, its sensitivity to keywords reduces its retrieval efficiency. In order to improve its effectiveness, the proposed system has extracted a concept-space dictionary, using the UWN ontology, in order to be used as a semantic index of the VSM search system instead of the traditionally used term-space. The proposed system enables a conceptual representation of the document space, which in turn permits the semantic classification of them and thus obtaining the semantic search benefits. Moreover, we introduced a new incidence measurement to calculate the semantic significance degree of the concept in a document instead of the traditional term frequency. Furthermore, we introduce a new formula for calculating the semantic weight of the concept to be used in determining the semantic similarity of two vectors. The system's experimental results showed an enhancement of the F-measure value to 81.62% using the semantic conceptual indexing instead of 63.04% using the standard syntactic one.

Still, the model's implementation suffers from some limitations. Consequently, the presented results will certainly be improved if those limitations are overcome. Therefore, as a future work, we have to solve the ambiguity problem by discriminating the meaning contextually. Also, we may work on refining the processing of the multiword expression expansions. That will improve the results noticeably since 12%[10] of the semantic expansions are in the form of multiword expressions. We also will try to expand the named entities in order to use them in the environment of the linked data. Moreover, the improvement of the Arabic knowledge representation in the UWN will help to overcome its limitations that directly affects the search results. Another open research area is to solve the problems of the Arabic language morphological analysis in order to prevent the consequent errors occurred in the indexing process, and hence, the construction of the search dictionary. We also may try to use Google Translation API with the UWN in order to find results for these terms that have results in languages other than Arabic.

References

1. Khaled, S.: A Survey of Arabic Named Entity Recognition and Classification. Computational Linguistics 40(2), 469–510 (2014), doi:10.1162/COLIa00178.
2. Saleh, L.M.B., Al-Khalifa, H.S.: AraTation: An Arabic Semantic Annotation Tool. In: The 11th International Conference on Information Integration and Web-based Applications & Services (2009)
3. Tazit, N., Bouyakhf, E.H., Sabri, S., Yousfi, A., Bouzouba, K.: Semantic internet search engine with focus on Arabic language. In: The 1st International Symposium on Computers and Arabic Language & Exhibition © KACST & SCS (2007)
4. Cardoso, J.: Semantic Web services: Theory, tools, and applications. IGI Global (March 30, 2007) ISBN-13: 978-1599040455
5. Hepp, M., De Leenheer, P., de Moor, A.: Ontology management: Semantic web, semantic web services, and business applications. Springer (2008) ISBN: 978-0-387-698899-1
6. Kashyap, V., Bussler, C., Moran, M.: The Semantic Web: Semantics for Data and Services on the Web (Data-Centric Systems and Applications). Springer (August 15, 2008) ISBN-13: 978-3540764519
7. Panigrahi, S., Biswas, S.: Next Generation Semantic Web and Its Application. IJCSI International Journal of Computer Science Issues 8(2) (March 2011)
8. Unni, M., Baskaran, K.: Overview of Approaches to Semantic Web Search. International Journal of Computer Science and Communication 2(2), 345–349 (2011)
9. Renteria-Agualimpia, W., López-Pellicer, F.J., Muro-Medrano, P.R., Nogueras-Iso, J., Zarazaga-Soria, F.J.: Exploring the Advances in Semantic Search Engines. In: de Leon F. de Carvalho, A.P., Rodríguez-González, S., De Paz Santana, J.F., Rodríguez, J.M.C. (eds.) Distrib. Computing & Artif. Intell., AISC, vol. 79, pp. 613–620. Springer, Heidelberg (2010)
10. Kassim, J.M., Rahmany, M.: Introduction to Semantic Search Engine. In: International Conference on Electrical Engineering and Informatics, ICEEI 2009 (2009)
11. Habash, N.Y.: Introduction to Arabic Natural Language Processing. Association for Computational Linguistics 30 (August 2010) ISBN 978-1-59829-795-9
12. Al-Zoghby, A.M., Eldin, A., Hamza, T.T.: Arabic Semantic Web Applications: A Survey. Journal of Emerging Technologies in Web Intelligence, 52–69 (2013)

[10] This percentage is computed from the retrieved UWN expanding results.

13. Elkateb, S., Black, W., Vossen, P., Farwell, D., Pease, A., Fellbaum, C.: Arabic WordNet and the challenges of Arabic: The Challenge of Arabic for NLP/MT. In: International Conference at the British Computer Society (October 23, 2006)
14. Al-Khalifa, H.S., Al-Wabil, A.S.: The Arabic Language and the Semantic Web: Challenges and Opportunities. In: International Symposium on Computers and the Arabic Language, Riyadh, Saudi Arabia (November 2007)
15. Omar, D.: Arabic Ontology and Semantic Web. al-Mu'tamar al-duwali lil-lughah. al-lughat al-'Arabiyah bayn al-inqirad wa al-tatawwur, tahaddiyat wa tawqi'at, Jakarta, Indonesia July 22-25 (2010),
-الأنطولوجيا العربية و الويب الدلالي: المؤتمر الدولي للغة العربية (اللغة العربية بين الانقراض والتطور- التحديات والتوقعات) ، جاكرتا, إندونيسيا: 22- 25 يوليو 2010.
16 Zhao, Y.-H., Shi, X.-F.: Shi: The Application of Vector Space Model in the Information Retrieval System. Software Engineering and Knowledge Engineering: Theory and Practice 162, 43–49 (2012)
17 de Melo, G., Weikum, G.: UWN: A Large Multilingual Lexical Knowledge Base. In: Annual Meeting of the Association of Computational Linguistics (2012)
18 Oudah, M.M., Shaalan, K.: A Pipeline Arabic Named Entity Recognition Using a Hybrid Approach. In: Proceedings of the 24th International Conference on Computational Linguistics (COLING 2012) (2012)
19 Al-Zoghby, A.M., Eldin Ahmed, A.S., Hamza, T.: Utilizing Conceptual Indexing to Enhance the Effectiveness of Vector Space Model. International Journal of Information Technology and Computer Science (IJITCS) 5(11) (2074) (October 2013) ISSN: 2074-9007

Experiments with Query Expansion for Entity Finding

Fawaz Alarfaj, Udo Kruschwitz, and Chris Fox

University of Essex,
School of Computer Science and Electronic Engineering,
Wivenhoe Park, Colchester, CO43SQ
{falarf,udo,foxcj}@essex.ac.uk

Abstract. Query expansion techniques have proved to have an impact on retrieval performance across many retrieval tasks. This paper reports research on query expansion in the entity finding domain. We used a number of methods for query formulation including thesaurus-based, relevance feedback, and exploiting NLP structure. We incorporated the query expansion component as part of our entity finding pipeline and report the results of the aforementioned models on the CERC collection.

1 Introduction

Information retrieval (IR) systems are designed to meet users' information needs efficiently and effectively. Most of these systems, including Web search engines, perform this task by retrieving a set of documents, which are ranked by how relevant they are to any given query. Nevertheless, information needs are not always best represented by documents. In fact, many query types are best answered with a set of entities. For example, if a user is looking for a list of chefs who have shows on the TV Food Network, he or she must manually extract the list from the set of documents retrieved by the search engine, which can be a time-consuming and tedious process involving long documents. Modern search engines have recently started to recognise this need by identifying some well-known entities. Figure 1 shows an example of Google search engine results for the query "Boeing 747". Google recognises the entity and generates a slider on the right-hand side with some basic information about this entity.

Such capabilities are limited to only a number of entity types and to simple entity queries. Thus, entity search represents an active area of research with many unsolved challenges. To this end, the Text REtrieval Conference (TREC) established the Enterprise and Entity tracks to research this topic. A special type of entity finding, which is known as expert finding, is where the task is to find relevant experts with knowledge about the given topic. The focus of this research will be the expert finding task, and we will use the CERC data set and test collection to evaluate different setups for a more accurate system.

A. Gelbukh (Ed.): CICLing 2015, Part II, LNCS 9042, pp. 417–426, 2015.
DOI: 10.1007/978-3-319-18117-2_31

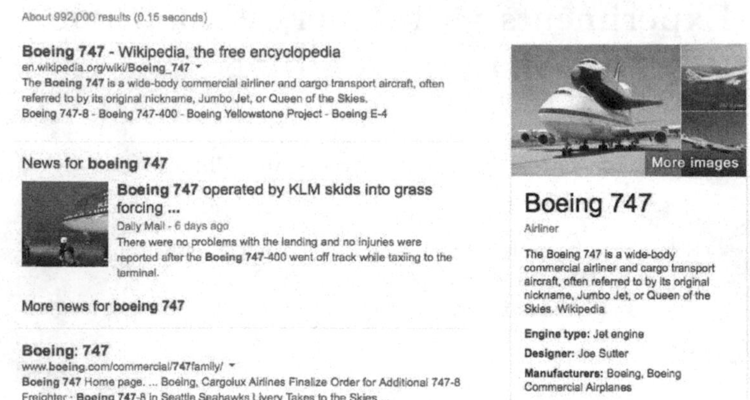

Fig. 1. Boing 747 query results in Google

2 Background

With the establishment of the expert finding task, a general platform was created to develop and test expert finding algorithms using a number of datasets and evaluation benchmarks. This has helped foster research in this field and has resulted in the introduction of a number of expert finding algorithms, including the statistical models [1,2], voting model [3], and graph-based models [4], among others. Expert finding continues to attract the attention of many researchers, most recently in the study of social media websites like Facebook [5] and Twitter [6,7] and community question-answering websites such as Yahoo! Answers [8] and Stack Overflow [9]. Among the successful approaches developed for this task are those which incorporate proximity in their models. For instance, the work of Petkova and Croft [10] directly addresses the use of different *kernel* functions that capture the proximity of candidates and query term occurrences. Other works that examine proximity in expert finding include [11,12,13,14]. Many of the systems that use proximity consider the distance between entities within a fixed window or multiple windows of text within the document. Some researchers consider a number of pre-processing steps to enhance the overall system, including query expansion [15,1], the underlying document retrieval method [16], and document structure [17].

Alarfaj et al. [18] proposed the application of an adaptive window size by creating a unique window size for each document, instead of a fixed window size across the whole collection. The size of the proposed window is a function of various document features. The idea is to use these features to set the window size and, thereby, improve the overall ranking function. The four features suggested by Alarfaj et al. are: document length, the number of entities in a document, the average sentence length, and the readability index. This method is used in this study.

The expert finding system, like any information retrieval application can be seen as a kind of pipeline, in which the accuracy of the overall application depends on the performance of each single component in the pipeline. The aim of this study is to test two components, namely, the query expansion method and the document retrieval algorithm. We compared four query expansion methods including: using thesaurus to expand the main query, removing stop words, extracting collocations, and pseudo-relevance feedback. Each method has been tested with three document retrieval algorithms: the Vector Space Model, Language Modeling, and BM25.

3 Experimental Setup

The CSIRO Enterprise Research Collection (CERC) represents some real-world search activity within the CSIRO organisation, in which the topics and the relevant judgments are made by the internal staff. Each test topic includes a keyword query and a narrative, which provide extra detail about the information needs of the query. Additionally, the query topics in 2007 contained examples of web pages related to the query. In 2008, the topics were requests for information received by enquiries staffers at CSIRO. These topics have been extracted from a log of real email enquiries. Selected emails were used after removing any identifying information and greetings, etc. Figure 2 shows example topics from the 2007 and 2008 TREC Enterprise track.

TREC topics usually have the following main fields:

NUM is the number of the query.
TITLE is a short query. Generally, this field is used as the query topic.
NARR is the narrative part. This part contains more information about the query. This part is used in query expansion, see Section 5.
PAGE Examples of key pages: Lists the URLs of a few pages that would be relevant. This part is only in the 2007 queries.

```
<top>
<num> CE-013</num>
<query>human clinical trials</query>
<narr>
Overview of the Human Nutrition Clinic
facility and the latest clinical nutrition
trials which are recruiting volunteers.
</narr>
<page>CSIRO145-08477954</page>
<page>CSIRO145-11110519</page>
<page>CSIRO142-02910122</page>
</top>
```

2007 topic

```
<top>
<num> CE-055</num>
<query>case moth identification</query>
<narr>
I have photographs of several different
Case Moths and have been endeavouring
to find photographs or drawings that could
assist in identification.
</narr>
</top>
```

2008 topic

Fig. 2. Example of the topics used in the TREC expert finding task in 2007 and 2008

The CERC judgments should be more accurate when compared to the TREC's 2005 and 2006 judgments, because they were made with first-hand knowledge. Furthermore, the set of relevant experts to each query is small compared to W3C[1]. This simulates reality, because users are usually interested in seeing the main contacts for their query rather than a long list of people with various levels of knowledge about the topic.

Although there is no master list of candidates in the CERC test collection, the process for generating the list of candidates was straightforward because all the email addresses conforms to a standard format (i.e.`firstname.lastname@csiro.au`), which can largely facilitate the recognition process. By locating the email addresses, we could also find the candidates' full names. After locating all email occurrences in the collection, we removed the duplicates and non-personal emails, which resulted in a total of 3,480 candidates.

Table 1. Statistics of the CERC Enterprise research test collections

	TREC 2007	TREC 2008
# docs	370,715	370,715
# People	3,480	3,480
Avg. Doc Length	354.8	354.8
# Queries	50	77⋆
# qrels	152	2,710
Avg. Experts/Topic	3.06	10.38

⋆ Only 55 queries have relevant judgements, which have been used in the study.

4 Methodology

Unlike other approaches, which depend on external sources, we ranked candidates based on two matrices: co-occurrence and proximity. It could be argued that a candidate is an expert in a certain topic when he/she appears together with the query term more often and if they appear within a closer proximity to each other. A similar approach, provided by Alarfaj [18] is easy to replicate for the current problem. On this ground, we implemented the adaptive window size approach using four document features. Thus, the value of the candidate *ca* in document *d* is calculated as follows:

$$p(ca|d,q) = \frac{P_{occu}(ca|d) + P_{kernel}(ca|d)}{\zeta} \quad (1)$$

where $P_{occu}(ca|d)$ represents the first assumption (i.e., the co-occurrence of entities given evidence of their relation), and $P_{kernel}(ca|d)$ represents the second assumption (i.e., the close distance between entities given further evidence of

[1] W3C has an average of 40 experts per topic.

their relation). The value of the constant ζ is chosen to ensure that $p(c|d)$ is a probability measure. The value of ζ is computed as follows:

$$\zeta = \sum_{i=1}^{N} (P_{occu}(ca_i|d) + P_{kernel}(ca_i|d)), \tag{2}$$

where N is the total number of candidates in the document d.

For the co-occurrence function, a $TF - IDF$ weighting scheme is applied to capture the candidate importance at the current document, while at the same time discriminating against it in terms of the general importance:

$$P_{occu}(ca|d) = \frac{n(ca,d)}{\sum_i n(ca_i,d)} \cdot \log \frac{|D|}{|\{d' : n(ca,d') > 0\}|} \tag{3}$$

where $n(ca,d)$ is the frequency with which a candidate appears in the document. $\sum_i n(ca_i,d)$ is the sum of all candidates in that document. $|D|$ is the number of documents in the collection. $d' : n(ca,d') > 0$ is the number of documents where the candidate appears. The proximity part, $P_{kernel}(ca|d)$, is defined as follows:

$$P_{kernel}(ca|d) = \frac{k(t,c)}{\sum_{i=1}^{N} k(t,ca_i)} \tag{4}$$

A non-uniform Gaussian kernel function has been used to calculate the candidate's proximity:

$$k(t,c) = \frac{1}{\sqrt{2\pi\sigma^2}} \exp(\frac{-u^2}{2\sigma^2}), \quad u = \begin{cases} |c-t|, & \text{if } |c-t| \leq \mathrm{w} \\ \infty, & \text{otherwise} \end{cases} \tag{5}$$

where c is the candidate position in the document, t is the main topic position, and w is the window size for the current document. As mentioned earlier, the proximity in this work is considered using the adaptive window size approach, meaning that each document has its own unique window. The window size is defined using document features including: Document Length, Candidate Frequency, Average Sentence Length, and Readability Index. The aforementioned features are combined in the following equation:

$$\begin{aligned} \mathrm{WindowSize} = \frac{\sigma}{4} * (\log(\frac{1}{\mathrm{DocLength}}) * \beta_1 + \mathrm{CanFreq} * \beta_2 \\ + \mathrm{AvgSentSize} * \beta_3 + \mathrm{ReadabilityIndex} * \beta_4) \end{aligned} \tag{6}$$

In equation 6, the σ variable allows us to scale the window size, whereas the weighting factors β determine each feature's contribution in the equation. After identifying the window size for the current document, it was applied to every instance of the query entity found within the document. This enables the extraction of any candidate entities accompanying the query entity. Each candidate is given a weight in the window depending upon its proximity to the query entity. The proximity weight is calculated using a Gaussian kernel function, which according to previous work [10], produces the best results in this context.

5 Query Expansion

In this part, we detail the query expansion techniques that were compared in this study. Three main sources have been used for query expansion: the original query, the narrative part, and the retrieved documents as detailed in Table 2.

Table 2. Query expansion methods used in this study

Tag	Query expansion method	Data sources
QE1	Thesaurus-based	Query term
QE2	Stop words removal	Narrative Part
QE3	Collocations	Narrative Part
QE4	Relevance feedback	Retrieved documents

Using different query expansion algorithms may have produced disparate influences on the overall performance of expert finding systems. In some cases, applying query expansion can result in a worse performance as in [19]. One of the main problems that may have an impact on query expansion for expert finding is *topical drift* as discussed in [15]. The query expansion methods used in this study are as follows:

Relevance Feedback
In relevance feedback, users are involved in the retrieval process by giving a feedback on the relevance of the documents and, thereby, help revise results sets. However, in some scenarios, as in our applications, user involvements tend to be very limited. In such cases, we resort to pseudo relevance feedback, which automates the process of relevance feedback by assuming that the top k ranked documents[2] are relevant. According to this assumption, one can revise the initial results set. In this work, we used the Rocchio algorithm [20] implementation for Lucene search engine[3].

Thesaurus Expansion
For thesaurus expansion, we used WordNet[4] to automatically extract synonyms. In this method, we normalised each query by removing stop words and punctuation. Next, we expanded each term in the query with synonyms and related words.

Stop Words Removal
This approach is very common among TREC runs [21,22]. In this study, we deleted all the stop words from the query's narrative field (see Figure 2) and used it to expand the query.

[2] We have used the default value of $k = 10$. It could be possible to further improve the performance of this method by varying k, but we have left this for future work.

[3] `http://lucene-qe.sourceforge.net`

[4] `http://wordnet.princeton.edu`

Collocations

According to [23], "a collocation is an expression consisting of two or more words that correspond to some conventional way of saying things." In this work, we used tag patterns to represent collocations. Table 3 outlines the tag patterns suggested by Justeson and Katz [24], which were used in this study. Figure 3 gives an example for query expansion using collocations patterns and removing stop words methods.

Table 3. Part of speech tag patterns as suggested by [24] where A is an adjective, N is a lexical noun (i.e. not a pronoun); and P is a preposition

Tag Pattern	Example
A N	linear function
N N	regression coefficients
A A N	Gaussian random variable
A N N	cumulative distribution function
N A N	mean squared error
N N N	class probability function
N P N	degrees of freedom

Narrative:

Different techniques for slag granulation, use of waste heat, use of different materials to replace traditional Portland cement (e.g. geopolymers)

Collocations:	-Stop words:
- *"slag granulation"* - *"waste heat"* - *"traditional Portland cement"* - *"Portland cement"*	*different techniques slag granulation waste heat different materials traditional Portland cement geopolymer*

Fig. 3. Comparison between using collocations patterns and removing stop words of the narrative part of query CE-017

6 Results

We evaluated the expert finding system with different query expansion and document retrieval methods. The results were evaluated using two standard metrics: mean average precision (MAP) and mean reciprocal rank (MRR). Details on the performance of the expert finding system are given in Table 4. We tested three document retrieval algorithms: the Vector Space Model (VSM), Language Modelling (LM), and BM25. Results showed that using different query expansion algorithms appears to have an influence on the overall performance. For

Table 4. Results of the expert finding system with different query expansion techniques. Best scores are in boldface. Significance for each document retrieval algorithmis is tested against the baseline using a two-tailed paired t-test. We indicate $p < 0.01$ using $\triangle$.

		TREC 2007		TREC 2008	
		MAP	MRR	MAP	MRR
VSM	$QE0$ (baseline)	0.5022	0.7313	0.4464	0.7956
	$QE1$	0.4804	0.7145	0.4351	0.7740
	$QE2$	0.5102	0.7741	0.4618	0.8434
	$QE3$	0.5242	0.7957△	0.4776△	0.8495△
	$QE4$	0.5137	0.7807△	0.4724	0.8411△
LM	$QE0$ (baseline)	0.5089	0.7336	0.4514	0.7976
	$QE1$	0.4907	0.7117	0.4441	0.7726
	$QE2$	0.5252	0.8071△	0.4639	0.8517△
	$QE3$	**0.5937**△	**0.8766**△	**0.5140**△	0.9228△
	$QE4$	0.5819△	0.8583△	0.5084	0.9137△
BM25	$QE0$ (baseline)	0.5210	0.7634	0.4573	0.8161
	$QE1$	0.4975	0.6810	0.4455	0.7553
	$QE2$	0.5535	0.7440	0.4763	0.8656△
	$QE3$	0.5662△	0.8157△	0.5022	**0.9376**△
	$QE4$	0.5605	0.7978	0.4962	0.9293△

instance, in $QE1$, the performance declined by about 4% to 10% compared to the baseline, $QE0$, in which we did not use any query expansion. Among the query expansion techniques used in this study, $QE3$ proved to have the highest impact with an improvement of up to 14.8%. On the other hand, it is clear that the document retrieval algorithms had a slight effect, which may indicate the stability of the retrieval model. However, if we to compare the three document retrieval algorithms, the improvement over the baseline for the results obtained using language-modelling algorithm are marginally greater than using others algorithms. The results also suggest that generally the percentage improvements in MRR results are slightly higher than the percentage improvements in MAP results. With regard to the test collections, results seems to be consistent with no clear pattern about which one has more sensitivity regarding the query expansion methods.

7 Conclusions

In this paper, we have presented experiments on the role of query expansion in entity finding. We compared four query expansion methods. Each method was tested with three document retrieval algorithms, Vector Space Model, Language Modelling, and BM25. Results show that the best performance was achieved with collocation query expansion and with language modelling document retrieval algorithms. Expert finding is a specific application, and, in future work, we plan to test the aforementioned methods with more general application of entity finding, where the task will be to retrieve any type of entity, not just experts.

References

1. Fang, H., Zhai, C.: Probabilistic models for expert finding. In: Amati, G., Carpineto, C., Romano, G. (eds.) ECIR 2007. LNCS, vol. 4425, pp. 418–430. Springer, Heidelberg (2007)
2. Balog, K., Azzopardi, L., de Rijke, M.: Formal models for expert finding in enterprise corpora. In: SIGIR 2006, pp. 43–50. ACM (2006)
3. Macdonald, C., Ounis, I.: Voting for candidates: Adapting data fusion techniques for an expert search task. In: CIKM 2006, pp. 387–396. ACM (2006)
4. Serdyukov, P., Hiemstra, D.: Modeling documents as mixtures of persons for expert finding. In: Macdonald, C., Ounis, I., Plachouras, V., Ruthven, I., White, R.W. (eds.) ECIR 2008. LNCS, vol. 4956, pp. 309–320. Springer, Heidelberg (2008)
5. Hecht, B., Teevan, J., Morris, M.R., Liebling, D.J.: Searchbuddies: Bringing search engines into the conversation. In: ICWSM 2012, pp. 138–145 (2012)
6. Ghosh, S., Sharma, N., Benevenuto, F., Ganguly, N., Gummadi, K.: Cognos: crowdsourcing search for topic experts in microblogs. In: SIGIR 2012, pp. 575–590. ACM (2012)
7. Cheng, Z., Caverlee, J., Barthwal, H., Bachani, V.: Finding local experts on twitter. In: WWW 2014, pp. 241–242 (2014)
8. Aslay, Ç., O'Hare, N., Aiello, L.M., Jaimes, A.: Competition-based networks for expert finding. In: SIGIR 2013, pp. 1033–1036. ACM (2013)
9. Yang, L., Qiu, M., Gottipati, S., Zhu, F., Jiang, J., Sun, H., Chen, Z.: CQArank: Jointly model topics and expertise in community question answering. In: CIKM 2013, pp. 99–108. ACM (2013)
10. Petkova, D., Croft, W.B.: Proximity-based document representation for named entity retrieval. In: CIKM 2007, pp. 731–740. ACM (2007)
11. Petkova, D., Croft, W.B.: Hierarchical language models for expert finding in enterprise corpora. International Journal on Artificial Intelligence Tools 17, 5–18 (2008)
12. Fu, Y., Xiang, R., Liu, Y., Zhang, M., Ma, S.: A CDD-based formal model for expert finding. In: CIKM 2007, pp. 881–884. ACM (2007)
13. Macdonald, C., Hannah, D., Ounis, I.: High quality expertise evidence for expert search. In: Macdonald, C., Ounis, I., Plachouras, V., Ruthven, I., White, R.W. (eds.) ECIR 2008. LNCS, vol. 4956, pp. 283–295. Springer, Heidelberg (2008)
14. Alarfaj, F., Kruschwitz, U., Fox, C.: An adaptive window-size approach for expert-finding. In: Proceedings of DIR 2013 Dutch-Belgian Information Retrieval Workshop, Delft, the Netherlands, pp. 4–7 (2013)
15. Macdonald, C., Ounis, I.: Expertise drift and query expansion in expert search. In: Proceedings of CIKM 2007, pp. 341–350. ACM (2007)
16. Macdonald, C., Ounis, I.: The influence of the document ranking in expert search. Information Processing & Management 47, 376–390 (2011)
17. Zhu, J., Song, D., Rüger, S.: Integrating multiple windows and document features for expert finding. Journal of the American Society for Information Science and Technology 60, 694–715 (2009)

18. Alarfaj, F., Kruschwitz, U., Fox, C.: Exploring adaptive window sizes for entity retrieval. In: de Rijke, M., Kenter, T., de Vries, A.P., Zhai, C., de Jong, F., Radinsky, K., Hofmann, K. (eds.) ECIR 2014. LNCS, vol. 8416, pp. 573–578. Springer, Heidelberg (2014)
19. Wu, M., Scholer, F., Garcia, S.: RMIT university at TREC 2008: Enterprise track. In: Proceedings of TREC-17 (2008)
20. Rocchio, J.J.: Relevance feedback in information retrieval. Prentice-Hall, Englewood Cliffs (1971)
21. Duan, H., Zhou, Q., Lu, Z., Jin, O., Bao, S., Cao, Y., Yu, Y.: Research on Enterprise Track of TREC 2007 at SJTU APEX Lab. In: Proceedings of TREC-16 (2008)
22. SanJuan, E., Flavier, N., Ibekwe-SanJuan, F., Bellot, P.: Universities of Avignon & Lyon III at TREC 2008: Enterprise track. In: Proceedings of TREC-17 (2008)
23. Manning, C.D., Schütze, H.: Foundations of statistical natural language processing. MIT Press (1999)
24. Justeson, J.S., Katz, S.M.: Technical terminology: some linguistic properties and an algorithm for identification in text. Natural Language Engineering 1, 9–27 (1995)

Mixed Language Arabic-English Information Retrieval

Mohammed Mustafa[1] and Hussein Suleman[2]

[1]Department of Information Technology, University of Tabuk, Tabuk,
Kingdom of Saudi Arabia
mohd_must2001@hotmail.com

[2]Department of Computer Science, University of Cape Town, Cape Town,
Republic of South Africa
Hussein@cs.uct.ac.za

Abstract. For many non-English languages in developing countries (such as Arabic), text switching/mixing (e.g. between Arabic and English) is very prevalent, especially in scientific domains, due to the fact that most technical terms are borrowed from English and/or they are neither included in the native (non-English) languages nor have a precise translation/transliteration in these native languages. This makes it difficult to search only in a non-English (native) language because either non-English-speaking users, such as Arabic speakers, are not able to express terminology in their native languages or the concepts need to be expanded using context. This results in mixed queries and documents in the non-English speaking world (the Arabic world in particular). Mixed-language querying is a challenging problem and does not attained major attention in IR community. Current search engines and traditional CLIR systems did not handle mixed-language querying adequately and did not exploit this natural human tendency. This paper attempts to address the problem of mixed querying in CLIR. It proposes mixed-language (language-aware) IR solution, in terms of cross-lingual re-weighting model, in which mixed queries are used to retrieve most relevant documents, regardless of their languages. For the purpose of conducting the experiments, a new multilingual and mixed Arabic-English corpus on the computer science domain is therefore created. Test results showed that the proposed cross-lingual re-weighting model could yield statistically significant better results, with respect to mixed-language queries and it achieved more than 94% of monolingual baseline effectiveness.

1 Motivation

Multilingualism (mixing languages together in talking, for example) is a natural human tendency that is very common in multilingual communities, in which natives use more than one language in their daily business lives and everyday demands (such as teaching, economy and business). In such communities, natives are always able to express some keywords in languages other than their native tongue (Gey, et al., 2005). For examples, the typical Arabic speaker speaks a mixture of tightly-integrated words in both English and Arabic (and various slang variants). Talk/Text mixing in scientific Arabic references/books/lectures/forums is very prevalent in the Arabic-speaking world. Hong Kong

A. Gelbukh (Ed.): CICLing 2015, Part II, LNCS 9042, pp. 427–447, 2015.
DOI: 10.1007/978-3-319-18117-2_32

speakers typically speak Cantonese with many scattered English words. It is noticed, however, that in most of these bilingual/multilingual communities, the common factor in their mixed-language tendency is the use of English as a pivot/second language. This is especially true in technical domains, in which most terminology is borrowed from English. This mixed-language trend, known also as code-switching (Cheung and Fung, 2004), has been one of the major focus researches in linguistics and sociology.

With the growth of the Internet, especially in the few last decades, the mixed-language feature has begun to spread on the Web and gradually non-English natives (i.e. Arabic speakers), who are bilingual, begin to search the Web in a mixture of languages - mostly with English on the fringes but not at the core. They often do this in order to approximate their information needs more accurately, rather than using monolingual queries written in their native-tongue languages in searching. This new type of search can be identified as mixed or multilingual querying. It is also referred to as the bilingual query (Mustafa & Suleman, 2011; Mustafa, 2013; Reih and Reih, 2005). From Information Retrieval (IR) perspective, a mixed query is a query written in more than one language – usually bilingual. For instance, the query 'مفهوم ال polymorphism', (meaning: concept of polymorphism) is a mixed query that is expressed in two languages (Arabic and English). English portions in mixed queries are often the most significant keywords. In the same context, a mixed or a multilingual document can be defined as any document that is written in more than one language (Mustafa, 2013; Fung et al., 1999). Although the primary language in mixed documents is the non-English one (Arabic in this case), the English parts are mostly significant terms and are expected to be good candidates for search, e.g., technical terms or proper nouns. Two forms of text mixing in mixed documents are present in the Web. In the first form, text is mixed in form of text-switching or in tightly-integrated terms/portions/snippets/phrases between two languages, e.g., 'شرح ال polymorphism' (meaning: explain polymorphism). The second form of mixed-language text, which is the most common, consists of similar text (terms/phrases/snippets) description in both non-English and English languages, e.g., 'Indexing الفهرسة'. Probably in such a case, the scientific non-English term is accompanied by its corresponding translation in English so as to refine non-English terms - Arabic term in this example. This feature of co-occurring terms in non-English documents is interesting and has been widely used (Nie, 2010; Peters, 2012). For example, Zhang and Vines (2004) stated that in Chinese Web pages, English terms are very likely to be the translations of their immediately preceding Chinese terms and, hence, the feature was intensively used in mining to extract translations of a great number of terms in queries.

Current search engines and traditional IR systems do not handle mixed-language queries and documents adequately because, in most cases, the search result is often biased towards documents that exactly contain the same terms that are present in the mixed query, regardless of its constituent languages. Hence, the list is often dominated by mixed documents, especially at the top (mostly with one language as a primary), rather than by the most relevant documents. Figure 1 shows an example of a mixed Arabic-English query 'مقدمة في ال threading' (meaning: introduction to threading), submitted to the Google Web search engine. The search was conducted in January 2015. As shown in the figure, the search result is often biased towards mixed documents and thus, many monolingual and highly relevant documents, which are mostly written in English, will be ranked at the lower level of the retrieved list and thus can be easily missed by users – even if they are the best matching documents.

This paper proposes a new re-weighting model to handle the mixed-language feature in both queries and documents. The major idea is to re-weight mixed documents so as to make them comparable to monolingual documents and thus, most documents, regardless of their languages, can be retrieved according to relevance, rather than according to exact matching. The paper focuses on common computer science vocabulary with special attention on Arabic/English bilingual querying and writing (as the feature of mixing languages is very common in scientific Arabic writing) but the techniques can be implemented in other languages.

Fig. 1. An example of mixed Arabic-English query submitted to Google

2 Related Work

In his study to analyze Web users' behaviours, Lu, et al. (2006) tackled the reasons behind using multilingual trends of querying. The findings, which were extracted from the analysis of a query log of a search engine and more than 77,000 multilingual queries, showed that mixed query searching between Chinese and English was primarily caused by the use of computer technologies, and the facts that some Chinese words do not have a popular translation; and the culture, such as in Hong Kong, of using both Chinese and English in speaking and writing.

In their attempt to build a language-aware IR system, the studies of Mustafa and Suleman, (Mustafa & Suleman, 2011; Mustafa, et al., 2011a) are ones of the earlier works to introduce the problem of the mixed-language querying and writing to IR. It was shown in these studies that the phenomenon of mixed-querying and writing is very prevalent in non-speaking world in developing countries and is very common in non-English technical domains, especially in Arabic, which are always dominated by terminology borrowed from English.

Very recently Gupta, et al, (2014) and his colleagues' researchers from Microsoft stated that for many languages, including Arabic, there is a large number of transliterated contents on the Web in the Romanized script and the phenomenon is non-trivial for current search engines. Gupta showed that the problem, which they called Mixed-Script IR, is challenging because queries need to search in documents written in both scripts. The analysis of Gupta and his team was performed on the search log of the Bing. The researchers found that among 13 billions queries, 6% of them were containing one or more transliterated words.

In their work to build a language identifier for query words in code-mixed queries, Bhat, et al., (2014), stated that code-mixing is a socio-linguistic phenomenon that is prominent among multilingual speakers who always switch back and forth between two languages and the phenomenon is suddenly risen due to increasing interest in social networking.

Except the very recent work, neither mixed-language queries nor searches for mixed-language documents have yet been adequately studied. This is because the grounding belief is that the Cross-Language Information Retrieval (CLIR) task, in which users are able to retrieve documents in a language that is different form query language, is a translation followed by a monolingual retrieval, and, thus, most algorithms are strongly optimized for monolingual queries, rather than for mixed queries. Some CLIR studies investigated the use of a hybrid bi-directional approach that merges both document translations, from one direction, with query translation, from the other direction. Some studies tested this approach at word levels (Nie and Simard, 2001; Chen and Gey, 2004; Wang and Oard, 2006; Nie, 2010). Most results showed that such a combination is very useful, but the query sets in them were essentially monolinguals with a grounding base that the test collection is monolingual (in a single and a different language from query's language) and the major aim in these studies is to disambiguate translation, rather than handling the mixed-language feature in queries and documents. From that perspective, most approaches have focused on developing effective translation techniques by re-weighting translations when several alternative candidates are available for a certain source query term.

One of these techniques, which has been widely used and accepted, is the Structured Query Model (SQM) (Pirkola, 1998; Darwish and Oard, 2003). The key idea behind the SQM is to treat all the listed translations that are obtained from a monolingual translation resource as if they are synonyms. Formally, SQM estimates Term Frequency (TF), Document Frequency (DF) and document length as follows:

$$\mathrm{TF}\,(q_i,d_k) = \sum_{\{t|t\in T(q_i)\}} TF(t,d_k) \tag{1}$$

$$\mathrm{DF}\,(q_i) = \left| \bigcup_{\{t|t\in T(q_i)\}} \{d_t\} \right| \tag{2}$$

$$L'_{d_k} = l_{d_k} \tag{3}$$

Where TF (qi,dk) is the term frequency of the query term qi, in document dk and DF(qi) is the number of documents in which the term qi occurs, $TF(t,d_k)$ is the term frequency of the translation t in document d_k, T (qi) is the set of the known translations, with t representing elements in this set, d_t is the set of documents containing the translation t , L'_{d_k} and l_{d_k} is the length of the document d_k (the length of a document is kept in the SQM). The symbols are derived from Levow, et al. (2005).

SQM has been widely accepted and many studies, later, derived several variants using the same technique. Stated in (Darwish and Oard, 2003), Kwok presented a variant to structured query by substituting the union operator with a sum:

$$\mathrm{DF}\ (q_i) = \sum_{\{t|t \in T(q_i)\}} DF(t) \tag{4}$$

Other approaches, like the study of Darwish and Oard (2003) as an example, incorporate translation probabilities within TF and DF estimation with an assumption that using statistical models usually result in translations with both strong and weak probabilities (Darwish and Oard, 2003; Wang and Oard, 2006). In all these approaches, however, a single index, in which all documents are put together, in a monolingual language with a monolingual query set, was used.

But, in mixed-queries and documents, more than one language are present (multilingual and mixed documents) in document collection. In such a case, the task becomes of Multilingual Information Retrieval (MLIR) (Peters, et al., 2012). As in CLIR, one of the major architecture in MLIR is the use of a centralized architecture, which considers putting all documents into a single index pool. Queries are often translated into all the target languages (document collection languages) and then merged together to formulate another big queries, which are those submitted to the single index. For example, if the source query contains only the word 'deadlock' then its translation(s), e.g., 'الجمود' is merged to the original query, resulting in a mixed query 'deadlock الجمود', which will be submitted to the single index. Other approaches attempts also to solve the problem in terms of what is known as 'result fusion' (Peters, et al., 2012), in which each query (source or target) is used to retrieve documents from the big single index and then the results are merged together using result merging methods.

Gupta, et al., (2014) stated that the practical method for handling mixed documents can be to dividing these documents into corresponding sub-documents and/or sub-collections, but, this will make the documents useless. From that perspective, the use of a single centralized index is the most appropriate approach to the work presented here for two reasons. First, their document collection is often multilingual (probably with some mixed documents). Second, both approaches (mixed-language problem and centralized architecture of MLIR) utilize mixed queries.

But, the use of a centralized index, whenever there are several monolingual sub-collections, has been shown to have a major drawback, which is overweighting (Lin and Chen, 2003; Peters, 2012). Overweighting means that weights of documents in small collections are often preferred (due to low DF and to increase in the total number of documents). In this paper, this type of overweighting is called *traditional overweighting*.

With respect to mixed-language problems, other limitations can be also identified. In mixed-language queries and documents, term frequencies of similar terms across languages are assigned and computed independently- as if they are different or in a single monolingual language , despite the fact that these terms (the source and its translation) are akin to each other, but cross-lingually (note that query words in centralized architecture are often merged with their translation to formulate a big and mixed query like 'Inheritance الوراثة ' in which each word is a translation of the other). As a result for this drawback, mixed documents may dominate the top of the retrieved list as their scores will be computed from the entire query. In this paper, this drawback is called *biased term frequency*. The same arguments (independent computations of terms that are similar) also apply for Document Frequency (DF) computation. This would likely skew the final list or the term may suppress the impact of its translation(s) (or vice versa). In that perspective, the term with low document frequency would likely overweight, even if its translation(s) is of low importance (high DF). This skew in DF is called in this thesis *biased document frequency,* whereas its consequent overweighting in terms is called *overweighting due to mixture of texts.*

On the other hand, the feature of co-occurrence of terms (terms that are accompanied with their translations, e.g., the co-occurrence of 'الإقفال deadlock' together but in two different languages, would likely conspire to increase the scores of mixed documents and cause them to earn extra weights that are not really part of their weighting values. Consider the two documents that follow. The example is typically taken from a real Arabic textbook. The first document D_1 is a mixed document with Arabic as a primary language and English as a secondary language, whereas the second document D_2 is a monolingual English document:

D_1: "تؤدي عملية التطبيع *normalization* لإنشاء مجموعة جدول *tables* ذات..."
D_2: *"The process of normalization leads to the creation of tables, whose..."*

D1 is the exact translation of D2. But, since D2 is in Arabic, as a primary language, the translated English term 'normalization' co-occurs with its Arabic equivalent 'التطبيع'. In a centralized architecture, if the query is the word 'normalization', for example, the big merged query may probably contain something like 'التطبيع normalization', in which the translation 'التطبيع' is concatenated to source query. Such a query would cause the mixed document D1 to be ranked ahead of document D2 because the Arabic term 'التطبيع' tends to co-occur with its equivalent English term in D1. Thus, the document earns extra weights, as the TF was computed twice. This is a key problem in mixed-language queries and documents. Obviously, the attributes of such weighting in this example is not desirable. Accordingly, managing such shortcomings is crucial to improve the accuracy of term weighting in mixed documents. A complete example for the different types of drawbacks is provided by the authors in (Mustafa & Suleman, 2011; Mustafa, 2013).

3 Why Mixed Arabic-English Querying/Writing

Four major reasons were identified by the authors (Mustaf & Suleman, 2011; Mustafa 2013) for the mixed-language trend in speaking, writing and querying in the Arabic world:

Firstly: it is not always possible for Arabic speakers to provide precise Arabic translations for newly added terms, as most scientific terminology is borrowed from English, or/and not always feasible for those users to directly express their concepts in medicine and technology, for example. Secondly: the translation/transliteration, if any, of newly added terms to Arabic (referred to as Arabicization), is not usually performed on a regular basis (The Academy of Arabic Language, 2011). This is a significant problem because it makes the Arabic language limited in its vocabulary of up-to-date terminology and, thus, Arabic speakers are unable to express some keywords in their native tongue and English technical terms are instead utilized. Thirdly: one of the most significant problems with the Arabicization process, when it is performed, is that scientists who execute the process do not usually invite the experts and scientists in a given scientific domain to participate (The Academy of Arabic Language, 2011). This is a wide-spread problem in the Arabic world and it results in making translated/transliterated Arabic terms, in most cases, ambiguous, chaotic and are almost not understood by Arabic speakers. Fourthly: in order to avoid missing some valuable documents due to its regional variation, Arabic users prefer to express terminology in English, rather than in Arabic. In fact, Arabic documents that cover particular topics in technical domains are usually regionally variants. The problem of regional variation in Arabic, especially in scientific domains, is crucial. This is especially true when considering the Arabic-speaking world. The region has 22 countries, many of them with their own academy for the development of the language (Mirkin, 2010). Each academy translates/transliterates/Arabicizes new terminology individually, without a well-established coordination in most cases with its peers across the Arabic-speaking world (The Academy of Arabic Language, 2011). As a result, scientific modern terms in Arabic Gulf countries may be totally different from those in Levantine countries. Table 1 shows some samples of these regional variations, gathered from the Web in the computer science domain. The regional variation problem affects mixed-language queries solely because in most cases it has an impact on translation of mixed queries; for example, many candidate translations would be produced.

Table 1. Some regional variations in Arabic collected from the Web

English Term	Arabic Term	English Term	Arabic Term
Hardware	العتاد	Hashing	التشتت
	المكونات المادية		البعثرة
	المكونات الفيزيائية		الفرم
Linked List	القائمة المتصلة	Symmetric key	المفتاح المتناظر
	القائمة المتسلسلة		
	اللائحة المترابطة		المفتاح المتماثل
	السلسلة المتصلة		

4 Proposed Cross-Lingual Structured Query Model

The major idea behind the proposed re-weighting is to utilize the features of the mixed documents to extract reasonable weights in terms of TF, DF and document length for terms in mixed queries. On one hand, since a term may frequently appear in the same mixed document but in multiple languages then it is reasonable to handle these terms as synonymous but cross-lingually. For a query that requests the source term, a document with just one or more of its translation(s) along with any source term should be retrieved. The major subsequent result to this general assumption is that weights of such synonymous terms across languages would likely be computed together as if these terms are a single similar term, rather than decomposing their computations (in terms of TF and DF) individually. Thus far the mixed document can be viewed as if it is in a single monolingual language and resulting in making monolingual documents comparable and more competitive to those mixed ones. Using such a paradigm makes the weight cross-lingual, instead of monolingual.

On the other hand, since the technical term may be accompanied by its translation, then its term frequency and document frequency components should be adjusted. This is the basic notion of the proposed approach, which is called in this paper the cross-lingual structured query model and which adjusts TF and DF components, resulting in what is called in this paper cross-lingual TF and cross-lingual DF.

According to the above assumption, if the source query language term q or its translation A_i appears in the mixed document D, these terms are treated as if the query term q occurs in the document D and hence, both the translation A_i and the term q are considered as synonyms but in different languages. Formally, in the proposed model, the cross-lingual TF of the source query term can be expressed as:

$$\grave{TF}_{q,D} = TF_{q,D} + \sum_{\{i|i \in T_q\}} TF_{i,D} \tag{5}$$

Where, $\grave{TF}_{q,D}$ is the new computed frequency of occurrences of the source term q (this is the joint TF of synonyms across languages), D is a mixed document in more than one language, the $TF_{q,D}$ is the frequency of occurrence of the source term q in the document D, T_q is the set of the translations of the term q in the target (document) language and $TF_{i,D}$ is the number of occurrences of a given element in the translation set T_q that appears in the document D. Thus, $TF_{T,D}$ is the number of occurrences of the terms in the translation set T_q that occurs in document D. Symbols are derived from Levow, et al. (2005).

The second step of the proposed weighting aims to circumvent the impact of the TF component when a source query term is accompanied by its corresponding translation(s), e.g., 'deadlock الإقفال'. The premise made here is that since a source term tends to co-occur with its equivalent translation or vice versa, it is unlikely to compute weights for each of the co-occurring terms (a weight for the term in Arabic and another weight for its translation in English). In such a case, it is reasonable to apply a decaying factor for the joint TF of cross-lingual synonyms. Intuitively, the decay factor can be estimated based on how frequently terms in queries co-occur together in mixed documents, that is when a source term co-occurs with its translation(s), the cross-lingual joint TF, as in equation 5, is rebalanced by decreasing the frequency of the source query term by 1. Bilingual terms across languages are considered as 'co-occurred terms' in this paper if they appear together in a window of size 5. Such an assumption seems reasonable for two situations. Firstly, terms don't need to be in the same order in texts, e.g., 'deadlock الإقفال' and ' الأقفال deadlock' are similar. Secondly, individual terms in phrases did not need be exactly neighbouring to each other, e.g., the cross-lingual phrase 'mutual exclusion الاحتكار المتناوب'.

At this point, let's assume that the source language query term q is placed in a set Q and its translations are placed in another set T_q, where $T_q = \{a1, a2, \ldots, an\}$. The Cartesian product between these two sets, denoted below as C_q, will generate possible pair combinations between each source query term in the source query language with one of its translations, e.g., (q, a_1), $(q, a_2)\ldots$, (q, a_n). Hence, the decaying factor of the TF of the source query (synonyms across languages) can be computed as follows:

$$C_q = \mathrm{Q} \times Tq \tag{6}$$

$$T\grave{F}_{q,D} = TF_{q,D} + \sum_{\{i|i \in T_q\}} TF_{i,D} - \sum_{\{n|n \in C_q\}} TF_{n,D} \tag{7}$$

In which $TF_{n,D}$ is the frequency of occurrences of each element (pair) in the set C_q in document D. Other terms are previously defined. Summing up the number of occurrences of all pairs in the set C_q represents the decaying factor. Thus, the decaying factor suppresses the impact of the biased TF problem and handle the co-occurring terms in different languages.

Document length is an essential component in similarity computations. This is because the longer the documents the more terms paired with distinguished terms are assumed to be found and consequently leading such documents to have higher TF as well as increasing the likelihood of containing terms that match the user's query. Modification of TF in equation 7 has a potential consequent impact, even if it is low, on the document length as the latter needs to be updated by decreasing it by 1 for each co-occurring of cross-lingual terms in mixed documents. Formally:

$$\grave{L}_D = L_D - \sum_{\{n|n \in C_q\}} TF_{n,D} \tag{8}$$

Where $\grave{L}_D$ is the new length of document D and L_D is the original number of terms in the same document D and other terms in the formulae are previously defined.

Document frequency estimation in the proposed model depends on how frequently a certain query term or one of its corresponding translations occur in all documents, regardless of their languages. Assuming that terms are synonyms across languages, it is reasonable to count every document that contains each of these synonymous terms in the DF statistics. Accordingly, if a document D includes at least one translation a_i, that document can be handled as if it contains the query term q and vice-versa. This would minimize the problem of biased document frequency because the document frequency will be computed across all documents (those in Arabic, English and mixed documents). Thus, if the source query term, for example, appears in many documents, whereas one of its translations occurs only few times (high weight), the result list will not be skewed towards that translation, as the document frequency will be computed as a joint document frequency containing all documents that include the source term or one of its translation(s). Formally, cross-lingual joint DF is computed as:

$$D\grave{F}_q = DF_q \cup_{\{i|i \in T_q\}} DF_i \tag{9}$$

Where DF_q is the set of documents which contain the source language term q in monolingual and mixed documents, $D\grave{F}_q$ is the new computed document frequency of the source term q in all documents in the collection regardless of the language(s) present in these documents (this is the joint DF of synonyms across languages), DF_i is the set of document which contain any translation ai in documents, thus, the $DF_{T_q,D}$ is the set of documents in which one or more terms in the set T_q in the document collection occur and other terms are defined above.

If the Kwok formula (see equation 4), which alters the union operator (∪) to a normal summation operator (+) for simplicity, is used - as in this paper, equation 9 would become:

$$D\grave{F}_q = DF_q + \sum_{\{i|i \in T_q\}} DF_i \quad (10)$$

It is noted that the proposed cross-lingual SQ model didn't make use of translation probabilities, which have been shown to be important in CLIR. One might ask why not use such approaches of translation probabilities in the proposed weighting. The answer stems from the differences between searching and retrieval in a technical/specialized domain against searching and retrieval in general- domain news stories. For instance, developed techniques using new-genre domains are not set to be directly applied to scientific domains, especially in multilingual scientific Arabic collections. In such cases, there is a likelihood of poor retrieval. For example, the words 'object' and 'Oracle' might have valid entries with different meanings/ alternatives, each of which can be probabilistically estimated in general-purpose dictionaries. However, the same words are very specific if the searched domain is in common computer science. Therefore, for the news domain it might be suitable to retrieve documents that contain the most probable translation in the target corpus, rather than including all of them, when a set of synonymous translations are present. However, this can be considered as an undesirable behaviour in technical jargon as this criterion of choosing the most probable translation does not hold. Particularly, the converse is quite accurate, especially for a language with several regional variations – as in the Arabic-speaking world. In such cases there may be a very highly relevant document that contains relatively infrequent translations, which is pertaining to a specific region/dialect but it does not for others. Consider the Arabic translations for the technical English phrase 'object oriented programming' when the target corpus is in common vocabulary of computer science in Arabic. The translations are: ' البرمجة الشئية', ' البرمجة كائنية التوجه', ' البرمجة موجهة الأهداف' and 'البرمجة كائنية المنحى'. All these alternative translations can be used in scientific Arabic documents, but according to the dialect/tongue of the writer. Technical topics in Arabic computer science domain exhibit this specific behavior. The same arguments also apply to Table 1. Thus, the appearance of what seems to be a superfluous translation like 'البرمجة الشئية' in documents does not make such a translation as an undesirable or irrelevant.

5 Experimental Design

There are inherent difficulties when experiments in this paper were conducted. The major difficulty was the lack of a suitable mixed and multilingual Arabic-English test collection. It is true that many ad-hoc text collections were developed, but most of them, and almost the majority of the Arabic/English CLIR collections, are monolingual or consist of several monolingual sub-collections, rather than containing documents with mixed languages. Furthermore, query sets in these collections are essentially monolingual. It is also noted that existing test collections are primarily focused upon general domain news stories whose vocabulary is almost monolingual. Arabic is also rare among standard scientific test collections and they are almost not synchronic (regionally variant). Regional variation in technical Arabic domains is principal. For these reasons, a primarily Web-based multilingual and mixed Arabic-English test collection, with approximately 70,000 documents and 42 million words, on common computer science vocabulary was therefore developed. The collection, which has been named MULMIXEAC (MULtilingual and MIXed English Arabic Collection) contains also a set of 47 topics (and thus, 47 mixed queries) with a sufficient set of relevance judgments. The choice of the domain (computer science) was mainly governed by the spread of both multilingualism and Arabic regional variants in this domain. The collection had been gathered using different approaches. Students and tutors at two Arabic universities participated also in building both the document collection and the query set. Table 2 provides brief statistics for the collection. The last row in the table shows the total of each heading in the first row.

Table 2. Statistics for the MULMIXEAC collection. Figures are provided without stemming

Number of documents			Number of words		Number of distinct words	
English	Arabic	Mixed	English	Arabic	English	Arabic
51,217	483	17,484	37,169,213	4,683,724	512,976	162,032
69,184			41,852,937		675,008	

All documents in the collection are in HTML format. A language tag attribute "*lang*" inside a paragraph tag (<*p lang* ="*ar*"> or <*p lang* ="*en*">) was added to documents –depending on how much a certain document is mixed (or monolingual). So, if the document contains two portions, for example, in the two languages, then two paragraph tags, besides those that already appear, with different values for attribute '*lang*' will occur in that document. This is essential for preparing the texts for indexing as it would help to identify the correct stemmer. A complete analysis for the test collection (how it was collected, processed, cleaned and statistically tested) is described by the authors in (Mustafa, et al., 2011b; Mustafa, 2013).

Queries in the collection have been collected by asking potential users (students and tutors). In order to implement the blindness, the choice of the query language was deliberately avoided and, hence, participants could show their natural searching behaviours. Before submitting the queries, participants were only shown the categories of the MULMIXEAC corpus, but without any pre-knowledge about the corpus itself. To this point, a raw set, which was semi-blindly produced, consisting of 500 queries was obtained. In this acquired set, more than 48% (240 queries) of the created set were expressed in multilingual forms. Monolingual English queries represented also a higher proportion (about 42%), whereas Arabic the query proportion was only 10% of the total number of the submitted queries in the set. Users' behaviours in querying had been investigated and details are provided in (Mustafa, 2013). A complete analysis also for how the process was performed, including also relevance judgment, is provided in (Mustafa, 2013). The collected set of 47 queries, which results from different steps and it is described in (Mustafa, 2013), was put into similar formats as the TREC *topics* (queries), e.g., *Title*, *Narrative*, etc. However, extra fields were added also. The most important one is the *originalQuery* field, which stores the original mixed-language query in both Arabic and English, as it was submitted by the potential user. Queries were also numbered (DLIB01- DLIB47) for referencing. Some sample queries in the created set are listed in Table 2. More details and analysis about document collection, query set creation and relevance judgments can be found in (Mustafa, et al., 2011b; Mustafa, 2013).

Table 3. Examples of some mixed queries (DLIB01-DLIB07) in the created query set

Query #	Query	Counterpart in English
DLIB01	مفهوم الـ Deadlock	Concept of deadlock
DLIB02	ماذا نعني بالـ Secure Socket Layer	What is meant by Secure Socket Layer
DLIB03	الفرق بين الـInterpreter و الـ Assembler	Difference between interpreter and assembler
DLIB04	شرح Polymorphism في الجافا	Explain polymorphism in Java
DLIB05	مثال في Entity Relationship Model	Entity and Relationship Model, Example
DLIB06	تقنيات Data Mining	Data Mining techniques
DLIB07	تمارين Synchronized Methods جافا	Tutorials on synchronized methods in Java

As the retrieval search task for this work was to emphasize highly relevant documents and whether they are ranked at higher position, multiple levels of relevance (graded relevance) were employed. The assessment had been done on a six-point scale (0-5) with 5= High relevance, 4= Suitable relevance, 3= Partial relevance, 2= Marginal/very low relevance, 1= Possibly not- relevant and 0= Irrelevant. The relevance judgment was done for top 10 documents only due to resource limitations.

Indexing, Stemming and Query Translation

All documents regardless of their languages were placed together into a single pool index. Four logical field types were utilized to populate text during the indexing stage. These fields were *<TITLE-Arabic>*, *<CONTENTS-Arabic>*, *<TITLE-English>* and *<CONTENTS-English>*. Thus, depending on the language(s) of documents, some or all fields may be used. During indexing, documents were normalized and processed, For the Arabic texts in both monolingual Arabic the mixed documents, the prior-to-indexing step begins with processing the *kasheeda* and removing diacritical marks. Following this, a letter normalization process for the Arabic texts was also executed so as to render some different forms of some letters with a single Unicode representation, e.g., replacing the letters HAMZA (أ،إ) and MADDA (آ) with bare

ALIF (ا). English documents and English parts in mixed documents were also normalized in terms of case-folding. As there are many fields, during indexing, language dependent processing for stemming and stopwords removal was applied. The Lucene IR system with some developed components to apply the proposed cross-lingual Kwok's approximation, was used in all experiments. Arabic words were lightly stemmed using the LIGHT-10 stemmer, whereas the English words were stemmed by the SNOWBALL stemmer. In the experiments, an extension of the BM25 weighting scheme in terms of multiple weighted fields (Robertson, et al., 2004) has been implemented. That extension, which was developed by Robertson, is based on refraining from doing linear combination of scores obtained from scoring every field in documents. Instead, the proposed alternative is to calculate a single score for the linear combination of term frequencies and document frequencies of terms in the different fields, but weighted by the corresponding weighted fields. The tuning parameters values that were set in the Okapi implementation for experiments were 2 and 0.5 for K_1 and b, respectively.

The *OriginalQuery* field in all topics was used as a source query. Since each source query is originally mixed in two languages, two directions for translations had been identified, from Arabic to English and vice-versa. Since English portions in mixed queries are assumed to be technical terms, their translations were first looked up in a special English-to-Arabic computer-based dictionary, word-by-word. This is an in-house dictionary. If there is more than one translation present, all of them are retained and used. This is necessary in technical Arabic domains due to regional variations. If the English word is not found in the dictionary, Google Translator is used with only one sense returned. The same source, which is Google, was used to translate Arabic words, which are often taken from general purpose vocabulary, into English with at most one sense returned.

6 Experiments

Five runs were conducted. Due to nature of the experiments, three of these runs are baselines and the other two are the official runs:

- he first experiment, which was called $\mathbf{b_{IR}}$, is a monolingual English run, in which the original mixed queries were translated manually by human experts to English to generate monolingual English Queries. The b_{IR} experiment was conducted to upper-bound retrieval effectiveness for CLIR experiments reported in this paper.
- The second experiment, which was called $\mathbf{b_{IRengine}}$, arose from searching capabilities of existing search engines and how they handle mixed queries. Thus, the $b_{IRengine}$ run mimics, and thereby exploits, retrieval of search engines (search-engine-retrieval-like), in which mixed queries are posted to the MULMIXEAC collection index as they were submitted by originators and with performing any translation process.
- The third experiment was the CLIR lower baseline - named as $\mathbf{b_{CLIR}}$. It combines the centralized approach of indexing with the traditional structured query model(s). On one hand, the centralized architecture is the most similar approach to the work presented here and it is a widely reported baseline in MLIR. On the other hand,

SQM is another widely reported baseline that is able to cluster monolingual synonyms (regional variants in this case). Accordingly, in the methodology of this run, the two monolingual queries, which were obtained by bi-directional translations, as described above, were merged together to form another big and yet mixed query, in the two languages. Following this, the Arabic translation of the English technical terms are grouped together using Kwok's approximation. For example, in a mixed query like ' مفهوم الـ deadlock, the two bi-directional translated queries are merged to produce another big query 'مفهوم الجمود الاستعصاء التوقف التام الإغلاق الإقفال concept deadlock', in which words presented in the same colour are translations of each other. Next, the Arabic translations of the word 'deadlock' were structured together as many alternatives are found (equations 1 and 4 for estimating TF and DF, respectively). Thus, the final big query can be represented in SMART notation as: #SUM(concept deadlock # SYN(الجمود الاستعصاء التوقف التام الإغلاق الإقفال) مفهوم). Through these experiments, both the lower baseline and the search-engine-retrieval-like run ($\mathbf{b_{CLIR}}$ and $\mathbf{b_{IRengine}}$) are referred to as the *mixed-query baselines*, whenever comparisons referred to both of them together.

- The fourth run, which was called CRSQM-NODECAY, investigates the impact of the proposed cross-lingual SQM but the neighboring feature of co-occurring terms, as in 'deadlock الأقفال', was not considered. Its methodology is similar to the $\mathbf{b_{CLIR}}$ run but the cross-lingual structuring here is performed cross-lingually and according to equations 5 and 10 for estimating TF and DF, respectively. Note that the proposed cross-lingual weighting is only applied if the translations of the source query term are obtained from the technical dictionary.
- The fifth experiment, which was called CRSQM-DECAY, tested the proposed re-weighting scheme after using a damping factor for bilingual paired terms that tend to co-occur together. Its methodology is similar to the CRSQM-NODECAY run but the structuring here is performed cross-lingually with a damping factor for the TF. Thus, the TF, DF and document length were estimated as in equations 7, 10 and 8, respectively.

In all runs, the Discounted Cumulative Gain (DCG) was used to measure performance and it was computed for the top 10 documents for each query used for retrieval, as the retrieval task emphasizes highly relevant documents. The DCG values across all the 47 queries were averaged and the statistical Student's t-test measure was used.

Figure 2 depicts retrieval effectiveness of all runs presented. The retrieval effectiveness was assessed by the average DCG over 10 points (@ top k documents, k=1..10). Table 4 reports the same retrieval effectiveness of all runs presented in a tabular form. Values in the table are chosen for some document cut-off levels from 1 to 10, due to space consideration.

7 Results

As can be seen in the figure, approaches that make use of the proposed weighting algorithms produced more effective performance that was consistently higher than both mixed-query baselines (lower and search-engine-retrieval-like baselines).

The difference in effectiveness of the two runs (CRSQM-NODECAY and CRSQM-DECAY), compared to the two mixed-query baselines, begins with small values at the 1st top document (when k=1) and increases gradually as more documents were accumulated.

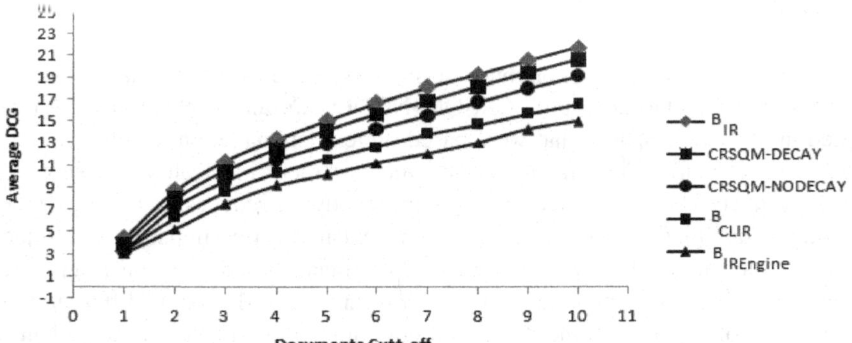

Fig. 2. The average DCG of all runs at document cut-off values [1..10] for the different runs

Table 4. Results of different runs in terms of average DCG for some document cut-off levels

Measure	Average DCG @							
Run	1	2	4	5	6	8	9	10
b_{IR}	4.447	8.745	13.424	15.174	16.771	19.409	20.671	21.882
b_{CLIR}	3.340	6.277	10.357	11.640	12.759	14.783	15.770	16.641
$b_{IRengine}$	3.040	5.250	9.249	10.255	11.298	13.125	14.335	15.064
CRSQM-NODECAY	3.319	7.277	11.559	12.961	14.380	16.868	18.089	19.249
CRSQM-DECAY	3.915	8.020	12.540	14.217	15.731	18.292	19.561	20.746

Table 5 lists the *p*-values of significance tests of both the CRSQM-NODECAY and CRSQM-DECAY runs, compared to the lower baseline b_{CLIR} at document cut-off levels: 2, 3, 4, 6, 8 and 10. White cells in the table indicate statistically and significantly better results than the baseline, while grey cells indicate that there is a difference, but it is statistically insignificant.

Table 5. P-values using the Student's t-test of both the CRSQM-NODECAY and CRSQM-DECAY runs against lower baseline (bCLIR)

Measure	Average DCG @					
Run	2	3	4	6	8	10
CRSQM-NODECAY	0.513	0.192	0.064	0.007	0.000	0.000
CRSQM-DECAY	0.297	0.079	0.020	0.002	0.000	0.000

This improvement in performance for the two different re-weighting schemes was attributed to the use of the proposed cross-lingual structured model, in which technical terms in queries are cross-lingually structured, regardless of their languages (by handling them as synonyms across languages and as if they are in a single language). This cross-lingual structuring resulted in that the weight of each technical source term, mostly in English, was calculated as a single weight consisting of re-estimating both the term frequency and the document frequency of the same source term with those in its all cross-lingual synonymous terms and regardless of their languages. This cross-lingual computation in both CRSQM-NODECAY and CRSQM-DECAY runs resulted in different impacts, that were based on text language, on documents (mixed versus monolingual). While in mixed documents structuring technical terms cross-lingually reduces their estimated scores significantly, it reduces, also the scores of monolingual documents, but with a slower rate. Such different impacts on documents stemmed from the different effects of the cross-lingual structuring on English term weights versus Arabic term weights, which in turn were reflected as different effects on the scores of mixed documents versus monolingual documents. It is evident that that Arabic terms were over-weighted in mixed-query baselines, due to the low number of documents in their corresponding sub-collection. But, when technical terms were appropriately and cross-lingually structured in the CRSQM-DECAY and the CRSQM-NODECAY runs, the document frequencies of Arabic technical terms, which are essentially technical translations, would increase significantly (note that the document frequency of the English technical term, which was relatively high, was added and the collection is dominated by English). Such increase in document frequencies of Arabic technical terms would probably have a reduction effect on their weights and moderates the overweighting problem. As a result, mixed documents, which mainly obtained their higher scores from these over-weighted Arabic translated terms, may re-weighted into lower weights, depending on their cross-lingual TF and DF statistics. Likewise, document frequencies for English technical terms, instead of Arabic terms, using the cross-lingual structuring, were also reduced, as structuring query terms across languages causes such English terms to expand their weight computations to include their synonymous terms in the Arabic language. But, this increase in the document frequencies of English terms was small because the Arabic sub-collection size was also small. Thus, their weights were not affected too much and consequently the scores of the monolingual English documents were not reduced too much. In this way, the overweighting problem in the CRSQM-DECAY and the CRSQM-NODECAY runs was moderated. Hence, the IDF factor of the cross-lingual structuring was used to make a difference in the weights of Arabic terms (mixed documents mainly) versus English terms (monolingual English documents). The consequent result of this re-weighting in the both CRSQM-DECAY and CRSQM-NODECAY runs was that many monolingual English documents, which were mostly more relevant, would probably be ranked ahead of mixed documents and thus, the dominance of mixed documents on top was broken, although some of these mixed documents were still placed at higher ranks due to their high term frequencies.

The term frequency component in the cross-lingual structuring was another reason for the better performance of the proposed re-weighting schemes. This is because structuring technical terms cross-lingually makes the term frequency component in mixed documents versus monolingual documents more comparable because the term frequency of terms would be counted regardless of their languages. The use of the cross-lingual structuring for term frequency suppresses impact of the biased TF and consequently causes an improvement in the proposed re-weighting scheme runs.

Contrary to such cross-lingual computations for estimating term frequency and document frequency components in the proposed approaches, the mixed-query baselines assigned weights of technical terms independently from weights of their translation(s), although these translations were monolingually structured, thus resulting in the deterioration of performance. Furthermore, the over-weighting problem makes both the lower baseline and the search-engine-retrieval-like runs yield significantly worse results.

From the results of the CRSQM-DECAY run, it is clear that extending the term frequency statistics of the cross-lingual structuring and re-weighting to consider the phenomenon of any two similar and bilingual terms that co-occurred together in documents (i.e. 'deadlock الإقفال', in which the term 'deadlock' co-occurs with its Arabic translation), suggests that such neighbouring terms in different languages, even within a predefined window, can have a substantial effect on retrieval performance. This is obvious when the average DCG values of the CRSQM-NODECAY weighting are compared with those in the CRSQM-DECAY weighting. The improvement in retrieval performance between the runs at top 10 was distinguishable and statistically significant (p-value < 0.000012). This moderate improvement in CRSQM-DECAY derived from the fact that both the prior well-established cross-lingual weighting in CRSQM-NODECAY (and the lower baseline as well) may cause some terms, even when they are cross-lingually structured, in the mixed merged queries to earn somewhat double weights, due to the co-occurrence of the same term in multiple languages in document. In the CRSQM-DECAY run the cross-lingual term frequency suppresses the impact of such co-occurred pairs into different languages.

However, the difference in performance in both the CRSQM-DECAY and the CRSQM-NODECAY runs was not consistent through all queries. Figure 3 illustrates a query-by-query comparison for some of the top queries using the CRSQM-NODECAY approach versus the CROSS-DECAY method. The majority of the queries

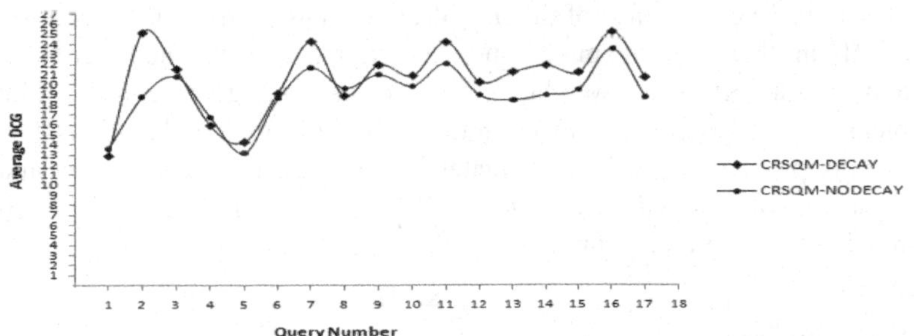

Fig. 3. Query-by-query comparisons, in terms of average DCGs, for the (CRSQM-DECAY) and the (CRSQM-NODECAY) runs, respectively

in the figure show better results, in most cases, or sustain the same performance, for CRSQM-DECAY over the CRSQM-NODECAY run. This was primarily derived from the artifact that a considerable number of technical terms in many mixed documents contain different snippets/phrases/terms but into two different languages. Note that the tuning parameter value of k_1 in the used okapi BM25 model in all experiments was set to the value 2. In figure 3 it is shown also that the performance, in terms of an average DCG, was indistinguishable and almost similar for some queries, especially for some of the first queries, but it was significant for others. There are only few cases in which the CRSQM -NODECAY run outperformed the CRSQM -DECAY (3 queries in the plotted graph). With respect to these queries, this was mainly because some English keywords, e.g., the word 'system', can be found in many documents but in different topics. Thus, if the term frequency of such a word is high in some irrelevant documents, this may probably result in the inclusion of these documents on the retrieval list, possibly at the top, but depending on their frequencies.

The findings of the proposed re-weighting schemes imply major observation. There is a considerable number of mixed documents that include terms that were written into two different languages. Handling these terms reasonably can result in a significant effect on performance.

Comparing baselines together, including the upper one, the best results are related to the upper baseline run (b_{IR}). This is because the retrieval of the upper baseline was performed using manually translated queries with experts and, hence, no noisy translations could affect retrieval. It can also be seen in the figure that the retrieval efficiency for the lower baseline (b_{CLIR}) run goes down and reaches a declining percentage of approximately 24% (average DCG at rank 10 was 16.641, whereas it was 21.882 for the b_{IR} run), compared to the full efficiency of the upper baseline run at top 10 ranked documents. A similar drop in the effectiveness of the naive search-engine-retrieval-like baseline ($b_{IRengine}$) also occurred. In particular, at rank position 10, the performance of $b_{IRengine}$ run falls to a low minimum of %31, compared to upper baseline and %10, compared to lower baseline. This decline in performance of the mixed-query runs (b_{CLIR} and $b_{IRengine}$), was mainly caused by that these runs were often attempting to perform exact matching between queries and documents, but with no sufficient analysis for the type of the submitted query (monolingual or mixed) and regardless of the language presented in each. Furthermore, the overweighting problem contributed to bad performance of the mixed-query runs. Particularly, it causes many terms, mostly in Arabic, in the mixed queries (original or mixed and merged after bi-directional translation) to overweight, as their corresponding sub-collection/language size, typically the Arabic one, included in the big multilingual collection is small. The overweighting problem and the exact matching of terms between queries and documents regardless of their languages makes the result lists bias towards mixed documents in the two mixed-query runs.

8 Conclusion

Although mixing languages together in talking/writing is a widespread human tendency in non-English speaking world, Arabic as example, but the problem has attained very little attention in IR community. In this paper, the mixed-language IR problem is introduced and tackled to show its importance and why the phenomenon should be considered in future search engines, which should allow multilingual users (and their multilingual queries) to retrieve relevant information created by other multilingual users and well studied. It was shown that existing search engines and traditional CLIR systems do not handle mixed language queries and documents adequately. This is because the majority of the existing algorithms, and also test collections, are optimized for monolingual weighting and queries, even if they are translated. The mixed-language problem in this paper has been studied through a corpus that had been created for this purpose. The corpus is multilingual and mixed in both Arabic and English, synchronic and specialized in common computer science vocabulary.

To meet the primary goal of building language-aware algorithms and/or mixed-language IR, the main focus of the paper was to explore weighting components. In particular, a cross-lingual re-weighting method (cross-lingual structured query model) was proposed. Thus, for any technical source query term, regardless of its language, it can be language-aware by obtaining its translations firstly and then all the candidate translations are grouped together with the source term itself, resulting in cross-lingual synonyms. This was done while taking into consideration different forms in which texts in different languages are mixed, e.g., bilingual co-occurring terms in mixed documents, and their impact on retrieval performance. Thus, based on such a mixed-language feature, term frequency, document frequency and document length components were adjusted using the cross-lingual re-weighted model. The experiments suggest that the use of the proposed model could yield statistically better results compared to traditional approaches. Furthermore, the model could suppress the impact of the independent computations of terms that are similar across languages in mixed queries and documents. The results lead to the conclusion that indeed it is possible to develop an IR system that can handle mixed queries and mixed documents effectively. However, it is important to validate the results in the future by using NDCG instead of DCG and the distribution of the relevant documents in each sub-collection should be also provided in the future work.

A number of potential directions are also worthy to be explored in the future. It is firstly being planned to extend the size of the MULMIXEAC corpus. The same Arabic and English languages are still the focus for this extension. During this stage also, a mixed language identifier will be developed. Language identifiers are important to mixed-language IR system. Another direction for future investigation is the BM25 Okapi field weighting space. It was shown that some approaches extended the model to multiple weighted fields. Such simple extension to multiple weighted fields was shown to be effective, but yet fields of documents are still in a monolingual language. In particular, in the future work, it is aimed to explore whether an estimated probability of how much a document is mixed can be incorporated in field weights (note that fields are in several languages).

References

Bhat, I., Mujadia, V., Tammewar, A., Bhat, R., Shrivastava, M.: IIIT-H System Submission for FIRE2014 Shared Task on Transliterated Search (2014), http://www.isical.ac.in/~fire/working-notes/2014/MSR/FIRE2014_IIITH.pdf

Chen, A., Gey, F.C.: Combining query translation and document translation in cross-language retrieval. In: Peters, C., Gonzalo, J., Braschler, M., Kluck, M. (eds.) CLEF 2003. LNCS, vol. 3237, pp. 108–121. Springer, Heidelberg (2004)

Cheung, P., Fung, P.: Translation disambiguation in mixed language queries. Machine Translation 18, 251–273 (2004)

Croft, W.B., Metzler, D., Strohman, T.: Search engines: Information retrieval in practice. Addison-Wesley (2010)

Darwish, K., Oard, D.W.: Probabilistic structured query methods. In: Proceedings of the 26th Annual International ACM SIGIR Conference on Research and Development in Information Retrieval, pp. 338–344. ACM (2003a)

Gey, F.C., Kando, N., Peters, C.: Cross-language information retrieval: The way ahead. Information Processing & Management 41, 415–431 (2005)

Gupta, P., Bali, K., Banchs, R., Choudhury, M., Rosso, P.: Query expansion for mixed-script information retrieval. In: Proc. of SIGIR, pp. 677–686. ACM Association for Computing Machinery (2014)

Levow, G.A., Oard, D.W., Resnik, P.: Dictionary-based techniques for cross-language information retrieval. Information Processing & Management 41, 523–547 (2005)

Lin, W.C., Chen, H.H.: Merging mechanisms in multilingual information retrieval. In: Peters, C., Braschler, M., Gonzalo, J. (eds.) CLEF 2002. LNCS, vol. 2785, pp. 175–186. Springer, Heidelberg (2003)

Lu, Y., Chau, M., Fang, X., Yang, C.C.: Analysis of the Bilingual Queries in a Chinese Web Search Engine. In: Proceedings of the Fifth Workshop on E-Business. Citeseer, Milwaukee (2006)

Mccarley, J.S.: Should we translate the documents or the queries in cross-language information retrieval? In: Proceedings of the 37th Annual Meeting of the Association for Computational Linguistics on Computational Linguistics, pp. 208–214. Association for Computational Linguistics (1999)

Mirkin, B.: Population levels, trends and policies in the Arab region: Challenges and opportunities. Arab Human Development, Report Paper, 1 (2010)

Mustafa, M., Suleman, H.: Multilingual Querying. In: Proceedings of the Arabic Language Technology International Conference (ALTIC), Alexandria, Egypt (2011)

Mustafa, M., Osman, I.M., Suleman, H.: Building a Multilingual and Mixed Documents Corpus. In: Proceedings of the Arabic Language Technology International Conference (ALTIC), Alexandria, Egypt (2011a)

Mustafa, M., Osman, I.M., Suleman, H.: Indexing and Weighting of Multilingual and Mixed Documents. In: Proceedings of the Annual Conference of the South African Institute of Computer Scientists and Information Technology (SAICSIT), vol. 52, pp. 110–120. ACM, Cape Town (2011b)

Mustafa, M.: Lost in Cyberspace: Mixed Queries and Mixed Documents. In: Proceedings of the 3rd international Conference on Linguistic and Cultural Diversity in Cyberspace, Yakutsk, Russia (2014)

Mustafa, M.: Mixed-Language Arabic English Information Retrieval, PhD Thesis, University of Cape Town, South Africa (2013)

Nie, J.Y.: Cross-language information retrieval. Synthesis Lectures on Human Language Technologies 3, 1–125 (2010)

Nie, J.-Y., Simard, M.: Using Statistical Translation Models for Bilingual IR. In: Peters, C., Braschler, M., Gonzalo, J., Kluck, M. (eds.) CLEF 2001. LNCS, vol. 2406, pp. 137–150. Springer, Heidelberg (2002)

Peters, C., Braschler, M., Clough, P.: Multilingual Information Retrieval: From Research to Practice. Springer, Heidelberg (2012) ISBN 978-3-642-23007-3

Pirkola, A.: The effects of query structure and dictionary setups in dictionary-based cross-language information retrieval. In: Proceedings of the 21st Annual International ACM SIGIR Conference on Research and Development in Information Retrieval, pp. 55–63. ACM (1998)

Rieh, H., Rieh, S.Y.: Web searching across languages: Preference and behavior of bilingual academic users in Korea. Library & Information Science Research 27, 249–263 (2005)

Robertson, S., Zaragoza, H., Taylor, M.: Simple BM25 extension to multiple weighted fields. In: Proceedings of the Thirteenth ACM International Conference on Information and Knowledge Management, pp. 42–49. ACM (2004)

The Academy of the Arabic Language, Sudan Office (2011)

Wang, J., Oard, D.W.: Combining bidirectional translation and synonymy for cross-language information retrieval. In: Proceedings of the 29th Annual International ACM SIGIR Conference on Research and Development in Information Retrieval, pp. 202–209. ACM (2006)

Zhang, Y., Vines, P.: Using the web for automated translation extraction in cross-language information retrieval. In: Proceedings of the 27th Annual International ACM SIGIR Conference on Research and Development in Information Retrieval, pp. 162–169. ACM (2004)

Improving Cross Language Information Retrieval Using Corpus Based Query Suggestion Approach

Rajendra Prasath[1,2], Sudeshna Sarkar[1], and Philip O'Reilly[2]

[1] Department of Computer Science and Engineering
Indian Institute of Technology, Kharagpur - 721 302, India
{drrprasath,shudeshna}@gmail.com
[2] Dept of Business Information Systems, University College Cork, Cork, Ireland
Philip.OReilly@ucc.ie

Abstract. Users seeking information may not find relevant information pertaining to their information need in a specific language. But information may be available in a language different from their own, but users may not know that language. Thus users may experience difficulty in accessing the information present in different languages. Since the retrieval process depends on the translation of the user query, there are many issues in getting the right translation of the user query. For a pair of languages chosen by a user, resources, like incomplete dictionary, inaccurate machine translation system may exist. These resources may be insufficient to map the query terms in one language to its equivalent terms in another language. Also for a given query, there might exist multiple correct translations. The underlying corpus evidence may suggest a clue to select a probable set of translations that could eventually perform a better information retrieval. In this paper, we present a cross language information retrieval approach to effectively retrieve information present in a language other than the language of the user query using the corpus driven query suggestion approach. The idea is to utilize the corpus based evidence of one language to improve the retrieval and re-ranking of news documents in the another language. We use FIRE corpora - Tamil and English news collections - in our experiments and illustrate the effectiveness of the proposed cross language information retrieval approach.

Keywords: Query Suggestion, Corpus Statistics, Cross-Lingual Document Retrieval, Retrieval Efficiency.

1 Introduction

With the advent of the world wide web, Internet users, speaking a language other than English, are steadily growing. These users create and share information on various topics in their own language and thus the documents in multiple languages grow rapidly over the world wide web. Users cannot access the information written in a language different from their own and hence require a cross language information retrieval(CLIR) system to access information in different languages. In such cross language information retrieval systems, a user

A. Gelbukh (Ed.): CICLing 2015, Part II, LNCS 9042, pp. 448–457, 2015.
DOI: 10.1007/978-3-319-18117-2_33

may query in a source language (known language to the user) and it has to be translated into the target language (unknown language to the user). Then the cross language information retrieval system has to retrieve information, from the unknown language collection, pertaining to the user query in the known language. Since the retrieval process depends on the translation of the user query, getting the correct translation(s) of the user query is of great interest. There could be many issues in getting the right translations.

For a pair of languages chosen by a user, resources, like incomplete dictionary, inaccurate machine translation system, and insufficient tools that could map the term contexts in one language to the similar term contexts in another language, may exist. With these insufficient resources, we have to find a mapping of user queries given in one language to its equivalent query in another language. Also for a given query, there might exist multiple translations. The right translation pertaining to user information needs has to be identified from multiple translations. The underlying corpus evidence may suggest a clue on selecting a suitable query that could eventually perform better document retrieval. To do this, we plan to develop a cross language information retrieval approach based on the corpus driven query suggestion approach. The idea is to use corpus statistics across news documents in different Indian languages and English and then propose a general methodology to utilize the corpus statistics of one language to improve the retrieval of news documents in the other language.

This paper is organized as follows: The next section presents a comprehensive review of literature related to various strategies in cross lingual information retrieval. Section 3 presents motivations and objectives of this research work. Then we describe the underlying cross lingual information retrieval problem and the issues associated with CLIR systems in Section 4. Then in Section 5, we describe our proposed CLIR approach in the context of Indian language pairs. We proceed by presenting our experimental results in Section 6. Finally Section 7 concludes the paper.

2 Existing Work

Capstick *et* al. [1] presented a fully implemented system MULINEX that supports cross-lingual search of the world wide web. This system uses dictionary-based query translation, multilingual document classification and automatic translation of summaries and documents. This system supports *English* and two European languages: *French* and *German*. Gelbukh [2] presented a thesaurus-based information retrieval system that enriches the query with the whole set of the equivalent forms. Their approach considers enriching the query only with the selected forms that really appear in the document base and thereby providing a greater flexibility. Zhou *et* al. [3] presented a survey of various translation techniques used in free text cross-language information retrieval. Ballesteros and Croft [4] illustrated the use of pre- and post-translation query expansion via pseudo relevance feedback and reported a significant increase in cross language information retrieval effectiveness over the actual (unexpanded) queries.

McNamee and Mayfield [5] also ensured these findings and showed that pre-translation led to the remarkable increase in retrieval effectiveness where as post-translation expansion was still useful in detecting poor translations. Shin *et* al. [6] presented a query expansion strategy for information retrieval in MEDLINE through automatic relevance feedback. In this approach, greater weights are assigned to the MeSH terms (that are classified for each document into major MeSH terms describing the main topics of the document and minor MeSH terms describing additional details on the topic of the document), with different modulation for major and minor MeSH terms' weights.Levow *et* al. [7] described the key issues in dictionary-based cross language information retrieval and developed unified frameworks, for term selection and the translation of terms, that identify and explain a previously unseen dependence of pre- and post-translation expansion. This process helps to explain the utility of structured query methods for better information retrieval.

3 Objectives

User seeking information may not find relevant information pertaining to his / her information need in a specific language. But information may be available in a different language for his / her information needs, but the user may not know that language. Thus the user may not be able to access the information present in a language that is different from his / her own. To support users to access information present in a different language, cross language document retrieval systems are necessary for different language pairs. In such systems, user query given in a source language has to be translated into the target language and then the cross language retrieval has to be performed.

Since the retrieval process depends on the translation of the user query, getting the correct translation of the user query is of great interest. There could be many issues in getting the right translation. For a pair of languages chosen by a user, resources, like incomplete dictionary, inaccurate machine translation system, and insufficient tools that could map the term contexts in one language to the similar term contexts in another language, may exist. With these insufficient resources, we have to find a mapping of user queries given in one language to its equivalent query in another language. Also for a given query, there might exist multiple translations. The right translation pertaining to user information needs has to be identified from multiple translations output. The underlying corpus evidence may suggest a clue on selecting a suitable query that could eventually perform better document retrieval. In order to do this, we plan to develop a cross language document retrieval system using a corpus driven query suggestion approach.

4 Cross Language Information Retrieval

In this section, we describe the working principle of a cross language information retrieval system. Users search for some information in a language of their choice and this language is considered as the source language. Users look for information to be retrieved and presented either in their own choice of the language or in a different language which we consider as the target language. Some cross language IR systems first perform the translation of the user query given in the source language and translates it into the target language. Then using the translated query, the CLIR system performs the document retrieval in that target language and translates the retrieved documents in the source language, so that the users can get the relevant information in a language that is different from their own.

4.1 Issues in CLIR Systems

We list below a few important issues in CLIR systems:

Query Translation: The main issue in CLIR is to develop tools to match terms in different languages that describe the same or similar meaning. In this process, a user is allowed to choose the language of interest and inputs a query to the CLIR system. Then the CLIR system translates this query into the desired language(s).

Document Translation: Often query translation suffers from certain ambiguities in the translation process, and this problem is amplified when queries are short and under-specified. In these queries, the actual context of the user is hard to capture and results in translation ambiguity. From this perspective, document translation appears to be more capable of producing more precise translation due to richer contexts.

Document Ranking: Once documents are retrieved and translated back into the source language, a ranked list has to be presented based on their relevance to the actual user query in the source language. So ranking of documents in source and / or target language is essential in cross language information retrieval.

5 The Proposed CLIR System

We present an approach to improve the cross lingual document retrieval using a corpus driven query suggestion (CLIR-CQS) approach. We have approached this problem from enhancing the query translation process in the cross language information retrieval by accumulating the corpus evidence and use the formulate query for better information retrieval. Here we assumed that a pair of languages: (s, t) is chosen and an *incomplete dictionary* (the translation of many terms in the language t may be missing) is given for this pair of languages.

5.1 Identifying Missing / Incorrect Translations

Any query translation system (either based on the dictionary based approach or statistics / example based approach) translates the user query given in the source language s into the target language t. Since the dictionary is incomplete and has limited number of entries, we may have missing or incorrect translation of the user query in language t. We present an approach that handles the missing or incorrect translation of the user query and to improve the retrieval of information in the target language t.

Let q_i^t be the partially correct translation of q^s. In this case, some query terms are translated into the target language and some are not. In case of missing translations, we use the co-occurrence statistics of query terms in language s and their translated terms in language t to identify the probable terms for missing translations of query terms that could result in better retrieval of cross lingual information retrieval (CLIR).

5.2 Corpus Driven Query Suggestion Approach

In this section, we describe the Corpus driven Query Suggestion(CQS) approach for the missing / incorrect translations. Let q_i^t be the translation (may be a correct or partially correct or incorrect translation) of q^s.

In this section, we consider the case in which some query terms are translated into the target language and some are not. In case of missing translations, we use the co-occurrence statistics of query terms in language s and their translated terms in language t to identify the probable terms for missing translations of query terms that could result in better retrieval of cross lingual information retrieval (CLIR). The proposed approach is given in Algorithm. 1.

First we identify the query terms for which the correct translation exists and find the set of co-occurring terms of these query terms. Then we perform weighting of these co-occurring terms. Then we present our procedure to identify the probable terms for missing translations in the actual user query by creating a connected graph using the actual query terms; the co-occurring terms in language and their available translations in the target language.

Weighting of Query Terms: Using corpus statistics, we compute the weight of the terms that co-occur with the query terms as given in Algorithm 2. We consider the initial set of top n documents retrieved for the user query in the source language s.

Scoring Candidate Terms: We perform the scoring of the co-occurring terms of correct translations in the target language as given in Algorithm 3. This generates a list of candidate terms for missing translations in the target language.

5.3 Document Ranking

We have used Okapi BM25 [8,9] as our ranking function. BM25 retrieval function ranks a set of documents based on the query terms appearing in each document, regardless of the inter-relationship between the query terms within a document. Given a query Q, containing keywords $q_1, q_2, \cdots, q_n$, the BM25 score of a document D is computed as:

$$score(Q, D) = \sum_{i}^{n} idf(q_i) \cdot \frac{tf(q_i, D) \cdot (k_1 + 1)}{tf(q_i, D) + k_1 \cdot (1 - b + b \cdot \frac{|D|}{\text{avgdoclength}})}$$

Algorithm 1. CLIR using probable terms for missing translations

Require: A machine translation system for query translation

Input: Query q^s having a sequence of keywords in language s

Description:

1. **Initial Set**: Input the user query to the search engine and retrieve the initial set of top n documents in language s: $D = \{d_1^s, d_2^s, \cdots, d_n^s\}$
2. **Co-occurring Terms**: Identify the list of terms that co-occur with each of the query terms. Let these terms list be SET_s
3. **Identify Correct Translation**: Using incomplete dictionary, identify the list of query terms for which the correct translation exists. Then for each correct translation, identify co-occurring terms in the target collection. Let SET_t be the list of these terms.
4. Compute weights of the co-occurring terms in top n documents using corpus statistics using the steps given in Algorithm 2
5. **Identify Probable Terms for Missing Translations of Terms**: For each term in SET_s and SET_t, create a bipartite network using incomplete dictionary as follows: if each term w^s in SET_s has a correct translation w^t in SET_t then draw a link from w^s to w^t. Repeat this for all terms in SET_s.
6. Let PT be the list of probable terms; Initialize the list $PT \leftarrow 0$
7. Compute $tscore(w_p)$ using the procedure given in 3
8. Sort terms in SET_t in decreasing order of $tscore(w_p)$, $1 \leq p \leq |SET_t|$. Choose $l\times$ (# terms for which no translation exists) and add then to the PT, where l denotes the number of aspects a user is interested in.
9. **Query Formulation**: Using the terms in PT, formulate the query by choosing $tscore(w_c)$, $1 \leq c \leq |PT|$ as their weights.
10. **Retrieve**: Now using the formulated query, retrieve the documents in the target collection and sort the documents in decreasing order of their similarity scores.
11. **return** top k documents ($k \leq n$) as the ranked list of documents

Output: The ranked list of top $k \leq n$ documents

Algorithm 2. Weighting of co-occurring terms

Input: SET_s - list of terms that co-occur with each of the query terms

1. Using corpus statistics, compute the weight of each co-occurring terms in top n documents as follows:
2. **for** each co-occurring term ct_j, $(1 \leq j \leq SET_s)$ **do**
3. Compute

$$termWeight(ct_j) = idf(ct_j) \times \frac{\sum_{i=1}^{n} tf(ct_j)}{max_{1 \leq j \leq |SET_s|}(\sum_{i=1}^{n} tf(ct_j))} \tag{1}$$

4. where $idf(ct_j)$ denotes inverse document frequency of the term ct_j.
5. **end for**

Algorithm 3. Scoring candidate terms

Input: SET_s - the list of terms that co-occur with each of the query terms;
SET_t - list of co-occurring terms(of correctly translated terms) in the target language.

1. **for** each term w_p $1 \leq p \leq |SET_t|$ **do**
2. compute $d = \#$terms in SET_s that have outlinks to w_p

$$tscore(w_p) = d + \frac{\sum_{i=1}^{l} termWeight(w_p)}{max_{1 \leq l \leq r}(\sum_{i=1}^{l} termWeight(w_p))} \quad (2)$$

where r denotes the number of terms having inlinks from SET_s

3. **end for**

where $tf(q_i, D)$ is the term frequency of q_i in the document D; $|D|$ is the length of the document D and *avgdoclength* is the average document length in the text collection; k_1, $k_1 \in \{1.2, 2.0\}$ and b, $b = 0.75$ are parameters; and $idf(q_i)$ is the inverse document frequency of the query term q_i.

The inverse document frequency $idf(q_i)$ is computed as:

$$idf(q_i) = \log \frac{N - df(q_i) + 0.5}{df(q_i) + 0.5}$$

where N is the total number of documents and $df(q_i)$ is the number of documents containing the term q_i.

6 Experimental Results

In this section, we present the experimental results of the proposed cross language information retrieval approach on the selected language pairs: *Tamil* and *English.* We have used the multi-lingual adhoc news documents collection of FIRE[1] datasets for our experiments. More specifically, we have used English and Tamil corpus of FIRE 2011 dataset and analyzed the effects of the proposed approach.

We have considered a set of 10 queries in the language: *Tamil* and for each query in Tamil, we consider the machine translated query in English using Google between the period 30 Jan - 09 Feb 2015 and the manual reference translation in English. The queries are listed in table 1. We have used an incomplete Tamil - English dictionary with 44,000 entries in which there are 20,778 unique entries and 21,135 terms have more than one meaning. We have used this dictionary for translating query terms and also to map the terms co-occurring with the correctly translated pairs. Since we use Lucene [2] as the indexing and retrieval system with BM25 ranking system. Since we retrieve top 20 documents for each query and perform the scoring of candidate terms. The average access time for terms set in Tamil

[1] Forum for Information Retrieval and Evaluation - `http://www.isical.ac.in/~fire/`

[2] `Lucene:www.apache.org/dist/lucene/java/`

Table 1. List of queries

No	Queries in Tamil	Google Translation	Reference Translation
1	வேங்கை மரங்கள் கடத்தல்	Leopard trees trafficking	vengai trees smuggling
2	தூசு படிந்த மரச்சட்டம்	Grime maraccattam	dirt ingrained wooden frame
3	மேற்கில் ஞாயிறு மறைவு	The death Sunday in the West	Sun sets in west
4	சேலம் வீரபாண்டி சிறையில் கலாட்டா	Salem Veerapandi booed in prison	outbreak in Salem Veerapandi prison
5	சசிகலா ஆதிமுக கட்சியில் இருந்து நீக்கம்	Shashikala alimuka removal from the party	Saskala expelled from ADMK party
6	தமிழக மீனவர்கள் போராட்டம்	Fishermen struggle	Tamilnadu fishermen struggle
7	சம்பா பயிர்கள் தண்ணீர் இன்றி வாட்டம்	Samba crops without water gradient	samba crops fade out without water
8	ஊட்டியில் மலர் கண்காட்சி நிறைவு விழா	Ooty flower show at the closing ceremony	closing ceremony of flower exhibition in Ooty
9	கோவையில் முக்கிய பிரமுகர் கைது	The main figure arrested in Coimbatore	important person arrested in Coimbatore
10	வெள்ளி முளைக்கும் நேரம்	Silver germination time	Moon rising time

Table 2. The selected queries in Tamil; the equivalent translations in English and the retrieval efficiency in Tamil monolingual retrieval

QID	Query in TAmil	Translated Query in ENglish Google Translate / (Derieved Query terms)	User Info Need	p@5	p@10
1	வேங்கை கடத்தல்	Wang conduction / (வேங்கை tree[273] smuggling[110] cut[88] sandle wood[71] tiger[70] வனத்துறையினர்[62] near[50] people[50], steps[45] area[44])	Info about smuggling of Venghai (tree)	0.8	0.65
2	தூசு படிந்த மரச்சட்டம்	Dust-stained maraccattam (dust[128] stained[115] wood[95] coated[75] glass[72] frame[61] time[58] police[52] road[50] people[49] நடவடிக்கை[38])	Info about the dust stained wooden frame	0.7	0.6
3	மேற்கில் ஞாயிறு மறைவு	Sunday on the west side (west[210] india[111] power[106] bengal[105] side[107] sets[101] indies[95] மறைவு[51] ஞாயிறு[48] இரங்கல்[31])	Sun sets on the west	0.6	0.55
4	சேலம் வீரபாண்டி சிறையில் கலாட்டா	Create virapanti Salem in jail (jail[802] வீரபாண்டி[499] ஆறுமுகம்[287] former[149] திமுக[144] court[102] central[98] police[79] authorities[74] prison[70])	Issues made by Salem Veerapandi in prison	0.7	0.5
5	சசிகலா ஆதிமுக கட்சியில் இருந்து நீக்கம்	Athimuka Shashikala from the disposal (சசிகலா[230] அதிமுக[211] party[192] court[166] ஜெயலலிதா[128] disposal[127] chief[118] state[83] minister[82] cases[81])	News about the Sasikala's suspension in ADMK party	0.65	0.6

* Calcutta and Telegraph are the most frequent terms occur in most of the documents. So these terms are not included in our derived query terms

is 765.3 milliseconds and 97.8 milliseconds. Since the retrieval of the initial set of documents, and finding co-occurrence terms from this initial set of documents take very neglibile amount of time (less than 2 seconds even for top 50 documents), we did not consider the retrieval time comparison in this work.

Table 3 presents the details of our experiments done in CLIR with machine translation of user queries with Google translation [3] and CLIA with the proposed corpus based query selection approach. We used Google translation to translate the user query given in Tamil language into English language. For every query term, we may either get one or more terms with correct meaning. Now the given

[3] https://translate.google.com

Table 3. Comparison of retrieval efficiency of top 10 search results: CLIR with Machine Translation (Google) vs CLIR with the proposed corpus based query suggestion approach

QID	Precision @ top 5		Precision @ top 10	
	CLIR-MT	CLIR-CQS	CLIR-MT	CLIR-CQS
1	0.10	0.40	0.25	0.40
2	0.15	0.25	0.20	0.45
3	0.10	0.35	0.25	0.35
4	0.20	0.40	0.35	0.55
5	0.10	0.45	0.40	0.50
6	0.15	0.50	0.35	0.60
7	0.10	0.35	0.25	0.40
8	0.20	0.50	0.40	0.55
9	0.10	0.25	0.20	0.35
10	0.15	0.35	0.25	0.45

query terms span over multiple queries with the permutation of the matching query terms (of different meaning). We use corpus statistics to score each of the queries. Then we have considered the top k queries to perform the formulation of a single weighted query. Then using this formulated query, we have performed cross language information retrieval with Tamil-English documents collection. Consider the query ID: 1. In this query, there are 3 tamil query terms: { *Vengai*, *Marangal*, *Kadaththal* }. The term *Vengai* may refer to two variations: *Vengai* - type of a tree whose botanical name is *Pterocarpus marsupium*, *leopard* - animal; *Marangal* - trees - the correct translation; and finally *Kadaththal* - may refer to at least 3 variations: *trafficking* or *smuggling* or *stealing.* This would give 2 x 1 x 3 = 6 different queries. We identify a set of terms that boosts these query variations and then choose the top k terms to form the single weighted query using query terms weighting approach.

During the evaluation of the proposed approach, we have used 3-points scale for making relevant judgments. We have considered top 10 documents for each query and manually evaluted the retrieved results using the metric: *precision @ top k* documents. The preliminary results show that the proposed approach is better in disambiguating the query intent when query terms that have multiple meanings are given by the users.

7 Conclusion

We have presented a document retrieval approach using corpus driven query suggestion approach. In this work, we have used corpus statistics that could provide a clue on selecting the right queries when translation of a specific query term is missing or incorrect. Then we rank the set of the derieved queries and select the top ranked queries to perform query formulation. Using the re-formulated weighted query, cross language information retrieval is performed. We have presented the comparison results of CLIR with Google translation of the user queries and CLIR with the proposed corpus based query suggestion. The preliminary results show that the proposed approach seems to be promising and we are exploring this further with a graph based approach that could unfold the hidden relationships between query terms in a given pair of languages.

References

1. Capstick, J., Diagne, A.K., Erbach, G., Uszkoreit, H., Leisenberg, A., Leisenberg, M.: A system for supporting cross-lingual information retrieval. Inf. Process. Manage. 36(2), 275–289 (2000)
2. Gelbukh, A.: Lazy query enrichment: A method for indexing large specialized document bases with morphology and concept hierarchy. In: Ibrahim, M., Küng, J., Revell, N. (eds.) DEXA 2000. LNCS, vol. 1873, pp. 526–535. Springer, Heidelberg (2000)
3. Zhou, D., Truran, M., Brailsford, T., Wade, V., Ashman, H.: Translation techniques in cross-language information retrieval. ACM Comput. Surv. 45(1), 1:1–1:44 (2012)
4. Ballesteros, L., Croft, W.B.: Phrasal translation and query expansion techniques for cross-language information retrieval. In: Proceedings of the 20th Annual International ACM SIGIR Conference on Research and Development in Information Retrieval, SIGIR 1997, pp. 84–91. ACM, New York (1997)
5. McNamee, P., Mayfield, J.: Comparing cross-language query expansion techniques by degrading translation resources. In: Proceedings of the 25th Annual International ACM SIGIR Conference on Research and Development in Information Retrieval, SIGIR 2002, pp. 159–166. ACM, New York (2002)
6. Shin, K., Han, S.-Y., Gelbukh, A., Park, J.: Advanced relevance feedback query expansion strategy for information retrieval in MEDLINE. In: Sanfeliu, A., Martínez Trinidad, J.F., Carrasco Ochoa, J.A. (eds.) CIARP 2004. LNCS, vol. 3287, pp. 425–431. Springer, Heidelberg (2004)
7. Levow, G.A., Oard, D.W., Resnik, P.: Dictionary-based techniques for cross-language information retrieval. Inf. Process. Manage. 41(3), 523–547 (2005)
8. Robertson, S.E., Walker, S.: Some simple effective approximations to the 2-poisson model for probabilistic weighted retrieval. In: Proc. of the 17th ACM SIGIR Conference on Research and Development in IR, SIGIR 1994, pp. 232–241. Springer-Verlag New York, Inc., New York (1994)
9. Robertson, S., Zaragoza, H.: The probabilistic relevance framework: Bm25 and beyond. Found. Trends Inf. Retr. 3(4), 333–389 (2009)

Search Personalization via Aggregation of Multidimensional Evidence About User Interests

Yu Xu, M. Rami Ghorab, and Séamus Lawless

CNGL Centre for Global Intelligent Content, Knowledge and Data Engineering Group
School of Computer Science and Statistics, Trinity College Dublin, Ireland
{xuyu,rami.ghorab,seamus.lawless}@scss.tcd.ie

Abstract. A core aspect of search personalization is inferring the user's search interests. Different approaches may consider different aspects of user information and may have different interpretations of the notion of *interest*. This may lead to learning disparate characteristics of a user. Although search engines collect a variety of information about their users, the following question remains unanswered: to what extent can personalized search systems harness these information sources to capture multiple views of the user's interests, and adapt the search accordingly? To answer this question, this paper proposes a hybrid approach for search personalization. The advantage of this approach is that it can flexibly combine multiple sources of user information, and incorporate multiple aspects of user interests. Experimental results demonstrate the effectiveness of the proposed approach for search personalization.

1 Introduction

Search engines have become the most used means for seeking information on the Web [1]. However, studies have shown that a large proportion of user queries are short and underspecified, which may fail to effectively communicate the user's information need [2]. Users may have completely different intentions for the same query under different scenarios [3, 4]. For example, a violinist may issue the query "string" to look for the information about violin strings, while a programmer may use the same query to find the information about a data type. Personalized search techniques [3, 4, 5, 6, 7] offer a solution to this issue. They infer users' search interests from usage information and adapt search results accordingly.

Search engines store a range of information about the user, such as submitted queries, clicked results, etc. In addition, information like the content of the clicked pages and links between pages, can also be used to infer the user's interests. Each one of these sources constitutes evidence of the user's interests but reflects a different aspect of the user [5]. However, the majority of previous approaches focus on just part of the available information. In addition, different approaches may have different interpretations of the concept *interest*. For example, in [6], the user's interests and information needs were inferred based on click-through data where the relationship between users, queries and Web pages was analyzed and represented by a quadruple <user, query, visited pages, weight>. The authors in [7] consider a different source of evidence for

A. Gelbukh (Ed.): CICLing 2015, Part II, LNCS 9042, pp. 458–467, 2015.
DOI: 10.1007/978-3-319-18117-2_34

inferring the user's domain interest; they use category information of Web pages and users' click interactions to analyze users' domain preference represented by an *n*-dimensional vector.

Each one of the abovementioned approaches only gives a limited view of the user's interests, i.e., only captures certain aspects/dimensions of the user's interests and preferences. This paper argues that taking into consideration multi-dimensional interests of the user may lead to better search personalization. This raises a number of research challenges: how can we achieve better performance in search adaptation by considering multiple perspectives of interests based on a range of user information? How can we combine these aspects and devise an appropriate hybrid approach?

A number of studies have attempted combining multiple information sources to discover users' potential interests. For example, in [8] click-through statistics, dwell time, title and snippet information of the page are exploited to compute a number of feature scores for each result. The search results are then re-ranked by aggregating these feature scores of each page. In [9], a user's historic search interactions and current search session are combined with topic information of each page for deriving the users' long-term and short-term interests. These two aspects of interests then together influence the ranking of the search results. Although these approaches have shown their effectiveness by combining more information about the user, the exploited information is still limited in each single approach. More importantly, they still confine the user interests to a particular dimension, e.g., degree of preference for each page or topic of interest in the above two works.

In order to answer above questions, this paper focuses on personalized re-ranking and proposes a general hybrid approach. It adapts search to users' multi-dimensional interests by merging different personalized rankings for each query. These rankings can be produced from any personalized approach based on user information. Specifically, the contributions of this paper are as follows: (1) Three algorithms for search personalization are proposed. Each algorithm infers the user's interests from a different perspective; (2) An appropriate aggregation strategy is selected to investigate the merging of the personalized rankings, which leads to different hybrid approaches; (3) This work evaluates the performance of the three approaches and the hybrid approaches using a large set of search logs from a major commercial Web search engine. The experimental results show the hybrid approaches outperform each single approach and the hybrid approach considering the all three perspectives of user interests performs the best.

2 Background

In recent years, researchers have discovered various information sources for search personalization, among which the users' search logs are an essential, such as the queries the user submitted, dwell time on clicked pages etc. The user's search history is the most direct evidence which reflects their preferences [4]. More recently, researchers have begun to explore external resources (e.g., social tags of web pages, social activities of users) and to incorporate them into user search history for search adaptation [10, 11]. This work considers the usage of users' search logs in the preliminary stage of the research.

In addition to the information gathering, there are two primary strategies to personalize search: Query Expansion and Result Re-ranking. The former performs the adaptation by modifying the query the user input in order to retrieve results that more relevant to the underlying information need [12]. It involves operations like augmenting the query with additional terms and altering the weights on each query term. The latter is the strategy most commonly used [13]. It reevaluates the relevance of the results to the user based upon her personal needs and re-orders them accordingly. The key to personalized ranking approaches is how to score the pages given certain information sources. This work focuses on the latter.

3 Personalized Approaches Based on User Search History

This section first describes the three proposed personalization algorithms, which capture users' interests from three perspectives: web pages, web domains and queries. It then discusses the aggregation strategy that is used to combine the approaches.

3.1 Click Frequency and Time Based Ranking

Re-finding is a common occurrence when using search engines, where the user issues a query just in order to find a web page that she has already read previously [14]. Therefore, for each query, this approach (denoted as FT-Click) assigns personalized scores to those pages clicked by the user among return results. Specifically, for a page *p* returned by a query *q* of user *u*, the personalized score can be computed by:

$$S^{FT-Click}(p,q,u) = \frac{\sum_{C \in Clicks(p,u)} DTime(C)}{|Clicks(u)| \cdot \alpha} \qquad (1)$$

where *Clicks(p, u)* is the clicks that *u* has previously made on Web page *p*; |*Clicks(u)*| is the total number of clicks made by *u*; *DTime(C)* is the dwell time on page *p* by click *C* of *u*; α is a smoothing parameter used for normalization.

FT-Click re-ranks the top *k* pages in a ranked list according to their personalized scores, generating a personalized ranking list, where *k* is a configurable parameter. But it only applies to the pages whose FT-Click scores exceed a configurable threshold parameter *TH*. This approach can help users quickly re-find pages, and enhance search effectiveness as proven in our experiments. But this approach has a disadvantage, i.e., it will have no impact when the user has not previously visited any page in the result list.

3.2 Web Domain Preference Based Ranking

A website usually has a main subject domain associated with it (e.g. *nba.com* : basketball). Users are typically interested in a, often small, subset of all subject domains.

Therefore, it can be assumed that a user often issues queries in these domains, and likely find the appropriate result from a limited number of websites [15].

Based on this, this paper proposes a personalized approach (denoted as P-Domain) that puts more emphasis on pages that belong to the user's favorite web domains. It represents a user's web domain preference by a m-tuple $D(u) = [(d_1, ID(u, d_1)), \ldots, (d_m, ID(u, d_m))]$, in which m is number of web domains used to represent users' domain preference; d_m is a certain web domain; $ID(u, d_m)$ represents the degree of interest of user u in web domain d_m. The m elements in every tuple are in descending order of ID value. The interest degree of user u on a certain web domain d is calculated by:

$$ID(u,d) = \frac{\sum_{C \in Clicks(d,u)} DTime(C)}{|Clicks(u)| \cdot \beta} \tag{2}$$

where $Clicks(d, u)$ is the click set that u made on those pages belong to web domain d; β is the normalization parameter. P-Domain takes the m web domains among $Clicks(d, u)$ with the largest interest degree as u's preference domains, and builds the corresponding web domain preference vector (DPV) $D(u)$. This approach assigns ID score as the personalized score of the top k pages in a returned list for each query.

3.3 Identical Query Based Ranking

The above two methods mainly exploit wisdom of individual click history. However, this approach also takes advantage of the search experience of other users to improve search. In popular commercial search engines, the previously issued queries indicate the most frequently-used combinations of keywords. Therefore, when a user issues a query it is possible to identify other users who issued the same queries in the past for the same or similar information need. In this paper, these users are called query-similar users of the current user for the current query. The challenge here is, to identify query-similar users from all those users issued the same query, and to make use of the click records of these users to improve search personalization.

It is reasonable to assume that a query issuer is more likely to have the same or similar information need as those who share similar interests for a given query. In addition, the user more likely prefers pages that were more frequently clicked by the query-similar users. Based on the above analysis, this paper proposes an identical query based re-ranking approach (denoted as I-Query). It adopts improved Pearson's correlation coefficient to compute the similarity of any two users u_i and u_j :

$$Sim(u_i, u_j) = \frac{\sum_{d \in d_{ij}} ID(u_i,d) \cdot ID(u_j,d)}{\sqrt{\sum_{d \in d_{ij}} ID(u_i,d)^2} \sqrt{\sum_{d \in d_{ij}} ID(u_j,d)^2}} \frac{|d_{u_i} \cap d_{u_j}|}{|d_{u_i} \cup d_{u_j}|} \tag{3}$$

where d_{ij} is the common domains in the DPVs of u_i and u_j; $|d_{u_i} \cap d_{u_j}|$ is the element number of the intersection of domains in DPVs of u_i and u_j. On a current query q issued by user u, $U(q)$ represents the set of users who issued q. The query-similar users of u on q, denoted as $SU(q, u)$, are defined as those users among $U(q)$ whose similarities with u are greater than the average similarity between u and users in $U(q)$:

$$SU(q,u) = \{su|su \in U(q) \text{ and } Sim(u,su) > AveSim(u,U(q))\} \quad (4)$$

Provided the query-similar users, the personalized score of a return page p for query q of user u can be computed by:

$$S^{I-Query}(p,q,u) = \frac{\sum_{su \in SU(q,u)} Sim(u,su) \cdot |Clicks(p,q,su)|}{|Clicks(p,q,SU(q,u))| \cdot \gamma} \quad (5)$$

in which $|Clicks(p, q, su)|$ is the click number of su on page p by query q in history; $|Clicks(p, q, SU(q, u))|$ is the total click number of all users in $SU(q, u)$ on page p by query q; γ is the smoothing parameter.

3.4 Aggregation of Personalized Rankings

The above three approaches are based on different user history data and infer the users' interests from different standpoints. They consider their interests in Web pages, domains and the Web pages the similar users like respectively. The assumption under investigation in this paper is that each approach adapts search results to different characteristics of the user, generating different personalized rankings, and therefore the aggregation of these rankings for a query may reflect users' real needs in a more complete manner. It is significant to investigate whether this aggregation can further improve search performance in practice.

The personalized results are in the form of a ranked list of web pages. The preferences reflected from multiple lists may be competing. So it is necessary to select an appropriate ranking aggregation strategy and take diverse preferences of the user into account for the final ranked list. Many ranking aggregation techniques have been proposed, e.g., Mean, Median, Stability Selection, Exponential Weighting Ranking Aggregation [16]. Since the personalized lists reflect different dimensions of user interests, the personalized scores of pages are not directly comparable. Therefore, having trialed a number of aggregation strategies, the most appropriate one, Weighted Borda-Fuse (WBF), was selected. WBF only considers the rank order of pages in merging the ranking lists and is widely used in the field of Information Retrieval (IR) [17]. Specifically, the WBF simulates the aggregation process as a voting election, in which each voter (personalized approach) ranks a fixed set of candidates (returned pages for a query) in the order of preference. Each page gets a point based on positions in all ranked lists and the weight of the voters. They are then re-ranked according to these points.

Equation (6) shows an example of merging two lists L_1, L_2 using WBF, in which p is a page in either L_1 or L_2; $r(p, L_1)$ is the rank of p in L_1 and a is the weight that satisfies $0 \leq a \leq 1$. The WBF generates a new personalized score for each page by aggregating the two lists. Then pages are ranked in ascending order of their aggregated scores.

$$S^{F-Borda}(p, L_1, L_2) = a \cdot r(p, L_1) + (1-a) \cdot r(p, L_2) \quad (6)$$

Given a query q issued by u, the search engine returns a default ordered list L_0 generated by a global ranking function. The three approaches can also generate three personalized ranking lists $L_{FT\text{-}Click}$, $L_{P\text{-}Domain}$, $L_{I\text{-}Query}$ by re-ranking L_0. For a single approach, the final results are generally presented by a merging of L_0 and a pure personalized ranking list (e.g., $L_{FT\text{-}Click}$). The following section describes an evaluation of the various combinations of L_0, $L_{FT\text{-}Click}$, $L_{P\text{-}Domain}$, $L_{I\text{-}Query}$ using WBF. In the experiment, we adjust the weights on each approach to seek the most effective combinations.

4 Evaluation

4.1 Dataset Description

The experiments reported in this paper are carried out on the dataset released for the Yandex[1] "Personalized Web Search Challenge"[2]. This dataset extracted from Yandex contains 30 days of search activity of more than 5 million users who are sampled from a large city. The dataset is grouped by search session. For each session, the session id, user id and session time are recorded. For a query within a session, the query terms, ordered URLs returned by search engine, the web domain of each URL, the submitted time of the query, and the click actions within the query are all extracted. Furthermore, the timestamp of each click is also recorded. Table 1 provides descriptive statistics of the dataset; more detailed information about the dataset is available on the competition's homepage[2]. The dataset is split into a training set (the first 27 days) and a test set (the last 3 days).

Table 1. Statistics of dataset

Item	Number
Unique Queries	21,073,569
Unique URLs	70,348,426
Unique Users	5,736,333
Training Sessions	34,573,630
Test Sessions	797,867
Clicks in the Training Data	64,693,054
Total Records in the Log	167,413,039

4.2 Evaluation Metric

The evaluation metric used in this paper is the Normalized Discounted Cumulative Gain (NDCG) measure, which is a widely used evaluation metric in the IR community. In this paper, we use the satisfaction grade of the user on a page to represent the true relevance (gold standard) for each test query, which is detailed below. We only take the top 10 pages for each query, so k is equal to 10. In addition, the optimal order of the k pages is based on the satisfaction grade each page gained. In our experiments, we calculate NDCG for each test query and then average all the NDCG values as the final evaluation value.

---

[1] www.yandex.com

[2] www.kaggle.com/c/yandex-personalized-web-search-challenge

The dwell time of each click is the time that passed between this click and the next click or the next query. It is well correlated with the probability of the user to satisfy her information need with the clicked page. The clicks with dwell time longer than a threshold are called satisfied clicks. The competition organizer proposed the use of three grades 0, 1, 2 to represent the satisfaction degree of a user on a clicked page:

- 0 grade (irrelevant) corresponds to pages with no clicks and clicks with dwell time less than 50 time units;
- 1 grade (relevant) corresponds to pages with clicks and dwell time between 50 and 399 time units;
- 2 grade (highly relevant) corresponds to pages with clicks and dwell time not shorter than 400 time units, and pages associated with last click of a session.

4.3 Experimental Results

In our experiments, we set α with 400 (time unit) and *TH* with the average personalized score of all clicked pages in training set for FT-Click; set β with 400 (time unit), *m* with 10 for P-Domain; set γ with 1 in I-Query; and *k* with 10.

4.3.1 Single Approach

Firstly, we experiment with each approach separately in order to validate their effectiveness. For each test query we directly merge the re-ranked list ($L_{FT\text{-}Click}$, $L_{P\text{-}Domain}$ or $L_{I\text{-}Query}$) with the default list L_0 using Equation (6), generating the final personalized ranking list. Fig.1 presents the average NDCG values of each approach on the test dataset. We vary the value of the weight *a* on L_0 in the range [0, 0.9] to find the best performance of each approach. The average NDCG value of default rankings of all test queries is the baseline marked using a straight line in Fig.1. It shows the three approaches are all effective in personalizing the user's search, outperforming the default ranking of the search engine with a proper *a*. The FT-Click, P-Domain and I-Query get the best performance (NDCG = 0.79647, 0.79258, 0.79301) when *a* is taken as 0, 0.3 and 0.4 respectively.

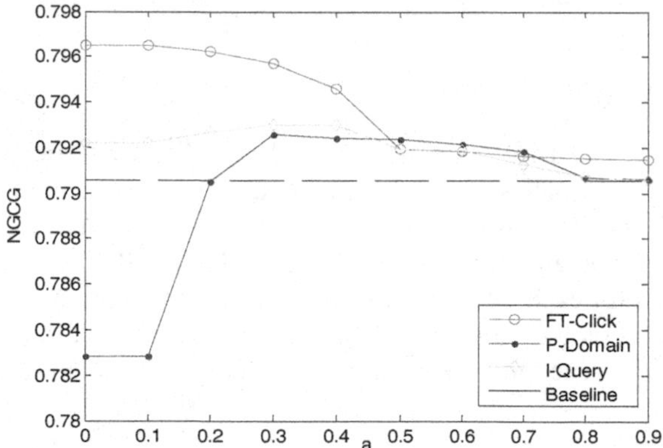

Fig. 1. Performance of Single Personalized Approach

4.3.2 Combinations of Two Approaches

In this experiment, we merge any two personalized ranking lists (e.g. $L_{FT\text{-}Click}$ and $L_{P\text{-}Domain}$) for each test query with WBF to observe their performance. Fig.2 presents the NDCG performance of three different combinations. In each combination, we vary the weight a on the first personalized approach (according to the order of title in each sub-graph of Fig.2) to find its best performance, and we compare its results with a baseline, the best NDCG value from two individual approaches forming the combination, which is marked with the straight line. For example, the baseline for the combination of FT-Click and P-Domain is 0.79647. The experimental results show that the three combinations outperform any single approach forming the combination with a proper a. They get the best performance (NDCG = 0.79692, 0.79674, 0.79499) when a is taken as 0.5, 0.7 and 0.4 respectively according to the order of Fig.2.

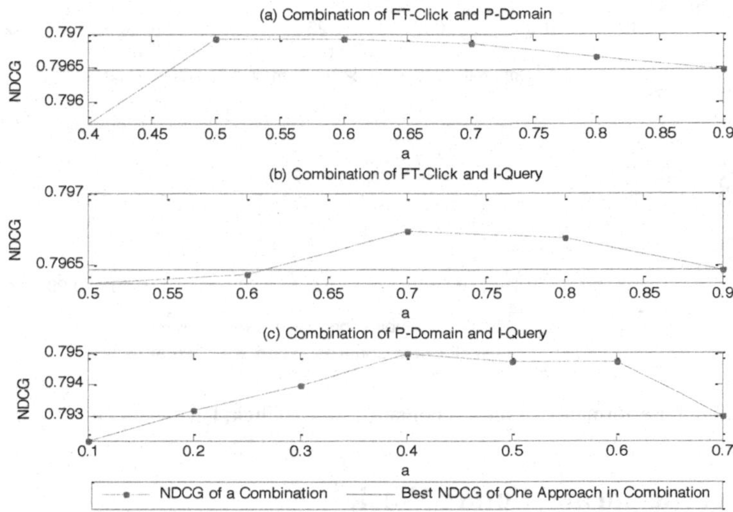

Fig. 2. Performance of the Combination of Two Personalized Approaches

4.3.3 The Combination of Three Approaches

The above experimental results indicate that the pure FT-Click personalized ranking performs best without the involvement of L_O when a is taken as 0. So we just study the aggregations of three personalized rankings ($L_{FT\text{-}Click}$, $L_{P\text{-}Domain}$ and $L_{I\text{-}Query}$) without considering L_O in the combination of three approaches. The ranking lists of the three approaches are aggregated using an extended WBF as shown in Equation (7), in which a, b, c are the weights assigned to each list; $0 \leq a, b, c \leq 1$ and $a + b + c = 1$.

$$S^{F-Borda}(p, L_1, L_2, L_3) = a \cdot r(p, L_1) + b \cdot r(p, L_2) + c \cdot r(p, L_3) \quad (7)$$

The above experiments also show that the combined approaches where FT-Click is assigned a larger weight achieve the best performance. This proves that FT-Click is more effective and should be assigned higher weight. Therefore in this experiment, based on the observations made, we set the weight *a* on FT-Click as 0.5 and 0.6 respectively, and adjusted the weight *b* on P-Domain to observe the performance of the combined approach. Fig. 3 shows the experimental results on which the best NDCG value obtained in above experiments (0.79692) is marked using a straight line as a baseline. We find the combined approach improves further on the basis of two-approach combinations. It gets a highest NDCG 0.79730 when $a = 0.5$ and $b = 0.3$.

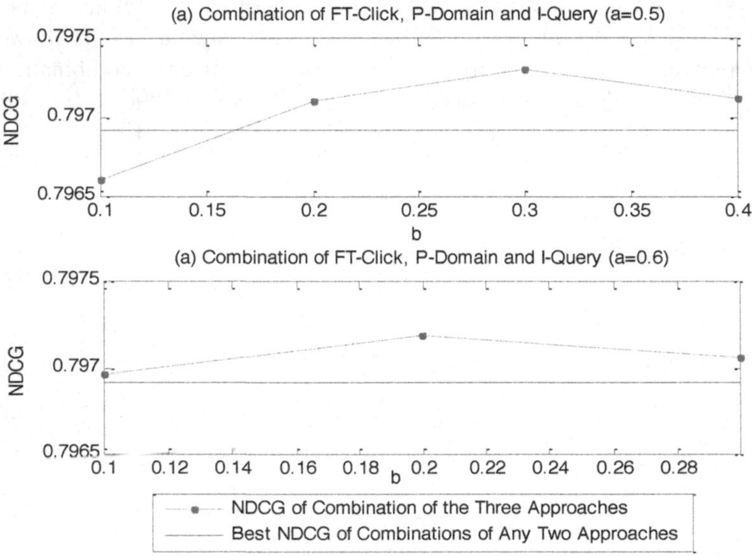

Fig. 3. Performance of the Combination of FT-Click, P-Domain and I-Query

5 Conclusion and Future Work

This paper explored the improvement of personalized search by combining various sources of information about the user. Three algorithms were proposed that utilized search logs to infer different perspectives of the user's interests. In addition, this paper chose an aggregation strategy that adapted search based on multiple perspectives of user interests. Experiments were conducted on a large scale real-world dataset with abundant user behavior logs and experimental results showed the effectiveness of the proposed approaches and aggregation strategy. This study demonstrates that search engines can further improve the relevance of search results with respect to a user, if personalization decisions are based on a holistic view of user interests.

Future work will involve investigating how the model parameters can be automatically fine-tuned. Besides, the study of query-sensitive aggregation methods will be a focus of our future work.

Acknowledgements. This work was supported by Science Foundation Ireland (Grand 12/CE/I2267) as part of CNGL (www.cngl.ie) at Trinity College Dublin.

References

1. Sanderson, M., Croft, W.B.: The history of information retrieval research. Proceedings of the IEEE 100, 1444–1451 (2012)
2. Dou, Z., Song, R., et al.: Evaluating the effectiveness of personalized web search. IEEE Transactions on Knowledge and Data Engineering 21(8), 1178–1190 (2009)
3. Teevan, J., Dumais, S.T., Liebling, D.J.: To personalize or not to personalize: modeling queries with variation in user intent. In: SIGIR 2008, pp. 163–170. ACM (2008)
4. Shen, X., Tan, B., Zhai, C.X.: Implicit user modeling for personalized search. In: CIKM 2005, pp. 824-831 (2005)
5. Micarelli, A., Gasparetti, F., Sciarrone, F., Gauch, S.: Personalized search on the world wide web. In: Brusilovsky, P., Kobsa, A., Nejdl, W., et al. (eds.) Adaptive Web 2007. LNCS, vol. 4321, pp. 195–230. Springer, Heidelberg (2007)
6. Sun, J.T., Zeng, H.J., Liu, H., et al.: Cubesvd: A novel approach to personalized web search. In: WWW 2005, pp. 382–390. ACM (2005)
7. Chirita, P.A., Nejdl, W., Paiu, R., et al.: Using ODP metadata to personalize search. In: SIGIR 2005, pp. 178–185. ACM (2005)
8. Agichtein, E., Brill, E., Dumais, S.: Improving web search ranking by incorporating user behavior information. In: SIGIR 2006, pp. 19–26. ACM (2006)
9. Bennett, P.N., White, R.W., Chu, W., Dumais, S.T., et al.: Modeling the impact of short- and long-term behavior on search personalization. In: SIGIR 2012, pp. 185–194. ACM (2012)
10. Kashyap, A., Amini, R., Hristidis, V.: SonetRank: leveraging social networks to personalize search. In: CIKM 2012, pp. 2045–2049. ACM (2012)
11. Zhou, D., Lawless, S., Wade, V.: Web search personalization using social data. In: Zaphiris, P., Buchanan, G., Rasmussen, E., Loizides, F. (eds.) TPDL 2012. LNCS, vol. 7489, pp. 298–310. Springer, Heidelberg (2012)
12. Ghorab, M.R., Zhou, D., O'Connor, A., et al.: Personalised information retrieval: survey and classification. User Modeling and User-Adapted Interaction 23(4), 381–443 (2013)
13. Xu, S., Bao, S., Fei, B., et al.: Exploring folksonomy for personalized search. In: SIGIR 2008, pp. 155–162 (2008)
14. Badesh, H., Blustein, J.: VDMs for finding and re-finding web search results. In: Proceedings of the 2012 iConference, pp. 419–420. ACM (2012)
15. Qiu, F., Cho, J.: Automatic identification of user interest for personalized search. In: WWW 2006, 727–736 (2006)
16. Dwork, C., Kumar, R., et al.: Rank aggregation methods for the web. In: WWW 2001, pp. 613–622. ACM (2001)
17. Aslam, J.A., Montague, M.: Models for metasearch. In: SIGIR'01, pp. 276–284. ACM (2001)

Question Analysis for a Closed Domain Question Answering System

Caner Derici[1], Kerem Çelik[1], Ekrem Kutbay[2], Yiğit Aydın[2], Tunga Güngör[1], Arzucan Özgür[1], and Günizi Kartal[2]

[1] Boğaziçi University
Computer Engineering Department
[2] Educational Technology
Bebek, Istanbul, 34342
{caner.derici,kerem.celik2,ekrem.kutbay,yigit.aydin, gungort,arzucan.ozgur,gunizi.kartal}@boun.edu.tr

Abstract. This study describes and evaluates the techniques we developed for the question analysis module of a closed domain Question Answering (QA) system that is intended for high-school students to support their education. Question analysis, which involves analyzing the questions to extract the necessary information for determining what is being asked and how to approach answering it, is one of the most crucial vcomponents of a QA system. Therefore, we propose novel methods for two major problems in question analysis, namely focus extraction and question classification, based on integrating a rule-based and a Hidden Markov Model (HMM) based sequence classification approach, both of which make use of the dependency relations among the words in the question. Comparisons of these solutions with baseline models are also provided. This study also offers a manually collected and annotated vgold standard data set for further research in this area.

1 Introduction

Question Answering (QA) systems aim to produce automatically generated answers for questions stated in natural languages. The drastic improvements in the Natural Language Processing (NLP) and Information Retrieval (IR) techniques in the past decade have led to the development of prominent QA systems, some of which are available for public use, such as AnswerMachine[1] and WolframAlpha[2]. It has even been possible to develop a QA system that can compete on a TV show against human opponents [8].

Building a fully capable QA system, however, has difficulties mostly due to numerous challanging sub-problems that need to be solved, such as question analysis (involving pre-processing and classification of questions), information retrieval, cross linguality and answer generation (involving answer extraction and formulation), along with some lower level subtasks, such as paraphrasing, common sense implication or reference resolution. In addition, the architecture of a QA system, as well as the techniques employed usually

[1] http://theanswermachine.tripod.com/

[2] http://www.wolframalpha.com/

A. Gelbukh (Ed.): CICLing 2015, Part II, LNCS 9042, pp. 468–482, 2015.
DOI: 10.1007/978-3-319-18117-2_35

depend on factors such as question domain and language. Many researchers have tackled the individual problems involved in such systems separately. While some are considered to be solved, the majority of the problems are still open to further research [9,1].

This study attempts to solve the first problem of a QA system, question analysis. The overall system is developed for Turkish-speaking high-school students to enable them to query in their natural language any question chosen from their course of study. Note that there is virtually no upper bound in the number of possible query frequency, as the system is intended for use by virtually all high schools in Turkey. Therefore in order for the system to be practically usable, besides accuracy, the overall architecture should be carefully designed, where each module is comprehensively analysed and evaluated individually. In this study, we present the development and evaluation of the first module, namely, question analysis in the pipeline of our system, intended for use in the prototype domain of Geography. The primary concern in question analysis is to extract useful information from a given question to be used in subsequent modules to finally generate a correct response. In particular, the information that indicates a certain type or a central property of the entity being asked, along with a classification of the question into pre-determined classes from the domain helps to reduce significantly the size of the work space of the further stages in the system such as information retrieval or candidate answer generation.

In the following example, the information indicating that the name of a plain is asked, which we refer to as the *focus*, and the classification *ENTITY.PLAIN* helps us to navigate around these concepts in the knowledge base, searching the answer.

"Türkiye'nin en büyük [*ovasının adı*] *nedir?"*

"What is the [*name of the*] *largest* [*plain*] *in Turkey?"*

For focus extraction, we developed a rule-based model, along with a Hidden Markov Model (HMM) based statistical model. We investigate the accuracy of the combination of these two in focus extraction. Additionally, for question classification, we show that a rule-based model is more successful in finding coarse classes than a tf-idf based bag-of-words baseline model that utilizes the frequencies of the words in a question.

Developing such a question analysis module, let alone a QA system for Turkish is especially challenging because it is an agglutinative language with a morphologically rich and derivational structure. For this reason, we pre-process the questions by performing morphological analysis and disambiguation, as well as dependency parsing using the Turkish NLP Pipeline [16,6,15]. Morphological analysis and disambiguation produces the root forms of the words and their part-of-speech (POS) tags. Dependency parsing produces the dependency relations among the words in a given sentence. The tags that are used by the dependency parser are defined in the Turkish Dependency TreeBank, which includes tags such as *SUBJECT*, *OBJECT*, *SENTENCE*, *MODIFIER*, *CLASSIFIER*, *POSSESOR*, and etc [6,7].

We propose a novel approach for question classification and focus detection, based on integrating a rule-based method with an HMM-based sequence classification method, for a closed-domain QA system. Additionally, we contribute the first manually collected and annotated gold standard question analysis data set for Turkish. The implementation codes and the gold standard Turkish question data will be publicly available for reproducability and further research[3].

2 Related Work

A fundamental task in a QA system is determining the type of the answer, and its properties and possible constraints. Given a query stated in a natural language, a QA system often extracts some immediate information such as the question class (e.g. *what, who, when, etc.*) based on the pre-determined answer types [4]. Recent state-of-the-art techniques for question classification often involve statistical approaches [12,13]. Additionally, some QA systems are in general more semantics oriented, and construct a knowledge base directly from raw question texts [10]. However, these systems determine only the type of a given question. They do not further determine, for example what type of entity is being asked, which would narrow down the search space significantly.

One approach in simulating a question analysis is to use general purpose search engines. One of the earliest studies that employs such a strategy, is an open-domain QA system, AnswerBus [19]. AnswerBus employs a bag-of-words strategy, where search engines are scored based on the number of hits they returned for each word. The total score of a search engine for a particular question is the sum of the hits returned for each word in the question. Based on their total scores, the best search engine is determined as the most appropriate knowledge source for answering the question. However, AnswerBus does not use any semantic information, nor does it extract any information to build a more competent answering strategy.

The first successful Turkish factoid QA system used a hybrid approach (both rule-based and statistical), not for question analysis, but for providing a direct answer by matching surface level question and answer patterns [5]. It doesn't employ any explicit question analysis, other than extracting the predefined question and answer patterns.

Inspired by its significant success, our system adapts its strategies for question analysis among the ones that are employed in one of the most powerful QA systems, IBM's Watson [11]. For analysing a given question (i.e. clue), Watson extracts firstly a part of the clue that is a reference to the answer (*focus*); second, it extracts the terms that denote the type of the entity asked *(lexical answer type, LAT)*; third, the class of the clue (*QClass*); and finally some additional elements of the clue (*QSection*) should it need special handling. Lally et al. [11] extensively evaluate the significance of distilling such information to produce correct answers. To extract these information, Watson mostly uses regular expression based rules combined with statistical classifiers to assess the learned reliability of the rules. Note that, the sole purpose of Watson is to win the Jeopardy! game, a well-known television quiz show where the quiz questions are presented as free formatted "clues", rather than complete question statements, rendering Watson's analysis methods specific to the Jeopardy! game. In a closed-domain QA system, on the other hand, it is sufficient to extract only *LAT* and *QClass* in order to analyse a complete question, since in a complete question sentence, what Watson refers to as the *focus* is often the question word (e.g. "What" in the example in Section 1). Therefore, the real

[3] `https://github.com/cderici/hazircevap`

focus of a question, what we refer to as the *focus* is closest in reality to what Watson refers to as *LAT*. In this regard, our definition of the focus is:

"the terms in the question that indicate what type of entity is being asked for".

A study most relevant for our question analysis is conducted by [3], where rule-based and statistical methods are utilized together to extract the question focus in an open-domain QA system. In this study, a binary classification using Support Vector Machines (SVM) is performed on words in the English questions that are parsed using a constituency parser. Further, experts with manually tailored rules are used to identify the different features which are then deployed in the SVM. In contrast, our analysis separately uses both rule-based and statistical models to extract the focus. It also performs question classification for Turkish questions that are parsed using a dependency parser. Additionally, a sequence classification is performed using a Hidden Markov Model (HMM) based algorithm, whose results are then combined with the results of the rule-based experts to produce the final focus. Unfortunately, our study is incompatible for comparison with this study. Firstly, because the definition of the focus in [3] depends on a constituency parser and a coreference resolver, which currently do not exist for Turkish. Therefore, it is neither possible to define equivalent rules for the English dataset, nor to apply the techniques proposed in [3] to the Turkish dataset.

3 System Architecture

Although the main technical contribution of this study is the methodology (i.e. the combination of the rule-based and statistical models), one of the tenets of this paper is to be an introduction of the QA system, upon which this analysis module resides and to be a starting point for the development of the subsequent modules. Consequently, this section introduces the general architecture of the system, as well as the way in which the question analysis module connects to it.

The overall architecture of the system is designed in concordance with the DeepQA technology, introduced in [8]. The primary principle in DeepQA is to have parallel units with multiple sub-modules that produce different candidate results for each sub-problem, which are then scored according to the evidence collected by trained machine learning models. Then the most likely candidate is returned as the final answer.

After question analysis, the extracted *focus* is used in the *Information Retrieval* module to fetch the relevant knowledge units[4] that are pruned and refined by the *QClass*. These relevant units are then fed to the *Candidate Answer Generation* module that has multiple different information retrieval algorithms to produce all possible relevant answer units. For each candidate answer unit, syntactic and semantic evidence units are collected, which are then used to score the candidate answers, the ones having low scores are pruned. Finally, the strong candidates are synthesized into the final answer set, where the most likely answer is fed to the answer generation module along with the other top k answers for providing optionality.

[4] We refrain from referring to these units as "documents", as we do not limit the format in which the knowledge is represented.

3.1 Question Analysis Module

The Question Analysis module consists of three parallel sub-modules, the *Distiller*, *HMM-Glasses* and *ClassRules*, illustrated in Figure 1. The first two modules are for extracting the question's *focus*, whereas the third module is for determining the most likely classification of the question *(QClass)* into the pre-defined question classes.

The *focus* indicates what exactly the given question is asking, and what type of entity it is. In the example in Section 1, the focus is the collection of these parts[5] of the question: "ovasının adı" *(name of a specific plain)*, since the question asks for a name. In particular, it asks the name of a plain. Therefore, the phrase "ova adı" *(name of a plain)* can be constructed even syntactically from the phrase "ovasının adı" *(name of a specific plain)*, since we already have the morphological roots attached to the question parts. Because "ova"*(plain)* is the root, and "sı" and "nın" are possessive suffixes which together mean: *"a name of a plain of"*. The *QClass* for this question is *ENTITY* (see Table 2).

In the following example, the *focus* is the parts "denizci kimdir" *(Who is the sailor)*, and the *QClass* is *HUMAN.INDIVIDUAL*. The rationale for the *focus* is that the question asks for a person's name, and it is known that the person is a sailor. Observe that we omit the distinctive properties of the entity in the question (e.g. the first sailor), because at this point, we are mostly interested in "is a" and "part of" relations that indicate a certain type of the entity. The remaining properties are used by the subsequent modules of the system to semantically prune both the *relevant knowledge units* and the *candidate answers*.

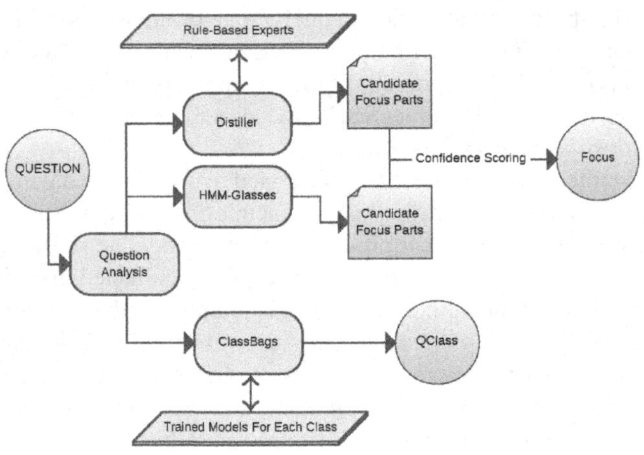

Fig. 1. Queestion Analysis Module

[5] Note that, we refer to each word of the question as a "part". A part represents a single word in the question that has been annotated with extra information such as its morphological root, part-of-speech, dependency tag, and etc.

"Dünyayı dolaşan ilk [*denizci kimdir*] *?"*

" [*Who is*] *the* [*sailor*] *that first circumnavigated the Earth?"*

4 Methodology

For *focus extraction*, we have a fastidious rule based focus extractor, the *Distiller*, with specifically tailored rules over the dependency trees for almost all types of factual questions in the Geography domain, and an HMM based classifier, *HMM-Glasses*, which uses a variation of the Viterbi algorithm [17] that essentially renders it somewhat more liberal than the Distiller to a certain extent. Other than one common trait, that is operating on the dependency relations among the words in the question, their approaches to the main problem (i.e. to extract the focus) are based on completely different principles in different levels of resolution. This distinction is critical to our methodology, since it provides the necessary insight for the model to efficiently handle languages with rich derivational structure, such as Turkish. At this point, a delicate balance is required for the combination of these models. For this purpose, we take into account the individual confidences of both the Distiller and HMM-Glasses, rendered through their individual performances over the training dataset. Additionally, for the classification of the question into predetermined classes from a certain domain (Geography in our case), we have a rule-based classifier, which extracts the coarse class by manually constructed phrase-based rules.

4.1 Focus Extraction

Distiller. We observed that in our selected domain of Geography, there are certain patterns of question statements (based on the predicate), common to the majority of the questions. We identified each such pattern (*question type*) and defined manually sets of rules (*experts*) for the extraction of the focus from the dependency parse tree of each question. We call this sets of rules together, *The Distiller*.

Currently we have seven rule-based experts, along with a generic expert that handles less frequent cases by using a single generic rule. The primary reason of the inclusion of a generic expert is data scarcity. However, we prefer to make it optional, because having a specific general expert along with a finite number of experts may result in a penalized precision as opposed to more or less increased recall, depending on the data set size, which may not always be a desirable option in practice. All experts and their question frequencies in the data set are given in Table 1.

The rules contain instructions to navigate through the dependency tree of a given question. For example, the rule for the "nedir"*(what is ...)* expert, and the rule for the "verilir"*(... is given ...)* expert, as well as the generic rule are as follows (examples provided in Figure 2).

nedir:*(what is ...)*

- Grab the *SENTENCE* in the question
- Grab and traceback from the *SUBJECT*, and collect only *POSSESSOR* and *CLASSIFIER*

***verilir:** (... is given ...)*

- Grab the *SUBJECT* of the *SENTENCE* in the question
- Grab and traceback from the first degree *DATIVE.ADJUNCT* of the *SENTENCE*, and collect only the firstdegree *MODIFIER*

generic:

- Grab the *SUBJECT* of the *SENTENCE* in the question
- Traceback from the *SUBJECT*, and collect the first degree *POSSESSOR* and/or *CLASSIFIER*, along only with their *POSSESSOR* and/or *CLASSIFIER*

Every rule-based expert has a confidence score based on its performance for extracting the correct focus parts from the questions belonging to its expertise. This score is used to indicate the reliability of the expert's judgement later when combining its result with the *HMM-Glasses*. The confidence scores, along with the focus parts of a question $\mathbb{Q}$ are reported by both the *Distiller* and the *HMM-Glasses* in the format of triplets:

$$\langle fpt, fpd, fpc \rangle_n$$

where $n \in \{1..|\mathbb{Q}|\}$ [6], fpt stands for *focus part text*, fpd is *focus part dependency tag* and fpc denotes *focus part confidence* score. Both models produce such triplets for

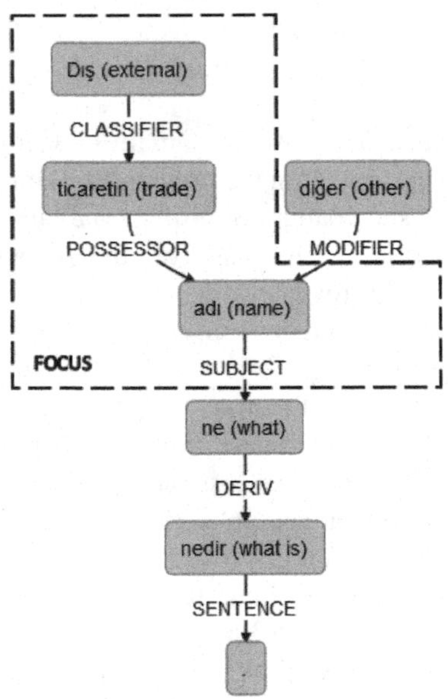

Dış ticaretin diğer adı nedir?
What is the other name of external trade?

Fig. 2. *nedir* expert tells that the focus of this question is "a name of external trade"

[6] $|\mathbb{Q}|$ denotes the number of words in the question $\mathbb{Q}$.

Table 1. Experts and their question frequencies in the training data

Expert Type	***Frequency (%)***
generic	25.6
hangi *(... which ...)*	19.5
nedir *(what is ...)*	15.0
denir *(... referred to as ...)*	9.6
kaç *(how many ...)*	9.6
verilir *(... is given ...)*	7.2
hangisidir *(which one is it ...)*	7.2
kadardır *(how much ...)*	6.3

each focus part that they extracted. However, there is a significant distinction in the way that the confidences are reported for each part of the extracted focus between the rule-based and the statistical models. As explained in detail in Section 4.1, *HMM-Glasses* work on individual parts of the question, while the *Distiller* extracts sub-trees from the dependency tree of the question. Therefore, the *Distiller*'s resolution is not big enough to consider the individual probabilities for each part to be in the focus. Thus the *Distiller* produces a collection of parts as the *focus*, along with a single confidence score (*total confidence score*) reported by the expert in charge, this is mapped to fpc scores of all parts, rendering all parts in the focus equal from the *Distiller*'s perspective.

HMM-Glasses. *HMM-Glasses* models the focus extraction as a HMM and performs a sequencial classification on the words in the question using the Viterbi algorithm. Having only two hidden states, namely *FOC* (i.e. the observed part is a focus part) and *NON* (i.e. the observed part is not a focus part), it treats each question part as an observation, and decides whether the observed part is a part of the focus of the question.

We first serialize the dependency tree of the question and feed the algorithm the serialized tree. Serialization (or encoding) of a tree is to systematically produce a sequential representation of it, which is mostly employed in the fields of applied mathematics, databases and networks [18,14]. Evidently the method with which the tree is serialized has an observable influence on the characteristics of the algorithm's results. We investigated this effect with two general serialization approaches, and empirically tested it (see Section 6). Common approaches in tree serialization try to efficiently serialize the tree within the information theoretical resource bounds (in terms of time and space), while taking into account also the deserialization process [2]. On the other hand, we are only concerned with the coherency of the tree structure. In other words, the dependency relations should be consistent among all the serialization methods. Therefore, we considered the simplest possible methods, *forward mode* and *backward mode*.

Forward and Backward Modes. While constructing the sequence from the dependency tree in forward mode, left children (according to the reverse visualization of the dependency tree) take precedence over the right children to be taken into the sequence.

Therefore, the left-most branch is taken first, then the branch on its immediate right is taken, and so on. Finally the parent is added. Backward mode is simply the other way around, where the right children take precedence over the left children. Any difference in serialization changes the whole learning process, thereby renders the learned features unique to a particular serialization. This therefore provides a noticeable diversity in the characteristics of the learning, depending on the serialization method. Below are the serializations of the question in Figure 2. Recall that we only consider the morphemes of the words (i.e. stripped from all the adjuncts).

***forward serialization** (->)*

Dış	ticaret	diğer	ad	ne
(external)	*(trade)*	*(other)*	*(name)*	*(what)*
FOC	FOC	NON	FOC	NON

***backward serialization** (<-)*

ne	ad	diğer	ticaret	Dış
(what)	*(name)*	*(other)*	*(trade)*	*(external)*
NON	FOC	NON	FOC	FOC

Essentially, forward mode serialization corresponds to reading the question from left to right (or start to end), while backward mode corresponds to reading it from end to start. Different serialization approaches potentially allow ensembles of various kinds of models, handling different parts of the question as they have learned different features of the data while training. Therefore, a more complex model can be obtained by combining multiple *HMM-Glasses* having different serialization approaches.

We model the focus extraction problem as a HMM by firstly computing the prior probabilities of our hidden states (i.e. *FOC* and *NON*), and secondly learning the probabilities from the given set of serialized questions as follows

$$a_{jk} = P(t^j|t^k) \qquad b_{ij} = P(w_i|t^j)$$

where a_{jk} represents the probability of being in state t^j given the previous state is t^k, and b_{ij} indicates the probability that the current observation is the word w_i given that the current state is t^j. Decoding is performed using the Viterbi algorithm, where the states correspond to the nodes in the produced Viterbi path indicating the most likely judgements for each part to be a focus part of the question. Further, the observation probabilities b_{ij} are used as confidence scores (i.e. fpc) in the triplets. Recall that all results are reported as triplets (see Section 4.1).

Dependency Tags vs. Word Texts. In all parts of the question analysis, taking advantage of the dependency relations among the words in the question whenever possible has prominent benefits, compared to mere syntactic approaches for languages with a rich derivational structure, where for instance possible long distance relationships in the question statement can easily be determined. Therefore, the very first design of the *HMM-Glasses* was planned to learn and evaluate the dependency tag sequence of a question, which essentially corresponds to learning the tree shape, rather than the sequence of words. However, this approach mislead the model, as there are some tags that occur more frequently in questions than others, as for example a question often has only one *SENTENCE* tag, while it has lots of *MODIFIER* tags. More importantly, the focus is often a small part of the question. Thus, for example, the judgement of whether a *MODIFIER* part is a focus part is strongly biased by the fact that the number of cases a *MODIFIER* is a NON will be orders of magnitude higher than otherwise. Furthermore, working with the normalized frequencies requires a lot of training data for the model to have a statistically significant learning experience. Therefore, *HMM-Glasses* currently learns the probabilities of the part texts (i.e. words) in the question. This leaves the model with no dependency relation information at hand. However, it is compensated by the *Distiller* as the experts use by definition only the dependency rules for extraction.

Combination of the Distiller and HMM-Glasses. Recall that the *Distiller* outputs the focus parts with a single total confidence score of the expert that produced the results. In addition with the part-wise confidences that *HMM-Glasses* produces, we have:

$$
\begin{array}{cc}
\text{HMM} & \text{Distiller} \\
\left\{\begin{array}{l} \langle \mathrm{fpn}_1, fpt_1, fpc_1 \rangle \\ \langle \mathrm{fpn}_2, fpt_2, fpc_2 \rangle \\ \quad\vdots \\ \langle \mathrm{fpn}_p, fpt_p, fpc_p \rangle \end{array}\right. & \left.\begin{array}{l} \langle \mathrm{fpn}_1, fpt_1, fpc \rangle \\ \langle \mathrm{fpn}_2, fpt_2, fpc \rangle \\ \quad\vdots \\ \langle \mathrm{fpn}_q, fpt_q, fpc \rangle \end{array}\right\}
\end{array}
$$

Combination of the candidate focus parts produced by different models is performed in a part-wise manner. In other words, models try to convince each other about each part being among the final focus parts. To do this, we make use of the fpc scores, weight them with the models' individual f-scores over the training data and grab the maximum. Note that, if a part is determined as a candidate focus part by only one of the models M_1 (i.e. the other model M_2 predicts that this part is not a focus part), then we compute the confidence score of M_1 as described above and compare it with the f-score of M_2. If the confidence score of M_1 is greater than that of M_2, the word is classified as a focus part, otherwise it is excluded from the focus.

4.2 Class Extraction

For question classification, we manually pre-determined two types of classes, namely coarse and fine classes, adapted from [12,13], with different semantic resolutions. A question's fine class establishes a strong link to the specific domain at hand, while its coarse class essentially introduces a generality into the model that would render the classification applicable in domains other than Geography.

Currently we have seven coarse classes (see Table 2), along with a total of 57 fine classes. In this study, we only concentrated on coarse classes. We plan to perform classification of fine classes using statistical approaches, which requires comprehensive number of questions in each fine class.

Table 2. Coarse Classes for the Geography Domain

Question Class	*Frequency (%)*
DESCRIPTION	25.2
NUMERIC	24.2
ENTITY	19.6
TEMPORAL	12.4
LOCATION	11.9
ABBREVIATION	3.8
HUMAN	2.4

In order to classify a given question into one of the coarse classes, we devised a set of common phrases for each class unique to that class. For example, for the class NUMERIC, we have two phrases: "kaç"*(how many/how much)* and "kadardır"*(this much/that many)*. The classifier searches for these patterns in a given question and classifies accordingly.

We additionally implement a statistical classifier that employs a tf-idf based weighted bag-of-words strategy, as a baseline model to compare with the rule-based approach. In baseline model the weight of a word w for a class c is computed as follows.

$$\text{tf-idf}_{w,c} = tf_{w,c} \times idf_w$$

where $\text{tf}_{w,c}$ indicates the number of times word w occurs in class c, and idf_w is computed as shown below.

$$idf_w = \log \frac{\text{\# of classes}}{\text{\# of classes containing w}}$$

Then, for a given question Q, we assign it to the class that maximizes the sum of the tf-idf scores of the words in the question:

$$\arg\max_{c} \sum_{w \in Q} \text{tf-idf}_{w,c}$$

5 Data

One of the major contributions of this study is to provide a gold standard, diverse set of Turkish questions from the prototype domain of Geography, manually annotated by human experts. The data set contains 977 instances in the following format: {QuestionText | FocusPartTexts | CoarseClass | FineClass }.

Approximately 30 percent of the dataset consisted of actual questions posed by teachers, collected from Geography-related textbooks and online materials. The rest were generated by three of the researchers, who are educational technologists, based on actual Geography texts used in grades 9 – 12 in high schools in Turkey.

Inter-Annotator Agreement. We made use of two strategies in data annotation: focus annotation and QClass annotation. Three researchers (two of whom are educational technologists) manually identified the focus in each question, while two researchers (one educational technologist) annotated the questions for QClass. The evaluations were later compared to the developer's judgment. The inter-annotator agreement scores for focus was 82%, and for QClass was 92%.

6 Evaluation and Results

One of the major challenges we face was not having a suitable baseline (from previous studies etc.) to indicate the actual hardness of the problem and the actual efficiency of our solutions. Therefore, we implemented a baseline model for focus extraction that identifies the words adjacent to a question keyword for certain proximity as focus parts. The proximity model has slightly worse than, but similar results with the tf.idf model. We chose to include only the baseline with the best results (i.e. tf.idf) for a clear comparison. Note that the baseline models are intentionally designed to be rather simple, because there is no prior study on statistical question analysis on Turkish. Therefore, the baselines are kept simple in order to set the lower bounds of the problem. Moreover, a tf-idf based statistical baseline model that employs a bag-of-words strategy is implemented for question classification as well. All the results are reported as comparisons to these baseline models in Table 3 and Table 4.

Table 3. Evaluation Results of All Models for Focus Extraction

Model	*Precision*	*Recall*	*F-Score*
Baseline (tf.idf model)	**0.769**	**0.197**	**0.290**
Distiller (Generic Enabled)	0.714	0.751	0.732
Distiller (Generic Disabled)	0.816	0.623	0.706
HMM-Glasses (Backward Mode)	0.839	0.443	0.580
HMM-Glasses (Forward Mode)	0.847	0.495	0.625
HMM-Glasses (Forward and Backward Mode)	0.821	0.515	0.633
Combined (Generic Enabled, Backward)	0.734	0.841	0.784
Combined (Generic Enabled, Forward)	0.732	0.846	0.785
Combined (Generic Enabled, Forward & Backward)	**0.721**	**0.851**	**0.781**
Combined (Generic Disabled, Backward)	0.821	0.759	0.789
Combined (Generic Disabled, Forward)	0.818	0.765	0.791
Combined (Generic Disabled, Forward & Backward)	**0.802**	**0.776**	**0.788**

Table 4. QClass Classification Results. Upper section is baseline tf-idf based model, and lower section is rule-based model.

Classses	*Precision*	*Recall*	*F-Score*
Description	0.662	0.908	0.764
Temporal	0.767	0.618	0.670
Numeric	0.801	0.758	0.776
Entity	0.100	0.025	0.040
Abbreviation	0.933	0.766	0.823
Location	0.759	0.212	0.312
Human	0.600	0.600	0.600
Tf.Idf Overall	**0.660**	**0.555**	**0.569**
Description	0.874	0.732	0.797
Temporal	1.000	1.000	1.000
Numeric	0.995	0.911	0.951
Entity	0.603	0.817	0.694
Abbreviation	0.871	0.894	0.883
Location	0.944	0.880	0.911
Human	0.869	0.833	0.851
Rule-based Overall	**0.879**	**0.867**	**0.869**

Since the data on which our models are evaluated have been prepared in this course of study, we build our strategy of evaluation around the concept of hygene, where we ensure two fundamental principles. Firstly, at any point and for each model, scores are obtained from the results produced for questions with which the model *never* crossed before. Secondly, for a reasonable comparison between the models, same scores are computed for different models with different settings using the *same* questions at each iteration of the evaluation.

To evaluate the *Distiller*, the rule-based experts are developed by using only the first 107 questions, that we had at the beginning. Therefore, the remaining questions are safely treated as test data, as there were no modifications done after having a larger number of questions.

Evaluations for all the models are performed using stratified 10-fold cross-validation over all the questions. The final results (i.e. precision, recall and f-score) for focus extraction are obtained by macro-averaging the individual results.

Recall that the *Distiller* has the option to enable and disable the generic expert, while the *HMM-Glasses* has *forward*, *backward* and *forward & backward* modes that calibrate the serialization of the dependency tree. All the different combinations of these settings for each model are seperately evaluated both individually and in combination, in each iteration of the folding process. The results for focus extraction and question classification are shown in Tables 3 and 4, respectively.

6.1 Focus Extraction Results

Individual evaluation of the *Distiller* resulted in comparable precision scores along with lower recall scores (compared to the combined models). A noticable outcome of the

Distiller evaluations is the behavior of the generic expert. Results indicate that generic expert lowers the accuracy of the retrieved results (i.e. precision), while increasing the coverage (i.e. recall) of the model. However, the two effects do not compensate, as the results show that f-score of the *Distiller* with the generic expert enabled is higher than the one with the generic expert disabled.

Distinct evaluation of the effect of the serialization methods indicates that for forward and backward modes, the forward mode is slightly better than the backward mode considering the f-scores. Backward mode seems to increase the recall of any model to which it is included, however, f-scores indicate that this increase in recall is not useful, because it in fact lowers the performance of the combined models whenever it is included.

In general, although the individual accuracies of the models are reasonable enough, the increase in the coverage (recall) for all combined models, having both the *Distiller* and *HMM-Glasses*, compared to the individual recall scores indicate that the combination is useful, as it does not sacrifice the precision scores that we observe in individual evaluations, thereby increasing also the f-scores. Therefore, we can conclude that the models complement each other nicely.

6.2 ClassRules Results

Results show that exploiting the domain knowledge resulted in a significant success that a statistical baseline model could not get near. However, manually crafted set of rules are a big problem when changing the domain. Therefore, a statistical learner that will automatically learn these domain specific phrases is planned for further development, since it requires significant amount of instances for each class. This scarcity is also the reason we leave the identification of fine classes for a future study. Table 4 shows the macro-averaged precision, recall and f-score of coarse class identification of the rule-based classifier, along with the results of the tf-idf based baseline classification.

7 Conclusion

In this study, we presented a novel combination of rule-based and statistical approaches to question analysis, employed in a closed-domain question answering system for an agglutinative language, such as Turkish. Our question analysis consists of focus extraction and question classification. For focus extraction, we have multiple rule-based experts for most frequent question types in Turkish. Additionally, we described a HMM-based novel sequence classification approach for focus extraction, along with combining the results of both rule-based and statistical models according to the individual confidence scores of each model. For question classification, we employed a rule-based classifier which uses pattern phrases uniqe to each class. We implemented baseline models for both problems, and have reported here the comparisons. In addition to the methodology offered, we also provide a set of manually annotated questions for both reproducability and further research.

Acknowledgments. This work was supported by The Scientific and Technological Research Council of Turkey (TÜBİTAK) under grant number 113E036.

References

1. Allam, A.M.N., Haggag, M.H.: The question answering systems: A survey. International Journal of Research and Reviews in Information Sciences (IJRRIS) 2 (2012)
2. Benoit, D., Demaine, E.D., Munro, J.I., Raman, V.: Representing trees of higher degree. In: Dehne, F., Gupta, A., Sack, J.-R., Tamassia, R. (eds.) WADS 1999. LNCS, vol. 1663, pp. 169–180. Springer, Heidelberg (1999)
3. Bunescu, R., Huang, Y.: Towards a general model of answer typing: Question focus identification. In: International Conference on Intelligent Text Processing and Computational Linguistics (CICLING) (2010)
4. Dominguez-Sal, D., Surdeanu, M.: A machine learning approach for factoid question answering. Procesamiento de Lenguaje Natural (2006)
5. Er, N.P., Çiçekli: A factoid question answering system using answer pattern matching. In: International Joint Coneference on Natural Langauge Processing, pp. 854–858 (2013)
6. Eryiğit, G.: The impact of automatic morphological analysis & disambiguation on dependency parsing of turkish. In: Proceedings of the Eighth International Conference on Language Resources and Evaluation (LREC), Istanbul, Turkey (2012)
7. Eryiğit, G., Nivre, J., Oflazer, K.: Dependency parsing of turkish. Computational Linguistics 34, 357–389 (2008)
8. Ferrucci, D.A.: Introduction to "this is watson". IBM Journal of Research and Development 56, 1–15 (2012)
9. Gupta, P., Gupta, V.: A survey of text question answering techniques. International Journal of Computer Applications 53, 1–8 (2012)
10. Katz, B.: Annotating the world wide web using natural language. In: Proceedings of the 5th RIAO Conference on Computer Assisted Information Searching on the Internet, pp. 136–159 (1997)
11. Lally, A., Prager, J.M., McCord, M.C., Boguraev, B.K., Patwardhan, S., Fan, J., Fodor, P., Chu-Caroll, J.: Question analysis: How watson reads a clue. IBM Journal of Research and Development 56, 2:1–14 (2012)
12. Li, X., Roth, D.: Learning question classifiers: the role of semantic information. Natural Language Engineering 12, 229–249 (2006)
13. Metzler, D., Croft, B.W.: Analysis of statistical question classification for fact-based questions. Information Retrieval 8, 481–504 (2005)
14. Munro, J.I., Raman, V.: Succinct representation of balanced parentheses and static trees. SIAM J. Comput. 31, 762–776 (2002)
15. Nivre, J., Hall, J., Nilsson, J., Chanev, A., Eryiğit, G., Kübler, S., Marinov, S., Marsi, E.: Maltparser: A language-independent system for data-driven dependency parsing. Natural Language Engineering Journal 13, 99–135 (2007)
16. Şahin, M., Sulubacak, U., Eryiğit, G.: Redefinition of turkish morphology using flag diacritics. In: Proceedings of The Tenth Symposium on Natural Language Processing (SNLP 2013) (2013)
17. Viterbi, A.: Error bounds for convolutional codes and an asymptotically optimum decoding algorithm. IEEE Transactions on Information Theory 13 (1967)
18. Wen, L., Amagasa, T., Kitagawa, H.: An approach for XML similarity join using tree serialization. In: Haritsa, J.R., Kotagiri, R., Pudi, V. (eds.) DASFAA 2008. LNCS, vol. 4947, pp. 562–570. Springer, Heidelberg (2008)
19. Zheng, Z.: Answerbus question answering system. In: Proceedings of the Second International Conference on Human Language Technology Research (HLT), pp. 399–404 (2002)

Information Extraction with Active Learning: A Case Study in Legal Text

Cristian Cardellino[1], Serena Villata[2], Laura Alonso Alemany[1], and Elena Cabrio[2]

[1] Universidad Nacional de Córdoba, Argentina
crscardellino@gmail.com, alemany@famaf.unc.edu.ar
[2] INRIA Sophia Antipolis, France
firstname.lastname@inria.fr

Abstract. Active learning has been successfully applied to a number of NLP tasks. In this paper, we present a study on Information Extraction for natural language licenses that need to be translated to RDF. The final purpose of our work is to automatically extract from a natural language document specifying a certain license a machine-readable description of the terms of use and reuse identified in such license. This task presents some peculiarities that make it specially interesting to study: highly repetitive text, few annotated or unannotated examples available, and very fine precision needed.

In this paper we compare different active learning settings for this particular application. We show that the most straightforward approach to instance selection, uncertainty sampling, does not provide a good performance in this setting, performing even worse than passive learning. Density-based methods are the usual alternative to uncertainty sampling, in contexts with very few labelled instances. We show that we can obtain a similar effect to that of density-based methods using uncertainty sampling, by just reversing the ranking criterion, and choosing the *most certain* instead of the *most uncertain* instances.

Keywords: Active Learning, Ontology-based Information Extraction.

1 Introduction and Motivation

Licenses and data rights are becoming a crucial issue in the Linked (Open) Data scenario, where information about the use and reuse of the data published on the Web need to be specified and associated to the data. In this context, the legal texts describing the licenses need to be translated into machine-readable ones to allow for automated processing, verification, etc.

Such machine-readable formulation of the licenses requires a high degree of reliability. For example, if the original license states that action A is forbidden and this prohibition is not reported in the RDF version of the license then this could lead to misuses of the data associated to that machine-readable license. For this reason, we need highly accurate performance in the task, to guarantee highly reliable outputs.

A. Gelbukh (Ed.): CICLing 2015, Part II, LNCS 9042, pp. 483–494, 2015.
DOI: 10.1007/978-3-319-18117-2_36

In this scenario, human intervention is unavoidable, to establish or validate the correspondence between concepts in ontologies and expressions in natural language. In this paper, we propose to ease this dependency by optimizing human intervention through an active learning approach. Active learning techniques [13] aim to get powerful insights on the inner workings of automated classifiers and resort to human experts to analyze examples that will most improve their performance. We show the boost in performances introduced by different improvements on a classical, machine learning approach to information extraction.

More precisely, in the experimental evaluation of our framework, we show that active learning produces the best learning curve, reaching the final performance of the system with fewer annotated examples than passive learning. However, the standard active learning setting does not provide an improvement in our study case, where very few examples are available. Indeed, if we choose to annotate first those instances where the classifier shows more uncertainty, the performance of the system does not improve quickly, and, in some cases, it improves more slowly than if instances are added at random. In contrast, selecting for annotation those instances where the classifier is most certain (*reversed uncertainty sampling*) does provide a clear improvement over the passive learning approach. It is well-known that uncertainty sampling does not work well with skewed distributions or with few examples, in those cases, density estimation methods work best. We show that using *reversed uncertainty sampling* in this particular context yields results in the lines of density estimation methods.

The rest of the paper is organized as follows: in Section 2 we discuss the general features of the active learning approach and related work, Section 3 presents our approach to ontology-based IE for licenses; in Section 4 we describe how we apply active learning techniques to this kind of problems. Experimental results comparing the different approaches are discussed in Section 5.

2 Relevant Work

Active learning [13] is a more “intelligent” approach to machine learning, whose objective is to optimize the learning process. This optimization is obtained by choosing examples to be manually labelled, by following some given metric or indicator to maximize the performance of a machine learning algorithm, instead of choosing them randomly from a sample. This capability is specially valuable in the context of knowledge-intensive Information Extraction, where very obtaining examples is costly and therefore optimizing examples becomes crucial.

The process works as follows: the algorithm inspects a set of unlabeled examples, and ranks them by how much they could improve the algorithm's performance if they were labelled. Then, a human annotator (the so-called “oracle”) annotates the highest ranking examples, which are then added to the starting set of training examples from which the algorithm infers its classification model, and the loop begins again. In some active learning approaches, the oracle may annotate features describing instances, and not (only) instances themselves. This latter approach provides even faster learning in some cases [6,12,10,15].

Different strategies have been applied to determine the most useful instances to be annotated by the oracle, including expected model change, expected error reduction or density-weighted methods [11]. The most intuitive and popular strategy is *uncertainty sampling* [9], which chooses those instances or features where the algorithm is most uncertain. This strategy has been successfully applied to Information Extraction tasks [3,14]. Uncertainty can be calculated by different methods depending on the learning algorithm. The simplest methods exploit directly the certainty that the classifier provides for each instance that is classified automatically. This is the information that we are exploiting.

However, we did not only use uncertainty sampling, but also the exact opposite. We explored both prioritizing items with highest certainty and with lowest certainty. We followed the intuition that, when a model is very small, based on very few data, it can be improved faster by providing evidence that consolidates the core of the model. This is achieved by choosing items with highest certainty, because they also provide the lowest entropy with respect to the model, and can help to redirect wrong assumptions that a model with very few data can easily make. When the core of the model is consolidated, items with highest uncertainty should provide a higher improvement in performance by effectively delimiting with more precision the decision frontier of the model. This phenomenon, which lies at the heart of well-known semi-supervised learning techniques like self-training (or *bootstrapping*), has also been noted by approaches combining density estimation methods when very few examples are available, and uncertainty sampling when the training dataset has grown [5,17].

Other approaches have been applied to fight the problem of learning with few examples, by finding the optimal seed examples to build a training set [4,7]. However, these approaches are complex and difficult to implement, thus lie beyond the capacities of the regular NLP practitioner. In contrast, the approach presented here is conceptually simple and easy to implement, as it is a wrapper method over your best-know classifier.

We developed an active learning tool inspired on Dualist [12]. As in Dualist, we provide a graphical user interface for the human oracle to answer the queries of the active learning algorithm. The base machine learning algorithm is also a Multinomial Naïve Bayes, but our method for ranking instances is uncertainty/certainty sampling based on the confidence of the classifier. Features can also be labelled, using Information Gain to select them, but sequentially with respect to instances, not simultaneously as in Dualist. As an addition, our approach allows for multiclass labeling, that is, an instance can be labelled with more than one class. Our active learning framework source together with the dataset is available at `https://github.com/crscardellino/nll2rdf-active-learner`.

3 Passive Learning IE System for Textual Licenses

As a base to our system, we used NLL2RDF, an Information Extraction system for licenses expressed in English, based on a (passive) machine learning approach [1]. The final goal of the system is to identify fragments of text that allow to identify a *prohibition*, a *permission* or an *obligation* (or *duty*) expressed by a license. When these fragments are identified, they are converted into an RDF machine-readable specification of the license itself. Section 3.1 provides a general overview of the system describing the representation of licensing information we selected, and Section 3.2 presents the machine learning approach adopted whithin the system, as well as the performances of the basic setting.

3.1 Overview of the System

The architecture of the system is based on a machine learning core, with an SVM classifier that learns from examples. Examples are manually assigned to one of a predefined set of classes associated to the licenses ontology. Many vocabularies exist to model licensing information. Some examples include LiMO[1], L4LOD[2], ODRS[3] and the well known Creative Commons Rights Expression Language (CC REL) Ontology[4]. So far the Linked Data community has mainly used the CC REL vocabulary, the standard recommended by Creative Commons, for machine-readable expression of licensing terms.

However, more complex licenses information can be defined using the Open Digital Rights Language (ODRL) Ontology[5], that allows to declare rights and permissions using the terms as defined in the Rights Data Dictionary[6]. This vocabulary, in particular, has not been specifically conceived for the Web of Data scenario, but it intends to provide flexible mechanisms to support transparent and innovative use of digital content in publishing, distribution and consumption of digital media across all sectors. ODRL allows to specify fine grained licensing terms both for data (thus satisfying the Web of Data scenario) and for all other digital media. The ODRL vocabulary defines the classes to which each text fragment needs to be translated by the system. It specifies different kinds of Policies (i.e., Agreement, Offer, Privacy, Request, Set and Ticket). We adopt `Set`, a policy expression that consists in entities from the complete model. Permissions, prohibitions and duties (i.e., the requirements specified in CC REL) are specified in terms of an `action`. For instance, we may have the action of attributing an `asset` (anything which can be subject to a policy), i.e., `odrl: action odrl: attribute`. For more details about the ODRL vocabulary, refer to the ODRL Community group.[7]

3.2 Machine Learning Core

The core of the system is based on passive machine learning. Given some manually annotated instances, a classifier is trained to assign each text fragment to

[1] `http://data.opendataday.it/LiMo`
[2] `http://ns.inria.fr/l4lod/`
[3] `http://schema.theodi.org/odrs/`
[4] `http://creativecommons.org/ns`
[5] `http://www.w3.org/ns/odrl/2/`
[6] `http://www.w3.org/community/odrl/`
[7] `http://www.w3.org/community/odrl/`

one or more of the given ontological classes, including the class of instances that is not associated to any meaning in the reference ontology (i.e., ODRL in this case), which is the case for the majority of sentences in any given license.

In the first approach, a Support Vector Machine classifier was used. Texts were characterized by the unigrams, bigrams and trigrams of lemmas, obtaining an f-measure that ranged from 0.3 to 0.78 depending on the class, with 0.5 average. Later on we included bigrams and trigrams of words that co-occur in a window of three to five words. This last feature is aimed to capture slight variations in form that convey essentially the same meaning.

These additional features increased the average accuracy of the system to 76%, kappa coefficient of .7. Although the performance of the system was fairly acceptable in general, it was not acceptable considering that we are dealing with legal information, and that an error in the system could cause an actual misuse of the data. Moreover, we found that it was difficult to improve such performances given the complexity of the task. Finally, we wanted to make it easier to port this system to other domains (i.e., other kind of legal documents like contracts, or policies), and to do that it was crucial to optimize the annotation effort (only 37 licenses where considered and annotated). For all these reasons, we decide to adopt an active learning setting.

In the active learning setting, we decide to use a different classifier that allowed easy manipulation of its inner workings, so that we could implement active learning tweaks easily. As in [12], a Multinomial Naïve Bayes (MNB) classifier was the classifier of choice.

As a baseline to assess the improvement provided by the active learning approach to the problem, we assess the performance of the MNB in a Passive Learning setting. The performance of the MNB by itself was quite below that of SVMs, of 63% (kappa coefficient of .6). Since it is well-known that bayesian methods are more sensitive to noise than SVMs, we applied Feature Selection techniques as a preprocessing to this classifier. We calculated the IG of each feature with respect to the classes, and kept only the 50 features with most IG, as long as they all had an IG over 0.001, those with IG below that threshold were discarded. Feature Selection yields an important improvement in performances, reaching an accuracy of 72%. This performance, however, is still below that of SVMs, and that is why we study a third improvement: one vs. all classification.

As pointed out above, MNB is highly sensitive to noise, which seems specially acute in this setting where we have only very few examples of many of the classes. To obtain better models in this context, we applied a one vs. all approach, where a different classifier is trained to distinguish each individual class from all the rest. This, combined with a separate Feature Selection preprocess for each of the classifiers yields a significant improvement in performances, reaching an accuracy of 83%, with a kappa coefficient of .8. This allows us to use MNB as a base classifier for active learning, without sacrificing loss in performance with respect to the SVM baseline. Results are summarized in Table 1.

Table 1. Accuracy of two passive learning classifiers with different configurations

	plain	with FS	one vs. all	one vs. all & FS	one vs. all & class-specific FS
SVM	76	76	71	73	73
MNB	63	72	60	78	**83**

4 Licenses IE within an Active Learning Loop

The benefits of active learning, as discussed before, are a faster learning curve and an optimization of the human effort needed to train an automatic classifier. We want to assess the impact of an active learning approach in the task of License Information Extraction.

We apply uncertainty sampling to assess the utility of instances and IG to assess the utility of features for a given model. We then explored the effects of ranking instances either by highest or lowest uncertainty.

We implemented a system to apply active learning to the kind of annotation that we aim to develop, with functionalities similar to those of Dualist [12]. The architecture of the system is visualized in Figure 1. The system is provided with an annotated and an unannotated dataset. A model is learnt from the annotated dataset, applying MNB in a one-vs-all setting with separated feature selection for each classifier. Then, the model is applied to an unannotated dataset, and instances in this dataset are ranked according to the certainty of the model to label them, ranking highest those with most certainty or with most uncertainty. The highest ranking instances are presented to the oracle, who annotates them, associating each instance to one or more of the classes defined by the ODRL ontology or the class "null" if none of the available classes apply for the instance.

Then the oracle is given the possibility to annotate features that she finds as clearly indicative of a given class. For each class, the list of features with highest IG with the class is provided, and the oracle selects those that she finds are indicative of the class. If the user chooses not to annotate features, they are selected by the automated feature selection technique, that is, the system keeps for each one-vs.-all classifier only the top 50 features with highest IG with the class or only those features with more than 0.001 IG with the class, whichever condition produces the biggest set. If the user chooses to annotate features, these are added to the pool of features selected with the default method.

Finally, the system is trained again with the annotated corpus, now enhanced with the newly annotated examples and possibly newly annotated features.

The system is built as a hybrid application with Perl[8], Scala[9] and Java[10]. It uses the libraries within Weka [16], including LibSVM [2], for the training and evaluation of the classifier and has a hybrid web application (that uses both Perl and Scala), to use in the local machine, created with `Play!`[11], a Model-View-Controller (MVC)

[8] `https://www.perl.org/`

[9] `http://www.scala-lang.org/`

[10] `https://java.com/`

[11] `https://www.playframework.com/documentation/2.3.x/Home`

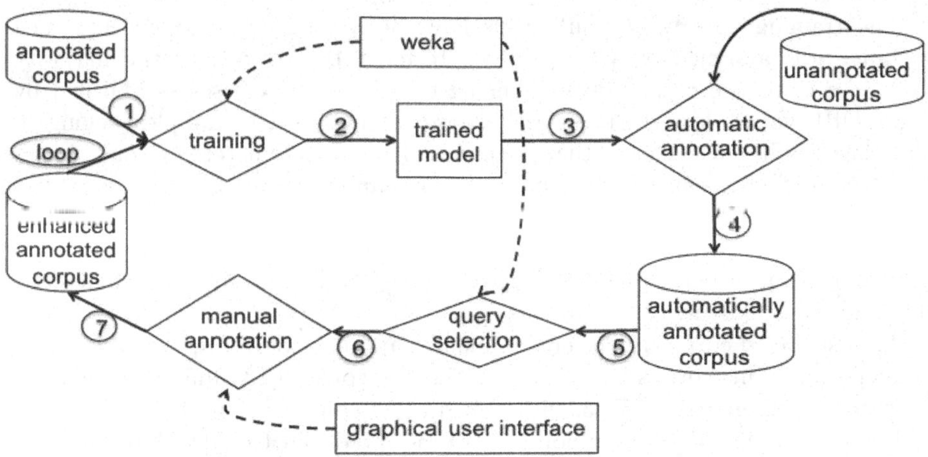

Fig. 1. Architecture of our basic License IE system, with the active learning component

Web application framework for the Scala language, in order to use HTML for a graphic interface.

The system is prepared for multi-class annotation, because more than one property may be expressed in a single sentence.

5 Experimental Setting

The goal of the experiments is to assess whether active learning strategies provide a faster learning curve than traditional methods, to alleviate portability of our approach to different domains. To that end, we compared active learning, with different parametrizations, with passive learning.

5.1 Dataset

We evaluated the performance of different learning methods using a manually annotated dataset of licenses. The corpus consists of the original labelled set of 37 licenses, and an unlabeled set of 396 licenses. It is composed of software licenses, source code licenses, data licenses, and content licenses; they are public as well as private domain licenses.

The labeled corpus has a total of 41,340 words, 2,660 of them unique. The mean of words per license is 1117.30, with a median of 400. This corpus has a total of 162 labelled instances, with a mean of 12.46 instances per class. The class with the most instances is *permission-to-distribute*, with a total of 33 instances, while there are three classes with just one instance: *permission-to-read, prohibition-to-derive and requirement-to-attach-source*. Classes with very few instances are known to provide for very poor learned models, so we discarded classes with less than 5 labelled instances.

The training and evaluation corpus have been tagged previously and each instance was assigned to a single class. It must be noted that the majority of sentences in the corpus do not belong to any of the classes established by the ODRL vocabulary. In the classification setting, these examples belong to the class "null", which is actually composed of several heterogeneous classes with very different semantics, with the only common factor that their semantics are not captured by the ODRL vocabulary. The fact that this heterogeneous majority class is always present seems a good explanation for why the one-vs-all approach is more performant: it is easier to define one single class than some heterogeneous classes.

The unlabeled corpus is gathered manually, and has no overlap with the annotated corpus. This corpus has a total of 482,259 words, 8,134 unique. The mean of words per license is 1217.83, with a median of 505.50.

For the manual dataset annotation we adopted the CONLL IOB format., The B and I tags are suffixed with the chunk type according to our annotation task, e.g. B-PERMISSION, I-PERMISSION. We first tokenized the sentences using Stanford Parser [8], and we then added two columns, the first one for the annotation of the relation, and the second one for the value The Stanford Parser is also used to parse the instances of the unannotated corpus. From the unannotated corpus, sentences are taken as instances to be annotated by the automated classifier or the oracle.

5.2 Evaluation Methods and Metrics

The evaluation task is done with an automated simulation of the active learning loop on the annotated corpus. In this simulation, from the 156 original instances on the corpus, we started with an initial random set of 20 instances (roughly 12% of the annotated corpus). From this initial set the first model was learned, using the Multinomial Naïve Bayes approach. After that, the model was evaluated using 10-fold cross-validation.

With this initial model, we proceed to use the rest of the annotated instances as the unannotated corpus. With the data from the first model we carry out the selection of the queries from this "unannotated corpus" for manual annotation. In our experiments we try with three different approaches: queries of automatically annotated instances where the classifier is most certain sample, queries of instances where the classifier is most uncertain, and random selection (passive learning). The selected queries are then annotated using the provided information (as these queries are, in fact, from the annotated corpus) and added to the annotated corpus as new instances.

Once again the annotated corpus is used in a second iteration for creation and evaluation of a new model. The process is repeated until all the "unannotated" instances are assigned their label. The number of newly annotated instances per iteration in our experiments is: 1, 3, 5 and 10.

The goal of this simulation is to show the steep of the curves in each one of the query selection methods in comparison to each other, with the highest slope being the best query selection strategy.

6 Analysis of Results

In Figure 2 we can see the learning curves of our active learning approach, obtained as described in Section 5.2. We can see that the "most certain" strategy performs consistently better than the passive and most uncertain strategies, improving performance with fewer instances. The other two perform comparably if the number of instances added at each iteration is high, and the "most uncertain" approach performs even worse than the passive approach (random) if instances are added one at a time for each iteration. These results confirm our hypothesis that, for models inferred from very few training examples, maximizing the entropy of examples is not useful, while providing more evidence to define the core of the classes does provide an improvement in performance.

In an error analysis we can indeed see that the classes with most error are the smallest classes.This shows the benefit of growing the set of annotated examples,

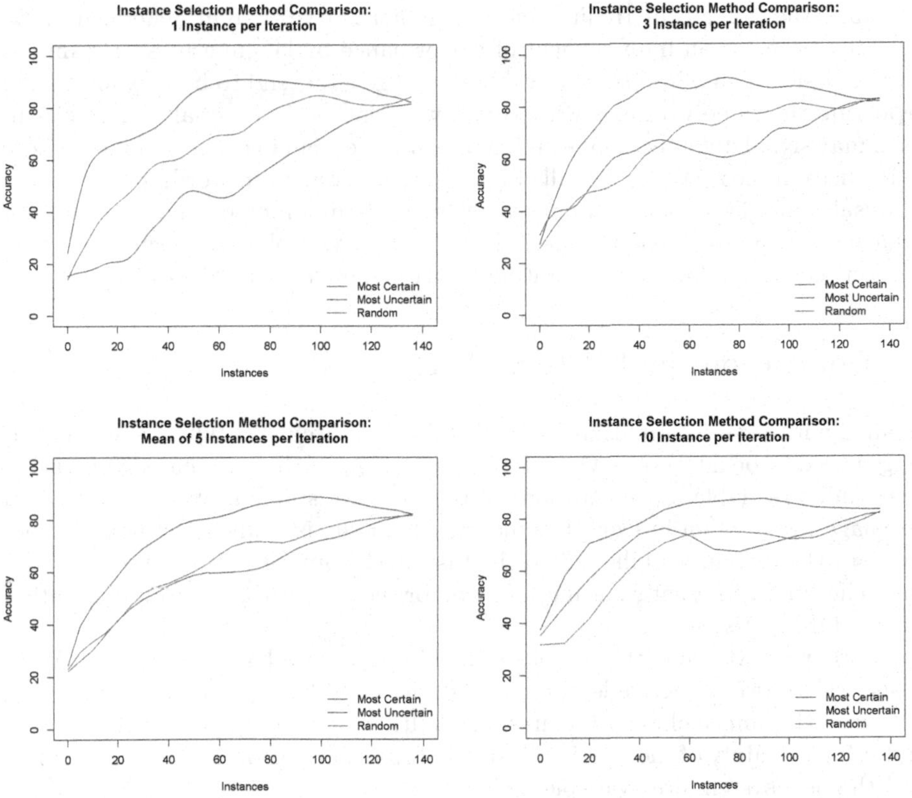

Fig. 2. Learning curves of active learning approaches with different policies for instance selection. In the y axis we depict accuracy, in the x axis, the number of instances added to training, and the different lines represent different strategies to pick the instances to be added in each iteration to the training corpus: random, ranked by most uncertain or by most uncertain.

and thus the utility of an active learning approach for this task. The best strategy to grow from a very small dataset, with classes with very few instances, seems to be by choosing instances that are very similar to those already labelled, which provides a faster improvement in the performance of the classifier.

When examples are selected applying the "most uncertain" strategy are, they mostly belong to the "null" class, that is, they do not signal any of the classes relevant for the problem. Most of the sentences in licenses do not belong to any of the classes defined by the ODRL vocabulary and are classified as "null".

Providing examples for the class "null" is specially harmful for the resulting model for two main reasons. First, it grows the majority class, while small classes are kept with the same few examples, thus adding the problem of having an imbalanced dataset to the problem of having small classes with few instances. Second, the class "null" is composed by many heterogeneous classes that are not included in the ODRL vocabulary, and therefore its characterization is difficult and may be misleading.

Besides this configuration of classes, which can be found in very different domains, the domain of IE in licenses and normative text in general may be specially prone to an improvement of performance by labeling most certain examples first, because licenses and legal texts in general tend to be very formulaic, repeating the same wordings with very few variations, and small differences in form may signal differences in meaning, much more than in other domains, where differences in meaning are signalled by bigger differences in wordings.

Results for the best performances achieved by different passive learning approaches are summarized in Table 1. Those results were obtained using the whole dataset, corresponding to the rightmost extreme in the graphics of Figure 2.

7 Conclusions and Future Work

Dealing with legal information in the Web is a research area where several challenges need to be addressed. One of these challenges is the automated generation of machine-readable representations of the licenses, starting from their natural language formulation in legal documents. This issue is challenging not only because of the difficulties inherent to the task itself, but also due to the fact that the generated representation must be conformant with the starting regulation to avoid data misuse.

In order to overcome this problem, in this paper, we have developed a Web-based framework for active learning of instances and features, much in the spirit of Settles [12], but including features like multiclass labeling for a single instance and using certainty of the classifier instead of Information Gain. Both the dataset and the new system are available online.

We have shown that, for the problem of inferring a classifier for normative text, where few labelled instances are available, active learning does provide a faster learning curve than traditional machine learning approaches, and it is thus an effective strategy to optimize human resources. It must be noted that in this specific setting, active learning is useful only if the most certain examples are selected to be hand-tagged, in contrast with the most frequent approach in active learning, called *uncertainty sampling*, where human annotators are given to annotate examples that the classifier is most uncertain about. This is caused by the fact that normative text is very formulaic thus tending to repetitions, but also to the fact that in this domain, slight differences in formulation tend to signal actual differences in meaning.

Several open issues have to be considered as future work. First, given the complexity of the task, our system provides an RDF representation of licenses considering their basic deontic components only, i.e., we model *permissions*, *prohibitions*, and *duties*. However, we plan to consider as future work further constraints expressed by the licenses, e.g., about time, payment information, and sub-licensing. Second, from the experimental setting perspective, we will explore some configurations that are left unexplored in this paper, like those classes with less than 5 labelled instances, with the most certain strategy to begin with. We will evaluate the performances of the system also using feature labeling by itself and in combination with instance labeling. We are currently exploring if there is a point in the development of the training set where it is more useful to switch from certainty sampling to uncertainty sampling, probably in correspondence with the different distributions of features in annotated and unannotated corpora. Finally, the system can be extended to a multilingual scenario (as far as the NLP pre-processing is available for the targeted languages), to provide machine readable versions of licenses published by national institutions, or licenses published in different languages.

References

1. Cabrio, E., Aprosio, A.P., Villata, S.: These are your rights - A natural language processing approach to automated RDF licenses generation. In: Proceedings of the SemanticWeb: Trends and Challenges - 11th International Conference, ESWC 2014, Anissaras, Crete, Greece, May 25-29, pp. 255–269 (2008)
2. Chang, C.-C., Lin, C.-J.: Libsvm - A library for support vector machines, The Weka classifier works with version 2.82 of LIBSVM (2001)
3. Culotta, A., McCallum, A.: Reducing labeling effort for structured prediction tasks. In: Proceedings of the 20th National Conference on Artificial Intelligence, AAAI 2005, vol. 2, pp. 746–751. AAAI Press (2005)
4. Dligach, D., Palmer, M.: Good seed makes a good crop: Accelerating active learning using language modeling. In: Proceedings of the 49th Annual Meeting of the Association for Computational Linguistics: Human Language Technologies: Short Papers, HLT 2011, vol. 2, pp. 6–10. Association for Computational Linguistics, Stroudsburg (2011)
5. Donmez, P., Carbonell, J.G., Bennett, P.N.: Dual strategy active learning. In: Kok, J.N., Koronacki, J., Lopez de Mantaras, R., Matwin, S., Mladenič, D., Skowron, A. (eds.) ECML 2007. LNCS (LNAI), vol. 4701, pp. 116–127. Springer, Heidelberg (2007)
6. Druck, G., Settles, B., McCallum, A.: Active learning by labeling features. In: Proceedings of the Conference on Empirical Methods in Natural Language Processing, pp. 81–90. ACL (2009)

7. Kearns, M.: Efficient noise-tolerant learning from statistical queries. J. ACM 45(6), 983–1006 (1998)
8. Klein, D., Manning, C.D.: Accurate unlexicalized parsing. In: Proceedings of the 41st Annual Meeting on Association for Computational Linguistics, ACL 2003, vol. 1, pp. 423–430. Association for Computational Linguistics, Stroudsburg (2003)
9. Lewis, D.D., Catlett, J.: Heterogeneous uncertainty sampling for supervised learning. In: Proceedings of the Eleventh International Conference on Machine Learning, pp. 148–156. Morgan Kaufmann (1994)
10. Pujara, J., London, B., Getoor, L.: Reducing label cost by combining feature labels and crowdsourcing. In: ICML Workshop on Combining Learning Strategies to Reduce Label Cost (2011)
11. Settles, B.: Active learning literature survey. Computer Sciences Technical Report 1648, University of Wisconsin–Madison (2009)
12. Settles, B.: Closing the loop: Fast, interactive semi-supervised annotation with queries on features and instances. In: Proceedings of the Conference on Empirical Methods in Natural Language Processing, pp. 1467–1478. ACL (2011)
13. Settles, B.: Active Learning. In: Synthesis Lectures on Artificial Intelligence and Machine Learning. Morgan Kaufmann (2012)
14. Settles, B., Craven, M.: An analysis of active learning strategies for sequence labeling tasks. In: Proceedings of the Conference on Empirical Methods in Natural Language Processing, pp. 1069–1078. ACL (2008)
15. Symons, C.T., Arel, I.: Multi-View Budgeted Learning under Label and Feature Constraints Using Label-Guided Graph-Based Regularization (2011)
16. Witten, I.H., Frank, E.: Data Mining: Practical machine learning tools and techniques. Morgan Kaufmann (2005)
17. Zhu, J., Wang, H., Yao, T., Tsou, B.K.: Active learning with sampling by uncertainty and density for word sense disambiguation and text classification. In: Proceedings of the 22nd International Conference on Computational Linguistics, vol. 1, pp. 1137–1144. Association for Computational Linguistics (2008)

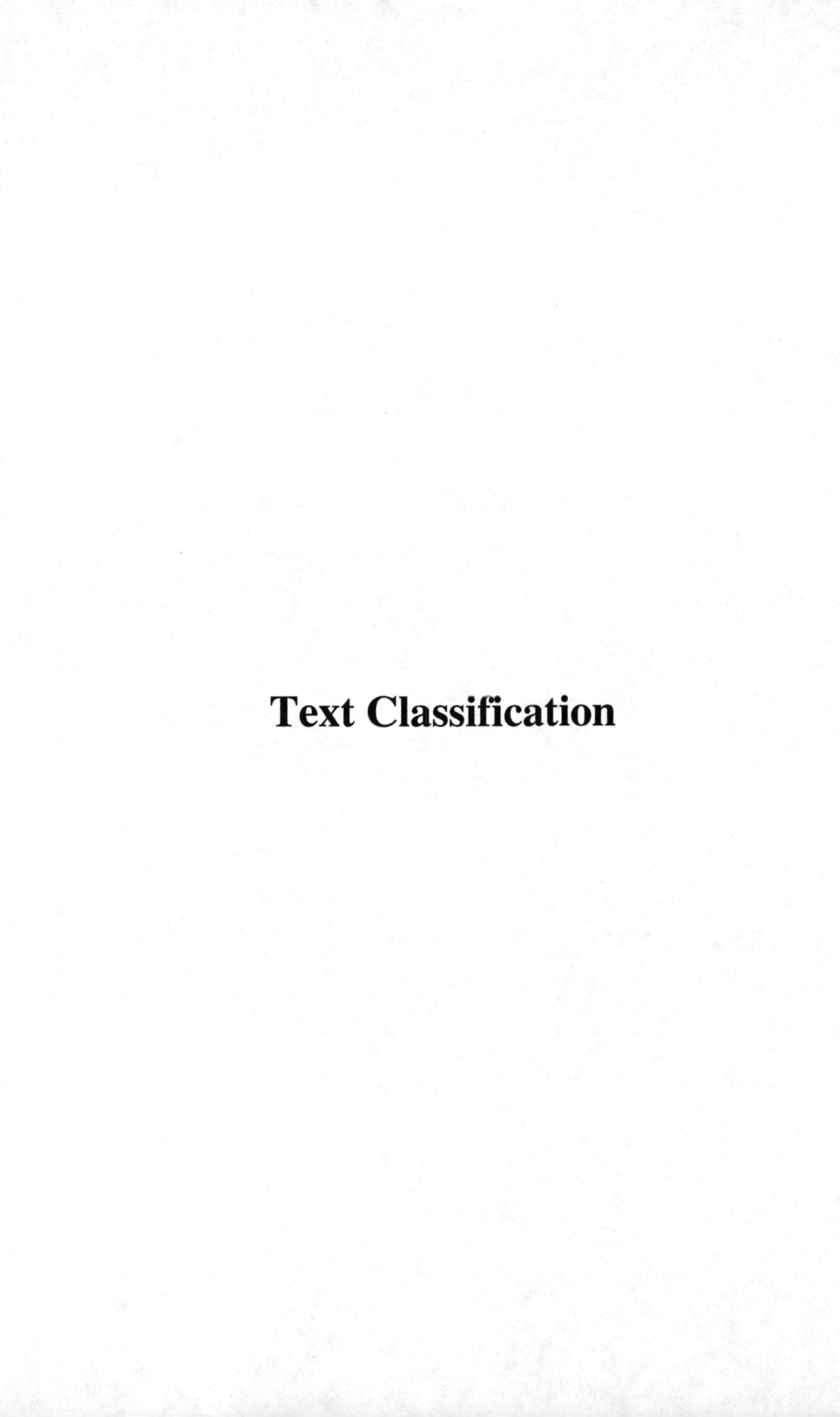

Text Classification

Term Network Approach for Transductive Classification

Rafael Geraldeli Rossi, Solange Oliveira Rezende, and Alneu de Andrade Lopes

Institute of Mathematics and Computer Science
University of So Paulo, So Carlos, SP, Brazil
{ragero,solange,alneu}@icmc.usp.br

Abstract. Transductive classification is a useful way to classify texts when just few labeled examples are available. Transductive classification algorithms rely on term frequency to directly classify texts represented in vector space model or to build networks and perform label propagation. Related terms tend to belong to the same class and this information can be used to assign relevance scores of terms for classes and consequently the labels of documents. In this paper we propose the use of term networks to model term relations and perform transductive classification. In order to do so, we propose (i) different ways to generate term networks, (ii) how to assign initial relevance scores for terms, (iii) how to propagate the relevance scores among terms, and (iv) how to use the relevance scores of terms in order to classify documents. We demonstrate that transductive classification based on term networks can surpass the accuracies obtained by transductive classification considering texts represented in other types of networks or vector space model, or even the accuracies obtained by inductive classification. We also demonstrated that we can decrease the size of term networks through feature selection while keeping classification accuracy and decreasing computational complexity.

1 Introduction

Text automatic classification (TAC) is useful to organize text collections, filtering e-mails, retrieving documents, and generating metadata, to cite few [30,24]. TAC has become an important research topic due to the huge number of applications and the proliferation of texts to disseminate information.

Generally TAC is applied using supervised inductive learning algorithms [30,24], which induce a classification model to classify new/unseen texts. Usually a large number of labeled documents are necessary to induce an accurate classification model. However, obtaining a high number of labeled documents is costly and time consuming.

TAC can also be performed by transductive learning. Transductive learning directly estimates the labels of unlabeled documents without creating a classification model. Transductive learning is performed considering texts represented

A. Gelbukh (Ed.): CICLing 2015, Part II, LNCS 9042, pp. 497–515, 2015.
DOI: 10.1007/978-3-319-18117-2_37

in a vector space model (VSM), such as Self-Training [32], Co-Training [3], Transductive SVM [10], and Expectation Maximization [16], or considering texts represented by networks, such as Gaussian Field and Harmonic Functions [34] or Learning with local and Global Consistency [33].

Due to the assumptions and drawbacks of algorithms based on VSM, transductive classification based on networks has been demonstrated to be a useful approach for transductive learning [35]. In such case, the dataset is modeled as a network and the labels of labeled documents are propagated to the unlabeled documents through the network connections. Label propagation using just few labeled documents may achieve higher accuracy than supervised inductive learning using a large number of labeled documents for TAC [20]. In this paper we focus on transductive learning since this approach is able to make use of the plenty of unlabeled texts available to perform and improve the performance of TAC, and saves user time necessary for labeling a large number of documents.

As documents and terms are presented in any text collections, we can always represent them by (i) **document networks**, in which documents are network objects and they are linked through hyperlinks or links representing interdependence between documents such as similarity; (ii) **term networks**, which terms are network objects and they are linked considering similarity, order of occurrence or syntactic/semantic relationship; or (iii) **bipartite networks**, in which documents and terms are network objects and they are linked when a term occurs in a document. The different types of networks models different patterns of the text collections, leading to different results.

Existing transductive classification algorithms rely on term frequency to directly classify texts represented in vector space model or to build document and bipartite networks to perform label propagation. They ignore relations among terms, which can be modeled by term networks. However, this type of information can improve the relevance scores of terms, which is useful to classify documents [19,20,30], since related terms tend to belong to the same class. Furthermore, improving the relevance scores of terms for classes consequently improves the classification performance.

Performing transductive classification through the setting of the relevance scores of terms modeled in a term network requires answering five questions: 1) what type of measure is appropriate to assign weights for relations between terms? 2) what relations should be considered? 3) how to assign initial relevance scores for terms in a network? 4) how to propagate the relevance scores among related terms? and 5) how to use these relevance scores to classify documents?

In this paper we investigate and propose solutions for all the above mentioned points. An evaluation carried out using 15 text collections from different domains shows that the classification accuracies obtained by transductive classification using term networks can surpass the accuracies obtained by algorithms based on VSM, document and bipartite networks, or even supervised inductive classification. Moreover, the proposed approach allows decreasing the network size through feature selection. We show that this procedure speeds up classification and keeps classification accuracy.

The remainder of this paper is organized as follows. Section 2 presents related works about transductive classification, representation of text as networks, and existing approaches for TAC considering term networks. We also present in this section the notations, concepts and technical details that are used in this paper. In Section 3 we describe our term network approach for transductive classification of texts. Section 4 presents the details of the experimental evaluation and the results. Finally, Section 5 presents the conclusions and points to future work.

2 Related Work, Background and Notations

The first researches about transductive learning for text classification considers text collections represented in vector space model [32,3,10,16]. Usually a bag-of-words is used to represent the text collection, in which each document is represented by a vector and each dimension of the vector corresponds to a term. The values in the vectors are based on the frequency of a term in a document, such as binary weights, term frequency (*tf*) or term frequency - inverse document frequency (*tf-idf*) [30].

Traditional and state-of-the-art transductive algorithms based on vector space model are [35]: Self-Training [32], Co-Training [3], Expectation Maximization (EM) [16], and Transductive Support Vector Machines (TSVM) [10]. There are also some combinations/variations of these algorithms to perform transductive learning. These algorithms have strong assumptions about the data properties/distribution. For instance, Self-training considers that the most confident classifications are correct and retrain a classification model iteratively considering the most confident classifications as labeled examples. EM considers that the texts are generated by generative model and TSVM has the assumption that the classes are well-separated, such that the hyperplane with maximal margin falls into a low density region. These assumptions are frequently violated in practice and the classification performance is degraded when they do not hold [35].

Network-based representation is a natural and direct way to represent textual data for different tasks. Different types of objects (vertices) and different relations (links) can be used to generate network-based representations. Formally, a network is defined as $N = \langle \mathcal{O}, \mathcal{R}, \mathcal{W} \rangle$, in which $\mathcal{O}$ represents the set of objects, $\mathcal{R}$ the set of relations among objects, and $\mathcal{W}$ the set of weights of the relations. A network is called homogeneous network if $\mathcal{O}$ consists of a single type of object, and heterogeneous network if $\mathcal{O}$ consists of h different types of objects ($h \geq 2$), i.e., $\mathcal{O} = \mathcal{O}_1 \cup \ldots \cup \mathcal{O}_h$ [8].

Performing transductive classification on networks requires modeling a text collection in a single network to allow the propagation of labels through the entire collections and consequently label all documents [20]. In order to do so, we can model text collection as document, term, or bipartite networks.

In a document network, $\mathcal{O} = \mathcal{D}$, in which $\mathcal{D} = \{d_1, d_2, \ldots, d_n\}$ represents the documents of a collection. $\mathcal{D}$ can be composed by labeled ($\mathcal{D}^L$) or unlabeled ($\mathcal{D}^U$) documents, i.e., $\mathcal{D} = \mathcal{D}^L \cup \mathcal{D}^U$. Documents are connected according to (i) "explicit" relations such as hyperlinks or citations [17], or (ii) considering similarity [2,34,33]. Here we focus on similarity-based document network, since it models any text collection as networks and provide better results than document networks based on explicit relations [2]. Usually two approaches are used to generate similarity-based document networks [35]: (i) fully connected-network or (ii) nearest neighbor network. In this paper we consider the most representative type of each approach: (i) Exp network and (ii) Mutual k Nearest Neighbors (kNN) network. In an Exp network, the weight of the relation between a document d_i and a document d_j (w_{d_i,d_j}) is given by a Gaussian function, i.e., $w_{d_i,d_j} = \exp(-dist(d_i, d_j)^2/\sigma^2)$, in which $dist(d_i, d_j)$ is the distance between the documents d_i and d_j, and σ controls the bandwidth of the Gaussian function. In mutual kNN network, an object d_i and an object d_j are connected if d_j is one of the k nearest neighbors of d_i and d_i is one of the k nearest neighbors of d_j.

In a term network, $\mathcal{O} = \mathcal{T}$, in which $\mathcal{T} = \{t_1, t_2, \ldots, t_m\}$ represents the terms of a collection. Terms are connected if (i) they precede or succeed each other in a text [1,12], (ii) they co-occur in pieces of texts as sentences/windows [25,14] or in the text collection [29,28,13] (also called similarity), or (iii) they present syntactic/semantic relationship [25,26].

In a bipartite network, $\mathcal{O} = \{\mathcal{D} \cup \mathcal{T}\}$ [20,4]. Thus, this is a heterogeneous network. $d_i \in \mathcal{D}$ and $t_j \in \mathcal{T}$ are wired if t_j occurs in d_i and the relation weight between them (w_{t_i,d_j}) is the frequency of t_j in d_i. Thus, just the terms and their frequencies in the documents are necessary to generate the bipartite network.

Regardless of the network representation, the computational structures to perform transductive classification are the same. Let $\mathcal{C} = \{c_1, c_2, \ldots, c_l\}$ be the set of label classes, and let $\mathbf{f}_{o_i} = \{f_{c_1}, f_{c_2}, \ldots, f_{c_{|\mathcal{C}|}}\}^T$ be the weight vector of an object o_i, which determines its weight or relevance score for each class[1]. All weight vectors are stored in a matrix $\mathbf{F}$. The predefined label of an object o_i, i.e., the label informed by a domain specialist/users or assigned to an object at the begging of the classification, is stored in a vector $\mathbf{y}_{o_i} = \{y_{c_1}, y_{c_2}, \ldots, y_{c_{|\mathcal{C}|}}\}^T$. In the case of labels assigned by domain specialists/users, the value 1 is assigned to the corresponding class position and 0 to the others. The weights of connections among objects are stored in a matrix $\mathbf{W}$.

Document and bipartite networks has been used for transductive classification of texts [34,33,20]. To the best of our knowledge, term networks have been used exclusively in supervised inductive learning for TAC. Usually, term networks are used to generate a vector-space representation considering edges [1] or subgraphs [9] as features. In this case, traditional supervised inductive learning algorithms based on vector space model can be used to perform TAC. However, the conversion from network representation to vector space model may lead to a extremely high dimensionality, which difficult their application in practical situations.

An alternative is to use term networks to induce a classification model. [23] and [15] build a term network for each document and perform graph matching to classify documents considering the classes of the most similar term networks. Both approaches present high computational cost due the graph matching and high memory consumption due to the need to keep all the term networks in

[1] A weight vector will be also treated by class information when referred to the weight vector of documents and relevance scores to the weight vector of terms.

memory. Moreover, the term networks presented [23] and [15] are generated considering HTML sections and controlled biomedical vocabulary respectively, which does not allow their application in any text collection.

[29] avoids the high consumption of memory and computation presented by graph matching approach. In such case, a term network is generated for each class. Terms are connected if they co-occur frequently in the same class. Then, the terms are ranked using PageRank algorithm [17] and these rankings are used as a classification model. A new document is labeled according to the most correlated ranking between its ranked terms and the ranked terms of each class. Despite speeding up the classification time and decrease the memory consumption, the classification model does not provide good classification accuracy.

3 Proposal: Term Network Approach for Transductive Classification of Texts

The term network approach for transductive classification of texts proposed here, named TCTN (Transductive Classification through Term Networks), has four main steps: (i) term network generation, (ii) initial relevance score setting, (iii) relevance score propagation, and (iv) text classification. In the next sections we present the details of these four steps.

3.1 Term Network Generation

A text collection is composed by "generic" terms, which occur in documents from several classes, and "specific" terms, which are most likely to appear in one or few classes. Specific terms tend to be strongly related among them and weakly related with generic terms. These characteristics can be useful to propagate the relevance scores among terms. Since there is no "explicit" information about the relations among terms in a text collection, we need to extract the relations analyzing the text collection. Besides, we need measures which assign high weights for relations among specific terms and low weights for relations among generic terms.

We can employ similarity (also called quality, interestingness or association) measures to compute relations among terms [7,27]. Different measures calculate the similarity between terms t_i and t_j ($\Omega(t_i, t_j)$) considering the information contained in the contingency matrix presented in Table 1. The contingency matrix contains the probability of occurrence of each term ($p(t_i)$ and $p(t_j)$), probability of no occurrence ($p(\neg t_i)$ and $p(\neg t_j)$), the joint probability of two terms co-occur ($p(t_i, t_j)$) and not co-occur ($p(\neg t_i, \neg t_j)$), and the probability of one term occurs without other term ($p(t_i, \neg t_j)$ and $p(\neg t_i, t_j)$).

We selected measures which comply with different characteristics and properties [7,27]. The selected measures were: Support, Yule's Q, Mutual Information, Kappa, and Piatetsky-Shapiro. Table 2 presents the selected similarity measures, their formulas and range of values.

In this paper we consider two approaches to wire terms considering their similarity: (i) `Threshold` approach, in which two terms are wired if their similarity is

Table 1. Contingency matrix for terms t_i and t_j

	t_j	$\neg t_j$	Total
t_i	$p(t_i, t_j)$	$p(t_i, \neg t_j)$	$p(t_i)$
$\neg t_i$	$p(\neg t_i, t_j)$	$p(\neg t_i, \neg t_j)$	$p(\neg t_i)$
Total	$p(t_j)$	$p(\neg t_j)$	1

Table 2. Formulas of the selected similarity measures [7,27]

Measure	Formula	
Support	$P(t_i, t_j)$	[0,1]
Yule's Q	$\frac{P(t_i,t_j)P(\neg t_i,\neg t_j)-P(t_i,\neg t_j)P(\neg t_i,t_j)}{P(t_i,t_j)(P\neg t_i,\neg t_j)+P(t_i,\neg t_j)P(\neg t_i,t_j)}$	[-1,1]
Mutual Information	$P(t_i, t_j)log_2\left(\frac{P(t_i,t_j)}{P(t_i)P(t_j)}\right)+$ $P(t_i, \neg t_j)log_2\left(\frac{P(t_i,\neg t_j)}{P(t_i)P(\neg t_j)}\right)+$ $P(\neg t_i, t_j)log_2\left(\frac{P(\neg t_i,t_j)}{P(\neg t_i)P(t_j)}\right)+$ $P(\neg t_i, \neg t_j)log_2\left(\frac{P(\neg t_i,\neg t_j)}{P(\neg t_i)P(\neg t_j)}\right)$	[-1,1]
Kappa	$\frac{P(t_i,t_j)+P(\neg t_i,\neg t_j)-P(t_i)P(t_j)-P(\neg t_i)P(\neg t_j)}{1-P(t_i)P(t_j)-P(\neg t_i)P(\neg t_j)}$	[-1,1]
Piatetsky-Shapiro	$P(t_i, t_j) - P(t_i)P(t_j)$	[-0.25,0.25]

above a threshold, and (ii) TopK approach, in which a term is wired to its k most mutual similar terms, i.e., a term t_i is wired to a term t_j if t_j is one of the most similar terms of t_i and t_i is one of the most similar terms of t_j. At the end of the wiring process, if two terms present a negative relation weight, all the other network relations are increased with the module of the most negative weight. This is performed to ensure the correct functioning of the relevance score propagation algorithm (Section 3.3). The computation of similarity measures considers the occurrence of terms in both labeled and unlabeled documents.

3.2 Initial Relevance Score Setting

Labeling terms requires knowledge about the classes present in the collection and a notion about the occurrence of terms for all classes. On the other hand, labeling documents is an easier task. The user can define the class of a document based on its content. Moreover, just few labeled documents can provide initial relevance scores for several terms. Thus, a mechanism to infer initial relevance scores (or predefined class information as presented in the previous section) to terms considering the labeled documents of a collection is useful and necessary.

The proposed initial relevance score of a term t_i for a class c_j is given by

$$y_{t_i,c_j} = \sum_{d_k \in \mathcal{D}^L} freq(d_k, t_i) y_{d_k c_j} / \sum_{d_k \in \mathcal{D}^L} freq(d_k, t_i), \quad (1)$$

where $freq(d_k, t_i)$ is the frequency of term t_i in document d_k and $y_{d_k c_j}$ is equal 1 if document d_k belongs to class c_j and 0 otherwise. This equation returns values close to 1 for terms that occur almost exclusively for one class and low values for terms that are equally distributed to several classes.

3.3 Relevance Score Propagation

The goal of relevance score propagation is to set the relevance scores of related terms through the network connections. We have 3 assumptions to set the relevance scores of terms. (i) the relevance scores of neighboring terms must be close; (ii) the final relevance scores of labeled terms must be close to their initial relevance scores; and (iii) terms with high number of relations should not dominate relevance score propagation. These three assumptions are enforced through the terms of the following objective function:

$$Q(\mathbf{F}(\mathcal{T})) = \frac{1}{2}\sum_{t_i,t_j \in \mathcal{T}} w_{t_i,t_j} \left\| \frac{\mathbf{f}_{t_i}}{\sqrt{\sum\limits_{t_k \in \mathcal{T}} w_{t_i,t_k}}} - \frac{\mathbf{f}_{t_j}}{\sqrt{\sum\limits_{t_k \in \mathcal{T}} w_{t_j,t_k}}} \right\|^2 + \mu \sum_{t_i \in \mathcal{T}} ||\mathbf{f}_{t_i} - \mathbf{y}_{t_i}||^2, \tag{2}$$

in which μ controls the importance of each term of Equation 2.

Equation 2 corresponds to objective function minimized by Global Consistency (LLGC) algorithm [33]. Algorithm 1 presents an iterative procedure to minimize Equation 2. Line 1 computes the degree (sum of relation weights) of each term. The degree is used in Line 2 to compute a normalized symmetric matrix, which is necessary for the convergence of relevance score propagation. Lines 3-5 performs relevance score propagation, in wich the class information of each term is set by the class information of neighboring terms (first term of Line 4) and the initial class information (second term of Line 4). The class information of neighbors and the initial class information are weighted by α and $(1 - \alpha)$ respectively. The iterative procedure is called "label propagation" [35] and in our case will propagate the relevance scores among terms.

Algorithm 1. Relevance Score Propagation

Input : $\mathcal{T}$,$\mathbf{W}$,$\mathbf{Y}$,
α - parameter to attenuate differences of the predefined relevance scores of network objects in consecutive iterations ($0 < \alpha < 1$)
Output: $\mathbf{F}$

1 $\mathbf{D} = diag(\mathbf{W} \cdot \mathbf{I}_{|\mathcal{T}|})$ `/* diag(...) is the diagonal matrix operator */`
2 $\mathbf{S} = \mathbf{D}^{-1/2} \cdot \mathbf{W} \cdot \mathbf{D}^{-1/2}$
3 **repeat**
4 | $\mathbf{F} \leftarrow \alpha \cdot \mathbf{S} \cdot \mathbf{F} - (1 - \alpha) \cdot \mathbf{Y}$
5 **until** *relevance scores of terms remains the same or fixed number of iterations*

3.4 Text Classification

The relevance scores assigned to terms through relevance score propagation are used for text classification. To do so we consider the relevance scores of terms for classes and their frequency in the documents. The class information of an unlabeled document $d_i \in \mathcal{D}^U$ for a class c_j is given by the weighted linear function

$$f_{d_i,c_j} = \sum_{t_k \in \mathcal{T}} freq(d_i, t_k) f_{t_k,c_j}. \quad (3)$$

The class or label of document d_i is given by the arg-max value of $\mathbf{f}_{d_i}$, i.e., $class(d_i) = \arg\max_{c_j \in \mathcal{C}} f_{d_i,c_j}$.

4 Experimental Evaluation

This section presents the textual document collections used in the experiments, experimental configuration, evaluation criteria, results and discussions.

4.1 Document Collection

We used 15 textual document collections from the following domains: scientific document (SD), web pages (WP), news articles (NA), sentiment analysis (SA), and medical documents (MD). The collections have different characteristics. The number of documents ($|\mathcal{D}|$) ranges from 299 to 11162, the number of terms ($|\mathcal{T}|$) from 1726 to 22927, the average number of terms per document ($|\overline{\mathcal{T}}|$) from 6.65 to 205.06, the number of classes ($|\mathcal{C}|$) from 2 to 16, the standard deviation considering the class percentages in each collection ($dev(\mathcal{C})$) from 0 to 18.89, and the percentage of the majority class ($m(\mathcal{C})$) from 7.69 to 51.12. Tables 3 presents the characteristics of the 10 collections.

For the collections La1, Oh0, Oh10, Oh15, Oh5, Ohscal, and Re0 [5] no preprocessing was performed since these collections were already preprocessed. For the others, single words were considered as terms, stopwords were removed, terms were stemmed using Porter's algorithm [18], HTML tags were removed, and only terms with document frequency ≥ 2 were considered. The collections are available at `http://sites.labic.icmc.usp.br/text_collections/`. More details about the collections are presented at [21].

4.2 Experiment Configuration and Evaluation Criteria

In the experimental evaluation we analyze: (i) if the proposed approach for transductive classification using term networks provides better accuracies than algorithms based on VSM, document and bipartite networks, or even supervised inductive classification; (ii) if we can reduce the term network size and keep classification accuracy; (iii) the level of disagreement among classifications provided by document, bipartite and term networks to validate the hypothesis that different networks extracts different patterns; and (iv) what similarity measure and way to wire terms generate term networks provide the highest accuracies.

We considered the Multinomial Nave Bayes (MNB) as supervised inductive classification algorithm since it is a parameter-free algorithm and is accurate for text classification [20]. We used the Weka's implementation of MNB [31].

Table 3. Characteristics of the textual document collections

Collection	$\lvert\mathcal{D}\rvert$	$\lvert\mathcal{T}\rvert$	$\lvert\overline{\mathcal{T}}\rvert$	$\lvert\mathcal{C}\rvert$	$dev(\mathcal{C})$	$\max(\mathcal{C})$
CSTR (SD)	299	1726	54.27	4	18.89	42.81
Dmoz-Health-500 (WP)	6500	4217	12.40	13	0.00	7.69
Dmoz-Science-500 (WP)	6000	4821	11.52	12	0.00	9.63
IrishSent (SA)	1660	8659	112.65	3	6.83	39.46
[illegible]	[illegible]	[illegible]	[illegible]	[illegible]	[illegible]	[illegible]
MultiDomainSent (SA)	8000	13360	42.36	2	0.00	50.00
NFS (SD)	10524	3888	6.65	16	3.82	13.39
Oh0 (MD)	1003	3183	52.50	10	5.33	19.34
Oh15 (MD)	913	3101	59.30	10	4.27	17.20
Oh5 (MD)	918	3013	54.43	10	3.72	16.23
Ohscal (MD)	11162	11466	60.39	10	2.66	14.52
Polarity (SA)	2000	15698	205.06	2	0.00	50.00
Re0 (NA)	1504	2887	51.73	13	11.56	40.43
Re8 (NA)	7674	8901	35.31	8	18.24	51.12
Reviews (NA)	4069	22927	183.10	5	12.80	34.11
WebKb (WP)	8282	22892	89.78	7	15.19	45.45

Self-Training[2], EM[3] and TSVM[4] were considered as transductive algorithms based on vector space model. We used the MNB as the learning algorithms for Self-Training. The Self-Training model was incremented using 5, 10, 15 and 20 most confident classifications in each iteration. We disregard Co-Training since it requires collections with two independent views. We are able to generate two views splitting the feature set but we empirically verified that this approach does not outperform Self-Training.

For EM we considered 1, 2, 5 and 10 components per class and 0.1, 0.3, 0.5, 0.7, and 0.9 as weights for unlabeled documents to set the probabilities of terms occur in the classes [16]. For TSVM we used $C = \{10^{-5}, 10^{-4}, 10^{-3}, 10^{-2}, 10^{-1}, 10^{-0}, 10^{1}\}$ and a linear kernel. We run TSVM with and without the function proposed in [10] to maintain the same class proportion of labeled documents in the classification of unlabeled documents.

[2] Self-training implementation used in this paper is available at `http://sites.labic.icmc.usp.br/ragero/cicling_2015/text_categorization_tool/TCT/TransductiveClassification_SelfTraining.java`

[3] EM implementation used in this paper is available at `http://sites.labic.icmc.usp.br/ragero/cicling_2015/text_categorization_tool/TCTAlgorithms/Transductive/ExpectationMaximization_Transductive.java`

[4] TSVM implementation used in this paper is available at `http://sites.labic.icmc.usp.br/ragero/cicling_2015/text_categorization_tool/TCTAlgorithms/Transductive/TSVM_Balanced_Transductive.java` and `http://sites.labic.icmc.usp.br/ragero/cicling_2015/text_categorization_tool/TCTAlgorithms/Transductive/TSVM_Unbalanced_Transductive.java`

We considered the LLGC[5] algorithm for transductive classification in document networks since it is the basis for relevance score propagation in the proposed approach. We used $\alpha = \{0.1, 0.3, 0.5, 0.7, 0.9\}$. We generate document networks considering `Exp` networks with $\sigma \in \{0.05, 0.2, 0.35, 0.5\}$ and mutual KNN networks with $k \in \{1, 7, 17, 37, 57\}$[6].

We used an heterogeneous version of LLGC, called GNetMine[7] [8], for transductive classification in bipartite networks. The classification procedure using bipartite networks was the same presented in [20]. We used $\alpha = \{0.1, 0.3, 0.5, 0.7, 0.9\}$.

For the proposed approach (TCTN[8]), we also considered $\alpha = \{0.1, 0.3, 0.5, 0.7, 0.9\}$. Term networks[9] were generated considering the five similarity measures (Support, Mutual Information, Kappa, Yule's Q, and Piatetsky-Shapiro) and the two approaches to connect terms (`Threshold` (ϵ) and `TopK` (κ)) presented in Section 3.1. We used $\kappa \in \{1, 7, 17, 37, 57\}$ for all similarity measures in `TopK` approach. The thresholds are defined according to

$$threshold = (\Omega_{max} - \Omega_{min}) * \epsilon + \Omega_{min}, \tag{4}$$

in which Ω_{min} is the minimum similarity value between two terms, Ω_{max} is the maximum similarity value between two terms, and we used $\epsilon \in \{0.00, 0.25, 0.50, 0.75\}$. With this we divide the range of similarity values in four equal intervals and consequently we consider all `term-term` connections, 25%, 50% and 75% of the most significant similarity values.

The iterative solutions proposed by the respective authors of the LLGC, GNetMine, EM and the proposed approach were used. The maximum number of iterations was set to 1000 [20]. The metric used for comparison was the accuracy, i.e., the percentage of correctly classified documents. The accuracies were obtained considering the average accuracies over 10 runs. In each run we randomly selected N documents from each class as labeled documents. We carried out experiments using $N = \{1, 10, 20, 30, 40, 50\}$. We start with the minimum number

5 LLGC implementation used in this paper is available at `http://sites.labic.icmc.usp.br/ragero/cicling_2015/text_categorization_tool/TCTAlgorithms/Transductive/`

6 The source code to generate document networks is available at `http://sites.labic.icmc.usp.br/ragero/cicling_2015/text_categorization_tool/TCTNetworkGeneration/DocumentNetworkGeneration_ID.java`

7 GNetMine implementation used in this paper is available at `http://sites.labic.icmc.usp.br/ragero/cicling_2015/text_categorization_tool/TCTAlgorithms/Transductive/GNetMine_DocTerm_Transductive.java`

8 TCTN implementation used in this paper is available at `http://sites.labic.icmc.usp.br/ragero/cicling_2015/text_categorization_tool/TCTAlgorithms/Transductive/TCTN_Transductive.java`

9 The source code to generate term networks is available at `http://sites.labic.icmc.usp.br/ragero/cicling_2015/text_categorization_tool/TCTNetworkGeneration/TermNetworkGeneration.java`

of labeled document per class and vary by factor of ten from 10 to 50. This variation in the number of labeled documents allows us to better demonstrate the behavior of the algorithms for different number of labeled documents, the trade-off between the number of labeled documents and classification performance, and the differences among supervised inductive learning algorithms and transductive learning algorithms as we increase the number of labeled documents. The remaining $|\mathcal{D}| - (N * |\mathcal{C}|)$ documents were used to evaluate the classification.

4.3 Results

In our first analysis we compared the proposed approach with the algorithms presented in Section 4.2. Figures 1 and 2 present the best accuracies obtained by the algorithms used for comparison and the proposed approach[10]. The proposed approach (black straight line) obtained the highest or close to the highest accuracy for all text collections. Document networks presented a similar behavior but term networks surpass them in most cases.

In general, term and document networks obtained higher accuracies than algorithms based on VSM or bipartite networks. Moreover, transductive algorithms based on VSM presented lower accuracies than MNB (supervised inductive learning algorithms) for most of the text collections.

We submitted the data presented on Figures 1 and 2 to Friedman test and Li's post-hoc test with 95% of confidence level to assess statistically significant differences among the classification algorithms[11]. This is an advisable statistically significant difference test to use when there is a control algorithm (the proposed one) and results from multiple datasets [6]. The proposed approach obtained the best average ranking when using 10 or more labeled documents per class. There were also statistically significant differences in comparison with GNetMine, EM, and TSVM when using 10 or more labeled document per class and statistically significant differences in comparison with Self-Training when using 1, 10 and 50 labeled documents per class.

Besides the proposed approach obtains a better average ranking in comparison with document networks, i.e., the use of term networks provided better classification accuracies than document networks in most cases, there is also another advantage of term networks: we can decrease the networks size (number of objects and number of relations) through feature selection while keeping classification accuracy. This is not possible in document networks since if we discard documents from the network they would not be classified.

[10] All generated results are available at `http://sites.labic.icmc.usp.br/ragero/cicling_2015/complete_results/`

[11] The Friedman test is a non-parametric test based on average ranking differences. It ranks the algorithms for each text collection individually, in which the algorithm with highest performance have the rank of 1, the second best performance 2, and so on. In the case of ties average ranks are assigned. Then the average ranking is computed for each algorithm considering the ranks in each text collection. Once there are statistically significant differences on the rankings, the Li's post test is used to find pairs of algorithms which produce differences.

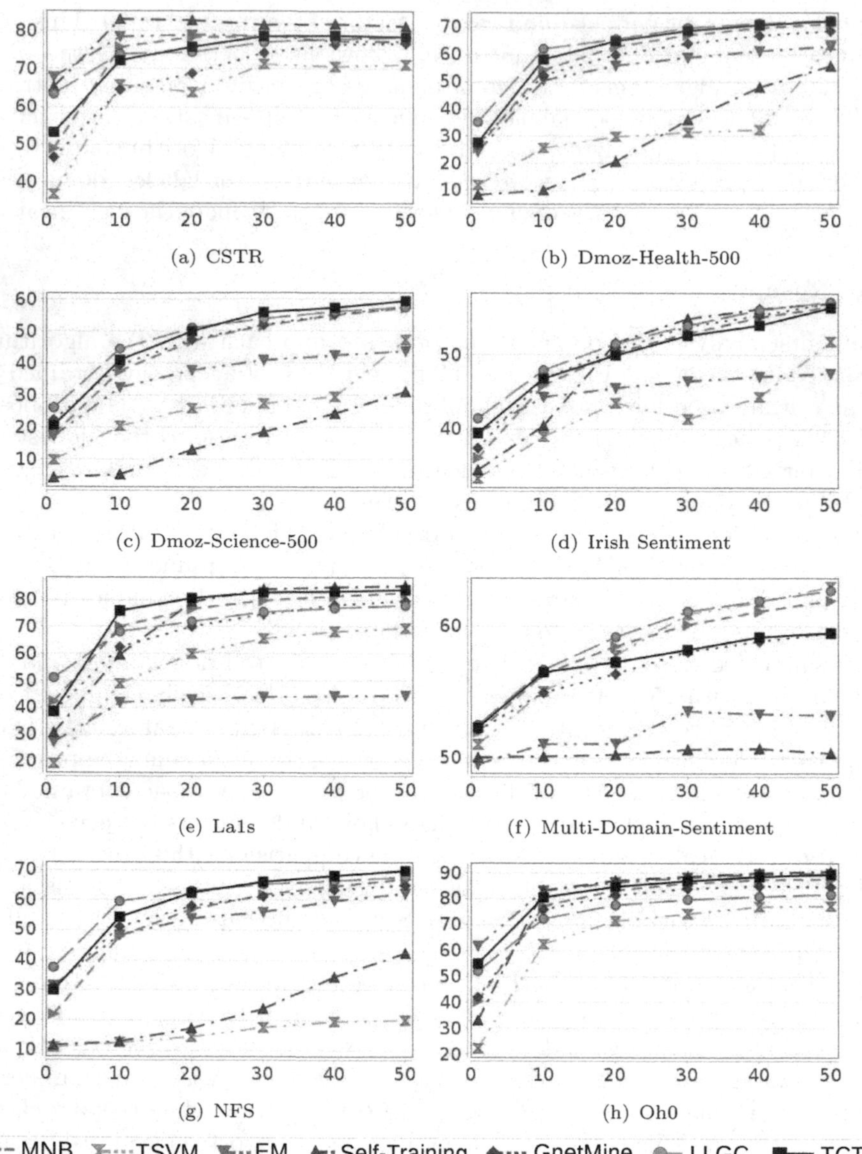

Fig. 1. Best classification accuracies obtained by the algorithms used in the experimental evaluation. X-axis presents the number of labeled documents per class and y-axis presents accuracy values.

We select 25% and 50% of top ranked term according to the sum of TF-IDF [22] to illustrate our assumption. In Figure 3 we present the best classification accuracies considering all terms, 25% and 50% of the terms. This figure shows that the accuracies obtained with smaller term networks were close to the accuracies obtained by term networks considering all terms of the text collections.

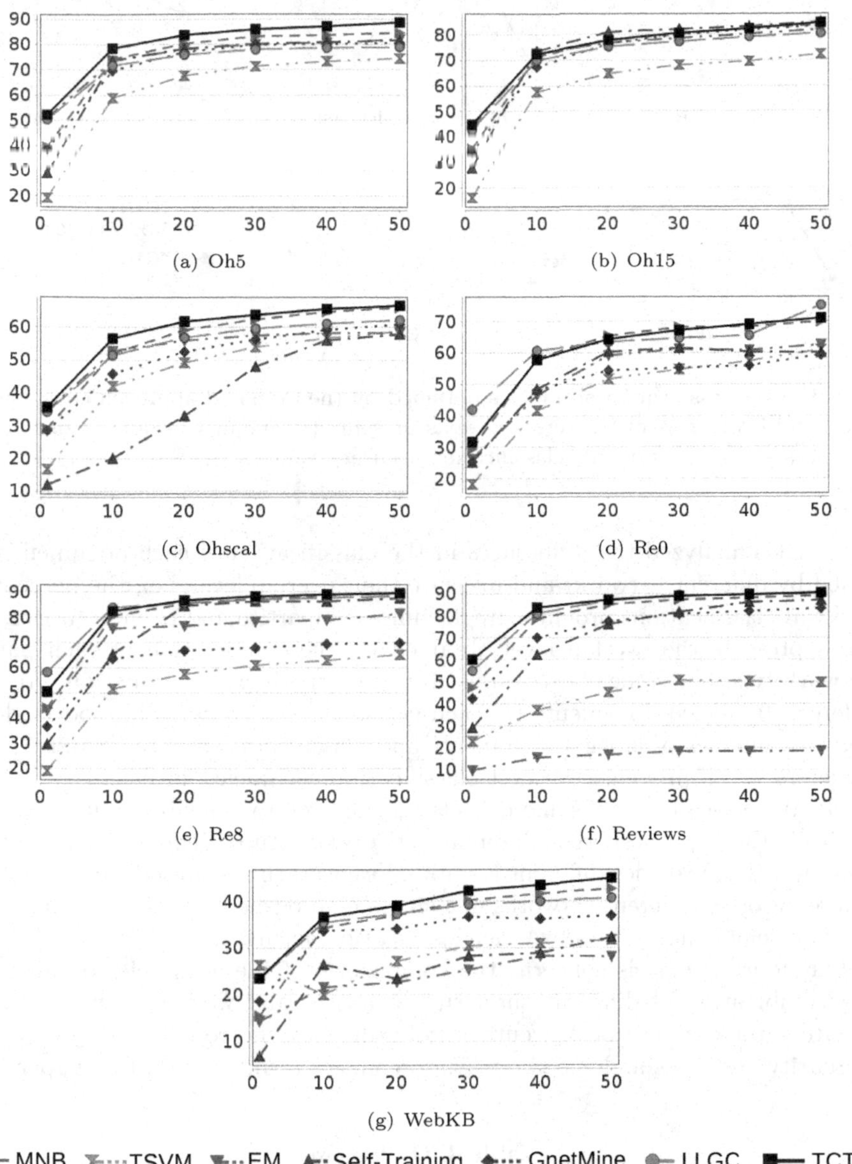

(a) Oh5 (b) Oh15 (c) Ohscal (d) Re0 (e) Re8 (f) Reviews (g) WebKB

Fig. 2. Best classification accuracies obtained by the algorithms used in the experimental evaluation. X-axis presents the number of labeled documents per class and y-axis presents accuracy values.

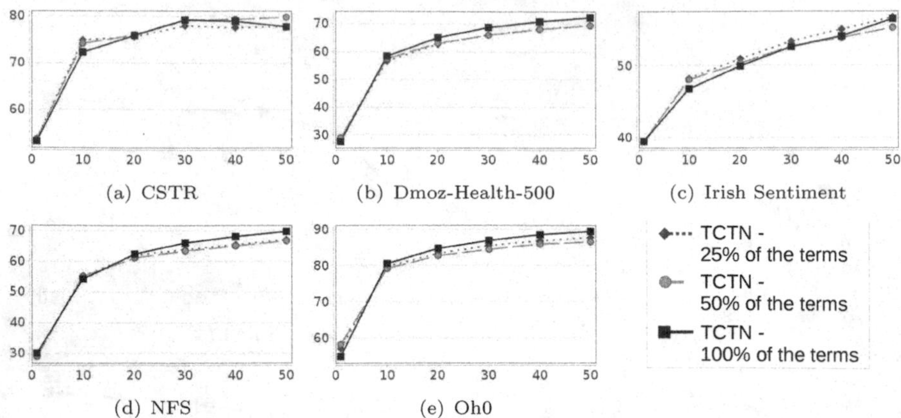

Fig. 3. Best classification accuracies obtained by the proposed approach considenring 100%, 50% and 25% of the terms. X-axis presents the number of labeled documents per class and y-axis presents classification accuracy values.

We also analyzed the differences in the classification of each document provided by different networks and measure their overlap level, i.e., the percentage of equal classifications provided by document, bipartite and term networks. Table 4 presents the overlap level obtained by the combination of each pair of network types for some collection used in the experimental evaluation. We considered 10 labeled documents for each class, since it provided a good trade-off between number of labeled documents and classification accuracy, and the parameters which provided the best classification accuracies for each type of network. We notice that for some collections the overlap reaches about 50%. This confirms the hypothesis that different networks capture different aspects of the data and thus provides different results. Moreover, this is an indicative that the combination of different networks in a single heterogeneous network or use them as an ensemble might improve the classification accuracy.

Our fourth analysis concerns the comparison of the accuracies obtained by the combination of different similarity measures and ways to connect terms. Figure 4 presents the best accuracies obtained by the combination of the five similarity measures and the two ways to connect terms used in this paper. The

Table 4. Ovelap level

Collection	Bipartite-Document	Bipartite-Term	Document-Term
Dmoz-Science-500	52.38%	69.03%	53.73%
IrishSent	47.69%	56.38%	91.12%
La1s	69,52%	80,16%	68,94%
NFS	60.95%	76.64%	63.84%
Oh15	75.46%	86.68%	74.75%

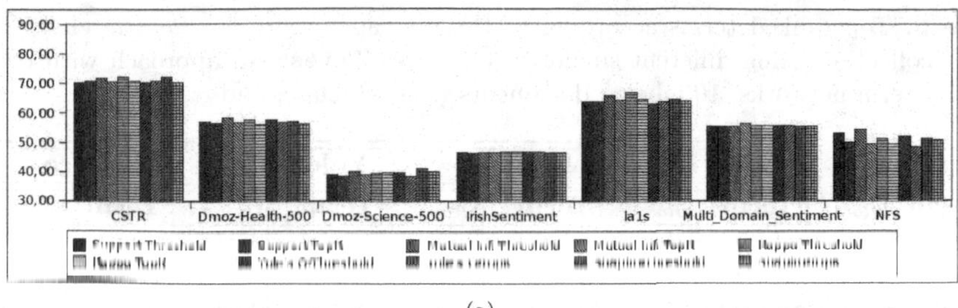

(a)

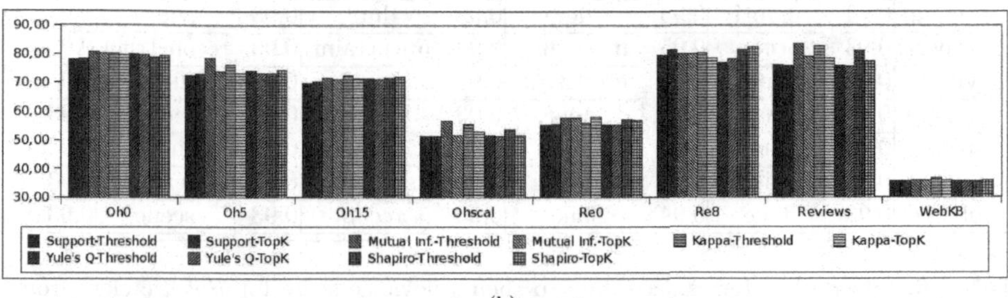

(b)

Fig. 4. Accuracies obtained by different similarity measures and ways to connect terms

use of Mutual Information, Kappa and Shapiro as similarity measure and the `Threshold` approach to connect terms presented better accuracies than other term networks. In general, the use of `Threshold` approach presented slightly better accuracies than `TopK` approach.

Despite the accuracies obtained by different similarity measures and ways to connect terms were close, the final relevance scores of terms were different for different term networks. To illustrate this, we ran the proposed term network approach for CSTR (Computer Science Technical Reports) collection [21], which is composed by technical reports about Systems, Theory, Robotics and Artificial Intelligence published in the Department of Computer Science at University of Rochester. We used $\alpha = 0.1$, 10 labeled documents for each class and the `Threshold` approach with $\epsilon = 0$. Table 5 presents the top 10 ranked terms for Robotics class considering term networks generated by different similarity measures. We notice that all top ranked terms are related to robotics. Moreover, some important terms to distinguish documents about robotics, as *image* and *camera*, are present in all top ranked terms of the networks generated by different similarity measures. However, we also notice a set of different terms, showing that different similarity measures generate different term networks and provide different relevance scores for terms.

The approaches to connect terms also have impact on the top relevant terms for classes. Table 6 presents the final relevance scores considering Support as similarity measure. We consider `Threshold` approach with $\epsilon \in \{0, 0.25\}$ and `TopK` approach with $\kappa \in \{7, 57\}$. Term networks with $\epsilon = 0$ and $\kappa = 57$, i.e., two approaches which produces more connections among terms, shared half of the

Table 5. Top ranked terms according to the relavance score for *Robotic* class from CSTR collection using different similarity measures, `Threshold` approach with $\epsilon = 0$ to build term networks, 10 labeled documents for each class, and $\alpha = 0.1$

Support		Mutual Inf.		Kappa		Yule's Q		Shapiro	
Term	$\mathbf{f}_{Rob}$	**Term**	$\mathbf{f}_{Rob}$	**Term**	$\mathbf{f}_{Rob}$	**Term**	$\mathbf{f}_{Rob}$	**Term**	$\mathbf{f}_{Rob}$
imag	0.95	hull	0.95	convei	0.93	convei	0.94	imag	0.93
view	0.94	convei	0.95	imag	0.93	fire	0.93	convei	0.93
scene	0.94	compactli	0.95	camera	0.93	spike	0.93	camera	0.93
recognit	0.94	empti	0.95	view	0.93	hull	0.93	view	0.93
properti	0.93	occup	0.95	uncertainti	0.93	uncertainti	0.93	reconstruct	0.93
visual	0.93	tightest	0.95	reconstruct	0.93	imag	0.93	nois	0.93
camera	0.93	imag	0.95	nois	0.93	true	0.93	accuraci	0.93
point	0.93	fire	0.94	hull	0.93	nois	0.93	random	0.93
dynam	0.93	spike	0.94	fire	0.93	shannon	0.93	uncertainti	0.93
observ	0.93	true	0.94	spike	0.93	recept	0.93	scene	0.93

Table 6. Top ranked terms according to their relevance score for *Robotic* class from CSTR collection using Support as similarity measure and different approaches to connect terms

Support $\epsilon = 0$		Support $\epsilon = 0.15$		Support $\kappa = 7$		Support $\kappa = 57$	
Term	$\mathbf{f}_{Rob}$	**Term**	$\mathbf{f}_{Rob}$	**Term**	$\mathbf{f}_{Rob}$	**Term**	$\mathbf{f}_{Rob}$
imag	0.95	absenc	0.90	realiti	1.17	level	1.20
view	0.94	absolut	0.90	calibr	1.07	realiti	1.14
scene	0.94	account	0.90	planner	1.06	camera	1.10
recognit	0.94	accuraci	0.90	kalman	1.06	recognit	1.09
properti	0.93	activ	0.90	sens	1.06	properti	1.08
visual	0.93	anatomi	0.90	point	1.06	view	1.07
camera	0.93	arbitrarili	0.90	cognit	1.06	kalman	1.06
point	0.93	artifact	0.90	sensori	1.05	point	1.06
dynam	0.93	assign	0.90	probabilist	1.05	imag	1.05
observ	0.93	beach	0.90	configur	1.05	configur	1.04

top ranked terms. They present a very different set of top ranked terms compared with term networks with lesser number of relations ($\epsilon = 0.25$ and $\kappa = 7$).

5 Discussion, Conclusions, and Future Work

In this paper we present a term network approach for transductive classification of texts. The proposed approach performs relevance score propagation in a term network to set the relevance scores of terms for classes. These relevance scores are then used to classify unlabeled documents. The proposed approach can be applied to any text collection since it does not depend of controlled vocabulary or specific structures in a text to generate term networks. Moreover, the proposed approach avoids the drawbacks of other existing classification algorithms based on term networks, such as computing similarities among networks, which has high computational cost, and mapping the network representation to a vector-space representation or generates term networks for each class or document, which has high memory consumption.

The transductive classification based on term networks proposed in this article surpasses the classification accuracy obtained by transductive algorithms based on vector space model, document and bipartite networks, or ever supervised inductive classification, for most of the evaluated text collections. We demonstrated that the use of Mutual Information, Kappa and Piatetsky-Shapiro and the `Threshold` approach generated term networks which provided the best classification accuracies. We also demonstrated that term-based networks capture different patterns in comparison to other types of network, which indicates that their combination can improve the classification accuracy, as we intend to do in future work.

We highlight that the number of terms of a text collection converges to a constant number as the number of documents increases [11]. Considering that the complexity to generate document networks is $O(|\mathcal{D}|^2 * |\mathcal{T}|)$ and to generate term networks is $O(|\mathcal{D}| * |\mathcal{T}|^2)$, the proposed approach is useful for for huge text collections or when the number of documents is higher than the number of terms. Moreover, we can decrease the term network size through feature selection, which speeds up classification time and keeps classification accuracy.

Acknowledgements. Grants 2011/12823-6, 2011/22749-8, and 2014/08996-0, Sao Paulo Research Foundation (FAPESP).

References

1. Aggarwal, C.C., Zhao, P.: Towards graphical models for text processing. Knowledge & Information Systems 36(1), 1–21 (2013)
2. Angelova, R., Weikum, G.: Graph-based text classification: learn from your neighbors. In: Proc. Special Interest Group on Information Retrieval Conference, pp. 485–492. ACM (2006)
3. Blum, A., Mitchell, T.: Combining labeled and unlabeled data with co-training. In: Proc. Conf. Computational Learning Theory, pp. 92–100. ACM (1998)
4. Dhillon, I.S.: Co-clustering documents and words using bipartite spectral graph partitioning. In: Proc. Int. Conf. Knowledge Discovery and Data Mining, pp. 269–274. ACM (2001)
5. Forman, G.: 19MclassTextWc dataset (2006), `http://sourceforge.net/projects/weka/files/datasets/text-datasets/19MclassTextWc.zip/download`
6. García, S., Fernández, A., Luengo, J., Herrera, F.: Advanced nonparametric tests for multiple comparisons in the design of experiments in computational intelligence and data mining: Experimental analysis of power. Information Sciences 180(10), 2044 (2010)
7. Geng, L., Hamilton, H.J.: Interestingness measures for data mining: A survey. ACM Computing Surveys 38(3) (2006)

8. Ji, M., Sun, Y., Danilevsky, M., Han, J., Gao, J.: Graph regularized transductive classification on heterogeneous information networks. In: Balcázar, J.L., Bonchi, F., Gionis, A., Sebag, M. (eds.) ECML PKDD 2010, Part I. LNCS, vol. 6321, pp. 570–586. Springer, Heidelberg (2010)
9. Jiang, C., Coenen, F., Sanderson, R., Zito, M.: Text classification using graph mining-based feature extraction. Knowledge-Based Systems 23(4), 302–308 (2010)
10. Joachims, T.: Transductive inference for text classification using support vector machines. In: Proc. Int. Conf. on Machine Learning, pp. 200–209 (1999)
11. Marcacini, R.M., Rezende, S.O.: Incremental construction of topic hierarchies using hierarchical term clustering. In: Int. Conf. Software Engineering & Knowledge Engineering, pp. 553–558 (2010)
12. Markov, A., Last, M., Kandel, A.: Model-based classification of web documents represented by graphs. In: Proc. Workshop on Web Mining and Web Usage Analysis, pp. 1–8 (2006)
13. Matsuo, Y., Sakaki, T., Uchiyama, K., Ishizuka, M.: Graph-based word clustering using a web search engine. In: Prof. Conf. on Empirical Methods in Natural Language Processing, pp. 542–550. ACL (2006)
14. Mihalcea, R., Tarau, P.: TextRank: Bringing order into texts. In: Proc. Conf. Empirical Methods in Natural Language Processing (2004)
15. Mishra, M., Huan, J., Bleik, S., Song, M.: Biomedical text categorization with concept graph representations using a controlled vocabulary. In: Proc. Int. Workshop on Data Mining in Bioinformatics, pp. 26–32. ACM (2012)
16. Nigam, K., McCallum, A.K., Thrun, S., Mitchell, T.: Text classification from labeled and unlabeled documents using EM. Machine Learning 39(2/3), 103–134 (2000)
17. Page, L., Brin, S., Motwani, R., Winograd, T.: The PageRank citation ranking: Bringing order to the web. Technical Report 1999-66, Stanford University (November 1998), http://ilpubs.stanford.edu:8090/422/
18. Porter, M.F.: An algorithm for suffix stripping. Readings in Information Retrieval 14(3), 130–137 (1980)
19. Rossi, R.G., de Andrade Lopes, A., de Paulo Faleiros, T., Rezende, S.O.: Inductive model generation for text classification using a bipartite heterogeneous network. Journal of Computer Science and Technology 3(29), 361–375 (2014)
20. Rossi, R.G., Lopes, A.A., Rezende, S.O.: A parameter-free label propagation algorithm using bipartite heterogeneous networks for text classification. In: Proc. Symposium on Applied Computing. ACM (2014)
21. Rossi, R.G., Marcacini, R.M., Rezende, S.O.: Benchmarking text collections for classification and clustering tasks. Tech. Rep. 395, Institute of Mathematics and Computer Sciences - University of Sao Paulo (2013), http://www.icmc.usp.br/CMS/Arquivos/arquivos_enviados/BIBLIOTECA_113_RT_395.pdf
22. Salton, G.: Automatic text processing: the transformation, analysis, and retrieval of information by computer. Addison-Wesley Longman Publishing Co., Inc. (1989)
23. Schenker, A., Last, M., Bunke, H., Kandel, A.: Classification of web documents using a graph model. In: Proc. Int. Conf. Document Analysis and Recognition, pp. 240–244. IEEE Computer Society (2003)
24. Sebastiani, F.: Machine learning in automated text categorization. ACM Computing Surveys 34(1), 1–47 (2002)
25. Solé, R.V., Corominas-Murtra, B., Valverde, S., Steels, L.: Language networks: their structure, function, and evolution. Complexity 15(6), 20–26 (2010)

26. Steyvers, M., Tenenbaum, J.B.: The large-scale structure of semantic networks: Statistical analyses and a model of semantic growth. Cognitive Science 29, 41–78 (2005)
27. Tan, P.-N., Kumar, V., Srivastava, J.: Selecting the right interestingness measure for association patterns. In: Proc. Int. Conf. Knowledge Discovery and Data Mining, pp. 32–41. ACM (2002)
28. Tseng, Y.-H., Ho, Z.-P., Yang, K.-S., Chen, C.-C.: Mining term networks from text collections for crime investigation. Expert Systems with Applications 39(11), 10082–10090 (2012)
29. Wang, W., Do, D.B., Lin, X.: Term graph model for text classification. In: Li, X., Wang, S., Dong, Z.Y. (eds.) ADMA 2005. LNCS (LNAI), vol. 3584, pp. 19–30. Springer, Heidelberg (2005)
30. Weiss, S.M., Indurkhya, N., Zhang, T.: Fundamentals of Predictive Text Mining. Springer London Ltd. (2010)
31. Witten, I.H., Frank, E.: Data Mining: Practical machine learning tools and techniques, 2nd edn. Morgan Kaufmann (2005)
32. Yarowsky, D.: Unsupervised word sense disambiguation rivaling supervised methods. In: Proc. Annual Meeting on Association for Computational Linguistics, pp. 189–196. Association for Computational Linguistics (1995)
33. Zhou, D., Bousquet, O., Lal, T.N., Weston, J., Schölkopf, B.: Learning with local and global consistency. In: Advances in Neural Information Processing Systems, vol. 16, pp. 321–328 (2004)
34. Zhu, X., Ghahramani, Z., Lafferty, J.: Semi-supervised learning using gaussian fields and harmonic functions. In: Proc. Int. Conf. Machine Learning, pp. 912–919. AAAI Press (2003)
35. Zhu, X., Goldberg, A.B., Brachman, R., Dietterich, T.: Introduction to Semi-Supervised Learning. Morgan and Claypool Publishers (2009)

Calculation of Textual Similarity Using Semantic Relatedness Functions

Ammar Riadh Kairaldeen[1] and Gonenc Ercan[2]

[1]University of Baghdad, Baghdad, Iraq
eng_ammar81@yahoo.com
[2] Department of Informatics, Hacettepe University, Turkey
gonenc@cs.hacettepe.edu.tr

Abstract. Semantic similarity between two sentences is concerned with measuring how much two sentences share the same or related meaning. Two methods in the literature for measuring sentence similarity are cosine similarity and overall similarity. In this work we investigate if it is possible to improve the performance of these methods by integrating different word level semantic relatedness methods. Four different word relatedness methods are compared using four different data sets compiled from different domains, providing a testbed formed of various range of writing expressions to challenge the selected methods. Results show that the use of corpus-based word semantic similarity function has significantly outperformed that of WordNet-based word semantic similarity function in sentence similarity methods. Moreover, we propose a new sentence similarity measure method by modifying an existing method which incorporates word order and lexical similarity called as overall similarity. Furthermore, the results show that the proposed method has significantly improved the performance of the overall method. All the selected methods are tested and compared with other state-of-the-art methods.

1 Introduction

In Natural Language Processing (NLP), determining the similarity between two sentences is a crucial task due to expressional variations in natural language, and has direct applications in different tasks in NLP and similar research fields.

The techniques used for detecting the similarity between two long texts are different than those used for short texts. Long text techniques rely on analyzing the shared words between two texts, which cannot be used in the short text techniques where shared words can be rare or even an empty set. Thus, similarity measures should take into account the syntactic and semantic structure of the sentences.

The techniques to measure the semantic similarity are applied and developed in different fields [1, 2]. For instance, in information retrieval (IR) it is used to solve the problem of measuring the similarity to assign a ranking score between a query and texts in a corpus [3]. In text summarization, sentence semantic similarity is used to cluster similar sentences [4]. In web page retrieval, sentence similarity can be effectively enhanced by calculating the page title similarity [5]. These are only a few

A. Gelbukh (Ed.): CICLing 2015, Part II, LNCS 9042, pp. 516–524, 2015.
DOI: 10.1007/978-3-319-18117-2_38

examples of the applications of sentence semantic similarity. Therefore, it is important to pursue research and development to improve the benefits of using similarity measures and lexical semantic resources in a wide range of applications.

In this work, we perform a comparative analysis of different word level semantic relatedness measures in sentence similarity task. Furthermore we show that it is possible to improve sentence similarity metrics by integrating word level semantic relatedness measures

2 Semantic Relatedness of Words

Different methods attempt to calculate word-to-word relatedness, some methods are based on information derived from a large corpus, while others are based on relationships between words defined in WordNet. In this work, four different methods are selected, one is corpus-based and three others are WordNet-based, as shown in Fig. 1.

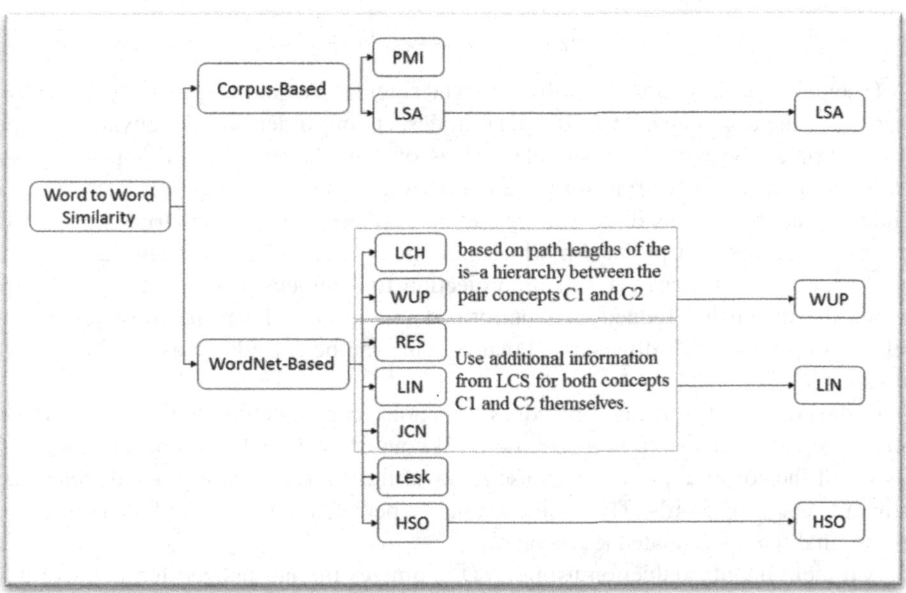

Fig. 1. Corpus-based and Wordnet-based Semantic Relatedness Functions

2.1 Corpus-Based Measures

Corpora based semantic relatedness measures depend on the distributional semantic hypothesis, which states that words occurring in similar contexts are semantically related. A context can be realized as a document, paragraph or sentence in a computational model. A computational model estimated from infinitely many discourse observations can represent semantic relatedness of words accurately. For this reason a method that is able to accommodate large datasets should be preferred. When compared to Wordnet or other knowledge based methods, corpora based methods are more flexible and does not require maintenance. Furthermore it is possible to adapt a domain specific corpora to achieve better performance in a certain domain.

Latent Semantic Analysis (LSA)

In this work one Corpus-based measure of semantic similarity is used [6]. In LSA, corpus term co-occurrences are collected from a large general domain corpora like Wikipedia and using dimensionality reduction the semantic relatedness is measured. The dimensionality reduction is typically done by truncating the matrix using Singular Value Decomposition (SVD). In this work we will be using term-by-term matrix for corpus representation.

The co-occurrence matrix A can directly be used with full dimensions as a semantic relatedness function, however both efficiency and effectiveness significantly improves by reducing the number of dimensions in A with Singular Value Decomposition (SVD). SVD decomposes the original matrix to three components as follows.

$$A_{m\times n} = U_{m\times k}\Sigma_{k\times k}V_{k\times n}^{T}$$

U and V matrices store the left and right singular vectors respectively. Σ matrix stores the singular values in its diagonal in descending order. The eigenvalues of the AA^T matrix are square of the singular values of A. Only retaining the top k singular values and vectors, and truncating the matrices to k yields a projection from $|V|$ to k dimensional space. This dimension reduction is referred to as SVD truncation. SVD truncation produces a projection which minimizes the Frobenius norm distance between the original matrix A and its projection to k-dimensional space $A_k^{'}$. In other words the dimension reduction is performed in a way to keep the most prominent relationships in the full-dimensional matrix. This can be considered as a noise removal strategy.

In the context of semantic relatedness this noise can be common, as words that are related to each other only in a specific local context will tend to co-occur in a small subset of the corpora. Following noise removal the matrix $A_k^{'}$ stores a k-dimensional semantic space of words. The cosine similarity between the vectors of $A_k^{'}$ is returned as the final semantic relatedness scores.

A dimensionality reduction using SVD optimizes the normalized least square differences between the original full dimensional vectors and the new projected vectors. This ensures that the removed information is mostly noise, keeping the most important relationships intact.

Furthermore this dimension reduction establishes second or even higher order relationships between words, as two words occurring in similar contexts are projected to similar subspaces. After dimension reduction the similarity between the projected vectors are used to calculate the semantic relatedness between the words.

In our implementation we use the English Wikipedia articles to form a semantic space using a method similar to Rapp [6] and Ercan [7]. This method applies an Entropy based weighting function to term-by-term matrix. This semantic space is able to correctly identify correct synonyms of 93% of TOEFL synonymy questions [7]. The co-occurrence matrix A is weighted using the formula below.

$$A_{ij} = log(1 + C_{ij}) - \sum_{k=0}^{|V|} p(w_i, w_k)\, log(p(w_i, w_k))$$

where, C_{ij} is the number of times i^{th} and j^{th} words co occur with each other. The term $p(w_i, w_k)$ is the probability of observing both w_i and w_k.

2.2 Wordnet-Based Measures

WordNet-based similarity functions are also used to measure the semantic relatedness of given two words. This is useful in different natural language tasks, as if a pair of sentences shares many similar words, we can guess that the meaning conveyed in the sentences are related in general.

First, it is important to clarify that when using Wordnet based algorithms the similarities are measured between concepts or word senses, instead of words. However due to polysemy a word may have more than one sense. In order to measure the similarity of such words, the similarity between all possible combinations of word senses is calculated and the maximum similarity is considered as the similarity of the words.

In most of Wordnet based measures the information in the IS-A hierarchy is exploited. A human would consider 'car' and 'boat' to be more similar rather than the pair 'boat' and 'tree' since 'car' and 'boat' have a more specific common ancestor, namely the 'vehicle' concept.

WordNet contains separate IS-A hierarchies for nouns and verbs. In such methods, similarities can only be found when both words are in one of these categories. For example, the nouns 'dog' and 'cat', and the verbs 'run' and 'walk.' However, adjectives and adverbs are not organized into IS-A hierarchies so these methods are not applicable.

Other relationship types are also used in Wordnet based semantic relatedness measures. These include part-of relationships ('tree' and 'garden'), as well as opposites ('dark' and 'light') and so on.

It is possible to make use of non-hierarchal information in WordNet in relatedness measures, applicable to different concept pairs including words with different parts of the speech. For instance "weapon" and "murder".

The Lowest Common Superconcept (LCS) for the two concepts C1 and C2 is the most specific concept that is an ancestor of both C1 and C2 (e.g LCS between cat and dog is pet). In the following, word-to-word semantic similarity functions are used to calculate sentence similarity. Both LCH and WUP [8, 9] are based on path lengths in the IS–A hierarchy between the concept pairs C1 and C2.

RES [10] uses additional information to LCS by including summation of the information content for both concepts C1 and C2. The LIN [11] scales the information content of the LCS by the summation, while JCN [12] takes the difference of the calculation and the information content from LCS.

Lesk [13, 14] incorporates information from WordNet glosses, and HSO [15] classifies relations in WordNet based on the direction, and classifies relations in WordNet as having direction.

3 Computing Sentence to Sentence Similarity Method

3.1 Cosine Similarity Method

Cosine similarity is a popular vector based similarity measure in both information retrieval and text mining. In this approach compared strings are transformed into vectors in a high-dimensional space so that the cosine of the angle between the vectors can be used to calculate the similarity [16].

$$Similarity = \cos(\theta) = \frac{\vec{A}.\vec{B}}{|\vec{A}||\vec{B}|}$$

3.2 Semantic Similarity Matrix Based Method

All the similarity scores between all word pairs in the sentences are taken into account. In this approach sentences are represented with a binary vector (with elements equal to 1 if a word is present and 0 otherwise), $\vec{a}$ and $\vec{b}$. The formula below shows how the similarity between these sentences can be computed [17].

$$sim\left(\vec{a},\vec{b}\right) = \frac{\vec{a}\,W\,\overrightarrow{b^T}}{|\vec{a}||\vec{b}|}$$

Where the W is a semantic similarity matrix containing information about the similarity of word pairs.

3.3 Overall Sentence Similarity Method

It is a combination of both cosine similarity and word ordering information between two sentences to find the preferable similarity measure in one formula as below.

$$S_{overall}(S_1, S_2) = \lambda\,\frac{\vec{a}\,.\,\overrightarrow{b^T}}{|\vec{a}||\vec{b}|} + (1-\lambda)\frac{||r_1 - r_2||}{||r_1 + r_2||}$$

The coefficient λ determines the importance of the two parts of the formula in the overall similarity calculation, wherever $\lambda \leq 1$ also the sentence structure is major in the processing text, then the value of λ should be greater than 0.5, (i.e. $\lambda \in (0.5\,,1]$) [18].

3.4 Enhanced Overall Sentence Similarity Method

A new method is proposed by enhancing the similarity formula explained above by changing the cosine similarity part, by the semantic similarity matrix part, as shown in the formula below.

$$SE_{overall}(S_1, S_2) = \lambda \frac{\vec{u}\, W \,.\, \overrightarrow{v^T}}{|\vec{a}||\vec{b}|} + (1 - \lambda) \frac{\| r_1 - r_2 \|}{\| r_1 + r_2 \|}$$

Where SE represents the enhanced formula using the semantic similarity formula instead of using the cosine similarity.

4 Corpora

Our data sets are part of Semantic Evaluation (SemEval). Four data sets are selected then used to apply the selected method to find sentence similarity. (Headlines, image, OnWN and MSR-Video) In each data set there are 750 sentence pairs. Table 1 summarizes our corpora in terms of number of sentences and words.

The biggest advantage of these data sets than the others is the gold standard it assembled using mechanical Turk, it contains a score between 0 and 5 for each pair of sentences. After applying the similarity methods in the data sets, Pearson correlation is used to compare the methods' similarity results. Also, statistical tests between two correlations are used to check if the differences are significant [19].

Table 1. Number of Sentences, Words and Unique Words in the Corpora

	Headlines	**images**	**OnWN**	**MSRvid**
NO of words with stop words	11228	13689	11617	9945
Average sentence length	7.5	9.1	7.7	6.6
No of words without stop words	8308	7464	5518	5065
Approximate sentence length without stop words	5.5	5	3.7	3.4
NO of sentences	1500	1500	1500	1500
NO of compared sentence pairs	750	750	750	750
Inter-tagger correlation percentage	79.4	83.6	67.2	88

5 Results

First we compared the methods by using a combination of all the corpora yielding a 6000 sentence test-bed. Table 2 shows the results of these experiments. The results show that using semantic matrix with word semantic similarity function with LSA obtains the highest Pearson correlation than all other methods. On the other hand, all the enhanced semantic similarity methods using WordNet-based word semantic similarity functions obtained higher Pearson correlation than the semantic matrix method using the same function.

Table 2. Pearson Correlation Rank for all Corpora

	All corpuses	Method Rank
Semantic Matrix – LSA	0.719	1
Overall Similarity – LSA	0.709	2
Cosine Similarity	0.664	3
Overall Similarity	0.661	4
Overall Similarity – LIN	0.5079	5
Semantic Matrix – LIN	0.504	6
Overall Similarity -WUP	0.2975	7
Semantic Matrix - WUP	0.287	8
Overall Similarity - HSO	0.1920	9
Semantic Matrix – HSO	0.187	10

The results indicates using semantic matrix method with corpus-based word semantic similarity function LSA achieves a significant performance gain compared to using the same method with WordNet-based word semantic similarity functions (i.e. LIN, WUP and HSO). This shows that adding the word semantic relatedness functions to cosine similarity significantly improves the performance. On the other hand, adding the word order similarity to the semantic function methods achieved a significant difference only when WordNet-based word semantic similarity function was used.

The results show adding word semantic similarity function into the introduced overall similarity method achieves a significant performance compared with the overall similarity method that does not use the word semantic similarity function. Also, the results show using Corpus-Based word semantic relatedness function significantly improves the similarity result compared to using WordNet-based word semantic relatedness function.

Table 3 summarizes the results of our experiments, to compare this work with other researchers in the same datasets but with different methods, we report the best and worst correlation values obtained in the SemEval competition. Different results were obtained by other researchers who used different techniques to find the sentence similarity and consider additional factors like using grammatical functions, supervised learning in their calculations [17].

Table 3. Compare with Other Results that use the Same Data Sets

Metrics names		Headlines	Images	OnWN	MSRvid
Other SemEval Participants Range	Max.	0.7837	0.8214	0.8745	0.8803
	Min.	0.0177	0.3243	0.3607	0.0057
Cosine Similarity		0.636	0.733	0.6364	0.714
Corpus-Based Word function and Semantic Matric	LSA	0.573	0.708	0.779	0.800
WordNet-Based Word function and Semantic Matric	LIN	0.550	0.463	0.600	0.603
	WUP	0.301	0.136	0.496	0.312
	HSO	0.155	0.131	0.306	0.142
Overall Similarity		0.622	0.727	0.6359	0.707
Enhances-Overall Similarity (LSA)		0.571	0.708	0.766	0.786
Enhances-Overall Similarity (LIN)		0.543	0.478	0.504	0.601
Enhances-Overall Similarity (WUP)		0.313	0.154	0.504	0.329
Enhances-Overall Similarity (HSO)		0.162	0.136	0.309	0.148

6 Conclusion

We have performed a comparative analysis of using word level semantic relatedness measures in semantic textual similarity problems. Our results indicate that Corpora based methods are significantly better than Wordnet based measures. These experiments show that adding word level semantic relatedness measures improves both cosine similarity and overall similarity methods defined in the literature. To the best of our knowledge the LSA based semantic relatedness measure which achieves good results in used in our experiments defined by Rapp [6] was not previously tested in semantic textual similarity task.

References

1. Cilibrasi, R., Vitányi, P.: The Google Similarity Distance. IEEE Trans. Know Data Engineering (2006)
2. Batet, M.: Ontology-Based Semantic Clustering. AI Communication 24 (2011)
3. Jones, K., Walker, S., Robertson, S.: A Probabilistic Model of Information Retrieval: Development and Comparative Experiments. Part. In: Information Processing and Management (2000)
4. Barzilay, R., Elhadad, M.: Using Lexical Chains for Text Summarization. In: Proceedings of the ACL Workshop on Intelligent Scalable Text Summarization (1997)

5. Mehran, S., Timothy, H.: A Web-Based Kernel Function for Measuring the Similarity of Short Text Snippets. In: WWW 2006. ACM Press (2006)
6. Rapp, R.: Discovering the Senses of an Ambiguous Word by Clustering its Local Contexts. In: Proc. 28th Annu. Conf. Gesellschaft für Klassif, pp. 521–528 (2004)
7. Ercan, G.: Lexical Cohesion Analysis for Topic Segmentation, Summarization and Keyphrase Extraction. Phd. Dissertation. Bilkent University (2012)
8. Leacock, C., Chodorow, M., Miller, G.: Using Semantics and WordNet Relation for Sense Identification. Association for Computational Linguistics (1998)
9. Wu, Z., Palmer, M.: Verb semantics and Lexical Selection. In: Proceedings of the Annual Meeting of the Association for Computational Linguistics (1994)
10. Resnik, P.: Using Information Content to Evaluate Semantic Similarity in a Taxonomy. In: Proceedings of the 14th International Joint Conference on Artificial Intelligence (1995)
11. Francis, W., Henry, K.: Frequency Analysis of English Usage. Lexicon and Grammar. Houghton Mifflin, Boston (1982)
12. Lin, D.: An Information-Theoretic Definition of Similarity. In: Proceedings of the International Conference on Machine Learning (1998)
13. Jay, J., David, W.: Semantic Similarity Based on Corpus Statistics and Lexical Taxonomy. In: Proceedings of International Conference Research on Computational Linguistics (ROCLING X), Taiwan (1997)
14. Choueka, Y., Lusignan, S.: Disambiguation by Short Contexts Computers and the Humanities (1985)
15. Satanjeev, B., Ted, P.: Extended Gloss Overlaps as a Measure of Semantic Rrelatedness. In: Proceedings of the Eighteenth International Joint Conference on Artificial Intelligence (2003)
16. Zaka, B.: Theory and Applications of Similarity Detection Techniques., Institute for Information Systems and Computer Media (IICM) Graz University of Technology A-8010 Graz, Austria (2009)
17. Samuel, F., Stevenson, M.: A Semantic Similarity Approach to Paraphrase Detection (2007)
18. Yuhua, L., Zuhair, B., David, M., James, O.: A Method for Measuring Sentence Similarity and its Application to Conversational Agents. IEEE Transactions on Knowledge and Data Engineering (2006)
19. Li, J., Bandar, Z., McLean, D., Shea, O.: A Method for Measuring Sentence Similarity and its Application to Conversational Agents. In: 17th International Florida Artificial Intelligence Research Society Conference, Miami Beach. AAAI Press (2004)

Confidence Measure for Czech Document Classification

Pavel Král[1,2] and Ladislav Lenc[1,2]

[1] Dept. of Computer Science & Engineering,
Faculty of Applied Sciences,
University of West Bohemia,
Plzeň, Czech Republic
[2] NTIS - New Technologies for the Information Society,
Faculty of Applied Sciences,
University of West Bohemia,
Plzeň, Czech Republic
{pkral,llenc}@kiv.zcu.cz

Abstract. This paper deals with automatic document classification in the context of a real application for the Czech News Agency (ČTK). The accuracy our classifier is high, however it is still important to improve the classification results. The main goal of this paper is thus to propose novel confidence measure approaches in order to detect and remove incorrectly classified samples. Two proposed methods are based on the *posterior* class probability and the third one is a supervised approach which uses another classifier to determine if the result is correct. The methods are evaluated on a Czech newspaper corpus. We experimentally show that it is beneficial to integrate the novel approaches into the document classification task because they significantly improve the classification accuracy.

1 Introduction

Automatic document classification is extremely important for information organization, storage and retrieval because the amount of electronic text documents is growing extremely rapidly. Multi-label document classification becomes currently significantly more important than the single-label classification because it usually corresponds better to the requirements of real applications.

Previously, we have developed an experimental multi-label document classification system for the Czech News Agency (ČTK)[1] based on the Maximum entropy classifier. The main goal of this system is to replace the manual annotation of the newspaper documents which is very expensive and time consuming. The resulting F-measure value of this system is higher than 80%, however this value is still far from perfect.

Therefore, in this paper, we propose a way how to detect incorrectly classified examples in order to improve the final classification score. Three novel Confidence Measure (CM) approaches are proposed, compared and evaluated for this task. The first two confidence measures are based on the *posterior* class probability. Then, we propose a supervised CM approach that combines these two methods by a classifier.

[1] http://www.ctk.eu

A. Gelbukh (Ed.): CICLing 2015, Part II, LNCS 9042, pp. 525–534, 2015.
DOI: 10.1007/978-3-319-18117-2_39

It is worthy of attention, that the confidence measure was never previously integrated to the Czech document classification. Moreover, to the best of our knowledge, no similar confidence measure approach in multi-label document classification field exists.

Section 2 is a short overview of the document classification and confidence measure approaches. Section 3 describes our document classification and confidence measure methods. Section 4 deals with the realized experiments on the ČTK corpus. We also discuss here the obtained results. In the last section, we conclude the research results and propose some future research directions.

2 Related Work

This section is composed of two parts. The document classification is described in the first one, while the second one is focused on the confidence measure task itself.

2.1 Document Classification

Document classification is usually based on supervised machine learning methods that exploit an annotated corpus to train a classifier which then assigns the classes of unlabelled documents. The most of works use Vector Space Model (VSM), which usually represents each document with a vector of all word occurrences weighted by their Term Frequency-Inverse Document Frequency (TF-IDF).

The main issue of this task is that the feature space in the VSM is highly dimensional which decreases the accuracy of the classifier. Numerous feature selection/reduction approaches have been introduced [1–3] to solve this problem.

Furthermore, a better document representation should help to decrease the feature vector dimension, e.g. using lexical and syntactic features as shown in [4]. Chandrasekar et al. further show in [5] that it is beneficial to use POS-tag filtration in order to represent a document more accurately. The authors of [6] and [7] use a set of linguistic features, however they do not improve the document classification accuracy.

More recently, some interesting approaches based on Latent Dirichlet Allocation (L-LDA) [8, 9] have been introduced. Another method exploits partial labels to discover latent topics [10]. Principal Component Analysis (PCA) [11] incorporating semantic concepts [12] has been also used for the document classification. Semi-supervised approaches, which progressively augment labelled corpus with unlabelled documents [13], have also been proposed.

The most of the proposed approaches is focused on English and only few works deal with Czech language. Hrala et al. use in [14] lemmatization and Part-Of-Speech (POS) filtering for a precise representation of Czech documents. In [15], three different multi-label classification approaches are compared and evaluated. The other recent works propose novel features based on the named entities [16] or on the unsupervised machine learning [9].

2.2 Confidence Measure

Confidence measure is used as a post-processing of the recognition/classification to determine whether a result is correct or not. The incorrectly recognized samples should be removed from the resulting set or another processing (e.g. manual correction) can be further realized.

This technique is mainly used in the automatic speech processing field [17–20] and is mostly based on the *posterior* class probability. However, it can be successfully used in another research areas as shown in [21] for genome maps construction, in [22] for stereo vision, in [23] for handwriting sentence recognition or in [24] for automatic face recognition.

Another approach related to the confidence measure is proposed by Proedrou et al. in the pattern recognition task [25]. The authors use a classifier based on the nearest neighbours algorithm. Their confidence measure is based on the algorithmic theory of randomness and on transductive learning.

The confidence measures are mostly used in the single-label classification. But the nature of many real-world classification problems is multi-label. One approach using confidence measures in the multi-label setting is proposed in [26]. The authors use semi-supervised learning algorithms and include a confidence parameter when assigning the labels. Two methods for the confidence value computation are proposed.

Another possibility how to deal with the confidence measures is to use a so called Conformal Predictor (CP) [27]. CP assigns a reliable measure of confidence and is used as a complement of machine learning algorithms. Author of [28] proposes to use a modification called Cross-Conformal Predictor (CCP) to handle the multi-label classification task. He states that this modification is more suitable for this task because of its lower computational costs.

The above mentioned approaches apply the confidence measures on other types of the data. Moreover, to the best of our knowledge, no similar confidence measure approach in multi-label document classification field exists.

3 Document Classification with Confidence Measure

The following sections are focused on our feature set, multi-label document classification approach and particularly on the proposed confidence measure methods.

3.1 Feature Set & Classification

The feature set is created according to Brychcín et al. [9]. They are used because the authors experimentally proved that the additional unsupervised features significantly improve classification results.

- **Words** – Occurrence of a word in a document. Tf-idf weighting is used.
- **Stems** – Occurrence of a stem in a document. Tf-idf weighting is used.
- **LDA** – LDA topic probabilities for a document.
- **S-LDA** – S-LDA topic probabilities for a document.
- **HAL** – Occurrence of a HAL cluster in a document. Tf-idf weighting is used.
- **COALS** – Occurrence of a COALS cluster in a document. Tf-idf weighting is used.

For multi-label classification, we use an efficient approach presented by Tsoumakas et al. in [29]. This method employs n binary classifiers $C_{i=1}^{n} : d \rightarrow l, \neg l$ (i.e. each binary classifier assigns the document d to the label l iff the label is included in the document, $\neg l$ otherwise). The classification result is given by the following equation:

$$C(d) = \cup_{i=1}^{n}: C_i(d) \tag{1}$$

The Maximum Entropy (ME) [30] model is used for classification.

3.2 Confidence Measure

Posterior Class Probability Approaches. The output of an individual binary classifier C_i is the posterior probability $P(L|F)$, where $L \in \{l, \neg l\}$ represents a binary class and F represents the feature vector created from the text document d.

We use two different approaches. The first approach, called ***absolute confidence value***, assumes that higher recognition score confirms the classification result. For the correct classification $\hat{L}$ the following two equations must be satisfied:

$$\hat{L} = \arg\max_{L}(P(L|F)) \tag{2}$$

$$P(\hat{L}|F) > T1 \tag{3}$$

The second approach, called ***relative confidence value***, computes the difference between the l score and the $\neg l$ score by the following equation:

$$\Delta P = abs(P(l|F) - P(\neg l|F)) \tag{4}$$

Only the classification results with $\Delta P > T2$ are accepted. This approach assumes that the significant difference between l and $\neg l$ classification scores confirms the classification result.

$T1$ and $T2$ are the acceptance thresholds and their optimal values are set experimentally.

Composed Supervised Approach. Let R_{abs} and R_{rel} be the scores obtained by the *absolute confidence value* and *relative confidence value* methods, respectively. Let variable H determine whether the document is classified correctly or not. A Multi-Layer Perceptron (MLP) classifier which models the *posterior* probability $P(H|R_{abs}, R_{rel})$ is used to combine the two partial measures in a supervised way.

In order to identify the best performing topology, several MLP configurations are built and evaluated. The MLP topologies will be described in detail in the experimental section.

4 Experiments

4.1 Tools and Corpus

For implementation of the multi-label classifier we used Brainy [31] implementation of Maximum entropy classifier. It has been chosen mainly because of our experience with this tool.

As already stated, the results of this work shall be used by the ČTK. Therefore, for the following experiments we used the Czech text documents provided by the ČTK. This corpus contains 2,974,040 words belonging to 11,955 documents annotated from a set of 37 categories. Figure 1 illustrates the distribution of the documents depending on the number of labels. This corpus is freely available for research purposes at http://home.zcu.cz/~pkral/sw/.

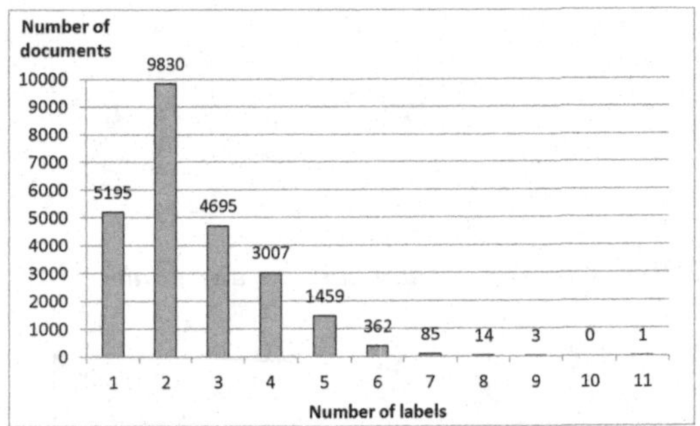

Fig. 1. Distribution of the documents depending on the number of labels

We use the five-folds cross validation procedure for all following experiments, where 20% of the corpus is reserved for testing and the remaining part for training of our models. For evaluation of the document classification accuracy, we use the standard Precision, Recall and F-measure (*F-mes*), also called F1-score, metrics [32]. The confidence interval of the experimental results is 0.6% at a confidence level of 0.95.

4.2 Experimental Results

ROC Curves of the Proposed Approaches. As in many other articles in the confidence measure field, we will use the Receiver Operating Characteristic (ROC) curve [33] for evaluation of our CM methods. This curve clearly shows the relationship between the true positive and the false positive rate for different values of the *acceptance* threshold.

Figure 2 depicts the performance of the *absolute confidence value* method, while the results of the *relative confidence value* approach are given in Figure 3.These figures demonstrate that both approaches are suitable for our task in order to identify incorrectly classified documents. These figures further show, that the *relative confidence value* method slightly outperforms the *absolute confidence value* approach.

Better accuracy of this approach can be explained by the fact that the significantly higher difference in the *posterior* probabilities (between l and $\neg l$ classes) is a better metrics than the simple absolute value of this probability.

Note that this evaluation can be done only for the first two proposed methods which depend on the acceptance threshold. The third approach will be evaluated directly by the F-measure metrics.

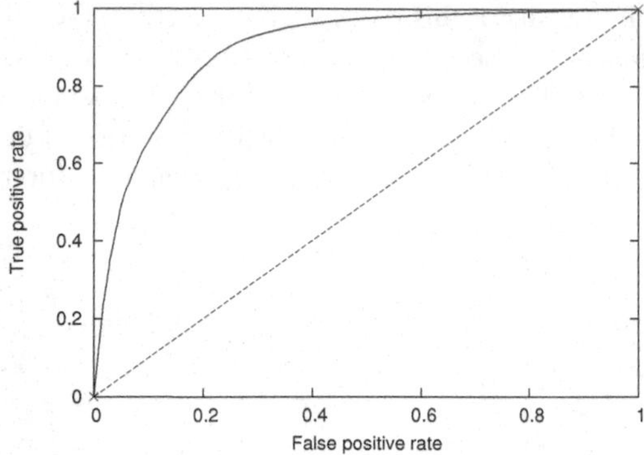

Fig. 2. ROC curve for the *absolute confidence value* method

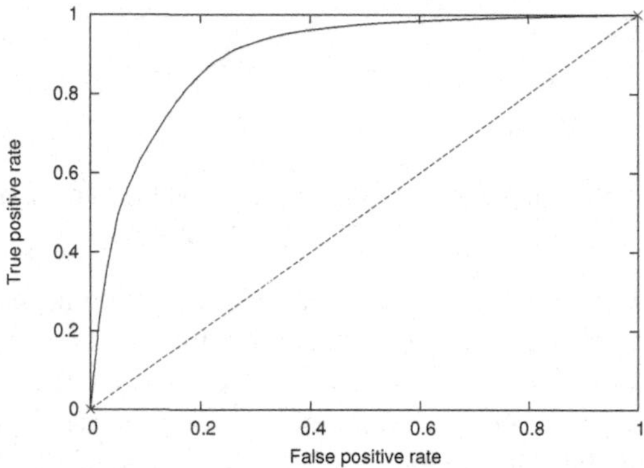

Fig. 3. ROC curve for the *relative confidence value* method

Dependency of the F-measure on the Acceptance Threshold. We deal with the multi-label classification task. The proposed confidence measure approaches thus significantly influence the resulting F-measure score. In this experiment, we would like to identify optimal acceptance thresholds for both CM methods.

Figure 4 shows the dependency of the F-measure value on the acceptance threshold for the *absolute confidence value* method, while the Figure 5 depicts the same dependency for the *relative confidence value* approach. These curves show that both optimal threshold values are close to 1. We can conclude that the correct classification must be associated with the significantly high level of the *posterior* probability (or significantly high difference between $P(l|F)$ and $P(\neg l|F)$ probability values).

Similarly as in the previous experiment, this evaluation is realized only for two first CM methods.

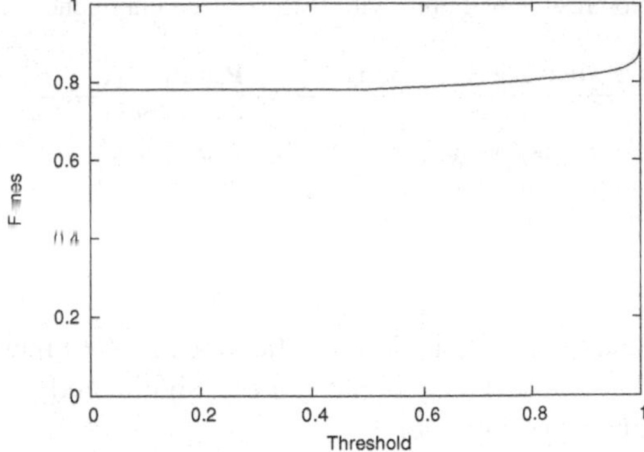

Fig. 4. Dependency of the F-measure on the acceptancce threshold for the *absolute confidence value* method

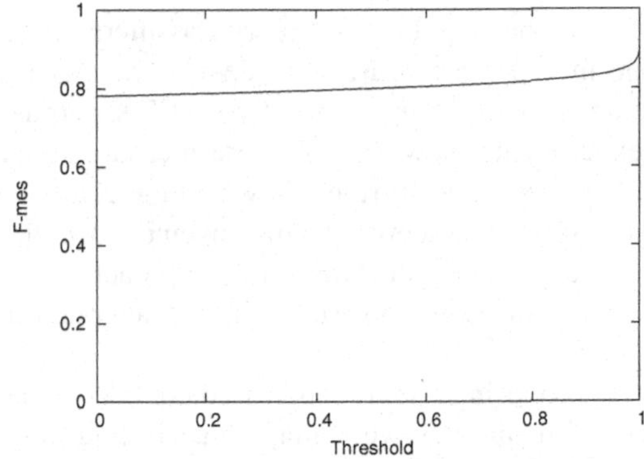

Fig. 5. Dependency of the F-measure on the acceptancce threshold for the *relative confidence value* method

Classification Results without and with the Proposed Confidence Measure Approaches. In the last experiment, we show the document classification scores in two cases: without and with the confidence measure. We also evaluate and compare the performance of the proposed confidence measure methods. As already stated, we use the standard Precision (*Prec*), Recall (*Rec*) and F-measure (*F-mes*) metrics for evaluation.

The results of this experiments are given by Table 1. The first line shows the classification scores without any confidence measure (the *baseline*). The two following lines depict the results of the *absolute* and *relative confidence value* methods. The optimal values of the thresholds $T1$ and $T2$ are based on the results of the previous experiment and are set in both cases to 0.99. The last line shows the results of the composed supervised approach which uses an MLP classifier. We set experimentally the following MLP topology as the best one: two input nodes (R_{abs} and R_{rel}), ten nodes in the hidden layer and two output nodes (classes *correct / not correct*).

Table 1. Classification results without / with the proposed confidence measures [in %]

Confidence Measure Approach	Prec	Rec	F-mes
-	89.0	75.6	81.7
Absolute confidence value	93.8	78.3	85.3
Relative confidence value	94.3	79.4	86.2
Composed supervised approach (MLP)	97.4	99.3	98.3

It is clearly visible that every individual confidence measure method improves the classification results. The improvement is then further significantly increased when the MLP is used to combine the two measures.

5 Conclusions and Future Work

In this paper, we proposed three confidence measure methods and integrated them into multi-label document classification scenario. The first two measures are based on the *posterior* class probability of the output of our binary classifiers, while the third method is a supervised one and incorporates an MLP to decide whether the classification is correct or not. The methods are evaluated on the Czech ČTK corpus of the newspaper text documents. The experiments show that all these measures improve significantly the classification results. Moreover, we further show that the composed supervised CM approach gives the best classification score. The improvement over the baseline (no CM used) reaches 16.6% in the absolute value when this approach is used. Therefore, we conclude that the confidence measure approach will be integrated into our document classification system.

The first perspective is proposing a semi-supervised confidence measure. In this approach, the CM model will be progressively adapted according to the processed data. We will further integrate other suitable individual measures into our composed MLP approach (use for example the so called *predictor* features [19]). The last perspective consists in evaluation of our proposed methods on different languages and language families.

Acknowledgements. This work has been partly supported by the European Regional Development Fund (ERDF), project "NTIS - New Technologies for Information Society", European Centre of Excellence, CZ.1.05/1.1.00/02.0090. We also would like to thank Czech New Agency (CTK) for support and for providing the data.

References

1. Forman, G.: An extensive empirical study of feature selection metrics for text classification. The Journal of Machine Learning Research 3, 1289–1305 (2003)
2. Yang, Y., Pedersen, J.O.: A comparative study on feature selection in text categorization. In: Proceedings of the Fourteenth International Conference on Machine Learning, ICML 1997, pp. 412–420. Morgan Kaufmann Publishers Inc., San Francisco (1997)

3. Lamirel, J.C., Cuxac, P., Chivukula, A.S., Hajlaoui, K.: Optimizing text classification through efficient feature selection based on quality metric. Journal of Intelligent Information Systems, 1–18 (2014)
4. Lim, C.S., Lee, K.J., Kim, G.C.: Multiple sets of features for automatic genre classification of web documents. Information Processing and Management 41, 1263–1276 (2005)
5. Chandrasekar, R., Srinivas, B.: Using syntactic information in document filtering: A comparative study of part of speech tagging and supertagging (1996)
6. Moschitti, A., Basili, R.: Complex linguistic features for text classification: A comprehensive study. In: McDonald, S., Tait, J.I. (eds.) ECIR 2004. LNCS, vol. 2997, pp. 181–196. Springer, Heidelberg (2004)
7. Wong, A.K., Lee, J.W., Yeung, D.S.: Using complex linguistic features in context-sensitive text classification techniques. In: Proceedings of 2005 International Conference on Machine Learning and Cybernetics, vol. 5, pp. 3183–3188. IEEE (2005)
8. Ramage, D., Hall, D., Nallapati, R., Manning, C.D.: Labeled lda: A supervised topic model for credit attribution in multi-labeled corpora. In: Proceedings of the 2009 Conference on Empirical Methods in Natural Language Processing, EMNLP 2009, vol. 1, pp. 248–256. Association for Computational Linguistics, Stroudsburg (2009)
9. Brychcín, T., Král, P.: Novel unsupervised features for czech multi-label document classification. In: Gelbukh, A., Espinoza, F.C., Galicia-Haro, S.N. (eds.) MICAI 2014, Part I. LNCS, vol. 8856, pp. 70–79. Springer, Heidelberg (2014)
10. Ramage, D., Manning, C.D., Dumais, S.: Partially labeled topic models for interpretable text mining. In: Proceedings of the 17th ACM SIGKDD International Conference on Knowledge Discovery and Data Mining, KDD 2011, pp. 457–465. ACM, New York (2011)
11. Gomez, J.C., Moens, M.-F.: Pca document reconstruction for email classification. Computer Statistics and Data Analysis 56(3), 741–751 (2012)
12. Yun, J., Jing, L., Yu, J., Huang, H.: A multi-layer text classification framework based on two-level representation model. Expert Systems with Applications 39, 2035–2046 (2012)
13. Nigam, K., McCallum, A.K., Thrun, S., Mitchell, T.: Text Classification from Labeled and Unlabeled Documents Using EM. Mach. Learn. 39, 103–134 (2000)
14. Hrala, M., Král, P.: Evaluation of the document classification approaches. In: Burduk, R., Jackowski, K., Kurzynski, M., Wozniak, M., Zolnierek, A. (eds.) CORES 2013. AISC, vol. 226, pp. 877–885. Springer, Heidelberg (2013)
15. Hrala, M., Král, P.: Multi-label document classification in czech. In: Habernal, I., Matousek, V. (eds.) TSD 2013. LNCS, vol. 8082, pp. 343–351. Springer, Heidelberg (2013)
16. Král, P.: Named entities as new features for czech document classification. In: Gelbukh, A. (ed.) CICLing 2014, Part II. LNCS, vol. 8404, pp. 417–427. Springer, Heidelberg (2014)
17. Senay, G., Linares, G., Lecouteux, B.: A segment-level confidence measure for spoken document retrieval. In: 2011 IEEE International Conference on Acoustics, Speech and Signal Processing (ICASSP), pp. 5548–5551. IEEE (2011)
18. Senay, G., Linares, G.: Confidence measure for speech indexing based on latent dirichlet allocation. In: INTERSPEECH (2012)
19. Jiang, H.: Confidence measures for speech recognition: A survey. Speech Communication 45, 455–470 (2005)
20. Wessel, F., Schluter, R., Macherey, K., Ney, H.: Confidence measures for large vocabulary continuous speech recognition. IEEE Transactions on Speech and Audio Processing 9, 288–298 (2001)
21. Servin, B., de Givry, S., Faraut, T.: Statistical confidence measures for genome maps: application to the validation of genome assemblies. Bioinformatics 26, 3035–3042 (2010)
22. Hu, X., Mordohai, P.: A quantitative evaluation of confidence measures for stereo vision. IEEE Transactions on Pattern Analysis and Machine Intelligence 34, 2121–2133 (2012)

23. Marukatat, S., Artières, T., Gallinari, P., Dorizzi, B.: Rejection measures for handwriting sentence recognition. In: Proceedings of Eighth International Workshop on Frontiers in Handwriting Recognition, pp. 24–29. IEEE (2002)
24. Li, F., Wechsler, H.: Open world face recognition with credibility and confidence measures. In: Kittler, J., Nixon, M.S. (eds.) AVBPA 2003. LNCS, vol. 2688, pp. 462–469. Springer, Heidelberg (2003)
25. Proedrou, K., Nouretdinov, I., Vovk, V., Gammerman, A.: Transductive confidence machines for pattern recognition. In: Elomaa, T., Mannila, H., Toivonen, H. (eds.) ECML 2002. LNCS (LNAI), vol. 2430, pp. 381–390. Springer, Heidelberg (2002)
26. Rodrigues, F.M., de M Santos, A., Canuto, A.M.: Using confidence values in multi-label classification problems with semi-supervised learning. In: The 2013 International Joint Conference on Neural Networks (IJCNN), pp. 1–8. IEEE (2013)
27. Nouretdinov, I., Costafreda, S.G., Gammerman, A., Chervonenkis, A., Vovk, V., Vapnik, V., Fu, C.H.: Machine learning classification with confidence: application of transductive conformal predictors to mri-based diagnostic and prognostic markers in depression. Neuroimage 56(2), 809–813 (2011)
28. Papadopoulos, H.: A cross-conformal predictor for multi-label classification. In: Iliadis, L., Maglogiannis, I., Papadopoulos, H., Sioutas, S., Makris, C. (eds.) Artificial Intelligence Applications and Innovations. IFIP AICT, vol. 437, pp. 241–250. Springer, Heidelberg (2014)
29. Tsoumakas, G., Katakis, I.: Multi-label classification: An overview. International Journal of Data Warehousing and Mining (IJDWM) 3, 1–13 (2007)
30. Berger, A.L., Pietra, V.J.D., Pietra, S.A.D.: A maximum entropy approach to natural language processing. Computational Linguistics 22, 39–71 (1996)
31. Konkol, M.: Brainy: A machine learning library. In: Rutkowski, L., Korytkowski, M., Scherer, R., Tadeusiewicz, R., Zadeh, L.A., Zurada, J.M. (eds.) ICAISC 2014, Part II. LNCS, vol. 8468, pp. 490–499. Springer, Heidelberg (2014)
32. Powers, D.: Evaluation: From precision, recall and f-measure to roc., informedness, markedness & correlation. Journal of Machine Learning Technologies 2, 37–63 (2011)
33. Brown, C.D., Davis, H.T.: Receiver operating characteristics curves and related decision measures: A tutorial. Chemometrics and Intelligent Laboratory Systems 80(1), 24–38 (2006)

An Approach to Tweets Categorization by Using Machine Learning Classifiers in Oil Business

Hanaa Aldahawi and Stuart Allen

School of Computer Science and Informatics, Cardiff University, Cardiff, UK
{Aldahawiha,Stuart.M.Allen}@cs.cardiff.ac.uk

Abstract. The rapid growth in social media data has motivated the development of a real time framework to understand and extract the meaning of the data. Text categorization is a well-known method for understanding text. Text categorization can be applied in many forms, such as authorship detection and text mining by extracting useful information from documents to sort a set of documents automatically into predefined categories. Here, we propose a method for identifying those who posted the tweets into categories. The task is performed by extracting key features from tweets and subjecting them to a machine learning classifier. The research shows that this multi-classification task is very difficult, in particular the building of a domain-independent machine learning classifier. Our problem specifically concerned tweets about oil companies, most of which were noisy enough to affect the accuracy. The analytical technique used here provided structured and valuable information for oil companies.

Keywords: Twitter, machine learning, text categorization, feature extraction, Oil business.

1 Introduction

Twitter is a micro-blogging network site which allows users to broadcast real-time messages of 140 characters called "tweets". Twitter was launched in 2006 and its users since then have vastly increased; today it has more than 284 million active users tweeting approximately 500 million tweets daily [16]. In response to these impressive numbers, many companies created an official accounts on Twitter to engage and communicate with their customers and stakeholders [7]. Analysing the content of tweets, even if they are re-tweets (re-sent tweets) or mentions (which mention the @companyname) is very important for companies which want to understand more about their customers. This paper focuses on the various machine learning classifiers that oil companies use for text categorization.

Text categorization is the process of automatically assigning one or more predefined categories to text documents [5]. In the present study, the terms "documents" and "tweets" refer to a similar concept. We can treat each tweet as a document and use text-categorization concepts such as tokenization, stemming, term-frequency and document-frequency [8] to encapsulate a flexible representation of the problem, making it easy for the text categorization algorithm to be efficiently applied to this

A. Gelbukh (Ed.): CICLing 2015, Part II, LNCS 9042, pp. 535–546, 2015.
DOI: 10.1007/978-3-319-18117-2_40

problem. However, the machine-learning community has considered using other concepts as documents in this regard. For example, the textual features of movies (e.g. genre, actor/actress, comments, plot, etc.) have been considered as documents on which to build a movie-recommender system [8]. Similarly, items of textual information about books have been treated as documents in while building a book recommender system [13].

The aim of this work is to categorize incoming tweets automatically into a number of pre-defined classes. Hence, this project can be termed a multi-class categorization problem [5]. The term multi-class refers to the machine-learning problem where the input instances/documents can be classified into more than two classes/categories. Multi-class categorization is difficult than binary-class classification (with only two output classes/categories). Certain tricks are available for converting multi-class problems into a series of binary-class problems and then predicting the output of classes by means of a voting scheme. One-versus-one (1v1), one-versus-all (1vR) and Directed Acyclic Graph (DAG) are typical methods [17]. Multi-class classification problems suffer from class imbalance, whereby a class which has more training data is more likely to be predicted as an output class also.

We used Weka library [11] for solving multiclass problem, which employs 1v1. It is, as used in [8]. We also assigned different prior weights inversely proportional to the training data to overcome the class-imbalanced problem.

2 Related Work

Text Categorization (TC) is a powerful building block in several kinds of information framework and methods of data management. Its purpose is to sort a set of documents into a predefined set of categories[15]. This task has multiple applications such as the automated indexing of scientific feature on the basis of the information database of technical descriptions, patent applications, the specific distribution of data to consumers, hierarchical clusters, spam recognition, document types, assigning authorship, etc.

Automated text categorization is attractive because it provides liberty from manually curating document databases, which, as the number of documents increases, can be time-consuming and inefficient. In addition, automated text classification includes information retrieval (IR) technology and machine learning (ML) technology, which is more accurate than manual optimization. Texts are mainly assigned to a specific category by comparison with a bag-of-words model of documents. However, during this process the linguistic features such as micro-text, semantic and syntax recognition are still ignored in the automated learning. Spam recognition and filter are a widely used way of applying text categorization wherein received emails are automatically categorized as spam/junk or non-spam [12].

Applying text classification in practice is an interesting topic for research when so much text-based data is generated every day. A deep understanding of text classification gives researchers the chance to develop new applications, for they can easily obtain data including emails and micro-text which requires classification. Earlier techniques used in text categorization were built up from linear classifiers, which focused on efficiency. Other aspects of text categorization include, for example, leveraging cross-category dependencies, ways of "borrowing" training examples surrounded by mutually dependent categories and ways of discovering latent structures in a functional space for the joint modelling of dependent categories [20, 21]. Current research focuses on classifying data according to topic; other types of class are also interesting, for example, classifying data by sentiment: or determining whether a review is positive or negative [14] or when texts are being classified, whether a text is misleading or not. Nevertheless, the models and procedure for topic categorization are also significant in these problems and some remarkable deliberations over the qualities of the categorization seem to be the best guides for improving performance.

Nowadays researchers use text mining techniques to predict the rising and falling of the stock market. For instance, [6] used the classifier ANN on Twitter data to understand users' moods in relation to the stock market and on this basis to predict its fluctuations. [21] predicted the results of stock market indicators such as the Dow Jones, NASDAQ and S&P 500 by analysing Twitter posts. Commonsense knowledge based approaches for text analysis has also gathered a lot of buzz in this field of research. This approach has been proved to work outstanding in the area of emotion detection and sentiment analysis [27][28] and in purely pattern recognition method such as music genre classification [29].

3 Proposed Approach

This section gives an overview of the methods of data collection and the way in which we built our experiments.

3.1 Datasets

The experiments in this work were conducted on BP America and Saudi Aramco, two of the greatest oil companies in the world. Oil trading is a controversial sector and Twitter is a good platform for displaying the honest opinions of every tweeter. Our results can help these companies to know and deal with those who mention them, by revealing the categories into which they fall. In this experiment we used two datasets extracted from Twitter between November 2012 and August 2014. The datasets of both companies contain 6000 mentions 3000 of each (@BP_America, @Saudi_Aramco).

3.2 Primary Analysis of Datasets

As sentiment has a crucial impact on categorising tweets and all tweets carry sentiment, we used this sentiment as a feature by which to categorize the tweets. Alchemy API (an automated sentiment analysis tool) [1] was used to classify the sentiment in their content through natural language processing. The possibilities of assessment were positive (0+), neutral (0) or negative (-0). Ideally, automated sentiment analysis tool can be used for big data but, for a range of reasons, does not always give accurate answers [4]. Hence we introduced the same dataset of BP America and Saudi Aramco to the Amazon Mechanical Turk platform (AMT) [2], which served as our human sentiment analysis tool. Then to identify the main user group we used AMT to categorize the users who mentioned a company name in their tweets. Each tweet content and user was classified by three workers and we incorporated the average sentiment score and user category agreement in our results. As a requester, we placed the HITs (Human Intelligent Tasks) on Mechanical Turk. Each HIT was displayed with such instructions as: *In this task you are asked to decide which category a Twitter user belongs to from the 8 options, then read the tweet and select whether it is positive, neutral or negative.* The wage on offer was $0.10 per HIT. All the AMT workers selected for this task had been vetted and verified by AMT as competent in the task we gave them. Moreover, in order to ensure classification accuracy, all AMT workers who participated were native English speakers based in the U.S.A. Their eligibility was reaffirmed by the results sheets which contained their i.d., country, region, city and IP.

3.3 Feature Extraction

Feature extraction techniques aim to find the specific pieces of data in natural language documents [5], which are used for building (training) classifiers. As an example, take the following tweet (from the BP dataset) and apply to it the concepts of feature extraction, one by one.

> *"RT @BP_America: Did you know the first service stations opened around 1910? Self-service stations did not become the norm until 1970s: http://t.co/4obuMaCS"*

We performed the following steps to extract the features from the given tweets:

3.3.1 Pre-Processing

The documents usually consist of string characters. Machine learning algorithms (e.g. Text Categorization algorithms) cannot work with these strings. We have to convert them into a format which is suitable for the machine learning classifiers.

There is sequencing of steps that we perform to crry out this task is, as follows:

1. Convert the documents into tokens—sequences of letters and digits.
2. Perform the following modifications
   - Remove HTML and other tags (e.g. Author tag (@), hash tag (#)
   - Remove URLs
   - Remove stop words
   - Perform stemming

Stop words are frequently occurring words that carry no (or very little) information; it is usual in machine learning to remove them before feeding the data to any learning algorithms. Hashtags and URLs should be removed, because they can confuse the classifier with irrelevant information. Stemming eliminates the case and inflection information from a word and maps them into the same stem. For example, the words *categorization*, *categorized* and categories all map into the same root stem 'category'.

The given tweet is processed and after each step yields the following:

3.3.2 Processing

After tokenization
"RT @BP_America Did you know the first service stations opened around 1910 Self-service stations did not become the norm until 1970s http://t.co/4obuMaCS"
After removing HTML and other tags
"RT Did you know the first service stations opened around 1910 Self-service stations did not become the norm until 1970s http://t.co/4obuMaCS"
After removing URLs
"RT Did you know the first service station opened around 1910 Self-service stations did not become the norm until 1970s"
After removing stop words
"RT Did know first service station opened around 1910 Self-service stations become norm until 1970s "
After performing stemming
"RT Did know first service ***station*** *opened around 1910 Self-service* ***station*** *become norm until 1970"*

4 Features and Categorization Labels

In this section we present the features/details that were used in the experiments that shown in Table 1. There are different reasons to chose these featuressuch as: they are easy to extract, simple but salient and intuitive and any machine learning classifier can be trained over them.

4.1 Feature Set Employed by Text Classification Algorithms

The following features are used for categorizing the tweets into the categories described in section 4.2, below

4.2 Output Categories

We used 8 categories of users and one of these categories was manually assigned as to each tweet user. The classification model trained on the following categories: general public, media, environmentalists, politicians, business analysts, oil company employees, government organisations and non-government organisations.

Table 1. Details of The features

No.	Feature name	Type	Feature details
1	**Tweet content**	N-grams	Sequence of the words in the tweet
2	**Tweet content**	String	Content of the tweet itself
3	**Automated sentiment**	Positive, Negative, Neutral	Sentiment of the tweet marked automatically
4	**Manual sentiment**	Positive, Negative, Neutral	Sentiment of the tweet marked manually
5	**Number of followers**	Any non-negative Integer value	Number of followers of the user who tweeted
6	**User description**	String	Description of the user who tweeted
7	**Re-Tweet (RT)**	Boolean (Yes, No)	If the tweet is original or has been re-tweeted
8	**Tweet length**	Discrete	Length of the tweet
9	**User URLs**	Boolean (Yes, No)	Does the user description have a URL?
10	**Tweet URL**	Boolean (Yes, No)	Does the tweet content have a URL?
11	**Tweet hashtags**	Boolean (Yes, No)	Does the tweet have a Hashtag?

5 Experiment Results and Discussion

5.1 Partitioning the Data Set into Testing and Training Set

We used the 5-fold cross validation scheme to portion the given data files into a testing and training set. We reported the average (accuracy) results obtained over the 5 folds. We used the 5-fold cross validation approach to partition the dataset, since it has been the preferred approach in the machine learning literature for reporting results. Many researchers have used it, for example [8, 9].

For the same experiments, we randomly divided the dataset into 20% test set and 80% training set. It has been used in [10].

We used two different techniques to make the problem simple, as for few algorithms we proposed (e.g. Switching hybrid) it is not manageable in small time to have 2 test sets and 2 training sets for 2 different datasets. It will take a lot of time in tuning parameters etc.

5.2 Evaluation Metric

We have used the accuracy metric for measuring the performance of the classification approaches. Formally, it is defined as [3], [18]:

$$Accuracy = \frac{Number\ of\ correctly\ classified\ tweets}{Total\ number\ of\ classified\ tweets}$$

The objective is to increase the accuracy score [18], which corresponds to lowering the rate of classification error.

5.3 Results

Table 2 presents the prediction accuracy percentage of the 11 extracted features produced by 4 different machine learning classifiers, namely, the support vector machine (SVM), k-Nearest Neighbors (KNN), Naïve Byes (NB) and Decision tree (DT).

Table 2. Experiment Results with the BP America and Saudi Aramco datasets

	Datasets	**BP America**				**Saudi Aramco**			
No. of Features	**Features**	**SVM**	**KNN**	**NB**	**DT**	**SVM**	**KNN**	**NB**	**DT**
1	**All features**	38.37	35.81	42.56	73.21	79.56	78.22	57.56	68.89
2	**All - Tweet-content N-grams**	34.65	36.98	42.09	37.67	79.56	79.33	78.89	69.11
3	**All - Tweet content string**	36.98	35.17	42.79	38.84	79.56	78.89	57.56	67.78
4	**All - Automated sentiment**	38.84	35.81	42.56	38.14	79.56	78.89	57.56	70.22
5	**All - Manual sentiment**	36.74	34.19	42.56	38.37	79.56	79.11	57.56	68.67
6	**All - Number of followers**	40.23	37.44	42.09	39.07	79.56	79.11	57.56	68.22
7	**All - User description string**	39.07	36.28	42.33	39.30	78.44	78.89	57.56	78.22
8	**All - RT**	38.84	35.81	42.56	39.77	79.56	78.89	57.56	70.44
9	**All - Tweet length**	38.60	39.07	42.56	37.67	80.0	79.11	57.56	71.78
10	**All - URL in user description**	38.84	36.51	42.56	38.14	79.56	79.11	57.56	68.44
11	**All - URL in Tweet content**	37.44	34.42	40.47	37.44	79.56	78.67	57.56	69.56
12	**All - Hashtag in Tweet content**	39.07	37.21	42.33	38.84	79.56	78.89	57.56	76.89

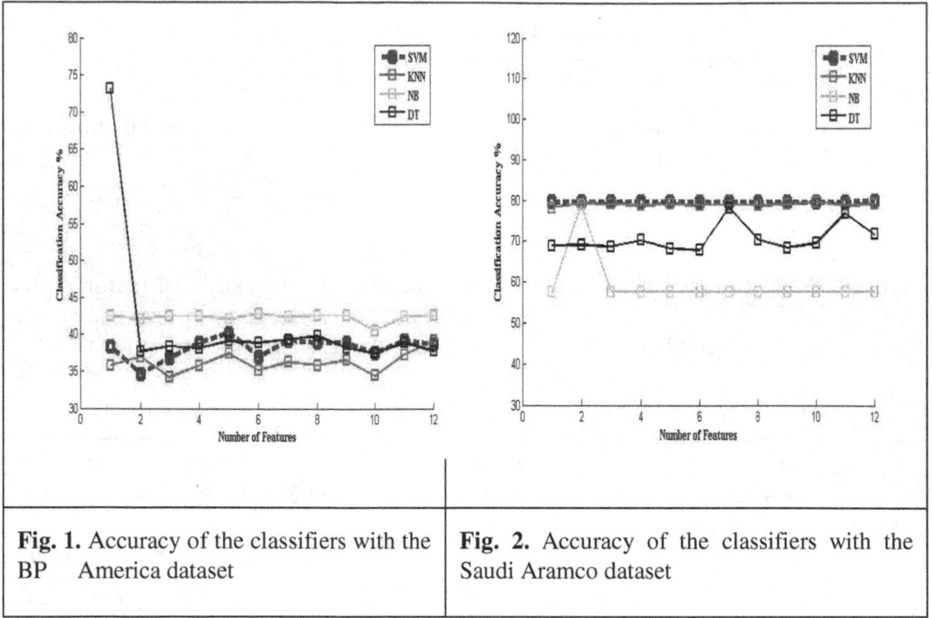

Fig. 1. Accuracy of the classifiers with the BP America dataset

Fig. 2. Accuracy of the classifiers with the Saudi Aramco dataset

For both of the datasets, we found that the N-gram is the most important feature, as can be seen in Table 2, below. The accuracy was low for the BP America dataset, because the tweets in this dataset are very noisy. Overall, we obtained 40.23% accuracy on the BP America dataset without even using the number of followers feature; indeed, we found that the number of followers feature reduced the accuracy. However, in both of the datasets, manual sentiment was found to be a more useful feature than automated sentiment was . On the Saudi Aramco dataset, the accuracy obtained was satisfactory. The length of the tweet was found to be a weak and redundant feature in both datasets. Figures 1 and 2 show the accuracy of the classifiers of both data sets. It can be seen that Naïve Byes proved a better classifier than SVM in the BP America dataset, while SVM, followed by the KNN classifier, performed better than the Saudi Aramco dataset. For other oil company datasets, or any controversial ones, we suggest using Naïve Byes or SVM classifiers, owing to their high rate of accuracy in classification.

5.4 Discussion

The results suggest that the Twitter data collected in our experiment are actually nonlinear in nature, i.e. it is very hard to classify the data with a linear classifier. However, SVM in particular performed best, because of its non-linear poly kernel function. Other classifiers suffered in their lack of a non-linear function as an error function. It is true that MLP has a feature for classifying non-separable data linearly but the sigmoid function used in MLP cannot classify the data well. The features are combined by vector wise concatenating of the features. We tried all feature combinations to evaluate the importance of each feature. Table 2 reported when the best performance was obtained.

The classifiers showed poor results with the BP America dataset but performed very well with the Saudi Armco dataset shown in Table 2. After observing both datasets, we found that the BP America dataset was very noisy and its tweets contain mainly URL. In the future, we aim to remove much of the noise to enhance the accuracy of classification.

6 Conclusion

We describe a text categorization framework based on 6000 Twitter posts mentioning oil companies. Our proposed approach is very similar to predicting the authorship of textual data. To the best of our knowledge, nobody so far had used Twitter data to predict authorship. The preliminary experiment showed satisfactory results. Our future work will focus on analysing the effect of other hybrid, ensemble or deep learning classifiers. Normalizing tweets and removing noise from them are also key techniques for enhancing accuracy. We aim to use more sentiment [23][25] and emotion [22] [24] [31] related features to enhance our system's capability. The proposed approach will also be enriched with more linguistic features such as the use of sentic patterns [23]. A novel fusion strategy will be developed in order to combine the features [26]. Use of concepts instead of words [32-40] will also be one of the major parts of the future work. We also aim to explore the use of contextual information carried by syntactic n-grams [41-44] and graph representation of text [45].

References

1. Alchemy API, AlchemyAPI,Inc. (2015), http://www.alchemyapi.com/
2. Amazon Mechanical Turk, https://www.mturk.com
3. Alag, S.: Collective intelligence in action. Manning, New York (2009)
4. Aldahawi, H., Allen, S.: Twitter Mining in the Oil Business: A Sentiment Analysis Approach. In: The 3rd International Conference on Cloud and Green Computing (CGC). IEEE (2013)
5. Billsus, D., Pazzani, M.: User Modeling for Adaptive News Access. User Modeling and User-Adapted Interaction 10(2-3), 147–180 (2000)
6. Bollen, J., Mao, H., Zeng, X.: Twitter Mood Predicts the Stock Market. Journal of Computational Science 2(1), 1–8 (2011)
7. Fournier, S., Avery, J.: The uninvited brand. Business Horizons 54(3), 193–207 (2011)
8. Ghazanfar, M.A.: Robust, Scalable, and Practical Algorithms for Recommender Systems, University of Southampto (2012)
9. Ghazanfar, M.A., Prügel-Bennett, A.: The Advantage of Careful Imputation Sources in Sparse Data-Environment of Recommender Systems: Generating Improved SVD-based Recommendations. Informatica (Slovenia) 37(1), 61–92 (2013)

10. Ghazanfar, M.A., Prügel-Bennett, A., Szedmak, S.: Kernel-Mapping Recommender System Algorithms. Information Sciences 208, 81–104 (2012)
11. Hall, M., Frank, E., Holmes, G., Pfahringer, B., Reutemann, P., Witten, I.H.: The WEKA Data Mining Software: an Update. ACM SIGKDD Explorations Newsletter 11(1), 10–18 (2009)
12. Jindal, N., Liu, B.: Review Spam Detection. In: The 16th International Conference on World Wide Web. ACM (2007)
13. Mooney, R.J., Roy, L.: Content-Based Book Recommending Using Learning for Text Categorisation. In: The 5th ACM Conference on Digital Libraries. ACM (2000)
14. Pang, B., Lee, L., Vaithyanathan, S.: Thumbs Up?: Sentiment Classification Using Machine Learning Techniques. In: The ACL 2002 Conference on Empirical Methods in Natural Language Processing, vol. 10. Association for Computational Linguistics (2002)
15. Sebastiani, F.: Machine learning in automated text categorization. ACM Computing Surveys (CSUR) 34(1), 1–47 (2002)
16. Twitter, Twitter,Inc. (2015), https://about.twitter.com/company
17. Witten, I.H., Frank, E.: Data Mining: Practical Machine Learning Tools and Techniques with Java Implementations. Morgan Kauffman, San Francisco (1999)
18. Witten, I.H., Frank, E.: Data Mining: Practical Machine Learning Tools and Techniques. Morgan Kaufmann, San Francisco (2005)
19. Zhang, J., Marszałek, M., Lazebnik, S., Schmid, C.: Local Features and Kernels for Classification of Texture and Object Categories: A Comprehensive Study. International Journal of Computer Vision 73(2), 213–238 (2007)
20. Zhang, T., Popescul, A., Dom, B.: Linear Prediction Models with Graph Regularization for Web-Page Categorisation. In: The 12th ACM SIGKDD International Conference on Knowledge Discovery and Data Mining. ACM (2006)
21. Zhang, X., Fuehres, H., Gloor, P.A.: Predicting Stock Market Indicators Through Twitter "I Hope It Is Not as Bad as I Fear". Procedia-Social and Behavioral Sciences 26, 55–62 (2011)
22. Poria, S., Gelbukh, A., Cambria, E., Yang, P., Hussain, A., Durrani, T.: Merging SenticNet and WordNet-Affect emotion lists for sentiment analysis. In: 2012 IEEE 11th International Conference on Signal Processing (ICSP), October 21-25, vol. 2, pp. 1251–1255 (2012)
23. Poria, S., Cambria, E., Winterstein, G., Huang, G.-B.: Sentic patterns: Dependency-based rules for concept-level sentiment analysis. Knowledge-Based Systems 69, 45–63 (2014), http://dx.doi.org/10.1016/j.knosys.2014.05.005 ISSN 0950-7051
24. Poria, S., Gelbukh, A., Das, D., Bandyopadhyay, S.: Fuzzy Clustering for Semi-supervised Learning–Case Study: Construction of an Emotion Lexicon. In: Batyrshin, I., González Mendoza, M. (eds.) MICAI 2012, Part I. LNCS, vol. 7629, pp. 73–86. Springer, Heidelberg (2013)
25. Cambria, E., Fu, J., Bisio, F., Poria, S.: AffectiveSpace 2: Enabling Affective Intuition for Concept-Level Sentiment Analysis. In: Twenty-ninth AAAI Conference on Artificial Intelligence (2015)
26. Poria, S., Cambria, E., Hussain, A., Huang, G.-B.: Towards an intelligent framework for multimodal affective data analysis. Neural Networks 63, 104–116 (2015), http://dx.doi.org/10.1016/j.neunet.2014.10.005 ISSN 0893-6080
27. Poria, S., Cambria, E., Ku, L.-W., Gui, C., Gelbukh, A.: A rule-based approach to aspect extraction from product reviews. In: SocialNLP 2014, vol. 28 (2014)

28. Poria, S., Gelbukh, A., Cambria, E., Das, D., Bandyopadhyay, S.: Enriching SenticNet polarity scores through semi-supervised fuzzy clustering. In: 2012 IEEE 12th International Conference on Data Mining Workshops (ICDMW), pp. 709–716. IEEE (2012)
29. Poria, S., Gelbukh, A., Hussain, A., Bandyopadhyay, S., Howard, N.: Music genre classification: A semi-supervised approach. In: Carrasco-Ochoa, J.A., Martínez-Trinidad, J.F., Rodríguez, J.S., di Baja, G.S. (eds.) MCPR 2012. LNCS, vol. 7914, pp. 254–263. Springer, Heidelberg (2013)
30. Poria, S., Gelbukh, A., Cambria, E., Hussain, A., Huang, G.-B.: EmoSenticSpace: A novel framework for affective common-sense reasoning. Knowledge-Based Systems 69, 108–123 (2014)
31. Poria, S., Gelbukh, A., Hussain, A., Howard, N., Das, D., Bandyopadhyay, S.: Enhanced SenticNet with Affective Labels for Concept-Based Opinion Mining. IEEE Intelligent Systems 28(2), 31–38 (2013), doi:10.1109/MIS.2013.4
32. Poria, S., Agarwal, B., Gelbukh, A., Hussain, A., Howard, N.: Dependency-based semantic parsing for concept-level text analysis. In: Gelbukh, A. (ed.) CICLing 2014, Part I. LNCS, vol. 8403, pp. 113–127. Springer, Heidelberg (2014)
33. Poria, S., Gelbukh, A., Agarwal, B., Cambria, E., Howard, N.: Common sense knowledge based personality recognition from text. In: Castro, F., Gelbukh, A., González, M. (eds.) MICAI 2013, Part II. LNCS, vol. 8266, pp. 484–496. Springer, Heidelberg (2013)
34. Cambria, E., Poria, S., Gelbukh, A., Kwok, K.: Sentic API: A common-sense based API for concept-level sentiment analysis. In: Proceedings of the 4th Workshop on Making Sense of Microposts (# Microposts2014), co-located with the 23rd International World Wide Web Conference (WWW 2014), Seoul, Korea. CEUR Workshop Proceedings, vol. 1141, pp. 19–24 (2014)
35. Agarwal, B., Poria, S., Mittal, N., Gelbukh, A., Hussain, A.: Concept-Level Sentiment Analysis with Dependency-Based Semantic Parsing: A Novel Approach. In: Cognitive Computation, pp. 1–13 (2015)
36. Poria, S., Cambria, E., Howard, N., Huang, G.-B., Hussain, A.: Fusing Audio, Visual and Textual Clues for Sentiment Analysis from Multimodal Content. Neurocomputing (2015)
37. Chikersal, P., Poria, S., Cambria, E.: SeNTU: Sentiment analysis of tweets by combining a rule-based classifier with supervised learning. In: Proceedings of the International Workshop on Semantic Evaluation, SemEval 2015 (2015)
38. Minhas, S., Poria, S., Hussain, A., Hussainey, K.: A review of artificial intelligence and biologically inspired computational approaches to solving issues in narrative financial disclosure. In: Liu, D., Alippi, C., Zhao, D., Hussain, A. (eds.) BICS 2013. LNCS, vol. 7888, pp. 317–327. Springer, Heidelberg (2013)
39. Pakray, P., Poria, S., Bandyopadhyay, S., Gelbukh, A.: Semantic textual entailment recognition using UNL. Polibits 43, 23–27 (2011)
40. Das, D., Poria, S., Bandyopadhyay, S.: A classifier based approach to emotion lexicon construction. In: Bouma, G., Ittoo, A., Métais, E., Wortmann, H. (eds.) NLDB 2012. LNCS, vol. 7337, pp. 320–326. Springer, Heidelberg (2012)
41. Sidorov, G.: Should syntactic n-grams contain names of syntactic relations. International Journal of Computational Linguistics and Applications 5(1), 139–158 (2014)
42. Sidorov, G., Gelbukh, A., Gómez-Adorno, H., Pinto, D.: Soft Similarity and Soft Cosine Measure: Similarity of Features in Vector Space Model. Computación y Sistemas 18(3) (2014)

43. Sidorov, G., Kobozeva, I., Zimmerling, A., Chanona-Hernández, L., Kolesnikova, O.: Modelo computacional del diálogo basado en reglas aplicado a un robot guía móvil. Polibits 50, 35–42 (2014)
44. Ben-Ami, Z., Feldman, R., Rosenfeld, B.: Using Multi-View Learning to Improve Detection of Investor Sentiments on Twitter. Computación y Sistemas 18(3) (2014)
45. Das, N., Ghosh, S., Gonçalves, T., Quaresma, P.: Comparison of Different Graph Distance Metrics for Semantic Text Based Classification. Polibits 49, 51–57 (2014)
46. Alonso-Rorís, V.M., Gago, J.M.S., Rodríguez, R.P., Costa, C.R., Carballa, M.A.G., Rifón, L.A.: Information Extraction in Semantic, Highly-Structured, and Semi-Structured Web Sources. Polibits 49, 69–75 (2014)

Speech Processing

Long-Distance Continuous Space Language Modeling for Speech Recognition

Mohamed Talaat, Sherif Abdou, and Mahmoud Shoman

Faculty of Computers and Information,
Cairo University. 5 Dr. Ahmed Zewail Street, 12613 Giza, Egypt
{mtalaat,s.abdou,m.essmael}@fci-cu.edu.eg
http://www.fci-cu.edu.eg

Abstract. The n-gram language models has been the most frequently used language model for a long time as they are easy to build models and require the minimum effort for integration in different NLP applications. Although of its popularity, n-gram models suffer from several drawbacks such as its ability to generalize for the unseen words in the training data, the adaptability to new domains, and the focus only on short distance word relations. To overcome the problems of the n-gram models the continuous parameter space LMs were introduced. In these models the words are treated as vectors of real numbers rather than of discrete entities. As a result, semantic relationships between the words could be quantified and can be integrated into the model. The infrequent words are modeled using the more frequent ones that are semantically similar. In this paper we present a long distance continuous language model based on a latent semantic analysis (LSA). In the LSA framework, the word-document co-occurrence matrix is commonly used to tell how many times a word occurs in a certain document. Also, the word-word co-occurrence matrix is used in many previous studies. In this research, we introduce a different representation for the text corpus, this by proposing long-distance word co-occurrence matrices. These matrices to represent the long range co-occurrences between different words on different distances in the corpus. By applying LSA to these matrices, words in the vocabulary are moved to the continuous vector space. We represent each word with a continuous vector that keeps the word order and position in the sentences. We use tied-mixture HMM modeling (TM-HMM) to robustly estimate the LM parameters and word probabilities. Experiments on the Arabic Gigaword corpus show improvements in the perplexity and the speech recognition results compared to the conventional n-gram.

Keywords: Language model, n-gram, Continuous space, Latent semantic analysis, Word co-occurrence matrix, Long distance, Tied-mixture model.

1 Introduction

N-gram model [1] [2] [3] is the most frequently used LM technique in many natural language processing applications. This is due to several reasons. First, n-gram models are easy to build; all what it requires is a plain text dataset. Second,

A. Gelbukh (Ed.): CICLing 2015, Part II, LNCS 9042, pp. 549–564, 2015.
DOI: 10.1007/978-3-319-18117-2_41

the computational overhead to build an n-gram model is virtually negligible given the amount of typically used data in many applications. Last, n-gram models are fast to use during decoding as it does not require any computation other than a table look-up. It defines the probability of an ordered sequence of n words by using an independence assumption that each word depends only on the last $n-1$ words. In case of trigram ($n = 3$), the probability for the word sequence $W = w_1, w_2, ..., w_N$ is:

$$P_{trigram}(W) = \prod_{i=1}^{N} P(w_i|w_{i-2}, w_{i-1}) . \tag{1}$$

In spite of their success, the n-gram models suffer from some major problems. One of the key problems in n-gram modeling is the inherent data sparseness of real training data. If the training corpus is not large enough, many actually possible word successions may not be well observed, leading to many extremely small probabilities. This is a serious problem and frequently occurs in many LMs. Usually some smoothing techniques are used to solve that problem by ensuring that some probabilities are greater than zero for words which do not occur or occur with very low frequency in the training corpus [4] [5] [6] [7] [8] [9] [10].

Another way to avoid data sparseness is by mapping words into classes in class-based LM resulting in a LM with less parameters. Class-based LM gives for infrequent words more confidence by relying on other more frequent words in the same class. The simplest class-based LM is known as class-based n-gram LM [11]. A common way to improve a class-based n-gram LM is by combining it with a word-based n-gram LM using interpolation method [12] [13]. Another approach is using a class-based n-gram LM to predict the unseen events, while the seen events are predicted by a word-based n-gram LM. This method is known as word-to-class back-off [14].

In addition to the data sparseness problem, the n-gram also suffers from the adaptability problem [15]. N-gram language model adaptation (to new domain, speaker, genre) is very difficult since it requires a huge amount of adaptation data to adapt the large number of the model parameters. The typical practice for this problem is to collect data in the target domain and build a domain specific language model. The domain specific language model is then interpolated with a generic language model trained on a larger domain independent data to achieve robustness [16].

Based on the Markov assumption, the word-based n-gram LMs are very powerful in modeling short-range dependencies but weak in modeling long-range dependencies Many attempts were made to capture long-range dependencies. The cache-based LM [17] used a longer word history (window) to increase the probability of re-occurring words. Also the trigger-based LM [18], which can be considered a generalization of the cache-based model where related words can increase the probability of the word that we predict. However, the training process (finding related word pairs) for such type of LMs is computationally expensive. There are also n-gram variants that models long range dependencies such as skip n-gram LM [19] [5] that skips over some intermediate words in the context, or variable-length n-gram LM [20] that uses extra context if it is considered to be more predictive.

To overcome the problems of the n-gram models the continuous parameter space LMs were introduced. In these models the words are treated as vectors of real numbers rather than of discrete entities. As a result, long-term semantic relationships between the words could be quantified and can be integrated into the model.

Bellegarda et al. [21] [22] [23] introduced latent semantic analysis (LSA) to language modeling. The concept of LSA was first introduced by Deerwester et al. [24] for information retrieval and since then there has been an explosion of research and application involving it. LSA maps words into a semantic space where two semantically related words are placed close to each other. Recently, LSA has been successfully used in language modeling to map discrete word into continuous vector space (LSA space). Bellegarda combines the global constraint given by LSA with the local constraint of n-gram language model. The same approach is used in [25] [26] [27] [28] [29] [30] but using neural network (NN) as an estimator. Gaussian mixture model (GMM) could also be trained on this LSA space [15]. Also, the tied-mixture LM (TMLM) was proposed in the LSA space [16]. Context dependent class (CDC) LM using word co-occurrence matrix was proposed in [31]. Instead of a word-document matrix, a word-phrase co-occurrence matrix is used in [32] as a representation of a corpus.

To apply the LSA, the text corpus must be represented by a mathematical entity called matrix. LSA is usually used together with the word-document matrix [33] to represent the corpus. Its cell contains the frequency of how many times a word occurs in a certain document in the corpus. Also, the word-word co-occurrence matrix was used in some previous studies, its cell a_{ij} contains the frequency of word sequence $w_j w_i$ in the corpus.

In this work we introduce a continuous parameter space LM based on LSA. We propose a different representation for the text corpus and taking into consideration the long range dependencies between words. We represent the text corpus by creating long-distance word co-occurrence matrices. These matrices represent the co-occurrences between different words on different distances in the corpus. Then LSA is applied to each one of these matrices separately. A tied-mixture HMM model is trained on the LSA results to estimate the LM parameters and word probabilities in the continuous vector space.

Rest of the paper is organized as follows. Section 2 gives a brief review about the continuous space language modeling. Section 3 includes the description of the LSA approach. Section 4 introduces our proposed long-distance matrices. Section 5 describes the tied-mixture modeling. Experiment results are presented in section 6. Finally section 7 includes the final discussion and conclusions.

2 Continuous Space Language Modeling

The underlying idea of the continuous space language modeling approach is to attack the data sparseness, adaptability and long range dependencies problems of the conventional n-gram models by performing the language model probability estimation in a continuous space. In the continuous space, words are not treated as discrete entities but rather vectors of real numbers.

To build a continuous space LM we need a mapping from the discrete word space to a representation in the continuous parameter space in the form of vectors of real numbers. Then we can train a statistical model to estimate the prediction probability of the next word given the mapped history in the continuous space.

As a result, long-term semantic relationships between the words could be quantified and can be integrated into the model, where in the continuous space we hope that there is some form of distance of similarity between histories such that histories not observed in the data for some word are smoothed by similar observed histories. This help to attack the data sparseness issue discussed above for n-gram LMs.

By moving to the continuous space, we can cast the language modeling problem as the acoustic modeling problem in the speech recognition application. In the acoustic modeling, large models can be efficiently adapted using a few utterances by exploiting the inherit structure in the model by techniques like maximum likelihood linear regression (MLLR) [34]. So, we can re-call the acoustic modeling adaptation tools to adapt language models in the continuous space. This addresses the adaptability issue discussed above for n-gram LMs [15] [16].

In this work we propose an approach to address the long range dependencies issue for the n-gram LMs with a practical implementation for key applications.

3 Latent Semantic Analysis

LSA extracts semantic relations from a corpus, and maps them to a low dimension vector space. The discrete indexed words are projected into LSA space by applying singular value decomposition (SVD) to a matrix that representing a corpus.

The first step is to represent the text as a matrix in which each row stands for a unique word and each column stands for a text passage or other context. Each cell contains the frequency with which the word of its row appears in the passage denoted by its column. In the original LSA, the representation matrix is a term-document co-occurrence matrix.

Next, LSA applies singular value decomposition (SVD) to the matrix. In SVD, a rectangular matrix is decomposed into the product of three other matrices. One component matrix describes the original row entities as vectors of derived orthogonal factor values, another describes the original column entities in the same way, and the third is a diagonal matrix containing scaling values such that when the three components are matrix-multiplied, the original matrix is reconstructed. For a matrix C with $M \times N$ dimension, SVD decomposes the matrix C as follows:

$$C \approx USV^T . \tag{2}$$

where U is a left singular matrix with row vectors and dimension $M \times R$, the matrix U is corresponding with the rows of matrix C. S is a diagonal matrix of singular values with dimension $R \times R$. V is a right singular matrix with row vectors and dimension $N \times R$, the matrix V is corresponding with the columns of matrix C. R is the order of the decomposition and $R \ll \min(M, N)$. These LSA matrices are then used to project the words into the reduced R-dimension LSA continuous vector space. In case of a term-document matrix used as a representation matrix, matrix U contains information about words while matrix V contains information about the documents. So, the matrix U is used to project words in the LSA space.

Assume we have a vocabulary of size V, each word $\{i \;\; 1 \leq i \leq V\}$ can be represented by an indicator discrete vector w_i having one at the ith position and zero in all other $V-1$ positions. This vector can be mapped to a lower dimension (R) vector u_i, using a projection matrix A of dimension $V \times R$ according to the following equation:

$$u_i = A^T w_i \;. \tag{3}$$

In other words, a continuous vector for word w_i is represented by the ith row vector of matrix A. So each word w_i has a continuous representation vector u_i.

Based on the word mapping in Equation 3, each history h consists of a set of $N-1$ words for an n-gram can be represented as a concatenation of the appropriate mapped words. The history vectors are of dimension $R(N-1)$. According to this mapping for word histories h, we can train a statistical parametric model on these histories continuous vectors and build a model to estimate the word probabilities $p(w|h)$ in the continuous LSA space.

4 Proposed Long-Distance Matrices

LSA starts from representing the corpus through a mathematical entity called a representation matrix. In the original LSA, the used representation matrix is a term-document co-occurrence matrix, where its cell $C(w_i, d_j)$ contains co-occurrence frequency of word w_i in document d_j. In [15] [16] [31], the word-word co-occurrence matrix is used to represent the corpus, where each cell $C(w_i, w_j)$ denotes the counts for which word w_i follow word w_j in the corpus.

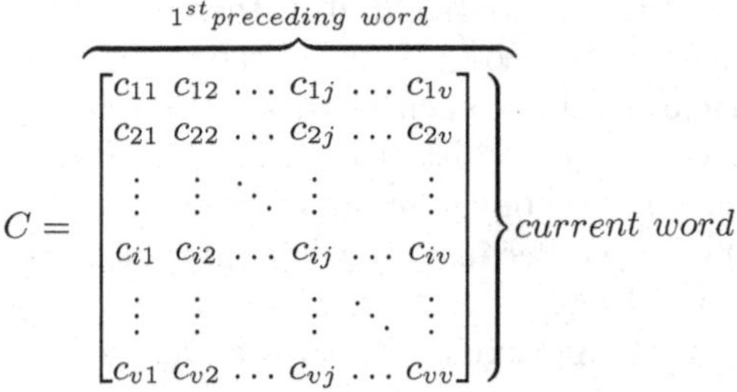

Fig. 1. Distance-One Word Co-occurrence Matrix

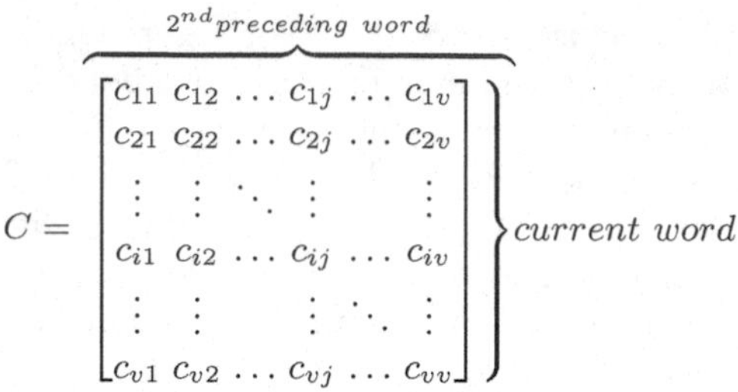

Fig. 2. Distance-Two Word Co-occurrence Matrix

In this work, we propose a representation for the corpus using distance based word-word co-occurrence matrices, where each matrix will represent the co-occurrence relation between each word and the previous words on different distances in the corpus as shown below.

The distance-one word co-occurrence matrix is a matrix representation where each row represents a current word w_i, and each column represents the 1st preceding word w_{i-1} as illustrated by Figure 1. Each cell $C(w_i, w_j)$ is a co-occurrence frequency of word sequence $w_j w_i$. This is a square matrix with dimension $V \times V$, where V is the vocabulary size. It represents the co-occurrence relations between each word and the first preceding words to that word appeared in the corpus.

The distance-two word co-occurrence matrix is a matrix representation where each row represents a current word w_i, and each column represents the 2nd preceding word w_{i-2} as illustrated by Figure 2. Each cell $C(w_i, w_j)$ is a co-occurrence frequency when the word w_j occurs as the 2nd preceding word of word w_i. This is a square matrix with dimension $V \times V$, where V is the vocabulary size. It represents the co-occurrence relations between each word and the 2nd preceding words to that word appeared in the corpus. And the same for distance-three matrix and so on.

In general, the distance-d word co-occurrence matrix is a matrix representation where each row represents a current word w_i, and each column represents the dth preceding word w_{i-d} as illustrated by Figure 3. Each cell $C(w_i, w_j)$ is a co-occurrence frequency when the word w_j occurs as the dth preceding word of word w_i. This is a square matrix with dimension V X V, where V is the vocabulary size. It represents the co-occurrence relations between each word and the words on distance d to that word appeared in the corpus.

By using these long-distance co-occurrence representation matrices, we hope to collect more information about each word and its relation with the previous words in the corpus on different distances, this may help to attack the long-dependencies problem of the conventional n-gram model. These matrices are large, but very sparse ones. Because of their large size and sparsity, we can apply the second step of LSA, by making SVD to each one of them separately to produce a reduced-rank approximation to each matrix of them.

$$C = \overbrace{\begin{bmatrix} c_{11} & c_{12} & \dots & c_{1j} & \dots & c_{1v} \\ c_{21} & c_{22} & \dots & c_{2j} & \dots & c_{2v} \\ \vdots & \vdots & \ddots & \vdots & & \vdots \\ c_{i1} & c_{i2} & \dots & c_{ij} & \dots & c_{iv} \\ \cdot & \cdot & & \cdot & \cdot & \cdot \\ \cdot & \cdot & & \cdot & & \cdot \\ c_{v1} & c_{v2} & \dots & c_{vj} & \dots & c_{vv} \end{bmatrix}}^{d^{th} preceding\ word} \left.\vphantom{\begin{bmatrix} 1\\1\\1\\1\\1\\1\\1 \end{bmatrix}}\right\} current\ word$$

Fig. 3. Distance-d Word Co-occurrence Matrix

Before proceeding in the SVD step, the entries of the co-occurrence matrices are smoothed according to Equation 4, this is because the co-occurrence matrices typically contain a small number of high frequency events and a large number of less frequent events, and the SVD derives a compact approximation of the co-occurrence matrix that is optimal in the least square sense, it best models these high frequency events, which may not be the most informative [16].

$$\hat{C}\left(w_i, w_j\right) = \log\left(C\left(w_i, w_j\right) + 1\right) . \tag{4}$$

Based on the SVD results from Equation 2, we construct a projection matrix A of dimension $V \times R$ corresponding to each word co-occurrence matrix by using the left singular matrix U, and the diagonal matrix S of the SVD results as follows:

$$A_{V \times R} = U_{V \times R} S_{R \times R} . \tag{5}$$

Now, we have a projection matrix A for each constructed long-distance word co-occurrence matrix. For example, if we create distance-one, distance-two, and distance-three word co-occurrences matrices then after the SVD for each one of them, we will have three projection related matrices. We can map the discrete word vector w_i $\{i \;\; 1 \leq i \leq V\}$ into the continuous space using Equation 3 to get a word vector u_i in the continuous space for each word from each projection matrix A.

As a result from the previous step, we have more than one mapped continuous vector u_i for each word w_i, then we concatenate these continuous vectors of each word to construct the final word vector that uniquely represent the word in the LSA continuous vector space. The final word vector will contain information about the relation of the word with the previous words appeared on different distances in the corpus. In the next section, we introduce the tied-mixture model used to estimate the word probabilities using these word vectors in the continuous space.

5 Tied-Mixture Modeling

Tied-Mixture Hidden Markov Model (TM-HMM) was proposed by Bellegarda et al. in [35] for continuous parameter modeling in speech recognition systems, which have a better decoupling between the number of Gaussians and the number of states compared to continuous density HMMs.

The TM-HHM is one in which all Gaussian components are stored in a pool and all state output distributions share this pool. Each state output distribution is defined by M mixture component weights and since all states share the same components, all of the state-specific discrimination is encapsulated within these weights. The output distribution for state w is defined as:

$$p\,(o|w) = \sum_{m=1}^{M} c_{w,m} \mathcal{N}_m(o, \mu_m, \Sigma_m)\,. \tag{6}$$

where o is the observation vector, M is number of mixture components, $c_{w,m}$ is the mixture weight related to state w for the mixture component m, μ_m, and Σ_m are the mean and covariance for the mth mixture component $\mathcal{N}_m$ respectively.

TM-HMM was used for language modeling in a Tied-Mixture Language Model (TMLM) [16], where each state represents a word w, and the model estimates the probability of observing the history vector (h) for a given word w, $p(h|w)$, and to calculate the posterior probability $p(w|h)$ of observing w as the next word given the history h, it uses the Bayes rule as follows:

$$p(w|h) = \frac{p(w)p(h/w)}{p(h)} = \frac{p(w)p(h/w)}{\sum_{v=1}^{V} p(v)p(h/v)}\,. \tag{7}$$

where $p(w)$ is the unigram probability estimate of word w, and V is the number of words in the vocabulary. Also it was shown that the unigram probabilities can be substituted for more accurate higher order n-gram probabilities. If this n-gram has an order that is equal to or greater than the one used in defining the continuous contexts h, then the TMLM can be viewed as performing a kind of smoothing of the original n-gram model as follows:

$$P_s(w|h) = \frac{P(w|h)p(h/w)}{\sum_{v=1}^{V} P(v/h)p(h/v)}\,. \tag{8}$$

where $P_s(w|h)$ and $P(w|h)$ are the smoothed and original n-grams.

The TM-HMM model parameters are estimated through an iterative procedure called the Baum-Welch, or forward-backward, algorithm [36], it maximizes the likelihood function via an iterative procedure. For the model estimation equations the readers are referred to [35].

In this work, we use the same TMLM proposed in [16] to estimates the probability $p(w|h)$ of observing w as the next word given the history h, but with our proposed word mapping and history representation h in the continuous vector space.

6 Experiments

In this work, we evaluated the developed long-distance continuous domain TMLM by integrating it in an Arabic speech recognition system, and we use the perplexity, and the word-error rate (WER) as an evaluation methods. The baseline system acoustic model was a standard triphone HMM model that was trained using 100 hours of broadcast news recordings of modern standard Arabic data. A tied $20K$ triphone states were trained with 16 Gaussians for each state cluster. The acoustic model was trained using diacritized data. Part of the acoustic model training data, 25 hours, was diacritized manually and the rest was diacritized using an Arabic diacritizer tool [37].

For the language model training data, we used part of the Arabic Gigaword data. We trained non-diacritized language models. We started with a small data set of two million words to investigate the feasibility of the proposed language model then used a larger data set of 50 million words.

In the first experiment, a statistical bigram language model using Modified Kneser-Ney smoothing has been built using SRILM toolkit [38], which is referred to as Word-2gr. The baseline bigram LM, the normal TMLM, and LD-TMLM are used to rescore the N-best list. We limited the rescoring to the 30-best (N=30) utterances in the list. We conducted the speech recognition language modeling experiments on a test set of 53 sentences.

The language model data has about $85K$ sentences comprising the $2M$ words. The vocabulary size is $91K$ unique words, (see table 1). First, we limit the construction of TMMs for words that occur 100 times or more, so we ended up in 2800 words including the beginning sentence and end sentence symbols. We mapped all the remaining words into one class, a sort of filter or unknown word, so the vocabulary size become 2801 words from the original $91K$ vocabulary.

Table 1. Small-Scale Language Model Data

LM Parameter	**Value**
No. of Sentences:	$85K$ Sentences
No. of Words:	$2M$ Words
Vocabulary Size:	$91K$ Words

The normal tied-mixture language model (TMLM) is trained by constructing the word co-occurrence matrix of dimensions 2801 × 2801. Each element $C(w_i, w_j)$ in the co-occurrence matrix is smoothed using Equation 4. Singular value decomposition (SVD) is performed on the resulting smoothed co-occurrence matrix. The SVDLIBC [1] toolkit with Lanczos method is used to compute the 50 (R=50) highest singular values and their corresponding singular vectors for the smoothed co-occurrence matrix. The resulting singular vectors are used to construct the projection to a 50-dimensional space. Each word in

[1] `http://tedlab.mit.edu/~dr/SVDLIBC/`

the vocabulary is represented by a vector of size 50. In another words, each bigram history is represented by a vector of size 50 representing that word. Then a TM-HMM is built and trained.

The proposed long-distance tied-mixture model (LD-TMM) with max distance (D=5) is trained as follows: we construct distance-one, distance-two, distance-three, distance-four, and distance-five word co-occurrence matrices, these are sparse matrices each of dimensions 2801×2801. Each element $C(w_i, w_j)$ in each co-occurrence matrix is smoothed using Equation 4. Singular value decomposition (SVD) is performed on each of the resulting smoothed co-occurrence matrices. The SVDLIBC toolkit with Lanczos method is used to compute the 10 (R=10) highest singular values and their corresponding singular vectors for each smoothed co-occurrence matrix. The resulting singular vectors of each matrix are used to construct the projection to a 10-dimensional space. Each word in the vocabulary is represented by a vector of size 50, this by concatenating the five word vectors of size 10 resulting from the SVD step for the five co-occurrence matrices. In another words, each bigram history is represented by a vector of size 50 representing that word. Thus, a document can be represented by a sequence of 50-dimensional vectors corresponding to the history of each of its constituent words. Then a TM-HMM is built and trained.

For the TMLM and LD-TMLM, we use the HTK toolkit [39] for building and training the TMM-HMM model, and the total number of the shared Gaussian densities (Gaussians pool) used is set to 200 . Also, when calculating the TMM score, the TMM likelihood probability generated by the model is divided by 40 to balance its dynamic range with that of the n-gram model.

Table 2 basically shows the speech recognition language model rescoring results for: the baseline word bigram (Word-2gr) with Modified Kneser-Ney smoothing, the normal TMLM , and the proposed long-distance TMLM (LD-TMLM) with max distance (D=5).

Table 2. Small-scale language model speech recognition rescoring results

Language Model (LM)	Word-Error Rate (WER)
Word-2gr	12.44 %
TMLM	11.47 %
LD-TMLM (D=5)	11.26 %

The first two rows in the table show the WER of the baseline bigram (Word-2gr) model and the WER of the tied-mixture continuous language model (TMLM), where the WER of the baseline Word-2gr model is 12.44% and the WER of the TMLM is 11.47%. The continuous TMLM shows improvement over the baseline Word-2gr model with 0.97%. The last row in the table shows that the proposed LD-TMLM improves the WER to 11.26%. From the results, the LD-TMLM provides improvements over the baseline bigram model by 1.18% and also provides improvements over the normal TMLM by 0.21%.

Table 3. Small-scale language models perplexity results

Language Model (LM)	**Log prob.**	**Perplexity(PP)**
Word-2gr	-6264.47	47.80
TMLM	-1474.40	2.48
LD-TMLM (D=5)	-1026.57	1.88
Word-2gr + LD-TMM (D=5)	-3645.52	9.50

Table 3 shows the log probabilities and perplexity results for: the baseline word bigram (Word-2gr) with Modified Kneser-Ney smoothing, the normal TMLM, the proposed long-distance bigram TMM (LD-TMM) with max distance (D=5), and an interpolation between the baseline bigram and the proposed LD-TMM. The interpolated LM uses uniform weights. These results for the same 53 reference sentences (3677 words) used in the rescoring results before.

The first two rows in the table show the perplexity of the baseline bigram (Word-2gr) model and the perplexity of the tied-mixture continuous language model (TMLM), where the perplexity of the baseline Word-2gr model is 47.80 and the perplexity of the TMLM is 2.48 . The continuous TMLM shows improvement in the perplexity results over the baseline Word-2gr model. The third row in the table shows that the proposed LD-TMLM (D=5) improves the perplexity to 1.88. An interpolation between the baseline bigram (Word-2gr) and the proposed LD-TMM (D=5) using uniform weights improves the perplexity results to 9.50 compared to the perplexity results of the baseline bigram (Word-2gr) model.

In the second experiment, we conducted the same rescoring and perplexity experiments on a large scale language model of training data that has about $2M$ sentences comprising about $50M$ words. The vocabulary size is $98K$ unique words, (see table 4). We limit the construction of TMMs for words that occur 100 times or more, so we ended up in 36842 words including the beginning sentence and end sentence symbols. We mapped all the remaining words into one class, a sort of filter or unknown word, so the vocabulary size become 36843 words from the original $98K$ vocabulary.

Table 4. Large-Scale Language Model Data

LM Parameter	**Value**
No. of Sentences:	$2M$ Sentences
No. of Words:	$50M$ Words
Vocabulary Size:	$98K$ Words

Similar to before, the word bigram (Word-2gr) with Modified Kneser-Ney smoothing is built as a baseline. Also, the normal TMLM is trained for the 36843 words where each word in the vocabulary is represented by a continuous vector of size 50. The proposed long-distance TMLM (LD-TMLM) with max distance (D=2) is trained where each word in the vocabulary is represented by a vector of size 50, this by concatenating the two word vectors of size 25 resulting from the SVD step for the two co-occurrence matrices.

Table 5 basically shows the speech recognition language model rescoring results for: the baseline word bigram (Word-2gr) with Modified Kneser-Ney smoothing, the normal TMLM , and the proposed long-distance TMLM (LD-TMLM) with max distance (D=2).

Table 5. Large-scale language model speech recognition rescoring results

Language Model (LM)	Word-Error Rate (WER)
Word-2gr	16.92 %
TMLM	16.06 %
LD-TMLM (D=2)	15.92 %

The first two rows in the table show the WER of the baseline bigram (Word-2gr) model and the WER of the tied-mixture continuous language model (TMLM), where the WER of the baseline Word-2gr model is 16.92% and the WER of the TMLM is 16.06%. The continuous TMLM shows improvement over the baseline Word-2gr model with 0.86%. The last row in the table shows that the proposed LD-TMLM improves the WER to 15.92%. From the results, the LD-TMLM provides improvements over the baseline bigram model by 1.0% and also provides improvements over the normal TMLM by 0.14%.

Table 6 shows the log probabilities and perplexity results for: the baseline word bigram (Word-2gr) with Modified Kneser-Ney smoothing, the normal TMLM, the proposed long-distance bigram TMM (LD-TMM) with max distance (D=2), and an interpolation between the baseline bigram and the proposed LD-TMM. The interpolated LM uses uniform weights. These results for the same 53 reference sentences (3677 words) used in the rescoring results before.

Table 6. Large-scale language models perplexity results

Language Model (LM)	**Log prob.**	**Perplexity(PP)**
Word-2gr	-9555.55	364.60
TMLM	-4341.68	14.58
LD-TMLM (D=2)	-6205.04	46.08
Word-2gr + LD-TMM (D=2)	-7880.295	129.62

The first two rows in the table show the perplexity of the baseline bigram (Word-2gr) model and the perplexity of the tied-mixture continuous language model (TMLM), where the perplexity of the baseline Word-2gr model is 364.60 and the perplexity of the TMLM is 14.58. The continuous TMLM shows improvement in the perplexity results over the baseline Word-2gr model. The third row in the table shows that the proposed LD-TMLM (D=2) improves the perplexity to 46.08 compared to the baseline bigram model (Word-2gr). An interpolation between the baseline bigram (Word-2gr) and the proposed LD-TMM (D=2) using uniform weights improves the perplexity results to 129.62 compared to the perplexity results of the baseline bigram (Word-2gr) model.

7 Conclusion

Continuous space language models have been shown to be a promising alternative approach to language modeling by means of their elegant property of treating words in a continuous space and the similarity between them is taken into account as the distance of their representations in this space. In this work we elaborated on continuous domain LMs to model the long distance relations between words. We proposed a continuous representation for the text corpus by constructing more than one word co-occurrence matrix that cover long distance dependencies between different words in the corpus. We used the tied-mixture HMM modeling to robustly estimate model parameters. Our initial experimental results validated the proposed approach with encouraging results compared to the conventional n-gram LM, significant and consistent improvements are observed when this type of model is applied to automatic speech recognition systems.

The computation cost of the TMLM is still expensive compared with the n-gram models. In our implementation we used several tricks of cashing, fast Gaussian computation and mixtures selection. This efforts managed to reduce the computation cost of the proposed model but still not fast enough to compete with n-grams. In our future effort we will investigate the integration of TMLM and n-grams to use the continuous domain model for the computation only of unseen words which would reduce the computation effort and make it a practical model.

References

[1] Markov, A.A.: An example of statistical investigation in the text of 'Eugene Onyegin' illustrating coupling of 'tests' in chains. In: Proceedings of the Academy of Sciences. VI, vol. 7 , St. Petersburg, pp. 153–162 (1913)

[2] Damerau, F.: Markov models and linguistic theory: an experimental study of a model for English, Janua linguarum: Series minor. Mouton (1971)

[3] Jelinek, F.: Statistical Methods for Speech Recognition. Language, Speech, & Communication: A Bradford Book. MIT Press (1997)

[4] Kneser, R., Ney, H.: Improved backing-off for m-gram language modeling. In: International Conference on Acoustics, Speech, and Signal Processing, ICASSP 1995, vol. 1, pp. 181–184 (1995)

[5] Ney, H., Essen, U., Kneser, R.: On structuring probabilistic dependencies in stochastic language modelling. Computer Speech and Language 8, 1–38 (1994)

[6] Good, I.J.: The population frequencies of species and the estimation of population parameters. Biometrika 40(3 and 4), 237–264 (1953)
[7] Jelinek, F., Mercer, R.L.: Interpolated estimation of markov source parameters from sparse data. In: Proceedings of the Workshop on Pattern Recognition in Practice, pp. 381–397. North-Holland, Amsterdam (1980)
[8] Katz, S.: Estimation of probabilities from sparse data for the language model component of a speech recognizer. IEEE Transactions on Acoustics, Speech and Signal Processing 35(3), 400–401 (1987)
[9] Lidstone, G.: Note on the general case of the Bayes–Laplace formula for inductive or a posteriori probabilities. Transactions of the Faculty of Actuaries 8, 182–192 (1920)
[10] Bell, T.C., Cleary, J.G., Witten, I.H.: Text Compression. Prentice-Hall, Inc., Upper Saddle River (1990)
[11] Brown, P.F., de Souza, P.V., Mercer, R.L., Pietra, V.J.D., Lai, J.C.: Class-based n-gram models of natural language. Comput. Linguist. 18(4), 467–479 (1992)
[12] Broman, S., Kurimo, M.: Methods for combining language models in speech recognition. In: Interspeech, pp. 1317–1320 (September 2005)
[13] Wada, Y., Kobayashi, N., Kobayashi, T.: Robust language modeling for a small corpus of target tasks using class-combined word statistics and selective use of a general corpus. Systems and Computers in Japan 34(12), 92–102 (2003)
[14] Niesler, T., Woodland, P.: Combination of word-based and category-based language models. In: Proceedings of the Fourth International Conference on Spoken Language, ICSLP 1996, vol. 1, pp. 220–223 (1996)
[15] Afify, M., Siohan, O., Sarikaya, R.: Gaussian mixture language models for speech recognition. In: IEEE International Conference on Acoustics, Speech and Signal Processing, ICASSP 2007, vol. 4, pp. IV–29–IV–32 (2007)
[16] Sarikaya, R., Afify, M., Kingsbury, B.: Tied-mixture language modeling in continuous space. In: Proceedings of Human Language Technologies: The 2009 Annual Conference of the North American Chapter of the Association for Computational Linguistics, NAACL 2009, pp. 459–467. Association for Computational Linguistics, Stroudsburg (2009)
[17] Kuhn, R., De Mori, R.: A cache-based natural language model for speech recognition. IEEE Transactions on Pattern Analysis and Machine Intelligence 12(6), 570–583 (1990)
[18] Rosenfeld, R.: A maximum entropy approach to adaptive statistical language modeling. Computer Speech and Language 10(3), 187–228 (1996)
[19] Nakagawa, S., Murase, I., Zhou, M.: Comparison of language models by stochastic context-free grammar, bigram and quasi-simplified-trigram (0300-1067). IEICE Transactions on Fundamentals of Electronics, Communications and Computer Sciences, 0300–1067 (2008)
[20] Niesler, T., Woodland, P.: A variable-length category-based n-gram language model. In: 1996 IEEE International Conference Proceedings on Acoustics, Speech, and Signal Processing, ICASSP 1996, vol. 1, pp. 164–167 (1996)
[21] Bellegarda, J., Butzberger, J., Chow, Y.-L., Coccaro, N., Naik, D.: A novel word clustering algorithm based on latent semantic analysis. In: 1996 Proceedings of the IEEE International Conference on Acoustics, Speech, and Signal Processing, ICASSP 1996, vol. 1, pp. 172–175 (1996)

[22] Bellegarda, J.: A multispan language modeling framework for large vocabulary speech recognition. IEEE Transactions on Speech and Audio Processing 6(5), 456–467 (1998)

[23] Bellegarda, J.: Latent semantic mapping (information retrieval). IEEE Signal Processing Magazine 22(5), 70–80 (2005)

[24] Deerwester, S., Dumais, S.T., Furnas, G.W., Landauer, T.K., Harshman, R.: Indexing by latent semantic analysis. Journal of the American Society for Information Science 41(6), 391–407 (1990)

[25] Bengio, Y., Ducharme, R., Vincent, P., Jauvin, C.: A neural probabilistic language model. Journal of Machine Learning Research 3, 1137–1155 (2003)

[26] Blat, F., Castro, M., Tortajada, S., Snchez, J.: A hybrid approach to statistical language modeling with multilayer perceptrons and unigrams. In: Matoušek, V., Mautner, P., Pavelka, T. (eds.) TSD 2005. LNCS (LNAI), vol. 3658, pp. 195–202. Springer, Heidelberg (2005)

[27] Emami, A., Xu, P., Jelinek, F.: Using a connectionist model in a syntactical based language model. In: 2003 IEEE Proceedings of the International Conference on Acoustics, Speech, and Signal Processing, ICASSP 2003, vol. 1, pp. I–372–I–375 (2003)

[28] Schwenk, H., Gauvain, J.: Connectionist language modeling for large vocabulary continuous speech recognition. In: 2002 IEEE International Conference on Acoustics, Speech, and Signal Processing (ICASSP), vol. 1, pp. I–765–I–768 (2002)

[29] Schwenk, H., Gauvain, J.-L.: Neural network language models for conversational speech recognition. In: ICSLP (2004)

[30] Schwenk, H., Gauvain, J.-L.: Building continuous space language models for transcribing european languages. In: INTERSPEECH, pp. 737–740. ISCA (2005)

[31] Naptali, W., Tsuchiya, M., Nakagawa, S.: Language model based on word order sensitive matrix representation in latent semantic analysis for speech recognition. In: 2009 WRI World Congress on Computer Science and Information Engineering, vol. 7, pp. 252–256 (2009)

[32] Fumitada: A linear space representation of language probability through SVD of n-gram matrix. Electronics and Communications in Japan (Part III: Fundamental Electronic Science) 86(8), 61–70 (2003)

[33] Rishel, T., Perkins, A.L., Yenduri, S., Zand, F., Iyengar, S.S.: Augmentation of a term/document matrix with part-of-speech tags to improve accuracy of latent semantic analysis. In: Proceedings of the 5th WSEAS International Conference on Applied Computer Science, ACOS 2006, pp. 573–578. World Scientific and Engineering Academy and Society (WSEAS), Stevens Point (2006)

[34] Leggetter, C., Woodland, P.: Maximum likelihood linear regression for speaker adaptation of continuous density hidden markov models. Computer Speech and Language 9(2), 171–185 (1995)

[35] Bellegarda, J.R., Nahamoo, D.: Tied mixture continuous parameter modeling for speech recognition. IEEE Transactions on Acoustics, Speech and Signal Processing 38(12), 2033–2045 (1990)

[36] Baum, L.E., Petrie, T., Soules, G., Weiss, N.: A maximization technique occurring in the statistical analysis of probabilistic functions of markov chains. The Annals of Mathematical Statistics 41(1), 164–171 (1970), doi:10.2307/2239727

[37] Rashwan, M., Al-Badrashiny, M., Attia, M., Abdou, S., Rafea, A.: A stochastic arabic diacritizer based on a hybrid of factorized and unfactorized textual features. IEEE Transactions on Audio, Speech, and Language Processing 19(1), 166–175 (2011)

[38] Stolcke, A.: SRILM – an extensible language modeling toolkit. In: Proceedings of ICSLP, vol. 2, Denver, USA, pp. 901–904 (2002)

[39] Young, S.J., Evermann, G., Gales, M.J.F., Hain, T., Kershaw, D., Moore, G., Odell, J., Ollason, D., Povey, D., Valtchev, V., Woodland, P.C.: The HTK book, version 3.4. In: Cambridge University Engineering Department, Cambridge, UK (2006)

A Supervised Phrase Selection Strategy for Phonetically Balanced Standard Yorùbá Corpus

Adeyanju Sosimi[1,*], Tunde Adegbola[2], and Omotayo Fakinlede[1]

[1]University of Lagos, Akoka-Lagos, Nigeria
azeezadeyanju@gmail.com, oafak@unilag.edu.ng
[2]Africa Language Technology Initiative, Bodija, Ibadan, Nigeria
taintransit@hotmail.com

Abstract. This paper presents a scheme for the development of speech corpus for Standard Yorùbá (SY). The problem herein is the non-availability of phonetically balanced corpus in most resource-scarce languages such as SY. The proposed solution herein is hinged on the development and implementation of a supervised phrase selection using Rule-Based Corpus Optimization Model (RBCOM) to obtain phonetically balanced SY corpus. This was in turn compared with the random phrase selection procedure. The concept of Exploitative Data Analysis (EDA), which is premised on frequency distribution models, was further deployed to evaluate the distribution of allophones of selected phrases. The goodness of fit of the frequency distributions was studied using: Kolmogorov Smirnov, Andersen Darling and Chi-Squared tests while comparative studies were respectively carried out among other techniques. The sample skewness result was used to establish the normality behavior of the data. The results obtained confirmed the efficacy of the supervised phrase selection against the random phrase selection.

Keywords: Standard Yorùbá, Corpus, Rule-Based Corpus Optimization Model, Phrase selection, Automatic Speech Recognition.

1 Introduction

Human Language Technology (HLT) applications currently exist for a vast majority of languages of the industrialized nations but this is not the case with most African languages such as the Standard Yoruba (SY) language. One of the impediments to such development for these Africa languages is the relatively lack of language corpus. Unlike the case of most African languages, language corpus have over the years been fully developed other continents as seen in the Thai language, Chinese and French languages amongst others which have different corpus from the English language. HLT development is premised on availability of phonetically rich and balanced digital speech corpus of the target language [1]. However, only few African languages have speech corpus required for HLT development, which in recent times have been given a higher priority as a result of global technological influence. SY is one of the few indigenous Nigerian languages that have benefitted from the platform of HLT.

A. Gelbukh (Ed.): CICLing 2015, Part II, LNCS 9042, pp. 565–582, 2015.
DOI: 10.1007/978-3-319-18117-2_42

[2] reported that SY is the native language of more than 30 million people within and outside Nigeria. With respect to speech recognition, the Yorùbá language bears a challenging characteristic in the usage of tones to discriminate meaning. Yorùbá speech is constructed by appropriate combination of elements from a phonological alphabet of three lexically contractive vocal gestures, namely consonants, vowels, and tones. According to [3], the three distinct tones, therefore, widen the scope of an x- syllable Yorùbá word to 3^x possible permutations while [4] considered two systems for SY ASR: oral vowels using fuzzy logic (FL) and artificial neural network (ANN) based models. [5] considered additional feature extraction methods to evaluate the effect of voice activity detection in an isolated Yorùbá word recognition system. However, in the case of continuous speech, the signal is affected by many factors such as sentence prosody, co-articulation, speaker's emotion, gesticulation, etc. [6]

To accomplish the target for the SY language, there is a need for efficient and effective continuous speech corpus development as presented by [7], where it was stated that the construction of phonetically rich and balanced speech corpus is based on the selection of a set of phrases. In the literature, various techniques have been proposed for such selection, and a major constraint with this is the cost development. An approach for such selection is to ensure that allophone interaction and distribution of phrases have equal parity without loss of information and also, not undermining language syntactic rules. [8] reported that uniform distribution and frequency of occurrence of phones appears to be the dominant paradigm in assessing allophones optimality. The adopted strategy for new language resource development is dependent on the task; the first scheme is particularly used for training a Text-to-Speech (TTS) system while the second type is better adapted for the development of ASR systems [8]. Generally, optimal distribution of allophones is of significance when developing corpora for resource-scarce language [1].

[9] reported that the use of prompts produces quality corpus for training the ASR system. [10] discussed the design of three large databases to cope with the challenges of the ASR system. [8], [11], [12], [13], and [14] considered the design and development of phonetically balanced corpus for an under-resourced language. These authors premised their work on the understanding that corpus are word- and sentence-based. However, corpus development using the sentence or phrase selection technique presents some challenges, which include how to harvest the selected phrase in the target language, how to preserve context integrity and how to classify a phonetically rich and balanced phrase [15].

This research seeks to address the challenges facing the SY corpus development. The focus is to develop and implement a Supervised Prompt Selection Strategy for the generation of a phonetically balanced SY speech corpus. Furthermore is the integration of a tone explicit model for addressing the tone characteristics for SY and test a speech recognizer using the SY corpus.

The following sections of this paper includes: Section 2, which presents the proposed methodology and implementation procedures while section 3 presents the evaluation SY speaker independent ASR based on the RBCOM corpus. Finally, Section 4 gives a brief conclusion and likely extensions required to enhance the performance of the Rule-Based Corpus Optimization Model (RBCOM).

2 Methodology

2.1 SY Corpus Development

The methodology of the SY corpus development was carried out based on the overall system design as illustrated in Figure 1. Once the text data is captured, a formatting process is carried out after which an optional set of text are generated followed by selection and evaluation of n-gram after which a prompt selection process is effected.

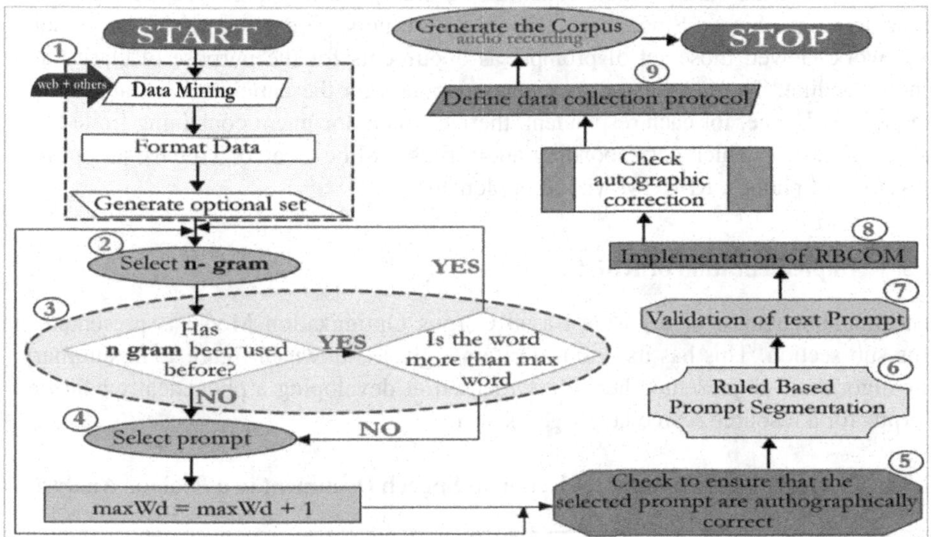

Fig. 1. Corpus Development Approach

2.2 Rule-Based Prompt Segmentation

The rules utilized herein, are adapted from the existing SY literature. Rule-based prompt segmentation was integrated into the general approach for pragmatic evaluation of phone interaction with a view to having optimal coverage. The RBCOM is based on the three possible syllable structures of SY; these include consonant-vowel CV, vowel V and syllabic nasal N. The rules for Yorùbá Phoneme Interaction are as follows:

Phonemes (Consonants)

1. The following phonemes co-occur with any other phonemes at the word initial position, middle, or final word position to have a CV syllable b, t, d, j, k, g, p, gb, f, s, ṣ and m.
2. 'h', 'r', 'l', 'y' and 'w' co-occur with oral and nasalized vowels (phonemes). Any of the consonants come before any of the vowels.
3. The consonant 'n' co-occurs with oral vowels (phonemes). It comes before any of the oral vowels in order to form a CV or CVV syllable structure.

Phonemes (Oral Vowels)

1. All the oral vowels and three nasalized vowels '-in-', '-ẹn-' and '-un-' co-occur with any other phonemes at the word initial, middle and final positions to form VV, CV or VC syllable structure.
2. The nasalized vowels '-an-' and '-ẹn-' co-occur with some consonants (phonemes) to form a CV syllabic structure. The vowel follows the consonants.

These rule sets were used for the segmentation of SY corpus into basic syllabic units. In this work, each valid syllable constitutes a class, and members of each class were considered as objects. Since a specific set of prompts were given to the respondent, this work viewed those set of prompts as documents for the purpose of clarity and understanding. The number of documents depends on the anticipated number of respondents. Hence, for each respondent, there exists a document containing finite lines of sentences for which corresponding audio files will be created. To have an optimal coverage of phones, RBCOM will be implemented.

2.3 Implementation of RBCOM

The implementation of the Rule Based Corpus Optimization Model is presented in this sub-section. This has its link from item eight as shown in figure 1. In summary, the algorithms as presented herein are focused at developing a phonetically balanced corpus for a resource scarce language like SY.

2.3.1 Algorithm 1: Random Selection of Speech Document in a Pool for Analysis

Algorithm 1
For $k = k\ to\ K;\ \ k = 1, \ldots, K$ /* $k =$ document
For $i = 1\ to\ I;\ \ i = 1, \ldots, \boldsymbol{N_T}$ /* $\boldsymbol{N_T} = Total\ No\ of\ sentence$ in R
/* $\boldsymbol{R}$ = *Respository*
Return Rnd /*return a random number between 0 and 1
$f_r = [\mathrm{Rnd} * i]$
Select $S_{ik} = \lceil f_r \rceil$
Let sentence r be sentence $i\ in\ document\ k$
where $[\mathrm{r} \in f_r]$
Next i
Next k

Section 2.3.1 presents the psuedocode of Algorithm 1 whose task is to select sentences from the repository pool for document composition. To determine sentence $\boldsymbol{i}$ in a document $\boldsymbol{k}$, a random number $\boldsymbol{Rnd}$ is generated $\ni \mathbf{0} \leq \boldsymbol{Rnd} \leq \mathbf{1}$ and multiplied by $\boldsymbol{N_T}$. Here, $\boldsymbol{S_{ik}} = \lceil \boldsymbol{f_r} \rceil$ i.e. sentence $\boldsymbol{i}$ in a document $\boldsymbol{k}$. This selection process continues until the finite number of lines of sentences are achieved for the respondents $\ni \mathbf{1} \leq \boldsymbol{k} \leq \boldsymbol{K}$. Sentences $\boldsymbol{i}$ as contained in $\boldsymbol{k}$ are then analyzed based on the algorithmic design described in Algorithm 2.

2.3.2 Algorithm 2: Population of Investigatory Array

```
Algorithm 2
Read {D_k}      k = 1, ..., K            /* create an array of for documents
 Read {S_nk}    n = 1, ..., N            /* S_nk = sentences n in documents k
Read {L_j}  j = 1,2, ......, J           /* syllables

 Do while k ≤ K                          /* read each document
   k = k + 1
   For n = 1 to N                        /* read each sentence in a document
   n = n + 1
  Read S_nk                              /* read sentence n in document k
   Flag = false
   Do while Flag = false for j = 1 to J  /* for syllable types
     For each L ϵ S_nk                   /* Read syllable in sentence
     if L ϵ C_j                          /* is syllable in class j of syllables?
     C_jk = C_jk + 1                     /* count j in S_nk and store
  else
       C_jk = C_jk                       /* syllable occurrence frequency
      Next j
        Test_ End_ of Sentence S_nk
     If End_of Sentence S_nk
       Flag = True
  Loop
    Next n
      Loop
```

In Algorithm 2, the investigatory array is populated based on SY syllable space as obtained in **Section 2.2**. For each document $\boldsymbol{k}$, select and read sentence $\boldsymbol{n}$ where $\boldsymbol{n} = 1,2,\ldots\ldots \boldsymbol{N}$. Furthermore, within each sentence $\boldsymbol{n}$, read syllable, classify and count number of syllable classes $\boldsymbol{j}$ where $\boldsymbol{j} = \mathbf{1}, \mathbf{2}, \ldots\ldots \boldsymbol{J}$ as $\boldsymbol{C_{jk}}$ until the search gets to the end of the sentence. Thus, $\{\boldsymbol{C_{jk}}\}$ for syllable class $\boldsymbol{j} = \mathbf{1}, \mathbf{2}, \ldots\ldots \boldsymbol{J}$ and document $\boldsymbol{k} = \mathbf{1}, \mathbf{2}, \ldots\ldots \boldsymbol{K}$ constitute an array $\boldsymbol{J} \times \boldsymbol{K}$, with non-negative entries (i.e. including zero entries where a syllable class is not found within a document) $\boldsymbol{C_{jk}}$ representing the number of times each syllable $\boldsymbol{j}$ is found in document $\boldsymbol{k}$. Syllable $\boldsymbol{j}$ with $\boldsymbol{C_{jk}} = \mathbf{0}$ are identified and strategies for the replacement of $\boldsymbol{S_{ik}}$ are described in Algorithm 3.

2.3.3 Algorithm 3: Identification of Array Entries and Replacement Strategies

Algorithm 3

Step I: Identification of array entries with zero and replacement

```
Read {C_jk}                                /* Read investigation array
Do while zero_content> 0
For j = 1 to J
For k = 1 to K
if C_jk > 0 ;
goto TRAY
elseif C_jk = 0                            /* if array entry is zero
```

Step II: Search the repository for viable replacement

```
Read Repository R
Do while m ≤ N_T            /*N_T is the total number of sentences in repository
    m = m + 1
RFlag = false
Do while RFlag = False
Read  r_m ∈ R               /* read sentence m in R
     For j = 1 to J
if   L_y = L_j              /* Syllable y is same as syllable in class j
         P1 = P1 + 1        /* No of syllable of type j seen in a sentence m ∈ M
else
Next j
End of Sentence = True
RFlag = True                Rflag is Repository flag
Loop
if P1 > P2
C_S = Label of Candidate_ Sentence = m          /* Replacement candidate
P2 = P1
P1 = 0
Select fc_s                 /* Sentence for replacement
Loop
Read {D_jk} ≡ {C_jk}
```

Step III: Ranking Syllables in Investigation Array

```
Q = 0                          /* copy array {C_ik}
for v = 1, ... ..., V
      w = 1, ... ..., W
if d_wk > d_vk
d_vk = Q    d_vk = d_wk ; d_wk = Q
h_v = w                        /* keep track of original label of syllable now ranked
g_w = v
Next w
Next v
```

Step IV: Determination of sentence to replace in Investigation pool

```
For i = 1 to 20
/* top twenty syllable will be considered as candidate for removal
q = q + 1
Do while n ≤ N
n = n + 1
For z = z to Z
g = h_z                        /* identify the original label
Flag = false
Do while Flag = False
Read S_nk                      /* read sentence
if L_q ∈ S_nk                  /* if syllable q is in sentence n
X_qn = X_qn + 1                /* increment number of syllable q in sentence
Next i
if End of statement n then
Flag = True
Loop
```

Algorithm 3 was executed in four steps. The first step is to search each document $\boldsymbol{k}$ for syllable $\boldsymbol{j}$ and the number of occurrence of syllable $\boldsymbol{j}$ is counted. Where zero entry is found in array $\{\boldsymbol{C_{jk}}\}$, i.e. $\boldsymbol{C_{jk}} = 0$, for each sentence $\boldsymbol{m}$, where $\boldsymbol{m} = \mathbf{1}$, **2,…..,** $\boldsymbol{N_T}$ in the repository is read as described in step two. In step three, the occurrence of syllable $\boldsymbol{j}$ in the repository is also counted and the sentence index is noted. The sentence with highest occurrence of the syllable $\boldsymbol{j}$ with $\boldsymbol{C_{jk}} = \mathbf{0}$ and syllable $\boldsymbol{i}$ with least occurrence in document $\boldsymbol{k}$ is selected. If in a document the array content $\boldsymbol{C_{jk}}$ is zero, the repository is searched for viable replacement. The highest and least occurrence of syllable $\boldsymbol{j}$ are noted. Finally, step four is the determination sentence $\boldsymbol{S_{nk}}$ to replace. All $\boldsymbol{S_{nk}}$ were read and the count of highest occurrence of syllable $\boldsymbol{j}$ are noted. The sentence $\boldsymbol{i}$ with $\boldsymbol{j}$ space syllables having the highest number of counts is selected for replacement and the array ratings obtained from Algorithm 3. This was in-turn evaluated using Algorithm 4 as presented in Section 2.3.4.

2.3.4 Algorithm 4: Evaluation Model of Array Ratings

The evaluation procedure of both the random and RBCOM generated corpus is presented in Section 2.3.4.

Algorithm 4

```
Read {C_jk}
for j = 1,2, ……..J                /*Column
for k = 1,2, ……..K                /* Row
CTs = CTs + C_jk                  /*Aggregate of k over all j
CTj = CTs/K                       /* Average count of syllables for each document k
Next k
    CTs = 0
Next j
for k = 1,2, …….K
for j = 1,2, ……..J
RT_k = RTs + C_jk                 /*Aggregate of j over all k
RTs = RTs/J                       /* Average count of syllables for each class j
Next j
  RTs = 0
Next k
for k = 1, …….K
for j = 1, ……..J
Obj_Sum = Obj_Sum + [|RT_j − C_jk| + |CT_k − C_jk|]
/* Model objective function
   Next j
Next k
                P4 = M            /* M is a large number
for j = 1 to W
if
    P3 < P4
  P4 = P3
Replace array {C_jk}
   else
Next w
```

Algorithm 4 depicts the basic steps involved in the Random and RBCOM corpus evaluation. To ensure optimal syllable distribution within and across all documents for **k = 1, 2, … K**, the objective of the evaluation model as encapsulated in algorithm 4 is to minimize the difference in syllable counts within and across documents without interfering with the language syntactic rules. In this algorithm the function $\boldsymbol{f(j,k)}$ is expressed as

$$f(j,k) = \sum_{j=1}^{J}\sum_{k=1}^{K}[|RT_j - C_{jk}| + |CT_k - C_{jk}|]$$

$\boldsymbol{RT_j}$ = Aggregate of j over all document $\boldsymbol{k}$
$\boldsymbol{C_{jk}}$ = Number of counts of particular syllable j in document $\boldsymbol{k}$
$\boldsymbol{CT_k}$ = Average count of syllable for each document $\boldsymbol{k}$

The processes from Algorithms 1 to 4 are iteratively repeated and values of $\boldsymbol{f(j,k)}$ compared until convergence got to the minimum. The RBCOM text prompt obtained on the implementation of Algorithms 1 to 4 were validated, based on schemes described in Section 2.4.

2.4 Validation of Text Prompt

The proposed text prompt validation model from the SY corpus development is as presented in this sub-section: Firstly, the modeling procedure for the Explorative Data Analysis (EDA) is premised on frequency distribution models, followed by the determination of the goodness of fit of the frequency distribution anchored on: (1) Kolmogorov Smirnov, (2) Andersen Darlin and (3) Chi-Squared test criteria. The evaluation procedure for the skewness of samples is based on the moment coefficient of skewness as shown in (1), (2), (3) and (4) below to evaluate m_3 and m_2 respectively.

$$\text{skewness: } g_1 = \frac{m_3}{m_2^{3/2}} \tag{1.0}$$

$$m_3 = \sum(x - \bar{x})^3 f/n \tag{2.0}$$

$$m_2 = \sum(x - \bar{x})^2 f/n \tag{3.0}$$

$$G_1 = \frac{\sqrt{n(n-1)}}{n-2} g_1 \tag{4.0}$$

Where,
$\bar{x}$ = mean,
f = frequency
n = sample size
m_3= third moment of the data set
m_2= variance
g_1= skewness
G_1 = sample skewness

3 Results and Discussion

The summary of SY corpus development is presented in this section. It covers the following: SY data and experiment, evaluation of random and RBCOM generated corpus.

3.1 Data and Experiment

The texts of this corpus were selected from various data sources which includes: newspapers, magazines, journals, books, letters, handwritten texts, movie scripts and extracts from the television. This corpus is a complete set of SY contemporary texts. The texts are about different subjects, including politics, arts, culture, economics, sports, stories, etc. The SY harvested data contains a total of 206,444 words with 5,689 distinct words.

In order to achieve maximal lexeme diversity, an n-gram prompt selection was used to generate the prompt list. At the end of evaluation stage, a selection of sentences was done based on the established protocol. The process ensures that each respondent has at least 10-15 similar prompts and 185-190 randomly selected prompts. Sentences and word were verified based on the context of language and grammar. The quality of corpora greatly affects the performance of ASR system; therefore, before recording, a syllable segment of words in prompt list was analyzed. The syllable segmentation was carried out based on the rules defined in Section 2.2.

Graphs I-V as contained in Figure 2 below represents an illustration of EDA plot for syllables in a randomly selected text prompt. In testing for normality, the text prompt was seen to be negatively skewed, with skewness value less than -1 signifying a highly skewed syllable distribution.

Furthermore, a general trend of syllable occurrence and variants of distribution based on goodness of fit is presented in Table 1. The results of the expert model mining of allophone occurrence indicate the Generalized Pareto (GP), Dagnum, Johnson SB and Error are the best performing models for Kolmogorov Smirnov test. For Andersen Darling test, the best performing models are Dagnum, Gen. Pareto, Wakeby and Gumbel-Max. For, Chi-Squared test, Wakeby and Dagnum appear to be the best representation. This in addition, further reinforce the skewness of the allophones.

Having established the limitations of the randomly selected prompt, the initial results for the implementation of the RBCOM are as shown in Figures 3 and 4. A state of stability was attained for the objective function from the 50th iteration upwards. The new set of data generated from the point of stability was analyzed and validated for optimality as presented in Figure 5.

Graphs VI-IX as contained in Figure 5 represents an illustration of EDA plot for syllables obtained from the point of stability of the RBCOM generated prompt. The output as seen in the Figure 5 shows that the data normality and skewness is between -1 and +1 which represents an improvement in the distribution of allophones.

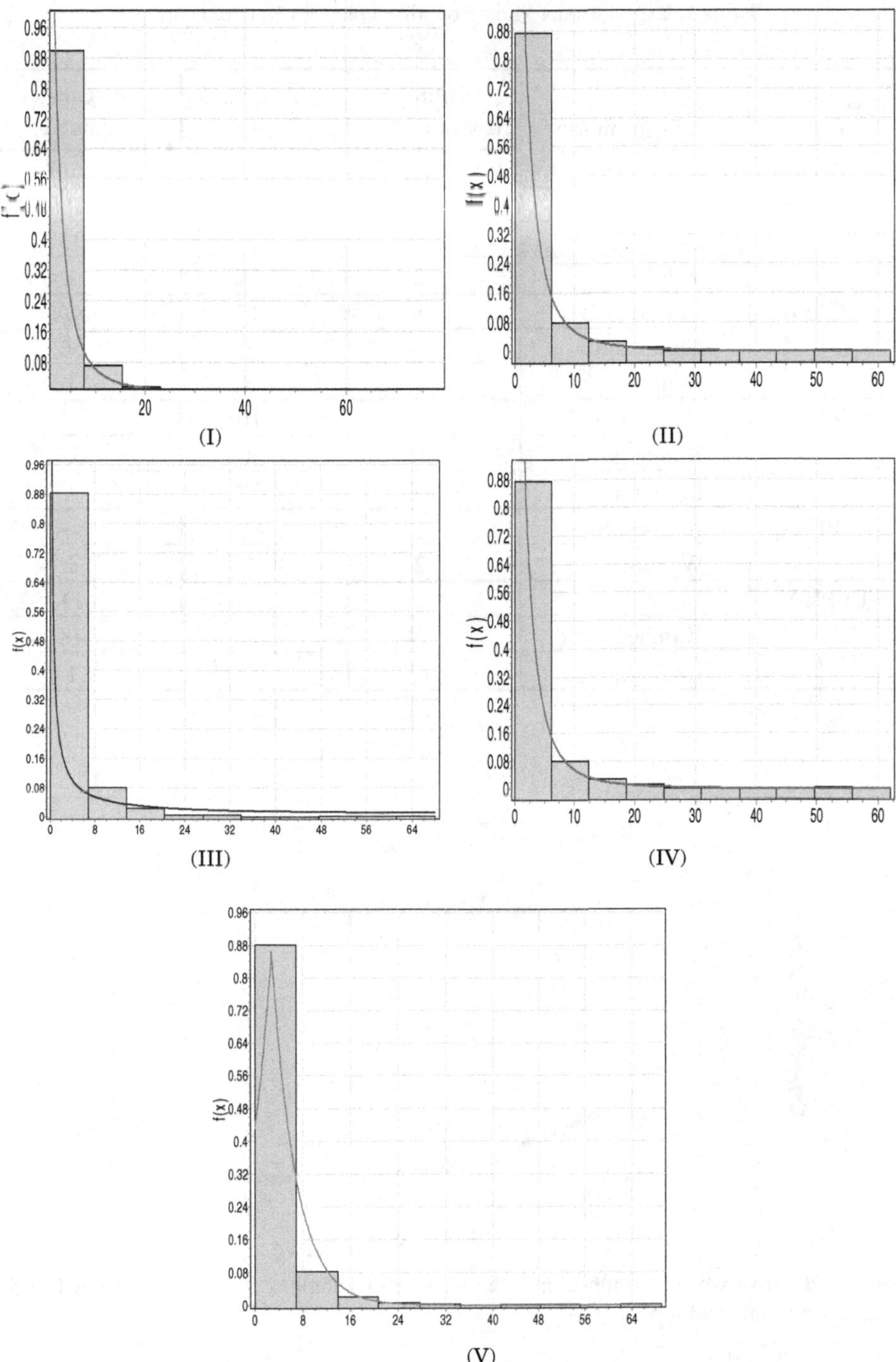

Fig. 2. SY syllable frequency trend for randomly selected prompts

Table 1. Expect Model Mining of Allophones for Random corpus

S/No	Distribution	Kolmogo-rov Smirnov	Andersen Darling	Chi-Squared
		Rank	Rank	Rank
TYPE I	Gen. Pareto	1	2	2
	Dagnum	28	1	33
	Wakeby	2	3	1
TYPE II	Johnson SB	1	48	NA
	Gen. Pareto	3	1	36
	Dagnum	2	8	1
TYPE III	Gen. Pareto	1	10	3
	Kumaraswamy	27	1	35
	Dagnum	13	22	1
TYPE IV	Johnson SB	1	49	NA
	Wakeby	2	1	1
TYPE V	Error	1	19	11
	Gumbel. Max	34	1	48
	Dagnum	13	17	1

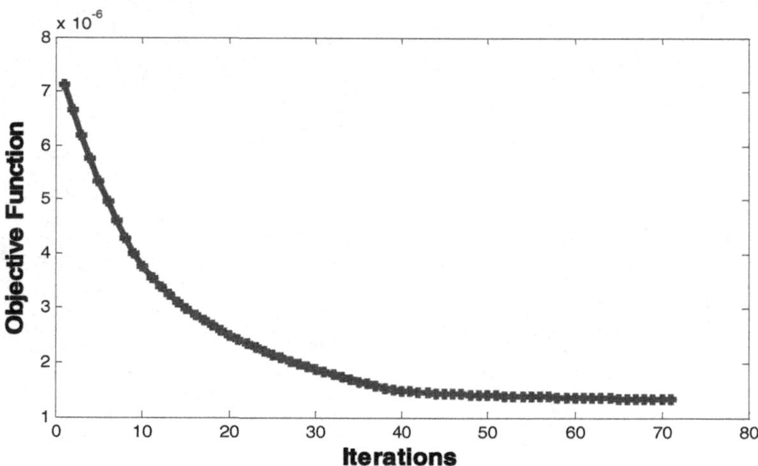

Fig. 3. Objective value as a function of the number of iterations for RBCOM applied to SY Corpus Development Problem (CDP)

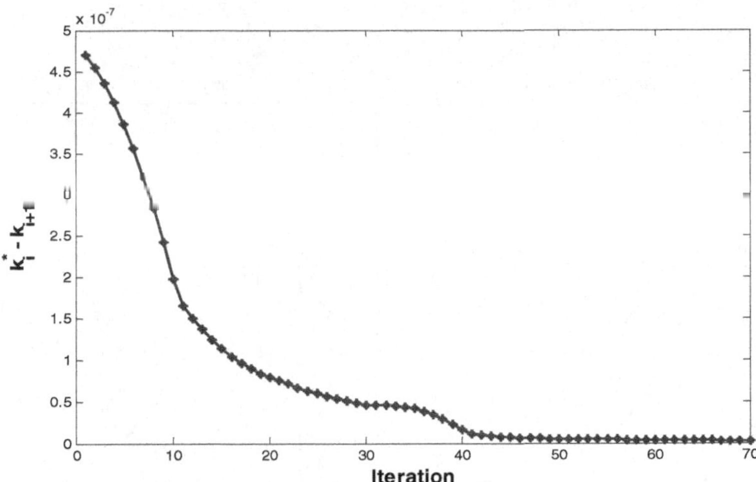

Fig. 4. Algorithm convergence as a function of the number of iterations for RBCOM applied to SY Corpus Development Problem (CDP)

A general trend of syllable occurrence and variants of distribution based on goodness of fit is presented in Table 2. The results of the expert model mining of allophone occurrence indicates that the Log Pearson 3, Gen. Extreme Value, Weibull, Burr, Lognormal, Pearson 5 (3P) and Log Logistic are the best performing models for Kolmogorov Smirnov, Andersen Darling and Chi-Squared tests.

3.2 Results of Random and RBCOM Generated Corpus

Figure 6 illustrates the syllable frequency for random and RBCOM corpus; the approximate syllable frequency range achieved for RBCOM prompt spans from 57 to 1200 while for the random scheme, the range spans from 0 to 6200. From the results, it is evident that random scheme cannot guarantee optimal allophone (syllable) coverage. Some syllables have a zero frequency of occurrence with very large frequency bandwidth. The RBCOM represents an improvement over the random scheme with low deviation of frequency distribution of syllable. This justifies the results presented in Figure 5 and Table 2.

A further analysis of Random and RBCOM corpus based on vague-linguistic terms are as presented in Graphs I-VI of Figure 7. The schemes were assessed based on three vague linguistic categories, namely low, middle and high range frequency distributions. Graphs I-VI of Figure 7 represents a performance profile of both Random and RBCOM corpus for low, middle and high range syllable frequency. For the low syllable frequency as depicted on Graphs I and II, the following outputs depicts the performance of both the Random and RBCOM corpus:16 syllables with zero frequency for random scheme, 35 syllables within 1-9 frequency range, and 109 syllables within 10-70 frequency range. The RBCOM schemes herein have their syllables frequency ranging from 155 to 162.

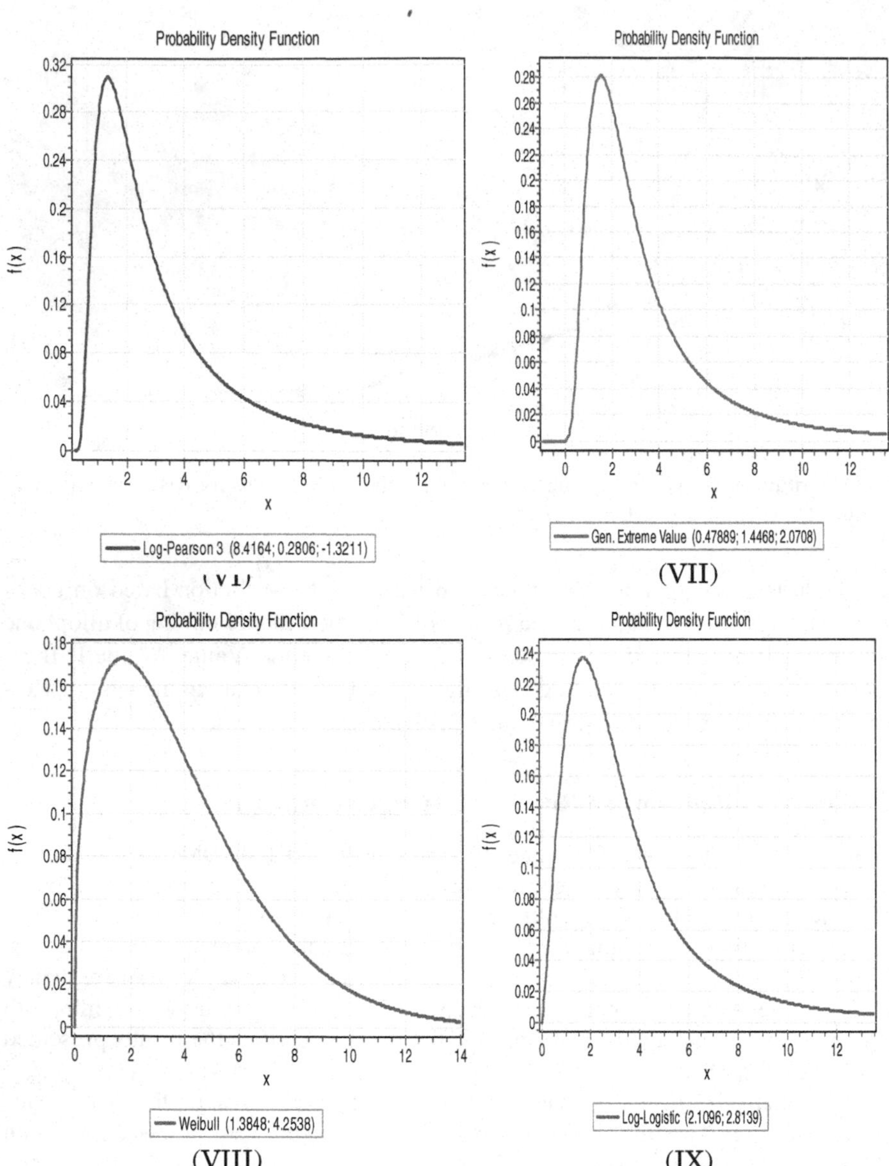

Fig. 5. SY syllable frequency trend for RBCOM prompt

Table 2. Expect Model Mining of Allophones for stable RBCOM corpus

S/No	Distribution	Kolmogorov Smirnov Rank	Andersen Dar-ling Rank	Chi-Squared Rank
TYPE I	Log Pearson 3	1	3	9
	Gen. Extreme Value	2	1	6
	Log. Logistics	6	8	1
TYPE II	Gen. Extreme Value	1	48	31
	Weibull	19	1	30
	Log Logistics	26	26	1
TYPE III	Weibull	1	10	3
	Burr	3	1	16
	Frechet	57	47	1
TYPE IV	Log Logistics	1	50	24
	Lognormal	8	1	7
	Pearson 5 (3P)	10	9	1

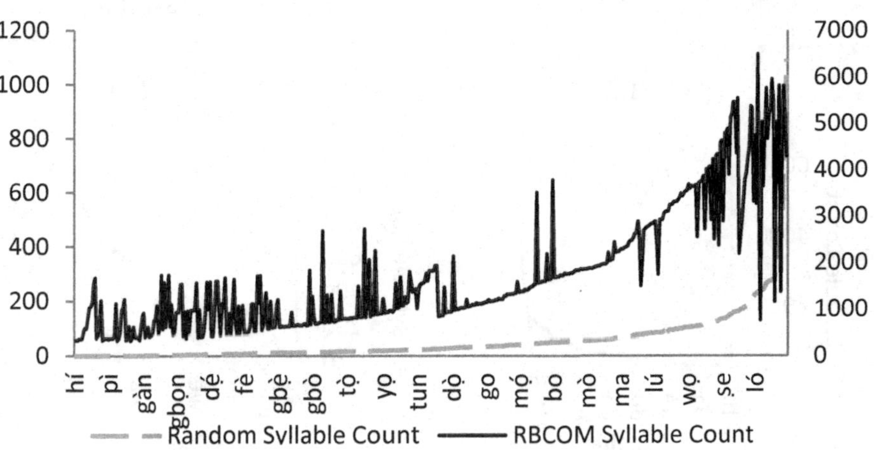

Fig. 6. Syllable Frequency Distribution

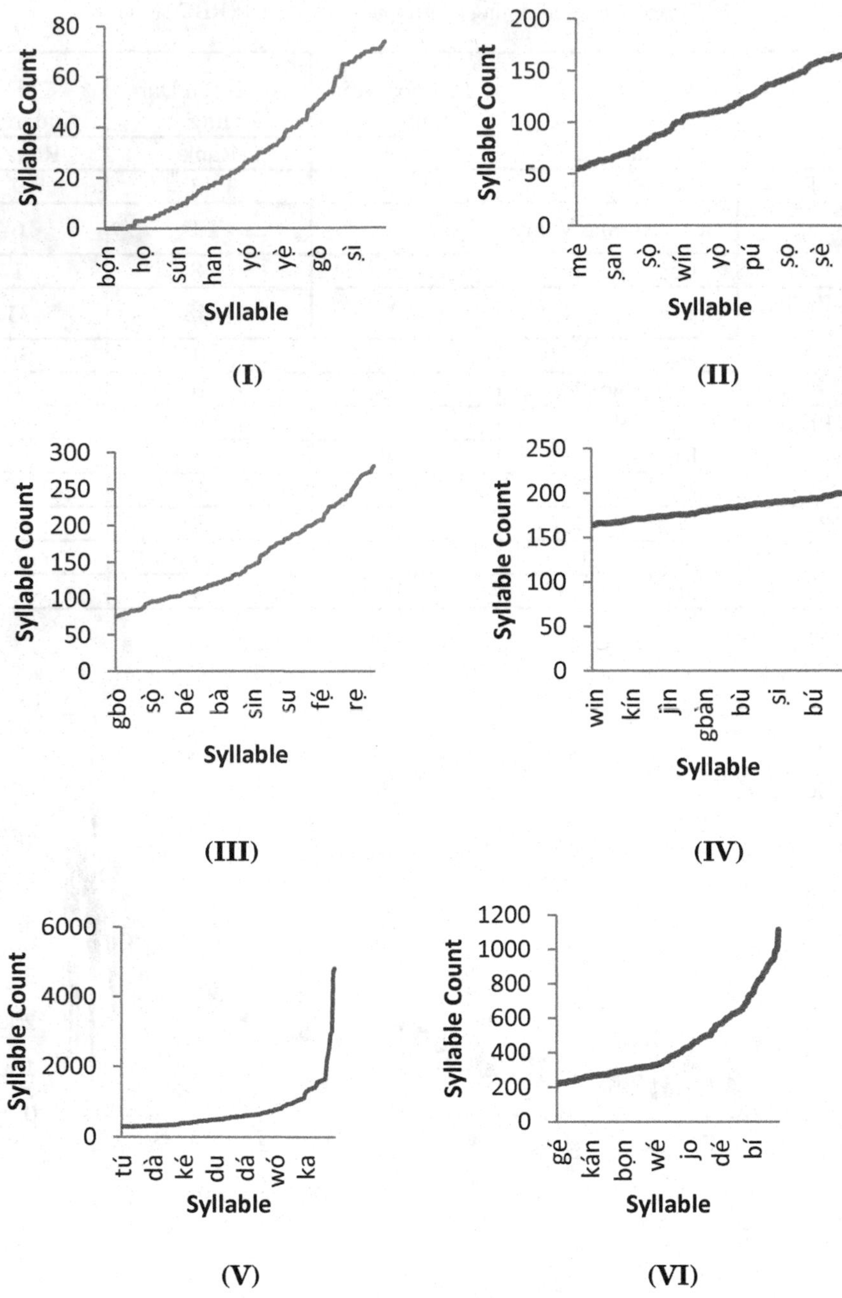

Fig. 7. Syllable Frequency against Syllable for Random and RBCOM Corpus Selection Strategy

Furthermore, in the case of the middle syllable frequency as obtained in Graphs III and IV of Figure 7, the performance of the Random and RBCOM corpus is premised on the following outcomes: availability of 120 syllables for the random scheme, frequency range of syllable is between 76 and 270 and RBCOM schemes have 61 syllables with syllable frequency range from 164-200. Also, Graphs V and VI of Figure 7 depicts the high syllable frequency which corresponds to the performance of both the Random and RBCOM corpus. The corresponding associated outputs includes: availability of 101 syllables for the random scheme, extension of the syllable frequency range from 282 to 6327 and availability of 241 syllables for the RBCOM schemes with syllable frequency range of 220-1113. These results have further stressed the initial assertion that corpus generated using the Random scheme are not phonetically balanced and also demonstrates the robustness of the RBCOM prompt selection strategy, in particular, for the moderate size corpus development of a resource-scarce language.

4 Conclusion

This paper has presented a new methodology for development of ASR corpus. The proposed technique which is premised on the use of Rule Based Corpus Optimization Model (RBCOM) has shown a higher prospect of results with respect to effectiveness. On comparison with the conventional randomized approach, the RBCOM demonstrated some level of superiority especially as displayed in the skewness of the distributed experimental data. While the random approach provided a skewness of less than -1, the RBCOM technique gave a skew distribution ranging from -1 to +1.

Furthermore, in classifying the allophone frequencies based on vague linguistic terms, the proposed RBCOM showed a better frequency range for each considered category of data unlike the random approach that generated a wider range for each class of allophones. For a future work, an allophone threshold for each genre of corpora is proposed.

References

1. Abushariah, M.A.A.M., Ainon, R.N., Zainuddin, R., Alqudah, A.A.M., Elshafei Ahmed, M., Khalifa, O.O.: Modern standard Arabic speech corpus for implementing and evaluating automatic continuous speech recognition systems. Journal of the Franklin Institute 349(7), 2215–2242 (2012)
2. Odéjobí, O.À.: A Quantitative Model of Yoruba Speech Intonation Using Stem-ML. INFOCOMP Journal of Computer Science 6(3), 47–55 (2007)
3. Adegbola, T., Owolabi, K., Odejobi, T.: Localising for Yorùbá: Experience, challenges and future direction. In: Proceedings of Conference on Human Language Technology for Development, pp. 7–10 (2011)
4. Àkànbí, L.A., Odéjobí, O.À.: Automatic recognition of oral vowels in tone language: Experiments with fuzzy logic and neural net-work models. Appl. Soft Comput. 11, 1467–1480 (2011)
5. Aibinu, A.M., Salami, M.J.E., Najeeb, A.R., Azeez, J.F., Rajin, S.M.A.K.: Evaluating the effect of voice activity detection in isolated Yorùbá word recognition system. In: 2011 4th International Conference on Mechatronics (ICOM), pp. 1–5. IEEE (May 2011)

6. Chomphan, S., Kobayashi, T.: Implementation and evaluation of an HMM-based Thai speech synthesis system. In: Proc. Interspeech, pp. 2849–2852 (August 2007)
7. Hoogeveen, D., Pauw, D.: CorpusCollie: a web corpus mining tool for resource-scarce languages (2011)
8. Cucu, H., Buzo, A., Burileanu, C.: ASR for low-resourced languages: Building a phonetically balanced Romanian speech corpus. In: 2012 Proceedings of the 20th European Signal Processing Conference (EUSIPCO), pp. 2060–2064. IEEE (August 2012)
9. Lecouteux, B., Linares, G.: Using prompts to produce quality corpus for training automatic speech recognition systems. In: MELECON 2008 - The 14th IEEE Mediterranean Electrotechnical Conference, pp. 841–846 (2008)
10. Nakamura, A., Matsunaga, S., Shimizu, T., Tonomura, M., Sagisaka, Y.: Japanese speech databases for robust speech recognition. In: Proceedings of the Fourth International Conference on Spoken Language, ICSLP 1996, vol. 4, pp. 2199–2202. IEEE (October 1996)
11. Metze, F., Barnard, E., Davel, M., Van Heerden, C., Anguera, X., Gravier, G., Rajput. N.: The Spoken Web Search Task. In: MediaEval 2012 Workshop, Pisa, Italy, October 4-5 (2012)
12. Lee, T., Lo, W.K., Ching, P.C., Meng, H.: Spoken language resources for Cantonese speech processing. Speech Communication 36(3), 327–342 (2002)
13. Abate, S.T., Menzel, W.: Automatic Speech Recognition for an Under-Resourced Language – Amharic. In: Proceedings of INTERSPEECH, pp. 1541–1544 (2007)
14. Raza, A.A., Hussain, S., Sarfraz, H., Ullah, I., Sarfraz, Z.: Design and development of phonetically rich Urdu speech corpus. In: 2009 Oriental COCOSDA International Conference on Speech Database and Assessments, pp. 38–43. IEEE (August 2009)
15. Wu, T., Yang, Y., Wu, Z., Li, D.: Masc: A speech corpus in mandarin for emotion analysis and affective speaker recognition. In: IEEE Odyssey 2006 on Speaker and Language Recognition Workshop, pp. 1–5. IEEE (June 2006)

Semantic Role Labeling of Speech Transcripts

Niraj Shrestha, Ivan Vulić, and Marie-Francine Moens

Department of Computer Science, KU Leuven, Leuven, Belgium
{niraj.shrestha,ivan.vulic,marie-francine.moens}@cs.kuleuven.be

Abstract. Speech data has been established as an extremely rich and important source of information. However, we still lack suitable methods for the semantic annotation of speech that has been transcribed by automated speech recognition (ASR) systems . For instance, the semantic role labeling (SRL) task for ASR data is still an unsolved problem, and the achieved results are significantly lower than with regular text data. SRL for ASR data is a difficult and complex task due to the absence of sentence boundaries, punctuation, grammar errors, words that are wrongly transcribed, and word deletions and insertions. In this paper we propose a novel approach to SRL for ASR data based on the following idea: (1) combine evidence from different segmentations of the ASR data, (2) jointly select a good segmentation, (3) label it with the semantics of PropBank roles. Experiments with the OntoNotes corpus show improvements compared to the state-of-the-art SRL systems on the ASR data. As an additional contribution, we semi-automatically align the predicates found in the ASR data with the predicates in the gold standard data of OntoNotes which is a quite difficult and challenging task, but the result can serve as gold standard alignments for future research.

Keywords: Semantic role labeling, speech data, ProBank, OntoNotes.

1 Introduction

Semantic role labeling (SRL) regards the process of predicting the predicate argument structure of a natural language utterance by detecting the predicate and by detecting and classifying the arguments of the predicate according to their underlying semantic role. SRL reveals more information about the content than a syntactic analysis in the field of natural language processing (NLP) in order to better understand "who" did "what" to "whom", and "how", "when" and "where". For example, in the following two sentences:

Mary opened the door.
The door was opened by Mary.

Syntactically, the subjects and objects are different. "Mary" and "the door" are subject and object in the first sentence respectively, while their syntactic role is swapped in the second sentence. Semantically, in both sentences "Mary" is ARG0 and "the door" is ARG1, since Mary opened the door.

SRL has many key applications in NLP, such as question answering, machine translation, and dialogue systems. Many effective SRL systems have been developed to work with written data. However, when applying popular SRL systems such as ASSERT [1],

A. Gelbukh (Ed.): CICLing 2015, Part II, LNCS 9042, pp. 583–595, 2015.
DOI: 10.1007/978-3-319-18117-2_43

Lund University SRL [2], SWIRL [3], and Illinois SRL [4] on transcribed speech, which was processed by an automatic speech recognizer (ASR), many errors are made due to the specific nature of the ASR-transcribed data.

When a state-of-the art SRL system is applied to ASR data, its performance changes drastically [5] due to many automatic transcription errors such as the lack of sentence boundaries and punctuation, spelling mistakes and insertions, or deletions of words and phrases. The lack of sentence boundaries is another major problem. If a sentence boundary detection system correctly identifies sentence boundaries in the ASR data then the SRL system might produce acceptable results, but unfortunately correct sentence boundary detection in ASR data remains a difficult and error-prone task. In this paper, we investigate whether a correct sentence boundary detector is actually needed for SRL, and whether the recognition of a predicate and its semantic role arguments within a certain window of words would not be sufficient to recover the semantic frames in speech data.

Therefore, we focus on frame segmentation rather than sentence segmentation. A segment is named a *frame segment* when the system finds a predicate and its semantic roles. Frame segments from the ASR data are generated as follows. Taking a fixed window size of words, we generate all possible segments by moving the window slider by one word. Considering this segment as a *pseudo-sentence*, we apply an SRL system, which generates many possible combinations of arguments for a predicate since the same predicate may appear in multiple segments. The system finally chooses the best arguments for the predicate. In summary, in this paper we propose a novel approach to SRL for ASR data based on the following idea:

1. Combine the evidence from different segmentations of the ASR data;
2. Jointly select a good frame segmentation;
3. Label it with the semantics of PropBank roles;

Experiments with the OntoNotes corpus [6] show improvements compared to the state-of-the-art SRL systems on the ASR data. We are able to improve 4.5% and 1.69% in recall and F_1 measure respectively in predicate and semantic role pair evaluation compared to a state-of-the-art semantic role labeling system on the same speech/ASR data set [5]. Our novel approach to SRL for the ASR data is very promising, as it opens plenty of possibilities towards improving the frame detection in speech data without sentence boundary detection. As an additional contribution, we semi-automatically align the predicates found in the ASR data with the predicates in the gold standard data of OntoNotes which is a quite difficult and challenging task, but the result can serve as gold standard alignments for future research.

The following sections first review prior work, then describe the methodology of our approach and the experimental setup, and finally present our evaluation procedure and discuss the results.

2 Prior Work

Semantic role labeling or the task of recognizing basic semantic roles of sentence constituents is a well-established task in natural language processing [7, 8], due to the existence of annotated corpora such as PropBank [9], NomBank [10], FrameNet [11] and shared tasks (CoNLL). Current semantic role labeling systems (e.g., SWIRL: [3],

ASSERT: [1], Illinois SRL: [4], Lund University SRL: [2]) perform well if the model is applied on texts from domains similar to domains of the documents on which the model was trained. Performance for English on a standard collection such as the CoNLL dataset reaches F_1 scores higher than 85% [12] for supervised systems that rely on automatic linguistic processing up to the syntactic level.

On the other hand, semantic role labeling of (transcribed) speech data is very limited, perhaps due to the non-availability of benchmarking corpora. Nevertheless, several authors have stressed the importance of semantic role labeling of speech data, for instance, in the frame of question answering speech interfaces (e.g., [13,14]), speech understanding by robots (e.g., [15]), and speech understanding in general [16]. Favre [5] developed a system for joint dependency parsing and SRL of transcribed speech data in order to be able to handle speech recognition output with word errors and sentence segmentation errors. He uses a classifier for segmenting the sentences trained on sentence-spit ASR data taking into account sentence parse information, lexical features and pause duration. This work is used as a baseline system for our experiments. The performance of semantic role labellers drops significantly (F_1 scores decrease to 50.76% when applying the ASSERT SRL system on ASR data) due to the issues with transcribed speech discussed in introduction. A similar decrease in performance is also noticed when performing SRL on non-well formed texts such as tweets [17].

We hope that this paper will stir up interest of the research community in semantic processing of speech.

3 Methodology

The main objective of this work is to identify suitable ASR segments that represent a predicate with its semantic roles, in a task that we call *frame segmentation*. Frame segments from the ASR data are generated by taking a window of a fixed size, and moving it word-by-word. This way, all possible combinations of segments in which a predicate might appear are generated. Considering each segment as a (pseudo-)sentence, the SRL system generates many possible combinations of arguments for a predicate. Our system then chooses the best arguments for the predicate based on an *evidence-combining approach*. Figure 1 shows a snippet of the raw ASR data, while figure 2 shows the results after moving the fixed-size window word-by-word (brute force segments). After applying the SRL system on these brute force segments, we obtain the labels as shown in figure 3. It is clearly visible that the same predicate occurs in different segments, and also different argument types occur in different segments with different text spans.

To evaluate our approach, we use the OntoNotes corpus [6] annotated with gold standard ProbBank semantic roles [9] and its transcribed speech data.[1] The speech corpus is plain text without any information about time and pause duration. Each token in the corpus is given in its own line with an empty line serving as the sentence boundary mark. We have decided to convert the data into the original raw format which corresponds to the actual ASR output (i.e., no sentence boundary, punctuation marks) by merging all tokens into a single line. The final input corpus then resembles the format of the snippet from figure 1.

[1] The transcribed corpus is provided by [5] with the consent of SRI (`http://www.sri.com`).

a much better looking newsnight i might add as powerless on sits in for anderson and they're and they're both off of that that is not a replacement colin powell thank you for your faith larry and thank you for your graciousness will give and we're going to get started here good evening everybody welcome to newsnight as larry just told you i'm paulus on filling in for the two men anderson cooper and aaron brown he lost his life long ago but there's still something modern science can give him back his identity the mystery of the frozen ever and continues next up the lab and anger in the hood over a sign of the times has

Fig. 1. An example of raw ASR-transcribed speech data

a much better looking newsnight i might add as powerless on sits in for anderson
much better looking newsnight i might add as powerless on sits in for anderson and
better looking newsnight i might add as powerless on sits in for anderson and they
... ...
... ...
and continues next up the lab and anger in the hood over a sign of
continues next up the lab and anger in the hood over a sign of the
next up the lab and anger in the hood over a sign of the times
up the lab and anger in the hood over a sign of the times has

Fig. 2. Brute force segments of window size 15 generated from the raw ASR data

3.1 SRL on Sentence-Segmented ASR Data (Baseline Model)

We compare against a competitive baseline and state-of-the-art model from [5]. We use the same corpus as in [5] which is derived from OntoNotes and which is ASR-transcribed. For our baselinewe use the sentence boundaries as defined in [5]. An SRL system is then applied on the sentences provided in this corpus.

3.2 Longest and Shortest Text Span Selection for Arguments

For a given predicate, there might exist many possible arguments with different argument text spans (see figure 3 again). The first task is to select the optimal text span for

51: good evening everybody welcome to newsnight as larry just told you i 'm paulus on [TARGET filling]
52: evening everybody welcome to newsnight as larry just told you i 'm paulus on [TARGET filling] in
53: everybody welcome to newsnight as larry just told you i 'm paulus on [TARGET filling] in for
54: welcome to newsnight as larry just told you i 'm paulus on [TARGET filling] in [ARG1 for the]
55: to newsnight as larry just told you i 'm paulus on [TARGET filling] in [ARG2 for the two]
56: newsnight as larry just told you i 'm paulus on [TARGET filling] in [ARG1 for the two men]
57: as larry just told you i 'm paulus on [TARGET filling] in [ARG1 for the two men] anderson
58: larry just told you i 'm paulus on [TARGET filling] in [ARG2 for the two men] [ARGM-TMP anderson cooper]
59: just told you i 'm paulus on [TARGET filling] in [ARG1 for the two men] anderson cooper and
60: told you i 'm paulus on [TARGET filling] in [ARG1 for the two men anderson cooper and aaron]
61: you i 'm paulus on [TARGET filling] in [ARG2 for the two men] [ARG1 anderson cooper and aaron brown]
62: i 'm paulus on [TARGET filling] in [ARG2 for the two men] [ARG1 anderson cooper and aaron brown he]
63: paulus on [TARGET filling] in [ARG1 for the two men] anderson cooper and aaron brown he lost
64: on [TARGET filling] in [ARG1 for the two men] anderson cooper and aaron brown he lost his

Fig. 3. Output of the SRL system on brute force segments

each argument. There might occur cases when the text spans of an argument may subsume each other, then either the longest or the shortest text span is chosen. For example, as shown in figure 3, argument type ARG1 exhibits different text spans, ranging from the shortest text span *for the* to the longest span *for the two men anderson cooper and aaron*. In addition, text spans of an argument might differ, and those text spans may not subsume each other. The text span is then selected based on the majority counts according to the occurrence of the two text spans in the corresponding segments, since a predicate cannot have two same argument types for the same dependent text span. Furthermore, text spans of different arguments may subsume each other. If that is the case, the longest or shortest text spans are selected.

Let us assume that the text spans for an argument are as follows:
$w_1 w_2 w_3 \ldots w_{i-1} w_i$
$w_1 w_2 w_3 \ldots w_{i-1} w_i w_{i+1} \ldots w_{j-1} w_j$
$w_1 w_2 w_3 \ldots w_{i-1} w_i w_{i+1} \ldots w_{j-1} w_j w_{j+1} \ldots w_{k-1} w_k$
In the *take-longest* span selection approach, text span $w_1 w_2 w_3 \ldots w_{i-1} w_i w_{i+1} \ldots w_{j-1}$ $w_j w_{j+1} \ldots w_{k-1} w_k$ is chosen. In the *take-shortest* approach text span $w_1 w_2 w_3 \ldots w_{i-1}$ w_i is chosen. There could be also the case where the argument type might have other text spans besides the above ones. Let us assume that there are additional two text spans:
$w_l w_{l+1} w_{l+2} \ldots w_m$
$w_l w_{l+1} w_{l+2} \ldots w_m w_{m+1} w_{m+2} \ldots w_{n-1} w_n$, where $l > k$ or $l < 1$.
Now, with the take-longest selection approach, we have two possible text spans: $w_1 w_2 w_3$ $\ldots w_{i-1} w_i w_{i+1} \ldots w_{j-1} w_j w_{j+1} \ldots w_{k-1} w_k$ and $w_l w_{l+1} w_{l+2} \ldots w_m w_{m+1} w_{m+2} \ldots$ w_n. Since the argument type can have only one text span, we then choose the first one since the text span $w_1 w_2 w_3 \ldots w_{i-1} w_i$ occurs more times (3 times) than $w_l w_{l+1} w_{l+2} \ldots$ w_m (2 times). The same heuristic is applied in the take-shortest selection approach. We label the take-longest selection approach as **win-n-L**, and the take-shortest approach as **win-n-S**, where the middle '**n**' represents the chosen window size.

3.3 Generating New Segments for a Predicate

Now, we explain a two-pass approach to generating new segments for a predicate. First, we use the output from the SRL system and the brute force approach discussed in 3.2 to detect the predicate. Following that, given this predicate, we identify new segments for the predicate and then again apply the SRL system. In this approach, the SRL system is applied on the brute force segments as discussed above. A predicate might appear in a sequence of segments. We select the first and the last segment of this sequence. These two segments are then merged using two different heuristics to generate two types of new segments. In the first approach, we simply merge the two segments by retaining overlapping tokens. We label this model as **newSeg-V1-win-n**.[2] In the second approach, the new segment starts from the first occurrence of a semantic role argument and ends at the last occurrence of the argument. This model is labeled as **newSeg-V2-win-n** Following that, we remove all the predicate and argument labels and re-run the SRL again on these two new segments.

[2] 'n' in each model is the chosen window size.

For example, given are the following two segments (the first and the last):
First segment: $w_1 w_2 [w_3] \dots [w_{i-1} w_i] w_{i+1} \dots w_{j-1} [w_j] w_{j+1} \dots [w_{k-1} w_k]$
Second segment: $[w_{k-1} w_k] w_{k+1} \dots w_{l-1} [w_l w_{l-1}] [w_{l+1}] \dots w_m$
where [] represents argument or predicate labels and tokens inside [] are argument or predicate values. When we generate a new segment with the first approach, we obtain the new segment as:
$w_1 w_2 [w_3] \dots [w_{i-1} w_i] w_{i+1} \dots w_{j-1} [w_j] w_{j+1} \dots [w_{k-1} w_k] w_{k+1} \dots w_{l-1} [w_l w_{l-1}]$
$[w_{l+1}] \dots w_m$
After removing the labels, the segment is:
$w_1 w_2 w_3 \dots w_{i-1} w_i w_{i+1} \dots w_{j-1} w_j w_{j+1} \dots w_{k-1} w_k w_{k+1} \dots w_{l-1} w_l w_{l-1} w_{l+1} \dots$
w_m
Using the second heuristic, we obtain the new segment as:
$[w_3] \dots [w_{i-1} w_i] w_{i+1} \dots w_{j-1} [w_j] w_{j+1} \dots [w_{k-1} w_k] w_{k+1} \dots w_{l-1} [w_l w_{l-1}] [w_{l+1}]$
After removing the labels, the new segment is:
$w_3 \dots w_{i-1} w_i w_{i+1} \dots w_{j-1} w_j w_{j+1} \dots w_{k-1} w_k w_{k+1} \dots w_{l-1} w_l w_{l-1} w_{l+1}$

4 Experimental Setup

We use the OntoNotes release 3 dataset which covers English broadcast and conversation news [6]. The data is annotated in constituent form.

We retain all the settings as described and used by [5]. We use the ASSERT SRL tool [1] for semantic role labeling which was one of the best SRL tools in the CoNLL-2005 SRL shared task[3] and also it outputs semantic roles in constituent form. We only use the sentence boundaries in the baseline state-of-the-art model. We investigate the performance of SRL on different windows sizes to predict the predicates with its semantic role arguments.

4.1 Evaluation

We evaluate different aspects related to the SRL process of identifying predicates and their arguments in the ASR data. Currently, we focus on the ability of the proposed system to predict predicates and arguments, and therefore evaluate the following aspects:

- Evaluation of predicates (**predEval**): We evaluate the ability of the system to predict solely the predicates. Here, we consider only a predicate which has arguments and we adhere to the same strategy in all subsequent evaluation settings. We evaluate the predicates considering different windows sizes. After the predicate identification step, we use these predicates in the subsequent evaluations.
- Evaluation of each argument type, that is, a semantic role (**argEval**): Here, we evaluate each argument type regardless of the found predicate. This evaluation is useful to identify different named entities like time, location, manner, etc.
- Evaluation of a predicate with each argument type (**predArgEval**): Here, we evaluate each argument type in relation to its found predicate. We are evaluating a pair comprising a predicate and its argument type.

[3] http://www.cs.upc.edu/~srlconll/st05/st05.html

– Evaluation of a predicate with all argument types (**predArgFullFrameEval**): In this evaluation, we evaluate the full frame for a predicate, that is, the predicate with all its argument types found in a segment. This evaluation is the most severe: if a predicate misses any of its arguments or classifies it wrongly, then the full frame is respectively not recognised or wrongly recognised.
– Evaluation of predicate-argument pairs, but now given a gold labeled correct predicate (**corrPredArgEval**): Here, we evaluate how well the system performs in identifying the correct arguments starting from a correct predicate.

4.2 Predicate Alignment between Gold Standard and Speech Data

Since the speech data does not contain any sentence boundaries and no alignment of its predicates with gold standard, it is necessary to align predicates between the gold standard and predicates identified by the system in speech data for the evaluation. If a predicate occurs once in both corpora, they are aligned using the one-to-one principle. But if a predicate appears more than once in either corpus then we have to align the predicates between two corpora. A predicate from speech data is aligned with a predicate in the gold standard, if three left and three right context words match. If the context words do not match, then we align the predicates manually. There are 41628 predicates in total contained in the gold standard, and the system has identified 25149 predicates out of which 13503 predicates have been aligned in the one-to-one fashion, 5447 predicates have been aligned relying on the left and right context matching, and we have manually aligned the remaining 6199 predicates. This constitutes our predicate alignment between the speech data and the gold standard.

4.3 Evaluation Metrics

Let FL be the final list retrieved by our system, and GL the complete ground truth list. To evaluate different evaluation aspects, we use standard precision (P), recall (R) and F_1 scores for evaluation.

$$P = \frac{|FL \cap GL|}{|FL|} \quad R = \frac{|FL \cap GL|}{|GL|} \quad F_1 = 2 \cdot \frac{P \cdot R}{P + R}$$

We also evaluate the overall system performance in terms of micro-averages and macro-averages for precision, recall and F_1. Suppose we have z arguments types. We then define the evaluation criteria as follows:

$$Micro\_avg(P) = \frac{\sum_{b=1}^{z} |FL \cap GL|}{\sum_{b=1}^{z} |FL|} \quad Micro\_avg(R) = \frac{\sum_{b=1}^{z} |FL \cap GL|}{\sum_{b=1}^{z} |GL|}$$

$$Micro\_avg(F_1) = \frac{2 \times Micro\_avg(P) \times Micro\_avg(R)}{Micro\_avg(P) + Micro\_avg(R)}$$

$$Macro\_avg(P) = \frac{1}{z} \sum_{b=1}^{z} P \quad Macro\_avg(R) = \frac{1}{z} \sum_{b=1}^{z} R$$

$$Macro\_avg(F_1) = \frac{2 \times Macro\_avg(P) \times Macro\_avg(R)}{Macro\_avg(P) + Macro\_avg(R)}$$

5 Results and Discussion

We have investigated the effects of the window size in identifying predicates and its semantic roles. The model **win-20-L** outperforms other variants of our system on a validation set, and also outperforms the baseline system in terms of F_1 measure by 0.94%, while recall is improved by 4.32%. In all models, including the baseline system, recall is lower than precision, and we have noticed that the SRL system is not able to identify the auxiliary verbs like *am, is, are, 're, 'm, has, have, etc.*, which occur many times in the test data. However, they are labeled as predicates with arguments in OntoNotes.

We use the predicates identified by **win-20-L** in other evaluation protocols as already hinted in 4.1. We again perform additional experiments with different windows sizes (5, 8, 10, 13, 15, 18, 20, 23, 25, 28, and 30). We show the results of all windows sizes in figures while the final best performing model **win-13-L** is shown in tables. We also generate new segments for every window size parameter setting, but report only the best results here, obtained by **newSeg-V1-win-5** and **newSeg-V2-win-5**.

Table 1 shows the comparison between the baseline system and **win-13-L** in argument based evaluations. Our system outperforms the baseline system when identifying almost all semantic roles. The results for the basic semantic role types likes: ARG0, ARG1, ARG3, ARGM-LOC, ARGM-MNR, ARGM-MOD seem quite satisfactory, with the F_1 score typically above 75%. On the other hand, the system does not perform well when identifying semantic role ARG2 compared to the other semantic roles. It was to be expected knowing that the identification of ARG2 is still a running problem in NLP SRL systems. From the table 1, it is also clear that our system is far better than the baseline system in predicting circumstantial argument roles like ARGM-LOC, ARGM-MNR, and ARGM-MOD which occur far from the predicate, and our system is able to correctly identify them because of the *take-longest* text span selection approach.

Figures 4(a), 4(b), and 4(c) show the argEval evaluations across all windows sizes with the longest and the shortest text span selection. In general, our models outperform the baseline system in terms of recall and F_1 measure. However, the models from **win-10-L** to **win-30-L** exhibit lower precision scores than the baseline system, which indicates that by increasing the window size, we add more noise in the evidence that is used by the system to detect correct arguments and text spans. The figures also reveal that, as we increase the windows size, the recall scores increase while precision scores decrease. We may also notice that the system is better in identifying correct semantic roles using the take-shortest approach to text span selection (when compared to take-longest) with larger window sizes, since very frequent semantic roles like ARG0 and ARG1 are typically composed of only a few words. However, this is not the case when the system evaluates the predicate-semantic role pair, as shown in figures 5(d), 5(e), and 5(f). In this evaluation, **Win-13-L** outperforms all other models as well as the baseline system. The results are further displayed in table 1. The model outperforms the baseline system when predicting the semantic roles ARG1, ARG2, ARG3, ARGM-LOC with their predicates but could not beat the baseline results when predicting ARG0, ARGM-MNR, ARGM-MOD. When we provide the correct predicate to our model, then our model outperforms the baseline system for semantic roles ARG0, ARGM-MNR as shown for the corrPredArgEval evaluation in table 1. This indicates that our SRL system

Table 1. A comparison of results obtained using the baseline model and **Win-13-L** in three different evaluation protocols (argEval, predArgEval and corrPredArgEval)

	argEval			predArgEval			corrPredArgEval		
	Precision	Recall	F_1	Precision	Recall	F_1	Precision	Recall	F_1
Baseline									
arg0	0.9889	0.7963	0.8822	0.7731	0.6225	0.6897	0.8644	0.6225	0.7238
arg1	0.9945	0.6805	0.8080	0.7479	0.5118	0.6077	0.8507	0.5118	0.6391
arg2	0.9929	0.3242	0.4888	0.6687	0.2184	0.3292	0.7386	0.2184	0.3371
arg3	0.9630	0.5219	0.6769	0.6498	0.3522	0.4568	0.7096	0.3522	0.4707
argm-loc	0.9506	0.4711	0.6300	0.5236	0.2595	0.3470	0.5819	0.2595	0.3589
argm-mnr	0.9457	0.4587	0.6178	0.5127	0.2487	0.3349	0.5746	0.2487	0.3471
argm-mod	0.9910	0.6640	0.7952	0.8533	0.5717	0.6847	0.9192	0.5717	0.7049
Macro average PRF	0.5430	0.2525	0.3447	0.3051	0.1543	0.2050	0.3383	0.1543	0.2120
Micro average PRF	0.9849	0.5803	0.7303	0.7204	0.4244	0.5342	0.8075	0.4244	0.5564
win-13-L									
arg0	0.9613	0.9023	0.9309	0.7067	0.6633	0.6843	0.8230	0.6659	0.7362
arg1	0.9932	0.7594	0.8607	0.7384	0.5645	0.6399	0.8647	0.5675	0.6853
arg2	0.9848	0.4226	0.5914	0.6001	0.2575	0.3604	0.6784	0.2582	0.3740
arg3	0.9084	0.6515	0.7588	0.5903	0.4234	0.4931	0.6472	0.4252	0.5132
argm-loc	0.8655	0.7674	0.8135	0.3789	0.3360	0.3562	0.4307	0.3405	0.3803
argm-mnr	0.9056	0.7078	0.7946	0.3738	0.2922	0.3280	0.4277	0.2939	0.3484
argm-mod	0.9865	0.6869	0.8098	0.8125	0.5657	0.6670	0.8849	0.5669	0.6910
Macro average PRF	0.5341	0.3365	0.4129	0.2949	0.1773	0.2215	0.3231	0.1780	0.2295
Micro average PRF	0.9581	0.7009	0.8096	0.6388	0.4673	0.5397	0.7363	0.4694	0.5733

Table 2. A comparison of results obtained using the baseline model and our models in predicate and all its semantic roles evaluation (predArgFullFrameEval)

	Precision	Recall	F_1
baseline	**0.2646**	**0.1865**	**0.2188**
win-5-L	**0.2452**	**0.1825**	**0.2093**
win-8-L	0.2344	0.1775	0.2020
win-10-L	0.2266	0.1724	0.1958
win-13-L	0.2176	0.1659	0.1883
win-15-L	0.2139	0.1636	0.1854
win-17-L	0.2096	0.1603	0.1817
win-20-L	0.2062	0.1593	0.1797
win-23-L	0.2045	0.1565	0.1773
win-25-L	0.2024	0.1547	0.1754
win-28-L	0.2023	0.1544	0.1751
win-30-L	0.2008	0.1532	0.1738
newSeg-V1-win-5	0.1874	0.1671	0.1767
newSeg-V2-win-5	0.1683	0.1755	0.1718

outputs different semantic roles according to the selected segment length, and selecting optimal segment length is essential for the overall performance of the system.

However, our models are unable to improve over the baseline system in the full frame evaluation in terms of precision, recall and F_1, although the results of **win-5-L** comes very close to the results of the baseline system, and is on a par with only a 0.4% lower recall score (not significant at $p < 0.005$ using a two-tailed t-test) and a 1.94% lower precision score (not significant at $p < 0.005$ using the same significance test). This evaluation is very strict since one missed or wrongly classified argument respectively results in a non-recognised or wrongly recognised frame. We have also investigated whether a small window size is better than larger window sizes in order to predict the full frame of a predicate. From figure 5(a), 5(b), and 5(c), it is visible that the take-longest approach to text span selection with smaller windows produces better results than the models relying on the take-shortest heuristic. We hope that our novel modeling principles may lead to further developments and new approaches to identifying predicates with their semantic roles without the need of sentence boundary detection, which is still a major problem for ASR-transcribed speech data [5].

6 Conclusion and Future Work

We have proposed a novel approach to identify PropBank-style semantic predicates and their semantic role arguments in speech data. The specific problem that we tackle in this paper concerns the absence of any sentence boundaries in transcribed speech data. We have shown that even with a very simple, but robust segmentation approach we attain results that are competitive or even better than state-of-the-art results on the OntoNotes speech data set. We have analysed different approaches to selecting correct predicates, arguments and their text spans.

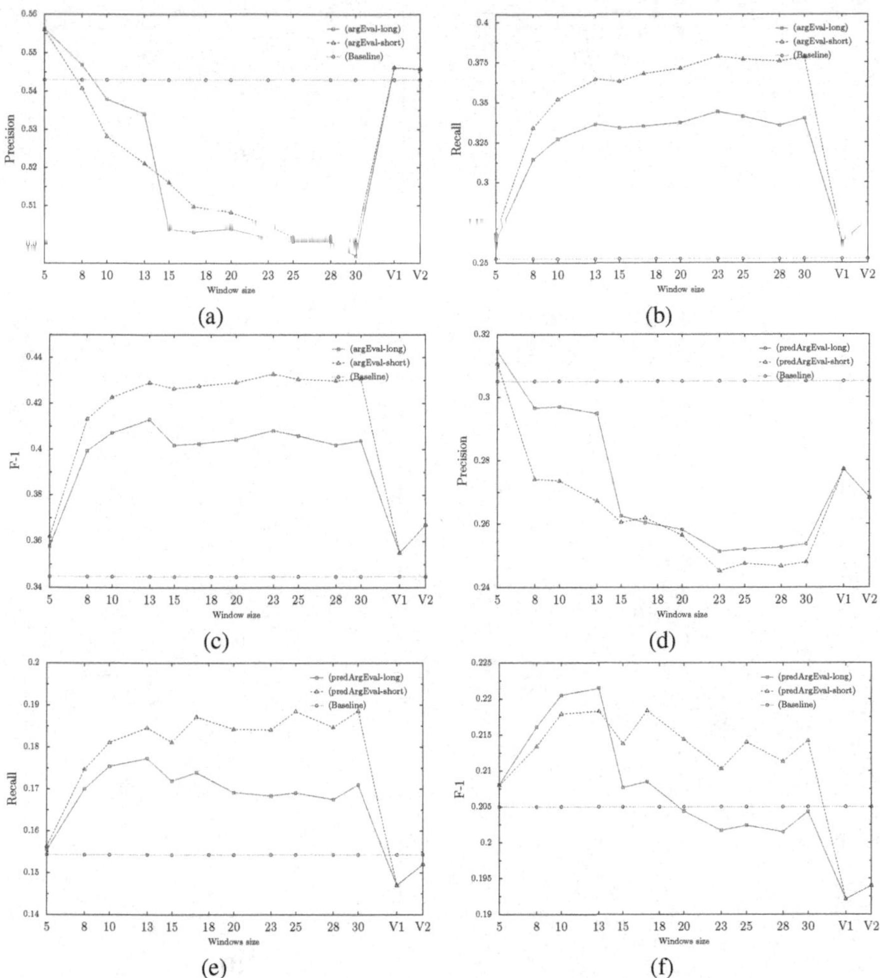

Fig. 4. Influence of windows size (when performing take-shortest and take-longest text span selection) on the overall results: (a) precision on argEval, (b) recall on argEval, (c) F_1 on argEval, (d) precision on predArgEval, (e) recall on predArgEval, and (f) F_1 on predArgEval. (In all figures, V1 and V2 are the best results obtained from **newSeg-V1-win-5** and **newSeg-V2-win-5** respectively.)

This work offers ample opportunities for further research. Currently, we do not employ any linguistic information in our models. The linguistic information will be exploited in future research in the form of language models and word embeddings trained on representative corpora, or in the form of shallow syntactic analyses of speech fragments (e.g., in the form of dependency information of phrases). We used an off-the-shelf SRL trained on written text data. We could investigate whether this SRL model could be transferred to speech data following the recent work from [18].

As another contribution we have built gold standard alignments between the predicates and arguments annotated in OntoNotes (which form the correct transcripts) and

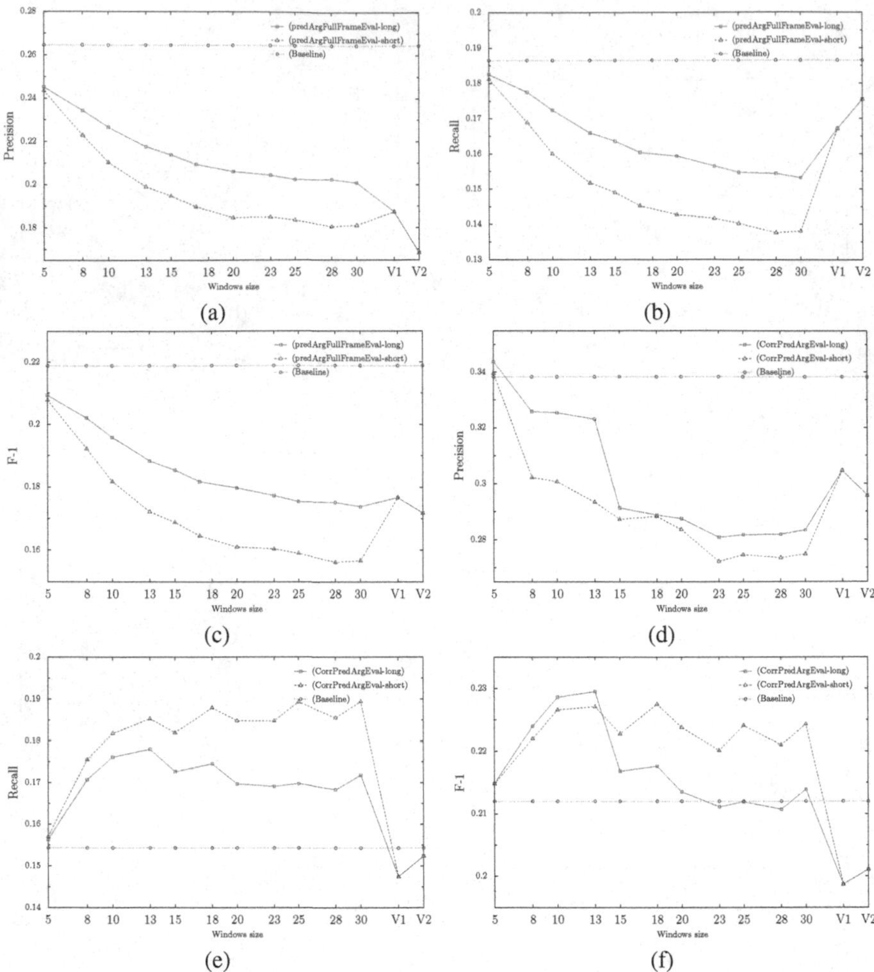

Fig. 5. Influence of windows size (when performing take-shortest and take-longest text span selection) on the overall results: (a) precision on predArgFullFrameEval, (b) recall on predArgFullFrameEval, (c) F_1 on predArgFullFrameEval, (d) precision on corrPredArgEval, (e) recall on corrPredArgEval, and (f) F_1 on corrPredArgEval. (In all figures, V1 and V2 are the best results obtained from **newSeg-V1-win-5** and **newSeg-V2-win-5** respectively.)

the original speech transcripts that were used in this and state-of-the-art research. This way we have produced an annotated corpus aligned to the original speech transcripts. We foresee that such a resource will be very valuable in future research on this topic.

References

1. Pradhan, S., Hacioglu, K., Ward, W., Martin, J.H., Jurafsky, D.: Semantic role chunking combining complementary syntactic views. In: Proceedings of the Ninth Conference on Computational Natural Language Learning, CoNLL 2005, Stroudsburg, PA, USA, pp. 217–220 (2005)

2. Johansson, R., Nugues, P.: Dependency-based semantic role labeling of PropBank. In: Proceedings of the Conference on Empirical Methods in Natural Language Processing, EMNLP 2008, pp. 69–78. Association for Computational Linguistics, Stroudsburg (2008)
3. Surdeanu, M., Turmo, J.: Semantic role labeling using complete syntactic analysis. In: Proceedings of the Ninth Conference on Computational Natural Language Learning, CoNLL 2005, pp. 221–224. Association for Computational Linguistics, Stroudsburg (2005)
4. Punyakanok, V., Roth, D., Yih, W.: The importance of syntactic parsing and inference in seamantic role labeling. Comput. Linguist. 34, 257–287 (2008)
5. Favre, B., Bohnet, B., Hakkani-Tür, D.: Evaluation of semantic role labeling and dependency parsing of automatic speech recognition output. In: Proceedings of the IEEE International Conference on Acoustics Speech and Signal Processing (ICASSP), pp. 5342–5345 (2010)
6. Hovy, E., Marcus, M., Palmer, M., Ramshaw, L., Weischedel, R.: OntoNotes: The 90% solution. In: Proceedings of the Human Language Technology Conference of the NAACL, Companion Volume: Short Papers. NAACL-Short 2006, pp. 57–60. Association for Computational Linguistics, Stroudsburg (2006)
7. Gildea, D., Jurafsky, D.: Automatic labeling of semantic roles. Comput. Linguist. 28, 245–288 (2002)
8. Màrquez, L., Carreras, X., Litkowski, K.C., Stevenson, S.: Semantic role labeling: An introduction to the special issue. Comput. Linguist. 34, 145–159 (2008)
9. Palmer, M., Gildea, D., Kingsbury, P.: The proposition bank: An annotated corpus of semantic roles. Comput. Linguist. 31(1), 71–106 (2005)
10. Meyers, A., Reeves, R., Macleod, C., Szekely, R., Zielinska, V., Young, B., Grishman, R.: The NomBank project: An interim report. In: Proceedings of the NAACL/HLT Workshop on Frontiers in Corpus Annotation (2004)
11. Baker, C.F., Fillmore, C.J., Lowe, J.B.: The Berkeley FrameNet project. In: Proceedings of the 36th Annual Meeting of the Association for Computational Linguistics and 17th International Conference on Computational Linguistics, ACL 1998, vol. 1, pp. 86–90. Association for Computational Linguistics, Stroudsburg (1998)
12. Zhao, H., Chen, W., Kit, C., Zhou, G.: Multilingual dependency learning: A huge feature engineering method to semantic dependency parsing. In: Proceedings of the Thirteenth Conference on Computational Natural Language Learning: Shared Task, CoNLL 2009, Boulder, Colorado, USA, June 4, pp. 55–60 (2009)
13. Stenchikova, S., Hakkani-Tür, D., Tür, G.: QASR: question answering using semantic roles for speech interface. In: Proceedings of INTERSPEECH (2006)
14. Kolomiyets, O., Moens, M.F.: A survey on question answering technology from an information retrieval perspective. Inf. Sci. 181, 5412–5434 (2011)
15. Hüwel, S., Wrede, B.: Situated speech understanding for robust multi-modal human-robot communication (2006)
16. Huang, X., Baker, J., Reddy, R.: A historical perspective of speech recognition. Commun. ACM 57, 94–103 (2014)
17. Mohammad, S., Zhu, X., Martin, J.: Semantic role labeling of emotions in Tweets. In: Proceedings of the 5th Workshop on Computational Approaches to Subjectivity, Sentiment and Social Media Analysis, pp. 32–41. Association for Computational Linguistics, Baltimore (2014)
18. Ngoc Thi Do, Q., Bethard, S., Moens, M.F.: Text mining for open domain semi-supervised semantic role labeling. In: Proceedings of the First International Workshop on Interactions Between Data Mining and Natural Language Processing, pp. 33–48 (2014)

Latent Topic Model Based Representations for a Robust Theme Identification of Highly Imperfect Automatic Transcriptions

Mohamed Morchid[1], Richard Dufour[1], Georges Linarès[1], and Youssef Hamadi[2]

[1] LIA - University of Avignon, France
firstname.lastname@univ-avignon.fr
[2] Microsoft Research, Cambridge, United Kingdom
youssefh@microsoft.com

Abstract. Speech analytics suffer from poor automatic transcription quality. To tackle this difficulty, a solution consists in mapping transcriptions into a space of hidden topics. This abstract representation allows to work around drawbacks of the ASR process. The well-known and commonly used one is the topic-based representation from a Latent Dirichlet Allocation (LDA). During the LDA learning process, distribution of words into each topic is estimated automatically. Nonetheless, in the context of a classification task, LDA model does not take into account the targeted classes. The supervised Latent Dirichlet Allocation (sLDA) model overcomes this weakness by considering the class, as a response, as well as the document content itself. In this paper, we propose to compare these two classical topic-based representations of a dialogue (LDA and sLDA), with a new one based not only on the dialogue content itself (words), but also on the theme related to the dialogue. This original Author-topic Latent Variables (ATLV) representation is based on the Author-topic (AT) model. The effectiveness of the proposed ATLV representation is evaluated on a classification task from automatic dialogue transcriptions of the Paris Transportation customer service call. Experiments confirmed that this ATLV approach outperforms by far the LDA and sLDA approaches, with a substantial gain of respectively 7.3 and 5.8 points in terms of correctly labeled conversations.

1 Introduction

Automatic Speech Recognition (ASR) systems frequently fail on noisy conditions and high Word Error Rates (WERs) make difficult the analysis of the automatic transcriptions. Solutions generally consist in improving the ASR robustness or/and the tolerance of speech analytic systems to ASR errors. In the context of telephone conversations, the automatic processing of these human-human interactions encounters many difficulties, especially due to the speech recognition step required to transcribe the speech content. Indeed, the speaker behavior may be unexpected and the train/test mismatch may be very large, while speech signal may be strongly impacted by various sources of variability: environment and channel noises, acquisition devices...

A. Gelbukh (Ed.): CICLing 2015, Part II, LNCS 9042, pp. 596–605, 2015.
DOI: 10.1007/978-3-319-18117-2_44

One purpose of the telephone conversation application is to identify the main theme related to the reason why the customers called. In this considered application, 8 classes corresponding to customer requests are considered (*lost and founds, traffic state, timelines...*). Additionally to the classical transcription problems in such adverse conditions, the theme identification system should deal with class (*i.e.* theme) proximity. For example, a *lost & found* request, considered as the main conversation theme, can also be related to itinerary (*where the object was lost?*) or timeline (*when?*). As a result, this particular multi-theme context makes identification of the main theme more difficult, ambiguities being introduced with the secondary themes.

An efficient way to tackle both ASR robustness and class ambiguity is to map dialogues into a topic space abstracting the ASR outputs. Dialogues classification will then be achieved in this topic space. Many unsupervised methods to estimate topic-spaces were proposed in the past. Latent Dirichlet Allocation (LDA) [1] was largely used in speech analytics applications [2]. During the LDA learning process, distribution of words into each topic is estimated automatically. Nonetheless, the class (or theme) associated to the dialogue is not directly taken into account in the topic model. Indeed, the classes are usually only used to train a classifier at the end of the process. As a result, such a system separately considers the document content (*i.e.* words), to learn a topic model, and the labels (*i.e.* classes) to train a classifier. Nonetheless, in the considered theme identification application, a relation between the document content (words) and the document label (class) exists.

This word/theme relation is crucial to efficiently label unseen dialogues. The supervised LDA [3] works around this drawback by considering the class belonging to a document, as a response during the learning process of the topic space. However, this representation could not substantially evaluate the relation between document content (words) and each theme. Indeed, these relations are evaluated through relations between topics and classes of the sLDA model. Moreover, these models (LDA and sLDA) need to infer an unseen document into each topic space to obtain a vectorial representation. The processing time during the inference phase as well as the difficult choice of an efficient number of iterations, do not allow us to evaluate effectively and quickly the best theme related to a given dialogue.

For these reasons, this paper is based on the work presented in [4] in which the authors proposed to use the Author-topic (AT) model [5] to represent a document instead of the classical LDA approach. The contribution of the paper is to go beyond this previous work by comparing the proposed AT model, called Author-topic Latent Variables (ATLV) representation, with the supervised LDA (sLDA) representation, which is an interesting alternative for classification tasks. For sake of comparison, results using the classical LDA topic-based representation [4] will also be reported. This robust ATLV representation takes into consideration all information contained into a document: the content itself (*i.e.* words), the label (*i.e.* class), and the relation between the distribution of words into the document and the label, considered as a latent relation. From this model, a vectorial representation in a continuous space is built for each dialogue. Then, a supervised classification approach, based on SVM [6], is applied. This method is evaluated in the application framework of the RATP call centre (Paris Public Transportation Authority), focusing on the theme identification task [7] and compared to LDA and sLDA approaches.

The rest of this paper is organized as follows. Topic model representations from document content are described in Section 2, by introducing LDA, sLDA and ATLV representations. Section 3 presents the experimental protocol and results obtained while finally, Section 4 concludes the work and gives some perspectives.

2 Topic-Model for Automatic Transcriptions

Dialogues, automatically transcribed using an Automatic Speech Recognition (ASR) system, contain many errors due to noisy recording conditions. An elegant way to tackle these errors is to map dialogues in a thematic space in order to abstract the document content. The most known and used one is the Latent Dirichlet Allocation (LDA) [1] model. The LDA approach represents documents as a mixture of latent topics. Nonetheless, this model does not code statistical relations between words contained into the document, and the label (*i.e.* class) that could be associated to it.

Authors in [3] proposed the supervised Latent Dirichlet Allocation (sLDA) model. This model introduces, in the LDA model, a response variable associated with each document contained into the training corpus. This variable is, in our considered context, the theme associated to a dialogue. In the sLDA model, the document and the theme are jointly modeled during the learning process, in order to find latent topics that will best predict the theme for an unlabeled dialogue of the validation data set. Although this model codes relation between the response variable and topics, this relation is not effective for theme identification task. Indeed, sLDA allows to relate a response (or theme) to a topic which is related itself to a set of words, and does not code strongly the relation between the document content (words occurrences) and the themes directly.

To go beyond LDA and, more importantly, sLDA [3] limits, an adapted Author-topic (AT) model is proposed here. The proposed Author-topic Latent Variables (ATLV) representation links both authors (here, the label) and documents content (words). The next sections describe LDA, sLDA, and ATLV based representations.

2.1 Latent Dirichlet Allocation (LDA)

In topic-based approches, such as Latent Dirichlet Allocation (LDA), documents are considered as a *bag-of-words* [8] where the word order is not taken into account. These methods demonstrated their performance on various tasks, such as sentence [9] or keyword [10] extraction. In opposition to a multinomial mixture model, LDA considers that a theme is associated to each occurrence of a word composing the document. Thereby, a document can change of topics from

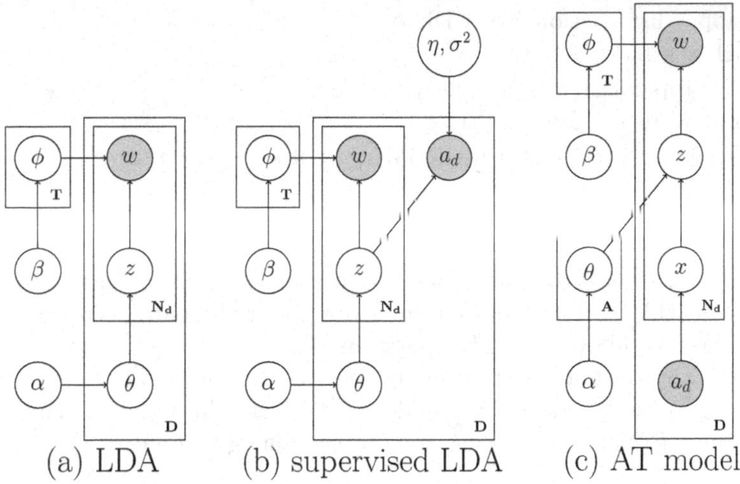

Fig. 1. Generative models for documents in plate notation for Latent Dirichlet Allocation (LDA) (a), supervised LDA (b) and Author-Topic (AT) (c) models

a word to another. However, the word occurrences are connected by a latent variable which controls the global respect of the distribution of the topics in the document. These latent topics are characterized by a distribution of word probabilities which are associated with them. LDA models have been shown to generally outperform other topic-based approches on information retrieval tasks [11].

The generative process corresponds to the hierarchical Bayesian model shown, using plate notation, in Figure 1 (a). Several techniques, such as Variational Methods [1], Expectation-propagation [12] or Gibbs Sampling [13], have been proposed to estimate the parameters describing a LDA hidden space. The Gibbs Sampling reported in [13], and detailed in [14], is used to estimate LDA parameters and to represent a new dialogue d with the rth topic space of size T. This model extracts a feature vector $V_d^{z^r}$ from the topic representation of d. The jth feature is:

$$V_d^{z_j^r} = \theta_{j,d}^r \,, \tag{1}$$

where $\theta_{j,d}^r = P(z_j^r|d)$ is the probability of topic z_j^r $(1 \leq j \leq T\ )$ generated by the unseen dialogue d in the rth topic space of size T.

2.2 Supervised LDA (sLDA)

Figure 1 (b) presents the sLDA model into its plate notation. Let $a \in R$ be the response (or theme in our theme identification context), and let fix the model parameters: T topics $\beta 1 : T$ (each β_k is a vector of term probabilities), the Dirichlet parameter α, and the response (theme) parameters η and σ^2. With the sLDA model, each document and response arises from the following generative process:

1. Draw topic distribution $\theta|\alpha \sim \text{Dir}(\alpha)$.
2. For each word
   (a) Draw topic assignment $z_n|\theta \sim \text{Mult}(\theta)$.
   (b) Draw word $w_n|z_n \sim \text{Mult}(\beta_{z_n})$.
3. Draw response variable (or theme) $a|z_{1:N_d}, \eta, \sigma^2 \sim \mathcal{N}\left(\eta^\top \overline{z}, \sigma^2\right)$.

with $\overline{z} = \frac{1}{N_d} \sum_{n=1}^{N_d} z_n$. The hyper-parameters of the sLDA model are estimated by performing a variational expectation-maximization (EM) procedure, also used in unsupervised LDA [1]. One can find out more about the parameters estimation or, more generally, about the sLDA itself, in [3].

The sLDA approach allows to directly estimate the probability for a theme a (or response) to be generated by a dialogue d. Then, the theme a which maximizes the prior $P(a|z_n, \eta, \sigma^2)$ is assigned to the dialogue d with:

$$C_{a,d} = \arg\max_{a \in A} \left\{P(a|d, z, \eta, \sigma^2)\right\} \tag{2}$$

Thus, each dialogue from the test or development set is labeled with the most likely theme given a sLDA model. This one does not require a classification method, which is not the case for LDA and ATLV representations.

2.3 Author-Topic Latent Variables (ATLV)

The Author-topic (AT) model, represented into its plate notation in Figure 1 (c), uses latent variables to model both the document content (words distribution) and the authors (authors distribution). For each word w contained into a document d, an author a is uniformly chosen at random. Then, a topic z is chosen from a distribution over topics specific to that author, and the word is generated from the chosen topic.

In our considered application, a document d is a conversation between an agent and a customer. The agent has to label this dialogue with one of the 8 defined themes, a theme being considered as an author. Thus, each dialogue d is composed with a set of words w and a theme a. In this model, x indicates the author (or theme) responsible for a given word, chosen from a_d. Each author is associated with a distribution over topics (θ), chosen from a symmetric Dirichlet prior ($\overrightarrow{\alpha}$) and a weighted mixture to select a topic z. A word is then generated according to the distribution ϕ corresponding to the topic z. This distribution ϕ is drawn from a Dirichlet ($\overrightarrow{\beta}$).

The parameters ϕ and θ are estimated from a straightforward algorithm based on Gibbs Sampling such as LDA hyper-parameters estimation method (see Section 2.1). One can find more about Gibbs Sampling and AT model in [5].

Each dialogue d is composed with a set of words w and a label (or theme) a considered as the author in the AT model. Thus, this model allows one to code statistical dependencies between dialogue content (words w) and label (theme a) through the distribution of the latent topics into the dialogue.

Gibbs Sampling allows us to estimate the AT model parameters, in order to represent an unseen dialogue d with the rth author topic space of size T. This method extracts a feature vector $V_d^{a_k^r} = P(a_k|d)$ from an unseen dialogue d with the rth author topic space Δ_r^n of size T. The kth $(1 \leq k \leq A)$ feature is:

$$V_d^{a_k^r} = \sum_{i=1}^{N_d} \sum_{j=1}^{T} \theta_{j,a_k}^r \phi_{j,i}^r \tag{3}$$

where A is the number of authors (or themes); $\theta_{j,a_k}^r = P(a_k|z_j^r)$ is the probability of author a_k to be generated by the topic z_j^r $(1 \leq j \leq T)$ in Δ_r^n. $\phi_{j,i}^r = P(w_i|z_j^r)$ is the probability of the word w_i (N_d is the vocabulary of d) to be generated by the topic z_j^r.

This representation, based on the AT latent variables, is called Author-topic Latent Variables (ATLV) representation in this work.

3 Experiments and Results

We propose to evaluate the effectiveness of the proposed approach in the application framework of the DECODA corpus [7,15,2].

3.1 Experimental protocol

The DECODA project [7] corpus, used to perform experiments on theme identification, is composed of human-human telephone conversations in the customer care service (CCS) of the RATP Paris transportation system. It is composed of 1,242 telephone conversations, corresponding to about 74 hours of signal, split into a train, development and test set, with respectively 740, 175 and 327 dialogues.

To extract textual content of dialogues, an Automatic Speech Recognition (ASR) system is needed. The LIA-Speeral ASR system [16] is used for the experiments. Acoustic model parameters were estimated from 150 hours of speech in telephone conditions. The vocabulary contains 5,782 words. A 3-gram language model (LM) was obtained by adapting a basic LM with the training set transcriptions. A "stop list" of 126 words[1] was used to remove unnecessary words (mainly function words) which results in a Word Error Rate (WER) of 33.8% on the training, 45.2% on the development, and 49.5% on the test. These high WER are mainly due to speech disfluencies and to adverse acoustic environments (for example, calls from noisy streets with mobile phones)

38 topic spaces are elaborated by varying the number of topics from 10 to 200 (step of 5 topics). The topic spaces are learned with a homemade implementation of LDA and AT models, while the version implemented by [17] is used for sLDA.

A classification approach based on Support Vector Machines (SVM) is performed using the *one-against-one* method with a linear kernel, to find out the main theme

[1] `http://code.google.com/p/stop-words/`

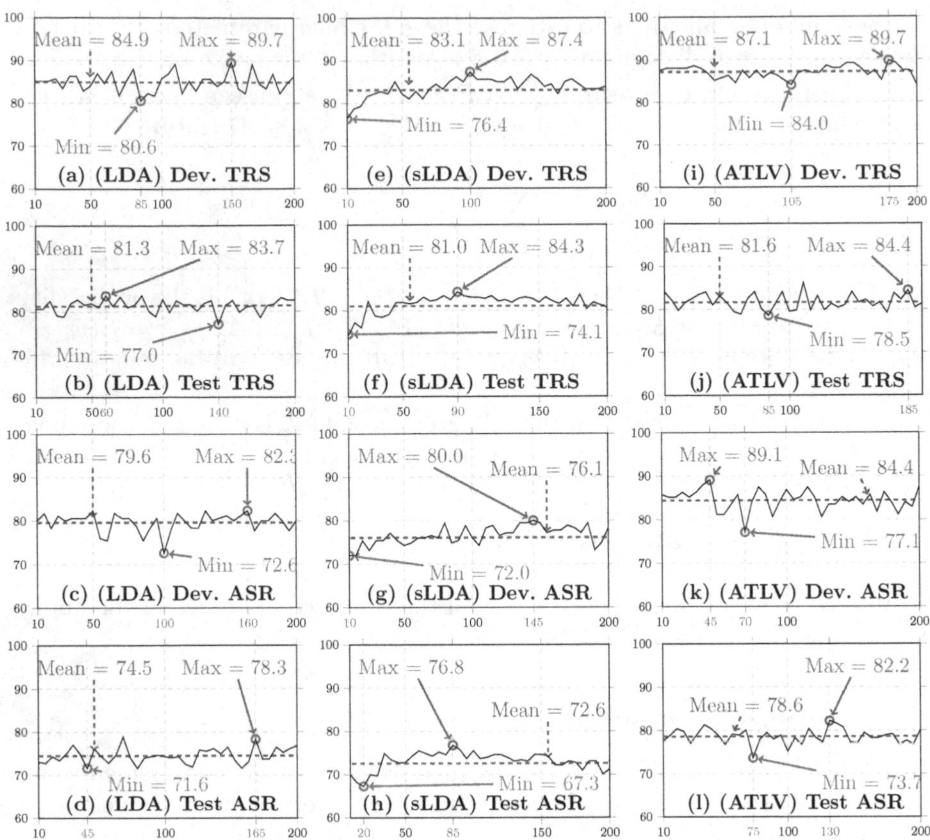

Fig. 2. Theme classification accuracies (%) using various LDA topic-based representations on the development and test sets with different experimental configurations. X-axis represents the number n of classes contained into the topic space ($10 \leq n \leq 200$).

of a given dialogue. This method gives a better accuracy than the *one-against-rest* [18]. SVM input is a vector representation of an unseen dialogue d [2].

For sake of comparison, experiments are conducted using the manual transcriptions only (TRS) and the automatic transcriptions only (ASR). The conditions indicated by the abbreviations between parentheses are considered for the development (Dev) and the test (Test) sets. Only homogenous conditions (TRS or ASR for both training and validations sets) are considered in this study. Authors in [2] notice that results collapse dramatically when heterogenous conditions are employed (TRS or TRS+ASR for training set and ASR for validation set).

[2] $V_d^{z_j^r}$ for a LDA representation and $V_d^{a_k^r}$ for an ATLV representation.

Table 1. Theme classification accuracies (%) for LDA, sLDA, and ATLV representations. **Best** corresponds to the best operating point obtained on the test data, while **Real** corresponds to the one estimated on the development set and applied to the test set.

Topic Model	**DATASET**		**Dev**		**Test**	
	Train	**Test**	**#topics**	**Best**	**Best**	**Real**
LDA	TRS	TRS	150	89.7	[illegible]	82.5
LDA	ASR	ASR	160	82.3	78.3	73.1
sLDA	TRS	TRS	100	87.4	84.3	83.1
sLDA	ASR	ASR	145	80.0	76.8	74.6
ATLV	TRS	TRS	175	89.7	84.4	**83.7**
ATLV	ASR	ASR	45	89.1	82.2	**80.4**

3.2 Results

The results obtained using manual (TRS) and automatic (ASR) transcriptions with respectively a topic-based representation from LDA, sLDA and ATLV, are presented in Figure 2. One can firstly point out that, for all dialogue representations (LDA, sLDA or ATLV), best results are obtained with manual transcriptions (TRS). Moreover, one can notice that the ATLV representation outperforms LDA, no matter the corpus (development or test) or conditions (TRS/ASR) studied.

In order to better compare performance obtained by all approaches (LDA / sLDA / ATLV), best results are reported in Table 1. Note that these results are given in **Best** and **Real** application condition, *i.e.* the **Real** configuration (number of topics contained into the topic space) being chosen with the **Best** operating point of the development set. As a result, a better operating point could exist in the test set, which could explain the performance difference between results reported in Table 1 and Figure 2.

With this real condition, we can note that the ATLV representation allows us to outperform both the LDA and sLDA approaches, with a respective gain of 1.2 and 0.6 points using the manual transcriptions (TRS), and of 7.3 and 5.8 points using the automatic transcriptions (ASR). This confirms the initial intuition that ATLV representation allows to better handle ASR errors than other topic-based ones. Improvements are mostly when ASR documents are used, and outcomes obtained for LDA and sLDA are quite close for both TRS and ASR configurations. One can point out that the results obtained for all topic-based representations in the TRS configuration are similar.

Another interesting point is the stability and robustness of the ATLV curve of the development set in TRS condition, comparatively to the LDA or sLDA representations. Indeed, the results are mainly close to the mean value (87.1%). The maximum achieved by both representations in TRS condition are the same. Thus, since dialogues are labeled (annotated) by an agent and a dialogue may contain more than one theme, this maximum represents the limit of a topic-based representation in a multi-theme context. Nonetheless, this remark is not applicable to the ASR condition.

4 Conclusion

Performance of ASR systems depends strongly to the recording environment. In this paper, an efficient way to deal with ASR errors by mapping a dialogue into a Author-topic Latent Variables (ATLV) representation space is presented. This high-level representation allows us to significantly improve the performance of the theme identification task. Experiments conducted on the DECODA corpus showed the effectiveness of the proposed ATLV representation in comparison to the use of the classic LDA representation or the more elaborated and adapted sLDA, with gains of at least 0.6 and 5.8 points (with the closest representation based on sLDA) respectively using manual and automatic transcriptions.

This representation suffers from the fact that theme distribution could not directly be estimated for an unseen document. Indeed, the proposed approach has to evaluate the probability $P(a_k|d)$ through the document content (words distribution) and the themes distribution. Thus, an interesting perspective is to add a new latent variable into the proposed model, to allow this model to infer effectively an unseen dialogue among all the authors.

References

1. Blei, D.M., Ng, A.Y., Jordan, M.I.: Latent dirichlet allocation. The Journal of Machine Learning Research 3, 993–1022 (2003)
2. Morchid, M., Dufour, R., Bousquet, P.M., Bouallegue, M., Linarès, G., De Mori, R.: Improving dialogue classification using a topic space representation and a gaussian classifier based on the decision rule. In: ICASSP (2014)
3. Mcauliffe, J.D., Blei, D.M.: Supervised topic models. In: Advances in Neural Information Processing Systems, pp. 121–128 (2008)
4. Morchid, M., Dufour, R., Bouallegue, M., Linarès, G.: Author-topic based representation of call-center conversations. In: 2014 IEEE International Spoken Language Technology Workshop (SLT) (2014)
5. Rosen-Zvi, M., Griffiths, T., Steyvers, M., Smyth, P.: The author-topic model for authors and documents. In: Proceedings of the 20th Conference on Uncertainty in Artificial Intelligence, pp. 487–494. AUAI Press (2004)
6. Vapnik, V.: Pattern recognition using generalized portrait method. Automation and Remote Control 24, 774–780 (1963)
7. Bechet, F., Maza, B., Bigouroux, N., Bazillon, T., El-Beze, M., De Mori, R., Arbillot, E.: Decoda: a call-centre human-human spoken conversation corpus. In: LREC 2012 (2012)
8. Salton, G.: Automatic text processing: the transformation. Analysis and Retrieval of Information by Computer (1989)
9. Bellegarda, J.: Exploiting latent semantic information in statistical language modeling. Proceedings of the IEEE 88, 1279–1296 (2000)
10. Suzuki, Y., Fukumoto, F., Sekiguchi, Y.: Keyword extraction using term-domain interdependence for dictation of radio news. In: 17th International Conference on Computational Linguistics, vol. 2, pp. 1272–1276. ACL (1998)

11. Hofmann, T.: Unsupervised learning by probabilistic latent semantic analysis. Machine Learning 42, 177–196 (2001)
12. Minka, T., Lafferty, J.: Expectation-propagation for the generative aspect model. In: Proceedings of the Eighteenth Conference on Uncertainty in Artificial Intelligence, pp. 352–359. Morgan Kaufmann Publishers Inc. (2002)
13. Griffiths, T.L., Steyvers, M.: Finding scientific topics. Proceedings of the National Academy of Sciences of the United States of America 101, 5228–5235 (2004)
14. Heinrich, G.: Parameter estimation for text analysis. Web: http://www.arbylon.net/publications/text-est.pdf (2005)
15. Morchid, M., Linarès, G., El-Beze, M., De Mori, R.: Theme identification in telephone service conversations using quaternions of speech features. In: INTERSPEECH (2013)
16. Linarès, G., Nocera, P., Massonié, D., Matrouf, D.: The lia speech recognition system: from 10xrt to 1xrt. In: Matoušek, V., Mautner, P. (eds.) TSD 2007. LNCS (LNAI), vol. 4629, pp. 302–308. Springer, Heidelberg (2007)
17. Wang, C., Blei, D., Li, F.F.: Simultaneous image classification and annotation. In: IEEE Conference on Computer Vision and Pattern Recognition, CVPR, pp. 1903–1910. IEEE (2009)
18. Yuan, G.X., Ho, C.H., Lin, C.J.: Recent advances of large-scale linear classification 100, 2584–2603 (2012)

Probabilistic Approach for Detection of Vocal Pathologies in the Arabic Speech

Naim Terbeh[1], Mohsen Maraoui[2], and Mounir Zrigui[1]

[1] LaTICE Laboratory, Monastir unit, Monastir-Tunisia
naim.terbeh@gmail.com,
mounir.zrigui@fsm.rnu.tn
[2] Computational mathematics Laboratory, University of Monastir, Monastir-Tunisia
maraoui.mohsen@gmail.com

Abstract. There are different methods for vocal pathology detection. These methods usually have three steps which are feature extraction, feature reduction and speech classification. The first and second steps present obstacles to attain high performance and accuracy of the classification system [20]. Indeed, feature reduction can create a loss of data. In this paper, we present an initial study of Arabic speech classification based on probabilistic approach and distance between reference speeches and speech to classify. The first step in our approach is dedicated to generate a standard distance (phonetic distance) between different healthy speech bases. In the second stage we will determine the distance between speech to classify and reference speeches (phonetic model proper to speaker and a reference phonetic model). Comparing these two distances (distance between speech to classify and reference speeches & standard distance), in the third step, we can classify the input speech to healthy or pathological. The proposed method is able to classify Arabic speeches with an accuracy of 96.25%, and we attain 100% by concatenation falsely classified sequences. Results of our method provide insights that can guide biologists and computer scientists to design high performance systems of vocal pathology detection.

Keywords: Arabic Speech Classification, Acoustic Model, Phonetic Transcription, Phonetic Model, Phonetic Distance, Healthy Speech, Pathological Speech.

1 Introduction and Stat of the Art

The speech signal produced by a speaker may contain characteristics that distinguish him from another signal: pathological or healthy speech. Our project focuses on the detection of falsely pronunciations in Arabic speech. The phonetic transcription can be used to generate a phonetic model of Arabic speech: the percentage of occurrence of each bi-phoneme in Arabic spoken language. The comparison between the standard phonetic model (numerical model [21]) of Arabic speech and that specific to the speaker can classify the speech produced by the concerned speaker to pathological or healthy.

A. Gelbukh (Ed.): CICLing 2015, Part II, LNCS 9042, pp. 606–616, 2015.
DOI: 10.1007/978-3-319-18117-2_45

There are 4 main types of pronunciation mistakes:

- The suppression: in this type of pronunciation defects the speaker removes one sound of the word and not the totality. The removal can exceed that one sound, the fact of increasing the difficulty of comprehension of produced speech.
- The substitution: this type of pronunciation defects summarizes the production of one sound instead of that intended. For example, to produce the sound " ث " instead of the sound " س ". This type of pronunciation defects is more known in children speech than in adult speech. Substitution poses comprehension problems of produced speech when it occurs frequently.
- The distortion: we talk about distorted phoneme when the speaker produces the sound by a bad manner but the new sound is similar to that desirable. For example, the pronunciation of the sound "ت" instead of the "ط". This type of pronunciation defects is known in adult speech than in children speech. Despite that this type of pronunciation defects occurs, the new pronounced sound is comprehensible in human speech.
- The addition: this type of pronunciation defects feels the addition of one or more sounds to the correct pronunciation. This is the least known type.

In the literature, there are several studies that treat human speech to detect pronunciation pathologies. There are several approaches that are based on the features contained in a speech signal.

Among these works, we can mention:

- The work of Vahid and al. [1] is to classify the speech signal to pathological or healthy. This work is based on artificial neural networks (ANN) [10]. The proposed method can be summarized in these three steps:
    - Extraction of MFCC [8] Coefficients Vector, and using this vector to create the characteristic vector (a vector which contains 139 features).
    - Use the method of principal component analysis (PCA) [8,9] to reduce the size of feature vector.
    - Based on the artificial neural networks (ANN), the speech signal in input will be classified to pathological or healthy.

The classification rate achieved is 91% based on 130 speech sequences (pathological and healthy).

- The work of Little and al. [2] which reports a simple nonlinear approach to online acoustic speech pathology detection for automatic screening purposes. Such an approach which combines between linear classification and biophysics of speech production (not linear) achieves an overall healthy/pathological detection performance of 91%.
- The work of Vahid and al. [3] that addressing the speech classification to pathological or normal (healthy), based on hidden Markov model (HMM) [4] and the LBG algorithm [5]. The classification procedure summarizes in three steps:

- Extraction of MFCC vector of speech to be classified.
- Extraction of the quantization vector [6,7] using the LBG algorithm.
- Using a HMM model to classify the speech in input to pathological or healthy.

This approach achieves a classification rate of 93%, using a base of 80 speech sequences (40 healthy and 40 pathological).

- The work of Kukharchik and Al. [11]; the main idea is to use the change of wavelet characteristics [14] and Support Vector Machines (SVM) [12] to classify a speech sequence to pathological or healthy. The proposed method account four steps:
  - Extraction of characteristics of speech to be classified using the KWT algorithm [13].
  - Delete frames corresponding to silence periods using the total energy criterion of frame [15].
  - Extracting only the vowels in the speech segment.
  - The classifier based on SVM uses pretreated data to classify the speech in input to pathological or healthy.

This approach achieves an accuracy rate of 99% using a base of 90 hours of segmented and indexed records (70 Hours of pathological speech, 20 Hours of healthy speech).

- The work of Martinez and al. [16] that use acoustic analysis of speech in different domains to implement a system for automatic detection of laryngeal pathologies. Authors use different processing techniques of speech signal: cepstrum, mel-cepstrum, delta cepstrum, delta mel-cepstrum, and FFT. Two systems has been developed; one trained to distinguish between normal and pathological voices and another more complex, trained to classify normal, bicyclic and rough voice. The classification rate reaches 96% for the first system and 91% for the second system.
- The work of Plante and al. [17] that the aim is to detect phonatory disorders in children speech using signal processing methods. This detection system has been tested on a population of 89 control subjects, 34 insufficient velar and 88 children with various laryngeal pathologies. Proposed method achieves an accuracy rate of 95%.

Our contribution consists to introduce a new probabilistic approach based on a phonetic model and phonetic distance (distance that separates different phonetic models) to detect pronunciation defects in human Arabic speech.

2 Materials and Methodology

2.1 Introduction

Our work to detect pronunciation defects contained in Arabic continuous speech consist to achieve a phonetic reference model of the Arabic speech which for each new speech sequence to be classified, we compare this model with the proper model of concerned speaker and we generate the phonetic distance that separates these two models. This comparison can classify speech to healthy or pathological.

2.2 Acoustic Model

Generating the Arabic phonetic model of the Arabic speech requires an acoustic model and a large Arabic speech corpus. The speech base must be recorded by native speakers and containing healthy and pathological speeches. The following table summarizes the corpus used to train our acoustic model.

Table 1. Summarize of our speech base to train acoustic model

Objective	Speech base size (min)	
Training our acoustic model	360	135 minutes of pathological speech 225 minutes of healthy speech

2.3 Sphinx_align Tool

We use the Sphinx_align tool and our acoustic model to obtain the phonetic transcription that corresponds to speech sequence in input. The following figure summarizes this task:

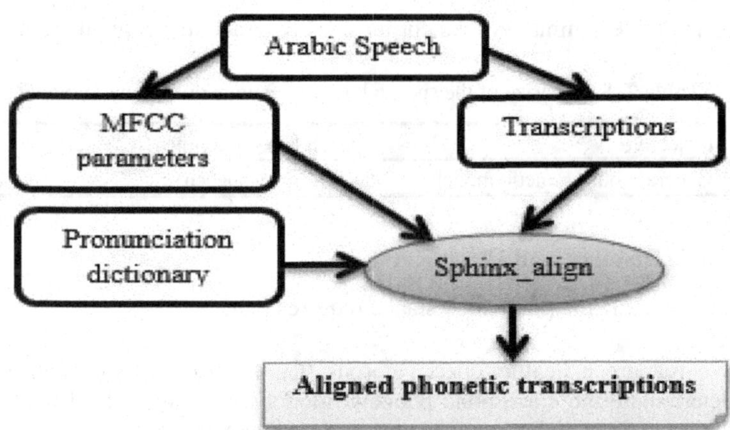

Fig. 1. Sphinx_align procedure

From the resulting phonetic transcription file (output of sphinx_align tool), we calculate probabilities of occurrence of each possible bi-phoneme in the Arabic speech corpus. The arrangement of probabilities of occurrences of bi-phonemes in a vector of 841 coefficients forms the Arabic phonetic model ($841=29^2$: Arabic language accounts 29 consonants; vowels don't pose pronunciation problems in our case). The following figure presents an extract from our Arabic phonetic model.

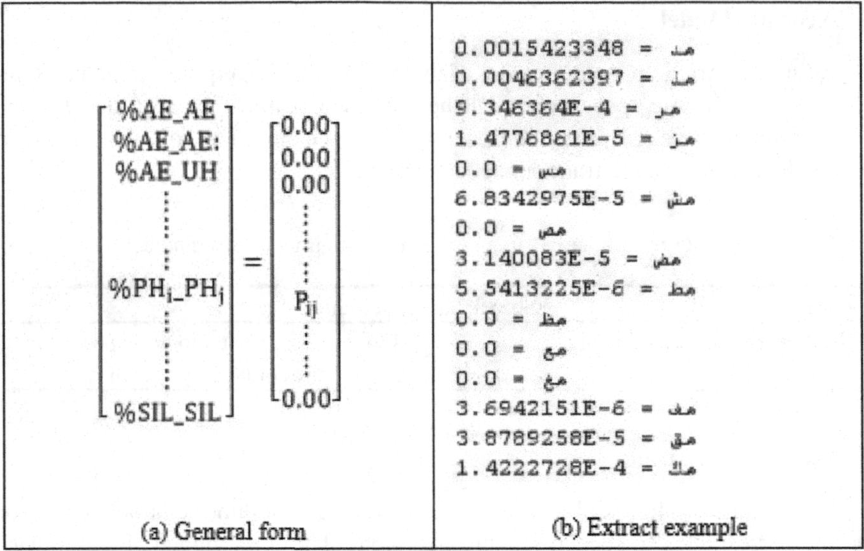

Fig. 2. Extract from the Arabic phonetic model

The following table summarizes the corpus used to generate our phonetic model.

Table 2. Summarize of the speech base to generate our phonetic model

Objective	Speech base size (min)
Generate our phonetic model	543 (healthy speech)

2.4 Phonetic Distance and Classification

This task to generate the phonetic distance requires that:

- We prepare n healthy speech corpus (C_i, $1 \leq i \leq n$), and for each corpus, we determinate the correspond phonetic model M_i, ($1 \leq i \leq n$). In this task, we use the acoustic model previously prepared and the Sphinx_align tool.

- We define $S=\{\alpha_{ij};\ 1\leq i,j\leq n \text{ and } i\neq j\}$ a set of angles that separate M_i and M_j ($\alpha_{ij}=\alpha_{ji}$ and $\alpha_{ii}=0$).
- We define the value Max=maximum{S}.
- We define the value δ=standard deviation{S}.
- We define the value Avg=average{S}.
- We calculate $\beta=Max+|Avg-\delta|$.

To calculate the set S, we follow these scalar product formulas:

$$M_i.M_j=\sum_{k=1}^{n} M_i[k]M_j[k] \quad (1)$$

$M_i.M_j=||M_i||.||M_j||.\cos(\alpha)$ with α is the angle that separates between M_i and M_j. (2)

$$\mathrm{Cos}(\alpha)=M_i.M_j/||M_i||.||M_j|| \quad (3)$$

For each new speaker, we use a speech sequence recorded by his voice and we follow the same previous procedure (section 2.3) to generate its proper phonetic model. We calculate the angle θ that separates this model and that reference to the Arabic speech. We distinguish tow cases:

- If $\theta\leq\beta$, then the speech in input (pronounced by the concerned speaker) is heathy
- Else ($\theta>\beta$) the speech in input is pathological

2.5 Classification Procedure

The proposed method to classify Arabic speech can be summarized in these following steps:

- Generation of n phonetic models of the Arabic speech (using one corpus for each phonetic model) to calculate the maximum distance between the Arabic phonetic models (phonetic distance),
- Generation of the phonetic reference model (the average of n previous models),
- For each new sequence to be classified, we generate the phonetic model proper to speaker (speaker can be normal, native, with disability, …),
- Compare these two models and classify the speech in input to healthy or pathological.

We can schematize our Arabic speech classification system as following:

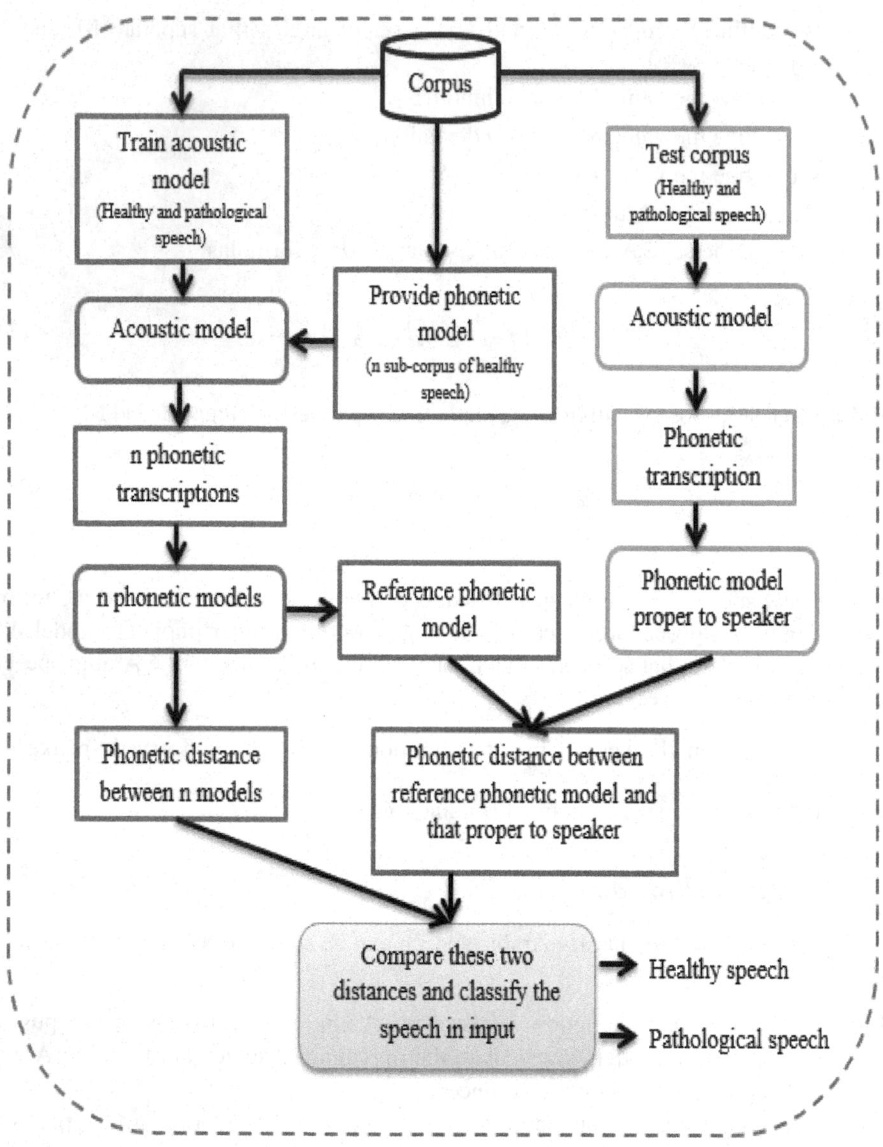

Fig. 3. General Form of Proposed Classification System

3 3 Tests and Results

3.1 Tests Conditions

The test is done under the following conditions:

- To train our acoustic model, a corpus of six hours of Arabic healthy and pathological speech (*.wav format) in mono speaker mode has been prepared.

- To generate our phonetic model, a base of nine hours healthy Arabic speech has been prepared (on *.wav format and in mono speaker mode).
- This healthy Arabic speech base is divided into four sub-corpuses, and for each one we determine its phonetic model.
- The test database was created with the help of a speech therapist. It counts 80 Arabic speech sequences which 60 are pathological and 20 are healthy. All records are on *.wav format and in mono speaker mode.

The following table summarizes test conditions:

Table 3. Summarize of our Arabic speech base

Corpus	Size	Prepared by	Speaker number	Age (year)	Objective
1st Corpus	80 records	Speech Therapist	80 Speakers	Between 13 and 45	Test
2nd Corpus	6 hours	Healthy Peoples	6 Speakers	Between 17 and 35	Training acoustic model
3rd Corpus	9 hours	Healthy and pathological peoples	12 speakers	Between 21 and 51	Generate the phonetic model

3.2 Experiment Results

The first step is to calculate the maximum distance between the four phonetic models prepared. Results are summarized in the following table:

Table 4. Distances between phonetic models

Model	M1	M2	M3	M4
M1	0	0.1306207571312359°	0.16295310606493837°	0.11036318831226814°
M2	0.1306207571312359°	0	0.1506320448535047°	0.1181644845403591°
M3	0.16295310606493837°	0.1506320448535047°	0	0.09816562350665492°
M4	0.11036318831226814°	0.1181644845403591°	0.09816562350665492°	0

From the previous table, we calculate different values of S, Max, Avg and δ as shown in the following table.

Table 5. Different values of S, Max, Avg and δ

S	Max	Avg	δ
0.1306207571312359°, 0.16295310606493837°, 0.11036318831226814°, 0.1506320448535047°, 0.1181644845403591°, 0.09816562350665492°	0.16295310606493837°	0.12848320073482686°	0.06439212422353331°

The previously table gives the value:

β= 0.16295310606493837° + |0.12848320073482686°-0.06439212422353331°| = 0.22704418257643192°.

The following table summarizes the pathological speech detection rate:

Table 6. Results of Pathology Detection

Test Corpus	Test results
60 pathological records	57 pathological records and 3 healthy records
20 healthy records	20 healthy records

The sixth table presents that three sequences from sixty are falsely classified. To identify the reason of this false classification, we try to classify sequences that combine two sequences from these three falsely classified. The following table summarizes classification results of combined sequences:

Table 7. Classification Results of Combined Records

Sequence	Class
Combination between the 1st and the 2nd sequences	pathological
Combination between the 1st and the 3rd sequences	pathological
Combination between the 2nd and the 3rd sequences	pathological

Last results (seventh table) present the impact of the sequence size in the classification procedure. Indeed, when we use a large sequence of speech we maximize the probability to see all possibilities of bi-phoneme combination in such spoken sequence; then detection of vocal pathology becomes easy. Against, if we use a short speech sequence, highly probable we don't have all combination possibilities between Arabic phonemes in such record; so detection of vocal pathology becomes more difficult.

In the following figure, we express different results obtained using our approach:

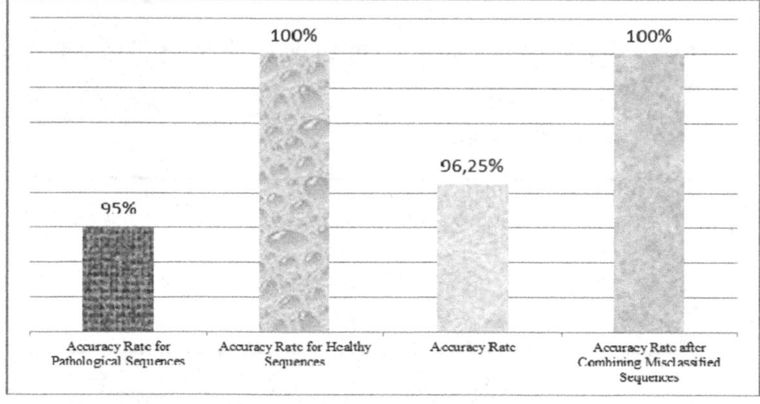

Fig. 3. Pathology detection rate

4 Conclusion and Future Works

Acoustic analysis presents the proper method in spoken language diagnostics to detect vocal pathologies. In this article, our acoustic model has been trained for healthy and pathological Arabic speech. For this purpose, a large corpus of Arabic healthy and pathological speech has been prepared. Also we test this acoustic model by another Arabic healthy corpus to generate the reference phonetic model and the reference phonetic distance that separates between two different phonetic models of Arabic speech. And finally, after generate the phonetic model proper to speaker, we calculate the distance that separates between this model and that reference. This distance will be comparted to the reference phonetic distance to classify the speech in input to healthy or pathological. Experiment results show that the proposed approach has high classification accuracy; indeed we attain a classification rate of 96.25%, and 100% after combining falsely classified sequences.

It may be possible to try to build an automatic system for detection of phonemes that pose pronunciation problems for native peoples suffering from language disabilities or foreign speakers that learn Arabic language.

Also it may be possible to benefit from this work to tray to elaborate an automatic speech correction system for peoples suffering from pronunciation problems [18, 19].

Acknowledgments. First, I would like to express my deepest regards and thanks to all the members of the evaluation research committee within CICLing conference. I extend also my advance thanks to my supervisor M. Mounir ZRIGUI for his valuable advice and encouragement.

References

1. Majidnezhad, V., Kheidorov, I.: An ANN-based Method for Detecting Vocal Fold Pathology. International Journal of Computer Applications 62(7) (January 2013)
2. Little, M., McSharry, P., Moroz, I., Roberts, S.: Nonlinear, Biophysically-Informed Speech Pathology Detection. In: ICASSP (2006)
3. Majidnezhad, V., Kheidorov, I.: A HMM-Based Method for V ocal Fold Pathology Diagnosis. IJCSI International Journal of Computer Science Issues 9(6(2)) (November 2012)
4. Bréhilin, L., Gascuel, O.: Modèles de Markov caches et apprentissage de sequences
5. Patane, G., Russo, M.: The enhanced LBG Algorithm. Neural Networks 14(9) (November 2001)
6. Makhoul, J., Roucos, S., Gish, H.: Vector Quantization in Speech Coding
7. Räsänen, O.J., Laine, U.K., Altosaar, T.: Self-learning Vector Quantization for Pattern Discovery from Speech. In: InterSpeech 2009, September 6-10, Brighton, United Kingdom (2009)
8. Ittichaichareon, C., Suksri, S., Yingthawornsuk, T.: Speech Recognition using MFCC. In: International Conference on Computer Graphics, Simulation and Modeling, ICGSM 2012, July 28-29, Pattaya, Thailand (2012)
9. Jolliffe, I.T.: Principal Component Analysis. 2nd edn. (2008)
10. Paquet, P.: L'utilisation des réseaux de neurones artificiels en finance. Document de Recherche n° 1997-1 (1997)

11. Kukharchik, P., Martynov, D., Kheidorov, I., Kotov, O.: Vocal Fold Pathology Detection using Modified Wavelet-like Features and Support Vector Machines. In: 15th European Signal Processing Conference, EUSIPCO 2007, September 3-7, Poznan, Poland, (2007); copyright by EURASIP
12. Archaux, C., Laanaya, H., Martin, A., Khenchaf, A.: An SVM based Churn Detector in Prepaid Mobile Telephony (2004)
13. Aguira-Conraria, L., Soares, M.J.: The Contnuous Wavelet Transform: A Primer (July 12, 2011)
14. Damerval, C.: Ondelettes pour la détection de caractéristiques en traitement d'images. Doctoral thesis (Mai 2008)
15. Hamila, R., Astola, J., Cheikh, F.A., Gabbouj, M., Renfors, M.: Teager energy and the ambiguity function. IEEE Transactions on Signal Processing 47(1) (1999)
16. Martinez César, E., Rufiner Hugo, L.: Acoustic analysis of speech for detection of laryngeal pathologies. In: 22nd Annual EMBS International Conference, July 23-28, Chicago Il (2000)
17. Fabrice, P., Christian, B.-V.: Détection acoustique des pathologies phonatoires chez l'enfant. Doctoral thesis (1993)
18. Terbeh, N., Labidi, M., Zrigui, M.: Automatic speech correction: A step to speech recognition for people with disabilities. In: ICTA 2013, October 23-26, Hammamet, Tunisia (2013)
19. Terbeh, N., Zrigui, M.: Vers la Correction Automatique de la Parole Arabe. In: Citala 2014, November 26-27, Oujda, Morocco (2014)
20. Zrigui, M., Ayadi, R., Mars, M., Maraoui, M.: Arabic Text Classification Framework Based on Latent Dirichlet Allocation. CIT 20(2), 125–140 (2012)
21. Zouaghi, A., Zrigui, M., Antoniadis, G.: Automatic Understanding of Spontaneous Arabic Speech - A Numerical Model. TAL 49(1), 141–166 (2008)

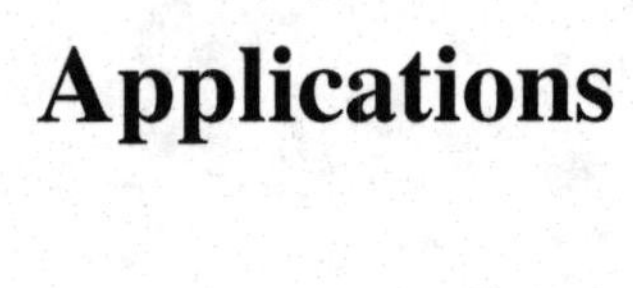

Applications

Clustering Relevant Terms and Identifying Types of Statements in Clinical Records

Borbála Siklósi

Pázmány Péter Catholic University,
Faculty of Information Technology and Bionics,
50/a Práter street, 1083 Budapest, Hungary
{siklosi.borbala}@itk.ppke.hu

Abstract. The automatic processing of clinical documents created at clinical settings has become a focus of research in natural language processing. However, standard tools developed for general texts are not applicable or perform poorly on this type of documents, especially in the case of less-resourced languages. In order to be able to create a formal representation of knowledge in the clinical records, a normalized representation of concepts needs to be defined. This can be done by mapping each record to an external ontology or other semantic resources. In the case of languages, where no such resources exist, it is reasonable to create a representational schema from the texts themselves. In this paper, we show that, based on the pairwise distributional similarities of words and multiword terms, a conceptual hierarchy can be built from the raw documents. In order to create the hierarchy, we applied an agglomerative clustering algorithm on the most frequent terms. Having such an initial system of knowledge extracted from the documents, a domain expert can then check the results and build a system of concepts that is in accordance with the documents the system is applied to. Moreover, we propose a method for classifying various types of statements and parts of clinical documents by annotating the texts with cluster identifiers and extracting relevant patterns.

Keywords: clinical documents, clustering, ontology construction,less-resourced languages.

1 Introduction

Clinical documents are created in clinical settings by doctors and assistants. Depending on local regulations, these types of texts are very different in quality, but even the best ones stay far behind proofread, general texts. Moreover, the content of such documents is from a very narrow domain, where general language use is usually suppressed by unique patterns of syntax and semantics. This results in a sublanguage, where standard tools developed for general texts are not applicable or perform poorly, especially in the case of less-resourced and morphologically complex languages, such as Hungarian. However, clinical documents contain valuable information, worth finding, even if the only 'resource' we have is just the collection of the documents themselves.

A. Gelbukh (Ed.): CICLing 2015, Part II, LNCS 9042, pp. 619–630, 2015.
DOI: 10.1007/978-3-319-18117-2_46

In order to be able to create a formal representation of knowledge in clinical records, first, a normalized representation of concepts should be defined. This can be done by mapping each record to an external ontology or other semantic resources. Even if such a resource was available, Zhang showed that there is a significant difference between the representation of concepts in such an artificial system and the cognitive behaviour of knowledge ([24]). Moreover, in the case of clinical documents, the representation of medical concepts by doctors and patients are also different ([12]). Thus, it is more reasonable to create a representational schema from the texts themselves rather than enforcing these documents to adjust to a predefined ontology of concepts, which, by the way, does not even exist for less-resourced languages. Having such an initial system of knowledge extracted from the documents, a domain expert can then check the results and build a system of concepts that is in accordance with the documents the system is applied to.

Thus, in our research, we aim at transforming the content of Hungarian ophthalmology records to different representations, which are useful to describe the content of the documents from different aspects, i.e. either a conceptual hierarchy or semantic patterns. We applied distributional semantic models, which capture the meaning of terms based on their distribution in different contexts. As Cohen et al. state in [2], such models are applicable to the medical domain, since the constraints regarding the meaning of words and phrases are stricter than in general language. In [13], it has been shown that in the medical domain distributional methods outperform the similarity measures of ontology-based semantic relatedness. Using this similarity metric, we propose a method for building a hierarchical system of concepts. These representations are then used as a basis of the manual construction of lexical thesauri of this domain, and also for an abstract representation of the documents. Based on such a representation, some semantic patterns can be defined, which can help the automatic identification of different types of statements found in the documents.

In the following section, we give a short review of related studies. Then, a description of our ophthalmology corpus follows, further explaining the differences between general language and the clinical sublanguage. In Section 4, the theory of distributional semantics and its application as a similarity measure between concepts are explained. It is followed by its use in a hierarchical clustering algorithm, resulting in a structured organization of concepts. Finally, an abstract representation is described based on this system of concepts, and the possible use of semantic patterns is demonstrated apllying them to the task of recognizing anamnesis statements.

2 Related Work

Semantics is needed to understand language. Even though most applications apply semantic models at a final stage of a language processing pipeline, in our case some basic semantic approaches were reasonable to apply as preprocessing steps. Still, our goal was not to create a fully functional semantic representation, thus the related literature was also investigated from this perspective.

There are two main approaches to automatic processing of semantic behaviour of words in free texts: mapping them to formal representations, such as ontologies as described in [7] and applying various models of distributional semantics ([3,14,9]) ranging from spatial models to probabilistic ones (for a complete review of empirical distributional models consult [2]). Handmade resources are very robust and precise, but are very expensive to create and often their representation of medical concepts does not correlate to real usage in clinical texts ([24,2]). On the other hand, the early pioneers of distributional semantics have shown that there is a correlation between distributional similarity and semantic similarity, which makes it possible to derive a representation of meaning from the corpus itself, without the expectation of precise formal definitions of concepts and relations ([2]).

Beside the various applications of distributional methods in the biomedical domain, there are approaches, where these are applied to texts from the clinical domain. Carroll et al. ([1]) create distributional thesauri from clinical texts by applying distributinal models in order to improve recall of their manually constructed word lists of symptoms and to quantify similarity of terms extracted. In their approach, the context of terms is considered as the surrounding words within a small window, but they do not include any grammatical information as opposed to our definition of features representing context. Still, they report satisfactory results for extracting candidates of thesaurus entries of nouns and adjectives, producing somewhat worse results in the latter case. However, the corpus used in their research was magnitudes larger than ours. As Sridharan et al. have shown, either a large corpus or a smaller one with high quality is needed for distributional models to perform well, emphasising the quality over size ([21]). This explains our slightly lower, but still satisfactory results. The similarity measure used in the research described in [1] was based on the one used in [8]. In this study, it is also applied to create thesauri from raw texts, however there it is done for general texts and is exploiting grammatical dependencies produced by high-quality syntactic parsers. A detailed overview of distributional semantic applications can be found in [2] and [22] and its application in the clinical domain is overviewed in [5].

3 The Hungarian Ophthalmology Corpus

In this research, anonymized clinical documents from the ophthalmology department of a Hungarian clinic were used. This corpus contains 334 546 tokens (34 432 sentences). The state of the corpus before processing was a structured high-level xml as described in [18]. It was also segmented into sentences and tokens and pos-tagged applying the methods of [10] and [11] respectively. Though such preprocessing tasks are considered to be solved for most languages in the case of general texts, they perform significantly worse in the case of clinical documents as discussed in the aforementioned publications. Still, this level of preprocessing was unavoidable. Furtheremore, multiword terms were also extracted from the corpus as described in [16].

When the ophthalmology notes are compared to a general Hungarian corpus, we find reasonable differences between the two domains. This explains some of the difficulties that prevent tools developed for general texts working in the clinical domain (there are significant differences between different departments as well, but in this paper 'clinical' is used to refer to 'ophthalmology' notes). These differences are not only present in the semantics of the content, but in the syntax and even in the surface form of the texts and fall into three main categories:

- **Syntactic behaviour:** Doctors tend to use shorter and rather incomplete and compact statements. This habit makes the creation of the notes faster, but the high frequency of ellipsis of crucial grammatical constituents makes most parsers fail when trying to process them. Regarding the distribution of part-of-speech (pos) in the two domains, there are also significant differences. While in the general corpus, the three most frequent types are nouns, verbs and adjectives, in the clinical domain nouns are followed by adjectives and numbers in the frequency ranking, while the number of verbs in this corpus is just one third of the number of the latter two. Another significant difference is that in the clinical domain, determiners, conjunctions, and pronouns are also ranked lower in the frequency list.
- **Spelling Errors:** Clinical documents are usually created in a rush without proofreading. Thus, the number of spelling errors is very high, and a wide variety of error types occur ([17]). These errors are not only due to the complexity of the Hungarian language and orthography, but also to characteristics typical of the medical domain and the situation in which the documents are created. The most frequent types of errors are: mistypings, lack or improper use of punctuation, grammatical errors, sentence fragments, Latin medical terminology not conforming to orthographical standards, and non-standard use of abbreviations.
- **Abbreviations:** The use of a kind of notational text is very common in clinical documents. This dense form of documentation contains a high ratio of standard or arbitrary abbreviations and symbols, some of which may be specific to a special domain or even to a doctor or administrator ([15]). These short forms might refer to clinically relevant concepts or to some common phrases that are very frequent in the specific domain. For the clinicians, the meaning of these common phrases is as trivial as the standard shortened forms of clinical concepts due to their expertise and familiarity with the context.

A detailed comparison of the ophthalmology corpus and a general Hungarian corpus can be found in [16] and [19].

4 Distributional Relatedness

The theory behind distributional semantics is that semantically similar words tend to occur in similar contexts ([4]) i. e. the similarity of two concepts is determined by their shared contexts. The context of a word is represented by a set of features, each feature consisting of a relation (r) and the related word (w'). For each word (w) the frequencies of all (w, r, w') triples are determined. In other studies, these relations are usually grammatical relations, however in the case of Hungarian ophthalmology texts, grammatical analysis performs poorly, resulting in a rather noisy model. Carroll et al. suggest using only the occurrences of surface word forms within a small window around the target word as features ([1]). In our research, a mixture of these ideas was used by applying the following relations to determine the features for a certain word:

- prev_1: the previous word
- prev_w: words preceding the target word within a distance of 2 to 4
- next_1: the following word
- next_w: words following the target word within a distance of 2 to 4
- pos: the part-of-speech tag of the actual word
- prev_pos: the part-of-speech tag of the preceding word
- next_pos: the part-of-speech tag of the following word

Words in this context are the lemmatized forms of the original words on both sides of the relations. To create the distributional model of words, a similarity measure needs to be defined over these features. Based on [8], the similarity measure we used was pointwise mutual information, which prefers less common values of features to more common ones, emphasising that the former characterize a word better than the latter. ([1])

First, each feature is associated with a frequency determined from the corpus. Then, the information contained in a triple of (w, r, w'), i.e. the mutual information between w and w' w.r.t. the relation r. ([6]) can be computed according to Formula 1:

$$I(w,r,w') = log\frac{||w,r,w'|| \times ||*,r,*||}{||w,r,*|| \times ||*,r,w'||} \tag{1}$$

While $||w,r,w'||$ corresponds to the frequency of the triple (w,r,w') determined from the corpus, when any member of the triple is a $*$, then the frequencies of all the triples corresponding the rest of the triple are summed over. For example, $||*, next\_1, szem||$ corresponds to the sum of the frequencies of words followed by the word *szem* 'eye'.

Then, the similarity between two words (w_1 and w_2) can be counted according to Formula 2

$$SIM(w_1,w_2) = \frac{\sum_{(r,w)\in T(w_1)\bigcap T(w_2)}(I(w_1,r,w)+I(w_2,r,w))}{\sum_{(r,w)\in T(w_1)} I(w_1,r,w) + \sum_{(r,w)\in T(w_2)} I(w_2,r,w)} \tag{2}$$

where $T(w)$ is the set of pairs (r,w') such that $I(w,r,w')$ is positive.

It should be noted that even though these models can be applied to all words in the raw text, it is reasonable to build separate models for words of different part-of-speech. Due to the relatively small size of our corpus and the distribution of part-of-speech as described in Section 3, we only dealt with nouns and nominal multiword terms that appear at least twice in the corpus.

Moreover, in order to avoid the complexity arising from applying this metric between multiword terms, these phrases were considered as single units, having the [N] tag when comparing them to each other or to single nouns. Figure 1 shows a heatmap where the pairwise similarities of terms found in a single ophthalmology document are shown. The lighter a square is, the more similar the two corresponding phrases are. As it can be seen on the map, the terms *"tiszta törőközeg"* ('clean refractive media') and *"békés elülső segmentum"* ('calm anterior segment') are similar with regard to their distributional behaviour, while for example the term *"neos mksz"* ('Neo-Synephrine to both eyes') is just slightly related to a few other terms in the particular document.

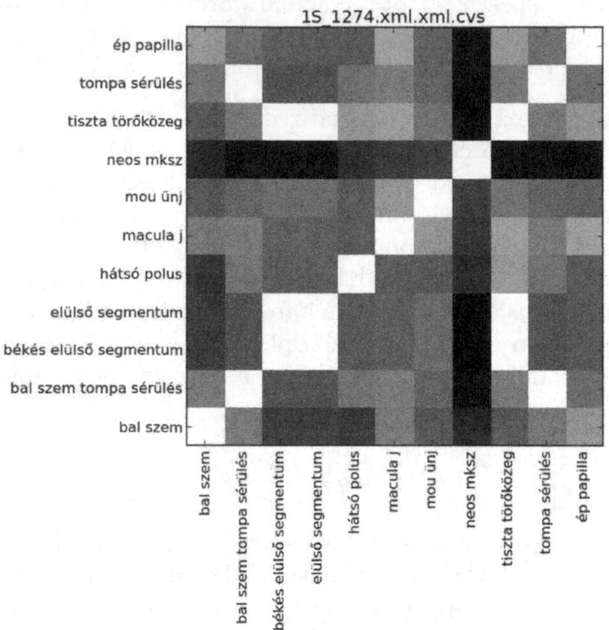

Fig. 1. The heatmap of pairwise similarities of terms extracted from a single document

The results show that the model does indeed indentify related terms, however due to the nature of distributional models, the semantic type of the relation may vary. These similarities are paradigmatic in nature, i.e. similar terms can be replaced by each other in their shared contexts. As this is true not only for synonyms, but also for hypernyms, hyponyms and even antonyms, such distinctions can not be made with this method. This shortcoming, however, does not prohibit the application of this measure of semantic relatedness when creating conceptual clusters as the basis of an ontology for the clinical domain. This can be done, because in this sublanguage the classification of terms should be based on their medical relevance, rather than on the common meaning of these words in every day language. Thus, no matter what their meaning might be, terms in the semantic group of e.g. 'signs and symptoms' are all related from the point of view of the semantics characterizing our specific clinical domain.

5 Conceptual Clusters

Based on the pairwise similarities of words and multiword terms, a conceptual hierarchy can be built. In order to create the hierarchy, we applied agglomerative clustering on the most frequent terms. Each term was represented by a feature vector containing its similarity to all the other terms. Formally, the c_i element of the vector $c(w)$ corresponding to term w is $SIM(w, w_i)$ as defined in Formula 2. The clustering algorithm was then applied on these vectors. The linkage method was chosen based on the cophenet correlation between the original data points and the resulting linkage matrix ([20]). The best correlation was achieved when using Ward's distance criteria ([23]) as the linkage method. This resulted in small and dense groups of terms at the lower level of the resulting dendrogram.

However, we needed not only the whole hierarchy, represented as a binary tree, but separate, compact groups of terms, i.e. well-separated subtrees of the dendrogram. The most intuitive way of defining these cutting points of the tree is to find large jumps in the clustering levels. To put it more formally, the height of each link in the cluster tree is to be compared with the heights of neighbouring links below it in a certain depth. If this difference is larger than a predefined threshold value (i.e. the link is inconsistent), then the link is a cutting point.

We applied this cutting method twice. First, we used a lower threshold value to create small and dense groups of terms. At this stage, the full hierarchy was kept and the nodes below the threshold were collapsed, having these groups of terms as leaf nodes (see for example node `1403` in Figure 2). In the second iteration, the hierarchy was divided into subtrees by using a higher threshold value. After this step, the hierarchy was only kept within these subtrees, but they were treated as single clusters. Each node in the tree was given a unique concept identifier.

Table 1 shows some examples of collapsed groups of terms. The resulting groups contain terms of either similar meaning, or ones having a semantically similar role (e.g. names of months or medicines, etc.), even containing some abbreviated variants as well (e.g. *"bes"*, *"békés elülső szegmens"*, *"békés elülső szegmentum"*, *"békés es"* – all standing for the term 'calm anterior segment'). Beside these semantically related groups of terms, there are some more abstract ones as well, which contain terms related to certain medical processes or phases of medical procedures. For example the term *"éhgyomor"* ('empty stomach') was grouped together with terms related to time and appointments, or *strab* ('strabism') and *párhuzamos szem* ('parallel eyes') were grouped together based on their medical relatedness. Figure 2 shows an example of a subtree with root identifier *2016* and the hierarchical organization of groups of terms in the leaf nodes.

6 Discovering Semantic Patterns

The clustering and ordering of terms extracted from clinical documents might be used directly as an initial point of a Hungarian medical (ophthalmological) ontology, containing phrases used by practitioners in their daily cases. However, since each group (and each node in the hierarchy) was given a unique identifier, words and phrases in the original text can be annotated or replaced by these

Table 1. Some example groups of terms as the result of the clustering algorithm

I1403	papilla, macula, cornea, lencse, jobb oldal, bal oldal, centrum, kör, szél, periféria, retina, szemhéj, elváltozás, vérzés, terület *papil, macula, cornea, lens, right side, left side, centre, circle, verge, periphery, retina, eyelid, change, blooding, area*
I1636	hely, kh, kötőhártya, szaru, conjunctiva, szemrés, szempilla, pilla, könnypont *place, cj, conjunctiva, white of the eye, conjunctiva, eyelid opening, eyelash, lash, punctum*
I1549	tbl, medrol, üveg, szemcsepp, gyógyszer *tbl, medrol, glass, eyedrop, medication*
I1551	folyamat, kivizsgálás, érték, idegentest, gyulladás, retinaleválás, látásromlás *process, examination, value, foreign body, imflamation, retinal detachment, worse vision*
I1551	ép papilla, halvány papilla, jó színű papilla, szűk ér, ép macula, fénytelen macula, kör fekvő retina, fekvő retina, rb, tiszta törőközeg, bes, békés elülső szegmentum, békés es *intact papil, vague papil, good colored papil, narrow vein, intact macula, dim macula, retina laying around, laying retina, ok, clean refractive media, cas, calm anterior segment, calm as*

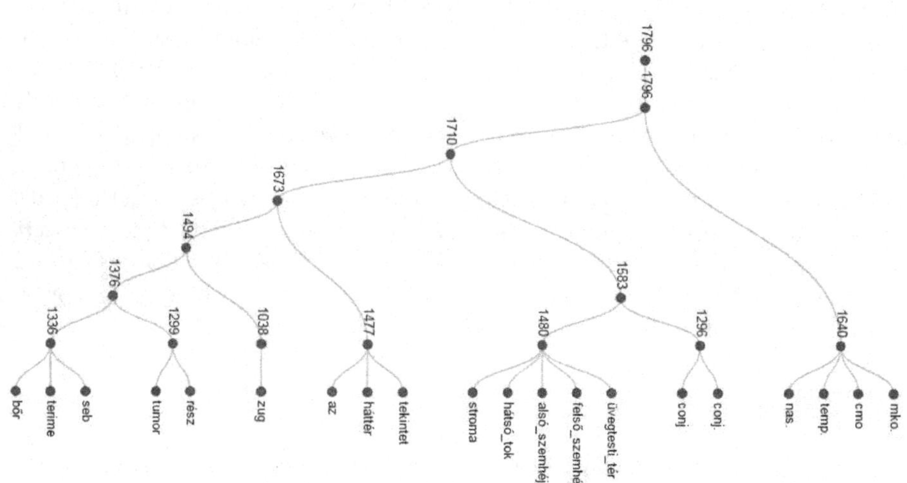

Fig. 2. A subtree cut out from the whole hierarchy containing groups of terms on the leaf nodes

concept ID's. Thus, a higher-level abstract representation can be created for the documents. Then, frequent patterns can easily be determined at this abstract level, no matter what the actual content and the frequency at actual realizations of the pattern are. Using cluster identifiers, the sentences not only became simpler, but these patterns were also easy to extract from them.

For example the pattern `1889|2139 1327 1627` characterizes expressions that contain some data about the state of one or both of the patient's eyes. The most frequent realizations of this pattern are: *"st. o. u."*, *"st. o. s."*, *"st. o. d."*, *"moct o. d."*, *rl. o. u."*, *"rl. o. sin."*, *"status o. s."*, *"távozáskor o. d." ('at leaving')*, *"b-scan o. d."*, etc. Another characteristic of this pattern is that it appears at the beginning of a sentence, thus this information can be used if this template is used as a separator between different statements in the documents.

These frequent patterns can also be used to classify sentences in each record. In Hungarian clinics, there are EHR (Electronic Health Record) systems, but there are no serious regulations about the use of such software. Thus, doctors or assistants, who are still used to the tradition of writing unstructured documents, tend to type all information into a single text box. Thus, various types of data that should appear under their own heading in the final document are mixed under a single (or no) heading. Thus, it is a separate task to find statements that should belong to different sections, such as anamnesis, lab tests, treatments, opinion, diagnosis, etc.

Thus, we have performed an experiment on defining patterns to find anamnesis sentences. In order to be able to exploit all the information we have already annotated our corpus with, we set up a simple search interface. Then, we were able to define patterns which include constraints applied to different annotation levels, such as the original text, the lemmas of words, part-of-speech, morphological analysis and cluster identifiers.

Anamnesis statements are the ones referring to the reason for a patient's visit to the doctor, their medical history or the aim of admission to hospital. We defined three patterns to cover such statements:

1. Including current complains and the reason of admitting the patient to the clinic. Such sentences include past tense singular verbs between any members of the cluster `1436` on both sides (not necessarily immediate neighbours). The pattern can be defined as `{I1436 .. VERB|Past|singular .. I1436}`. Sentences belonging to this group were like the one in the first example in Table 2.
2. Including some past events, or medical history. This pattern is similar to the previous one, differing only in the first part, i.e. there is no need for a preceding constraint before the verb resulting in the form of `{VERB|Past|singular .. I1436}`. An example is the second one in Table 2.
3. Including some conditional or wish statements. These are sentences describing some uncertain descriptions of the feelings of the patient or a certain wish. These are covered by the pattern `{I1436 .. VERB|Cond}`. But, since we found only few examples of such sentences, a simpler form of this pattern was also tried (i.e. including only the conditional verb), which produced

more results, but the precision was much lower in this case at the expense of recall. See the third example in Table 2 for a sentence covered by this pattern.

Table 2. Example sentences for each pattern describing anamnesis statements

1. `I1436..VERB|Past|singular..I1436`
'Our patient was admitted for the investigation of glaucoma.'
`Betegünk/N glaucomás/N kivizsgálás/N céljából/N került/V|Past felvételre/N.` `1436 1930 1551 1434 1436`
2. `VERB|Past|singular..I1436`
'His eyes are inflamed since yesterday.'
`Tegnap/N óta/P begyulladt/V|Past a/Det szeme/N.` `2249 1436`
3. `I1436..VERB|Cond`
'He feels as if there was something in his eyes.'
`Úgy/Adv érzi/V, mintha/C valami/N lenne/V|Cond a/Det szemében/N.` `2187 1436`

Table 3 shows the performance results of recognizing anamnesis sentences based on these patterns. ALL is the number of results, out of which TP (true positive) is the ratio of sentences classified correctly, and the rest are the ERRORS. Erroneously classified sentences fall into three categories. (1)FP (false positive) is the ratio of erroneous sentences classified incorrectly. (2)There were some sentences that were too ambiguous or malformed, so that we could not decide whether its content is anamnesis or not (UD). (3)There were also some sentences labelled as P/S, which contained some errors at a lower preprocessing level, such as a misspelling or bad pos-tag. However, only those errors were counted here, which caused the sentence to be erroneously classified. For example past participles are frequently mistagged as past tense verbs, which lead to the high ratio of errors in the second case.

Table 3. Results of recognizing anamnesis sentences based on multilevel patterns

PATTERN	ALL(#)	TP(%)	ERRORS(%)		
			FP	UD	P/S
1. `I1436..VERB|Past|singular..I1436`	147	0.972	0.4	0.4	0.2
2. `VERB|Past|singular..I1436`	192	0.961	0.16	0	0.84
3a. `I1436..VERB|Cond`	11	1.0	0	0	0
3b. `VERB|Cond`	145	0.889	0.75	0.187	0.062

7 Conclusion

In order to understand the meaning of the documents, or statements in clinical documents, a semantic model is needed. It is out of the scope of this work to create a complete model, however the foundations of such a semantic description have been laid.

The construction of hand-made semantic resources for a language is very expensive, requires language and domain-specific expertise and is not always in accordance with the cognitive representation of knowledge ([24]). While the domain-specific validation is unavoidable, the other two problems could partially be handled by applying unsupervised methods for ontology learning and recognition of semantic patterns of a sublanguage, such as medical language.

We have shown that the combination of distributional similarity measures and hierarchical clustering algorithms result in an organized system of concepts that is in accordance with the original corpus these models are derived from. This hierarchical structure and the abstract patterns relieved in the corpus after having the text annotated with concept identifiers, might be a valuable basis for the manual construction of domain-specific lexical resources and can be utilized for the automatic classification of statements.

Acknowledgements. The author would like to thank Attila Novák for the valuable and creative discussions, his linguistic support and the continuous encouragement he has provided throughout this research.

References

1. Carroll, J., Koeling, R., Puri, S.: Lexical acquisition for clinical text mining using distributional similarity. In: Gelbukh, A. (ed.) CICLing 2012, Part II. LNCS, vol. 7182, pp. 232–246. Springer, Heidelberg (2012)
2. Cohen, T., Widdows, D.: Empirical distributional semantics: Methods and biomedical applications. Journal of Biomedical Informatics 42(2), 390–405 (2009)
3. Deerwester, S., Dumais, S.T., Furnas, G.W., Landauer, T.K., Harshman, R.: Indexing by latent semantic analysis. Journal of the American Society for Information Science 41(6), 391–407 (1990)
4. Firth, J.R.: A synopsis of linguistic theory 1930-55. 1952-59, 1–32 (1957)
5. Henriksson, A.: Semantic spaces of clinical text: Leveraging distributional semantics for natural language processing of electronic health records (2013)
6. Hindle, D.: Noun classification from predicate-argument structures. In: Proceedings of the 28th Annual Meeting on Association for Computational Linguistics, ACL 1990, pp. 268–275, Stroudsburg, PA, USA (1990)
7. Jurafsky, D., Martin, J.H.: Speech and Language Processing: An Introduction to Natural Language Processing, Computational Linguistics, and Speech Recognition, 1st edn. Prentice Hall PTR, Upper Saddle River (2000)
8. Lin, D.: Automatic retrieval and clustering of similar words. In: Proceedings of the 17th International Conference on Computational Linguistics, COLING 1998, vol. 2, pp. 768–774, Stroudsburg, PA, USA (1998)
9. Lund, K., Burgess, C.: Producing high-dimensional semantic spaces from lexical co-occurrence. Behavior Research Methods, Instruments, and Computers 28(2), 203–208 (1996)

10. Orosz, G., Novák, A., Prószéky, G.: Hybrid text segmentation for Hungarian clinical records. In: Castro, F., Gelbukh, A., González, M. (eds.) MICAI 2013, Part I. LNCS (LNAI), vol. 8265, pp. 306–317. Springer, Heidelberg (2013)
11. Orosz, G., Novák, A., Prószéky, G.: Lessons learned from tagging clinical Hungarian. International Journal of Computational Linguistics and Applications 5(1), 159–176 (2014)
12. Patel, V.L., Arocha, J.F., Kushniruk, A.W.: Patients' and physicians' understanding of health and biomedical concepts: Relationship to the design of emr systems. J. of Biomedical Informatics 35(1), 8–16 (2002)
13. Pedersen, T., Pakhomov, S.V., Patwardhan, S., Chute, C.G.: Measures of semantic similarity and relatedness in the biomedical domain. Journal of Biomedical Informatics 40(3), 288–299 (2007)
14. Schütze, H.: Word space. In: Giles, L.C., Hanson, S.J., Cowan, J.D. (eds.) Advances in Neural Information Processing Systems 5, pp. 895–902. Morgan Kaufmann, San Francisco (1993)
15. Siklósi, B., Novák, A.: Detection and Expansion of Abbreviations in Hungarian Clinical Notes. In: Castro, F., Gelbukh, A., González, M. (eds.) MICAI 2013, Part I. LNCS (LNAI), vol. 8265, pp. 318–328. Springer, Heidelberg (2013)
16. Siklósi, B., Novák, A.: Identifying and Clustering Relevant Terms in Clinical Records Using Unsupervised Methods. In: Besacier, L., Dediu, A.-H., Martín-Vide, C. (eds.) SLSP 2014. LNCS (LNAI), vol. 8791, pp. 233–243. Springer, Heidelberg (2014)
17. Siklósi, B., Novák, A., Prószéky, G.: Context-aware correction of spelling errors in Hungarian medical documents. In: Dediu, A.-H., Martín-Vide, C., Mitkov, R., Truthe, B. (eds.) SLSP 2013. LNCS (LNAI), vol. 7978, pp. 248–259. Springer, Heidelberg (2013)
18. Siklósi, B., Orosz, G., Novák, A., Prószéky, G.: Automatic structuring and correction suggestion system for hungarian clinical records. In: 8th SaLTMiL Workshop on Creation and Use of Basic Lexical Resources for Lessresourced Languages, pp. 29–34 (2012)
19. Siklósi, B., Novák, A., Prószéky, G.: Context-aware correction of spelling errors in Hungarian medical documents. Computer Speech and Language (2014)
20. Sokal, R.R., Rohlf, F.J.: The comparison of dendrograms by objective methods. Taxon 11(2), 33–40 (1962)
21. Sridharan, S., Murphy, B.: Modeling Word Meaning: Distributional Semantics and the Corpus Quality-Quantity Trade-Off. In: Proceedings of the 3rd Workshop on Cognitive Aspects of the Lexicon, pp. 53–68. The COLING 2012 Organizing Committee (2012)
22. Turney, P.D., Pantel, P.: From frequency to meaning: Vector space models of semantics. J. Artif. Int. Res. 37(1), 141–188 (2010)
23. Ward, J.H.: Hierarchical grouping to optimize an objective function. Journal of the American Statistical Association 58(301), 236–244 (1963)
24. Zhang, J.: Representations of health concepts: a cognitive perspective. Journal of Biomedical Informatics 35(1), 17–24 (2002)

Medical Entities Tagging Using Distant Learning

Jorge Vivaldi[1] and Horacio Rodríguez[2]

[1] Universitat Pompeu Fabra, Roc Boronat 132, Barcelona, Spain
jorge.vivaldi@upf.edu
[2] Polytechnical University of Catalonia, Barcelona, Spain
horacio@lsi.upc.edu

Abstract. A semantic tagger aiming to detect relevant entities in medical documents and tagging them with their appropriate semantic class is presented. In the experiments described in this paper the tagset consists of the six most frequent classes in *SNOMED-CT* taxonomy (*SN*). The system uses six binary classifiers, and two combination mechanisms are presented for combining the results of the binary classifiers. Learning the classifiers is performed using three widely used knowledge sources, including one domain restricted and two domain independent resources. The system obtains state-of-the-art results.

1 Introduction

Semantic Tagging (*ST*) can be defined as the task of assigning to some linguistic units of a text a unique tag from a semantic tagset. It can be divided in two subtasks: detection and tagging. The first one is similar to term detection while the latter is closely related to Named Entity Classification, *NEC*.

Other *NLP* tasks related to ST are *Word Sense Disambiguation* (*WSD*), aiming to tag each word in a document with its correct sense from a senses repository, and *Entity Linking* (*EL*), aiming to map mentions in a document to entries in a Knowledge Base.

The key elements of the *ST* task are:

i) *the document*, or document genre, to be processed. We focus in this paper on domain restricted *ST*, specifically on the medical domain and the genre of documents treated are English Wikipedia[1] (*WP*) pages.
ii) *the linguistic units* to be tagged. There are two commonly followed approaches. Those that tag the entities occurring in the text and those that tag the mentions of these entities. Frequently, entities are represented by co-reference chains of mentions. Consider the following example (from the article "Asthma" of the English *WP*). "*Asthma* is thought to be caused by ... *Its* diagnosis is usually based on ... *The disease* is clinically classified ...". In these sentences there is an entity (*asthma*) referred three times, and, thus, forms a co-reference chain of three mentions. In this research, units to be tagged are terminological string found in WP.

[1] http://en.wikipedia.org/

A. Gelbukh (Ed.): CICLing 2015, Part II, LNCS 9042, pp. 631–642, 2015.
DOI: 10.1007/978-3-319-18117-2_47

iii) *the tagset.* A crucial point is its granularity (or size). The spectrum of tagset sizes is immense. In one extreme of the spectrum, fine-grained tagsets can consist of thousands (as is the case of *WSD* systems that use WordNet[2], *WN*, synsets as tags), or even millions (as is the case of wikifiers that use WP titles as tags). In the other extreme we can found coarse-grained tagsets. In the medical domain, for instance, in the *i2b2/VA* challenge, [1], the tagset consisted on three tags: *Medical Problem*, *Treatment*, and *Medical Test*. In the *Semeval-2013 task 9*, [2], focusing on drug-drug interation (*DDI*). Besides these task specific tagsets, subsets of Category sets in the most widely used medical resources (*MeSH®*, *SNOMED-CT*[3], *UMLS®*) are frequently used as tagsets. In this research we used a subset of the SN top categories.

In this work we deal with the task of *ST* of non-restricted documents in the medical domain using a tagset of the six most productive top semantic categories in *SNOMED-CT* (*SN*), (*Procedure, Body_structure, Substance, Pharmaceutical_or_biologic_product, Clinical_finding_disorder*, and, *Qualifier_value*). Our approach consists of learning six binary classifiers, one for each semantic class, whose results are combined by a simple meta-classifier. The cases to be classified are the mentions in the document corresponding to term candidates, i.e. sequence of words whose POS can form a valid terminological expression to refer to any of the concepts in the tagset. No co-reference resolution is attempted and, so, co-referring mentions could be tagged differently.

Most of the approaches to *ST* for small-sized tagsets, as our, use supervised Machine Learning, *ML*, techniques. The main problem found when applying these techniques is the lack of enough annotated corpora for learning. In our system we overcome this problem following a distant learning approach. Distant learning is a paradigm for relation extraction, initially proposed by [3], which uses supervised learning but with supervision not provided by manual annotation but obtained from the occurrence of positive training instances in a knowledge source or reference corpus. In the research reported here we have used *SN*, *WP*, and DBPEDIA[4], *DB*.

After this introduction, the organization of the article is as follows: In section 2 we sketch the state of the art of *ST* approaches. Section 3 presents the methodology followed. The experimental framework is described in section 4. Results are shown and discussed in section 5. Finally section 6 presents our conclusions and further work proposals.

2 Related Work

Within *ST*, the first faced problem and the one that has attracted more attention is *WSD*. [4] and [5] offer two excellent surveys on this issue. A more recent survey, covering many *ST* techniques and comparing them, can be found in [6]. [7] present an unified framework including *WSD* and *EL*.

[2] `http://wordnet.princeton.edu/`

[3] `http://ihtsdo.org/snomed-ct/`

[4] `http://wiki.dbpedia.org/`

Wikifiers proceed mostly into two steps: candidate detection and classification/ranking although facing the two tasks at a time has revealed some improvements. See [8] for an excellent, recent and comprehensive analysis.

Closely related to wikification is the task of *EL*. This task has got an explosive development starting with the *EL* challenge within the *TAC KBP* framework[5], from 2010. Overviews of the contests are the main sources of information: [9], [10], [11], and [12].

English is, by far, the most supported language for biomedical resources. The National Library of Medicine (NLM®) maintains the Unified Medical Language System (UMLS®) that groups an important set of resources to facilitate the development of computer systems to "understand" the meaning of the language of biomedicine and health. Please note, that only a fraction of such resources are available for other languages.

A relevant aspect of information extraction is the recognition and identification of biomedical entities (like *disease*, *genes*, *proteins* ...). Several named entity recognition (NER) techniques have been proposed to recognize such entities based on their morphology and context. NER can be used to recognize previously known names and also new names, but cannot be directly used to relate these names to specific biomedical entities found in external databases. For this identification task, a dictionary approach is necessary. A problem is that existing dictionaries often are incomplete and different variations may be found in the literature; therefore it is necessary to minimize this issue as much as possible.

There is a number of tools that take profit of the UMLS resources. Some the more relevant are:

- *Metamap* [13] is a pipeline that provides a mapping among concepts found in biomedical research English texts and those found in the UMLS Metathesaurus ®. For obtaining such link the input text undergoes a lexical/syntactic analysis and a number of mapping strategies. Metamap is highly configurable (it has data, output and processing options) and is being widely used since 1994 by many researchers for indexing biomedical literature.
- *Whatizit* [14] is also a pipeline for identifying biomedical entities. It includes a number of processes where each one is specialized in a type of task (*chemical entities*, *diseases*, *drugs*...). Each module processes and annotates text connecting to a publicly available specific databases (e.g. UniProtKb/Swiss-Prot, gene ontology, DrugBank....).

Keeping on the medical domain, an important source of information are the proceedings of the *2010 i2b2/VA challenge on concepts, assertions, and relations in clinical text*, [1]. The challenge included three sub-tasks, the first one, Concept Extraction, namely patient medical problems, treatments, and medical tests, corresponding to *ST*[6]. Almost all the participants followed a supervised approach. Regarding the first task, the one related to our system, final results

[5] `http://www.nist.gov/tac/2014/KBP/`

[6] The other two tasks were Assertion classification and Relation classification.

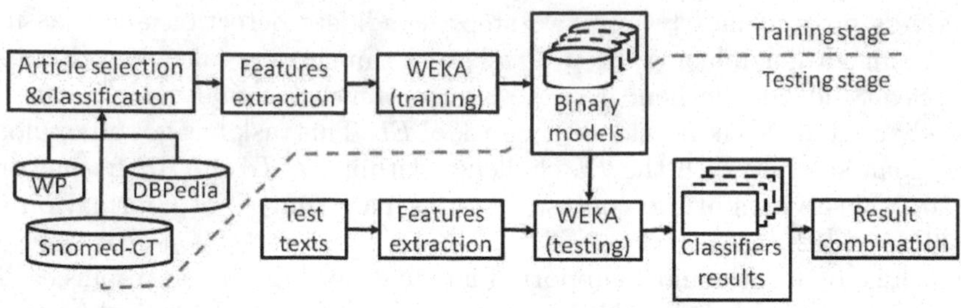

Fig. 1. Train and testing pipelines

(evaluated using F1 metric) range from 0.788 to 0.852 for exact matching and from 0.884 to 0.924 for the lenient inexact matching.

A more recent and interesting source of information is the DDI Extraction 2013 (task 9 of Semeval-2013, [2]. Focusing on a narrower domain, Drug-Drug interaction, the shared task included two challenges: i) Recognition and Classification of Pharmacological substances, and ii) Extraction of Drug-Drug interactions. The former is clearly a case of *ST*, in this case reduced to looking for mentions of drugs within biomedical texts, but with a finer granularity of the tagset, It included *drug*, *brand*, *group* (group of drug names) and *drug-n* (active substances not approved for human use). Regarding the first task, the overall results (using F1) range from 0.492 to 0.8. As DDI corpus was compiled from two very different sources, DrugBank definitions and Medline abstracts, the results are quite different depending on the source of the documents, for DrugBank, the results range from 0.508 to 0.827, while for Medline, obviously more challenging, the results range from 0.37 to 0.53.

3 Methodology

3.1 Outline

This paper, as most of the tools showed in section 2 proposes a machine learning solution to a tagging task. Therefore, it requires two main steps: train and annotatation (see Figure 1). The main drawback of this type of solutions is the dependency on annotated documents, which usually are hard to obtain. Our main target in this research is to train a classifier minimizing the impact of this issue and keeping good results. For such a purpose we use, within the distant learning paradigm, as learning examples, a set of seed words obtained with a minimal human supervision. Another main point is the tagset to be used. We decided to use as semantic classes the top level categories of the *SN* hierarchy. More specifically its six more frequent classes.

We obtain an instance-based classifier (upper section in Figure 1) for each semantic class using seed words extracted from three widely used knowledge sources (Section 3.2). The only form of human supervision is, as described below, the assignment of about two hundred *WP* categories to their appropriate *SN* semantic class. Later (lower section in Figure 1) such models are used to classify new instances.

3.2 Features Extraction

To obtain the seed terms needed for learning the classifiers, we proceed in three ways, using two different general purpose knowledge sources, *WP* and *DB*, and one, *SN*, specific for the medical domain (see, [15] and [16] for analysis of these and other resources).

WP, although being a general purpose resource, densely covers the medical domain; it contains terminological units from multiple medical thesauri and ontologies, such as Classification of Diseases and Related Health Problems (ICD-9, ICD-10), Medical Subject Headings (MeSH), and Gray's Anatomy, etc. *DB* is one of the central linked data dataset in *LOD*. It currently contains more than 3.5 million things, and 1 billion *RDF* triples with a nice coverage of the medical domain. *SN*, with more than 350,000 concepts, 950,000 English descriptions (concept names) and 1,300,000 relationships is the largest single vocabulary ever integrated into *UMLS*[7].

The main characteristics of the methods followed to obtain the seed terms from each of the above mentioned knoledge sources are the following:

a) *Wikipedia based seed terms.* Following the approach described in [17], that automatically extracts scored lists of terms from both *WP* pages titles and *WP* categories titles, we got the set of the most reliable *WP* categories[8]. This resulted on a set of 239 *WP* categories. We manually assigned to such categories a unique *SN* class (considering the full set of 19 classes). For each of these categories we obtained the full set of associated pages. For each page, we calculate a *purity factor*, i.e. a score (ranging in [0,1]), of the appropriateness of such page to a given *SN* class[9]. Finally, we only keep the most frequent *SN* classes (the set of 6 classes quoted before) and for such classes only the pages whose purity factor is above a threshold. In fact we have used only unambiguous pages, i.e. those having a purity of 1.
b) *SNOMED-CT based seed terms.* Based only in the six most frequent *SN* classes obtained as described above, we extracted for each class all the terms belonging to it that are present in both *SN* and *WP*. In this case some terms may be discarded due to its ambiguity[10].

---

7 `http://www.nlm.nih.gov/research/umls/`

8 See [17] for details about the way of obtaining such categories from *WP* resources. The system provides terms corresponding to both *WP* pages and categories, but we use here only the later.

9 A purity 1 means that all the *WP* categories attached to the page are mapped (directly or indirectly) into the same *SN* class, lower values of the purity may mean that the assignment of *WP* categories to *SN* classes is not unique or not exists.

10 It may happen that a given term from *SN* may be linked to two (or more) classes as for example the terms *inflammation* and *chlorhexidine*. The former is linked to both

c) *DBpedia based seed terms.* *DB* resulted the most productive way of obtaining seed terms, but the application of our extraction procedure, described below, was possible only for 3 of the 6 semantic classes, namely, *Clinical Finding/Disorder*, *Body_structure*, and *Pharmaceutical/biological product.* For accessing *DB* data we used the *DB Sparql endpoint*[11] that allows an efficient way of accessing the data. *DB* uses *RDF* as data model. So, the pieces of information are triples of the form <?s, ?p, ?o>[12]. The idea is starting with a set of seed terms, for each semantic tag, *t*, obtained from a knowledge source, *k*, ('sn' or 'wp'). Let S_t^k one of these sets, i.e. it contains the terms extracted from source *k* and mapped into the semantic class *t*. For instance, $S_{Body\_Structure}^{wp}$ is the set of seed words extracted from *WP* and tagged as *Body_structure.* For each $x \in S_t^k$ we collect all the <x, ?p, ?o> triples from *DB.* Let R_t^k the bag of predicates (?p) recovered. Some of the predicates are useless because its support (cardinality) is small. We denote the support of a predicate $p \in R_t^k$ as $support_t^{specific}(p)$[13], i.e. the count of all the triples involving whatever of the seed words of *t* and the predicate *p*. Obviously, many of the predicates, in spite of their support, are useless because they are not specific of the domain (for instance for the 318 seed terms in $S_{Body\_Structure}^{wp}$, the highest supported predicate (with a support of 412) was *http://www.w3.org/2000/01/rdfs-schema#label* obviously too general for being useful. The next step consists on collecting for each $p \in R_t^k$ all the <?s, p, ?o> triples from *DB*, i.e. the count of triples involving the predicate *p* without constraining the subject. We denote the support of a predicate $p \in R_t^k$ in this new searching as $support^{generic}(p)$. We compute then the specificity_ratio between both supports for the semantic class *t*:

$$specificity\_ratio_t(p) = \frac{support_t^{generic}(p)}{support_t^{specific}(p)}$$

From the predicates in R_t^k we remove those with small ratio (below a $threshold_{min}$) because they are not very productive and those with extremely high ratio (above a $threshold_{max}$) because they are too general and generate a lot of noise[14]. Additionally we have to remove those predicates selected for more than one semantic tag (for instance, the predicate *http://dbpedia.org/ontology/meshid* was selected as valuable predicate for both *Qualifier_Value* and *Clinical Finding/Disorder* and, so, has been removed from the two sets, because we are not able to decide the correct

qualifier value and *body structure* while the latter is linked to *pharmaceutical/biologic product* and *substance.* This issue may arise in all mentioned methods.

[11] http://dbpedia.org/sparql

[12] Often this notation stands for subject, predicate and object. The question mark prefix is used for identifying variables.

[13] The superindex *specific* refers to the search involving the seed terms, while the superindex *generic* refers to the unconstrained search.

[14] In the experiments reported here we set these two thresholds to 10 and 100 respectively.

assignment. With this strict criterion some of the semantic tags resulted on a null R_t^k.

In the first case we consider that the seed terms have been obtained with low human supervision while in the other two cases no supervision was needed. As can be noticed by the way of collecting the seed terms, above, in all the three cases terms have associated *WP* pages. The results, so, are sets of WP pages to be used for learning the classifiers.

Following [18], we generate training instances by automatically labelling each instance of a seed term with its designated semantic class. When we create feature vectors for the classifier, the seeds themselves are hidden and only contextual features are used to represent each training instance. Proceeding in this way the classifier is forced to generalize with limited overfitting.

3.3 ML Machinery

We created a suite of binary contextual classifiers, one for each semantic class. The classifiers are learned using, as in [18], SVM models using Weka toolkit [19]. Each classifier makes a weighted decision as to whether a term belongs or not to its semantic class. Examples for learning correspond to the mentions of the seed terms in the corresponding *WP* pages. Let $x_1, x_2, \ldots, x_n$ the seed terms for the semantic class t and knowledge source k, i.e. $x_i \in R_t^k$. For each x_i we obtain its *WP* page and we extract all the mentions of seed terms occurring in the page. Positive examples correspond to mentions of seed terms corresponding to semantic class t while negative examples correspond to seed terms from other semantic classes. Frequently, a positive example occurs within the text of the page but often many other positive and negative examples occur as well. Features are simply words occurring in the local context of mentions.

The corpus of each semantic class is divided into training and test sections. For processing the full corpus we use an in-house general purpose sentence segmenter and POS tagger to identify non empty words in each sentence and create feature vectors that represent each constituent in the sentence. For each example, the feature vector captures a context window of n words to its left and right[15] without surpassing sentence limits.

For evaluation we used *WP categories - SN classes* mappings as gold standard. We considered for each semantic class t a gold standard set including all the *WP* pages with purity 1, i.e. those pages unambiguously mapped to t. The accuracy of the corresponding classifier is measured against this gold standard set.

4 Experimental Framework

First, we proceed to collect the seed terms for each semantic class t and each knowledge source k. The results, depicted in Table 1 and Table 2, are discussed

[15] In the experiments reported here n was set to 3.

below. The processes for extracting seed words from to *WP* and *SN* are simple and no further explanations are needed. The third case, *DB*, however, is more complex and merits some attention. Let, for instance $k = DB$ and $t = Body\_structure$. We start the process with the 665 terms obtained from the knowledge source *WP*, see Table 2. Querying the *DB Sparql endpoint* with a set of queries <x, ?p, ?o>, for each of the values of $x \in S^{wp}_{Body\_Structure}$, we obtained 348 *rdf* triples involving 127 predicates. The most frequent predicate was *http://www.w3.org/2000/01/rdf-schema#label*, with 318 occurrences. We removed from the list predicates occurring less than 100 times resulting on a list of 28 predicates, named as $P_{Body\_Structure}$. We query again the *DB Sparql endpoint* now without instantiating the seed terms, i.e. we collect all the *rdf* triples involving the 28 predicates in $P_{Body\_Structure}$. The set of queries was in this case <?x, p, ?o>, for all the values of $p \in P_{Body\_Structure}$. We computed from the result the specificity-ratio and performed the filtering process described in section 3.2. Only the four predicates in Table 3 remained after the filtering process. Note that the specificity ratio (column 4) and generic count (column 3) assure a nice coverage.

Table 1 shows the global figures of the extraction process. In all the cases we found for each method a number of terms (row 2), a subset of them have the corresponding WP article (row 3) and the articles finally used (row 4). The reason to discard some WP articles are: i) only pages with a length greater that 100 words are accepted, ii) some pages has been discarded due to difficulties in extracting useful plain text (pages consisting mainly of itemized lists, formulas, links, and so) and iii) Only *WP* pages with a purity 1 have been selected. For the *DB* source, as discussed above, seed terms were extracted only for three of the semantic classes. In Table 2 we show the number of accepted terms splitted according the semantic class to which they belong.

Table 1. Terms effectively used for training

	WP only	SNOMED	DBpedia
WP categories	239	not applicable	not applicable
Total number of terms	10,802	74,107	15,761
terms found in WP	3,683	6,017	15,117
terms used	2,402	2,979	8,682

5 Results

As mentioned above the evaluation settings have been followed using *WP* categories / *SN* classes mappings as golden standard and *WP* pages as input documents. For each seed term we obtained its corresponding *WP* page and, after cleaning, POS tagging, and sentence segmenting, we extracted all the mentions. For each mention the vector of features is built and the 6 learned binary classifiers are applied to it. If none of the classifiers classify the instance as belonging

Table 2. Terms effectively used for training according to its SNOMED class

	WP only	SNOMED	DbPedia
Procedure	82	325	0
Substance	87	373	0
Body structure	665	143	1,952
Pharmaceutical/biological product	78	418	4,471
Qualifier value	973	108	0
Clinical Finding/Disorder	517	1,612	2,259
Total	2,402	2,979	8,682

Table 3. Predicates for collecting seed terms for the semantic class *body structure*

	specific terms	general terms	specificity ratio
http://dbpedia.org/ontology/grayPage	118	3,102	26.29
http://dbpedia.org/property/graysubject	118	3,118	26.42
http://dbpedia.org/property/graypage	118	3,110	26.36
http://dbpedia.org/ontology/graySubject	118	3,111	26.36

to the corresponding semantic class no answer is returned. If only one of the classifiers classifies positively the instance, the corresponding class is returned. Otherwise a combination step has to be carried out. For combining the results of the binary classifiers two methods have been implemented:

- *Best Result.* As results of binary classifiers are scored, this method simply returns the class of the best scored individual result.
- *Meta-classifier.* A SVM multiclass classifier is trained using as features the results of the basic binary classifiers together the context data already used in the basic classifiers. The resulting class is returned.

Table 4 depicts the global results got when applying both combination methods for the three knowledge sources extracted [16]. As it can be seen, using the metaclassifier slightly outperforms the best score method. Using *DB* as source of seed words consistently outperforms the other sources although, as said above, is only applicable to three of the six *SN* semantic classes, so for the other classes another source should be used.

It is difficult to compare our results with other state-of-the-art systems performing the same task because of the lack of gold standard dataset and the differences on used tagsets. A shallow comparison could be carried out with the

[16] Reported values are an average over the results for each *SN* class. Actual values, for the case of *SN* only seed terms, range among 73.0 to 94.8 (precision) and 67.1 to 93.6 (recall).

Table 4. Results obtained with different seed terms sources and corpus sections

Origin of the seed terms	Best result	using a meta-classifier	
		precision	recall
WP	87,4	89,7	89,4
SNOMED	87,4	88,9	88,7
DBpedia	94,0	94,9	94,9

Concept Extraction task of the *2010 i2b2/VA challenge on concepts, assertions, and relations in clinical text*, [1] and with the DDI Extraction 2013 (task 9 of Semeval-2013, [2], both sketched in section 2. Our results clearly outperform those obtained in these contests (specially if in the second case the comparison is performed with the results on Medline sources, closer to *WP* pages), although to be fair, and lacking a direct comparison, we simply say that our results can be considered state-of-the-art.

6 Conclusions and Further Work

We have presented a system that automatically detects and tags medical terms in general medical documents. The tagset used is derived from *SN* taxonomy. The results of the system, although not directly comparable with other systems, seem to reach at least state-of-the-art accuracy (compared with best systems in related contests). A relevant benefit of this approach is that the effort for obtaining positive/negative exemples for training has been reduced to a minimum.

The framework developed allows to perform additional experimenting changing several design parameters like the number of terms used for training, context width, features definition, etc. Some tests will be performed to optimize such parameters.

Some of the tools used in this experimentation are general purpose ones. Its performance may not be appropriate for some medical terms (ex. *1,3-difluoro-2-propanol* or *8-cyclopentyl-1,3-dipropylxanthine*, among others). We plan to introduce some improvement in our tools or use already existing/available specialised tools, such as Metamap, [13].

Although no direct comparison with other systems can be made, as justified in section 5, a shallow comparison with similar systems has been done and our results could be considered state-of-the-art.

Several lines of research and a pending work will be followed in the next future.

- As our results are based on three knowledge sources, an obvious way of possible improvement is the combination and/or the specialization of the resources for learning more accurate classifiers. The impossibility of applying the *DB* based approach, clearly the most productive one, to all the classes merits a deeper investigation.

- Using a finer grained tagset (for instance, replacing the general class *Pharmaceutical_or_biological_product* by the *DDI* tagset) could result on improvements in experiment replicating Semeval 2013 challenge.
- Moving from semantic tagging of medical entities to semantic tagging of relations between such entities is a highly exciting objective, in the line of recent challenges in the medical domain (and beyond).

Acknowledgements. This work was partially supported by the SKATER project (Spanish Ministerio de Economía y Competitividad, TIN2012-38584-C06-01 and TIN2012-38584-C06-05).

References

1. Uzuner, Ö., South, B.R., Shen, S., DuVall, S.L.: 2010 i2b2/va challenge on concepts, assertions, and relations in clinical text. J. Am. Med. Inform. Assoc. 18, 552–556 (2011)
2. Segura-Bedmar, I., Martínez, P., Zazo, M.H.: Lessons learnt from the ddi extraction-2013 shared task. Journal of Biomedical Informatics (2014), ISSN: 1532-0464
3. Mintz, M., Bills, S., Snow, R., Jurafsky, D.: Distant supervision for relation extraction without labeled data. In: Proceedings of the ACL, pp. 1003–1011 (2009)
4. Agirre, E., Edmonds, P.: Word sense disambiguation: Algorithms and applications. In: AAAI Workshop. Nancy Ide and Chris Welty (2006)
5. Navigli, R.: Word sense disambiguation: A survey. ACM Comput. 41 (2009)
6. Gerber, A., Gao, L., Hunte, J.: A scoping study of (who, what, when, where) semantic tagging services. In: Research Report, eResearch Lab. The University of Queensland (2011)
7. Moro, A., Roganato, A., Navigli, R.: Entity linking meets word sense disambiguation: A unified approach. Transactions of ACL (2014)
8. Roth, D., Ji, H., Chang, M.W., Cassidy, T.: Wikification and beyond: The challenges of entity and concept grounding. In: Tutorial at the 52nd Annual Meeting of the Association for Computational Linguistics (2014)
9. Ji, H., Grishman, R., Dang, H.T., Griffitt, K., Ellis, J.: Overview of the tac 2010 knowledge base population track. In: Text Analysis Conference, TAC (2010)
10. Ji, H., Grishman, R., Dang, H.T.: Overview of the tac 2011 knowledge base population track. In: Text Analysis Conference, TAC (2011)
11. Mayfield, J.: Javier Artiles, H.T.D.: Overview of the tac2012 knowledge base population track. In: Text Analysis Conference, TAC (2012)
12. Mayfield, J., Ellis, J., Getmana, J., Mott, J., Li, X., Griffitt, K., Strassel, S.M., Wright, J.: Overview of the kbp 2013 entity linking track. In: Text Analysis Conference, TAC (2013)
13. Aronson, A.R., Lang, F.M.: An overview of metamap: historical perspective and recent advances. JAMIA 17, 229–236 (2010)
14. Rebholz-Schuhmann, D., Arregui, M., Gaudan, S., Kirsch, H., Jimeno, A.: Text processing through web services: calling whatizit. Bioinformatics Applications Note 4, 296–298 (2008)
15. He, J., de Rijke, M., Sevenster, M., van Ommering, R., Qian, Y.: Generating Links to Background Knowledge: A Case Study Using Narrative Radiology Reports, Glasgow, Scotland, UK (2011)

16. Yeganova, L., Kim, W., Comeau, D., Wilbur, W.J.: Finding biomedical categories in medline ®. In: BMC Biomedical Semantics (2012)
17. Vivaldi, J., Rodríguez, H.: Using wikipedia for term extraction in the biomedical domain: first experience. Procesamiento del Lenguaje Natural 45, 251–254 (2010)
18. Huang, R., Riloff, E.: Inducing domain-specific semantic class taggers from (almost) nothing. In: Proceedings of the 48th Annual Meeting of the Association for Computational Linguistics, Uppsala, Sweden, pp. 275–285 (2010)
19. Hall, M., Frank, E., Holmes, G., Pfahringer, B., Reutemann, P., Witten, I.: The weka data mining software: An update. SIGKDD Explorations (2009)

Identification of Original Document by Using Textual Similarities

Prasha Shrestha and Thamar Solorio

University of Houston
Department of Computer Science
4800 Calhoun Rd. Houston, TX, 77004
pshrestha3@uh.edu, solorio@cs.uh.edu

Abstract. When there are two documents that share similar content, either accidentally or intentionally, the knowledge about which one of the two is the original source of the content is unknown in most cases. This knowledge can be crucial in order to charge or acquit someone of plagiarism, to establish the provenance of a document or in the case of sensitive information, to make sure that you can rely on the source of the information. Our system identifies the original document by using the idea that the pieces of text written by the same author have higher resemblance to each other than to those written by different authors. Given two pairs of documents with shared content, our system compares the shared part with the remaining text in both of the documents by treating them as bag of words. For cases when there is no reference text by one of the authors to compare against, our system makes predictions based on similarity of the shared content to just one of the documents.

Keywords: original document, bag-of-words, document provenance, plagiarism.

1 Introduction

When two documents have shared content, the first question that arises is whether it was the author of one document or the other that produced the original content. The answer to this question has important implications in terms of establishing provenance and authorship of the information in the shared content. When the presence of this shared content has been found by plagiarism detection systems, identifying the original document can help somebody to be exonerated of plagiarism. This will especially be useful in the academic scenario when two students are found to have similar content in their assignment. Usually both are held under blame. But in same cases, the student whose work was plagiarized might not even be aware of it. Another obvious use is when a person makes a claim of plagiarism of their work when is no information about which version of the document came first. In this case a system that finds the original document can help to settle the dispute.

Identifying the original document can also be a first step towards establishing the provenance of a document. Provenance is important because it has critical applications in security. There has been a lot of work in recording provenance for different types of data in e-science [1]. Several methods have also been proposed for developing automatic provenance recording systems in the cloud. There have been standards set on the

A. Gelbukh (Ed.): CICLing 2015, Part II, LNCS 9042, pp. 643–654, 2015.
DOI: 10.1007/978-3-319-18117-2_48

properties that these provenance recording systems must satisfy [2] and all the details that provenance information for cloud processes should contain [3]. There have even been work done for recording provenance of experimental workflows and even to establish provenance for art [4]. But document provenance has hardly had any research effort devoted to it. It would be easy to record document provenance but this is rarely done and in the instances of plagiarism, people are likely to try to hide this information rather than to document it. If the provenance has not been recorded, establishing provenance from a pool of documents is the only option left and it is a very hard problem. The problem is more tractable if the modifications on the document have been made by different authors. For cases when a document written by an author gets subsequently modified by other authors, our method can be useful to extract provenance. If the whole document or parts of the document has been modified by another author, our system can compare the modified section with other works from both authors to decide which version of the document is the original one. Our method can be applied for all the documents in question pairwise until the entire lineage is traced.

The problem we are dealing with and authorship attribution are also closely related. But one major difference is that in authorship attribution, the document or piece of text that we are trying to attribute to an author is untouched by any other author. It has been written solely by that author. But in our case, we have a piece of text that has been written by one author and in most cases, modified by another author to use in his own work. We are trying to attribute the text used by both authors to one of them. This adds a layer of complexity to our problem. Nonetheless, the ideas used in this work can also be applied in the scenario when the authorship of a piece of text is disputed between two authors. Given that text with disputed authorship and other documents written by these authors, one can use our system without modification in order to attribute the work to one of the authors.

We have used a simple yet effective method in order to solve the problem of finding the original document out of two documents. We first separate the content shared by them from both documents. We then divide the rest of the text in the documents into segments and create a bag of word representation of these segments and also of the segment with the shared content. We then extract the top most frequent words from each of these segments. The next step is to find the overlap between the top words from the shared content and the top words of all of the segments of both documents. The document whose segments have the higher average overlap with the shared content will be classified as the original document. Similarly, from the perspective of document provenance, the shared segment will have originated from this document and thus will be the predecessor of the other document.

This paper also deals with the case when between the two documents, in one of them all of the text is similar to parts of the other document. In this case, there is no additional reference text to compare against for one of the authors. Here, the prediction needs to be done only based upon the similarity or dissimilarity of the shared text to the text of only one of the authors. This scenario can happen in real life as well where all of the text written by an author has been fully lifted from one or more sources without adding any original content. This is a much harder problem and will generally have lower accuracy than when text from both authors is available. Our system can be used, although cautiously for this scenario as well.

2 Related Work

At the time of this writing, we were able to find only one previous work that deals with a similar problem as ours. Grozea and Popescu (2010), in their work, have proposed a solution for finding the direction of plagiarism [5]. The idea behind their approach is that the n-grams present in the plagiarized passage will repeat more throughout the original document than in the plagiarized document. This makes it very likely for these n-grams to occur much earlier in the source document than in the plagiarized document. They have used character 8-grams and only considered the first one of the n-gram matches between the plagiarized and the non-plagiarized sections. Then they plotted these matches and then found the asymmetry in the plots. Their work is a continuation of the system they submitted to the PAN 2009 External Plagiarism Detection Competition and they used the same data for this experiment as well. They were able to obtain an overall accuracy of 75.42% on this dataset.

The above work is the only one we could find that deals with the exact same problem as the one we are trying to solve. But the work on plagiarism detection: both intrinsic and extrinsic, problems dealing with authorship and the problem of anomaly detection are relevant to our task.

Our problem is very similar to the intrinsic plagiarism detection problem. In intrinsic plagiarism detection, the task is to figure out if a document has been plagiarized or not by using the text in just that document as the reference. So, in this problem as well as our problem requires the checking of how similar parts of a document are as compared to other parts of the same document. For the intrinsic plagiarism detection problem, Stamatatos (2009) proposed that the inconsistencies within the document, mainly stylistic, can point towards the plagiarized passage [6]. They use bag of character trigrams of automatically segmented passages in the document and use a sliding text window to compare the current text in the window to the whole document. They only deal with documents that have less than half plagiarized content because otherwise the style function will represent the style of the plagiarist and not of the true author. As in this method, most of the approaches to plagiarism detection, both intrinsic and extrinsic make use of n-grams. Barrón-Cedeño and Rosso (2009) have tried to investigate the best value for n when performing an n-gram comparison [7]. They used word n-grams in their method and found out that low values of n generally work better for n-gram based methods.

Intrinsic plagiarism detection can also be modeled as a one class classification problem, with the non-plagiarized text falling under the target class and all other plagiarized texts being the outliers [8]. Stein et al. (2011) used this approach along with a large number of lexical, syntactic and semantic features. In order to perform outlier identification, they assumed that the feature values of the outliers have uniform distribution and then use using maximum likelihood estimation. They also employ as a post-processing step, a technique called unmasking by Koppel and Schler (2004) [9]. This method works by removing the most discriminating features gradually such that, after a few iterations, the remaining features cannot properly discriminate between texts written by the same author but can still discriminate between texts from different authors.

Our problem as well as the problem of intrinsic plagiarism is similar to the problem of anomaly detection as well, since a plagiarized passage behaves like an anomaly. Guthrie et al. (2007) too have used a large variety of stylistic, sentiment and readability features in order to find an anomalous segment in a text [10]. The rank features used by them are particularly unique and they use rank correlation coefficient rather than similarity measures for the rank features. They rank a list of articles, prepositions, conjunctions, pronouns, POS bigrams and POS trigrams and then calculate the Spearman rank correlation coefficient. They found out that their accuracy improves as the segment size increases.

The problem of authorship attribution is related to our problem because both involve examining a text with undoubted authorship to check if another piece of text having unknown or dubious authorship is also written by the same author. In our problem, for the case when there is no reference text from one of the authors the problem becomes even more similar to authorship attribution, albeit on text written or changed by both authors. Stamatatos (2009) noted that although many kinds of lexical, character, syntactic and semantic features are used in authorship attribution, lexical features are the most prominently employed features in authorship attribution systems [11]. They also noticed that most systems considered the text as a bag of words with the stopwords being the most discriminating and most widely used features.

In order to determine if two documents have been written by the same author, rather than treating it as a one-class classification problem, Koppel and Winter (2014) have converted the problem to a many-candidates problem [12]. While in the one-class classification problem, we only have text written by the target author and we need to find out if a given document is written by this author vs any existing author. It is not possible to obtain text for every author in the world. So, they have created impostor documents and then tried to find out if the current document is more similar to a document written by the target author over any other impostor documents. So, the complexity of the problem is reduced from being a target vs outlier problem to a classification problem with a known set of classes.

3 Methodology

The input to our system is a pair of documents with known plagiarized content between them. In most real plagiarism cases, a single document might have passages taken from multiple source documents, which is also the case in the dataset we use. For this reason, we perform our classification on a per passage basis. Our system tries to attribute each one of the plagiarized passages to one of the documents separately. For example, if there are two documents containing similar passages, one of them will be the original document for this particular passage. But there might be several such passages inside a single document, originally appearing in several other documents. Thus, for each such passage, the original document might be different. For this reason, we perform our classification on a per passage basis. This shared or plagiarized content needs to be compared with only the text that has purely been written by the authors in question. For this reason, we also remove all other passages known to be shared with some other documents. After this, we are only left with the texts written by the two authors in question. In most of the cases, we have enough text from both authors in order to make a comparison. But for a few cases, we only have text from one of the authors to compare against. The method we used for the case when we have some amount of text from both authors is described in Section 3.1. For the few cases where we only have the text from one of the authors, we describe the method we used is described in Section 3.2.

3.1 Overlap between Words

Two pieces of text written by the same author are more likely to have similar word usage patterns. We use this idea in order to compare the shared content with the rest of each authors' text. We first find the most frequent word tokens in the shared, plagiarized part and in the non-plagiarized parts of both of the documents, and then use the overlap between these tokens in order to decide which document the plagiarized passage was originally taken from. This is a two class classification problem, but with a very limited amount of data that is representative of the two classes.

In any document, and especially in the long ones, the writing style and word usages of an author can change subtly throughout the document. The particular passage that has been copied by one author from another author's document may be similar to some parts of the text, but not so much to the others. For this reason, we first divide the unplagiarized passage of both documents into segments. In most of the cases in our dataset, the text purely written by a single author i.e. the non-plagiarized part of the text is longer than the plagiarized part. We chunk the non-plagiarized text into equal length segments in such a way that there are enough segments to compare the plagiarized text against, while also keeping the segments similar in length to the plagiarized text. But for cases when the non-plagiarized text is very short in comparison to the plagiarized text, the whole text comprises a single segment. We then tokenize the segments into words and retain everything, including stopwords.

After obtaining the segments, we proceed on to extract f frequent words from each of the segments, including the plagiarized ones. We set the value of f according to the segment size so that we will have enough words to compare for documents or segments of any size. We set f to one fourth of the segment size except for the case when one of the segments is smaller than this value. In this case, f will be equal to the size of the smaller segment. Thus, for large segments, we end up taking only the most frequent words. But for small segments, there will not be many words to compare against if we just take the most frequent ones. For cases where the segments are very small, either due to the plagiarized passage being small or the unplagiarized content in one of the documents being short, we use all of the tokens.

With these most frequent words in hand, the next step is to check how similar the plagiarized passage is to the two sets of non-plagiarized text. This similarity score is calculated as shown in Equation 1 below.

$$avg\_overlap(p,u) = \frac{\sum_{i=1}^{len(u)} |fw(p) \cap fw(u_i)|}{len(u)} \quad (1)$$

The value for $avg\_overlap$ is calculated between a plagiarized segment p and the set of non-plagiarized segments u of a document. $fw(x)$ represents the most frequent words in a segment x. This score will provide us with the extent of overlap between the plagiarized segment p and the set of non-plagiarized segments u in a document. The score calculates the overlap of the plagiarized passage with each of the non plagiarized segments of a document. It then computes the average overlap. This $avg\_overlap$ score is calculated for a plagiarized passage and the set of segments of both candidate documents. For a passage, the original document is taken as the document that produces a higher score with this passage.

As is the case in most plagiarized text, most of the plagiarized segments have been obfuscated. Due to this, we actually have two different versions of the same passage, in the two documents. We choose to make predictions for them individually and then combine the results of the predictions later. We calculate $avg\_overlap$ for the first version of the passage with both of the documents and obtain a prediction for that passage about the document it actually belongs to. We repeat the same for the other version of the passage. If both versions of the plagiarized segment predict the same document as the original, the final prediction is also the same document. But if they disagree, we go back to the $avg\_overlap$ values to make the final prediction. We have two $avg\_overlap$ scores for each version of the plagiarized passage, as calculated before. We take the higher of these scores for both passages and then again compare these two scores. Our system then uses the prediction for the version of the plagiarized segment that has higher $avg\_overlap$ score with its predicted original document.

3.2 Meta Learning for Predictions Using Single Documents

In cases where a document has been fully plagiarized, there is no reference text for one or both of the authors and the method described in the previous section becomes inapplicable. This scenario occurs in three cases. First, an entire document might be the product of content plagiarized from parts of another document. Second, the entire original document might be plagiarized into another document having some content of its own. The third and very rare case is when a document is fully plagiarized to form a new document and no extra content is added to it. For this last case, we do not have any reference to compare against in either of the two documents. This problem is nearly impossible to solve, and will require information outside of the two documents and is thus outside of the scope of our work.

For the other two cases, we have some reference text for one of the two authors. We make use of this author's text to perform a one class classification to decide whether the plagiarized text has also been written by the same author. The intuition behind this method is simple. A piece of text originally written by an author will resemble other content produced by the same author.

For the document having content additional to the plagiarized text, we first divide this non-plagiarized content into segments and then obtain the most frequent word unigrams in a similar way as the previous method. We also obtain these most frequent tokens for the plagiarized content in both documents in the same way. We then calculate the overlap score for both versions of the plagiarized passage with the set of segments obtained from the document having reference text. The scoring here also used the same formula as shown in Equation 1.

We will already have in hand the scores that we obtain for the case described in Section 3.1. We use these scores as our training data to train a logistic regression model. We then take the scores that we have just obtained for the one reference document and two versions of the plagiarized passage. We feed these to get the predictions from the model. Although there are only two classes that can be predicted for each of the passages, the documents that these two classes represent vary in every instance. This makes it a hard problem to get as good results as in the case discussed in the previous section.

4 Dataset

It is hard to find data for real cases of plagiarism or unintentional copying. For our experiments, we used the dataset from the text alignment subtask of PAN plagiarism detection task. This dataset consists of a set of documents with some content taken from one document and copied into another, either verbatim or with some changes. In this dataset, a document containing the plagiarized content is called a suspicious document and a document containing purely original content is called a source document. We removed the information about whether a document was source or suspicious and treated both documents equally in order to mimic the scenario of the problem we are trying to solve.

The plagiarism detection task at PAN has been taking place every year since 2009 and they have released a new or modified corpus in most years. We performed detailed experiments on the PAN 2009 corpus, in order to compare our results with Grozea and Popescu (2010), the only other known system dealing with the same problem [5]. But we also evaluated our system on all the other existing versions of the PAN dataset. In PAN 2009 corpus, the documents have been artificially plagiarized by using different methods of obfuscation as a human plagiarist would [13]. They have used replacement by synonyms, shuffling, text insertion and deletion. In the 2009 dataset, some documents have the plagiarized passages copied verbatim, while others have high or low levels of obfuscated plagiarized passages. In another form of obfuscation called translation obfuscation, they used a machine translator to translate English passages into a chain of other languages and then translated it back to English. In newer versions of the dataset, they have also used summary obfuscation. In summary obfuscation, the passages from one document are summarized before being inserted into the other document.

Apart from the PAN dataset, we also tested our system on a prominent case of plagiarism that had appeared in the media. Many works of a famous journalist works were alleged to have been plagiarized from other sources.[1] Those allegations were found to be true and the magazines where they were published issued statements expressing that those articles did not meet their standards and some even fired him. We only collected those news articles where he had plagiarized more than two sentences from another news article. We found three such cases among the plagiarism allegations against him.

[1] https://ourbadmedia.wordpress.com/2014/08/19/did-cnn-the-washington-post-and-time-actually-check-fareed-zakarias-work-for-plagiarism

4.1 Results and Analysis

The results we obtained for the PAN 2009 dataset are shown in Table 1. We obtained an accuracy of 85.56% on the overall test dataset. This is a lot higher than the 75.42% obtained by the only known previous work [5], who also used the same dataset. In the case of real plagiarism, the more obfuscated the text is, the more it deviates from the writing style of the original author and will reflect the writing style of the plagiarist. As expected, the results were better in the case of no obfuscation and the problem was harder for higher levels of obfuscation. When there is no obfuscation, our accuracy is 88.12% but it drops down to 79.54% for high obfuscation.

Table 1. Accuracy on the PAN 2009 dataset

Data Type	Number of Passages	Accuracy (%)
No Obfuscation	26855	88.12
Low Obfuscation	26628	86.04
High Obfuscation	13658	79.54
Translation Obfuscation	6381	85.72
Overall	73522	85.56
Grozea and Popescu (2010)	73522	75.42

The results for all of the PAN datasets are shown in Table 2. We obtained accuracy comparable to the PAN 2009 dataset for the PAN 2011 and 2012 datasets as well. But the accuracy on PAN 2013 data is notably lower than on all other datasets. To find the reason for this, we looked at the lengths of the documents in these datasets. We found that the length of documents in PAN 2013 dataset is significantly shorter than that of the other PAN datasets as shown in Table 3. When documents are short, it is harder to capture the writing style of an author given that small amount of information. Our segment size is also small and there are less segments to compare the plagiarized passage against. The top most frequent words obtained might not represent how the author truly writes for this case. This made our accuracy drop significantly. There might also be a bigger problem because in the older PAN datasets, the plagiarized passage and the document where it was inserted into to create simulated plagiarism were randomly chosen. As such, the plagiarized and non-plagiarized parts of the same document might have different topics. It is possible that in the experiments with the older datasets, our system might have been doing topic classification along with the detection of the original document. But even in this PAN 2013 dataset where the corpus creators have tried to stay within the same topic for both plagiarized and non-plagiarized text, our accuracy is fairly reasonable.

We performed an analysis of the effect of length on accuracy by using the PAN 2009 dataset. Since we are comparing plagiarized passages against non-plagiarized ones, both their lengths can affect our system. For example, if the non-plagiarized portion contains just five words while the plagiarized portion contains 5000 words, although the whole document will be long, our prediction might be hampered by the brevity of

Table 2. Accuracy on other PAN datsets

Dataset	Number of Passages	Accuracy (%)
PAN 2009	73522	85.56
PAN 2011	49621	82.14
PAN 2012	12495	85.98
PAN 2013	1007	74.07

Table 3. Average length of documents across PAN datasets

Dataset	Avg. # of words per document
PAN 2009	47653
PAN 2011	50582
PAN 2012	50315
PAN 2013	2462

the non-plagiarized passage. For this analysis, we considered the length of a document as the length of its shorter portion: either the plagiarized or the non-plagiarized part. We sorted the documents in ascending order by length and divided them into 10 buckets containing equal number of documents. The first bucket contains the shortest 10% of the documents, the second bucket contains the next shortest 10% and so on while the tenth bucket contains the longest 10% of the documents. We then looked at the accuracy for these buckets. Figure 1 shows the bucket index with the average number of words in the documents of that bucket and the accuracy obtained for that bucket of documents. The accuracy for the bucket with longer documents is considerably higher, 90 than that for the bucket with shorter documents, although the curve is not ascending uniformly. But the accuracy for the tenth bucket containing the longest documents is the highest at 90.99% while the accuracy for the first document is comparatively low at 82.65%. This also further shows that the length of the documents in the dataset plays a great role in the prediction accuracy.

On our data collected from real plagiarism case as describe in 4, we were able to predict the original document correctly for two cases out of the three. We believe that the size of the documents might have again played a role in the result. The one case where we designate the wrong document as the original one is where we have the least amount of text. Although this dataset is too small to draw any conclusions, our method does seem to work well for real cases, given that there is enough text to compare.

For a small minority of documents not having reference text by one of the authors, the results obtained by using the method described in Section 3.2 is shown in Table 4. The accuracy for this method is not comparable to the case when we have reference text for both documents. This problem of identifying whether a piece of text is written by a particular author or not, given very small samples of text written by that author is an inherently hard problem and is thus inclined to suffer from lower accuracy. But as seen in the same table, this situation occurs in less than 0.5% of the data for PAN 2009-12 datasets. In real plagiarism cases as well, the plagiarist is likely to plagiarize some parts

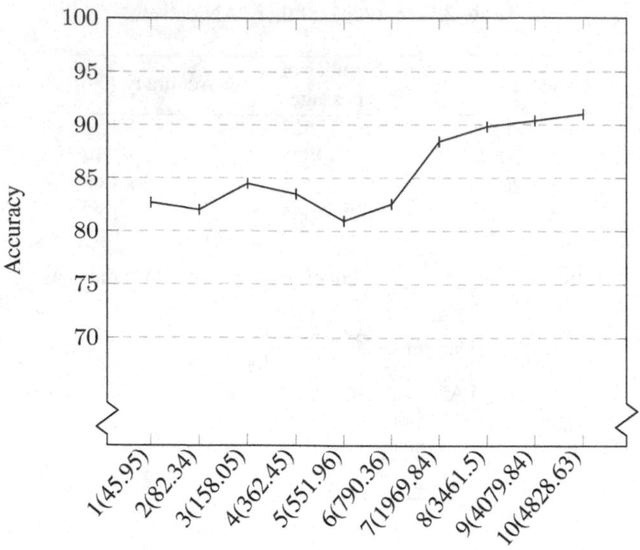

Fig. 1. Accuracy on buckets of documents sorted by length

of the document and also add his own content in order to not get caught. This was also the case for all of the real plagiarism data that we collected. The author had only copied certain parts of another article and inserted them into his own article. So, the situation where there is no other text in a document other than the plagiarized one is very less likely to occur.

Table 4. Accuracy for Meta Learning Method

Data Type	% of Total Documents	Accuracy (%)
PAN 2009	0.0054	52.65
PAN 2011	0.46	51.14
PAN 2012	0.46	43.10
PAN 2013	9.86	47.74

As is the case with systems dealing with authorship or author profiling that use word n-grams, stopwords were the most discriminating features. They were the most frequently occurring tokens in the overlap between the two segments. Apart from stopwords, the words that belong to the topic of the document also occurred in the overlap. For documents that were stories, there were also a lot of named entities present as the common words between the segments.

Apart from the documents for which we have results shown in the above tables, there are four document pairs in the dataset which belong to the third scenario as described in

Section 3.2. For these cases, neither of the methods gives a classification. This is a very hard problem and it occurred only in the 2009 dataset and that too in only four cases. A solution to this problem will surely require more information that what is available in the dataset as we will need to collect more data from the authors which out of the scope of our work.

5 Conclusions and Future Work

We have presented a method to identify the original document out of two when they have a piece of shared content between them. Our method is a good solution to the problem of finding the original document and it performs well across different datasets. As expected, the results were better for lower levels of obfuscation. Even for higher levels of obfuscation, our accuracy is close to 80%. Also, more correct predictions are made when there is sufficient text to capture the writing style of an author. As the text becomes shorter, the problem gets harder. But even for short documents, we obtain reasonable accuracy. Only in the case when one of the documents has been fully plagiarized, our accuracy is low. But since we are making predictions based on only one of the documents, lower accuracy is to be expected. There are rare cases where we cannot apply any of our methods due to neither of the two documents having any reference text. This is a problem that can have practical applications. This also relates to the problem of document provenance when there are only minor changes made to a document, as can happen when somebody is proofreading a document. This is an interesting problem and we would like to explore it in the future. We have also not dealt with the problem of self-plagiarism where an author reuses his/her own text. For this case, we cannot make use of the writing style of the author to determine which document is the original. This will require a completely different method. We also leave this for future work. But for now, we can surely say that when we have two different authors and text written by them to compare against, our method gives good performance.

Acknowledgments. We would like to thank the anonymous reviewers for providing us with the helpful feedback. This research is partially funded by The Office of Naval Research under grant N00014-12-1-0217.

References

1. Simmhan, Y.L., Plale, B., Gannon, D.: A survey of data provenance in e-Science. SIGMOD Rec. 34, 31–36 (2005)
2. Muniswamy-Reddy, K.K., Macko, P., Seltzer, M.: Provenance for the cloud. In: Proceedings of the 8th USENIX Conference on File and Storage Technologies, FAST 2010, pp. 15–14. USENIX Association, Berkeley (2010)
3. Muniswamy-Reddy, K.K., Holland, D.A., Braun, U., Seltzer, M.: Provenance-aware storage systems. In: Proceedings of the Annual Conference on USENIX 2006 Annual Technical Conference, ATEC 2006, p. 4. USENIX Association, Berkeley (2006)

4. Green, R.L., Watling, R.J.: Trace element fingerprinting of australian ocher using laser ablation inductively coupled plasma-mass spectrometry (LA-ICP-MS) for the provenance establishment and authentication of indigenous art*. Journal of Forensic Sciences 52, 851–859 (2007)
5. Grozea, C., Popescu, M.: Who's the thief? Automatic detection of the direction of plagiarism. In: Gelbukh, A. (ed.) CICLing 2010. LNCS, vol. 6008, pp. 700–710. Springer, Heidelberg (2010)
6. Stamatatos, E.: Intrinsic plagiarism detection using character n-gram profiles. In: 3rd PAN Workshop Uncovering Plagiarism, Authorship and Social Software Misuse, vol. 2, p. 38 (2009)
7. Barrón-Cedeño, A., Rosso, P.: On automatic plagiarism detection based on n-grams comparison. In: Boughanem, M., Berrut, C., Mothe, J., Soule-Dupuy, C. (eds.) ECIR 2009. LNCS, vol. 5478, pp. 696–700. Springer, Heidelberg (2009)
8. Stein, B., Lipka, N., Prettenhofer, P.: Intrinsic plagiarism analysis. Language Resources and Evaluation 45, 63–82 (2011)
9. Koppel, M., Schler, J.: Authorship verification as a one-class classification problem. In: Proceedings of the Twenty-First International Conference on Machine Learning, p. 62. ACM (2004)
10. Guthrie, D., Guthrie, L., Allison, B., Wilks, Y.: Unsupervised anomaly detection. In: Proceedings of the 20th International Joint Conference on Artifical Intelligence, IJCAI 2007, pp. 1624–1628. Morgan Kaufmann Publishers Inc., San Francisco (2007)
11. Stamatatos, E.: A survey of modern authorship attribution methods. Journal of the American Society for Information Science and Technology 60, 538–556 (2009)
12. Koppel, M., Winter, Y.: Determining if two documents are written by the same author. Journal of the Association for Information Science and Technology 65, 178–187 (2014)
13. Potthast, M., Stein, B., Barrón-Cedeño, A., Rosso, P.: An evaluation framework for plagiarism detection. In: Coling 2010: Posters, pp. 997–1005. Coling 2010 Organizing Committee, Beijing (2010)

Kalema: Digitizing Arabic Content for Accessibility Purposes Using Crowdsourcing

Gasser Akila, Mohamed El-Menisy, Omar Khaled, Nada Sharaf, Nada Tarhony, and Slim Abdennadher

Computer Science and Engineering Department,
The German University in Cairo
{gasser.akila,elmenisy,omar.saeed21,nadatarhony}@gmail.com,
{nada.hamed,slim.abdennadher}@guc.edu.eg

Abstract. In this paper, we present "Kalema", a system for digitizing Arabic scanned documents for the visually impaired such that it can be converted to audio format or Braille. This is done through a GWAP which offers a simple, challenging game that helps attract many volunteers for this cause. We show how such a tedious task can be achieved accurately and easily through the use of crowdsourcing.

Keywords: Crowdsourcing, GWAP, Arabic, Digitization, Accessibility.

1 Introduction

According to WHO [10], around 285 million people around the world suffer from visual impairment, with around 90% of them living in developing countries. The Arab visually impaired face a lot of difficulties since most of the Arabic content they need in schools and/or work is not yet digitized. It is thus hard to use such content on their own without using the help of well-sighted person(s).

In addition, due to the unavailability of digital copies of some Arabic text, valuable books have been unfortunately lost [1]. Much research has thus been conducted to enhance text digitization for Arabic [6]. The performance of Arabic Optical Character Recognition (OCR) systems for text with diacritics is unfortunately very poor. Diacritics are the signs placed above or below the text. They provide found above or under Arabic letters.

The aim of Kalema is to help digitize the Arabic content needed by the visually impaired through crowd-sourcing, by letting the crowd play a fun Game With A Purpose (GWAP) contributing to the digitization. Kalema also makes use of the social aspect by letting users sign in using their social network accounts in order to save their scores and be able to challenge their friends.

Through this paper, we will first discuss some of the related work in Section 2. Afterwards, in Section 3 we will go through the implementation process and the different modules that were implemented throughout the project. Lastly, we derive some conclusions and discuss the achieved results.

A. Gelbukh (Ed.): CICLing 2015, Part II, LNCS 9042, pp. 655–662, 2015.
DOI: 10.1007/978-3-319-18117-2_49

2 Related Work

Kalema provides a fully automated system where the admin uploads the scanned documents. Documents automatically get cut into words which can be used in a simple, yet challenging game to be played by the volunteers. This helps to eliminate the bottleneck regarding the collection and validation of users' input.

2.1 Limitations of Arabic OCR Systems

Despite of the availability of OCR systems for the Arabic language, their output is rather inaccurate [6,9,12]. The performance is unfortunately worse when it comes to Arabic text with diacritics. The output of some of the famous Arabic OCR systems such Google's Tesseract [7,11], and ABBYY [2] has poor accuracy for two main reasons. First, the Arabic calligraphy is connected and contains a lot of dots on different letters, which makes it hard for the OCR system to recognize as opposed to unconnected calligraphy like English. Furthermore, the quality of the scanned documents and different aging factors also contribute to the poor OCR output accuracy. After trying different Arabic OCR systems and confirming that their output is incomprehensible, we decided to let the crowd provide the digitization. Validation depends on how many users agree with the entered digitization.

One of the most well-known systems in the OCR correction field is Digitalkoot [8], which is a GWAP that uses crowdsourcing in order to correct the OCR output for scanned documents. The game asks users to validate whether the digital text accurately corresponds to the image of the word. The more words a user validates the higher their score gets. The main difference between Digitalkoot and Kalema is that, the OCR output for the English like scripts being used in Digitalkoot is far more comprehensible, which allows the volunteers to validate the OCR output rather than providing the digital text.

2.2 Why GWAPs?

Human Computation aims at making use of the ability of humans to solve different types of problems and perform tasks that computers are incapable of such as image recognition for example. The concept was modeled by Luis von Ahn in 2005 [13]. GWAPs [3] on the other hand apply the concept of human computation. They make use of the entertainment users get while playing a game to help in solving a problem as a side-effect. Such games could be used to collect large amounts of data instead of the rather costly traditional methods. The ESP game [14] is a very popular GWAP where users help identify the contents of an image. The two players get the same image and they have to supply different tags/labels to it. The more similar labels the two players enter, the higher their scores get. Recently, some work has been done to make use of the huge amount of Arabic-speaking Internet to digitize Arabic words that OCR programs where unable to digitize [5].

There are different social network campaigns and initiatives [4] that are concerned with digitizing Arabic content for the visually impaired. However, they are done through a manual process where an admin sends out scanned pages for volunteers who reply back with the digital Arabic text for these pages. This process is rather boring for volunteers since they need to read and type large amounts of text. Furthermore, it is tiring for the admins who need to manually collect the results, and validate them before being converted to audio format or Braille.

GWAPs can be used in different fields including computer vision, security, adult content filtering, OCR correction and digitization in general. The concept helps in achieving measurably accurate results based on crowdsourcing while still making the user motivated to perform the crowdsourced task as they are given with the feel and look of computer games. The concepts of GWAPs are suitable with our purpose as they would help digitize a large amount of Arabic content and yet, users will keep coming back to the game hoping to achieve better scores and constantly improving their typing speed and accuracy.

3 Approach

The implementation phase was divided into three major modules. The first module is the back-end server processing of the scanned pages. Such pages are uploaded through an admins interface. OCR related steps for processing the images are done through Tesseract. The second module is the actual GWAP implementation with the different gaming aspects like the gameplay, scoring, and ranking. Ruby on Rails was used to develop the application both for the admin side and the users side. The third and last module is involved in validating the users inputs and assigning scores to digitized words.

3.1 Admin Side and OCR module

An admin interface was developed to allow the system admin to create new books and upload scanned images of their pages.The uploaded pages are then cut to separate words as shown in 1 in order to be used by the players. Googles Tesseract OCR system was used to cut the scanned pages of the books into words. We installed its Linux package on our Ubuntu 12.04 server. Arabic training data was added to its training data folder to be able to detect Arabic text.

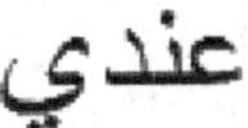

Fig. 1. Single word image produced by Tesseract

A Ruby interface with Tesseracts API was used to connect it with the application. After each page is uploaded, it is saved to the database and assigned to its appropriate book. Whenever the admin chooses to generate the words from the scanned pages, an API call to Tesseract is made which returns the coordinates and widths and heights of rectangles surrounding separate words. Such data is then used to crop the scanned image into different words which are also saved to the database and linked to their corresponding page. At this point, the generated images for separate words can be used in the gameplay.

3.2 User Interface and Gameplay

The user interface was designed to be as simple and as usable as possible. When a user first visits the game, they are greeted with a message explaining the aim of the game "to type in the words as quickly and as accurately as possible as they fall". Users are offered with the option of signing in using Facebook or Twitter so that they can keep their score and check their rankings later on. The user can then choose to proceed to play or sign in using their social network account which afterwards redirects them to the gameplay page as shown in Fig. 2.

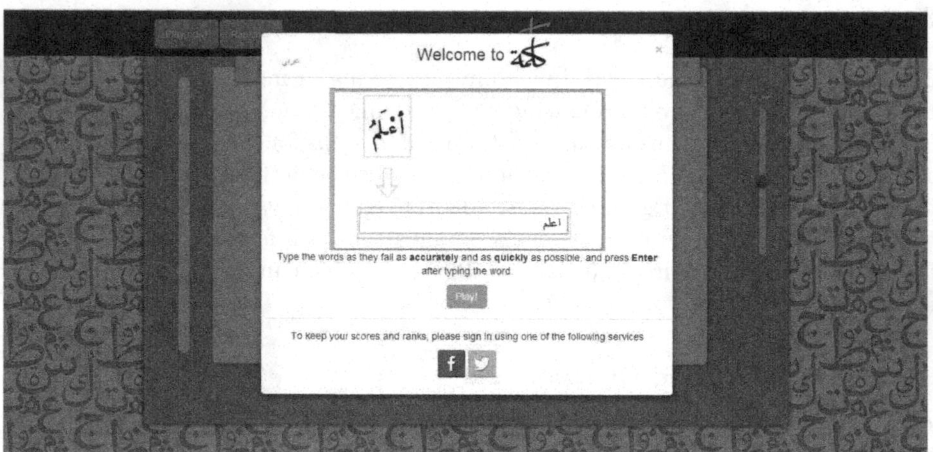

Fig. 2. Welcome screen: Showing the game instructions to users and providing the option to sign in using Facebook or Twitter

The game interface is shown in Fig. 3. Users can start the game by pressing the start button and the words will start falling. The text field where the user inputs the words get automatically focused when the game starts and the user should start to type in the words and press the Enter key after entering any word in order for the next word to appear and start falling. There is a progress bar that gets filled relative to the predefined number of words "currently 20 words".Users can also change the speed of the falling words using a slider. If the user does not press enter before the word reaches the bottom of the game container, their input is still recorded and the text field is cleared automatically to allow the new word to be typed in.

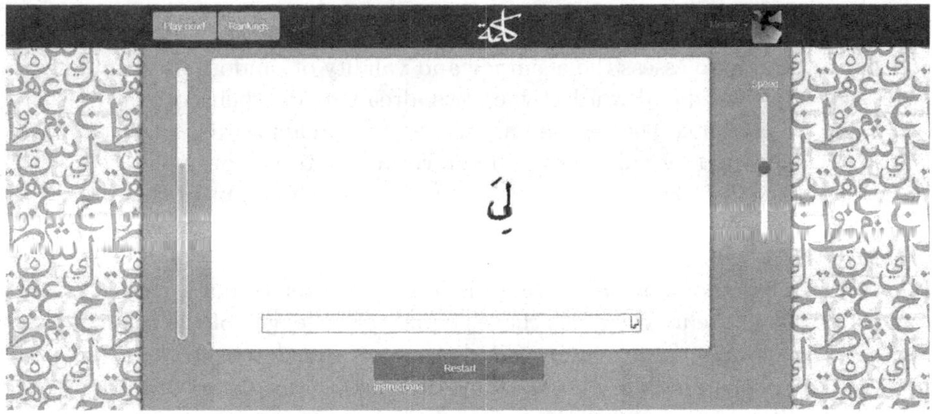

Fig. 3. Game interface: Users enter the word in the text field as they fall. The progress bar is on the left and speed settings on the right.

After the words to be entered are done, the score of the user gets calculated based on both accuracy and time. The scoring system will be discussed further in the following subsection. The user then has the option to share their score on Facebook or Twitter, which would help motivate their friends and followers to try the game and contribute to it.

The game makes the users feel challenged by having an all-time "Rankings" page as shown in Fig. 4. Such page displays the ranking of users who achieved the highest scores. It shows the number of games each user has played. This can significantly motivate users to play more games in order to achieve higher scores.

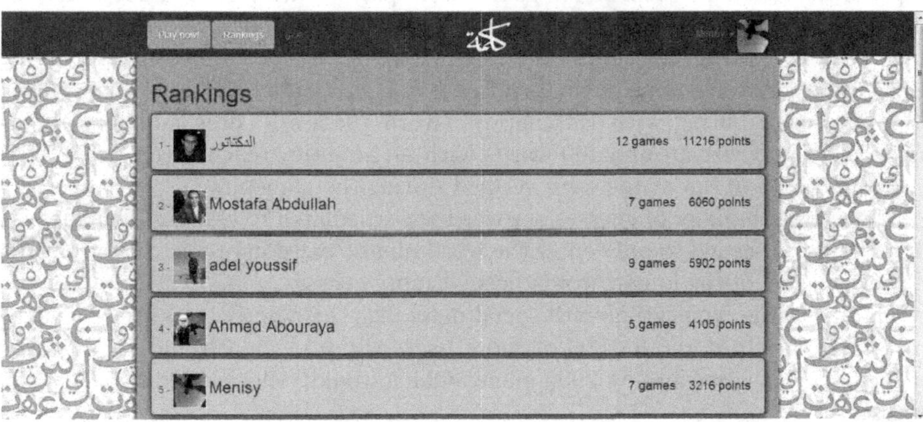

Fig. 4. Rankings page: Showing user rankings with the number of games played and the total score

3.3 Input Validation and Scoring System

In order to be able to assess the accuracy and validity of random users input, we used a set of pre-validated words "where we already know their correct digitization", and we presented them randomly within the game in order to check if the user is actually entering valid input. If the responses to the pre-validated word do not match with at least 60% accuracy, we discard the input of this user.

For every word that is not yet digitized, we initially have an empty array structure which will hold the different guesses to this word and a score assigned to each guess based on how many times is has been entered. For a new guess, or a new input from the user, a score that is equal to the length of the input is first assigned. Afterwards, Levenshtein edit distance is calculated between the new input and all of the previous guesses. This distance is subtracted from the score of the guess. If a guess matches the input yielding a distance of zero, we increase the score assigned to this guess by twice the length of the guess. However, if the guess is not found in the previous guesses it gets appended to them with its final score. The final score for the guess is then returned in order to be added to the scores of the other words in the game to calculate the users score with respect to the time taken in the game.

After enough guesses have been entered for a word, through redundancy, the scoring mechanism will converge and guesses with relatively high scores will be considered as the correctly digitized word. These validated digitized words can then be used again to validate the users' inputs in later games.

4 Results

Kalema release was on 15th of May 2014, and over 4 days "until the date this paper was written", a total of 448 games were played with a sum of 167 minutes of gameplay (2 hours and 47 minutes). Around 10960 inputs, or word guesses, have been saved from the users over this period.

22 users registered with their social networks accounts and played a total of 68 games contributing with 1360 inputs (word guesses). These numbers alone helped us to digitize around 400 words with an accuracy reaching up to 90%.

Around 33% of the words were verified during the gameplay.

The average number of guesses per word was calculated to be 2.32. This means that the users needed to only enter the word almost twice in order to be verified. However, some of the longer words needed more guesses.

Thanks to the integration with social networks, the game is expected to have an organic reach of about 1000 or more users within a month and an average of 2 games per user yielding 2000 games which would yield around 40000 word guesses.

The accuracy can be further improved using a validation game where users decide whether or not the digital text is correct compared to the image of the word.

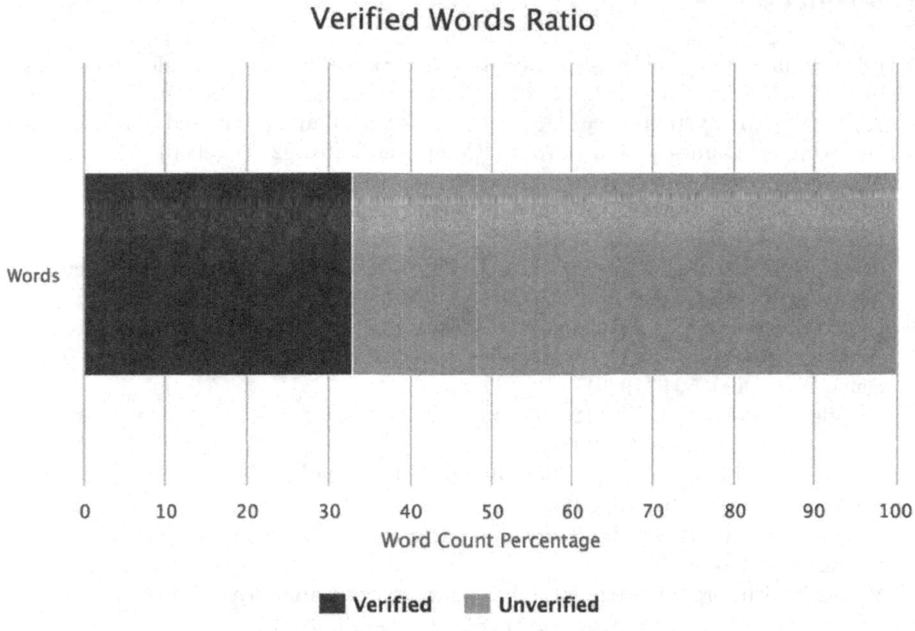

Fig. 5. Verified Words' Percentage

5 Conclusion

Arab visually impaired users find a hard time using non-digitized Arabic content. Kalema was developed to provide a platform that allows volunteers to play a game in which they type falling words quickly and accurately, these words come from scanned Arabic documents that need to be digitized.

The initial results show that crowdsourcing is indeed powerful when it comes to micro-tasks that need to be done by humans.It was also obvious that the social aspect is rather important regarding online games, as more users were attracted to play the game when they found their friends playing it.

The game interface can be improved further in order to be more attractive for users. Other game modes can also be implemented which would provide more gaming options suited for different types of volunteers. At a later stage and after collecting a significant amount of data, a validation game can be implemented where volunteers are asked to choose whether or not the digitized text corresponds to the image, it will be a simpler game meant for not yet validated users.

Kalema proved to be relatively successful relative to its short release period. Different users on social networks commented that the game is rather challenging and that it helps them improve their Arabic typing speed. By having more reach through social networks, Kalema would help digitize an enormous amount of Arabic content which would benefit thousands of visually impaired users around the Middle East.

References

1. Library fire in Egypt clashes destroys 'irreplaceable' 200-year-old documents, http://edition.cnn.com/2011/12/17/world/africa/egypt-unrest
2. ABBYY. OCR System, http://abbyy.com (2008) (Online; accessed May 19, 2014)
3. von Ahn, L.: Games with a Purpose. Computer 39(6), 92–94 (2006)
4. Alkafif, S.: Social initiative to digitize Arabic content (2011), https://www.facebook.com/sadik.alkafif (Online; accessed May 19, 2014)
5. Bakry, M., Khamis, M., Abdennadher, S.: Arecaptcha: Outsourcing arabic text digitization to native speakers. In: 11th IAPR International Workshop on Document Analysis Systems (DAS), pp. 304–308 (April 2014)
6. Bazzi, I., Schwartz, R., Makhoul, J.: An omnifont open-vocabulary ocr system for english and arabic. IEEE Transactions on Pattern Analysis and Machine Intelligence 21(6), 495–504 (1999)
7. Google. Tesseract OCR (2006), http://code.google.com/p/tesseract-ocr/ (Online; accessed May 19, 2014)
8. Finnish National Library. Digitalkoot (2011), http://www.digitalkoot.fi/ (Online; accessed May 19, 2014)
9. Märgner, V., El Abed, H.: Guide to OCR for Arabic Scripts. Springer-Verlag, London (2021)
10. World Health Organization. Visual impairment and blindness (2013), http://www.who.int/mediacentre/factsheets/fs282/en/ (Online; accessed May 19, 2014)
11. Smith, R.: An overview of the tesseract ocr engine. In: Proceedings of the Ninth International Conference on Document Analysis and Recognition, ICDAR 2007, vol. 02, pp. 629–633. IEEE Computer Society, Washington, DC (2007)
12. Smith, R., Antonova, D., Lee, D.-S.: Adapting the Tesseract Open Source OCR Engine for Multilingual OCR. In: Proceedings of the International Workshop on Multilingual OCR, MOCR 2009, pp. 1:1–1:8. ACM, New York (2009)
13. Von Ahn, L.: Human Computation. PhD thesis, Pittsburgh, PA, USA, AAI3205378 (2005)
14. von Ahn, L., Dabbish, L.: Labeling images with a computer game. In: Proceedings of the SIGCHI Conference on Human Factors in Computing Systems, CHI 2004, pp. 319–326. ACM, New York (2004)

An Enhanced Technique for Offline Arabic Handwritten Words Segmentation

Roqyiah M. Abdeen[1], Ahmed Afifi[2], and Ashraf B. El-Sisi[1]

[1] Computer Science dept.,
roqyiahabdeen@ymail.com, ashrafelsisim@yahoo.com
[2] Information technology dept.,Faculty of Computers and Information,
Menofia University, Egypt
ah.z.afifi@gmail.com

Abstract. The accuracy of handwritten word segmentation is essential for the recognition results; however, it is extremely complex task. In this work, an enhanced technique for Arabic handwriting segmentation is proposed. This technique is based on a recent technique which is dubbed in this work the base technique. It has two main stages: over-segmentation and neural-validation. Although the base technique gives promising results, it still suffers from many drawback such as the missed and bad segmentation-points(SPs). To alleviate these problems, two enhancements has been integrated in the first stage: word to sub-word segmentation and the thinned word restoration. Additionally, in the neural-validation stage an enhanced area concatenation technique is utilized to handle the segmentation of complex characters such as س. Both techniques were evaluated using the IFN/ENIT database. The results show that the bad and missed SPs have been significantly reduced and the overall performance of the system is increased.

Keywords: Arabic handwriting segmentation, character recognition, word thinning, neural network for character recognition.

1 Introduction

Handwriting recognition is a transformation process of the human handwritten scripts into a digital symbols that can be stored on a computer system [**1**].The concept of handwriting recognition can be divided into on-line and off-line recognition systems. In the on-line system, the required features are captured during the writing process such as speed, direction and the two-dimensional coordinates of successive points. The online systems are usually used in the tablet devices and smart phones[**2**]. Off-line handwriting recognition refers to the process of recognizing words that have been scanned from a paper or a book. After the document stored in a computer, it is become ready to perform further processing to allow superior recognition[**3**].

Although great efforts has been exerted in the area of handwriting recognition, the results of Arabic handwriting recognition and segmentation did not reach the required level of accuracy [**4**]. Due to the nature of the Arabic language especially the handwritten scripts, some characters can be connected to another character from both sides and some characters are not. The same character can be written in many shapes

A. Gelbukh (Ed.): CICLing 2015, Part II, LNCS 9042, pp. 663–681, 2015.
DOI: 10.1007/978-3-319-18117-2_50

according to its position in the word: isolated, at the end, on the middle or at the beginning of the word. In addition, the Arabic words use an external objects like "dots", "Hamza" that make the task of segmentation more complicated. Moreover, characters that do not touch each other but occupy the same horizontal space increase the difficulty of segmentation. Furthermore, different writers and the same writer under different conditions write some Arabic characters in completely different ways[**3,4**] as shown in **Figure 1.** One of the major problems in recognizing unconstrained cursive words is the segmentation process.

There are many techniques have been proposed in the field of Arabic segmentation. A hybrid segmentation technique is introduced in [**5**].This technique is based on two main stages: the heuristic stage and the artificial neural network (ANN) validation stage. In the heuristic stage, the connected block of characters (BCs) are extracted, the topographic features are generated and the pre-segmentation points are calculated. The neural validation stage verifies whether pre-segmentation points are valid or invalid. However, they could not achieve high accuracy because of the existence of external objects were affects the BCs extraction process and the segmentation point detection. In[**6**]**,** the authors proposed a segmentation technique based on extracting the features of the strokes that lie on the upper side of the word image. At first, a simple smoothing operation is utilized to eliminate the pixels located improperly around the contour of the image. Thereafter, the freeman chain coding scheme is used to find the coordinates of the pixels which lie on the contour. Then each two adjacent pixels coordinates are paired and purified and the slope between them are calculated. After that, an algorithm is utilized to allocate a sign (either + or -) for each purified pair, and the segmentation points are determined based on these signs. The algorithm showed high speed processing because there is no need for extracting further data like baseline, skeleton, or the V/H projection; however, it still suffer from some limitation like the segmenting of Arabic ligatures and (the vertical writing of many characters above each other). Another segmentation method was used in [**7**]. Firstly, an improved projection based method for baseline detection is employed. Then, the connected components (ccs) are extracted and the small parts like dots in the image are ignored. After that, the distances among different components are analyzed. The statistical distribution of this distance is then obtained to determine an optimal threshold for words segmentation. However, their system showed a failure in some cases due to the variation in handwriting, especially irregular spaces between sub-words and words, such as too large spaces between sub-words or too small spaces between words. Among the methods which are employed in the Arabic words segmentation is the new multi-agents segmentation technique which are introduced in [**8**]. The proposed approach utilizes seven agents which work together on the thinned image of the handwritten text to identify the regions where the setting of segmentation points is not permitted such as loops and cavities. Then, a specific set of topological rules are used on the remaining parts of the thinned image to determine the set of possible cut-points that will lead to a successful isolation of individual characters. However, these agents segment many characters in two parts like the characters "ض" and "ـم". The authors in [**9**], introduced a novel diacritics extraction and graphemes segmentation techniques. The strategy used in that segmentation technique lies in combining the local writing direction information and the neighbourhood geometric characteristics in a way that utilizes the nature of Arabic script, which has proven to achieve a promising segmentation performance. The proposed segmentation algorithm in [**10**] is start with segmenting the word into sub-words depends on the space between the characters and then the baseline of each sub-word is computed. The descenders of sub-words are then detected and deleted before computing the vertical projection to find correct candidate points for the skeleton of sub-word and they have achieved a good segmentation result.

The enhanced technique which is described in this paper integrates several recent techniques in a single technique in order to alleviate segmentation challenges and reach a satisfactory result. The main aim of this technique is to reduce the number of bad and missed SPs. The enhanced technique solves several problems which are mentioned in the previous works:

- The overlapping problem. Sub-word separation technique is employed to separate the overlapped character (the characters that share the same horizontal space) completely and correctly.
- The sub-word separation is not depend on the distance between the words or sub-words and do not use the vertical histogram as in the previous works. Nevertheless, it separates the sub-words by tracking each object to find the connected components and isolate them from each other based on the existence of the spaces between them. Consequently, the incorrect or inaccurate separation of the sub-words is never occurred.
- The dots and the external objects are removed before the sub-word separation and the segmentation process to avoid any wrong or inaccurate result.
- The segmentation of the Arabic ligatures: "ـحم","لحـ","بجـ", "لـ", "لى", "عمـ", "لحـ","لا" is not necessary because the neural network is trained to detect these shapes as a character and it can be translated into two characters in the recognition phase.
- The missed features due to the quick writing and the thinning process in some characters (specially the strokes of the character "ـسـ" and the characters which have the shape "ـبـ") is recovered using the image restoration algorithm as will be obtained in the proposed method.
- The segmentation process of the complex characters such as "س","ص" is improved by enhancing the neural validation stage.

The base and the enhanced techniques are evaluated using 500 words from the standard IFN/ENIT database.

The rest of the paper is organized as following. In Section 2, the proposed method is described. In Section 3, database and evaluation metrics is presented. The experimental results and analysis are discussed in Section 4. Finally, the conclusion and future work is put forward in Section 5.

2 Proposed Method

In this work, a technique for Arabic handwriting segmentation is proposed. This technique integrates the strengths of several recent techniques to alleviate their limitations. The base technique in this work is the one proposed in[**3**]. In this technique, all perspective segmentation points are detected in the first stage. In the second stage, the neural network is employed to validate these points and discards the wrong points.

Although that technique can produce reasonable segmentation results, two problems appears in that technique: the first is the overlapped characters (more than one character occupy the same horizontal space), which cannot be separated from each other in the first stage. **Figure2** shows an example of words containing overlapped characters. The second problem is that many segmentation points are missed after the thinning process because many important features are removed. **Figure 3** illustrate an example of this problem. To overcome these limitations, two additional stages are added. In the first stage, each word is divide into several sub-words according to characters connection [**10**]. This stage enhances the separation of overlapping characters. In the second stage a restoration technique is used to add more features to the characters which are affected by the thinning process, and accordingly the number of missed segmentation points can be reduced[**9**]. In the following sub-sections, the proposed enhanced technique will be described in details.

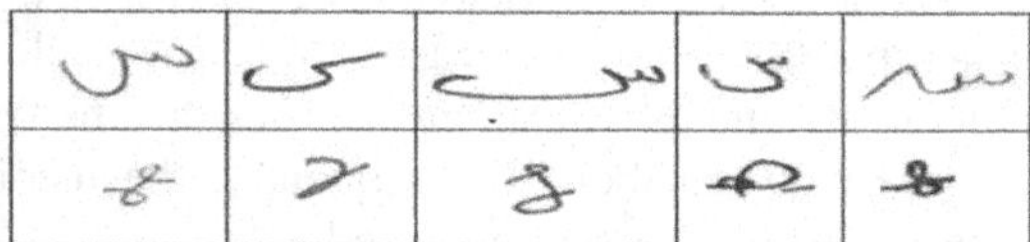

Fig. 1. Same characters written by different persons in different ways

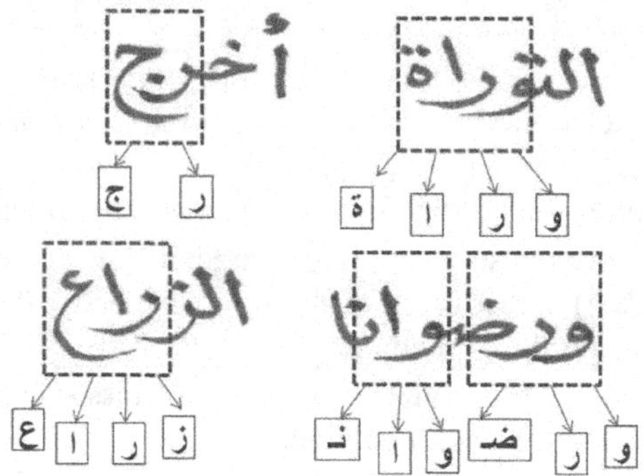

Fig. 2. Arabic Handwriting words contains overlapped characters

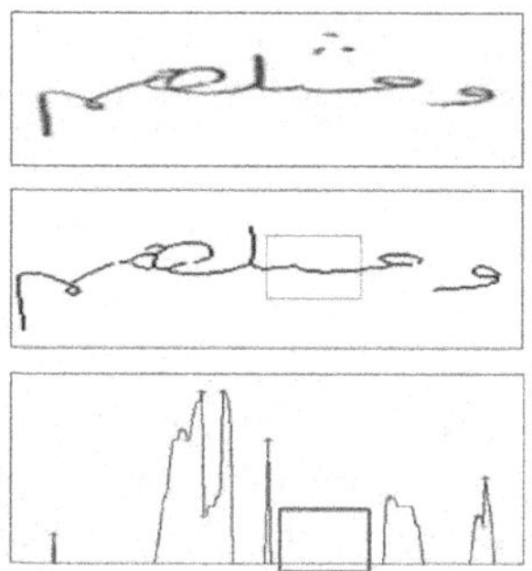

Fig. 3. Missed segmentation point after the thinning process

2.1 The Base Technique

The base technique in this work is based on neural network validation, it consists of two stages. The first stage is the Arabic Heuristic Segmenter (AHS) which is utilized to find the prospective segmentation points(PSP). In this stage, the image is converted into a binary image and filtered with an appropriate filter. After that, the dots and any punctuation marks are removed to find the SPs easily and correctly. To remove the dots, the average character density (ACD) of all Arabic script shapes are calculated. Then for any connected component, if its density is less than the (ACD), it will be removed. In the next step, the image is thinned. The middle region of the thinned image which contains the base line is then extracted using the horizontal histogram. To find the upper and lower baseline, the average of the local minima and maxima points above and below the largest density point is calculated to find the upper and lower baseline respectively as in **Figure 4**. A modified vertical histogram is then constructed by calculating the distance between the upper and lower foreground pixels for each column. The histogram is calculated for the middle region of the thinned image not on the original image to avoid the presence of excessive number of SPs. The local minima points on this histogram are considered as PSPs. The average character width approach is used to add or remove additional segmentation points. The average character width (ACW) is calculated by dividing the word image width on the number of local minima's. Then the distance between each two local minima points is calculated. If this distance is larger than the (ACW), a new SPs must added by finding another local minima for this region. If the distance between two local minima points is smaller than the (ACW), one of this two SPs should be removed. Finally, any SP passes more than one foreground pixel in the same vertical line with a white space between them means that this SP passes through an open or closed hole and it will be removed. **Figure 5.a** shows the word image after the modified histogram and **Figure 5.b** shows the image after applying the ACW approach and the holes detection.

In the second stage, the feed forward back propagation neural network classifier is utilized. This step aims to examine all PSPs to decide which one is invalid to remove it and which is valid to keep it as a correct segmentation point. Initially, the features of 60 shapes of Arabic characters for 20 writers are extracted. **Figure6** shows these 60 shapes. The feature extraction used in this work is the Modified Direction Feature (MDF)[**3**]. In the (MDF), each boundary pixel is assigned a number according to its direction; two for vertical, three for right diagonal, four for horizontal and five for left diagonal. After that, the direction values are normalized by replacing any value differ from its neighborhood according to one of its neighborhood values as shown in **Figure7**. Thereafter, the Location Transition (LT) and Direction Transition (DT) matrices are calculated. The LT matrices are calculated by finding location of the transitions from background to foreground pixel for each raw from left to right and from right to left, and for each column from top to bottom and vice versa. Only three transition are selected. If the transitions number is less than three then, zeros are added for this transitions. The LT matrices size which are used to find the transition on the rows =number of rows of image ×number of transitions (3 in this work) and the LT matrices size for columns = number of columns of image × 3. Finally, each value in the LT matrices is divided by the row number for the row matrices and by the column number for the column matrices.

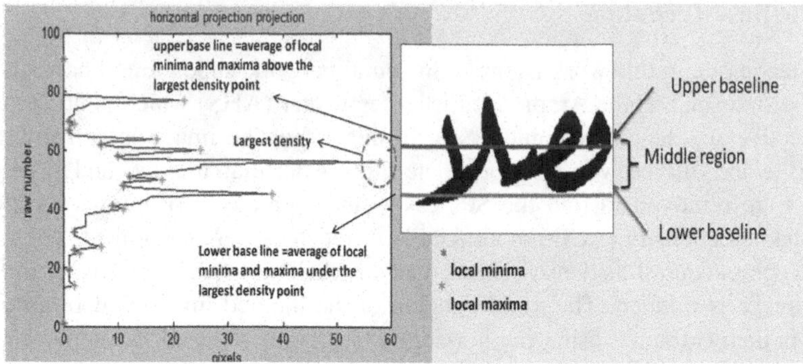

Fig. 4. Finding the middle region of the word image

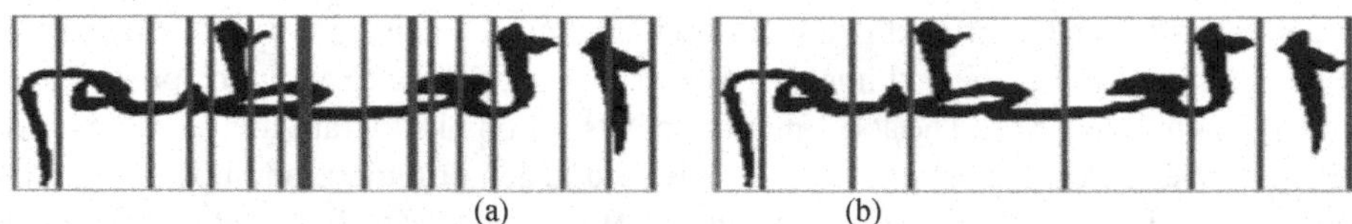

Fig. 5. a) Initial prospective points, b) Segmentation points after applying the ACW and open hole rules

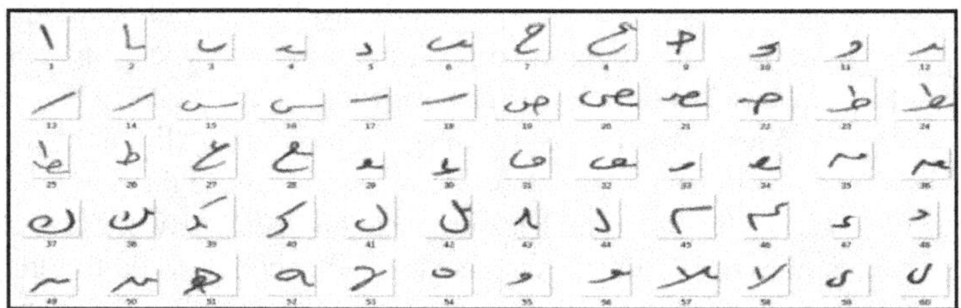

Fig. 6. The 60 input shapes used for training

The process for finding the DT matrix is the same but the direction value is used rather than the transition. Then, each value in the DT matrix is divided by 10 to simplify the numbers. After finding the LT (4 LT matrices, 2 for rows and 2 for columns) and DT matrices (4 DT matrices, 2 for rows and 2 for columns), a local averaging is used to resample all the matrices in the same size (in this work the re-sampled size is 5 ×3). The number of the features results from this operation is 120 features. The neural network is trained by 120 features extracted from the 60 shapes from 20 different writers, and also trained by incorrect characters. Another neural network is trained by the features of correct and incorrect segmentation Area (SA). The result of the first neural network is 61 values (60 values for the 60 shapes 'the largest value determining the correct character' and the last value for the incorrect character 'reject neuron'). The result for the second neural network is one value (more than 0.5 for the correct SA and less than 0.5 for the incorrect ones).

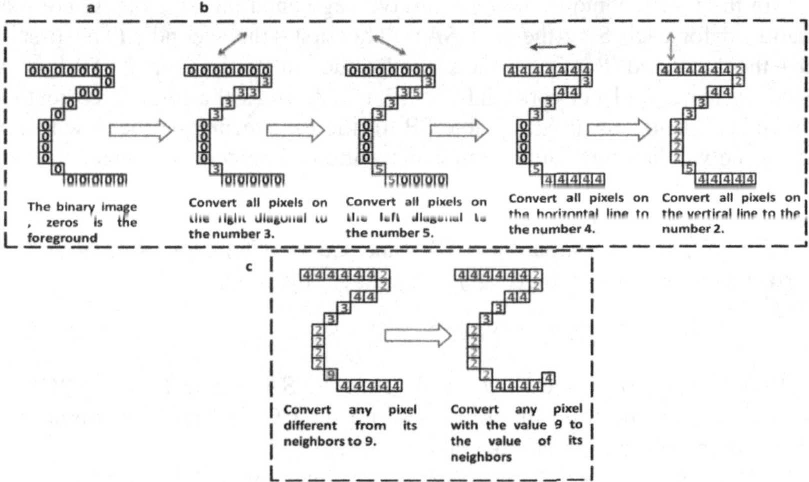

Fig. 7. a) The binary image, b) determining direction values, c) normalization steps

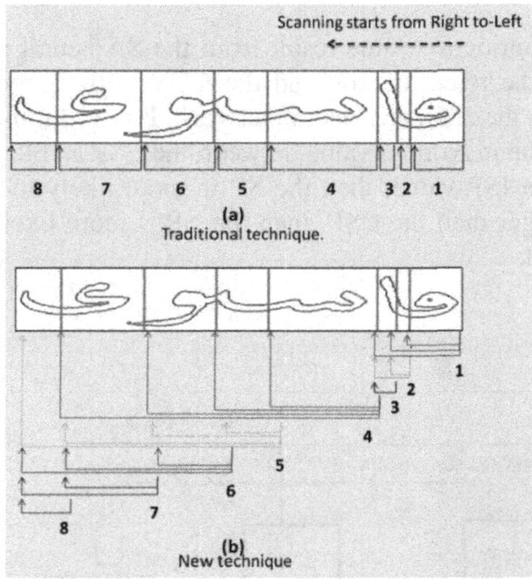

Fig. 8. The scanning process in the traditional and the new techniques

In the testing process, the word image is scanned from right to left. Then, for each SP the Right Character (RC), Center Character (CC) and the Segmentation Area (SA) are extracted. Thereafter, the 120 features for each part are calculated and tested by the two neural networks to find the confidence values. The scanning process of the segmentation areas to check the validity of each SP utilizes one of two scanning techniques: the traditional technique and the new technique which are proposed in[**3**]. Both techniques are explained in **Figure 8**. In both techniques the scanning process starts from right to left. In the traditional technique for each SP the RC, CC and SA are extracted and tested by the neural network to decide if the SP is accepted or rejected. In the new technique, four successive segmentation Areas (SA) are merged and examined for each SP :(the first SA), (The first +the second), (The first + the second + the third) and (the first + the second + the third + the fourth). All four Areas are tested by the neural network and the SP which gives the highest correct confidence value is selected as an acceptable SP. In the base technique the new technique is used for the validation because some handwritten characters are longer than others and this causes more bad segmentation points. The validity of the PSPs is determined based on the fusion of three neural confidence values: Segmentation Area Validation (SAV), Right Character Validation (RCV) and Center Character Validation (CCV). The Correct Segmentation Point (CSP) is calculated by Equation (1):

$$f(CSP) = SAV + RCV + (1 - CCV) \tag{1}$$

Where, SAV is the confidence value result from the SA neural network, RCV is the largest confidence value from the 60 values and the CCV is the value number 61 of the reject neuron for the character neural network.

The Incorrect Segmentation Point (ISP) is calculated by the Equation (2):

$$f(ISP) = -SAV + RCV + CCV \tag{2}$$

Where, SAV is the confidence value result from the SA neural network, RCV is the value number 61 of the reject neuron and the CCV is the largest confidence value from the 60 values for the character neural network. **Figure9** illustrate the steps of the testing step. Finally, the maximum value between the CSP and ISP is computed. If the CSP is larger than the ISP value, then the SP is more likely to be a correct SP and kept. If the ISP is larger than the CSP, then the SP is more likely to be incorrect SP and it will be removed.

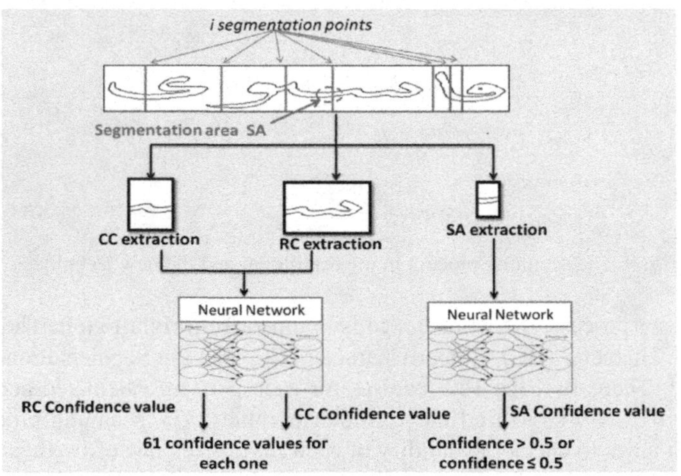

Fig. 9. The testing process

2.2 Sub-word Separation Technique

In Arabic language, most of the words can be divided into two or more sub-words before segmenting them into characters. This makes the word segmentation process easier and accurate. In the base technique, the vertical histogram is used to determine the segmentation points. Some writers write more than one character above each other. Therefore, the SPs cannot separate these characters completely and more than one character may be found in one segment. In **Figure 10.a** the character "ر" cannot be separated correctly from the character "ا" because both characters are written above each other, and also a part of the character "ع" is found in the same segment. To solve this problem, the word should be divided into sub-word images before determining the SPs as in **Figure 10.b**. To separate the sub-words from each other, initially a tracking algorithm is employed to track the boundary of each object on the image in order to find the connected components. Thereafter, each separated object is stored in an isolated image. After objects are isolated from each other, they are grouped again in one image to deal with it as one word and SPs are added between sub-words. These SPs will named the Verified SPs (VSPs) so that it cannot be removed during the neural network validation. **Figure 11** shows the steps of the sub-word separation process. The resulting images are then used to find the baseline and to determine the SPs.

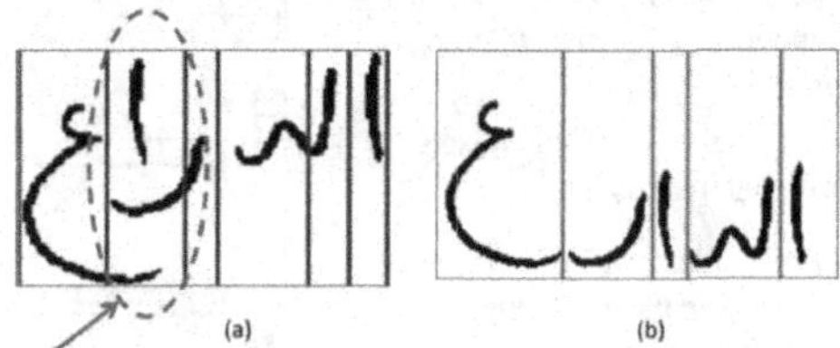

Fig. 10. a) The image before sub-word separation, b) after the sub-word separation

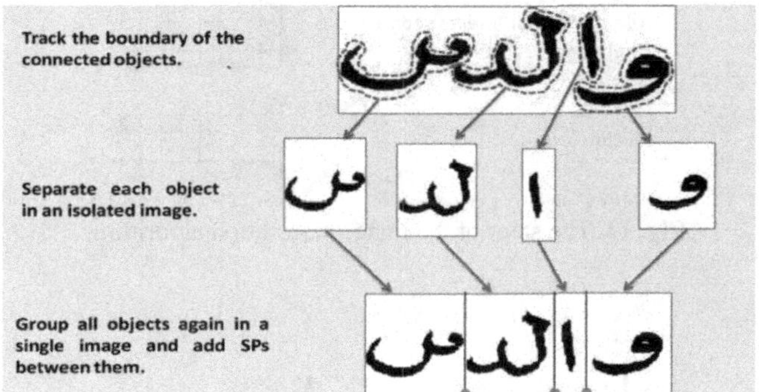

Fig. 11. The sub-words separation process

2.3 The Thinned Image Restoration

In the thinning process, the original image is converted to another image with less thickness. Some Arabic characters losses important features after the thinning process, which lead to a wrong characters segmentation and classification. In order to avoid this problem, an image restoration algorithm is applied on the image after thinning. Initially, a binary morphological template matching operation (hit-and-miss operation) is used to find the location of the intersection/end pixels in the thinned image. After determining their location, the second step is to add additional foreground pixels. First search all rows above the previous pixels in the thinned image until reach to a background pixel.

Once the location of this pixels are determined, the pixels with the same locations on the original image are replaced with a foreground pixel. Finally, the thinning process is applied again on the original image after adding the new pixels to give the restored thinned image. To avoid the connectivity between the unconnected characters, the word is firstly divided to sub-words before applying this algorithm. **Figure 12** shows a brief explanation of this algorithm and the **figure13** shows the sub-word "سا" after and before the utilization of image restoration algorithm.

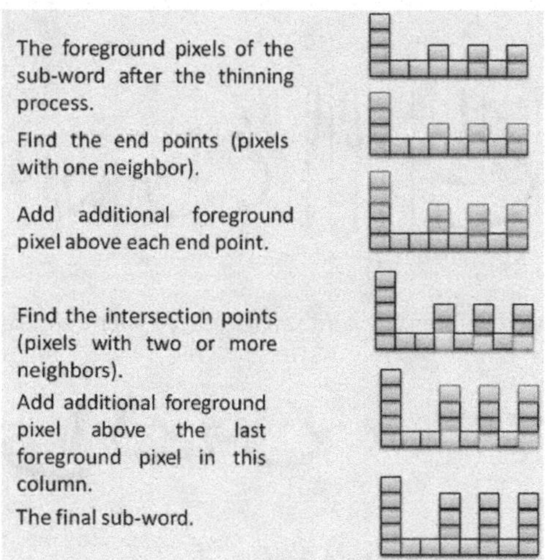

Fig. 12. The steps of the image restoration algorithm

Fig. 13. Sub-word before and after the image restoration algorithm

2.4 Enhanced Segmentation Area Scanning Technique

Although the new scanning algorithm used in the base technique outperforms the traditional scanning technique, it is also suffer from some limitations. In this subsection the proposed enhancement is explained. Some complex Arabic characters such as the "س" and "ص" are usually segmented into many correct characters (i.e. the "س" can recognized as "ـسـ","ـسـ", and "ـس" and all of them are a correct shapes).If the aforementioned new technique is applied on the character "س" the result will be as shown in the **Figure 14**, the character is over segmented and in some cases the problem of over segmenting the character "س" is increased. The proposed enhanced new technique contains a new step; if the checked region of the current SP is not correct or it is correct but it does not give the largest confidence value, it will be removed before checking the next SP. **Figure 15** shows the character "س" after applying the new enhanced technique. Many experiments were made on the previous techniques and the enhanced new technique gives the best result especially in the case of the characters "س" and "ص".This technique will be utilized in the proposed enhanced technique.

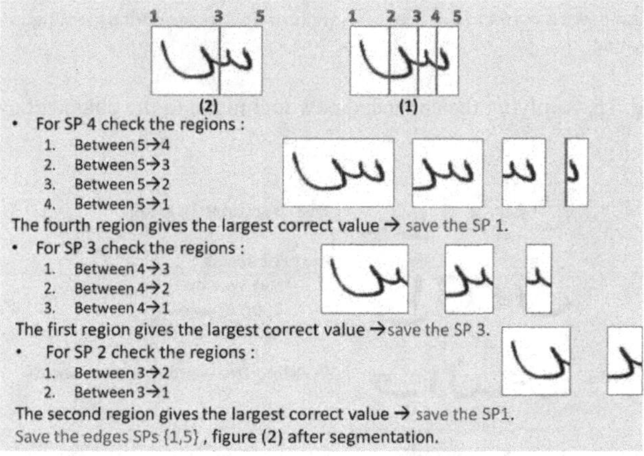

1 3 5
1 2 3 4 5
(2)
(1)
For SP 4 check the regions :
1. Between 5→4
2. Between 5→3
3. Between 5→2
4. Between 5→1
The fourth region gives the largest correct value → save the SP 1.
For SP 3 check the regions :
1. Between 4→3
2. Between 4→2
3. Between 4→1
The first region gives the largest correct value →save the SP 3.
For SP 2 check the regions :
1. Between 3→2
2. Between 3→1
The second region gives the largest correct value → save the SP1.
Save the edges SPs {1,5} , figure (2) after segmentation.

Fig. 14. Applying the new technique to the character س

2.5 The Enhanced Technique

In the enhanced technique, all the previous techniques and enhancements are joint together in a single technique. **Figure 16** shows the flowchart of the new enhanced technique.

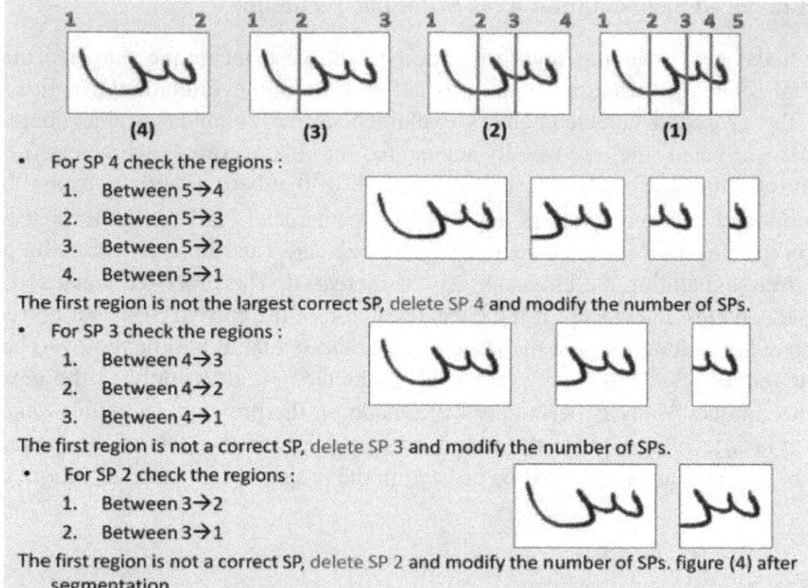

1 2 1 2 3 1 2 3 4 1 2 3 4 5
(4) (3) (2) (1)
• For SP 4 check the regions :
1. Between 5→4
2. Between 5→3
3. Between 5→2
4. Between 5→1
The first region is not the largest correct SP, delete SP 4 and modify the number of SPs.
• For SP 3 check the regions :
1. Between 4→3
2. Between 4→2
3. Between 4→1
The first region is not a correct SP, delete SP 3 and modify the number of SPs.
• For SP 2 check the regions :
1. Between 3→2
2. Between 3→1
The first region is not a correct SP, delete SP 2 and modify the number of SPs. figure (4) after segmentation.

Fig. 15. Applying the enhanced new technique to the character س

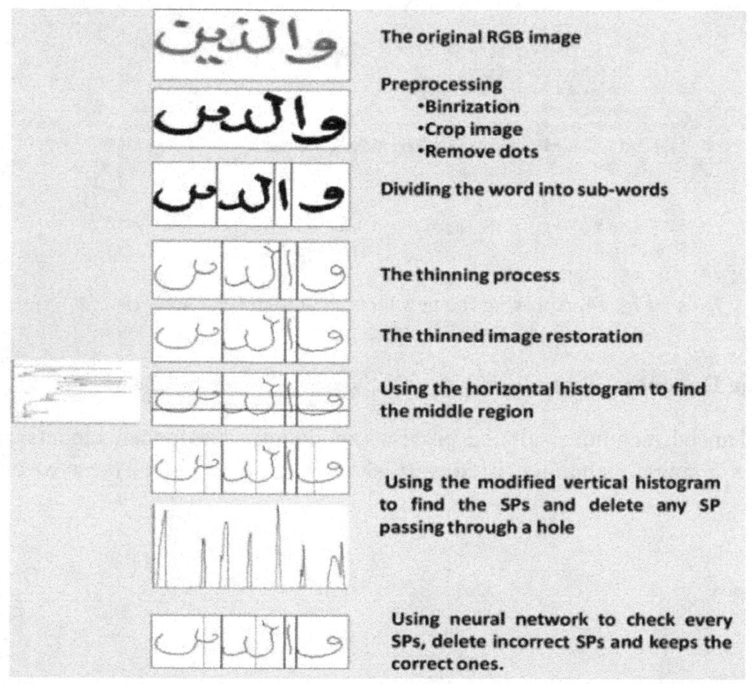

Fig. 16. The complete process of the enhanced technique

3 Database and Evaluation Metrics

The base technique and the enhanced technique are consisted of two stages: the over segmentation stage and the neural validation stage. The over segmentation stage in the enhanced technique contains the same steps of the base technique as well as two additional steps: the sub-word separation technique and the image restoration technique. Initially, the performance of both techniques are evaluated after the over segmentation stage. Then, they are evaluated again after the neural validation stage. The base technique and the enhanced technique were tested using 500 words which are extracted from the standard **IFN/ENIT** database. A random set of 500 words which are written by different writers were selected for the experiments. Thereafter, the 60 shapes were extracted manually from twenty different writers for the neural network training, so the size of the training set is 1200 characters (20 writers x 60 shapes). Initially, the number of SPs of all the words are calculated. The accuracy of both techniques is evaluated according to the number of correct SPs. The correct SPs are the points that split two touching characters completely and correctly. The incorrect SPs can be divided into three types: the "over segmentation points", the "bad segmentation points" and the "missed segmentation points". The over segmentation point occurs if the SP divides one character into two or more parts. The bad segmentation point splits two characters from each other in such a way that one of the characters or both characters are not separated correctly as shown in **Figure 10**. The last type of incorrect SPs is the missed SP which means that two connected characters are not separated at all.

4 Experimental Results and Discussion

The results presented here are divided into two sub-sections. The first sub-section discusses experiments results for the base technique. The second sub-section introduces the experiment results of the new enhanced technique.

4.1 The Performance of the Base Technique

The performance of the base technique is initially evaluated after the over segmentation stage**. Table 1** displays results after applying the over segmentation stage on 500 words from the standard database. The total number of segmentation points from the 500 words are 4436 SPs. The number of correct SPs is 3424 gave a percentage of 77.18 %. The percentage of total errors is 22.92 %.The over segmentation error is the greatest one and the bad and missed error are approximately equals. The best result is when the over segmentation error is more than the missed error because the goal of the over segmentation is to separate all characters from each other such that no two connected characters remain. Then the neural network is developed to remove any incorrect SP but not adding additional points. **Figure 17** shows an example of a correct segmented words taken from the standard database after the over segmentation stage. The most common error which are appear after the over segmentation stage is the over segmentation error. The neural network is developed to fix this type of errors. **Figure 18** provides samples of handwriting words contains over segmentation points after the over segmentation stage and the same words after the neural validation stage. As may be seen in **Table 2**, the over segmentation error in after the neural network validation stage is reduced from 14.17% as in **Table 1** to 9%. Therefore, the neural network performed well with this type of errors and remove the unneeded SPs. Some words still containing over segmentation error and the neural network failed to repair this error as in **Figure 19**. The second type of segmentation errors is the missed SPs. This type of errors may occurs if several characters are written above each other or if no local minima is detected in this region when calculating the vertical histogram. The neural network is not responsible for adding the missed SP and furthermore may be increase the number of missed SP. Note that the characters such as "حم","لح","بج", "لم", "لى", "عم","مح","لا" each of them is considered as a single character and the neural network were trained by a sample of these characters to recognize them as a correct characters. Consequently there is no need to divide them from each other. As may be seen in **Table 2**, the missed error are increased after the neural network validation from 4.5% as in **Table 1** to 7%. The last type of errors is the bad segmentation points in which the SP is not segment two touching characters correctly. This type of error is happens a lot for the writers who writes the characters above each other. The neural network can remove some but not all bad SPs. As may be seen in **Table 2**, the bad error is decreased from 4.12% as in **Table 1** to 4% which is a very slight difference. The last two types of errors can be alleviated using the proposed technique as we will explain in the next sub section.

Table 1. Over segmentation result of the base technique

words	SP	Correct segmentation	Over-segmentation error rates			
			Over-segmentation	Missed	Bad	Total
The 1st 100 words	819	654 79.85%	110 13.43%	36 4.39%	19 2.30%	165 20.14%
The 2nd 100 words	936	707 75.53%	174 18.58%	25 2.67%	30 3.20%	229 24.46%
The 3rd 100 words	848	657 77.47%	105 12.38%	35 4.12%	51 6%	191 22.52%
The 4th 100 words	903	686 75.96%	130 14.30%	49 5.42%	38 4.20%	217 24%
The 5th 100 words	930	720 77.41%	110 11.80%	55 5.91%	45 4.83%	210 22.58%
total	4436	**3424** **77.18%**	629 14.17%	200 4.50%	183 4.12%	**1017** **22.92%**

Fig. 17. Samples of successfully segmented word images after the over segmentation stage

Fig. 18. Samples of word images after the over segmentation stage and after the neural validation

Table 2. Neural Network validation of the base technique

words	SP	Correct segmentation	Over-segmentation error rates			
			Over segmentation	Missed	Bad	Total
The 1st 100 words	699	569 81.4 %	53 7.58 %	60 8.5 %	17 2.43 %	130 18.59 %
The 2nd 100 words	801	646 80.64 %	90 11.23 %	37 4.6 %	28 3.49 %	155 19.35 %
The 3rd100 words	741	588 79.35 %	55 7.4 %	58 7.8 %	40 5.39 %	153 20.6 %
The 4th 100 words	800	631 78.87 %	68 8.5 %	64 8 %	37 4.62 %	169 21.21 %
The 5th 100 words	860	679 78.95 %	88 10.23 %	56 6.5 %	37 4.3 %	181 21 %
total	3901	**3113** **79.8 %**	354 9 %	275 7 %	159 4 %	**788** **20.19 %**

Fig. 19. Samples of word images still containing over segmentation error

4.2 The Performance of The Enhanced Technique

In the proposed enhanced technique, two major problems are reduced: the missed SPs and the bad SPs. As previously mentioned, the main reason for the bad SP and the missed SP is when several characters are written above each other. Another reason for the missed SP is when no local minima is detected after the vertical histogram in this region and this may be due to the thinning process. **Table 3** shows the over segmentation results of the enhanced technique. Comparing this table with **Table 1**, it is noticed that the bad SPs percentage is reduced from 4.12% in the base technique to 1.54% in the enhanced technique. Furthermore, the missed SPs are reduced from 4.5% to 2.66% as in **Table 3**. **Figure 20** shows how the sub-word separation technique significantly improve the segmentation of some complex handwriting words. It solves the missed SPs problem as in the word "المحارزه" and the characters "ز" and "ر" are separated completely and correctly. Also the characters "ر", "ا" and "ع" in the word "ذراع بن زياد" are segmented correctly and the same in the rest of the words.

Table 3. Over segmentation result of the enhanced technique

words	SP	Correct segmentation	Over-segmentation error rates			
			Over-segmentation	Missed	Bad	Total
The 1st 100 words	810	660 81.48 %	115 14.19 %	25 3.08 %	10 1.23 %	165 20.14 %
The 2nd 100 words	918	716 77.99 %	170 18.5 %	20 2.17 %	12 1.3 %	202 22 %
The 3rd 100 words	788	666 84.51 %	98 12.43 %	16 2 %	8 1 %	122 15.48 %
The 4th 100 words	897	706 78.7 %	140 15.6 %	29 3.2 %	22 2.4 %	191 21.29%
The 5th 100 words	910	750 82.41 %	120 13.18 %	25 2.7 %	15 1.6 %	160 17.58 %
total	4323	**3498** **80.91 %**	643 14.87 %	115 2.66 %	67 1.54 %	**840** **19.43 %**

Fig. 20. Words shows the bad and missed SP after the base and enhanced technique

Moreover, the bad SPs problem are solved. For example, in the word "حي بو صفارة" the character "ر" is separated completely from the character "ة" and the same for the other words. The total segmentation accuracy after the over segmentation stage for the base technique is 77.18% and for the enhanced technique is 80.91% from the **Tables 2,3**. Another main reason for the missed SPs is that some writers does not write some letters clearly which lead to more missed or incorrect SPs. **Figure 21** shows some examples of handwriting words contains missed and bad SPs before and after the enhancements and the figure shows how the image restoration algorithm can solve this type of problems. For example the word "أمش" is detected as "امى-س" with three characters before applying the image restoration but it detected correctly as "س" by the neural network after applying the algorithm. Also the word "الكثير" have a missed SP but it is detected after applying the algorithm and the same with the remaining words. The **Figure 21** also shows how the neural network validation stage detect the character "س" correctly as in the word "أمس" and the other shape of this character "سـ" as in the words "التكسر" and "المسارح".

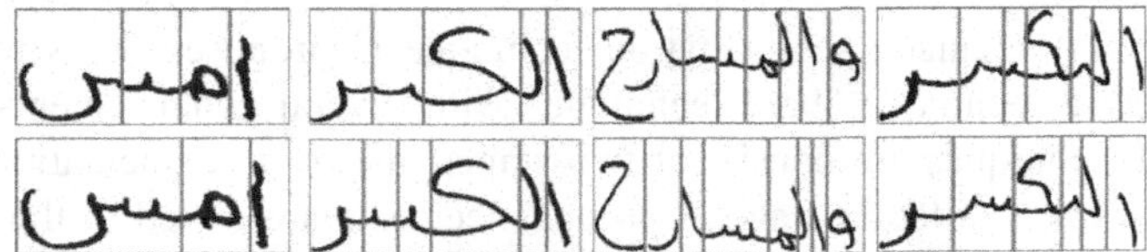

Fig. 21. Word samples shows the effect of the image restoration algorithm

Table 4. Neural Network validation result of the enhanced technique

writers	SP	Correct segmentation	Over-segmentation error rates			
			Over segmentation	Missed	Bad	Total
The 1st 100 words	706	588 83.28 %	70 9.9 %	44 6.23 %	4 0.5 %	118 16.7 %
The 2nd 100 words	784	658 83.92 %	76 9.69 %	50 6.37 %	0 0.0 %	126 16.07 %
The 3rd100 words	745	600 80.53 %	70 9.39 %	66 8.85 %	9 1.2 %	145 19.46 %
The 4th 100 words	810	640 79 %	99 12.2 %	64 7.9 %	7 0.86 %	170 20.98 %
The 5th 100 words	826	655 79.29 %	94 11.38 %	77 9.32 %	0 0.0 %	171 20.7 %
total	3871	**3141** **81.14 %**	409 10.56 %	301 7.77 %	20 0.5 %	**730** **18.85 %**

The results after the neural network validation stage are shown in **Tables 4**. As can be seen from **Table 4**,the neural network is succeeded in reducing the rate of the over and the bad SPs. The percentage of the over segmentation error after the neural validation are decreased from 14.87% as in the **Table 3** to 10.56% as in **Table 4** and the same are with the bad SPs. The bad SPs are reduced from 1.54% to 0.5%. In contrast to the bad and over segmentation error, the missed error percentage has increased after the neural validation from 2.66% to 7.77%. It also increased in comparison to the base technique as in **Table 2**. This problem is caused by the classifier and it will resolved in the future work by enhancing this classifier or using a new one. Likewise, the over segmentation error increased in comparison to the base technique and this because the characters are separated from each other and the word width is increased. Consequently, the ACW approach locates additional SPs in the word. The overall segmentation accuracy are improved compared with the result after the over segmentation.

The correct SPs percentage after neural validation are risen from 80.91% which is the percentage after the over segmentation to 81.14% after the neural validation. The overall segmentation accuracy are also increased after the enhanced technique in comparison with the base technique after the neural validation. After the base technique the correct SPs percentage was 79.8% and increased to 81.14% after the enhanced technique as mentioned previously.

5 Conclusion and Future Work

In this paper, an enhanced segmentation technique is proposed. This technique is based on a recent technique "base technique" with several enhancements. The base and enhanced techniques are consist of two stages: the over segmentation stage and the neural network validation stage. Two enhancements are added to the first stage: the word is divided into sub-words to solve the problem of the overlapped characters and a restoration algorithm is employed to add some lost features to the thinned word. Another enhancement added to the second stage to solve the problem of some complex characters such as "س" and "ص".

The base and the enhanced technique were evaluated using the standard IFN/ENIT database. The results show that the bad segmentation points are highly reduced in comparison to the base technique. Also, the missed SP is reduced after applying the enhancements but it increased again after applying the neural validation stage. In the future, we tend to employ different classifiers to improve the segmentation accuracy. Also, another feature extraction techniques will used to train the classifiers to improve the character recognition process.

References

1. Blumenstein, M.: Cursive character segmentation using neural network techniques, School of Information and Communication Technology, Griffith University, Gold Coast campus, PMB 50 Gold Coast Mail Centre, Queensland 9726, Australia
2. Al Hamad, H.A.: Use an Efficient Neural Network to Improve the Arabic Handwriting Recognition. In: Proceedings of the 2013 International Conference on Systems, Control, Signal Processing and Informatics (2013)
3. Al Hamad, H.A., Zitar, R.A.: Development of an efficient neural-based segmentation technique for Arabic handwriting recognition. Pattern Recognition 43, 2773–2798 (2010)
4. Elaiwat, S., AL-abed Abu-zanona, M., AL-Zawaideh, F.H.: A Three Stages Segmentation Model for a Higher Accurate off-line Arabic Handwriting Recognition. World of Computer Science and Information Technology Journal (WCSIT) 2(3), 98–104 (2012), ISSN: 2221-0741
5. Hamid, A., Haraty, R.: A Neuro-Heuristic Approach for Segmenting Handwritten Arabic Text. IEEE (2001)
6. Abdulla, S., Al-Nassiri, A., Salam, R.A.: Offline Arabic Handwriting Word Segmentation Using Rotational Invariant Segments Features. The International Arab Journal of Information Technology 5(2) (April 2008)

7. AlKhateeb, J.H., Jiang, J., Ren, J., Ipson, S.S.: Component-based Segmentation of Words from Handwritten Arabic Text. International Journal of Computer Systems Science and Engineering 5, 1 (2009)
8. Elnagar, A., Bentrcia, R.: A Multi-Agent Approach to Arabic Handwritten Text Segmentation. Journal of Intelligent Learning Systems and Applications (2012)
9. Eraqi, H.M., Abdelazeem, S.: A new Efficient Graphemes Segmentation Technique for Offline Arabic Handwriting. In: International Conference on Frontiers in Handwriting Recognition (2012)
10. Lawgali, A., Bouridane, A., Angelova, M., Ghassemlooy, Z.: Automatic segmentation for Arabic characters in handwriting documents. In: 18th IEEE International Conference on Image Processing (2011)

Erratum: Aspect-Based Sentiment Analysis Using Tree Kernel Based Relation Extraction

Thien Hai Nguyen and Kiyoaki Shirai

School of Information Science,
Japan Advanced Institute of Science and Technology
1-1 Asahidai, Nomi, Ishikawa 923-1292, Japan
{nhthien,kshirai}@jaist.ac.jp

A. Gelbukh (Ed.): CICLing 2015, Part II, LNCS 9042, pp. 114–125, 2015.
DOI: 10.1007/978-3-319-18117-2_9

DOI 10.1007/978-3-319-18117-2_51

In the originally published version, the 12th reference was wrong. It should read as follows:

12. Nguyen, T.H., Shirai, K.: Text classification of technical papers based on text segmentation. In: Métais, E., Meziane, F., Saraee, M., Sugumaran, V., Vadera, S. (eds.) NLDB 2013. LNCS, vol. 7934, pp. 278–284. Springer, Heidelberg (2013)

The original online version for this chapter can be found at
http://dx.doi.org/10.1007/978-3-319-18117-2_9

Author Index

Printed in the United States
By Bookmasters